Microbial Bioreactors for Industrial Molecules

Microbial Bioreactors for Industrial Molecules

Edited by

Sudhir P. Singh
Center of Innovative and Applied Bioprocessing (DBT-CIAB)
Mohali
India

Santosh Kumar Upadhyay
Department of Botany
Panjab University
Chandigarh
India

WILEY

This edition first published 2023

Registered Office(s)
John Wiley & Sons, Inc., 111 River Street, Hoboken, NJ 07030, USA
John Wiley & Sons Ltd, The Atrium, Southern Gate, Chichester, West Sussex, PO19 8SQ, UK

For details of our global editorial offices, customer services, and more information about Wiley products visit us at www.wiley.com.

Wiley also publishes its books in a variety of electronic formats and by print-on-demand. Some content that appears in standard print versions of this book may not be available in other formats.

Library of Congress Cataloging-in-Publication Data
Names: Singh, Sudhir P., editor. | Upadhyay, Santosh Kumar, editor.
Title: Microbial bioreactors for industrial molecules / edited by Sudhir P Singh, Santosh Kumar Upadhyay.
Description: Hoboken, NJ: Wiley, 2023. | Includes bibliographical references and index.
Identifiers: LCCN 2023000234 (print) | LCCN 2023000235 (ebook) | ISBN 9781119874065 (cloth) | ISBN 9781119874072 (adobe pdf) | ISBN 9781119874089 (epub)
Subjects: MESH: Bioreactors | Microbiological Phenomena | Molecular Biology–methods | Industrial Microbiology–methods
Classification: LCC QP517.M65 (print) | LCC QP517.M65 (ebook) | NLM QW 40 | DDC 572/.33–dc23/eng/20230331
LC record available at https://lccn.loc.gov/2023000234
LC ebook record available at https://lccn.loc.gov/2023000235

Cover Design: Wiley
Cover Image: © JUAN GAERTNER/SCIENCE PHOTO LIBRARY/Getty Images

Set in 9.5/12.5pt STIXTwoText by Straive, Pondicherry, India

Contents

List of Contributors

K. Abraham Peele
Department of Biotechnology, Vignan's Foundation for Science Technology & Research, Vadlamudi Andhra Pradesh, India

Carlos Alberto Acosta-Saldívar
Facultad de Ciencias Biológicas Universidad Autonoma de Coahuila, Torreón Coahuila, Mexico

Cristobal Noe Aguilar
Bioprocesses and Bioproducts Research Group, Food Research Department School of Chemistry, Autonomous University of Coahuila, Saltillo Coahuila, Mexico

Pedro Aguilar-Zárate
Engineering Department, Tecnológico Nacional de México/I. T. de Ciudad Valles Ciudad Valles San Luis Potosí, Mexico

Shayma Thyab Gddoa Al-Sahlany
Department of Food Science College of Agriculture University of Basrah Basra City, Iraq

Ayodeji Amobonye
Department of Biotechnology and Food Science, Faculty of Applied Sciences Durban University of Technology Durban, South Africa

Christiana Eleojo Aruwa
Department of Biotechnology and Food Science, Faculty of Applied Sciences Durban University of Technology, Durban South Africa and Department of Microbiology, School of Sciences Federal University of Technology Akure, Nigeria

Mubeen Ashraf
Department of Microbiology University of Central Punjab Lahore, Pakistan

Nagamani Balagurusamy
Facultad de Ciencias Biológicas Universidad Autonoma de Coahuila, Torreón Coahuila, Mexico

Victor Emmanuel Balderas-Hernández
División de Biología Molecular Instituto Potosino de Investigación Científica y Tecnológica, A. C. San Luis Potosí San Luis Potosí, Mexico

Deepika Baranwal
Department of Home Science Arya Mahila PG College Banaras Hindu University, Varanasi Uttar Pradesh, India

Shashi Kant Bhatia
Department of Biological Engineering College of Engineering Konkuk University Seoul, South Korea

Susana Calderón-Toledo
Laboratorio de Biología Molecular
Facultad de Farmacia y Bioquímica
Universidad Nacional Mayor de San Marcos
Lima, Peruz

Christian Iván Cano-Gómez
Tecnológico Nacional de México/IT de Ciudad Valles, Ciudad Valles
San Luis Potosí, Mexico

Iara Nobre Carmona
Department of Forest Sciences
University of São Paulo – Luiz de Queiroz College of Agriculture, USP – ESALQ, Piracicaba
Sao Paulo, Brazil

Tania Castillo
Departamento de Ingeniería Celular y Biocatálisis, Instituto de Biotecnología
Universidad Nacional Autónoma de México, Cuernavaca
Morelos, Mexico

Mónica L. Chávez-González
Bioprocesses and Bioproducts Research Group, Food Research Department
School of Chemistry, Autonomous University of Coahuila, Saltillo
Coahuila, Mexico

Jorge Alberto Vieira Costa
Laboratory of Biochemical Engineering
College of Chemistry and Food Engineering, Federal University of Rio Grande, Rio Grande
Rio Grande do Sul, Brazil

Whyara Karoline Almeida da Costa
Laboratory of Microbial Processes in Foods
Department of Food Engineering
Center of Technology, Federal University of Paraíba, João Pessoa
Paraíba, Brazil

Michele Greque de Morais
Laboratory of Microbiology and Biochemistry, College of Chemistry and Food Engineering, Federal University of Rio Grande, Rio Grande
Rio Grande do Sul, Brazil

Paulo Renato Souza de Oliveira
Department of Forest Sciences
University of São Paulo – Luiz de Queiroz College of Agriculture, USP – ESALQ, Piracicaba
Sao Paulo, Brazil

Claucia Fernanda Volken de Souza
Food Biotechnology Laboratory
University of Vale do Taquari – Univates
Lajeado, Rio Grande do Sul, Brazil and
Biotechnology Graduate Program
University of Vale do Taquari – Univates, Lajeado
Rio Grande do Sul, Brazil

Gauri Garg Dhingra
Department of Zoology
Kirori Mal College
University of Delhi
New Delhi, India

Ananias Francisco Dias Júnior
Department of Forestry and Wood Sciences
Federal University of Espírito Santo, UFES, Jerônimo Monteiro
Espírito Santo, Brazil

Alvaro Díaz-Barrera
Escuela de Ingeniería Bioquímica
Pontificia Universidad Católica del Valparaíso
Valparaíso, Chile

Ankita Dua
Department of Zoology, Shivaji College
University of Delhi, Raja Garden
New Delhi, India

Genesis Escobedo-Morales
Facultad de Ciencias Biológicas
Universidad Autonoma de Coahuila, Torreón
Coahuila, Mexico

Karina Maldonado-Ruiz Esparza
Department of Chemical, Industrial and Food Engineering, Universidad Iberoamericana, Lomas de Santa Fe
Mexico City, Mexico

Alexis García
School of Food Engineering
Faculty of Engineering
Universidad del Valle, Tuluá
Valle del Cauca, Colombia

Adriano Gennari
Food Biotechnology Laboratory
University of Vale do Taquari – Univates
Lajeado, Rio Grande do Sul, Brazil and
Biotechnology Graduate Program
University of Vale do Taquari – Univates, Lajeado
Rio Grande do Sul, Brazil

Javier Ulises Hernández-Beltrán
Facultad de Ciencias Biológicas
Universidad Autonoma de Coahuila, Torreón
Coahuila, Mexico

Nazim Hussain
Center for Applied Molecular Biology (CAMB)
University of the Punjab
Lahore, Pakistan

M. Indira
Department of Biotechnology
Vignan's Foundation for Science
Technology & Research, Vadlamudi
Andhra Pradesh, India

Nida Khaliq
Department of Microbiology
University of Central Punjab
Lahore, Pakistan

S. Krupanidhi
Department of Biotechnology
Vignan's Foundation for Science
Technology & Research, Vadlamudi
Andhra Pradesh, India

Vinay Kumar
Department of Physiology and Cell Biology
The Ohio State University Wexner Medical Center, Columbus
OH, USA

Roshan Kumar
Post-Graduate Department of Zoology
Magadh University, Bodh Gaya
Bihar, India

Daniel Neutzling Lehn
Food Biotechnology Laboratory
University of Vale do Taquari – Univates
Lajeado
Rio Grande do Sul, Brazil

Fernanda Leonhardt
Food Biotechnology Laboratory
University of Vale do Taquari – Univates, Lajeado
Rio Grande do Sul, Brazil

Liliana Londoño
BIOTICS Group, School of Basic Sciences
Technology and Engineering
Universidad Nacional Abierta y a Distancia – UNAD
Bogota, Colombia

Miriam Paulina Luévanos-Escareño
Facultad de Ciencias Biológicas
Universidad Autonoma de Coahuila, Torreón
Coahuila, Mexico

Marciane Magnani
Laboratory of Microbial Processes in Foods
Department of Food Engineering
Center of Technology, Federal University of Paraíba, João Pessoa
Paraíba, Brazil

Tehreem Mahmood
Department of Biotechnology
Quaid-i-Azam University
Islamabad, Pakistan

Arundhati Mehta
Department of Biotechnology
Guru Ghasidas Vishwavidyalaya, Bilaspur
Chhattisgarh, India

Alejandro Mendez-Zavala
Facultad de Ciencias Quimicas
Universidad Autonoma de Coahuila, Saltillo
Coahuila, Mexico

Mariela R. Michel
Engineering Department
Tecnológico Nacional de México/I. T. de Ciudad Valles, Ciudad Valles
San Luis Potosí, Mexico

Julio Montañez
Facultad de Ciencias Quimicas
Universidad Autonoma de Coahuila, Saltillo
Coahuila, Mexico

Lourdes Morales-Oyervides
Facultad de Ciencias Quimicas
Universidad Autonoma de Coahuila, Saltillo
Coahuila, Mexico

Juliana Botelho Moreira
Laboratory of Microbiology and Biochemistry
College of Chemistry and Food Engineering
Federal University of Rio Grande
Rio Grande
Rio Grande do Sul, Brazil

Diana B. Muñiz-Márquez
Facultad de Estudios Profesionales Zona Huasteca, Universidad Autónoma de San Luis Potosí, Ciudad Valles
San Luis Potosí, Mexico

Alaa Kareem Niamah
Department of Food Science
College of Agriculture, University of Basrah
Basra City, Iraq

Aeshna Nigam
Department of Zoology, Shivaji College
University of Delhi, Raja Garden
New Delhi, India

Graziela Barbosa Paludo
Food Biotechnology Laboratory
University of Vale do Taquari – Univates
Lajeado, Rio Grande do Sul
Brazil and Biotechnology Graduate Program
University of Vale do Taquari – Univates, Lajeado
Rio Grande do Sul, Brazil

Satya Narayan Patel
Center of Innovative and Applied Bioprocessing (DBT-CIAB), Mohali
Punjab, India

Ami R. Patel
Division of Dairy Microbiology
Mansinhbhai Institute of Dairy and Food Technology-MIDFT, Dudhsagar Dairy Campus, Mehsana
Gujarat, India

Lorena Pedraza-Segura
Department of Chemical
Industrial and Food Engineering
Universidad Iberoamericana, Lomas de Santa Fe
Mexico City, Mexico

Ruth Pedroza-Islas
Department of Chemical
Industrial and Food Engineering
Universidad Iberoamericana, Lomas de
Santa Fe
Mexico City, Mexico

Carlos Peña
Departamento de Ingeniería Celular y
Biocatálisis, Instituto de Biotecnología
Universidad Nacional Autónoma de
México, Cuernavaca
Morelos, Mexico

Allana Katiussya Silva Pereira
Department of Forest Sciences
University of São Paulo – Luiz de Queiroz
College of Agriculture, USP – ESALQ,
Piracicaba
Sao Paulo, Brazil

Adalberto Pessoa-Junior
Department of Biochemical and
Pharmaceutical Technology
School of Pharmaceutical Sciences
University of São Paulo
São Paulo, Brazil

Alejandra Pichardo
Department of Biotechnology
Universidad Autonoma Metropolitana-
Unidad Iztapalapa, Colonia Vicentina
Mexico City, Mexico

Santhosh Pillai
Department of Biotechnology and
Food Science
Faculty of Applied Sciences
Durban University of Technology
Durban, South Africa

Anna María Polanía
School of Food Engineering Faculty of
Engineering
Universidad del Valle, Tuluá
Valle del Cauca, Colombia

Belén Ponce
Escuela de Ingeniería Bioquímica
Pontificia Universidad Católica del Valparaíso
Valparaíso, Chile

Yashwant Kumar Ratre
Department of Biotechnology
Guru Ghasidas Vishwavidyalaya, Bilaspur
Chhattisgarh, India

Gaby Renard
Quatro G Pesquisa &
Desenvolvimento Ltda
TECNOPUC, Porto Alegre
Rio Grande do Sul, Brazil

Luis V. Rodríguez-Durán
Biochemical Engineering Department
UAM-Mante
Universidad Autónoma de Tamaulipas.
Ciudad Mante
Tamaulipas, Mexico

Iván Salmerón
School of Chemical Science
Autonomous University of Chihuahua
Chihuahua, Mexico

Anjali Saxena
Department of Zoology, Bhaskaracharya
College of Applied Sciences
University of Delhi, Dwarka
New Delhi, India

Daniel Segura
Departamento de Microbiología Molecular
Instituto de Biotecnología
Universidad Nacional Autónoma de
México, Cuernavaca
Morelos, Mexico

Nihir Shah
Division of Dairy Microbiology
Mansinhbhai Institute of Dairy and Food
Technology-MIDFT, Dudhsagar Dairy
Campus, Mehsana
Gujarat, India

Areej Shahbaz
Center for Applied Molecular Biology (CAMB)
University of the Punjab
Lahore, Pakistan

Sweety Sharma
Center of Innovative and Applied Bioprocessing (DBT-CIAB), Mohali
Punjab, India

Sapnita Shinde
Department of Biotechnology
Guru Ghasidas Vishwavidyalaya, Bilaspur
Chhattisgarh, India

Dhananjay Shukla
Department of Biotechnology
Guru Ghasidas Vishwavidyalaya, Bilaspur
Chhattisgarh, India

Ashish Kumar Singh
Center of Innovative and Applied Bioprocessing (DBT-CIAB), Mohali
Punjab, India

Smita Singh
Department of Allied Health Sciences
Chitkara School of Health Sciences
Chitkara University, Rajpura
Punjab, India

Sudhir P. Singh
Center of Innovative and Applied Bioprocessing (DBT-CIAB), Mohali
Punjab, India

Vibha Sinha
Department of Biotechnology
Guru Ghasidas Vishwavidyalaya, Bilaspur
Chhattisgarh, India

Vivek Kumar Soni
Department of Biotechnology
Guru Ghasidas Vishwavidyalaya, Bilaspur
Chhattisgarh, India

Subash Chandra Sonkar
Multidisciplinary Research Unit
Maulana Azad Medical College and Associated Hospitals
University of Delhi
New Delhi, India

Prem Prakash Srivastav
Agricultural and Food Engineering Department, Indian Institute of Technology Kharagpur, Kharagpur
West Bengal, India

Ana Luiza Machado Terra
Laboratory of Microbiology and Biochemistry
College of Chemistry and Food Engineering
Federal University of Rio Grande
Rio Grande
Rio Grande do Sul, Brazil

Mamta Thakur
Department of Food Technology
School of Sciences
ITM University, Gwalior
Madhya Pradesh, India

Neerja Thakur
Department of Biotechnology and Microbiology, RKMV, Shimla
Himachal Pradesh, India

Daniel Tobías-Soria
Facultad de Ciencias Quimicas
Universidad Autonoma de Coahuila, Saltillo
Coahuila, Mexico

Soubhagya Tripathy
Agricultural and Food Engineering Department, Indian Institute of Technology Kharagpur, Kharagpur
West Bengal, India

Santosh Kumar Upadhyay
Department of Botany
Panjab University, Chandigarh
India

Viviana Urtuvia
Escuela de Ingeniería Bioquímica
Pontificia Universidad Católica del Valparaíso
Valparaíso, Chile

Gemilang Lara Utama
Faculty of Agro-Industrial Technology
Universitas Padjadjaran, Sumedang
Indonesia and Center for Environment and Sustainability Science
Universitas Padjadjaran, Bandung
Indonesia

Fabiola Veana
Tecnológico Nacional de México/IT de Ciudad Valles, Ciudad Valles
San Luis Potosí, Mexico

T. C. Venkateswarulu
Department of Biotechnology
Vignan's Foundation for Science Technology & Research, Vadlamudi
Andhra Pradesh, India

Deepak Kumar Verma
Agricultural and Food Engineering Department, Indian Institute of Technology Kharagpur, Kharagpur
West Bengal, India

K. Vidya Prabhakar
Department of Biotechnology
Vikrama Simhapuri University, Nellore
Andhra Pradesh, India

Naveen Kumar Vishvakarma
Department of Biotechnology
Guru Ghasidas Vishwavidyalaya, Bilaspur
Chhattisgarh, India

Giandra Volpato
Federal Institute of Education
Science and Technology of Rio Grande do Sul, Porto Alegre
Rio Grande do Sul, Brazil

James Winterburn
Department of Chemical Engineering
The University of Manchester
Manchester, UK

Jorge Enrique Wong-Paz
Facultad de Estudios Profesionales Zona Huasteca, Universidad Autónoma de San Luis Potosí, Ciudad Valles
San Luis Potosí, Mexico

Amparo Iris Zavaleta
Laboratorio de Biología Molecular
Facultad de Farmacia y Bioquímica
Universidad Nacional Mayor de San Marcos
Lima, Peru

Preface

The presence of microorganisms is found virtually everywhere in the environment, as the unseen majority on earth. On the planet, any branch of science cannot be imagined to be unaffected by the dynamics of the natural microbial communities. Recent advances in interdisciplinary studies have helped in enhancing our understanding of the association between microbiomes and human beings. The potential of microbial bioresources has been realized in the advancement of various sectors, such as biotechnology, food technology, agricultural development, and health. The plentiful diversity in the microbiome of the earth's biosphere fosters many known and unknown solutions to the socio-economic issues. Many such microbial strains identified in research collections are required to be evaluated for the scope of technological value. The holistic approach to the maintenance and use of the earth's bioresource can facilitate the development of innovative bioreactors based on microbial wealth. Microorganisms are the source of a variety of biomolecules, such as enzymes, fatty acids, antibiotics, exopolysaccharides, biosurfactants, organic acids, rare sugars, functional metabolites, bioactive peptides, specialized metabolites, and nutraceuticals. The microbial enzymes are of enormous usage in food, pharmaceutical, cosmetic, and agricultural industries. The genomic resource of the microflora can be edited and/or engineered for continuous and upscale production of desirable biomolecules. Microbial cell systems can be developed into bio-factory for the production of high-value molecules. The scientific vision should be to exploit the basic and applied aspects of the strain metadata with environment safety and management. This book covers the diverse knowledge about industrial and innovative aspects of microbial cells and the derived biomolecules in numerous fields, including pharmaceuticals, nutraceuticals, food, biomass processing, etc. This book will act as a repository to get information on the application of microbial resources as bioreactors. This comprehensive wealth of information is useful for graduate students, academicians, researchers, and the general public.

Sudhir P. Singh,
Center of Innovative and Applied Bioprocessing (DBT-CIAB),
Mohali, Punjab, India
Santosh Kumar Upadhyay,
Department of Botany, Panjab University,
Chandigarh, India

1

Microbial Bioreactors: An Introduction

Ashish Kumar Singh[1], Santosh Kumar Upadhyay[2], and Sudhir P. Singh[1]

[1] *Center of Innovative and Applied Bioprocessing (DBT-CIAB), Mohali, Punjab, India*
[2] *Department of Botany, Panjab University, Chandigarh, India*

1.1 Microbial Bioresources

Organisms that are too small for the human eye and whose structure cannot be deciphered by the naked eye without a microscope are known as "microorganisms" or "microbes." All unicellular organisms are included in the group of microorganisms. Along with archaea and eubacteria, the term "microbes" is used for different members of algae, fungi, viruses, and protozoans [1]. Microbes are ubiquitous; some are beneficial, and some are harmful to human beings [2]. The diverse role of microorganisms on the planet makes the earth a greatly sustainable and inhabitable ecosystem. Microbial resources have good potential to produce a broad range of high-value compounds [3]. The microbial communities in different ecological niches are gaining more attention due to the increasing demands of various bioactive molecules for food, neutraceutical, and pharmaceutical industries [1,4,5]. The microbes present in traditional fermented products such as cheese, bread, and wine have also been broadly used in industries for the bulk production of different polymers, high-value chemicals, monomers, and biopharmaceuticals such as hormones, enzymes, vitamins, antibiotics, and vaccines [6, 7]. Together with the availability of complete genome sequencing data, progress in molecular biology techniques, recombinant DNA technology (RDT), CRISPR-Cas9 as a genome editing tool has allowed easy genetic manipulation of microbes to enhance or improve the production of different high-value biomolecules that could be carbohydrates, proteins, hormones, enzymes, lipids, etc. [6, 8–12]. These engineered or native microbial cells that act as biological devices for producing natural molecules as pharmaceuticals and industrial significance could be called as "microbial bioreactors." These microbial resources have the potential to make a variety of high-value chemicals, enzymes, bioactive peptides, secondary metabolites, etc. In addition, microbial systems are used to produce biofuel and biogas and for environmentally friendly bioremediation applications. A few specific examples have been discussed in this section; the upcoming chapters will go into greater detail on these topics.

Microbial Bioreactors for Industrial Molecules, First Edition.
Edited by Sudhir P. Singh and Santosh Kumar Upadhyay.

1.2 Microbial Bioresources for the Production of Enzymes

The ocean or marine environment is one of the most extensive untapped frontiers to human beings [13]. The largest aquatic ecosystem on the planet is the marine environment, which has the most critical biodiversity, including animals, plants, and microbes such as fungi, bacteria, and viruses [5, 14–20]. The ocean has moderate atmospheric pressure on the surface and massive pressure in the deepest ocean area. They also have zero sea ice temperature to extremely high temperatures above 300 °C in hydrothermal vents and low saline conditions to salt-saturated areas. This diverse range of environmental conditions is adapted by different life forms present in marine settings. They are metabolically diverse to produce various enzymes that can perform uniquely in industrial environments [13, 21]. As the ocean or marine contributes approximately half of the global primary production, they act as a vital nutritional source and a favorable alternative for food security. The marine environment has huge biological and ecological diversity, and this variability permits the production of several natural compounds used for humankind in agriculture, remediation, nutrition, health, etc. [21, 22]. Based on their ecological function and habitat, marine bacteria and fungi secret different novel enzymes and enzyme variants unique to nature [14]. The marine environment acts as a library for the various inimitable and potential enzymes such as lipase, chitinase, protease, pectinase, nucleases, and xylanase [22].

Many microbe-borne enzymes, viz., invertase, cellulase, xylanase, lipase, keratinase, amylase, lactase, and protease, have been industrially produced and commercialized in the past few decades due to their diverse vital role, eco-friendliness, cost-effectiveness, and economical feasibility [23, 24]. Pectinases have received significant attention worldwide as biological catalysts since they have wide applications in different industries like juice, paper, and food [25–27]. Pectinases have been most widely studied in plant origin, mainly from fruits, but their extraction and purification often need special conditions due to their thermolabile nature [25]. Therefore, the production of pectinases from the microbial origin is getting more attention nowadays as an alternative strategy due to its stability and easy extraction process.

The nitrile compounds or organic cyanides are carboxylic acids substituted by cyanide with the chemical formula R-CN, which are widely spread in the environment. Plant nitrile compounds in their natural state are cyanolipids, β-cyanoalanine, ricinine, cyanoglycosides, etc. [28–31]. Nitriles can also be found as metabolic intermediates in microorganisms. These compounds are essential for synthetic purposes and widely used at the industrial level to produce compounds such as carboxylic acids, amides, pharmaceutical products, polymers, heterocyclic compounds, and pesticides [28]. However, due to the presence of the cyano group, these are highly toxic, carcinogenic, and mutagenic [28, 30]. Therefore, the widespread usage of these substances could cause environmental issues [28]. Microorganisms can degrade many nitrile compounds by using the enzyme nitrilase and nitrile hydratase. These microbes use nitriles in the form of carbon or nitrogen source for their growth. In recent years, microbial-originated nitrilase enzymes have been used to convert nitriles into beneficial chemical compounds and clean up nitrile-contaminated soil and water [28]. Due to their ease of handling, manipulation, and culture under controlled conditions, microbes are attractive candidates for synthesizing economically significant enzymes.

1.3 Microbial Bioresources for Therapeutic Application

The age-old quote, "Let food be the medicine and medicine be the food," is given by Hippocrates, and it has become an ideology of the health-conscious population in today's lifestyle [32–34]. Afterwards, a Russian Nobel Prize winner, Eli Metchnikoff, recognized the beneficial role of some selected bacteria on the human gastrointestinal tract and proposed the "Theory of Longevity" [35, 36]. Several microorganisms used for the treatment of disease led to the development of the concept of "probiotics." In the year 1954, Ferdinand Vergin first gave the term "probiotika," i.e. probiotics [32, 36]. The probiotic history commenced with the early civilization when humans started consuming fermented foods in their diet. Elie Metchnikoff suggested that human health could be boosted after manipulating the gut microbiome with the help of good bacteria in the yoghurt [32, 35, 36]. The beneficial effect of undigestible food constituents such as fibers on the host's health is known as "prebiotics." Prebiotics generally modulate or enhance the growth of some selective bacteria, such as *Lactobacillus* and *Bifidobacteria*, in the colon [33, 34].

The term "synbiotics" was first introduced in 1995, which is less popular than prebiotics and probiotics. The combination of prebiotics and probiotics is known as synbiotics. Gibson and Roberfroid first proposed the term "synbiotics" in 1955. After several revisions, the International Scientific Association for Probiotics and Prebiotics (ISAPP) proposed the definition of synbiotics as "The mixture of live microorganisms and substrate, selectively utilized by host microorganisms that offer the health benefits on host health" [37]. The host microorganisms include the normal microflora of the host gastrointestinal tract and the externally cultured microorganisms taken in the form of probiotics [34–36, 38, 39]. Synbiotics have several health benefits, including immunomodulatory, antiallergenic, antimicrobial, antidiarrheal, hypoglycemic, anticarcinogenic, and hypolipidemic. They also increase minerals' absorption and act as an anti-osteoporotic activity [35].

Several enzymes have also been used as therapeutic drugs [40]. Among the enzymes, L-asparaginase has received substantial attention due to its prospective use as an oncological and acrylamide-decreasing agent in the food industry. In addition, L-asparaginase is also used in the pharmaceutical industry to treat various illnesses, including chronic lymphosarcoma, acute lymphoblastic leukemia, Hodgkin's disease, reticulosarcoma, and lymphocytic leukemia [41, 42]. Several microorganisms and some plants have been reported to have L-ASNase activity. However, due to the complex process of extraction and purification of enzymes from plants and animals, microorganisms act as a precious alternative for producing L-asparaginase [40–44]. Currently, industrial production of L-asparaginase has been carried out using the microbial strains of *Escherichia coli, Pseudomonas, Staphylococcus, Rouxiella, Pseudonocardia, Lactobacillus,* Acinetobacter, and *Erwinia chrysanthemi*, isolated from different environmental, clinical, and food samples [43].

Cancer has become a leading cause of mortality worldwide and is an essential barrier to improving life expectancy in both developed and developing countries [45]. According to World Health Organization (WHO) 2019, in 112 of 183 nations, cancer is the first or second major cause of death before the age of 70, and it ranks third or fourth in another 23 countries [45–47]. The International Agency for Research on Cancer (IARC) estimates that in 2020, cancer will account for more than 19.3 million new cases and 10 million mortality worldwide [45]. The key hurdles to managing cancer are aggressiveness, drug resistance, and cancer burden. Until recently, different types of traditional therapies, such as radiotherapy, hormonal therapy, chemotherapy, immunotherapy, and surgery, were used to treat all types of cancer [48, 49]. Vaccination is one of the most significant and successful disease prevention and control methods. Vaccines successfully eradicate harmful microorganisms and are employed as preventative and therapeutic strategies against diseases. Conventional vaccinations have high production costs, laborious purifying procedures, and biosafety concerns, necessitating time-consuming biosafety evaluations for commercial production. Molecular farming of vaccines, utilizing biomolecules' production in microorganisms or plant cells, offers several benefits compared to conventional systems, including simplicity in manufacture, storage, better yields, stability, and safety [50–52].

The microbial systems can be exploited for the biosynthesis of specialized metabolites, secondary products, pigments, toxins, and other substances that are helpful to the organism but are not involved in primary metabolism. Some of these items have the potential to be therapeutic medicinal agents. Microbial bioproduction has primarily met the ever-increasing need for medications made from natural resources, which has shown to be beneficial for the growing population. The many wear-and-tear processes continuously affect us, causing aging [53]. Skin beauty has been considered a crucial indicator of personal health throughout history and culture. Additionally, it influences social traits like behavior, attractiveness, and self-esteem [54]. New products called nutricosmetics and cosmeceuticals are currently being developed for the food and cosmetic industries [54]. Antiaging products have become more popular due to economic expansion, changing lifestyles, and improved health awareness. The most effective strategies for delaying aging and extending life include calorie/dietary restriction, genetic modification, and antiaging chemical therapy [55]. A chapter in this book focuses on the origin, bioproduction, and connections between antiaging chemicals from the microbial world and human health.

1.4 Microbial Bioresources for Biogenesis

Today, fossil fuel-derived conventional plastics are one of the most crucial materials in different fields: industrial, domestic, packaging, machinery frames, and furniture. Due to their versatile nature, such as strength, durability, degradation resistance, and lightness, they have almost replaced wood, glass, and metals in several cases [56, 57]. However, the excessive production and use of plastic have become an environmental concern because it is persistent and nonbiodegradable. As a result, it accumulates in the environment, posing a threat to life on earth [58]. Therefore, researchers are exploring biologically produced plastics, i.e. bioplastics, with ecofriendly and biodegradable properties [56]. These bioplastics include polyhydroxyalkanoates (PHAs), polylactic acid, polyesters, etc. [57, 58]. PHAs accumulate in microbial cells during unbalanced growth conditions as intracellular carbon and act as energy reserves in several microorganisms [57]. Therefore, it is essential to discuss a general overview of the upstream and downstream microbial biosynthesis of PHAs and their challenges.

Excessive use of petroleum or fossil energy sources for fuel production poses adverse environmental and socioeconomic effects [59]. It creates an energy crisis and boosts the search for new alternatives to mitigate fossil fuel energy consumption with negative environmental impact [59, 60]. Bioreactors provide a suitable environment for microbial biomass to carry out biochemical reactions and energy conversion [61]. Bioreactor technology is one of the most promising methods for microbial biomass production and energy conversion due to its simplicity, sustainability, moderate reaction condition, minimum carbon output, and low raw material utilization [59, 62].

The biogas plant is an appealing technology for sustainable renewable energy production. An intricate microbiological community converts organic wastes into biogas during anaerobic digestion [63]. As a result, this energy production and waste management method is an example of sustainability [63, 64]. The biomass used for digestion and the amount of microbial inoculum in plants controlled the quantity and quality of biogas, such as the composition of methane, carbon dioxide, and other gases produced [63, 65, 66]. The principal constituent of biogas is CH_4 (50–70%), CO_2 (30–50%), nitrogen (0–3%), and water vapor (95–10%), along with ammonia, hydrogen sulfide, hydrocarbons, and siloxanes [67].

Alginates are linear polysaccharides comprising different fractions of β-D-mannuronate (M) linked to α-L-guluronate (G) residues by β-1,4 bond [68–73]. Alginates are significant biopolymers employed as stabilizing, thickening, and gelling agents in the medical, industrial, and commercial sectors [68, 70]. In addition, alginate microspheres have been utilized therapeutically to release medicines, proteins, vaccines, and cells under controlled conditions. Brown algae are currently used to produce alginate. However, depending on the surrounding environment, the polymer's composition changes. Therefore, alginates should be biosynthesized with the specific physicochemical characteristics needed in specialized applications. As an exopolysaccharide, this polymer may be produced by *Pseudomonas* and *Azotobacter* [68, 70–72, 74]. Extensive research is going on for the production of alginate using microbial bioresources. A chapter in this book comprehensively describes the microbial biosynthesis of alginate and its genetic regulation, bacterial production of alginate at the bioreactor level, and different cultivation methods for enhancing alginate production at quality and quantity levels.

Around the world, plants and animals account for most oils and fats [75]. Lipids are a group of naturally occurring organic molecules, e.g. triacylglycerol, phospholipids, and glycolipids. They are classified according to their solubility in organic or nonpolar solvents like benzene, acetone, and chloroform [76]. Lipids, such as fats (solids) and oils (liquids), are classified as nutritional sources with a high level of metabolic energy [76, 77]. Lipids are significant in many biological processes, including cell signaling cascade, energy storage, and structural components of plasma membranes [76]. Microorganisms make up a significantly smaller fraction of the fat. Therefore, it is far more expensive to produce oils and fats from microorganisms than from plants [77, 78]. Animal fats were previously relatively inexpensive since they are frequently produced as byproducts or main products of the meat and dairy industries [75]. Biotechnological processes need to be explored to produce high-value oils and lipids at an economical cost [79, 80]. This book dedicates a chapter describing the wide range of microorganisms, such as algae, bacteria, fungi, and yeast, for oil, fat, and fatty acid sources.

1.5 Microbial Fermentation

From an historical point of view, lactic acid has a very long history. Swedish chemist Carl Wilhelm Scheele first discovered it in the year 1780 from sour milk in brown syrup. Based on its origin, it was given the name "Mjolksyra." Until 1857, it was considered a milk component, but later on, Louis Pasture suggested that lactic acid is a fermentation product of milk produced by certain microorganisms. After that, French scientist Fremy used fermentation to produce lactic acid. In 1881, this event contributed to the first industrial production of lactic acid in the United States using microbes [81, 82]. Lactic acid is a type of organic acid and is authorized as generally regarded as safe (GRAS) by the US Food and Drug Administration. Lactic acid has diverse roles in the food industry. It acts as a fermentation agent, food preservative, decontaminant, acidulant, flavor enhancer, viscosifier, cryoprotectant, etc. The chemical industry uses it as a pH regulator, mosquito repellent, green solvent, metal complexing agent, and neutralizer. It is also used in the cosmetic industry in the form of moisturizers, anti-acne agents, humectants, skin rejuvenating agents, etc. It is also helpful in the medicine or pharmaceuticals industry as dialysis solutions, surgical sutures, immune stimulants, controlled drug delivery systems, etc. [83–86]. Industrial synthesis of lactic acid is done through either chemical synthesis or microbial fermentation. However, microbial fermentation has some advantages; they are produced in the pure form, whereas synthesis of lactic acid via a chemical process always gives a racemic mixture [81]. Globally, the fermentation of carbohydrates through homolactic bacteria is used to produce lactic acid commercially. For example, different modified or optimized bacterial strains of lactobacilli are used to produce lactic acid. The industrial production of pure lactic acid can be done through microbial fermentation using different carbohydrates such as sucrose, maltose, starch, and glucose, derived from various feedstocks such as whey, barley malt, molasses, and beet sugar [81, 83–85].

For human consumption, milk can be derived from various animals, including cows, goats, sheep, buffalo, and humans [87]. However, the rich nutrient content of this milk – which contains proteins, lipids, carbohydrates, vitamins, minerals, and vital amino acids – provides a perfect habitat for the growth of numerous bacteria [87]. The enzyme, β-glycosidase, breaks down lactose in milk, producing lactose-free milk, which is sweeter than regular milk and suitable for lactose-intolerant people [88–92]. The food industry uses the lactose-breaking enzyme β-galactosidase to improve the flavor, sweetness, solubility, and ease of digestion of dairy products [91]. So successive book chapters describe a brief history of β-galactosidase, its structure, recombinant manufacture, and significant alterations made to the enzyme to enhance its functionality.

1.6 Microbial Biodegradation

One of the essential components of renewable bioresources on the earth is lignocellulosic biomass [93]. The lignocellulosic biomass comprises three major components, namely lignin (15–20%), hemicellulose (25–30%), and cellulose (40–50%) [94–97]. Lignin is a complex biopolymer consisting of polyphenols with a molecular weight of approximately 20,000 daltons and low biodegradability [94]. Due to the complex structure of lignin, its degradation becomes challenging compared to cellulose and hemicellulose [95]. Different chemical methods, such as treatment of aqueous ammonia, steam explosion, and acid hydrolysis, are used to degrade lignin, but this method generates toxic byproducts and requires high costs [98]. Therefore, biological processes of lignin degradation using different ligninolytic enzymes from microbial cell factories are gaining more attention nowadays. The biological methods of lignin degradation are more cost-effective and ecofriendly [94, 95]. The microbial enzymes of lignolytic functions have been discussed in the subsequent chapter.

A large spectrum of anthropogenic chemicals has been introduced into the soil, water, and air due to rising human activity in areas like agriculture, industry, and urbanization during the past few decades [99]. These harmful chemicals include a variety of organic substances such as petroleum hydrocarbons, xenobiotic substances, polycyclic aromatic hydrocarbons, halogenated substances, phenolic substances, volatile organic compounds (VOCs), pesticides, nitroaromatic substances, polychlorinated biphenyls (PCBs), and inorganic substances such as salts, nitrates, phosphates, and heavy metals such as arsenic (As) and copper (Cu). The growth and metabolic processes of plants, soil microbes, soil structure and fertility, aquatic species, and the biogeochemical cycling of elements are all negatively impacted by a contaminated ecosystem, which ultimately affects the ecosystem and human health [99–101]. Therefore, removing organic and inorganic pollutants from the contaminated region to support our society's sustainable growth is necessary [99]. The term "bioremediation" describes a collection of processes that uses biological systems to restore or purge damaged environments [102–104]. Bioremediation is an established method of decontaminating a polluted environment that is sustainable and kind to the environment. Of the microorganisms recovered from various environmental samples, only a tiny percentage are easily culturable [100]. We now better understand the bacteria that live in a given environment because of molecular techniques like metagenomics, transcriptomics, and fluxomics [100, 105].

Plastics are synthetic polymers that have a wide range of uses. Plastics are suitable for various applications due to their flexibility, strength, and erosion resistance [106]. Plastics made from petroleum offer a lot of good qualities. They are highly durable due to their small weight and extremely stable chemical and physical characteristics. They are produced in bulk and are well-established, resulting in meager costs. As a result, they are now commonplace in the global economy. However, because petro-plastic wastes are resistant to natural biodegradation processes, they significantly accumulate in the environment. Micro- and nano-sized plastic particles are already pervasive in terrestrial and aquatic environments due to their massive accumulation in municipal waste systems [105, 107, 108]. A large amount of waste is produced in which about 40% of plastics are used as single-use applications [105]. Numerous industrial and home uses have made considerable use of synthetic polymers, such as polyurethane (PUR), polyethylene terephthalate (PET), polypropylene (PP), polyethylene (PE), polystyrene (PS), and polyvinyl chloride (PVC) [100, 105–107, 109, 110]. Therefore, the biodegradation of plastics by different microorganisms and enzymes is a promising method for depolymerization reactions used for petrochemicals to turn them into monomers for recycling or mineralizing them into carbon dioxide, water, and fresh biomass with the concurrent creation of higher-value bioproducts [107]. Microbial bioresource for plastic-degrading enzymes has been discussed in this book.

1.7 Microbioresources for High-Value Metabolites

Sugars that have distinct physiological and structural characteristics are known as functional sugars. Due to their availability in traces in nature, they are also called "rare sugars" [111]. Functional sugars have various applications in pharmaceuticals, chemical, nutritional, and food industries [112, 113]. The low-calorie value and several health benefits make functional sugars preferable food ingredients [113]. However, as functional sugars are in trace amounts in honey and plant materials, their extraction from plants becomes very challenging. Also, the chemical synthesis of functional sugars creates difficult reaction conditions, limited product yield, several byproduct formations, the use of expensive chemicals, and environmental and safety issues [113, 114]. Exploring microbial bioresources for the synthesis and bioproduction of functional sugars is a necessity for developing feasible industrial processes.

The central ideology of science is to upgrade the quality of human life, and for several years, many people have concentrated on improving this quality [115]. The exploration of bioactive peptides is one of the promising approaches among the prior attempts. Bioactive peptides comprise short-chain amino acids usually 2–20 amino acids, derived from different plants, animals, and microbial sources [115–117]. These bioactive peptides have several known and unknown beneficial effects on animal and human health. These peptides act as antimicrobial, antidiabetic, antioxidant, antitumor, and antihypertensive agents [117]. Due to their distinctive qualities, bioactive peptides are used extensively in the pharmaceutical and food industries. However, its industrial production is still challenging, particularly regarding purity, cost, yields, and environmental sustainability [116, 117]. Chemical synthesis is the primary method used to produce bioactive peptides, which consume many solvents and increase residue production [115]. To overcome these obstacles and enable the large-scale bioproduction of these microbial peptides, it is crucial to research the metabolic engineering of the bacterial host to obtain bioactive peptides in bulk.

Antibiotics, pigments, growth hormones, anticancer drugs, and other microbial metabolites have been demonstrated as promising agents for improving human and animal health [118, 119]. Bacterial and fungal communities synthesize a wide range of aforementioned bioactive molecules with emerging benefactions to human health [118–120]. The secondary metabolites are typically produced during the microorganisms' late growth phase, and they are inhibited during the logarithmic phase [118]. Therefore, a well-designed bioreactor is necessary for producing secondary microbial metabolites in addition to nutrition. To enhance the production of the desired secondary metabolites, the bioreactor must provide microorganisms with the culture conditions required for the growth of microorganisms. The subsequent chapter describes different types of bioreactors, their design, and their impact on the production of secondary metabolites.

In conclusion this book compiles the global perspectives of microbes as bioreactors, crucial for the production of high-value biomolecules of emerging benefaction to human health and the environment.

Acknowledgments

The Department of Biotechnology (DBT), Govt. of India, is acknowledged for all kinds of support. AKS acknowledges ICMR fellowships.

References

1 Shintani, T., Upadhyay, S.K., and Singh, S.P. (2022). An introduction to microbial biodiversity and bioprospection. In: *Bioprospecting of Microorganism-Based Industrial Molecules*, 1e, 1–5. Wiley.

2 Abbas, A., Irfan, M., Khan, S. et al. (2021). Microbes: role in industries, medical field and impact on health. *Saudi J. Med. Pharm. Sci.* 7: 278–282. https://doi.org/10.36348/sjmps.2021.v07i06.010.

3 Singh, S.P. and Upadhyay, S.K. (2021). *Bioprospecting of Microorganism-Based Industrial Molecules*. https://doi.org/10.1002/9781119717317.

4 Sharma, M., Singh, D.P., Rangappa, K.S. et al. (2020). The biomolecular spectrum drives microbial biology and functions in agri-food-environments. *Biomolecules* 10: 1–8. https://doi.org/10.3390/biom10030401.

5 Singh, S.P. and Upadhyay, S.K. (eds.) (2021). *Bioprospecting of Microorganism-Based Industrial Molecules*. John Wiley & Sons Ltd. doi:10.1002/9781119717317.

6 Singh, V. (2006). *Microbial Cell Factories Engineering for Production of Biomolecules*. Academic Press, Elsevier. Stacy Masucci.

7 Shintani, T., Upadhyay, S.K., Singh, S.P. (2021). An introduction to microbial biodiversity and bioprospection. In: *Bioprospecting of Microorganism-Based Industrial Molecules* (ed. S.P. Singh and S.K. Upadhyay). John Wiley & Sons Ltd. https://doi.org/10.1002/9781119717317.ch1.

8 Nielsen, J., Tillegreen, C.B., and Petranovic, D. (2022). Innovation trends in industrial biotechnology. *Trends Biotechnol.* 40: 1–13. https://doi.org/10.1016/j.tibtech.2022.03.007.

9 Kalsoom, M., UR Rehman, F., Shafique, T. et al. (2020). Biological importance of microbes in agriculture, food and pharmaceutical industry: a review. *Innovare J. Life Sci.* 1–4. https://doi.org/10.22159/ijls.2020.v8i6.39845.

10 Upadhyay, S.K. (ed.) (2021). *Genome Engineering for Crop Improvement*. John Wiley & Sons Ltd. doi:10.1002/9781119672425.

11 Alok, A., Chauhan, H., Upadhyay, S.K. et al. (2021). Compendium of plant-specific CRISPR vectors and their technical advantages. *Life* 11: 1021. https://doi.org/10.3390/life11101021.

12 Sushmita, Kaur, G., Upadhyay, S.K. Verma, P.C. (2021). An overview of genome-engineering methods. In: *Genome Engineering for Crop Improvement* (ed. S.K. Upadhyay), 1–21. John Wiley & Sons Ltd. https://doi.org/10.1002/9781119672425.ch1.

13 Ferrer, M., Méndez-García, C., Bargiela, R. et al. (2019). Decoding the ocean's microbiological secrets for marine enzyme biodiscovery. *FEMS Microbiol. Lett.* 366: 1–7. https://doi.org/10.1093/femsle/fny285.

14 Rao, T.E., Imchen, M., and Kumavath, R. (2017). Marine enzymes: production and applications for human health. *Adv. Food Nutr. Res.* 80: 149–163. https://doi.org/10.1016/bs.afnr.2016.11.006.

15 Upadhyay, S.K. and Singh, S.P. (eds.) (2023). *Plants as Bioreactors for Industrial Molecules*. John Wiley & Sons Ltd. doi:10.1002/9781119875116.

16 Upadhyay, S.K. and Singh, S.P. (eds.) (2021). *Bioprospecting of Plant Biodiversity for Industrial Molecules*. John Wiley & Sons Ltd. doi:10.1002/9781119718017.

17 Krishnan, R., Singh, S.P., Upadhyay, S.K. (2021). An introduction to plant biodiversity and bioprospecting. In: *Bioprospecting of Plant Biodiversity for Industrial Molecules*

(ed. S.K. Upadhyay and S.P. Singh), 1–13. John Wiley & Sons Ltd. https://doi.org/10.1002/9781119718017.ch1.

18 Arya, S.K., Shiva, S., Upadhyay, S.K. (2021). Entomotoxic proteins from plant biodiversity to control the crop insect pests. In: *Bioprospecting of Plant Biodiversity for Industrial Molecules* (ed. S.K.Upadhyay and S.P. Singh), 15–52. John Wiley & Sons Ltd. doi.org/10.1002/9781119718017.ch2.

19 Dixit, S., Shukla, A., Singh, V., Upadhyay, S.K. (2021). Engineering of plant metabolic pathway for nutritional improvement; recent advances and challenges. In: *Genome Engineering for Crop Improvement* (ed. S.K.Upadhyay), 351–379. John Wiley & Sons Ltd. doi.org/10.1002/9781119672425.ch20.

20 Dixit, S., Shukla, A., Singh, V., Upadhyay, S.K. (2021). Bioprospecting of natural compounds for industrial and medical applications; current scenario and bottleneck. In: *Bioprospecting of Plant Biodiversity for Industrial Molecules* (ed. S.K. Upadhyay and S.P. Singh), 53–57. John Wiley & Sons Ltd. https://doi.org/10.1002/9781119718017.ch3.

21 Hosseini, H., Al-Jabri, H.M., Moheimani, N.R. et al. (2022). Marine microbial bioprospecting: exploitation of marine biodiversity towards biotechnological applications – a review. *J. Basic Microbiol.* https://doi.org/10.1002/jobm.202100504.

22 Beygmoradi, A. and Homaei, A. (2017). Marine microbes as a valuable resource for brand new industrial biocatalysts. *Biocatal. Agric. Biotechnol.* 11: 131–152. https://doi.org/10.1016/j.bcab.2017.06.013.

23 Satapathy, S., Rout, J.R., Kerry, R.G. et al. (2020). Biochemical prospects of various microbial pectinase and pectin: an approachable concept in pharmaceutical bioprocessing. *Front. Nutr.* 7: 1–17. https://doi.org/10.3389/fnut.2020.00117.

24 Thakur, P., Singh, A.K., Singh, M., and Mukherjee, G. (2022). Extracellular alkaline pectinases production: a review. *J. Microbiol. Biotechnol. Food Sci.* 11. https://doi.org/10.55251/jmbfs.3745.

25 Bhardwaj, N., Kumar, B., Agrawal, K., and Verma, P. (2021). Current perspective on production and applications of microbial cellulases: a review. *Bioresour. Bioprocess* 8. https://doi.org/10.1186/s40643-021-00447-6.

26 Shrestha, S., Rahman, M.S., and Qin, W. (2021). New insights in pectinase production development and industrial applications. *Appl. Microbiol. Biotechnol.* 105: 9069–9087. https://doi.org/10.1007/s00253-021-11705-0.

27 El, E.H.A., Elsayed, E.A., Suhaimi, N. et al. (2018). Bioprocess optimization for pectinase production using *Aspergillus niger* in a submerged cultivation system. *BMC Biotechnol.* 18: 71.

28 Sahu, R., Meghavarnam, A.K., and Janakiraman, S. (2019). A simple, efficient and rapid screening technique for differentiating nitrile hydratase and nitrilase producing bacteria. *Biotechnol. Rep.* 24: e00396. https://doi.org/10.1016/j.btre.2019.e00396.

29 Gong, J.S., Lu, Z.M., Li, H. et al. (2012). Nitrilases in nitrile biocatalysis: recent progress and forthcoming research. *Microb. Cell Factories* 11: 1–18. https://doi.org/10.1186/1475-2859-11-142.

30 Thuku, R.N., Brady, D., Benedik, M.J., and Sewell, B.T. (2009). Microbial nitrilases: versatile, spiral forming, industrial enzymes. *J. Appl. Microbiol.* 106: 703–727. https://doi.org/10.1111/j.1365-2672.2008.03941.x.

31 Vaishnav, A., Kumar, R., Singh, H.B., and Sarma, B.K. (2022). Extending the benefits of PGPR to bioremediation of nitrile pollution in crop lands for enhancing crop productivity. *Sci. Total Environ.* 826: 154170. https://doi.org/10.1016/j.scitotenv.2022.154170.

32 Mitropoulou, G., Nedovic, V., Goyal, A., and Kourkoutas, Y. (2013). Immobilization technologies in probiotic food production. *J Nutr. Metab.* 2013: https://doi.org/10.1155/2013/716861.

33 Kumar, V., Naik, B., Kumar, A. et al. (2022). Probiotics media: significance, challenges, and future perspective – a mini review. *Food Prod. Process. Nutr.* 4. https://doi.org/10.1186/s43014-022-00098-w.

34 Pandey, K.R., Naik, S.R., and Vakil, B.V. (2015). Probiotics, prebiotics and synbiotics – a review. *J. Food Sci. Technol.* 52: 7577–7587. https://doi.org/10.1007/s13197-015-1921-1.

35 Panesar, P.S. and Anal, A.K. (2022). *Probiotics, Prebiotics, and Synbiotics Technology Advancement Towards Safety and Industrial Applications*, 1e (ed. P.S. Panesar and A.K. Anal). Wiley https://doi.org/10.1007/10_2008_097.

36 Yadav, M.K., Kumari, I., Singh, B. et al. (2022). Probiotics, prebiotics and synbiotics: safe options for next-generation therapeutics. *Appl. Microbiol. Biotechnol.* 106: 505–521. https://doi.org/10.1007/s00253-021-11646-8.

37 Boyapati, R., Srikanth, C., Kumar, N.K. et al. (2017). Synbiotics- 'SYNC' ing together – a new and innovative approach. *Int. J. Periodontol. Implantol.* 2: 5–7.

38 Krumbeck, J.A., Walter, J., and Hutkins, R.W. (2018). Synbiotics for improved human health: recent developments, challenges, and opportunities. *Annu. Rev. Food Sci. Technol.* 9: 451–479. https://doi.org/10.1146/annurev-food-030117-012757.

39 Dahiya, D. and Nigam, P.S. (2022). Probiotics, prebiotics, synbiotics, and fermented foods as potential biotics in nutrition improving health via microbiome-gut-brain axis. *Fermentation* 8: 303.

40 Sindhu, R. and Manonmani, H.K. (2018). Expression and characterization of recombinant L-asparaginase from Pseudomonas fluorescens. *Protein Expr. Purif.* 143: 83–91. https://doi.org/10.1016/j.pep.2017.09.009.

41 El-Naggar, N.E.A. and El-Shweihy, N.M. (2020). Bioprocess development for L-asparaginase production by Streptomyces rochei, purification and in-vitro efficacy against various human carcinoma cell lines. *Sci. Rep.* 10: 1–21. https://doi.org/10.1038/s41598-020-64052-x.

42 Chand, S., Mahajan, R.V., Prasad, J.P. et al. (2020). A comprehensive review on microbial l-asparaginase: bioprocessing, characterization, and industrial applications. *Biotechnol. Appl. Biochem.* 67: 619–647. https://doi.org/10.1002/bab.1888.

43 Cachumba, J.J.M., Antunes, F.A.F., Peres, G.F.D. et al. (2016). Current applications and different approaches for microbial L-asparaginase production. *Braz. J. Microbiol.* 47: 77–85. https://doi.org/10.1016/j.bjm.2016.10.004.

44 Castro, D., Marques, A.S.C., Almeida, M.R. et al. (2021). L-asparaginase production review: bioprocess design and biochemical characteristics. *Appl. Microbiol. Biotechnol.* 105: 4515–4534. https://doi.org/10.1007/s00253-021-11359-y.

45 Sung, H., Ferlay, J., Siegel, R.L. et al. (2021). Global cancer statistics 2020: GLOBOCAN estimates of incidence and mortality worldwide for 36 cancers in 185 countries. *CA Cancer J. Clin.* 71: 209–249. https://doi.org/10.3322/caac.21660.

46 Schirrmacher, V. (2019). From chemotherapy to biological therapy: a review of novel concepts to reduce the side effects of systemic cancer treatment (review). *Int. J. Oncol.* 54: 407–419. https://doi.org/10.3892/ijo.2018.4661.

47 Robinson, H.L. and Amara, R.R. (2005). T cell vaccines for microbial infections. *Nat. Med.* 11: S25. https://doi.org/10.1038/nm1212.

48 Sedighi, M., Zahedi Bialvaei, A., Hamblin, M.R. et al. (2019). Therapeutic bacteria to combat cancer; current advances, challenges, and opportunities. *Cancer Med.* 8: 3167–3181. https://doi.org/10.1002/cam4.2148.

49 Housman, G., Byler, S., Heerboth, S. et al. (2014). Drug resistance in cancer: an overview. *Cancers (Basel)* 6: 1769–1792. https://doi.org/10.3390/cancers6031769.

50 Shanmugaraj, B., Bulaon, C.J.I., and Phoolcharoen, W. (2020). Plant molecular farming: a viable platform for recombinant biopharmaceutical production. *Plants* 9: 1–19. https://doi.org/10.3390/plants9070842.

51 Buyel, J.F. (2019). Plant molecular farming – integration and exploitation of side streams to achieve sustainable biomanufacturing. *Front. Plant Sci.* 9: 1–17. https://doi.org/10.3389/fpls.2018.01893.

52 Parvathy, S.T. (2020). Engineering plants as platforms for production of vaccines. *Am. J. Plant Sci.* 11: 707–735. https://doi.org/10.4236/ajps.2020.115052.

53 Baeshen, N.A., Sheikh, M.N.B.A., Bora, R.S. et al. (2014). Cell factories for insulin production. *Microb. Cell Factories* 13: 141. https://doi.org/10.1016/S0065-7743(08)60586-2.

54 Hernandez, D.F., Cervantes, E.L., Luna-vital, D.A. et al. (2021). Food-derived bioactive compounds with antiaging potential for nutricosmetic and cosmeceutical products. *Crit. Rev. Food Sci. Nutr.* 61: 3740–3755. https://doi.org/10.1080/10408398.2020.1805407.

55 Kapahi, P., Kaeberlein, M., and Hansen, M. (2017). Dietary restriction and lifespan: lessons from invertebrate models. *Ageing Res. Rev.* 39: 3–14. https://doi.org/10.1016/j.arr.2016.12.005.

56 Nduko, J.M. and Taguchi, S. (2021). Microbial production of biodegradable lactate-based polymers and oligomeric building blocks from renewable and waste resources. *Front. Bioeng. Biotechnol.* 8: 1–18. https://doi.org/10.3389/fbioe.2020.618077.

57 Surendran, A., Lakshmanan, M., Chee, J.Y. et al. (2020). Can polyhydroxyalkanoates be produced efficiently from waste plant and animal oils? *Front. Bioeng. Biotechnol.* 8: 1–15. https://doi.org/10.3389/fbioe.2020.00169.

58 Palmeiro-Sánchez, T., O'Flaherty, V., and Lens, P.N.L. (2022). Polyhydroxyalkanoate bio-production and its rise as biomaterial of the future. *J. Biotechnol.* 348: 10–25. https://doi.org/10.1016/j.jbiotec.2022.03.001.

59 Love, J. (2022). Microbial pathways for advanced biofuel production. *Biochem. Soc. Trans.* 50: 987–1001. https://doi.org/10.1042/BST20210764.

60 Adegboye, M.F., Ojuederie, O.B., Talia, P.M., and Babalola, O.O. (2021). Bioprospecting of microbial strains for biofuel production: metabolic engineering, applications, and challenges. *Biotechnol. Biofuels* 14: 1–21. https://doi.org/10.1186/s13068-020-01853-2.

61 Zhang, L., Zhang, B., Zhu, X. et al. (2018). Role of bioreactors in microbial biomass and energy conversion. *Green Energy Technol.* https://doi.org/10.1007/978-981-10-7677-0_2.

62 Al Makishah, N.H. (2017). Bioenergy: microbial biofuel production advancement. *Int. J. Pharm. Res. Sci.* 6: 93–106. https://ijpras.com/storage/models/article/W0JDdFW1M9hQDGGbNkMnli28TW9LBxcpGG4id2VAtG3Tib3XxjWqN3cy350S/bioenergy-microbial-biofuel-production-advancement.pdf.

63 Senés-Guerrero, C., Colón-Contreras, F.A., Reynoso-Lobo, J.F. et al. (2019). Biogas-producing microbial composition of an anaerobic digester and associated bovine residues. *MicrobiologyOpen* 8: 1–13. https://doi.org/10.1002/mbo3.854.

64 Kabeyi, M.J.B. and Olanrewaju, O.A. (2022). Biogas production and applications in the sustainable energy transition. *J. Energy* 2022: 1–43. https://doi.org/10.1155/2022/8750221.

65 Srivastava, N., Rohit Srivastava, K., Bantun, F. et al. (2022). Improved production of biogas via microbial digestion of pressmud using CuO/Cu2O based nanocatalyst prepared from pressmud and sugarcane bagasse waste. *Bioresour. Technol.* 362: 127814. https://doi.org/10.1016/j.biortech.2022.127814.

66 Harirchi, S., Wainaina, S., Sar, T. et al. (2022). Microbiological insights into anaerobic digestion for biogas, hydrogen or volatile fatty acids (VFAs): a review. *Bioengineered* 13: 6521–6557. https://doi.org/10.1080/21655979.2022.2035986.

67 Shabbir, M.N.A.M., Saif, H., Khan, S.H. et al. (2021). Microbial and biotechnological advancement in biogas production. In: *Environmental Microbiology and Biotechnology* (ed. A. Singh, S. Srivastava, and D.D.P. Rathore), 31–64. Singapore: Springer Nature Singapore Pte Ltd.

68 Núñez, C., López-Pliego, L., Ahumada-Manuel, C.L., and Castañeda, M. (2022). Genetic regulation of alginate production in *Azotobacter vinelandii* a bacterium of biotechnological interest: a mini-review. *Front. Microbiol.* 13. https://doi.org/10.3389/fmicb.2022.845473.

69 Zhang, L., Li, X., Zhang, X. et al. (2021). Bacterial alginate metabolism: an important pathway for bioconversion of brown algae. *Biotechnol. Biofuels* 14: 1–18. https://doi.org/10.1186/s13068-021-02007-8.

70 Urtuvia, V., Maturana, N., Acevedo, F. et al. (2017). Bacterial alginate production: an overview of its biosynthesis and potential industrial production. *World J. Microbiol. Biotechnol.* 33: 1–10. https://doi.org/10.1007/s11274-017-2363-x.

71 Dudun, A.A., Akoulina, E.A., Zhuikov, V.A. et al. (2022). Competitive biosynthesis of bacterial alginate using azotobacter vinelandii 12 for tissue engineering applications. *Polymers (Basel)* 14. https://doi.org/10.3390/polym14010131.

72 Hay, I.D., Rehman, Z.U., Moradali, M.F. et al. (2013). Microbial alginate production, modification and its applications. *Microb. Biotechnol.* 6: 637–650. https://doi.org/10.1111/1751-7915.12076.

73 Barzkar, N., Sheng, R., Sohail, M. et al. (2022). Alginate lyases from marine bacteria: an enzyme ocean for sustainable future. *Molecules* 27: 1–27. https://doi.org/10.3390/molecules27113375.

74 Yoshida, N., Takase, R., Sugahara, Y. et al. (2022). Direct production of polyhydroxybutyrate and alginate from crude glycerol by *Azotobacter vinelandii* using atmospheric nitrogen. *Sci. Rep.* 12: 1–9. https://doi.org/10.1038/s41598-022-11728-1.

75 Rattray, J.B.M. (1984). Biotechnology and the fats and oils industry – an overview. *J. Am. Oil Chem. Soc.* 61: 1701–1712. https://doi.org/10.1007/BF02582132.

76 Kannan, N., Rao, A.S., and Nair, A. (2021). Microbial production of omega-3 fatty acids: an overview. *J. Appl. Microbiol.* 131: 2114–2130. https://doi.org/10.1111/jam.15034.

77 Cho, I.J., Choi, K.R., and Lee, S.Y. (2020). Microbial production of fatty acids and derivative chemicals. *Curr. Opin. Biotechnol.* 65: 129–141. https://doi.org/10.1016/j.copbio.2020.02.006.

78 Boulton, C.A. (1988). The biotechnology of microbial oils and fats. *Resour. Appl. Biotechnol.* 131–140. https://doi.org/10.1007/978-1-349-09574-2_14.

79 Ochsenreither, K., Glück, C., Stressler, T. et al. (2016). Production strategies and applications of microbial single cell oils. *Front. Microbiol.* 7. https://doi.org/10.3389/fmicb.2016.01539.

80 Shah, A.M., Yang, W., Mohamed, H. et al. (2022). Microbes: a hidden treasure of polyunsaturated fatty acids. *Front. Nutr.* 9: 1–21. https://doi.org/10.3389/fnut.2022.827837.

81 Ghaffar, T., Irshad, M., Anwar, Z. et al. (2014). Recent trends in lactic acid biotechnology: a brief review on production to purification. *J. Radiat. Res. Appl. Sci.* 7: 222–229. https://doi.org/10.1016/j.jrras.2014.03.002.

82 Eiteman, M.A. and Ramalingam, S. (2015). Microbial production of lactic acid. *Biotechnol. Lett.* 37: 955–972. https://doi.org/10.1007/s10529-015-1769-5.

83 Abedi, E. and Hashemi, S.M.B. (2020). Lactic acid production – producing microorganisms and substrates sources-state of art. *Heliyon* 6: e04974. https://doi.org/10.1016/j.heliyon.2020.e04974.

84 Aliwarga, L., Wardani, A.K., Aryanti, P.T.P., and Wenten, I.G. (2019). Recent development of lactic acid production using membrane bioreactors. *IOP Conf. Ser. Mater. Sci. Eng.* 622: https://doi.org/10.1088/1757-899X/622/1/012023.

85 Miller, C., Fosmer, A., Rush, B. et al. (2019). Industrial production of lactic acid. *Compr Biotechnol.* 208–217. https://doi.org/10.1016/B978-0-12-809633-8.09142-1.

86 Tsapekos, P., Alvarado-Morales, M., Baladi, S. et al. (2020). Fermentative production of lactic acid as a sustainable approach to valorize household bio-waste. *Front. Sustain.* 1: 1–12. https://doi.org/10.3389/frsus.2020.00004.

87 Quigley, L., O'Sullivan, O., Stanton, C. et al. (2013). The complex microbiota of raw milk. *FEMS Microbiol. Rev.* 37: 664–698. https://doi.org/10.1111/1574-6976.12030.

88 Vera, C., Guerrero, C., and Illanes, A. (2022). Trends in lactose-derived bioactives: synthesis and purification. *Syst. Microbiol. Biomanufact.* 2: 393–412. https://doi.rg/10.1007/s43393-021-00068-2.

89 Widyastuti, Y., Rohmatussolihat, and Febrisiantosa, A. (2014). The role of lactic acid bacteria in milk fermentation. *Food Nutr. Sci.* 5: 435–442. https://doi.org/10.4236/fns.2014.54051.

90 Nivetha, A. and Mohanasrinivasan, V. (2017). Mini review on role of β-galactosidase in lactose intolerance. *IOP Conf. Ser. Mater. Sci. Eng.* 263: https://doi.org/10.1088/1757-899X/263/2/022046.

91 Saqib, S., Akram, A., Halim, S.A., and Tassaduq, R. (2017). Sources of β-galactosidase and its applications in food industry. *3 Biotech* 7: 1–7. https://doi.org/10.1007/s13205-017-0645-5.

92 Movahedpour, A., Ahmadi, N., Ghalamfarsa, F. et al. (2022). β-Galactosidase: from its source and applications to its recombinant form. *Biotechnol. Appl. Biochem.* 69: 612–628. https://doi.org/10.1002/bab.2137.

93 Janusz, G., Pawlik, A., Sulej, J. et al. (2017). Lignin degradation: microorganisms, enzymes involved, genomes analysis and evolution. *FEMS Microbiol. Rev.* 41: 941–962. https://doi.org/10.1093/femsre/fux049.

94 Kumar, A. and Chandra, R. (2020). Ligninolytic enzymes and its mechanisms for degradation of lignocellulosic waste in environment. *Heliyon* 6: e03170. https://doi.org/10.1016/j.heliyon.2020.e03170.

95 Plácido, J. and Capareda, S. (2015). Ligninolytic enzymes: a biotechnological alternative for bioethanol production. *Bioresour. Bioprocess* 2. https://doi.org/10.1186/s40643-015-0049-5.

96 Chukwuma, O.B., Rafatullah, M., Tajarudin, H.A., and Ismail, N. (2020). Lignocellulolytic enzymes in biotechnological and industrial processes: a review. *Sustainability* 12: 1–31. https://doi.org/10.3390/su12187282.

97 Ozcirak Ergun, S. and Ozturk, U.R. (2017). Production of ligninolytic enzymes by solid state fermentation using *Pleurotus ostreatus. Ann. Agrar. Sci.* 15: 273–277. https://doi.org/10.1016/j.aasci.2017.04.003.

98 Rahman, N.H.A., Rahman, N.A., Aziz, S.A., and Hassan, M.A. (2013). Production of ligninolytic enzymes by newly isolated bacteria from palm oil plantation soils. *Bioresources* 8: 6136–6150. https://doi.org/10.15376/biores.8.4.6136-6150.

99 Kumar, V., Shahi, S.K., Singh, I. (2018). Bioremediation: an eco-sustainable approach for restoration of contaminated sites. In: *Microbial Bioprospecting for Sustainable Development* (ed. J. Singh, D. Sharma, G. Kumar, N.R. Sharma), 115–136. Springer Nature Singapore Pte Ltd. https://doi.org/10.1007/978-981-13-0053-0

100 Sharma, P., Bano, A., Singh, S.P. et al. (2022). Recent advancements in microbial-assisted remediation strategies for toxic contaminants. *Cleaner Chemical Engineering* 2: 100020. https://doi.org/10.1016/j.clce.2022.100020.

101 Kumar, S., Thakur, N., Singh, A.K. et al. (2022). Aquatic macrophytes for environmental pollution control. In: *IN: Phytoremediation Technology for the Removal of Heavy Metals and Other Contaminants from Soil and Water*, 1ee, 291–315. Elsevier Inc.

102 Saxena, S., Suman, S., and Yadav, N. (2022). Technology through degradation of pollutants by microbiological processes. *Just Agriculture* 2.

103 Yadav, A.N., Suyal, D.C., Kour, D. et al. (2022). Bioremediation and waste management for environmental sustainability. *Journal of Applied Biology and Biotechnology* 10: 1–5. https://doi.org/10.7324/JABB.2022.10s201.

104 Bala, S., Garg, D., Thirumalesh, B.V. et al. (2022). Recent strategies for bioremediation of emerging pollutants: a review for a green and sustainable environment. *Toxics* 10: 484. https://doi.org/10.3390/toxics10080484.

105 Verschoor, J.A., Kusumawardhani, H., Ram, A.F.J., and de Winde, J.H. (2022). Toward microbial recycling and upcycling of plastics: prospects and challenges. *Frontiers in Microbiology* 13: 1–16. https://doi.org/10.3389/fmicb.2022.821629.

106 Shilpa, B.N. and Meena, S.S. (2022). Microbial biodegradation of plastics: challenges, opportunities, and a critical perspective. *Frontiers of Environmental Science and Engineering* 16: https://doi.org/10.1007/s11783-022-1596-6.

107 Mohanan, N., Montazer, Z., Sharma, P.K., and Levin, D.B. (2020). Microbial and enzymatic degradation of synthetic plastics. *Frontiers in Microbiology* 11. https://doi.org/10.3389/fmicb.2020.580709.

108 Temporiti, M.E.E., Nicola, L., Nielsen, E., and Tosi, S. (2022). Fungal enzymes involved in plastics biodegradation. *Microorganisms* 10: 1–27. https://doi.org/10.3390/microorganisms10061180.

109 Danso, D., Chow, J., and Streita, W.R. (2019). Plastics: environmental and biotechnological perspectives on microbial degradation. *Applied Environmental Microbiology* 85. https://doi.org/10.1128/AEM.01095-19.

110 Srikanth, M., Sandeep, T.S.R.S., Sucharitha, K., and Godi, S. (2022). Biodegradation of plastic polymers by fungi: a brief review. *Bioresources and Bioprocessing* 9. https://doi.org/10.1186/s40643-022-00532-4.

111 Mijailovic, N., Nesler, A., Perazzolli, M. et al. (2021). Rare sugars: recent advances and their potential role in sustainable crop protection. *Molecules* 26. https://doi.org/10.3390/molecules26061720.

112 Ahmed, A., Khan, T.A., Dan Ramdath, D. et al. (2022). Rare sugars and their health effects in humans: a systematic review and narrative synthesis of the evidence from human trials. *Nutrition Reviews* 80: 255–270. https://doi.org/10.1093/nutrit/nuab012.

113 Abbasi, A.R., Liu, J., Wang, Z. et al. (2021). Recent advances in producing sugar alcohols and functional sugars by engineering *Yarrowia lipolytica*. *Frontiers in Bioengineering and Biotechnology* 9: 1–9. https://doi.org/10.3389/fbioe.2021.648382.

114 Li, A., Cai, L., Chen, Z. et al. (2017). Recent advances in the synthesis of rare sugars using DHAP-dependent aldolases. *Carbohydrate Research* 452: 108–115. https://doi.org/10.1016/j.carres.2017.10.009.

115 Akbarian, M., Khani, A., Eghbalpour, S., and Uversky, V.N. (2022). Bioactive peptides: synthesis, sources, applications, and proposed mechanisms of action. *International Journal of Molecular Science* 23. https://doi.org/10.3390/ijms23031445.

116 Romero-Luna, H.E., Hernández-Mendoza, A., González-Córdova, A.F., and Peredo-Lovillo, A. (2022). Bioactive peptides produced by engineered probiotics and other food-grade bacteria: a review. *Food Chem X* 13. https://doi.org/10.1016/j.fochx.2021.100196.

117 Du, Z. and Li, Y. (2022). Review and perspective on bioactive peptides: a roadmap for research, development, and future opportunities. *Journal of Agriculture and Food Research* 9: 100353. https://doi.org/10.1016/j.jafr.2022.100353.

118 Singh, B.P., Rateb, M.E., Susana, R.-C. et al. (2019). Editorial: microbial secondary metabolites : recent developments and technological challenges. *Frontiers in Microbiology* 10: 914. https://doi.org/10.3390/md170.

119 Fouillaud, M. and Dufossé, L. (2022). Microbial secondary metabolism and biotechnology. *Microorganisms* 10: 123.

120 Hu, D., Lee, S.M., Li, K., and Mok, K.M. (2022). Exploration of secondary metabolite production potential in actinobacteria isolated from *Kandelia candel* Mangrove plant. *Frontiers in Marine Sci* 9: 700685. https://doi.org/10.3389/fmars.2022.700685.

2

Microbial Bioresource for the Production of Marine Enzymes

Lorena Pedraza-Segura, Karina Maldonado-Ruiz Esparza, and Ruth Pedroza-Islas

Department of Chemical, Industrial and Food Engineering, Universidad Iberoamericana, Lomas de Santa Fe, Mexico City, Mexico

2.1 Introduction

Marine biotechnology, or "blue" biotechnology, explores and exploits the biodiversity of the marine environment to obtain compounds that are applied in the food, cosmetic, pharmaceutical, and chemical industries in general. Among the vast marine diversity, microorganisms (archaea, bacteria, and fungi) isolated from both pelagic and benthic habitats, from sources such as marine sponges, algae, sediments, and coasts, are studied for various reasons, such as their adaptability to environmental conditions with adverse effects on most terrestrial microorganisms. They are halo- and osmotolerant and can be found in temperate, hot, and extremely cold climates, for example. Bioprospecting in marine environments allows knowing the large number of species of cultivable and non-cultivable microorganisms via either traditional techniques or metagenomics [1].

Among biomolecules, enzymes have a market size valued at US$10.69 billion in 2020 [2]; therefore, the search for new sources of enzymes or molecules with special properties is a preponderant activity [3]. Bioprospecting for marine enzymes has been carried out in a traditional way, starting from isolated and cultivated microorganisms, but also with the support of metagenomic analysis, since there are microorganisms that are non-cultivable or living in symbiosis with other organisms, in such a way that they cannot be isolated [4]. This has contributed to the knowledge of the variety of enzymes derived from microorganisms within the marine ecosystem, which are found within all the classes proposed by the Enzyme Commission (EC), as shown in Table 2.1.

2.2 Prokaryotes

Marine bacteria are good sources of enzymes as most of the nutrients they require are not easily accessible compared to their terrestrial counterparts [8, 9]. Marine niches are so varied that they offer unique enzymatic characteristics and are related to marine biogeochemical

Microbial Bioreactors for Industrial Molecules, First Edition.
Edited by Sudhir P. Singh and Santosh Kumar Upadhyay.

Table 2.1 Examples of marine enzymes.

EC class	Enzyme	Microorganism	Reference
1	Alcohol dehydrogenase	*Moraxella* sp.	[5]
2	Glutathione transferase	*Pseudoalteromonas* sp.	[5]
3	Cellulase	*Bacillus* sp.	[6]
4	Alginate lyase	*Microbulbifer* sp.	[7]
5	Triose phosphate isomerase	*Pseudomonas* sp.	[5]
6	DNA ligase	*Pseudoalteromonas haloplanktis*	[5]

cycles [10]. Marine bacteria use at an industrial level has been increasing, due to their hyperthermostability, halophilicity, barophilicity, cold adaptability, chemoselectivity, regioselectivity, degradability of recalcitrant molecules, stereoselectivity, and tolerance to solvents that is almost always shown in halophilic enzymes. In addition to this, marine bacteria are more stable than enzymes obtained from animals and plants [11, 12].

Marine bacteria associated with other organisms through biotic relationships can be found, and these interactions can occur in invertebrate organisms such as corals, marine sponges, sea cucumbers, crustaceans, or even vertebrates such as fish, and relationships can be beneficial for both or solely for one. In the marine environments of Antarctica, competition for resources is greater due to temperature, and bacteria are the most abundant life form, with adaptations at the genetic level that are not always reversible, thus giving them an advantage [13].

Low-temperature enzymes are of such interest that they have their own classification (psychrophilic); the temperatures they require to carry out enzymatic reactions make them special [14]; the cold environment is characterized by the challenging conditions that it has, in terms of salinity and temperature; and both characteristics affect the viscosity of the water, which makes chemical reactions difficult; therefore, it could be assumed that the density of the microorganism decreases under these conditions; however, in reality it is maintained. Reference [15] provides a comparative table between psychrophilic and mesophilic enzymes, showing the values of entropy, enthalpy, and free energy and noting the variations in these values since the enthalpy and entropy are greater for mesophilic enzymes. The action mechanism of psychrophilic enzymes serves to decrease the enthalpy of reaction. In a study by Marx et.al. [15] chitinases from *Serratia* and *Arthrobacter* were compared, and the results obtained show that the enthalpy of reaction is lower for the latter.

Psychrophilic enzymes have been of biotechnological interest for the following reasons: they have a better cost-effective relationship, which translates into a lesser quantity of enzyme at lower temperatures; they can catalyze reactions at temperatures where undesirable chemical reactions can be reduced; they can be deactivated at medium temperatures, which decreases energy expenditure and the use of chemical products; and they can catalyze reactions at temperatures where there is less bacterial contamination [5].

A large number of marine enzymes have been detected, isolated, characterized, and purified for industrial use, such as proteases, chitinases, keratinases, pullulanases, amylases, xylanases, agarases, lipases, peroxidases, tyrosinases, cellulases, glutaminases, and laccases [16]. One group of interest is antifouling enzymes produced by marine bacteria, since coatings commonly used for buildings and ships have shown negative effects on the environment. The function of these enzymes is the decomposition of the adhesive components and the production of repellents, and they can be oxidoreductases, lyases, ligases, among others [17].

2.2.1 Amylases

Among the most used enzymes in industry are amylases, due to their use in food to produce emulsifiers, in the textile industry for starch saccharification, among others. Among the amylase-producing bacteria are the actinomycetes, some examples of which are *Streptomyces griserorubens*, *Streptomyces rochet*, and *Streptomyces parvus*. Bukhari et al. [18] carried out an experimental design developed by Plackett Burman (PB), in which they observed that the maximum production of amylase is with yeast extract (YE) and $CaCl_2$, and the enzymatic activity varies depending on the strain, but in general they have 123 U/mL at 42 °C. Rathore et al. [19] reported that *Streptomyces lopnurensis* is capable of producing amylases and proteases simultaneously, obtaining 104 and 189 U/mL, respectively, in an optimized medium.

2.2.2 Proteases

Proteases have been widely studied in industries such as food, detergents, and cosmetics, and the proteases obtained from archaea have potential use because they can withstand high temperatures, osmotic pressure, and salinity, making them more resistant to detergent conditions and in some cases useful in salty foods [20].

Proteases are used to improve detergents by removing dirt from fabrics, and they can be used with biosurfactants and at different pH ranges. Added to foods, they eliminate turbidity in juices, make bread/pastry dough uniform, coagulate milk to make cheeses, modify flavors, etc.

2.2.3 Bactericide

Among the different types of proteases, fibrinolytic enzymes have been shown to be of medical interest for cardiovascular diseases such as thrombosis, which is characterized by the formation of fibrin coatings [21] and caused 31% of deaths worldwide in 2017, according to the World Health Organization (WHO) [22]. Marine bacteria have the ability to produce this type of enzyme, including *Streptomyces radiopugnans* [23], *Streptomyces parvulus* [24], *Marinobacter aquaeolei* [25], *Bacillus licheniformis* [26], *Bacillus velezensis* [27], among others, but those that have shown the best characteristics are of the genus *Bacillus*.

2.2.4 L-Asparaginase

Another important example of hydrolases is L-asparaginase, which is an enzyme capable of carrying out the hydrolysis of L-asparagine into aspartic acid and ammonium. Asparaginase can be divided into two main groups, according to the way it is obtained, which can be type I (intracellular) and type II (extracellular) [28], but the enzyme that is used commercially is type II, and within the marine microorganisms there are various bacteria and archaea capable of producing type II, such as *Pyrococcus furiosus*, *Thermococcus kodakaraensis*, *Thermococcus gammatolerans*, *Mesoflavibacter zeaxanthinifaciens*, *Enterobacter hormaechei*, and *B. velezensis* [29–34]. The interest in this enzyme lies in the fact that it is of interest in the medical area, where it has been used for more than 40 years and is in the "WHO Model List of Essential Medicines 2021," which means that it is on the list of minimum necessary medicines for the health system, generated by the WHO and renewed every two years. Another more recent area of application for this enzyme is in the area of food where it is used for the reduction of acrylamide, whose discovery in 2002 led to the development of regulations in some countries.

2.2.5 Carbohydrases

Glucanases can be used in animal feeds for increasing feed digestibility and improving nutritional value.

Carrageenan is one of the major sulfated polysaccharides and is a common food additive with about 80% usage in this area. Carrageenanase breaks down the polysaccharides into smaller molecules (oligosaccharides) with antiviral, anticoagulant, and antitumor activities [8]. *Cellulophaga flavobacteria* are producers of carrageenan enzymes. In a study by Howlader et al. [35], six different species of *Cellulophaga* were compared; the concentrations of NaCl and YE in the medium were 30 g/L and 3 g/L, respectively; the optimal growth temperature was 25 °C; and there was a period of 48 hours of incubation. The species show different characteristics in terms of growth conditions, but for enzymatic production the one with the best characteristics is *Cellulophaga algicola*, which is why the enzyme was characterized. Zhao et al. [36] reported the production of cold-adapted carrageenan obtained from *Pseudoalteromonas* sp. ZDY3, which grows in 5% NaCl and 0.5% YE and has a concentration of 0.453 U/mg of carrageenan. The enzyme showed to have good stability at temperatures below 35 °C, which makes it an important candidate for industrial applications.

Cellulase is used in the textile and paper industries, in cotton and linen processing, as a biofertilizer, and in biorefineries. Xylanase is applied to obtain high value-added products, such as xylitol, in the paper industry since it helps to solubilize lignin and thus less chlorine is used. Further, it can degrade polysaccharides in fruit/vegetable juices or beer, helping in clarification, etc. Some examples of xylanase-producing microorganisms are *Saccharophagus degradans*, *Microbulbifer* sp. [37], *Pantoea ananatis* [38], and *Bacillus aquimaris* [39], which is capable of tolerating solvents, for which it is considered ideal for industrial uses, and in addition its optimal temperature is 30 °C instead of 50 °C, which demonstrates a reduction in operating costs.

Xylanases can be used in the xylitol production process to improve the stability and texture of bread dough, in addition to modifying the flavors of some foods. It is used in the paper industry to provide more shine and resistance for paper.

2.3 Marine Archaea

Archaea belong to the Monera kingdom and prokaryotic domain, being of interest for their unique biochemical and physiological characteristics whose potential in the biotechnological area is wide, since they can function as factories for compounds of interest or for bioremediation because they are capable of degrading hydrocarbons, metals, and dehalogenating compounds [40].

Being tolerant to high temperatures and salt concentrations, the polymerases of hyperthermophilic archaea are of interest to the medical and research areas, since high temperatures are used in PCR to make copies of DNA, and some examples of microorganisms from which this polymerase can be obtained are: *T. gammatolerans* [41] and *Thermococcus onnurineus* [42], which have good polymerase activity at temperatures of 94 and 80 °C, respectively, making their use suitable in molecular biology.

Chitinases are hydrolytic enzymes that degrade chitin, an insoluble polymer, and the most abundant polysaccharide in nature, which is commonly found in the cytoskeletons of insects, crustaceans, and fungal cell walls [43]. Potential applications of chitinases range from their use in biorefineries, for treating the waste obtained from some industries such as food for the treatment of crustacean residues, to later be used to obtain some bioproducts of interest such as bioethanol, while in medicine its use can be to modify chitin for reuse in membranes, tissue engineering, and the direct use of the enzyme in the lysis of some cancer cell lines [44]. Archaea chitinases are less studied than bacterial chitinases, but they have the advantage of halotolerance and thermostability, and among the producing species is *Thermococcus chitonophagus* that is isolated from a deep-sea hydrothermal vent environment [45].

In several cases, the enzymes of marine and terrestrial archaea are produced by bacteria, from the expression of the genes of the former in microorganisms that are better known and manageable in mesophilic conditions, such as *Escherichia coli* and *Bacillus subtilis* [46].

Table 2.2 summarizes the most relevant enzymes of microorganisms of marine origin, grouped by the type of enzyme.

Table 2.2 Enzymes, microbial origin, and applications.

Enzyme	Microorganism	Applications	References
Hidrolases EC 3			
Glucanase	*Willopsis saturnus, Glaciozyma antarctica, Pichia anomala, Sulfolobus shibatae, Zobellia galactanivorans, Formosa agariphila, Bacillus lehensis*	Prebiotics	[47–53]
Alginate lyase	*Yarrowia lipolytica, Pseudoalteromonas carrageenovor, Cobetia* sp., *Agarivorans* sp., *Vibrio* sp., *Photobacterium* sp., *Microbulbifer* sp.	Biofuels and biochemical products	[54, 55]
Invertase	*Leucosporidium antarcticum*	HFS	[54]
Polygalacturonase	*Cryptococcus liquefaciens, Thalassospira frigidphilosprofundus, Fusarium moniliforme*		[47, 56]
Phytase	*Kodamaea ohmeri, Rhodotorula mucilaginosa, Penicillium polonicum*	Animal feed	[47, 49, 54, 57–59]
Xylanase	*Candida davisiana, Cryptococcus adeliensis, Guehomyces pullulans, Ochrovirga pacifica, Marinimicrobium* sp., *Cladosporium* sp., *Aspergillus niger*	Biorefinery, food and paper industries	[54, 60–62]

(Continued)

Table 2.2 (Continued)

Enzyme	Microorganism	Applications	References
Esterases	*Vibrio* sp., *Plexaura homomalla, Pseudoalteromonas haloplanktis, Bacillus licheniformis, Staphylothermus, Pyrodictium* sp., *Archaeglobus* sp., *Teredinibacter* sp., *Sulfolobus* sp., *Vibrio fischeri, Pelagibacterium halotolerans, Bacillus subtilis, Erythrobacter seohaensis SW-135, Thalassospira* sp., *Oleispira antarctica, Pseudomonas oryzihabitans HUP022, Vibrio fischeri, Pseudoalteromonas arctica, Bacillus* sp., *Pseudonocardia antitumoralis, Bacillus* sp., *Pseudoalteromas* sp. *strain*	Detergents	[10, 63–65]
L-Glutaminase	*Vibrio costicola, Bacillus velezensis, Providencia* sp., *Streptomyces olivochromogenes, Halomonas meridiana, Alcaligenes faecalis, Kosakonia radiciantans, Vibrio axureus, Brevundimonas diminuta, Pseudomonas aeruginosa, Streptomyces rimosus*	Medicine	[63–66]
Proteases	*Aureobasidium pullulans, Leucosporidium antarcticum, Metschnikova reukafii, Rhodotorula mucilaginosa, Yarrowia lipolytica, Bacillus halodurans, B. licheniformis*	Food, detergents, bactericide	[26, 49, 56, 67–69]
Inulinase	*Candida kefyr, Cryptococcus aureus, Debaryomyces cantarelli, Debaryomyces hansenii, Kluyveromyces fragilis, Kluyveromyces marxianus, Pichia guilliermondii, Yarrowia lipolytica, Alkalibacillus filiformis*		[14, 49, 54, 58]
β-Galactosidase	*Arthrobacter* sp., *Pseudoaletromonas haloplanktis, Enterobacter ludwigii, Alkalilactibacillus ikkense*	Food	[5, 63]
Cellulase	*Aureobasidium pullulans, Geotrichum candidum, Pichia salicaria, Streptomyces variabilis, Kocuria rosea, Stenotrophomonas maltophilia, Bartalinia robillardoides, Penicillium pinophilum*	Agriculture, biorefinery, food, paper industry, textiles, detergents	[49, 54, 67, 70–72]
Lipase	*Aureobasidium pullulans, Candida antarctica, Candida intermedia, Candida parapsilosis, Candida rugosa, Candida tropicalis, Geotrichum marinum, Leucosporidium scotia, Lodderomyces elongisporus, Pichia guilliermondii, Rhodotorula mucilaginosa, Yarrowia lipolytica, Oceanobacillus, Halobacillus truperi*	Bactericide, medicine, bioremediation, biorefinery, chemical industry	[47, 49, 56, 58]

Table 2.2 (Continued)

Enzyme	Microorganism	Applications	References
α-Amylase	*Aureobasidium pullulans, Cryptococcus antarctica, Engyodontium album, Penicillium* sp.	Food, pharmacy, chemical industry, bioremediation	[47, 49, 58, 62]
Oxidoreductases EC 1			
Alcohol dehydrogenase	*Flavobacterium frigidimaris, Moraxella* sp.	Biocatalysis	[5, 63]
Superoxide dismutase	*Cryptococcus* sp., *Debaryomyces hansenii, Rhodotorula* spp., *Udeniomyces* spp.		[47, 73]
Transferases EC2			
Polymerase	*Thermotoga maritima, Thermotoga neapolitana, Pyrococcus furiosus*	Medical applications (PCR)	[64]

2.4 Eukaryotes

2.4.1 Yeasts

If the terms "marine bacteria," "marine fungi," and "marine yeasts" are entered in a specialized search engine, it will be seen that references to yeasts are much less. Perhaps this is a reflection of the proportion of bacteria and yeasts existing in the ocean, in terms of gigatons of carbon (GtC). Bacteria represent 1.5 GtC vs. 0.3 for fungi (yeasts and molds), and few fungal species have been discovered and identified [74]. However, marine yeasts are investigated, among other purposes, for the production of enzymes for commercial applications because, like those of bacterial origin, they present resistance to extreme conditions of salinity, temperature, and pressure [54, 58].

Marine yeasts can be found in sediments, water columns, associated with algae, invertebrates, and mammals and belong to several genera such as *Kluyveromyces*, *Candida*, *Saccharomyces*, *Pichia*, *Geotrichum*, *Hanseniaspora*, *Torulopsis*, *Cryptococcus*, *Debaromyces*, *Yarrowia*, *Wickerhamomyces*, among others [54, 75]. Even in smaller numbers than bacteria, yeasts are an important source of enzymes and produce several of industrial importance, and there are already bioprocesses developed for the production of enzymes [47].

In the specialized body of literature, there are several works focused on different enzymes from marine yeasts, and most of them are hydrolases. For example, microorganisms isolated from cold environments receive special attention because their enzymes (amylases, proteases, lipases) can act at low temperatures and be applied in detergents, in the modification of starch, and processes in the food industry, since their reaction conditions are favorable [76]. Yeasts also stand out in the production of enzymes such as inulinases, which break the fructosidic bonds of inulin, to obtain oligosaccharides (prebiotics) and monosaccharides. They are also used in the production of high-fructose syrups (HFSs) [49, 54, 58, 77].

Apart from hydrolases, another enzyme of interest is superoxide dismutase (SOD), whose action allows marine microorganisms to survive in their habitat, facing oxidative stress, together with other enzymatic and non-enzymatic systems. This enzyme has important applications in the preservation of biologicals, reduction of damage caused by smoking, removal of the products of the Amadori and Maillard reactions, and protection from skin damage by oxidation, through cosmetics, among others [73]. SOD is common in organisms, but marine yeasts such as *Debaryomyces hansenii* produce it efficiently [47, 73].

In general, in the consulted literature, most of the research focuses on hydrolytic enzymes, probably because they have the largest market share, whether it is in detergents or various processes. The relevant information is also shown in Table 2.2.

2.4.2 Enzymes from Marine-Derived Fungi

Fungi are among the great diversity of microorganisms in the marine environment. It is estimated that they represent more than 1500 species [78, 79], although it may be an underestimated number [80]. They have been called marine-derived fungi because they have not been able to be classified as obligate or facultative microorganisms, hence the debate on this continues [81]. Currently, they are the object of attention due to the potential of compounds that can be obtained from them with medical, agricultural, and biotechnological applications that are still poorly explored [8, 78, 81–83].

Marine-derived fungi produce extracellular enzymes with remarkable properties of thermostability, tolerance to saline media and low temperatures, chemoselectivity, barophilicity, regioselectivity, and stereoselectivity [8]. Oxidoreductases, hydrolases, transferases, isomerases, among others, have been identified, and all are of great interest due to their potential applications [8, 47]. Information on the type of marine-derived fungi enzymes can be found in [78], the marine environments from which they have been isolated and the corresponding species in [80], and a classification of marine fungi in [84]. Among the enzymes of interest are amylase, xylanase, protease, glucosidase, chitinase, alginate lyase, and lipase, as shown in Table 2.2. In several cases, bioprocesses are already being developed to produce and purify them, as is the case with chitinase from *Penicillium janthinelum* [47].

Although they can be isolated from very diverse marine sources [75, 79, 81, 85], one of the environments with a great diversity of marine-derived fungi is the mangrove swamps [82]. In the work of De Paula et al. [86], fungi were isolated from the aerial roots of trees from a mangrove swamp in Brazil, and those researchers found that most of the species produced hydrolytic and ligninolytic enzymes, the former predominating; 85% produced cellulases and 78% produced pectinases. They found laccases and peroxidases among the species that produced ligninolytic enzymes. Some fungi had a wide enzymatic production, predominantly laccases (*Microsphaeropsis arundinis* as the best producer, reaching 1037.11 U/L of enzyme using seawater) and peroxidases; the most abundant genus was *Trichoderma* sp. Other genera identified were *Gliocladium* sp., *Microsphaeropsis* sp., *Geotrichum* sp., *Cryphonectria* sp., *Epicoccum* sp., and species of *M. arundinis*, *Trichoderma atroviride*, and *Trichoderma harzianum*, and in all cases the production of laccases was higher in the presence of seawater. The enzymatic extracts of *M. arundinis* and *Trichoderma villosa* were useful to decolorize real textile effluents containing LANASET® Yellow PA-4G, LANASET® Orange PA-2R, and ERIONYL® Red by 17%. Due to the importance of these enzymes in bioremediation and effluent treatment, among other applications, this part of the chapter focuses on laccases, due to their potential and the special characteristics of this group of enzymes, that are derived from marine fungi.

Regarding the treatment of dyes, there are works in which the discoloration of different dyes was tested (methyl orange, acid green 3, methylene blue, Remazol, brilliant blue, crystal violet, and congo red), from textiles, papers, and paints, using *Alternaria* sp. IA202, *Alternaria* sp. G55, *Cadophora* sp. AS21-1, *Cadophora luteo-olivacea*, and *Phoma* sp. 2 BRO-2013, isolated from the sediment of a coastal area and from Lake Cacalburnu in Turkey. The highest activities of laccases corresponded to *Alternaria* sp. and *C. luteo-olivacea*. With *Alternaria* sp., discolorations of 53–98% were achieved depending on the type of dye, after an exposure time of 144 hours [87]. Another research group tested the degradation potential of a textile dye using marine-derived fungi enzymes. A marine-derived basidiomycete, isolated from a sponge and identified as *Peniophora* sp., produced laccases, and their activity was compared with those of *Peniophora cinerea* of terrestrial origin. A semi-purified enzyme concentrate was obtained, with two isoforms for those of marine origin and five for those of terrestrial origin, with different sizes and molecular weights. In the activity tests, the laccases from *P. cinerea* saturated faster than those from *Peniophora* sp., which presented higher relative activity when the pH was between 6 and 7. The terrestrial laccases had higher activity than those of marine origin at temperatures between 30 and 40 °C. However, the thermal stability of both types of enzymes, after one hour, showed that the residual activity of the enzymes of the marine fungus was 20 and 10% for laccases of terrestrial origin. The percentage of textile blue discoloration was 67.38% for *P. cinerea* and 61.17% for *Peniophora* sp., and an increase in toxicity was reported when laccases of marine origin were used that, according to those authors, may be due to the generation of intermediate products during the degradation of the dye molecules, which may be more toxic. In this study, chemical mediators were not used to enhance the oxidation reactions. It is noteworthy that *P. cinerea* did not grow in a saline environment, which gives an advantage to the enzymatic production from the marine-derived fungi, which is able to use seawater for its massive production instead of freshwater [88].

Considering the advantages of enzymes from the marine-derived fungus *Peniophora* sp. with tolerance to saline stress, extreme pH values, a greater range of temperature, and that the purpose of investigating new sources of enzymes is for their industrial use, this study undertook the production of laccases from *Peniophora* sp. CBMAI 1063 in an air-lift bioreactor (ALBR; working volume 3.5 L), with artificial seawater and the addition of copper sulfate, and in a stirred tank reactor (3.5 L and 150 rpm) [89].

Wikee et al. [90], characterized the decolorization of dyes from two recombinant laccases of the marine-derived fungus *Pestalotiopsis* sp. KF079 isolated from marshes of the Baltic Sea, having *Aspergillus niger* as host of expression. The activity of both purified enzymes (PsLac1 and PsLac2) was evaluated. With PsLac1, a 70–100% decolorization was achieved in the presence of 1-hydroxybenzotriazole (HBT) and 24 hours of incubation for the commercial dyes azure blue, reactive black, acid yellow, and nitrosulfonazo III. Poly-R478 degradation reached 50% after 48 hours of incubation with PsLac1; regardless of the presence of HBT and bromocresol purple (BCP), degradation was 80% at 2 hours of incubation with both laccases. The activity mediated by salt concentration was higher for PsLac2, reaching 350% activity relative to 5% salt concentration.

These findings are relevant in the treatment of effluents from the textile or paper industry whose characteristics are extreme values of contaminants, pH, and salts. Even though there are few studies on the activity of laccases obtained from marine-derived fungi, where the largest group found in mangroves belongs to the *Ascomycete* class [82], it has been shown that they can degrade dyes in saline and alkaline media, opening new study perspectives for the use of marine-derived fungi [91]. Currently, fungal laccases of terrestrial origin are the most used in the degradation of phenolic and non-phenolic aromatic compounds, being the main producer *Basidiomycete* [82]. However, those of marine origin have significant potential due to their particular characteristics, hence there is a growing interest due to their adaptation to saline environments, and their use in applications to make processes more environmentally friendly.

The action of laccases in non-steroidal drugs whose presence has been detected in municipal waters is being studied, as a result of their action on aromatic compounds. The importance of degrading these and other drugs is related to their toxicity and stability, which means they can remain, not only in municipal water but also in the soil, for a long period of time. These types of compounds are considered emerging contaminants due to their possible consequences on human health and the environment [92] and, given their characteristics, their treatment with conventional methods (flocculation, activated sludge, ionization, etc.) has been inefficient. In their study to remove olsalazyne (5-aminosalicylic acid; 5-ASA), a drug used to treat ulcerative colitis, Bankole et al. [93] used a crude enzymatic extract of *Aspergillus aculeatus* strain bpo2 of marine origin obtained from a beach on the coast of Lagos, Nigeria. They evaluated the activity of laccases and tested the optimal concentration of oxidation–reduction mediators 2,2′-azino-bis (3-ethylbenzothiaz oline-6-sulfonic acid) (ABTS), HBT, and *p*-coumaric acid to improve degradation, reaching up to 99.5% contaminant degradation when ABTS (1 mM) was used as a laccase-catalyzed mediator.

Other persistent organic pollutants are highly stable polychlorinated biphenyls (PCBs), which are neurotoxic to humans as they bioaccumulate in tissues [94], act as estrogens [95], and induce cancer [96]. Various methods have been used for their degradation, such as incineration and bioremediation using marine-derived fungi. The degradation of PCB29 (2,4,5-trichlorobiphenyl) has been demonstrated with two laccases isolated from *Cladosporium* sp. TM138-S3, selected from 104 species for their high production of laccase and ability to degrade PCBs. The expression of laccase activity could be increased 2.6-fold by the addition of $CuSO_4$. Both enzymes showed thermal stability by retaining more than 65% of their activity in a temperature range of 30–65 °C. The maximum activity was at 50 °C and pH of 3. A removal by degradation of more than 71% of PCB29 was achieved with one of the laccases in the presence of ABTS as oxidation mediator [97].

Due to contamination in air, rivers, lakes, and oceans, polycyclic aromatic hydrocarbons (PAHs) are of great concern. They come mainly from the incomplete combustion of petroleum and its derivatives, tobacco, forest fires, grilled foods, among other sources. Undoubtedly, petroleum-related activities contribute the greatest proportion to the production of these pollutants compared to other activities. In particular, when there are petroleum-related accidents in marine environments, contamination with PAHs can represent close to 50% of the contaminants of this type that, due to their characteristics, reach various ecosystems where they settle and remain, constituting a threat of significant concern for the ecosystem from their toxic, mutagenic, and carcinogenic properties and resistance to biodegradation [98–100]. The US Environmental Protection Agency (EPA) has identified 16 priority PAHs [99] classifying as low molecular weight PAHs having three or fewer fused benzene rings with partial water solubilization properties, such as fluorene and naphthalene, and those with four or more, featuring high molecular weight, for example, pyrene and benzopyrene. The higher the molecular weight and the greater the number of benzene rings, the greater is the increase in hydrophobicity. For the degradation of this group of compounds, various types of bacteria and fungi with potential activity have been studied [101]. Although the first investigations on the degradation of hydrocarbons occurred 50 years ago [102], systematic research on the matter, with marine-derived fungi or their enzymes, is more recent. It is noteworthy that PAHs, especially those having four or more aromatic rings, can cause significant health risks and due to their chronic toxicity potential, they are a priority for the EPA, hence the importance of having biotechnological alternatives to degrade them [102, 103]. Bankole et al. [104] investigated the capacity of the filamentous fungus *Mucor irregularis* bp1 derived from the marine environment, for the degradation of fluorene, achieving a degradation of 81.5% at a pH of 7 and a temperature of 32.5 °C, when its concentration was 100 mg/L, using 2 g of dry weight of the fungal mycelium and five days of incubation. Proposing an optimization experiment using response surface methods, a degradation of 82.5% was reached. The presence of glucose and manganese ions enhanced the degradation. The main role for the degradation of fluorene to phenol was played by laccases, although the activities of lignin peroxidases (LiPs) and manganese peroxidases (MnPs) were also detected, presenting different induction times during degradation, which increased the enzymatic activities favoring the removal of fluorene.

To degrade PAHs, especially pyrene and benzopyrene, 3 strains of marine-derived fungi *Tinctoporellus* sp. CBMAI1061, *Marasmiellus* sp. CBMAI 1062, and *Peniophora* sp. CBMAI1063 were used. The 97.2% degradation of benzopyrene was achieved in seven days of incubation using *Marasmiellus* sp. CBM AI 1062, and for pyrene, 92.8% degradation was obtained. The proposed mechanism was by the cytochrome P450 (CYP) enzyme system and the activity of epoxy-hydrolases. The other fungi were not efficient for the degradation of the two substrates studied [105]. Vasconcelos et al. [106] also studied the degradation of pyrene and benzopyrene using marine-derived fungi, obtaining good results with the ascomycetes *Tolypocladium* sp. strain CBMAI 1346 and *Xylaria* sp. CBMAI 1464 isolated from marine sponges. With the first, a higher production of laccases, MnPs, and LiPs was observed than with the second, achieving pyrene degradation without the generation of intermediate toxic compounds after 21 days of incubation. Those authors also reported that when the fungi were exposed to pyrene, there was no absorption by the mycelium, but this was not the case when the contaminant was benzopyrene, where micellar absorption was 68.9% with *Tolypocladium* sp. strain CBMAI 13 and 83.06% with *Xylaria* sp. CBMAI 1464. Due to the higher production of ligninolytic enzymes and without showing pyrene absorption by the mycelium, the authors chose to experiment with *Tolypocladium* sp. strain CBMAI 13 to optimize pyrene degradation, achieving a removal of up to 94.17% in a medium with 35 ppm salinity, 0.350 g/50 mL malt extract, 0.150 g/50 mL peptone, 0.3 g/50 mL YE, 4 mM $MnSO_4$, and at pH 7. From transcriptomic analyses, it was determined that the degradation capacity of this ascomycete is mainly due to the action of enzymes of the cytochrome P450 (CYP) system; however, the action of laccases and MnPs cannot be dissociated.

Research in mycoremediation continues to have alternatives to the degradation of contaminating organic compounds, derived from human and industrial activity, which due to their chemical nature make their degradation difficult, and they are one of the biggest problems of contamination and alteration of the ecosystem in the world [107], as discussed above. Of interest for being areas of high contamination by PAHs, the environmental impact of seaports is studied to identify species of fungi with ligninolytic activity that have potential use for bioremediation. For example, Greco et al. [108] studied the port of Genoa, where they isolated 437 strains belonging to 12 genera and 23 species. The most recurrent genera were *Aspergillus* and *Penicillium* and the species *Penicillium solitum* and *Galactomyces geotrichum*. Their presence depended on the depth at which the samples were taken, highlighting that the presence of fungi was not detected on the surface of the water. Both can grow at temperatures in the range 25–30 °C. Their ability to degrade polymeric substances and, in the case of *G. geotrichum*, hydrocarbons have been reported. Given the presence of PAHs as contaminants, it can be assumed that they are adapted to the environment. Species of genus *Trichoderma* are also recurrently found. However, the authors conclude that the degradation capacity of the fungi found still needs to be studied for their use in the mycoremediation of port waters.

Plastics constitute another problem of environmental contamination that is a cause for concern, where polyethylene, being the majority contributor due to its high use as a packaging material, constitutes an important ecological problem, and it has been suggested that, through fungal-based biodegradation, a future solution could be found due to its extracellular production of oxidative enzymes, which has already been the object of study for several years. For the degradation of polyethylene, a marine-derived fungus, *Alternaria alternata* FB1, was used, demonstrating that it can reduce the weight of a polyethylene film by 95% in 120 days of treatment, opening an alternative to the biodegradation of these types of film materials by species isolated from marine environments. To verify the action of *A. alternata* FB1, the changes in the polyethylene film were detected using scanning electron microscopy (SEM), thus being able to verify the growth of the fungus on the surface of the film and the microdestruction of the structure, the decrease in crystallinity between 52 and 62.7%, and the depolymerization of polyethylene, using various analytical techniques. The predominant end product was diglycolamine, a four-carbon compound, although the route by which it could be produced as a result of polyethylene degradation has not yet been established. Based on gene transcription analysis, the involvement of 153 potential enzymes during polyethylene degradation by *A. alternata* was established, including laccases, peroxidases, oxidoreductases, hydrolases, dehydrogenases, oxidases, reductases, esterases, lipases, and cutinases, with the first three being the most relevant in the degradation process. The authors proposed that degradation followed a 3-step model: (i) colonization/corrosion, (ii) depolymerization, and (iii) assimilation/mineralization [109].

In addition to polyethylene, there are other widely used plastics such as polypropylene (PP), polystyrene (PS), polyvinyl chloride (PVC), and polyethylene terephthalate (PET). Due to its great use in containers for non-alcoholic beverages and bottled water, the latter has become an ecological problem, especially because it has a single use as a container [110]. Of the more than 300 million tons of plastic produced per year in the world, it is estimated that approximately 10% of the total ends up in the sea. During the process of manufacturing and using these materials, and also due to the degradation they suffer as environmental contaminants, microplastics are produced that, as they are modified by marine microorganisms (for example), can settle on the ocean floor and remain there for hundreds of years [111]. Other sources of microplastics derive from textiles in general and therefore from the manufacture of clothing, cosmetics, and paint [112]. In addition to the oceans and seas, these microplastics can reach freshwater bodies [113] and one of the mechanisms is from soil erosion, where contamination by these materials is also of great importance [114]. Microplastics are solid particles smaller than 5 mm, and nanoplastics are smaller than 1 mm [115]. Due to their size, they are likely to be ingested by the marine biota and reach humans through the food chain. Marine biota, by ingesting microplastics, suffer physical damage to the intestines, their filtering activity is altered, their digestive tract is affected, and even death can occur. Microplastics have been detected in more than 200 species, many of them edible, and in food products such as canned sardines, sugar, sea salt, honey, and bottled water [112]. However, the effects on humans are currently being studied, as the routes of access to these plastic microparticles may be, in addition to the food chain, inhalation and contact with the skin [116]. For both animals and humans, there is the aggravating fact that microplastics can form aggregates and, due to their chemical nature, absorb other contaminants such as pesticides and antibiotics and even pathogens, thus increasing their toxicity [117], hence their biodegradation is an alternative method to alleviate this serious pollution problem. The fungus *Zalerion maritimum*, obtained from the Portuguese coast, was used for the degradation of microplastics. High biomass growth rates were observed in the first seven days of treatment in a minimal culture medium (glucose/peptone/malt extract/sea salts). The removal of the plastic (polyethylene) occurred between 7 and 14 days, and the fungus was able to use polyethylene as a substrate, which opens the possibility of using this abundant microorganism as a bioremediation strategy (mycoremediation) [118]. The great potential of fungi in general for the degradation of petroleum-derived polymers due to their great ability to adapt and use them as a carbon source by producing surfactants (hydrophobins), and due to their intra- and extracellular enzymatic machinery, has been described in Sánchez [119].

Finally, a field that is yet to be explored for the use of marine-derived fungus enzymes is the food industry, where, due to their low specificity and high catalytic efficiency, laccases can degrade many different types of substrates, thus allowing their application in the food industry, its use is in the treatment of agroindustrial effluents to degrade lignin, and a wide variety of substrates such as polyphenols and other aromatic compounds, as aforementioned. It is interesting that in the food industry laccases have various applications. For example, they can be used as cross-linking agents for proteins and polysaccharides [120], to improve the properties of nanostructured collagen films for biodegradable packaging [121], to improve the viscoelastic properties of gluten in baking or to improve the behavior of gluten-free bread-product doughs [122], and in the fruit/vegetable juice industry to stabilize color [123]. The applications of these enzymes in the food industry and in other areas are wide ranged, and those derived from marine fungi expand their potential use due to their aforementioned special characteristics.

References

1 Lozada, M. and Dionisi, H.M. (2015). Microbial bioprospecting in marine environments. In: *Springer Handbook of Marine Biotechnology*, 307–326. Berlin, Heidelberg: Springer.

2 Grand View Research. (2022). Enzymes market size, share & trends analysis report by type (industrial, specialty), by product (carbohydrase, proteases), by source (microorganisms, animals), by region, and segment forecasts, 2021–2028. https://www.grandviewresearch.com/industry-analysis/enzymes-industry (accessed April 2022).

3 Singh, J., Sharma, D., Kumar, G., and Sharma, N.R. (ed.) (2018). *Microbial Bioprospecting for Sustainable Development*. Springer Singapore.

4 Madhavan, A., Sindhu, R., Parameswaran, B. et al. (2017). Metagenome analysis: a powerful tool for enzyme bioprospecting. *Applied Biochemistry and Biotechnology* 183 (2): 636–651.

5 Bruno, S., Coppola, D., di Prisco, G. et al. (2019). Enzymes from marine polar regions and their biotechnological applications. *Marine Drugs* 17 (10): 544.

6 Dos Santos, Y.Q., De Veras, B.O., De Franca, A.F.J. et al. (2018). A new salt-tolerant thermostable cellulase from a marine *Bacillus* sp. strain. *Journal of Microbiology and Biotechnology* 28 (7): 1078–1085.

7 Yang, J., Cui, D., Ma, S. et al. (2021). Characterization of a novel PL 17 family alginate lyase with exolytic and endolytic cleavage activity from marine bacterium *Microbulbifer* sp. SH-1. *International Journal of Biological Macromolecules* 169: 551–563.

8 Zhang, J., Jiang, L., Chen, X. et al. (2021). Recent advances in biotechnology for marine enzymes and molecules. *Current Opinion in Biotechnology* 69: 308–315.

9 Singh, S.P. and Upadhyay, S.K. (eds.) (2021). *Bioprospecting of Microorganism-Based Industrial Molecules*. John Wiley & Sons Ltd. doi:10.1002/9781119717317.

10 Barzkar, N., Sohail, M., Tamadoni Jahromi, S. et al. (2021). Marine bacterial esterases: emerging biocatalysts for industrial applications. *Applied Biochemistry and Biotechnology* 193 (4): 1187–1214.

11 Zhang, C. and Kim, S.K. (2010). Research and application of marine microbial enzymes: status and prospects. *Marine Drugs* 8 (6): 1920–1934.

12 Mohapatra, B.R., Bapuji, M., and Sree, A. (2003). Production of industrial enzymes (amylase, carboxymethylcellulase and protease) by bacteria isolated from marine sedentary organisms. *Acta Biotechnologica* 23 (1): 75–84.

13 Lo Giudice, A. and Rizzo, C. (2018). Bacteria associated with marine benthic invertebrates from polar environments: unexplored frontiers for biodiscovery? *Diversity* 10 (3): 80.

14 Loperena, L., Soria, V., Varela, H. et al. (2012). Extracellular enzymes produced by microorganisms isolated from maritime Antarctica. *World Journal of Microbiology and Biotechnology* 28 (5): 2249–2256.

15 Marx, J.C., Collins, T., D'Amico, S. et al. (2007). Cold-adapted enzymes from marine Antarctic microorganisms. *Marine Biotechnology* 9 (3): 293–304.

16 Dumorne, K. and Severe, R. (2018). Marine enzymes and their industrial and biotechnological applications. *Minerva Biotecnologica* 30 (4): 113–119.

17 Kristensen, J.B., Meyer, R.L., Laursen, B.S. et al. (2008). Antifouling enzymes and the biochemistry of marine settlement. *Biotechnology Advances* 26 (5): 471–481.

18 Bukhari, S.I., Al-Agamy, M.H., Kelany, M.S. et al. (2021). Production optimization using Plackett-Burman and Box-Behnken designs with partial characterization of amylase from marine actinomycetes. *BioMed Research International* 2021: 5289848.

19 Rathore, D.S. and Singh, S.P. (2021). Kinetics of growth and co-production of amylase and protease in novel marine actinomycete, *Streptomyces lopnurensis* KaM5. *Folia Microbiologica* 66 (3): 303–316.

20 Tavano, O.L., Berenguer-Murcia, A., Secundo, F., and Fernandez-Lafuente, R. (2018). Biotechnological applications of proteases in food technology. *Comprehensive Reviews in Food Science and Food Safety* 17 (2): 412–436.

21 Barzkar, N., Jahromi, S.T., and Vianello, F. (2022). Marine microbial fibrinolytic enzymes: an overview of source, production, biochemical properties and thrombolytic activity. *Marine Drugs* 20 (1): 46.

22 Altaf, F., Wu, S., and Kasim, V. (2021). Role of fibrinolytic enzymes in anti-thrombosis therapy. *Frontiers in Molecular Biosciences* 8: 476.

23 Dhamodharan, D. (2019). Novel fibrinolytic protease producing *Streptomyces radiopugnans* VITSD8 from marine sponges. *Marine Drugs* 17 (3): 164.

24 Alencar, V.N., Nascimento, M.C., Ferreira, J.V. et al. (2021). Purification and characterization of fibrinolytic protease from *Streptomyces parvulus* by polyethylene glycol-phosphate aqueous two-phase system. *Anais da Academia Brasileira de Ciências* 93: e20210335.

25 Barzkar, N., Khan, Z., Jahromi, S.T. et al. (2021). A critical review on marine serine protease and its inhibitors: a new wave of drugs? *International Journal of Biological Macromolecules* 170: 674–687.

26 Kumar, L.K., Samuel, M.K., Mooventhan, H. et al. (2021). Production of fibrinolytic protease from a halobacterium *Bacillus licheniformis* VITLMS isolated from marine Sponges of Rameshwaram Coast, India. *Current Bioactive Compounds* 17 (2): 165–173.

27 Lu, M., Gao, Z., Xing, S. et al. (2021). Purification, characterization, and chemical modification of *Bacillus velezensis* SN-14 fibrinolytic enzyme. *International Journal of Biological Macromolecules* 177: 601–609.

28 Gesto, D.S., Cerqueira, N.M., Fernandes, P.A., and Ramos, M.J. (2013). Unraveling the enigmatic mechanism of L-asparaginase II with QM/QM calculations. *Journal of the American Chemical Society* 135 (19): 7146–7158.

29 Bansal, S., Gnaneswari, P., Mishra, P., and Kundu, B. (2010). Structural stability and functional analysis of L-asparaginase from *Pyrococcus furiosus*. *Biochemistry (Moscow)* 75: 375–381.

30 Chohan, S.M. and Rashid, N. (2013). TK1656, a thermostable L-asparaginase from *Thermococcus kodakaraensis*, exhibiting highest ever reported enzyme activity. *Journal of Bioscience and Bioengineering* 116: 438–443.

31 Zuo, S., Xue, D., Zhang, T. et al. (2014). Biochemical characterization of an extremely thermostable L-asparaginase from *Thermococcus gammatolerans* EJ3. *Journal of Molecular Catalysis B: Enzymatic* 109: 122–129.

32 Lee, S.J., Lee, Y., Park, G.H. et al. (2016). A newly identified glutaminase-free L-Asparaginase (L-ASPG86) from the marine bacterium *Mesoflavibacter zeaxanthinifaciens*. *Journal of Microbiology and Biotechnology* 26 (6): 1115–1123.

33 Qeshmi, F.I., Homaei, A., Fernandes, P., and Javadpour, S. (2018). Marine microbial L-asparaginase: biochemistry, molecular approaches and applications in tumor therapy and in food industry. *Microbiological Research* 208: 99–112.

34 Mostafa, Y., Alrumman, S., Alamri, S. et al. (2019). Enhanced production of glutaminase-free L-asparaginase by marine *Bacillus velezensis* and cytotoxic activity against breast cancer cell lines. *Electronic Journal of Biotechnology* 42: 6–15.

35 Howlader, M.M., Molz, J., Sachse, N., and Tuvikene, R. (2021). Optimization of fermentation conditions for carrageenase production by *Cellulophaga* species: a comparative study. *Biology* 10 (10): 971.

36 Zhao, D., Jiang, B., Zhang, Y. et al. (2021). Purification and characterization of a cold-adapted κ-carrageenase from *Pseudoalteromonas* sp. ZDY3. *Protein Expression and Purification* 178: 105768.

37 Tanaka, D., Ohnishi, K.I., Watanabe, S., and Suzuki, S. (2021). Isolation of cellulase-producing *Microbulbifer* sp. from marine teleost blackfish (*Girella melanichthys*) intestine and the enzyme characterization. *The Journal of General and Applied Microbiology* 67 (2): 47–53.

38 Jatt, A.N., Tang, K., Liu, J. et al. (2015). Quorum sensing in marine snow and its possible influence on production of extracellular hydrolytic enzymes in marine snow bacterium *Pantoea ananatis* B9. *FEMS Microbiology Ecology* 91 (2): 1–13.

39 Lentini, V., Gugliandolo, C., and Maugeri, T.L. (2007). Identification of enzyme-producing thermophilic bacilli isolated from marine vents of Aeolian Islands (Italy). *Annals of Microbiology* 57 (3): 355–361.

40 Krzmarzick, M.J., Taylor, D.K., Fu, X., and McCutchan, A.L. (2018). Diversity and niche of archaea in bioremediation. *Archaea* 2018: 3194108.

41 Zhang, L., Jiang, D., Shi, H. et al. (2020). Characterization and application of a family B DNA polymerase from the hyperthermophilic and radioresistant euryarchaeon *Thermococcus gammatolerans*. *International Journal of Biological Macromolecules* 156: 217–224.

42 Cho, Y., Lee, H.S., Kim, Y.J. et al. (2007). Characterization of a dUTPase from the hyperthermophilic archaeon *Thermococcus onnurineus* NA1 and its application in polymerase chain reaction amplification. *Marine Biotechnology* 9 (4): 450–458.

43 Zikakis, J. (ed.) (2012). *Chitin, Chitosan, and Related Enzymes*. Elsevier.

44 Stoykov, Y.M., Pavlov, A.I., and Krastanov, A.I. (2015). Chitinase biotechnology: production, purification, and application. *Engineering in Life Sciences* 15 (1): 30–38.

45 Dukariya, G. and Kumar, A. (2020). Distribution and biotechnological applications of chitinase: a review. *International Journal of Biochemistry and Biophysics* 8: 17–29.

46 Alquéres, S.M.C., Almeida, R.V., Clementino, M.M. et al. (2007). Exploring the biotechnologial applications in the archaeal domain. *Brazilian Journal of Microbiology* 38: 398–405.

47 Bonugli-Santos, R.C., dos Santos Vasconcelos, M.R., Passarini, M.R. et al. (2015). Marine-derived fungi: diversity of enzymes and biotechnological applications. *Frontiers in Microbiology* 6: 269.

48 Mohammadi, S., Hashim, N.H.F., Mahadi, N.M., and Murad, A.M.A. (2021). The cold-active Endo-β-1, 3 (4)-Glucanase from a marine psychrophilic yeast, *Glaciozyma antarctica* PI12: heterologous expression, biochemical characterisation, and molecular modeling. *International Journal of Applied Biology and Pharmaceutical Technology* 12 (1): 279–300.

49 Chi, Z., Liu, G.L., Lu, Y. et al. (2016). Bio-products produced by marine yeasts and their potential applications. *Bioresource Technology* 202: 244–252.

50 Boyce, A. and Walsh, G. (2018). Expression and characterisation of a thermophilic endo-1, 4-β-glucanase from *Sulfolobus shibatae* of potential industrial application. *Molecular Biology Reports* 45 (6): 2201–2211.

51 Labourel, A., Jam, M., Jeudy, A. et al. (2014). The β-glucanase ZgLamA from *Zobellia galactanivorans* evolved a bent active site adapted for efficient degradation of algal laminarin. *Journal of Biological Chemistry* 289 (4): 2027–2042.

52 Belik, A.A., Rasin, A.B., Kusaykin, M.I., and Ermakova, S.P. (2022). Two GH16 Endo-1, 3-β-D-Glucanases from *Formosa agariphila* and *F. algae* bacteria have complete different modes of laminarin digestion. *Molecular Biotechnology* 64 (4): 434–446.

53 Jaafar, N.R., Khoiri, N.M., Ismail, N.F. et al. (2020). Functional characterisation and product specificity of Endo-β-1, 3-glucanase from alkalophilic bacterium, *Bacillus lehensis* G1. *Enzyme and Microbial Technology* 140: 109625.

54 Zaky, A.S., Tucker, G.A., Daw, Z.Y., and Du, C. (2014). Marine yeast isolation and industrial application. *FEMS Yeast Research* 14 (6): 813–825.

55 Barzkar, N., Sheng, R., Sohail, M. et al. (2022). Alginate lyases from marine bacteria: an enzyme ocean for sustainable future. *Molecules* 27 (11): 3375.

56 Navvabi, A., Razzaghi, M., Fernandes, P. et al. (2018). Novel lipases discovery specifically from marine organisms for industrial production and practical applications. *Process Biochemistry* 70: 61–70.

57 Adapa, V., Pulicherla, K., and Rao, K.S. (2019). Marine psychrophile-derived cold-active polygalacturonase: enhancement of productivity in *Thalassospira frigidphilosprofundus* S3BA12 by whole cell immobilization. *Biochemical Engineering Journal* 144: 135–147.

58 Sarkar, A. and Rao, K.B. (2016). Marine yeast: a potential candidate for biotechnological applications – a review. *Asian Journal of Microbiology, Biotechnology & Environmental Sciences* 18 (3): 627–634.

59 Kalkan, S.O., Bozcal, E., Hames Tuna, E.E., and Uzel, A. (2020). Characterisation of a thermostable and proteolysis resistant phytase from *Penicillium polonicum* MF82 associated with the marine sponge *Phorbas* sp. *Biocatalysis and Biotransformation* 38 (6): 469–479.

60 Hettiarachchi, S.A., Kwon, Y.K., Lee, Y. et al. (2019). Characterization of an acetyl xylan esterase from the marine bacterium *Ochrovirga pacifica* and its synergism with xylanase on beechwood xylan. *Microbial Cell Factories* 18 (1): 1–10.

61 Yu, H., Zhao, S., Fan, Y. et al. (2019). Cloning and heterologous expression of a novel halo/alkali-stable multi-domain xylanase (XylM18) from a marine bacterium *Marinimicrobium* sp. strain LS-A18. *Applied Microbiology and Biotechnology* 103 (21): 8899–8909.

62 Duarte, A.W.F., Barato, M.B., Nobre, F.S. et al. (2018). Production of cold-adapted enzymes by filamentous fungi from King George Island, Antarctica. *Polar Biology* 41 (12): 2511–2521.

63 Kennedy, J., Marchesi, J.R., and Dobson, A.D. (2008). Marine metagenomics: strategies for the discovery of novel enzymes with biotechnological applications from marine environments. *Microbial Cell Factories* 7 (1): 1–8.

64 Sarkar, S., Pramanik, A., Mitra, A., and Mukherjee, J. (2010). Bioprocessing data for the production of marine enzymes. *Marine Drugs* 8 (4): 1323–1372.

65 Fernandes, P. (2014). Marine enzymes and food industry: insight on existing and potential interactions. *Frontiers in Marine Science* 1: 46.

66 Barzkar, N., Sohail, M., Tamadoni Jahromi, S. et al. (2021). Marine microbial L-glutaminase: from pharmaceutical to food industry. *Applied Microbiology and Biotechnology* 105 (11): 4453–4466.

67 Barzkar, N., Homaei, A., Hemmati, R., and Patel, S. (2018). Thermostable marine microbial proteases for industrial applications: scopes and risks. *Extremophiles* 22 (3): 335–346.

68 Barzkar, N. (2020). Marine microbial alkaline protease: an efficient and essential tool for various industrial applications. *International Journal of Biological Macromolecules* 161: 1216–1229.

69 Balachandran, C., Vishali, A., Nagendran, N.A. et al. (2021). Optimización de la producción de proteasas a partir de *Bacillus halodurans* bajo fermentación en estado sólido utilizando residuos agrícolas. *Revista Saudita de Ciencias Biológicas* 28 (8): 4263–4269.

70 Yousefi-Mokri, M., Sharafi, A., Rezaei, S. et al. (2019). Enzymatic hydrolysis of inulin by an immobilized extremophilic inulinase from the halophile bacterium *Alkalibacillus filiformis*. *Carbohydrate Research* 483: 107746.

71 Trivedi, N., Reddy, C.R.K., and Lali, A.M. (2016). Marine microbes as a potential source of cellulolytic enzymes. *Advances in Food and Nutrition Research* 79: 27–41.

72 Barzkar, N. and Sohail, M. (2020). An overview on marine cellulolytic enzymes and their potential applications. *Applied Microbiology and Biotechnology* 104 (16): 6873–6892.

73 Zeinali, F., Homaei, A., and Kamrani, E. (2015). Sources of marine superoxide dismutases: characteristics and applications. *International Journal of Biological Macromolecules* 79: 627–637.

74 Bar-On, Y.M. and Milo, R. (2019). The biomass composition of the oceans: a blueprint of our blue planet. *Cell* 179 (7): 1451–1454.

75 Amend, A., Burgaud, G., Cunliffe, M. et al. (2019). Fungi in the marine environment: open questions and unsolved problems. *MBio* 10 (2): e01189–e01118.

76 Rosa, L.H. (ed.) (2019). *Fungi of Antarctica: Diversity, Ecology and Biotechnological Applications*. Springer.

77 Rawat, H.K., Soni, H., Treichel, H., and Kango, N. (2017). Biotechnological potential of microbial inulinases: recent perspective. *Critical Reviews in Food Science and Nutrition* 57 (18): 3818–3829.

78 Parte, S., Sirisha, V.L., and D'Souza, J.S. (2017). Biotechnological applications of marine enzymes from algae, bacteria, fungi, and sponges. In: *Advances in Food and Nutrition Research*, vol. 80, 75–106. Academic Press.

79 Gladfelter, A.S., James, T.Y., and Amend, A.S. (2019). Marine fungi. *Current Biology: CB* 29 (6): R191–R195. https://doi.org/10.1016/j.cub.2019.02.009.

80 Jones, E. (2011). Are there more marine fungi to be described? *Botanica Marina* 54 (4): 343–354. https://doi.org/10.1515/bot.2011.043.

81 Sette, L.D. and Santos, R.C.B. (2013). Ligninolytic enzymes from marine-derived fungi: production and applications. In: *Marine Enzymes for Biocatalysis*, 403–427. Woodhead Publishing.

82 Barone, G., Varrella, S., Tangherlini, M. et al. (2019). Marine fungi: biotechnological perspectives from deep-hypersaline anoxic basins. *Diversity* 11 (7): 113.

83 Birolli, W.G., Lima, R.N., and Porto, A.L. (2019). Applications of marine-derived microorganisms and their enzymes in biocatalysis and biotransformation, the underexplored potentials. *Frontiers in Microbiology* 10: 1453.

84 Jones, E.B., Suetrong, S., Sakayaroj, J. et al. (2015). Classification of marine ascomycota, basidiomycota, blastocladiomycota and chytridiomycota. *Fungal Diversity* 73 (1): 1–72.

85 Kwon, Y.M., Bae, S.S., Choi, G. et al. (2021). Marine-derived fungi in Korea. *Ocean Science Journal* 56 (1): 1–17.

86 De Paula, N.M., da Silva, K., Brugnari, T. et al. (2022). Biotechnological potential of fungi from a mangrove ecosystem: enzymes, salt tolerance and decolorization of a real textile effluent. *Microbiological Research* 254: 126899. https://doi.org/10.1016/j.micres.2021.126899.

87 Toker, S.K., Hüseyin Evlat, H., and Koçyiğit, A. (2021). Screening of newly isolated marine-derived fungi for their laccase production and decolorization of different dye types. *Regional Studies in Marine Science* 45: 101837. https://doi.org/10.1016/j.rsma.2021.101837.

88 De Jesus Fontes, B., Kleingesinds, E.K., Giovanella, P. et al. (2021). Laccases produced by *Peniophora* from marine and terrestrial origin: a comparative study. *Biocatalysis and Agricultural Biotechnology* 35: 102066.

89 Mainardi, P.H., Feitosa, V.A., de Paiva, L.B.B. et al. (2018). Laccase production in bioreactor scale under saline condition by the marine-derived basidiomycete *Peniophora* sp. CBMAI 1063. *Fungal Biology* 122 (5): 302–309.

90 Wikee, S., Hatton, J., Turbé-Doan, A. et al. (2019). Characterization and dye decolorization potential of two laccases from the marine-derived fungus *Pestalotiopsis* sp. *International Journal of Molecular Sciences* 20 (8): 1864.

91 Raghukumar, C. (2000). Fungi from marine habitats: an application in bioremediation. *Mycological Research* 104 (10): 1222–1226.

92 Taheran, M., Naghdi, M., Brar, S.K. et al. (2018). Emerging contaminants: here today, there tomorrow! *Environmental Nanotechnology, Monitoring and Management* 10: 122–126. https://doi.org/10.1016/j.enmm.2018.05.010.

93 Bankole, P.O., Semple, K.T., Jeon, B.H., and Govindwar, S.P. (2021). Impact of redox-mediators in the degradation of olsalazine by marine-derived fungus, *Aspergillus aculeatus* strain bpo2: response surface methodology, laccase stability and kinetics. *Ecotoxicology and Environmental Safety* 208: 111742.

94 Schantz, S.L., Widholm, J.J., and Rice, D.C. (2003). Effects of PCB exposure on neuropsychological function in children. *Environmental Health Perspectives* 111 (3): 357–576.

95 Kester, M.H., Bulduk, S., Tibboel, D. et al. (2000). Potent inhibition of estrogen sulfotransferase by hydroxylated PCB metabolites: a novel pathway explaining the estrogenic activity of PCBs. *Endocrinology* 141 (5): 1897–1900. https://doi.org/10.1210/endo.141.5.7530.

96 Loganathan, B.G. and Masunaga, S. (2009). PCBs, dioxins, and furans: human exposure and health effects. In: *Handbook of Toxicology of Chemical Warfare Agents* (ed. R.C. Gupta), 245–253. San Diego, CA: Academic Press.

97 Nikolaivits, E., Siaperas, R., Agrafiotis, A. et al. (2021). Functional and transcriptomic investigation of laccase activity in the presence of PCB29 identifies two novel enzymes and the multicopper oxidase repertoire of a marine-derived fungus. *Science of the Total Environment* 775: 145818. https://doi.org/10.1016/j.scitotenv.2021.145818.

98 Haritash, A.K. and Kaushik, C.P. (2009). Biodegradation aspects of polycyclic aromatic hydrocarbons (PAHs): a review. *Journal of Hazardous Materials* 169 (1–3): 1–15.

99 Sachaniya, B.K., Gosai, H.B., Panseriya, H.Z. et al. (2019). Polycyclic aromatic hydrocarbons (PAHs): occurrence and bioremediation in the marine environment. *Marine Pollution: Current Status, Impacts and Remedies* 1: 435–466.

100 Premnath, N., Mohanrasu, K., Rao, R.G.R. et al. (2021). A crucial review on polycyclic aromatic hydrocarbons-environmental occurrence and strategies for microbial degradation. *Chemosphere* 280: 130608.

101 Ghosal, D., Ghosh, S., Dutta, T.K., and Ahn, Y. (2016). Current state of knowledge in microbial degradation of polycyclic aromatic hydrocarbons (PAHs): a review. *Frontiers in Microbiology* 1369: https://doi.org/10.3389/fmicb.2016.01369.

102 Bhatt, J.K., Ghevariya, C.M., Dudhagara, D.R. et al. (2014). Application of response surface methodology for rapid chrysene biodegradation by newly isolated marine-derived fungus *Cochliobolus lunatus* strain CHR4D. *Journal of Microbiology* 52 (11): 908–917.

103 Passarini, M.R., Rodrigues, M.V., da Silva, M., and Sette, L.D. (2011). Marine-derived filamentous fungi and their potential application for polycyclic aromatic hydrocarbon bioremediation. *Marine Pollution Bulletin* 62 (2): 364–370.

104 Bankole, P.O., Semple, K.T., Jeon, B.H., and Govindwar, S.P. (2021). Biodegradation of fluorene by the newly isolated marine-derived fungus, *Mucor irregularis* strain bpo1 using response surface methodology. *Ecotoxicology and Environmental Safety* 208: 111619.

105 Vieira, G.A., Magrini, M.J., Bonugli-Santos, R.C. et al. (2018). Polycyclic aromatic hydrocarbons degradation by marine-derived basidiomycetes: optimization of the degradation process. *Brazilian Journal of Microbiology* 49: 749–756.

106 Vasconcelos, M.R., Vieira, G.A., Otero, I.V. et al. (2019). Pyrene degradation by marine-derived ascomycete: process optimization, toxicity, and metabolic analyses. *Environmental Science and Pollution Research* 26 (12): 12412–12424.

107 Revenga, C., Brunner, J., Henninger, N. et al. (2000). *Pilot Analysis of Global Ecosystems: Freshwater Systems*. Washington, DC: World Resources Institute.

108 Greco, G., Cutroneo, L., Di Piazza, S. et al. (2020). Trapping of marine-derived fungi on wooden baits to select species potentially usable in mycoremediation. *Italian Journal of Mycology* 49: 101–115.

109 Gao, R., Liu, R., and Sun, C. (2022). A marine fungus *Alternaria alternata* FB1 efficiently degrades polyethylene. *Journal of Hazardous Materials* 431: 128617.

110 Chen, Y., Awasthi, A.K., Wei, F. et al. (2021). Single-use plastics: production, usage, disposal, and adverse impacts. *Science of the Total Environment* 752: 141772.

111 Cole, M., Lindeque, P., Halsband, C., and Galloway, T.S. (2011). Microplastics as contaminants in the marine environment: a review. *Marine Pollution Bulletin* 62 (12): 2588–2597.

112 Toussaint, B., Raffael, B., Angers-Loustau, A. et al. (2019). Review of micro-and nanoplastic contamination in the food chain. *Food Additives & Contaminants: Part A* 36 (5): 639–673.

113 Wang, Z., Zhang, Y., Kang, S. et al. (2021). Research progresses of microplastic pollution in freshwater systems. *Science of the Total Environment* 795: 148888.

114 Rehm, R., Zeyer, T., Schmidt, A., and Fiener, P. (2021). Soil erosion as transport pathway of microplastic from agriculture soils to aquatic ecosystems. *Science of the Total Environment* 795: 148774.

115 GESAMP, S. (2015). Fate and effects of microplastics in the marine environment: a global assessment (ed. P.J. Kershaw) (No. 90, p. 96). IMO/FAO/UNESCO-IOC/UNIDO/WMO/IAEA/UN/UNEP/UNDP Joint Group of Experts on the Scientific Aspects of Marine Environmental Protection), Rep. Stud.–GESAMP.

116 Prata, J.C., da Costa, J.P., Lopes, I. et al. (2020). Environmental exposure to microplastics: an overview on possible human health effects. *Science of the Total Environment* 702: 134455.

117 Padervand, M., Lichtfouse, E., Robert, D., and Wang, C. (2020). Removal of microplastics from the environment. A review. *Environmental Chemistry Letters* 18 (3): 807–828.

118 Paço, A., Duarte, K., da Costa, J.P. et al. (2017). Biodegradation of polyethylene microplastics by the marine fungus *Zalerion maritimum*. *Science of the Total Environment* 586: 10–15.

119 Sánchez, C. (2020). Fungal potential for the degradation of petroleum-based polymers: an overview of macro-and microplastics biodegradation. *Biotechnology Advances* 40: 107501.

120 Li, X., Li, S., Liang, X. et al. (2020). Applications of oxidases in modification of food molecules and colloidal systems: laccase, peroxidase and tyrosinase. *Trends in Food Science & Technology* 103: 78–93.

121 Tian, X., Wang, Y., Duan, S. et al. (2021). Evaluation of a novel nano-size collagenous matrix film cross-linked with gallotannins catalyzed by laccase. *Food Chemistry* 351: 129335.

122 Manhivi, V.E., Amonsou, E.O., and Kudanga, T. (2018). Laccase-mediated crosslinking of gluten-free amadumbe flour improves rheological properties. *Food Chemistry* 264: 157–163. https://doi.org/10.1016/j.foodchem.2018.05.017.

123 Agrawal, K., Chaturvedi, V., and Verma, P. (2018). Fungal laccase discovered but yet undiscovered. *Bioresources and Bioprocessing* 5 (1): 1–12. https://doi.org/10.1186/s40643-018-0190-z.

3

Lactic Acid Production Using Microbial Bioreactors

Juliana Botelho Moreira[1], Ana Luiza Machado Terra[1], Whyara Karoline Almeida da Costa[2], Marciane Magnani[2], Michele Greque de Morais[1], and Jorge Alberto Vieira Costa[3]

[1] Laboratory of Microbiology and Biochemistry, College of Chemistry and Food Engineering, Federal University of Rio Grande, Rio Grande, Rio Grande do Sul, Brazil
[2] Laboratory of Microbial Processes in Foods, Department of Food Engineering, Center of Technology, Federal University of Paraíba, João Pessoa, Paraíba, Brazil
[3] Laboratory of Biochemical Engineering, College of Chemistry and Food Engineering, Federal University of Rio Grande, Rio Grande, Rio Grande do Sul, Brazil

3.1 Introduction

Lactic acid has been widely used in the food industry as a flavoring, acidulant, and preservative. This product has also been applied in the pharmaceutical and textile industries. Substrates with a high lactose content can be used in fermentation to produce lactic acid. Cheese whey, soy milk, corn, and potato stand out as potential substrates [1–4]. Lactic acid is applied as a monomer to obtain polylactic acid (PLA). The search for biodegradable polymers such as PLA is another reason for the rise in the lactic acid market [5, 6]. Lactic acid can be produced via fermentation using several alternative and low-cost substrates. In addition, the fermentation process helps to reduce the environmental impact and energy consumption using low temperatures [7, 8].

The production of lactic acid via fermentation can exhibit different levels of product yield according to the strain of microorganisms used. Bacterial strains, yeasts, fungi, and microalgae are microorganisms capable of producing lactic acid [2, 9]. For conducting the fermentation process, various operating parameters, inoculum size, nutritional requirement, and reactor configurations can favor lactic acid production [7]. Lactic acid fermentation can be performed by batch, fed-batch, repeated batch, and continuous modes. The reactors used in this process can have different configurations, such as stirred tank bioreactor, continuous fixed-bed bioreactor, cascade bioreactor, continuous chemostat cultivation, and membrane bioreactor [10]. However, the recovery of lactic acid is the main limiting factor of all reactor configurations tested for industrial production of lactic acid via fermentation [11, 12]. Improving the fermentation digester and the lactic acid recovery process and using low-cost biomass will contribute to the viability of the fermentation

Microbial Bioreactors for Industrial Molecules, First Edition.
Edited by Sudhir P. Singh and Santosh Kumar Upadhyay.

process for producing lactic acid. Future research should focus on developing a reactor that can improve product yield and optimize fermentation process conditions according to the bioreactor configuration [10].

In this context, this chapter provides approaches to lactic acid production using microbial reactors, emphasizing process parameters, lactic acid-producing microorganisms, and several alternative substrates. The ways of conducting the fermentation process and their main advantages and disadvantages, as well as the contribution of different configurations of bioreactors to improve the yield and productivity of lactic acid, are also mentioned, along with the main challenges facing the entire process of obtaining this product.

3.2 Microbial Lactic Acid Producers

Chemical synthesis and microbial fermentation are processes applied for industrially producing lactic acid. Fermentation is a clean process with the production of optically pure lactic acid (L-(+)-lactic acid) [13, 14]. Lactic acid bacteria (LAB), yeast, and filamentous fungi naturally produce lactic acid. Moreover, metabolic engineering allows microbial strains to use unconventional carbon for lactic acid production [2].

LAB species such as *Lactobacillus*, *Streptococcus*, *Leuconostoc*, and *Enterococcus* are used for lactic acid production [15]. Genetically modified *Saccharomyces cerevisiae* synthetized lactic acid in continuous mode [16]. *Rhizopus* is the principal fungi used for lactic acid production. *Rhizopus oryzae* and *Rhizopus microsporus* are producers of high amounts of lactic acid [17]. *Bacillus* sp. and *Escherichia coli* are also capable of producing both lactate isomers [18].

The lactic acid-producing microorganisms influence the characteristics of the produced lactic acid. They can be homofermentative and heterofermentative strains. A homofermentative strain synthesizes a single product (lactic acid). On the other hand, a heterofermentative strain can produce other products [19]. Furthermore, mixed culture of microorganisms has been proposed, where each performs a specific conversion. For this, at least two microorganisms must be compatible and have similarities in the aspects of nutritional and environmental requirements [2].

3.2.1 Bacteria

Homofermentative organisms are used in commercial lactic acid production processes. *Lactobacillus* and associated genera, *Streptococcus*, *Enterococcus*, and *Pediococcus*, stand out. The maximum productivity of these microorganisms is found at pH 5.5–6.5 [19]. Many have amylase activity originating from various plants and animals. However, there are limitations related to complex nutritional requirements and slightly lower temperatures in the fermentation process. These statements can lead to contamination, raise costs, and lower productivity due to early stage amylase production. Otherwise, they require partially hydrolyzed substrates [19, 20].

Lactobacillus spp. has demonstrated high fermentation capacity. The use of *Bacillus* spp. showed the potential to reduce the costs of fermentation processes [19]. Thermophilic

Bacillus coagulans strains are able to utilize sugars from lignocellulosic biomass to homofermentatively produce L-lactic acid under non-sterile conditions. However, it is necessary to expand research related to metabolic engineering to expand its industrial applications [2]. *Corynebacterium glutamicum* and *E. coli* showed high lactic acid productivity after genetic modification [19]. *Enterococcus faecalis* has been described as lactic acid producer from agricultural feedstock [21].

3.2.2 Fungi and Yeast

Different renewable carbon resources can be metabolized by *Rhizopus* spp. to produce high lactic acid content [19]. They also specify advantages over bacterial processes, such as (i) consuming a chemically defined medium, facilitating product recovery; (ii) consuming complex carbohydrates; (iii) producing high concentrations of L-lactic acid to metabolize glucose to manufacture polylactides. *R. oryzae* 2062 and *R. arrhizus* 36017 were able to produce lactic acid in a simultaneous saccharification and fermentation process [19, 22].

Pleissner et al. [23] evaluated a mutant of *R. oryzae* to obtain lactic acid. The strain provided lactic acid production about ten times greater than the original strain. In addition, yeasts can tolerate acidic conditions. Genetic manipulation can improve low lactic acid-producing yeasts [19]. However, there is a need for additional studies on approaches including fungi to produce lactic acid by fungi.

3.2.3 Microalgae

Microalgae are photosynthetic microorganisms widely recognized for their high carbon fixation capacity, which has contributed to minimizing the effects generated by greenhouse gases. These microorganisms use sunlight, carbon dioxide, water, and macronutrients and micronutrients to produce biomass rich in macromolecules interesting for obtaining bioproducts. In addition, microalgae can utilize alternative sources of nutrients, including different types of industrial waste and effluents [24]. Thus, microalgae are promising raw materials to produce lactic acid and reduce substrate costs for fermentation. Several microalgae genera can produce lactic acid, including *Nannochlorum*, *Nannochloropsis*, *Scenedesmus*, *Synechococcus*, and *Synechocystis* [19].

Research has focused on using microalgae mainly in biofuels production [24, 25], and the remaining residues from the lipid and carbohydrate extraction processes for obtaining biodiesel and bioethanol, respectively, have been neglected. Value-added chemicals can be obtained by microalgal biomass from carbohydrate components present. The catalytic transformation of algae into value-added products deserves further investigation, especially lactic acid. In this sense, Xia et al. [26] found a yield of 33.9% in the production of lactic acid from *Scenedesmus* biomass on Fe-Sn-Beta catalyst. The authors also investigated the conversion of macromolecules in the microalgae cell to lactic acid. The protein had a positive effect, promoting the production of lactic acid. On the other hand, the lipid component showed a strong inhibitory effect. Therefore, microalgae residue demonstrates high potential for the production of value-added chemical products, contributing to cost reduction and sustainable development of the environment.

3.3 Alternative Substrates for Lactic Acid Production

The biorefinery concept and the circular bioeconomy have motivated the search for promising and sustainable raw materials. The alternative and low-cost substrates can improve disposal and handling systems, reducing processing costs. Generally, new substrates investigated are byproducts or waste streams from other processes [27]. Different sources of carbohydrates are used as a substrate to produce lactic acid, including vegetable, agricultural, forestry, dairy residues [19, 28], and household waste [29].

Producing lactic acid from unconventional carbon sources requires the insertion of metabolic engineering of microbial strains. The production of D-lactic acid is a highlighted approach due to the demand for thermostable PLA production. However, its large-scale production to commercialization is still a challenge [2]. Algal biomass is a promising alternative to the production of lactic acid due to the absence of lignin and because it has a high content of carbohydrates and proteins [19].

Obtaining fermentable sugars from agricultural biomass is possible from pretreatment in the biomass. Thus, biomass presents potential to use in the production of lactic acid. The great challenge is to increase the quality of high-concentration sugars economically. Genetic engineering contributes to the resistance of strains to acidic environments and producing lactic acid at low pH. In addition, the use of neutralizing agents during fermentation is minimized. Thus, these aspects help to reduce the costs of lactic acid production [2].

3.4 Fermentation Process Parameters

Lactic acid yield and productivity depend on several factors, where pH is one of the critical parameters in the fermentation of microorganisms. The optimal pH for lactic acid production will depend on the microbial strain used [7]. Hassan et al. [30] produced lactic acid from organic waste using different isolates of *Enterococcus durans* BP130. The authors observed high stability and production at pH 8 and 9, obtaining 14.3 and 16.9 g/L of lactic acid, respectively, at 50 °C after 48 hours of fermentation. Trakarnpaiboon et al. [31] analyzed the effects of pH (5–7), using starch as a substrate, in the *R. microsporus* cultivation. The highest lactic acid production (83–84 g/L) was obtained in the pH range 5–6. The lowest production (54.8 g/L) was reported at pH 7.

Temperature is another crucial factor that can affect the growth kinetics of microorganisms, with a significant influence on the use of the substrate for obtaining lactic acid [7]. The optimal temperature for fermentation depends on the substrate and inoculum used [32, 33]. Zhang et al. [34] studied the quality of lactic acid fermentation at different incubation temperatures of alfalfa silage with *Lactobacillus plantarum* and *Lactobacillus casei*. The authors observed that 20 and 30 °C provided lower lactic acid production and a decrease in the count of coliform bacteria. However, when used at 40 °C, the silage treated with *L. casei* presented a lower coliform count and higher lactic acid content than the untreated and treated with *L. plantarum*.

Another factor that can influence the fermentation process is sterilization. Industrially used moist heat sterilization prevents contamination with microorganisms and the production of undesirable byproducts. However, this heat pretreatment raises production costs and can change the substrate composition. Operating costs for the fermentation process are estimated at up to 15% of the sterilization value [35]. High temperatures for thermophilic microorganisms, extreme thermophiles, and hyperthermophiles can be used in the fermentation process to increase productivity and reduce contamination [36]. The contamination probability during lactic acid production can be reduced in acidic or alkaline media. However, increasing the temperature and using acidic and alkaline conditions can compromise the economic viability of the fermentation process [33].

The inoculum size is also a crucial parameter in lactic acid production, as it determines the increase in microbial proliferation, yield, and productivity. Inoculums of 5–10% (v/v) prevent heterolactic fermentation and reduce the lag phase. However, the use of inoculum with a concentration greater than 5% (v/v) tends to increase the cost of the process [33]. Panesar et al. [37] obtained maximum lactic acid production (33.72 g/L) at a concentration of 2–4% (v/v) of *L. casei*. On the other hand, the lowest lactic acid production was observed using 1% (v/v) of the starter culture. The researchers concluded that the density of the starter culture interfered with the increase in lactic acid concentration [38]. However, a larger inoculum can cause nutrient depletion and interfere with cell growth [7, 33].

Regarding the carbon/nitrogen (C/N) ratio, carbon may be available in the form of sugars with high energy content. Nitrogen can be supplied through inorganic compounds, amino acids, and peptides [39]. The availability of nitrogen can interfere with the direction of energy obtained in catabolism. Under conditions with excess nitrogen, energy can be used for assimilation and cell growth. However, nitrogen limitation can prevent energy utilization for cell growth [40].

The hydraulic retention time (HRT) is an operational parameter that influences lactic acid production, where the organic loading rate (OLR) must be high and sufficient to provide a daily amount of carbon to the fermentative microorganisms. However, high OLR can produce a higher lactic acid concentration, although the bioreactor operation in this condition is unstable. Thus, the HRT must be long enough to allow the hydrolysis of complex organic matter. On the other hand, long HRT reduces the amount of manageable substrate per day [41]. In this context, Palomo-Briones et al. [42] showed that keeping HRT short can prevent the development of lactate-producing microorganisms in the bioreactor.

3.5 Mode Improvement of Lactic Acid and Reactor Configuration

The fermentation to obtain lactic acid is commonly operated by batch, fed-batch, continuous (Table 3.1), and repeated batch mode, with their respective advantages and disadvantages (Figure 3.1). In this sense, Pejin et al. [52] evaluated the effect of adding malt rootlets extract or soybean meal extract on brewer's spent grain hydrolysate in lactic acid fermentation in batch and fed-batch modes. In the batch fermentation, the use of 50% of malt root extract provided the highest concentration of lactic acid, yield, and volumetric productivity, with values of 25.73 g/L, 86.31%, and 0.95 g/L/h, respectively. With the addition of the same byproduct and the same concentration, there was a rise in lactic acid concentration (~58 g/L), a yield of approximately 90%, and volumetric productivity (~1.2 g/L/h) using fed-batch fermentation.

Table 3.1 Lactic acid production via fermentations of diverse producer microorganisms and substrates.

Producer microorganism	Substrate	Fermentation process	LA production (g/L)	LA yield	LA productivity	Reference
Lactobacillus rhamnosus	Waste from sweet potato	Batch fermentation	10	NF	NF	[43]
Enterococcus mundtii	Food waste and spent mushroom	Batch fermentation	59.04	0.72 g/g	1.27 g/L/h	[44]
Enterococcus mundtii	Xylose	Continuous fermentation	32.3	0.79 g/g	5.33 g/L/h	[45]
Enterococcus mundtii	Corn steep liquor	Continuous fermentation	41.0	1.01 g/g	6.15 g/L/h	[45]
Lactobacillus casei	*Sophora flavescens* residues and food waste	Batch fermentation	48.4	0.904 g/g	NF	[46]
None	Food waste and waste-activated sludge	Batch fermentation	13.18	0.52 g/g TCOD	NF	[47]
None	Food waste and waste-activated sludge	Batch fermentation	29.55	NF	7.39 g/L/d	[48]
Microbial consortium (*Clostridium sensustricto*, *Escherichia*, and *Enterococcus*)	Sugarcane Molasses and corn steep liquor powder	Batch fermentation	112.34	0.81 g/g	4.49 g/L/h	[49]
Enterococcus durans	Banana peels	Batch fermentation	28.8	0.85 g/g	0.60 g/L/h	[30]
Lactobacillus acidophilus and *Lactobacillus amylovorus*	Cassava bagasse and corn steep liquor	Batch fermentation	31.6	NF	0.11 g/L/h	[50]
Lactobacillus acidophilus and *Lactobacillus amylovorus*	Cassava bagasse and corn steep liquor	Fed-batch fermentation	66.9	NF	0.46	[50]
Lactobacillus plantarum	Glucose	Fed-batch fermentation	178.17	0.84 g/g	1.24 g/L/h	[51]
Lactobacillus plantarum	Microalgal hydrolysate	Batch fermentation	42.34	0.93 g/g	7.56 g/L/h	[51]
Lactobacillus plantarum	Microalgal hydrolysate	Continuous fermentation	39.72	0.99 g/g	9.93 g/L/h	[51]

LA, lactic acid; NF, not found; TCOD, total chemical oxygen demand.

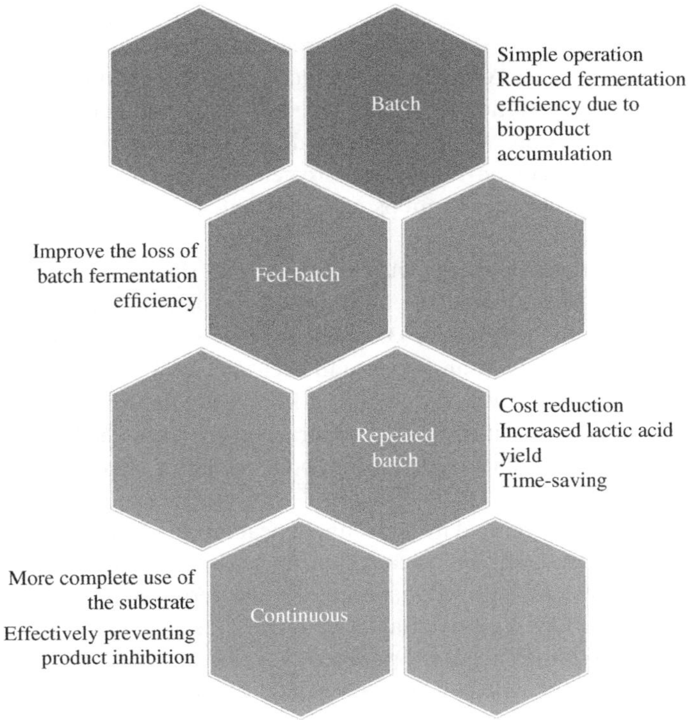

Figure 3.1 Schematic illustration of main characteristics of operation modes commonly used for the fermentation to obtain lactic acid.

Another study investigated the fermentation process with liquefied cassava starch in batch and fed-batch modes at pH 5.5. Lactic acid production was evaluated in the simultaneous saccharification and fermentation of liquefied cassava starch by *R. microsporus* DMKU 33. The bioreactor used was 5 L with agitation at 200 rpm and aeration at 0.75 vvm. The authors observed 91.8 g/L of lactic acid in 72 hours and productivity of 1.28 g/L/h in batch fermentation. The initial starch concentration, in this case, was 153.4 g/L. Furthermore, under these conditions, a yield of 0.76 g/g was achieved, which was lower than the yield found in fermentation with ~100 g/L of liquefied cassava starch (0.84 g/g). With the initial concentration of liquefied cassava starch of 102.7 g/L and the addition of 41.4 g/L and 36 hours, the fed-batch fermentation provided 105.3 g/L of lactic acid in 84 hours, with a total productivity of 1.25 g/L/h and yield of 0.93 g/g. Using the same substrate concentration, the fed-batch fermentation had significantly higher lactic acid yields than the batch fermentation [31].

Xu et al. [53] investigated the repeated-batch mode for lactic acid production while demonstrating the valorization of organic waste. The study also evaluated the stabilization of lactic acid production from a mixture of food waste and waste-activated sludge during long-term fermentation. The relative abundance of the main genera of LAB (*Alkaliphilus, Dysgonomonas, Enterococcus*, and *Bifidobacterium*) in the repeat batch reactor was stabilized (44.5%) and increased compared to the batch reactor (26.2%). This work demonstrated a high yield of lactic acid (0.72 g/g total chemical oxygen demand) through repeated batch fermentation. Furthermore, the lactic acid productivity rate improved by 0.11 g/L/h compared to the batch reactor. Another study used food waste for producing lactic acid using uncontrolled pH fermentation in batch, semicontinuous, and percolation systems. The selectivity values found were 93% for batch mode, 84% for semicontinuous process, and 75% for percolation system on a chemical oxygen demand (COD) basis. Moreover, the work reached lactic acid concentrations of 32, 16, and 15 $gCOD_{lactic\ acid}$/L, respectively [11].

Continuous fermentation works continuously with substrate addition and product removal [10]. Peinemann et al. [54] evaluated the addition of glucoamylase in batch operation. After 24 hours, the titer found was 50 g/L, 63% yield, and 2.93 g/L/h of productivity. With continuous fermentation, there was an increase in titer and yield (69 g/L, 86%, respectively). Although the productivity was lower (1.27 g/L/h) with the addition of glucoamylase, continuous fermentation utilizes the substrate more completely and comprehensively. Furthermore, the authors concluded that both modes of fermentation are economically profitable.

Reactors used for the production of lactic acid by biotechnological processes can involve different configurations, such as stirred tank bioreactor, continuous fixed-bed bioreactor, cascade bioreactor, continuous chemostat cultivation, and membrane bioreactor [10]. The stirred tank bioreactor system is widely used to obtain lactic acid [55, 56]. This reactor allows controlling fermentation conditions in a short production period and has a low risk of contamination. Furthermore, it can be switched between different production tasks with low retrofit costs. However, these bioreactors have high labor costs and have downtime related to sterilization, reinoculation, and cleaning [57].

The high concentration of lactic acid can compromise microbial action in the fermentation process. Thus, bioreactors for the in situ separation of lactic acid have been investigated. Matsumoto and Furuta [58] performed in situ separation of lactic acid by organic solvent extraction fermentation in an air-lift bioreactor. The authors found that lactic acid was obtained and extracted to an organic phase for 600 hours.

Membrane technology has been related as one of the most efficient and energy-saving processes for obtaining lactic acid. Furthermore, the combination of membrane and bioreactor reduces downstream processing [12]. Fan et al. [59] established an anaerobic membrane system for continuous lactic acid fermentation. A membrane module with a rotary vane pump pumped the fermentation broth. A cross-flow filtration system used full recycling mode when no fresh medium was fed. When starting the continuous fermentation, a multichannel peristaltic pump was used to feed the reactor with fresh medium. During continuous operation, the collected permeate was partially removed as a product with the peristaltic pump. The outlet flow was equal to the inlet flow (fresh medium), and the remaining permeate was then recycled back to the reactor (Figure 3.2). The study found that the lactic acid productivity of the membrane system was five times higher than in conventional batch processes [59].

In this context, Taleghani et al. [12] produced lactic acid from whey lactose in a membrane bioreactor and compared it with the performance of the conventional bioreactor. The study found maximum lactic acid yield (89%) at the dilution rate of 0.04 h^{-1} with the membrane bioreactor, while the conventional bioreactor was 47%. These results proved an

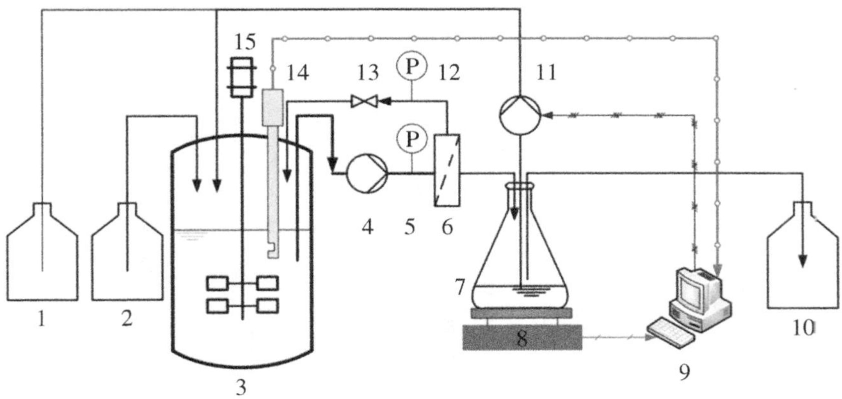

Figure 3.2 Membrane bioreactor system with online biomass monitoring using the optical sensor for continuous fermentation: (1) base, (2) culture medium, (3) stirred tank reactor, (4) rotary vane pump, (5,12) manometers, (6) membrane module, (7) intermediate reservoir, (8) electric balance, (9) control system, (10) product, (11) peristaltic pump, (13) valve, (14) optical sensor, and (15) motor. *Source:* Fan et al. [59] / MDPI / CC BY 4.0.

improvement in the performance of the bioreactor using a membrane. Furthermore, the maximum productivity of lactic acid obtained in the membrane reactor was 6.3 g/L/h, while the conventional reactor provided 3.4 g/L/h. The results showed that integrating membranes with fermentation contributed positively to lactic acid concentration and productivity.

Continuous fermentation through immobilized cells has improved fermentation efficiency, protecting it from toxic compounds [60]. Phenolic substances and associated aldehydes in the lignocellulosic substrate negatively affect microbial growth for lactic acid fermentation [61]. In this sense, a study used cross-linkable F127 bis-polyurethane methacrylate (F127-BUM/T15) to immobilize *Lactobacillus bulgaricus* cells for producing lactic acid. A continuous cell recycling fermentation system (Figure 3.3) was investigated, using glucose and corn stover hydrolysate as carbon sources. Total lactic acid production via immobilized cells was approximately 2000 g after 100 days of fermentation. The lactic acid yield obtained was higher (2.68 g/L/h) than that of free cells (0.625 g/L/h) [60]. On the other hand, this process still has limitations for industrial application due to the low efficiency of materials incorporation and low mechanical performance. Therefore, more research should focus on improving these characteristics so that the immobilization of cells in polymer hydrogels can be applied on a large scale to produce lactic acid with high efficiency.

3.6 Challenges

One of the main challenges in the fermentation process for lactic acid production is to provide the producing microorganism with the ideal condition, mainly of temperature and pH, while obtaining high concentrations, yields, and productivity of lactic acid [38]. The great challenge of this step occurs because as the fermentation progresses, the pH of the

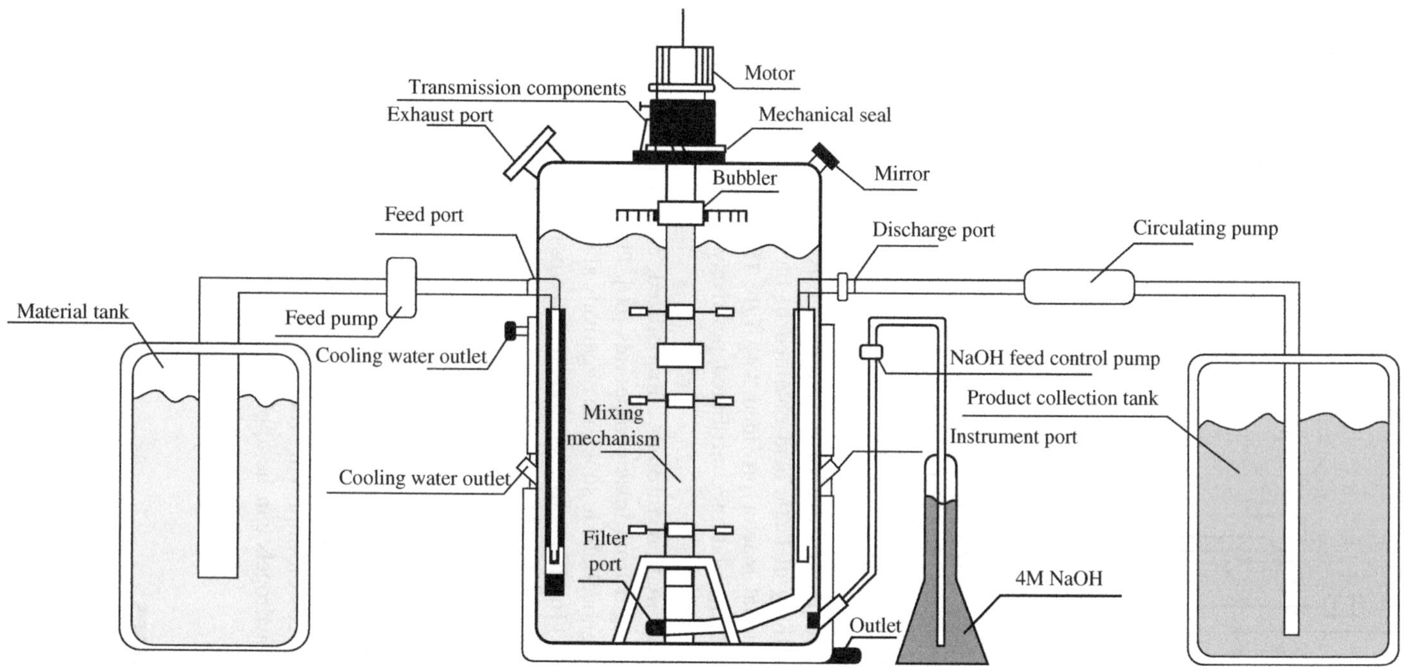

Figure 3.3 Schematic diagram of continuous fermentation by immobilized *L. bulgaricus* T15 cells. *Source:* Guo et al. [60] / MDPI / CC BY 4.0.

medium decreases, which can lead to inhibition and cellular inactivation of the producing microorganism [62, 63]. However, there are strains capable of obtaining lactic acid even in acidic medium [31]. Inhibition of the final product and substrate is another challenge in the context of the lactic acid production process, along with the formation of byproducts that can negatively interfere with product yield [64].

It is recognized that the initial concentration of the inoculum can help reduce process costs [33]. However, assessing the impact of the inoculum is challenging. The inoculum contains lactic acid and other interfering products on the initial pH. Lactic acid produced in a percolation system without pH control needs process optimization to reduce costs. Furthermore, lactic acid recovery can face problems related to ethanol contamination since its boiling point (78 °C) is different from that of lactic acid (122 °C) [11].

The cost of recovery and lactate concentration of the cultured broth is reported in the literature as up to 80% of the total cost of lactic acid production [12, 65, 66]. Thus, alternative systems for the separation process have been investigated. Another challenge in the production of lactic acid by the metabolic route is the costs of substrate and fermentation processes, which constitute about 40–70% of the total cost of the process [7, 33]. Lignocellulosic biomass, food waste, and microalgae have been indicated as alternative substrates to contribute to the economical production of lactic acid. Pretreatment of lignocellulosic materials to release fermentable sugars can release inhibitory compounds to microorganisms. Moreover, the complex composition of lignocellulosic biomass may limit its use for commercial production of lactic acid [7]. The food waste substrate needs pH control and a wide spectrum of fermentation products leading to increased lactic acid separation costs [11]. Therefore, using algae and microalgae for lactic acid production is a promising alternative to reduce production costs since these organisms can use carbon dioxide as a carbon source and other industrial off-gases and wastewater as nutrients [7].

3.7 Conclusions

Producing lactic acid from microbial reactors is a sustainable process with promising results. Fungi and yeasts are producer microorganisms for lactic acid fermentation, but LAB are the most commonly used microorganisms. Lignocellulosic biomass is a frequently mentioned substrate for fermentation. However, it needs pretreatment, which involves costs for the process. Microalgae are promissory substrates in this scenario since their use in fermentation minimizes costs, contributing to the treatment of industrial effluents and reducing the emission of greenhouse gases. The lactic acid production via fermentation can be conducted by different operational modes, with continuous and fed-batch being highlighted. Stirred tank reactors are commonly used to obtain lactic acid. On the other hand, membrane bioreactors have shown promising results concerning productivity and lactic acid yield compared to conventional production processes.

The main bottlenecks are end product inhibition, substrate inhibition, and byproduct formation. Furthermore, the cost associated with processing to recover lactic acid limits its commercial production via microbial fermentation. Using alternative low-cost substrates and economical bioreactors that provide better lactic acid yields and productivity contributes to lactic acid fermentation's economic viability. Besides, optimizing the fermentation process corresponding to the type of reactor in addition to developing new recovery processes are other crucial aspects for obtaining sustainable and economic high-purity lactic acid through a fermentation process.

Acknowledgments

This research was developed within the scope of the Capes-PrInt Program (Process # 88887.310848/2018-00). The authors also are grateful to the Coordenação de Aperfeiçoamento de Pessoal de Nível Superior – Brasil (CAPES) – Finance Code 001.

References

1 Aliwarga, L., Wardani, A.K., Aryanti, P.T.P. et al. (2019). Recent development of lactic acid production using membrane bioreactors. *IOP Conference Series: Materials Science and Engineering* 622: 012023.

2 Juturu, V. and Wu, J.C. (2016). Microbial production of lactic acid: the latest development. *Critical Reviews in Biotechnology* 36: 967–977.

3 Upadhyay, S.K. and Singh, S.P. (eds.) (2021). *Bioprospecting of Plant Biodiversity for Industrial Molecules.* John Wiley & Sons Ltd. doi: 10.1002/9781119718017.

4 Upadhyay, S.K. and Singh, S.P. (eds.) (2023). *Plants as Bioreactors for Industrial Molecules.* John Wiley & Sons Ltd. doi:10.1002/9781119875116.

5 Gupta, B., Revagade, N., Hilborn, J. et al. (2007). Poly (lactic acid) fiber: an overview. *Progress in Polymer Science* 32: 455–482.

6 Komesu, A., Oliveira, J.A.R.D., Martins, L.H.D.S. et al. (2017). Lactic acid production to purification: a review. *BioResources* 12: 4364–4383.

7 Ahmad, A., Banat, F., and Taher, H. (2020). A review on the lactic acid fermentation from low-cost renewable materials: recent developments and challenges. *Environmental Technology and Innovation* 20: 101138.

8 Zhao, Z., Xie, X., Wang, Z. et al. (2016). Immobilization of *Lactobacillus rhamnosus* in mesoporous silica-based material: an efficiency continuous cell-recycle fermentation system for lactic acid production. *Journal of Bioscience and Bioengineering* 121 (6): 645–651.

9 Singh, S.P. and Upadhyay, S.K. (eds.) (2021). *Bioprospecting of Microorganism-Based Industrial Molecules.* John Wiley & Sons Ltd. doi:10.1002/9781119717317

10 Song, L., Yang, D., Liu, R. et al. (2022). Microbial production of lactic acid from food waste: latest advances, limits, and perspectives. *Bioresource Technology* 345: 126052.

11 Bonk, F., Bastidas-Oyanedel, J.-R., Yousef, A.F. et al. (2017). Exploring the selective lactic acid production from food waste in uncontrolled pH mixed culture fermentations using different reactor configurations. *Bioresource Technology* 238: 416–424.

12 Taleghani, H.G., Ghoreyshi, A.A., and Najafpour, G.D. (2018). Thin film composite nanofiltration membrane for lactic acid production in membrane bioreactor. *Biochemical Engineering Journal* 132: 152–160.

13 Ghaffar, T., Irshad, M., Anwar, Z. et al. (2014). Recent trends in lactic acid biotechnology: a brief review on production to purification. *Journal of Radiation Research and Applied Science* 7: 222–229.

14 Narayanan, N., Roychoudhury, P.K., Srivastava, A. et al. (2004). L (+) lactic acid fermentation and its product polymerization. *Electronic Journal of Biotechnology* 7: 167–178.

15 Park, Y.H., Cho, K.M., Kim, H.W. et al. (2010). Method for producing lactic acid with high concentration and high yield using lactic acid bacteria. US Patent 7682814 B2. CJ Cheil Jedang Corporation, Seoul, Republic of Korea.

16 Mimitsuka, T., Sawai, K., Kobayashi, K. et al. (2015). Production of d-lactic acid in a continuous membrane integrated fermentation reactor by genetically modified *Saccharomyces cerevisiae*: enhancement in d-lactic acid carbon yield. *Journal of Bioscience and Bioengineering* 119: 65–71.

17 Kitpreechavanich, V., Maneeboon, T., Kayano, Y. et al. (2008). Comparative characterization of l-lactic acid-producing thermo tolerant *Rhizopus* fungi. *Journal of Bioscience and Bioengineering* 106: 541–546.

18 Qin, J., Zhao, B., Wang, X. et al. (2009). Non-sterilized fermentative production of polymer-grade L-lactic acid by a newly isolated thermophilic strain *Bacillus sp.* 2-6. *PLoS One* 4: e4359.

19 Abedi, E. and Hashemi, M.B. (2020). Lactic acid production – producing microorganisms and substrates sources-state. *Heliyon* 6: e04974.

20 Abdul-Abbas, S.J., Al Alnabi, D.I.B., Al-Hatim, R.R. et al. (2022). Effects of mixed strains of rhizopus oryzae and lactobacillus on corn meal fermentation. *Food Science and Technology* 42: e73621.

21 Reddy, L.V., Kim, Y.M., Yun, J.S. et al. (2015). L-lactic acid production by combined utilization of agricultural bioresources as renewable and economical substrates through batch and repeated-batch fermentation of *Enterococcus faecalis* RKY1. *Bioresource Technology* 209: 187–194.

22 Ghosh, B. and Ray, R.R. (2011). Current commercial perspective of *Rhizopus oryzae*: a review. *Journal of Applied Sciences* 11: 2470–2486.

23 Pleissner, D., Demichelis, F., Mariano, S. et al. (2017). Direct production of lactic acid based on simultaneous saccharification and fermentation of mixed restaurant food waste. *Journal of Cleaner Production* 143: 615–623.

24 Moreira, J.B., Santos, T.D., Duarte, J.H. et al. (2021). Role of microalgae in circular bioeconomy: from waste treatment to biofuel production. *Clean Technologies and Environmental Policy* 1–11.

25 Maia, J.L., Cardoso, J.S., Mastrantonio, D.J.S. et al. (2020). Microalgae starch: a promising raw material for the bioethanol production. *International Journal of Biological Macromolecules* 165: 2739–2749.

26 Xia, M., Shen, Z., Gu, M. et al. (2021). Efficient catalytic conversion of microalgae residue solid waste into lactic acid over a Fe-Sn-Beta catalyst. *Science of the Total Environment* 771: 144891.

27 Olszewska-Widdrat, A., Alexandri, M., López-Gómez, J.P. et al. (2020). Batch and continuous lactic acid fermentation based on a multi-substrate approach. *Microorganisms* 8 (7): 1084.

28 Mladenović, D., Pejin, J., Kocić-Tanackov, S. et al. (2018). Lactic acid production on molasses enriched potato stillage by *Lactobacillus paracasei* immobilized onto agro-industrial waste supports. *Industrial Crops and Products* 124: 142–148.

29 Tsapekos, P., Alvarado-Morales, M., Baladi, S. et al. (2020). Fermentative production of lactic acid as a sustainable approach to valorize household bio-waste. *Frontiers in Sustainability* 1: 1–12.

30 Hassan, S.E.-D., Abdel-Rahman, M.A., Roushdy, M.M. et al. (2019). Effective biorefinery approach for lactic acid production based on co-fermentation of mixed organic wastes by *Enterococcus durans* BP130. *Biocatalysis and Agricultural Biotechnology* 20: 101203.

31 Trakarnpaiboon, S., Srisuk, N., Piyachomkwan, K. et al. (2017). L-lactic acid production from liquefied cassava starch by thermotolerant *Rhizopus microsporus*: characterization and optimization. *Process Biochemistry* 63: 26–34.

32 Jan, C.P. and Daniel, P. (2020). Continuous pretreatment, hydrolysis, and fermentation of organic residues for the production of biochemicals. *Bioresource Technology* 295: 122256.

33 Rawoof, S.A.A., Kumar, P.S., Vo, D.-V.N. et al. (2020). Production of optically pure lactic acid by microbial fermentation: a review. *Environmental Chemistry Letters* 19: 539–556.

34 Zhang, Q., Yu, Z., Wang, X. et al. (2018). Effects of inoculants and environmental temperature on fermentation quality and bacterial diversity of alfalfa silage. *Animal Science Journal* 89 (8): 1085–1092.

35 Demichelis, F., Fiore, S., Pleissner, D. et al. (2018). Technical and economic assessment of food waste valorization through a biorefinery chain. *Renewable and Sustainable Energy Reviews* 94: 38–48.

36 Abdel-Rahman, M.A., Tashiro, Y., Zendo, T. et al. (2015). *Enterococcus faecium* QU 50: a novel thermophilic lactic acid bacterium for high-yield l-lactic acid production from xylose. *FEMS Microbiology Letters* 362: 1–7.

37 Panesar, P.S., Kennedy, J.F., Knill, C.J. et al. (2010). Production of L(+) lactic acid using *Lactobacillus casei* from whey. *Brazilian Archives of Biology and Technology* 53: 219–226.

38 Abdel-Rahman, M.A. and Sonomoto, K. (2016). Opportunities to overcome the current limitations and challenges for efficient microbial production of optically pure lactic acid. *Journal of Biotechnology* 236: 176–192.

39 Bouabidi, Z.B., El-Naas, M.H., and Zhang, Z. (2019). Immobilization of microbial cells for the biotreatment of wastewater: a review. *Environmental Chemistry Letters* 17: 241–257.

40 Ma, J., Li, Y., Jin, D. et al. (2021). Reasonable regulation of carbon/nitride ratio in carbon nitride for efficient photocatalytic reforming of biomass-derived feedstocks to lactic acid. *Applied Catalysis B: Environmental* 299: 120698.

41 Swiatkiewicz, J., Slezak, R., Krzystek, L. et al. (2021). Production of volatile fatty acids in a semi-continuous dark fermentation of kitchen waste: impact of organic loading rate and hydraulic retention time. *Energies* 14 (11): 2993.

42 Palomo-Briones, R., Razo-Flores, E., Bernet, N. et al. (2017). Dark-fermentative biohydrogen pathways and microbial networks in continuous stirred tank reactors: novel insights on their control. *Applied Energy* 198: 77–87.

43 Pagana, I., Morawicki, R., and Hager, T.J. (2014). Lactic acid production using waste generated from sweet potato processing. *International Journal of Food Science & Technology* 49 (2): 641–649.

44 Ma, X., Gao, M., Wang, N. et al. (2021). Lactic acid production from co-fermentation of food waste and spent mushroom substance with *Aspergillus niger* cellulase. *Bioresource Technology* 337: 125365–125365.

45 Abdel-Rahman, M.A., Tashiro, Y., Zendo, T. et al. (2016). Highly efficient l-lactic acid production from xylose in cell recycle continuous fermentation using *Enterococcus mundtii* QU 25. *RSC Advances* 6 (21): 17659–17668.

46 Zheng, J., Gao, M., Wang, Q.H. et al. (2017). Enhancement of L-lactic acid production via synergism in open co-fermentation of *Sophora flavescens* residues and food waste. *Bioresource Technology* 225: 159–164.

47 Li, X., Chen, Y.G., Zhao, S. et al. (2015). Efficient production of optically pure L-lactic acid from food waste at ambient temperature by regulating key enzyme activity. *Water Research* 70: 148–157.

48 Li, J., Zhang, W.J., Li, X. et al. (2018). Production of lactic acid from thermal pretreated food waste through the fermentation of waste activated sludge: effects of substrate and thermal pretreatment temperature. *Bioresource Technology* 247: 890–896.

49 Sun, Y., Xu, Z., Zheng, Y. et al. (2019). Efficient production of lactic acid from sugarcane molasses by a newly microbial consortium CEE-DL15. *Process Biochemistry* 81: 132–138.

50 Macedo, J.V.C., Rank, F.F.B., Escaramboni, B. et al. (2020). Cost-effective lactic acid production by fermentation of agro-industrial residues. *Biocatalysis and Agricultural Biotechnology* 27: 101706.

51 Chen, P.-T., Hong, Z.-S., Cheng, C.-L. et al. (2020). Exploring fermentation strategies for enhanced lactic acid production with polyvinyl alcohol-immobilized *Lactobacillus plantarum* 23 using microalgae as feedstock. *Bioresource Technology* 308: 123266.

52 Pejin, J., Radosavljević, M., Pribić, M. et al. (2018). Possibility of L-(+)-lactic acid fermentation using malting, brewing, and oil production by-products. *Waste Management* 79: 153–163.

53 Xu, X., Zhang, W., Gu, X. et al. (2020). Stabilizing lactate production through repeated batch fermentation of food waste and waste activated sludge. *Bioresource Technology* 300: 122709.

54 Peinemann, J.C., Demichelis, F., Fiore, S. et al. (2019). Techno-economic assessment of non-sterile batch and continuous production of lactic acid from food waste. *Bioresource Technology* 289: 121631.

55 Germec, M., Karhan, M., Bialka, K.L. et al. (2018). Mathematical modeling of lactic acid fermentation in bioreactor with carob extract. *Biocatalysis and Agricultural Biotechnology* 14: 254–263.

56 Neto, D.P.C., Pereira, G.V.M., Finco, A.M.O. et al. (2018). Efficient coffee beans mucilage layer removal using lactic acid fermentation in a stirred-tank bioreactor: kinetic, metabolic and sensorial studies. *Food Bioscience* 26: 80–87.

57 Villadsen, J., Nielsen, J., and Lidâen, G. (2011). *Bioreaction Engineering Principles*. New York: Springer.

58 Matsumoto, M. and Furuta, H. (2018). In situ extractive fermentation of lactic acid by *Rhizopus oryzae* in an air-lift bioreactor. *Chemical and Biochemical Engineering Quarterly* 32 (2): 275–280.

59 Fan, R., Ebrahimi, M., and Czermak, P. (2017). Anaerobic membrane bioreactor for continuous lactic acid fermentation. *Membranes* 7 (2): 26.

60 Guo, Y., Wang, G., Chen, H. et al. (2022). Continuous fermentation by *Lactobacillus bulgaricus* T15 cells immobilized in cross-linked F127 hydrogels to produce D-lactic acid. *Fermentation* 8 (8): 360.

61 Zhang, B., Wu, L., Wang, Y. et al. (2022). Re-examination of dilute acid hydrolysis of lignocellulose for production of cellulosic ethanol after de-bottlenecking the inhibitor barrier. *Journal of Biotechnology* 353: 36–43.

62 Lee, J.M., Jang, W.J., Lee, E.W. et al. (2020). β-Glucooligosaccharides derived from barley β-glucan promote growth of lactic acid bacteria and enhance nisin Z secretion by *Lactococcuslactis*. *LWT* 122: 109–114.

63 Lin, H.T.V., Huang, M.Y., Kao, T.Y. et al. (2020). Production of lactic acid from seaweed hydrolysates via lactic acid bacteria fermentation. *Fermentation* 6: 37.

64 Turner, T.L., Kim, E., Hwang, C. et al. (2017). Conversion of lactose and whey into lactic acid by engineered yeast. *Journal of Dairy Science* 100: 124–128.

65 Hulse, J.H. (2004). Biotechnologies: past history, present state and future prospects. *Trends in Food Science & Technology* 15: 3–18.

66 Hu, Y., Kwan, T.H., Daoud, W.A. et al. (2017). Continuous ultrasonic-mediated solvent extraction of lactic acid from fermentation broths. *Journal of Cleaner Production* 145: 142–150.

4

Advancement in the Research and Development of Synbiotic Products

Anna María Polanía[1], Alexis García[1], and Liliana Londoño[2]

[1] *School of Food Engineering, Faculty of Engineering, Universidad del Valle, Tuluá, Valle del Cauca, Colombia*
[2] *BIOTICS Group, School of Basic Sciences, Technology and Engineering, Universidad Nacional Abierta y a Distancia – UNAD, Bogota, Colombia*

4.1 Introduction

One of the ways to protect the host from invading microorganisms is to maintain a stable gut community. However, sometimes alterations occur when there are changes in the diet, infections, age, consumption of antibiotics, so that the intestinal microbiota is affected, presenting metabolic, pathogenic, or inflammatory conditions that can trigger intestinal diseases, inflammatory diseases, metabolic syndrome, atopy, or even colorectal cancer [1].

The concept of synbiotics was initially reported 25 years ago, it was simply the idea of combining nondigestible fermentable foods (prebiotics) with probiotics, and for this reason they were defined as "mixtures of probiotics and prebiotics that beneficially affect the host." The word is composed of the prefix Greek *'syn'* meaning *'together'* and the suffix *'biotic'* meaning *'pertaining to life'*. To establish a more adequate use of the word 'synbiotic' the International Scientific Association for Probiotics and Prebiotics (ISAPP) in 2019 established synbiotic as "a mixture comprising live microorganisms and substrate(s) selectively utilized by host microorganisms that confer a health benefit to the host" [2]. To date, two categories of synbiotics have been defined: complementary and synergistic. The first is constituted by a prebiotic and a probiotic that together provide health benefits but do not necessarily have to be co-dependent in their function; the amount to be used for each component should be a dose that has been shown to be effective individually. In the second group, the co-administered live microorganisms selectively utilize the substrate contained in the synergists [3].

It has been evidenced in various research that some microorganisms found in the gastrointestinal tract manage to play important roles in the maintenance of health. Synbiotics fall into this category and could be used as therapeutic strategies to improve

Microbial Bioreactors for Industrial Molecules, First Edition.
Edited by Sudhir P. Singh and Santosh Kumar Upadhyay.

human health in various clinical conditions. Their importance lies in the fact that they could help to address two targets in Goal 3 of Sustainable Development Goals (SDGs): prevention of premature mortality from noncommunicable diseases and promotion of mental health and well-being [4].

The purpose of this chapter is to provide a clear definition of synbiotics, its mechanisms of action, and possible effects on different health conditions, detailing the different types of bioreactors for its production, the encapsulation methods to improve its properties and stability, and its applications in different fields due to that can be used as therapeutic agents in the treatment of some skin diseases such as acne, melasma, and atopic dermatitis; in the same way they can be used in animal feed where can increase productivity, reduce infections by pathogens and improve the quality of the products where they are applied; however, its main application is in the development of functional foods due to the health benefits it brings to the consumer.

4.2 Probiotics, Prebiotics, and Synbiotics

4.2.1 Probiotics

Probiotic is derived from a Greek word meaning "for life" and its definition has undergone many modifications, it was introduced by Vergin, who proposed that these microorganisms were favorable for the gut microflora. The latest definition was jointly proposed by the Food and Drug Administration (FDA) and the World Health Organization (WHO), who state that probiotics are "live microorganisms that, when administered in adequate amounts, confer a health benefit on the host" [5,6]. The widely used probiotic microorganisms are of the genus *Lactobacillus* species *reuteri* and *rhamnosus*, *Bacillus* coagulans, *Escherichia coli strain Nissle 1917*, *bifidobacteria*, some strains of *Lactobacillus casei*, and acidophilus-group, also included the yeast *Saccharomyces boulardii* and some strains of *enterococcus* such as *Enterococcus faecium SF68* [7]. Some of the advantages of probiotics in the human organism is the effect it develops on the microbiota since it generates an adequate balance between pathogenic microorganisms and the bacteria needed for the proper functioning of the organism [8]. Some research shows that these microorganisms have health benefits such as improving intestinal transit, reducing the risk of obesity, stimulating mineral absorption, and lowering postprandial glucose levels [9–11]. These live microorganisms can be used to produce functional foods and the preservation of some types of products. Due to the positive effect that it generates, the probiotics can be useful to reestablish the natural microbiota, and for this reason, they are often used in the diet of people who have undergone therapies with antibiotics [12].

Probiotics are also regulated as they must be safe for human and animal health. In the United States, the regulatory organization is the FDA (Food and Drug Administration), which guarantees that the microorganisms used have Generally Regarded as Safe (GRAS) status. In Europe, the EFSA (European Food Safety Authority) oversees regulation and issues the concept of Qualified Presumption of Safety (QPS), which implies additional criteria for the safety evaluation of bacterial supplements, i.e. it includes the history of safe use and absence of risk of acquired resistance to antibiotics [13,14].

4.2.2 Requirements and Selection Criteria for Probiotic Strains

To optimize the use of probiotic strains, it is vital to perform safety evaluations, but this is not an easy activity to carry out [15]. The action of probiotics when used as microbial additives in animal feeds is not completely known, but when microorganisms adhere to the gastrointestinal tract, they can survive adverse conditions and generate beneficial effects on the protection and stability of the intestinal flora. They are also involved in digestive and metabolic processes and immune system response. Therefore, the characteristics of probiotics improve animal health, increase productivity, and improve host immunity [16]. When selecting these microorganisms, strains and even genera that have demonstrated beneficial or specific effects are included. The evaluation of probiotics is based on the safety and risk–benefit ratio established with the use of a specific probiotic strain. The probiotics used to produce probiotic formulas in animals are isolated from individuals belonging to the species for which they are destined, since the health benefits generated are specific to each species. Because of this, the biological material obtained can be adapted to various conditions found in the alimentary tract of the treated animal species [17].

Regarding the selection of layers for human use, a systematic approach should be followed to ensure that consumers have a safe probiotic. Some characteristics to consider are origin, genus identification, strain, stability in the gastrointestinal tract, and viability. For the evaluation of probiotics in foods, the guidelines provided by "ICMR-DBT" should be considered, which mention the steps to identify the functionality of probiotics. In the case of probiotics for human use, it is recommended that they be isolated from the human intestine or breast milk, since this way they adhere better to the intestinal wall and are safer. They can also be isolated from fermented products such as gundruk, sinki, khalpi, soidon, goyang, among others. To determine the probiotic potential of the microorganism, it is necessary to identify the strain down to the genus level. In addition to primary identification techniques, molecular and genetic techniques such as 16SrRNA DNA sequencing, fatty acid methyl ester (FAME), DNA-DNA hybridization technique, polymerase chain reaction (PCR) amplification and DNA and RNA hybridization should be performed [18]. The conditions that a microorganism must meet to be used as a probiotic are presented in Figure 4.1.

4.3 Prebiotics

A prebiotic can be defined as a "substrate that is selectively utilized by host microorganisms conferring a health benefit." Prebiotics are mainly nondigestible fibers that support the growth of some genus of microorganisms in the colon benefiting the health of the host, including some strains of *bifidobacteria* and *lactobacillus* [23–25]. Prebiotics can be used as an alternative option to probiotics or consumed in a complementary manner. It is important to highlight that some prebiotic strains can promote the growth of multiple indigenous intestinal bacteria. In addition, they have a great capacity to modify strains and species within the intestinal microbiota that are not easy to identify a priori [26]. Also, the prebiotics can exert benefits in the urogenital tract, oral cavity, and skin [27, 28].

Host-associated stress tolerance ability

When ingested the probiotic should be able to be stable in the different parts of the digestive tract and withstand the stress conditions of the human body, such as the action of some digestive enzymes like pepsin chemotrypsin, amylase, and lysozyme. Likewise, the strains must tolerate acid and bile, and thermal shocks resulting from changes in internal body temperature [18].

Adhesive properties

To define the probiotic potential of the strain, it is important to define its adhesion to the intestinal wall, as this ensures that the bacteria are not washed away, but self-aggregate to increase their cell density and biomass in the digestive tract. This also ensures that the strain has a better interaction with epithelial cells to generate host-associated functional effects [18].

Antimicrobial activity

The strains should have the ability to survive against potential pathogenic microorganisms present in the intestine. Probiotic strains should prevent the adhesion of pathogens to epithelial cells in the organism by secretion of antibodies, lactic acid, bacteriocin sakacin A, acticins, and alyteserin-1a [19].

Immune modulating response

The probiotic should produce metabolites that stimulate the maturation and function of immune cells. Bacteria stimulate immunoglobulin secretion and cytokine production. On the other hand, the strain should be selected according to the target host to enhance the systemic immune response [20].

Host-associated functional criteria

One of the main characteristics of probiotics is the health benefits it provides to the organism. Some of the properties that have been evidenced when consuming them are antidepressant, antidiabetic, antioxidant, antiobesity, anticolestrol, anticancer activity, in addition to reducing gastrointestinal diseases, diarrhea, gastroenteritis, and children's allergie. [21, 22].

Good technological properties

A probiotic can effectively offer benefits to the host if the strain can survive the storage period and remain efficient and viable. Ideally, probiotic bacteria should be able to grow in nutritional supplements (ideally inexpensive fermentation media), food matrices, and microaerophilic conditions. Another important aspect is that the strain can withstand different physical handling techniques during product processing without losing its efficiency and viability [18].

Figure 4.1 Characteristics of probiotics.

Much research has reported the effect of prebiotics in the prevention or delay of cardiovascular diseases with hypercholesterolemia, osteoporosis, obesity, diabetes, intestinal inflammation, and even gastrointestinal infections [24, 29]. Prebiotics are not easily degraded through the intestinal tract due to the absence of degrading microorganisms or digestive enzymes. However, when these microorganisms pass through the intestine and settle in the colon, the microbiota break them down and the food necessary for their maintenance is obtained, generating small molecules of carbohydrates and short-chain fatty acids, whose function is to provide energy to nearby bacteria [30].

Some foods that constitute a potential source of probiotics are fruits, cereals, vegetables, and other edible plants. These foods are artichokes, tomatoes, asparagus, bananas, garlic, green vegetables, onions, flax seeds, oats, wheat, and barley [26, 31]. The intake of these products prevents constipation and diarrhea problems, helps in the production of B vitamins and improves the immune system, reduces the probability of developing osteoporosis due to the increase in calcium absorption, reduces the symptoms of inflammatory bowel disease and therefore reduces the factors that lead to colon cancer, reduces the risk of diabetes, and improve the metabolism of carbohydrates [24, 29]. In addition to these benefits, the intake of products containing probiotics leads to a decrease in the population of pathogenic bacteria present in the intestinal tract, particularly *Campylobacter jejuni*, *Salmonella* spp., and *E. coli* [30]. The most recognized prebiotics are glucooligosaccharides (GuOS), fructooligosaccharides (FOS), inulin, xylooligosaccharides (XOS), galactooligosaccharides (GaOS), lactulose, maltooligosaccharides (MO), isomaltooligosaccharides (IMO), lactosaccharose, lactulosucrose, raffinose, fructans, maltodextrin, polydextrose, sorbitol, and gum arabic [25, 27, 32–36]. Lactulose accounts for about 40% of the oligosaccharides produced. Fructans such as oligofructose and inulin are widely used in connection with other probiotic species [26, 37]. In fact, FOS, inulin, and GaOS are commonly employed in many food products, include baby foods [27, 30].

4.3.1 Requirements and Selection Criteria for Prebiotic Strains

According to Wang [38], with the goal of classifying the foods as prebiotics is possible on establishing five criteria (Figure 4.2). The first one states that prebiotics are not digested, i.e. they are only partially digested in the upper tract. For this reason, prebiotics settle in the colon and are selectively fermented by bacteria that are beneficial (requirement of the second criterion) [39]. When the fermentative process is developed, there is an increase in the amount and/or production of short-chain fatty acids (SCFAs), the fecal mass increases, and the pH in the colon is reduced, leading to a reduction of fecal enzymes and nitroso products, thus strengthening the immune system [40] and benefiting the host (requirement of the third criterion). Another criterion that is often considered is the activity of bacteria in the gut and/or selective growth stimulation that are directly related to health protection. The last criterion states that prebiotics must have the ability to withstand food processing conditions without being altered, degraded, or chemically modified, as well as being accessible to bacterial metabolism in the gut [26].

One of the factors determining fermentative activity in prebiotics is the stimulation of the gut microbiota, which in turn influences the level of SCFA and confers health benefits to the host [41]. Likewise, prebiotics reduce intestinal pH, preserving the osmotic retention of water in the intestine. However, it is important not to over-consume probiotics because an excess of

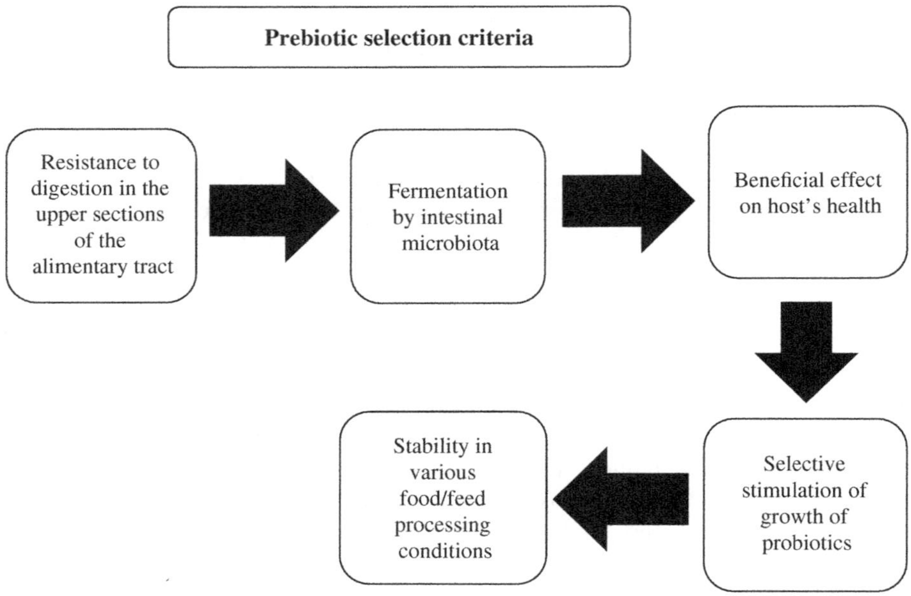

Figure 4.2 Requirements for potential prebiotics. *Source:* Markowiak and Śliżewska [26] / MDPI / CC BY 4.0.

probiotics can cause diarrhea and flatulence, an effect that is not generated when probiotics are consumed in excess [40]. Prebiotics can be ingested on a long-term basis and for the purpose of improving health. They have the advantage of not generating allergies and protect the intestinal flora when antibiotics are ingested. It is important to mention that the elimination of pathogens is better when antibiotic treatment is applied; however, these also manage to destroy part of the intestinal flora, for this reason although the use of prebiotics has less effect with pathogenic bacteria, it presents the advantages mentioned above [26, 42].

4.4 Synbiotics

Synbiotics can be defined as "a mixture composed of live microorganisms and substrate(s) used selectively by host microorganisms conferring a health benefit to the host" [3]. Synbiotics can be used to promote the survival of microorganisms that benefit the intestinal flora, when added to feed or food, and increase the production of specific bacterial strains of the intestinal tract [43]. Synbiotics are identified as healthy due to the beneficial effect they generate and are usually associated with the individual combination of prebiotics with probiotics [23]. As a consequence of the numerous combinations that can be obtained, a wide interest has been generated in recent years for the application of synbiotics due to the excellent modulation of the microbiota [26, 44].

Synbiotics can act in two ways: by improving the health of the host after ingesting a mixture of probiotic and prebiotic strains or by promoting autochthonous beneficial microflora, such as bifidobacteria, after ingesting prebiotics alone. The efficacy of synbiotics has been demonstrated in both animals and humans by demonstrating synergistic effects on host health, i.e. synbiotics can exert the following benefits [45]:

- Improve survival and implantation of probiotics in the colon.
- Stimulate the growth or activate the metabolism of beneficial bacteria for the colon (probiotics).
- Improve the microbial composition in the gastrointestinal tract.

Probiotic strains such as *Bifidobacteria* spp., *B. coagulans*, *Lactobacilli*, and *S. boulardii* are usually used in synbiotic formulations; and prebiotic strains include oligosaccharides such as GOS, FOS, inulin, prebiotics from natural sources such as yacón and chicory roots. Other benefits that the consumption of synbiotics can generate in humans are the increase of *lactobacilli* and *bifidobacteria* levels, leading to a balance of the intestinal microbiota, improving the immunomodulatory capacity, improving liver function in cirrhotic patients, reducing incidences of nosocomial infections in surgical patients, and preventing bacterial translocation [46, 47].

4.4.1 Synbiotic Selection Criteria

One of the first considerations when a synbiotic formula is to be developed is to select the right prebiotics and probiotics, i.e. that each when used individually should exert a beneficial effect on the health of the host [26]. "If a product contains a prebiotic and probiotic that has evidence that each microorganism has benefits individually, but not as a whole, it cannot be called synbiotic" [3]. When selecting synbiotic microorganisms, it must be clear which properties will be beneficial to probiotic metabolism. Formulations can be called synbiotic as long as a selective stimulation of the growth of other microorganisms can be assured [14].

Synbiotics must ensure consistent safety and performance. Microorganisms that are part of the synbiotic should have a publicly available annotation and genomic sequence, as well as be tested to determine that no genes present safety concerns (e.g. transferable antibiotic resistance or toxin production), and current taxonomic nomenclature should be used to name and designate a traceable strain. These strains should belong to internationally recognized culture collections and be available to the scientific community for research purposes [3].

4.4.2 Mechanism of Action of Synbiotics

Mixing probiotics with prebiotics generates a modulation of metabolic activity in the intestine and also preserves the intestinal biostructure, inhibiting the potential for pathogens in the gastrointestinal tract and the development of beneficial microbiota [23]. Synbiotics provide minors concentrations of undesirable metabolites, as well as inactivation of carcinogenic substances and nitrosamines. Therefore, using these microorganisms leads to an increase in carbon disulfides, SCFAs, methyl acetates, and ketones, thus generating a positive effect on the health of the host [48]. Regarding the therapeutic efficacy, synbiotics present anticancer, antiallergic, and antibacterial effects. Likewise, they counteract intestinal diseases and prevent constipation and diarrhea. They can also be effective in preventing osteoporosis, reducing blood fat and sugar levels, treating brain disorders associated with abnormal liver function and regulating the immune system [46]. In Figure 4.3, it is possible to appreciate the forms of action of synbiotics, according to their modification of the intestinal microbiota with appropriately selected probiotics and prebiotics.

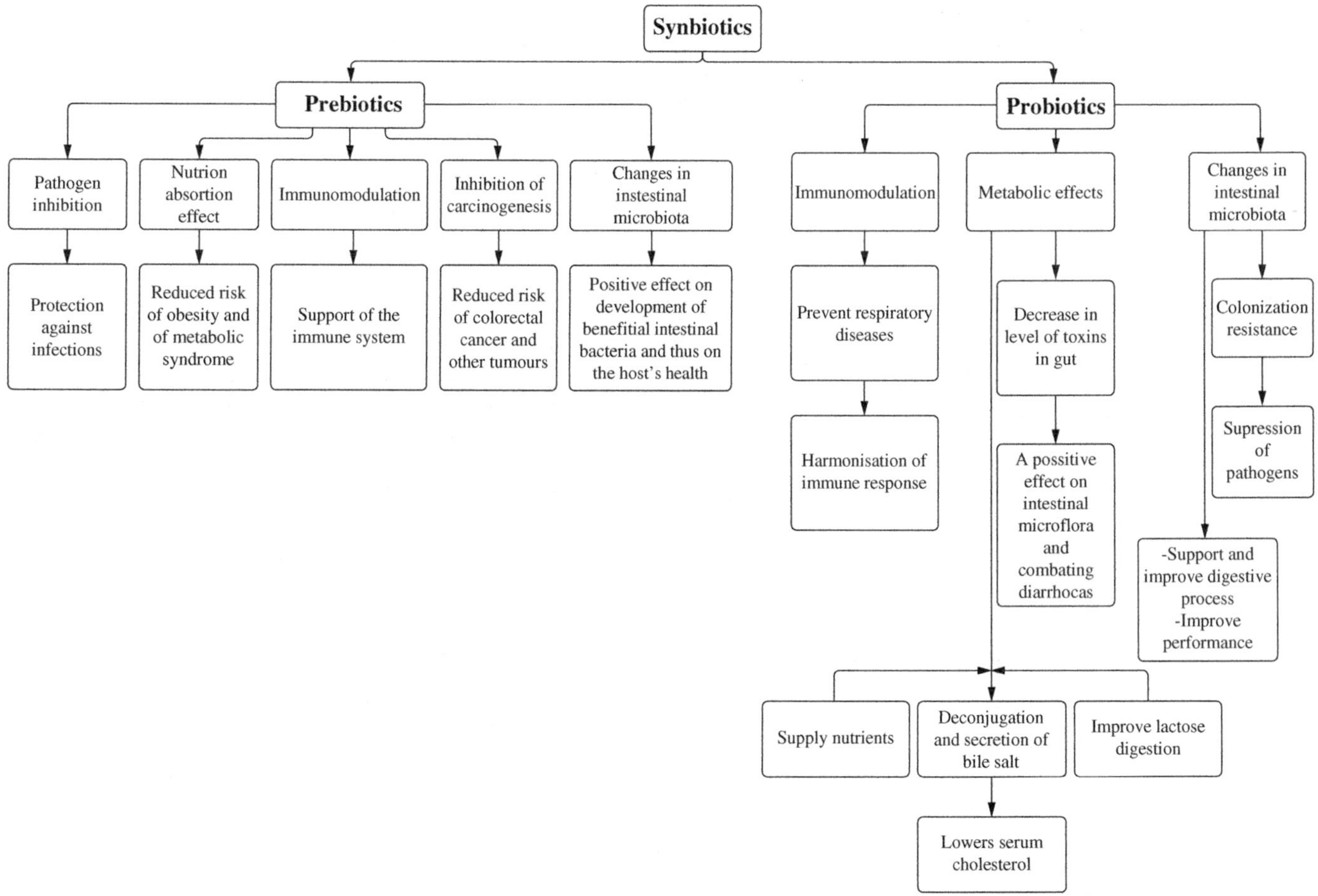

Figure 4.3 Mechanisms of action of synbiotics and their effects. *Source:* Markowiak and Śliżewska [26] / MDPI / CC BY 4.0.

4.5 Health Benefits from Synbiotics

The role of synbiotics on human health is a field of great interest and ongoing research. Multiple claims about the potential of synbiotics for improving different medical conditions have been made over the years, these include benefits on intestinal health, treatment, and prevention of different diseases (cardiac, biliary, liver, among others), and enhancement in the relationship between the gastrointestinal tract and the central nervous system, which can reduce the probability of presenting disorders such as depression, Alzheimer's disease, and Parkinson's disease [49].

One of the main challenges is the confirmation of the health benefits in the target host of complementary and synergistic synbiotics. Also, evidence of selective use by the co-administered live microorganism (synergistic synbiotic) or by the endogenous microbiota (complementary synbiotic) must also be generated.

Table 4.1 presents some recent studies of the effect of synbiotics over different health conditions. Concentration and composition of synbiotic (strains of probiotics and type pf prebiotic), characteristic of the subjects of the study, and statistical significance are considered.

Table 4.1 Health benefit claims and synbiotics.

Health benefit claim	Synbiotic used	Group study	Results	Reference
Reduction of infection in colorectal cancer post-surgery	Probiotics: *Lactobacillus acidophilus* NCFM (10^9) *Lactobacillus rhamnosus* HN001 (10^9) *Lactobacillus paracasei* LPC-37 (10^9) *Bifidobacterium lactis* HN019 (10^9) Prebiotics: Fructooligosaccharides (6 g)	Adults Intervention group (synbiotics) $N = 49$ Control group (placebo), $N = 42$	21.4% (9 out of 42 patients) in the control group presented surgical wound infection compared with only 2.0% (1 out of 49 patients) due to the administration of the synbiotic ($p = 0.002$)	[50]
Improvement of gastrointestinal discomfort and inflammatory status	Probiotic: *Bifidobacterium animalis lactis* Vesalius 002 (LMG P-28149) 5×10^9. Prebiotic: Fructo-oligosaccharides (FOS, ACTILIGHT, 4.95 g)	Middle-aged adults with transit disorders (85% women) Intervention group (synbiotics) $N = 13$ Control group (placebo) $N = 14$	After a 30-days intervention the number of days with gastrointestinal discomfort were reduced in the intervention group compared with the control group ($p < 0.05$)	[51]

(*Continued*)

Table 4.1 (Continued)

Health benefit claim	Synbiotic used	Group study	Results	Reference
Obesity treatment	Probiotics: *L. acidophilus, L. casei Bifidobacterium bifidum* (2×10^9 *CFU/g each)* Prebiotics: *Inulin (0.8 g)*	Overweight adults (body mass index between 25 and 35 kg/m^2) Intervention group (synbiotics) $N = 30$ Control group (placebo), $N = 29$	A significant reduction in body weight was found between the intervention group and control ($p = 0.03$) Changes in blood lipid profile between intervention and control group were observed. A statically significant difference in total cholesterol was found ($p = 0.01$)	[52]
Avoid post-partum complications	*Probiotic:* *L. acidophilus* strain T16 *(IBRC-M10785) L. casei* strain T2 *(IBRC-M10783)* *Bifidobacterium bifidum strain T1 (IBRC-M10771)* (2×10^9 *CFU/g each)* Prebiotic: *0.8 g inulin (HPX)*	Pregnant Women with Gestational Diabetes Intervention group (synbiotics) $N = 30$ Control group (placebo), $N = 30$	It was found that newborn hospitalization and hyperbilirubinemia decreased by synbiotic supplementation ($p = 0.006$) (1 case in the intervention group compared with 9 in the placebo)	[53]
Reduction of Stress related biomarkers	Probiotic: *5.0* $\times 10^9$ *CFU of L. paracasei HII01 and 5.0 × 109 CFU of Bifidobacterium animalis subsp. Lactis* (1×10^{10} *CFU/day)* Prebiotic: *5 g Galacto-oligosaccharides (GOS-700-P)* *5g Oligofructose (Orafti®P95)*	Non stressed subjects ($n = 12$) Stressed subjects ($n = 19$) Both groups received synbiotic preparation	It was found after a 12-week intervention, a significant reduction of cortisol, a hormone associated with stress states, in both the non-stressed ($p = 0.015$) and stressed group ($p = 0.033$) This was confirmed with an evaluation of the stressed state of the participants At the initial state of the study, 94% of the stressed group was classified in a mild stress sate after the intervention it reduced to 52%	[54]

Synbiotics effect over the postoperative recovery and reduction of possible infections is a subject of great interest as is shown in the study of [50]. During surgery the gut microbiota could be seriously affected, which could lead to proliferation of harmful bacteria, infections, and in some cases bacteremia (presence of bacteria in the bloodstream). Supplementation with synbiotic becomes a reliable strategy to contribute with a better recovery that promotes the growth of beneficial microorganism and its metabolites.

Modification of gut microbiota can also be helpful to treat non transmissible diseases as obesity or alleviating stress symptoms that worsen preexistent conditions or lead to stress-related disorders. For obesity, the manipulation of the gut microbiota could benefit weight loss due to its effect over metabolic variables such as cholesterol, blood lipid profile, insulin among others [55]. Stress leads to many forms of physical and mental problems; using synbiotic to reduce the biological response of biomarkers associated with stress is a valid form to improve the well-being of the patient [54]. It is possible to see the wide scope of possible application of synbiotic in health and the necessity for continuous research and verification of its actual benefits to develop efficient treatment alternatives for disease control.

4.6 Bioreactor Design for Synbiotic Production

According to the International Scientific Association for Probiotics and Prebiotics (ISAPP), there are three definitions for synbiotics: (i) synbiotic defined as "a mixture comprising live microorganisms and substrate(s) used selectively by host microorganisms that confers a health benefit to the host" with host microorganisms that are considered resident or autochthonous as probiotics; (ii) complementary symbiotic that is comprised of a prebiotic that assists the indigenous host microorganisms with a combined probiotic; finally (iii) synergistic synbiotic, in which the co-administered microorganisms selectively employ the substrate [3]. Regardless of the three definitions, for a synbiotic to provide a health benefit to the host, it must overcome all the barriers of the gastrointestinal tract, reaching the minimum number of viable cells capable of colonizing the intestine, so one of the key factors during the production of synbiotics is to achieve, among others, a high viability. Most synbiotics are produced through fermentation, so the selection of the type of fermenter, the definition of the operating conditions and the establishment of the culture medium are key variables to increase the efficiency of their production.

Most synbiotics are produced through fermentation processes. Fermentation is defined as a natural decomposition process in which complex organic substances are converted into simple compounds through chemical transformations by the action of biological catalysts [56] or industrially as a technology that using microorganisms or their parts can generate value-added products [57]. At a general level, fermentation is divided into two types of submerged fermentation (SmF) and solid-state fermentation (SSF). Solid fermentation is a process carried out on solid substrates in the absence of free water but with the necessary amount of water to allow the growth of microorganisms [58], while submerged fermentation involves the immersion of microorganisms in an aqueous solution containing all the nutrients necessary for growth [59]. Fermentation is carried out in fermenters or bioreactors, in which the appropriate conditions for the growth of microorganisms are guaranteed. During the fermentation process, different factors influence the growth of the microorganisms, which must be controlled to achieve an efficiency that allows industrial scale-up in the production of synbiotics. Among the limiting factors are mass transfer, temperature, pH, oxygen flow, among others, which can be modified by adding control systems to the bioreactors or by improving some physical systems such as agitation in the case of submerged fermentation [60, 61].

To obtain both probiotics and prebiotics through submerged fermentation, traditional bioreactors can be used, such as flask [62, 63], stirred tank fermenters [64, 65], tower fermenters, airlift fermenter, bubble column reactor, fluidized bed reactor, packed bed reactors, bioreactor [60, 66], or novel bioreactors such as the membrane-based [67, 68]. These bioreactors can operate in different modes of operation batch, continuous, repeated batch, and fed batch [68]. The selection of the fermentation mode depends on the ratio of substrate consumption to biomass and products [60]. For the selection of the type of bioreactor, the mode of operation, and the control variables, it is essential to know the reactions that take place during fermentation, mainly the behavior of the microbial cells, which can respond to different factors such as cell growth, substrate concentration, nutrient depletion, absorption rates, among others [61]. For example, during the production of fructo-oligosaccharide (FOS)-type prebiotics, sucrose is the main substrate in the medium for FOS synthesis, so its initial concentration is a process parameter. As a result of sucrose consumption and β-fructofuranosidase activity, glucose appears whose concentration can inhibit the fructosyl-transferring reaction and thus affect FOS synthesis. In this case, controlling the concentration of glucose in the medium can favor the yields of the process. In this sense, different strategies have been studied to improve the fermentation process, among which it has been found that the operation of the bioreactor in fed-batch mode can increase the yields in the production of FOS. The fed-batch operation is used in the industry to overcome the disadvantages derived from the inhibition of the reactions by substrate, thus increasing the concentrations of viable cells. During this one of the main control variables is the substrate feeding rate, which is directly related to the productivity [69].

In recent years, another bioreactor used in the production of synbiotics is the membrane bioreactor, which integrates a membrane module (microfiltration, ultrafiltration, nanofiltration, electrodialysis) to the traditional fermenter allowing simultaneous production and purification of a given product while maintaining cell growth. The membrane bioreactor can also operate in different modes such as batch, semi-continuous, continuous, immobilized, and membrane recycle [68]. Considering the advantages of this type of bioreactors, Burghardt [67] evaluated the enzymatic catalysis process to produce FOS, finding that the membrane reactor allows the constant removal of glucose that causes inhibition of the enzymatic reactions and maintains an adequate concentration of sucrose as the main substrate of the process. However, during the process the membrane pores were clogged possibly due to the combination of proteins with polysaccharides, so it is necessary to use additional systems that allow membrane cleaning.

In solid-state fermentation, one of the great challenges has been the design of bioreactors at industrial level to obtain better process yields. In this case, bioreactors can be divided into four groups: (i) bioreactors without forced aeration or occasional agitation, such as tray bioreactors; (ii) bioreactors with forced agitation, such as column bioreactors; (iii) bioreactors without forced agitation or with continuous slow agitation, such as packed bed, stirred drum, and rotary drum bioreactors; and (iv) bioreactors with slow continuous stirring and forced stirring bioreactors, among which are stirred-aerated bioreactors, gas-solid fluidized bed, and rocking drum [70, 71].

In the production of synbiotics through SSF, most of the bioreactors used in laboratory studies are flask and tray bioreactors due to the ease of control of the process conditions. Considering this Wu [72] used flask to ferment whole oats with *Rhizopus oryzae* and *Lactobillus plantarum* to obtain a synbiotic product through solid-state fermentation, finding that the process increased the antioxidant capacity and amino acid content, improving the nutritional quality of the oats.

4.7 Microencapsulation and Nanotechnology to Ensure Their Viability

Microencapsulation and nanotechnology allow the modification of the structure and size of materials, which leads to changes in the way that matter behaves in biological systems. One of the most interesting applications of these technologies is the controlled or sustained release of cells or bioactive substances at specific sites, resulting in better absorption and bioavailability, which can enhance the health benefits attributed to synbiotics.

Some challenges are presented to assure the stability and viability of synbiotics, being one of the main acidic and bile-rich environment of the stomach, thus requiring some sort of protection for the living cells that conform synbiotics. Microencapsulation could be a way to achieve this goal; in this technology a continuous coating is formed around the substance of interest (encapsulated material), stabilizing and protecting the synbiotics from the changing and harsh conditions of the gastrointestinal tract allowing them to reach the target site.

Proper selection of the encapsulation technique and wall material are crucial to the viability and proper release of the encapsulated symbiotics; different methodologies for microencapsulation are available among them are spray drying, extrusion, emulsification, and freeze drying. Some recent research on the use of microencapsulation in symbiotics are presented below.

Wu and Zhang [73] researched the role of the prebiotic part of microencapsulated synbiotics in the stability and survival rate of the living cells present in the product under simulated gastrointestinal conditions. Microcapsules formed from sodium alginate (SA), prebiotic arabinoxylan fiber (AX), and arabinoxylan oligosaccharides (AXOS) of different molecular weights were used to protect *L. plantarum* (LP). When low-molecular-weight AXOS (2.1×10^4) were employed for the formation of the microspheres, highest microencapsulation efficiency and survival rate were found ($85.45 \pm 2.36\%$ and $79.03 \pm 0.85\%$ respectively) this behavior was attributed to the formation of more porous structures in the microspheres increasing the space available for LP cells and a better fermentability of the low molecular weight AXOS promoting the viability of the probiotics. These results show the high importance for a proper selection of prebiotics to develop an ideal carrier for target delivery of probiotics and promote a synergistic relationship between the parts that form the synbiotic product.

Similarly, Silva et al. [74] researched the effect of FOS as prebiotic source and encapsulating material over the viability of *Lactobacillus acidophilus* in an alginate–gelatine matrix. It was found that under gastrointestinal tract simulated conditions, the survival of cells was higher for the microcapsules compared with free cells. Microcapsules with FOS presented a better protection maintaining its structure and allowing a sustained release of the probiotic during the intestinal phase. Once again it is possible to see that the prebiotic part of a synbiotic product play an important role over the stability and viability of these.

4.8 Nanoparticles

The use of prebiotic nanoparticles can improve the beneficial properties of synbiotics. Hong et al. [75] developed nanoparticles of pullulan, a polysaccharide of maltriose units with high gastric acid resistance and potential to enhance the growth of probiotics. The antimicrobial activity of *L. plantarum* with internalized pthalyl pullulan nanoparticles (PPN/LP) was compared with LP alone and LP with internalized pullulan. It was found that growth inhibition of pathogenic microorganism *E. coli* K99 and *Listeria monocytogenes* was better for PPN/LP. It was theorized that PPN apply some intracellular stress in the LP cells, which stimulate the expression of bioactive peptides (bacteriocins). These results show the high potential of nanoparticles in the development of better symbiotic products, more research is needed about the mechanism of action of nanoparticles in the production of bioactive substances, which can lead to industrialized cell factories for specific products.

Another application of prebiotic nanoparticles for the development of a synbiotic product was researched by Fayed et al. [76] who produced prebiotic nanoparticles of PLGA, a synthetic biodegradable polymer, loaded with inulin (a polysaccharide of fructose units). Nanoparticles were encapsulated with a mixture of probiotics from the *Bifidobacterium genus* (*Bifidobacterium bifidum, Bifidobacterium animalis*, and *Bifidobacterium lactis*) in double-layer microcapsules composed by alginate-arabic gum. It was found that the inclusion of prebiotic nanoparticles promoted the viability of probiotic even if the prebiotic was separately encapsulated in a PLGA matrix. This could be explained by the mechanism of inulin release that occurred in two phases: the first one in a rapid way (53% of inulin content released in 30 minutes, under simulated intestinal conditions) followed by a slow stage (87% of inulin content released in a period of 3 days); this behavior creates a sustained source of prebiotic that increase the survival of microorganism compared with cells without microencapsulation.

4.9 Applications in Various Fields such as Dermatological Diseases, Animal Feed, and Functional Foods

4.9.1 Dermatological Diseases

Some research with synbiotics has been reported, where the usefulness of these microorganisms in the treatment of these diseases is evidenced. For example, melasma is a disease whose exact incidence is unknown in many parts of the world and mainly affects

Hispanics, Latin Americans, and Asians living in areas that receive high-intensity ultraviolet radiation. It is a disorder of melanogenesis that causes localized, chronic, and acquired hypermelanosis of the skin [77]. It is characterized by the production of irregular light to dark macules on areas of the skin that are exposed to the sun, usually on the cheeks, chin, forehead, nose, and upper lip [78]. However, the use of natural products reduces the risk of side effects, reduces costs, and increases efficacy in chronic conditions, which is why it is of interest to study natural alternatives for this treatment. In this sense, it has been reported that some synbiotics present some advantages such as anti-inflammatory activities, inhibition of tyrosinase activity, antioxidant activity, ultraviolet protection, which makes them suitable to be included in the treatment of this disease. For these reasons, the researchers Piyavatin et al. [79] investigated whether the use of oral commercial synbiotic (TS6, conformed by probiotics of the genus *Lactococcus*, *Lactobacillus*, and *Bifidobacterium* with prebiotics such as fructooligosaccharides) could be useful in the treatment of this disease. Women aged 30–50 years with facial melasma were randomly put in two groups: Placebo or oral supplementation and treated for 12 weeks. A significant statistic difference was found in the patients treated with synbiotics for different melasma indicators as severity level of melasma (mMASI), melanin index crown among others. A reduction in the UV damage is possible promoted by the reduction of neutrophils who liberates enzyme related to skin aging. Therefore, it is possible to conclude that using synbiotics would be a viable alternative in the treatment of melasma.

Another skin disease that affects a large part of the population is acne. This chronic inflammatory skin disease is caused by *Cutibacterium acnes*. Kim et al. [80] evaluated the synergistic antibacterial activities of probiotic lactic acid bacteria (LAB) "*Lactobacillus acidophilus* A001F8" with *Curcuma longa rhizome extract* (CLE) as a synbiotic against *C. acnes*. The results evidenced that the synbiotic combination was effective to significantly increase the inhibition zone diameters against *C. acnes* compared to the use of *L. acidophilus* A001F8 or CLE alone. According to the above results, *L. acidophilus* together with the prebiotic *Curcuma longa rhizome* extract presented a synergistic effect of antibacterial activity against *C. acnes*, and therefore, it could be used in the development of cosmetics or drugs to treat acne. Al-Ghazzewi and Tester [81] studied the potential of synbiotics on *Propionibacterium acnes*. These authors evaluated *in vitro* the ability of probiotic bacteria and konjac glucomannan hydrolysates (GMH) to inhibit acne-inducing bacterium. The probiotic bacterial strains evaluated inhibited the growth of this bacterium. However, the inhibition increased significantly when the prebiotic (GMH) was added. According to these results, the use of synbiotics could be a biotherapeutic alternative for the treatment of this disease, eliminating the use of topical or systemic drugs that generate side effects in the organism.

Finally, another relevant dermatological disease is atopic dermatitis (AD). This disease is one of the most common dermatologic disorders affecting up to 20% of children as well as 1–3% of adults worldwide [82]. The research conducted by Noll et al. [83] compared the effectivity of control, prebiotic, and synbiotic baths to treat atopic dermatitis. The results showed a considerable decrease in the scoring of atopic dermatitis (SCORAD) in the evaluated times in those patients who underwent daily baths of synbiotics or prebiotics. After 14 days a significant statistic difference was found through the SCORAD in the patients treated with synbiotic bath (a mixture of different strains of acid lactic bacteria and prebiotics such as inulin and apple pectin) scoring the lowest value between the treatments. An improvement in itching and dryness of the skin and its bacterial microbiome was also observed in patients who took daily synbiotic baths; analysis of the composition of skin microbiome of the patients revealed that a modification of microorganism present could explain some of the beneficial effects observed. The time and the combination of type of bath and exposure was reported not significant. Therefore, these baths with synbiotics could be a promising alternative for cutaneous application in the treatment of AD. The effect of dietary and topical administration of synbiotics over skin injuries caused by AD was researched in an animal model by Kim et al. [84]. Mices were treated with a synbiotic product conformed by *L. plantarum*, *Bifidobacterium longum*, and *Pediococcus pentosaceus,* and β-*glucans* extracted from oats. It was found that combination of dietary and topical synbiotic treatment had a better effect than its individual administration. Mice treated topically + orally showed almost normal structure in its histological analysis and the extend of skin lesions were significantly minor compared with the negative control group and other treatments. β-Glucans and lactic acid bacteria promote a positive influence in the health of the test subject. However more research is needed to appropriately correlate the metabolic and immunologic paths who contributed to this behavior.

The use of synbiotic microorganisms is established as a viable alternative for the treatment of these dermatological diseases, which would make possible the elimination of antibiotics and topicals that are generally formulated for the treatment of these diseases that also trigger side effects.

4.9.2 Functional Foods

In recent years, the demand for healthy foods that improve the health of the consumer has been increasing. The main function of a food is to provide the nutrients (macro and micronutrients) necessary for the proper functioning of the body. However, it is recognized that there are foods that have functions beyond nutrition, which gives rise to a new category of foods, "functional foods." For many years, the main institutions that regulate food production and scientists worldwide have been debating the definition of functional foods, and although there is still no single definition, it can be generally said that they are foods or food ingredients that contain bioactive compounds that have a proven biological function beyond nutrition, improving the consumer's health and preventing the onset of diseases. [85]. As the definition states, the functionality must be demonstrated, and the compounds must be in adequate concentrations to be bioavailable and bio-accessible after consumption [86]. Due to the functionalities demonstrated, synbiotics are considered functional ingredients, which can be found naturally in fermented foods or can be incorporated into food matrices, and in both cases their application gives rise to functional foods.

Among the most recognized fermented foods with synbiotics are dairy products such as kefir, but there are also several examples with cereals, oilseeds, fruits, and vegetables [56, 87–89]. Zhang et al. [90] developed a functional food with oat flour by solid-state fermentation of *L. plantarum* TK9 and *B. animalis* subsp. lactis V9, with synbiotic potential due to lactic acid bacteria and the presence of β-glucans. One of the main characteristics of this food is the content of free nitrogen and the concentration of peptides with molecular weight lower than 6000 Da after fermentation, which suggests that these compounds could be better assimilated during their metabolism in the human body resulting in a positive effect on the consumer's health. Functional foods have also been obtained from soybeans, one of the most recognized worldwide for its functional properties is tempeh. Tempeh is a product obtained by solid-state fermentation of soybeans mainly with *Rhizopus oligosporus* or *R. oryzae* in which by the action of these, the proteins of the legume are more bio-assimilable, in addition to releasing a varied amount of bioactive compounds that are recognized for their antioxidant activity [91]. Beverages with synbiotics are also considered functional foods; Alvarado-Cóndor et al. [92] evaluated the process variables to obtain a fermented kefir-type beverage from whey, finding that at a fermentation temperature range of 28.5–29.7 °C and a concentration of 43.3% whey powder, it is possible to obtain a beverage with physicochemical characteristics similar to those defined for kefir. Fermented soy milk (or its by-products) is also considered functional because during the process the lactic acid bacteria such as *L. plantarum* make the components of the beverage more biodigestible, in addition to having a higher content of antioxidants, isoflavones, and other bioactive compounds [93]. Other traditional products such as yogurt can also be considered functional foods due to the functional potential of the lactic acid bacteria and prebiotic compounds they contain [42].

To develop functional foods, synbiotics can be incorporated into food matrices. In this case, Sáez-Orviz et al. [94] developed a synbiotic coating film with lactobionic acid and *L. plantarum* to coat a cottage cheese. During the study, the changes during storage and the survival of bacteria in simulated digestion were evaluated to affirm the prebiotic and probiotic characteristics of the film, finding that the synbiotic films have potential for the development of functional foods, among others, because they were able to maintain a high viable cell count to be considered probiotic.

4.9.3 Animal Feed

The use of synbiotics in animal production as an aid to increase productivity, reduction of pathogen infection and improvement on the quality of derivate products as meat, eggs, and milk are discussed in the following section.

Dai et al. [95] researched the physiological and intestinal microbiota changes of turbot fed with an extruded aquafeed supplemented with a synbiotic composed of stachyose + *L. Casei* (25.0 g/kg stachyose + 1×10^8 CFU/g, respectively). The results showed that the addition of synbiotic supplementation promote the growth of intestinal epithelial cells compared with the individual addition of the prebiotic or probiotic part in the feed. More intestinal cells produce a greater absorptive surface and led to a better digestive capacity and absorption of nutrients, demonstrating the synergistic properties of this feed additive. Synbiotic supplementation also promote the growth of lactobacilli reducing the viability of mycoplasma pathogens demonstrating the microbiota altering properties of the additive.

Hashem et al. [96] studied the use of Alginate-$CaCl_2$ nanoparticles as wall material for the encapsulation of a symbiotic product conformed by Moringa oleifera leaf extract (prebiotic) and *saccharomyces cerevisiae* yeast cells (probiotic) over the growth performance and gut health of rabbits during the growing phase. Statistically significant differences ($p < 0.05$) were found for weight gain between the control and the animals fed with encapsulated product (synbiotic concentration: 11×10^{12} CFU Yeast + 0.15 g moringa extract/kg diet), which can be explained by physiological changes in the length of the intestine (an increase of 16.2% with respect to the control), which facilitates the absorption and utilization of nutrients. In terms of intestinal health, the consumption of microencapsulated products affected the count of salmonella in the intestine, significantly reducing its population (approximately in a 42%) and favoring the growth of beneficial lactic acid bacteria.

According to the results reported, the use of symbiotics has a significant effect reducing the presence of pathogens in the intestinal tract of the animals studied, which is why they represent an interesting alternative to the conventional use of antibiotics, due to the food safety risks associated with their use and the danger posed by the development of resistant pathogens.

This potential was researched by [97] comparing the effect of additive type in diet (control-no additive, antibiotic, and synbiotic) over the cecal presence of pathogenic and beneficial microorganisms in broilers. The synbiotic used was a microencapsulated *L. plantarum (MLP)* (1.0×10^{10} CFU/g product) in combination with FOS, which were supplemented to the diet at a level of 0.1 g of MLP/kg of feed + 3 g of FOS/kg of feed.

The antibiotic used for comparison was *aureomycin* in a concentration of 0.15 g/kg of feed. *Aureomycin* is commonly used for prevention of infections caused by gram-positive and gram-negative bacteria and to promote growth. No statistically significant differences were found in the *E. coli* population of broilers fed with the antibiotic with respect to those fed with the symbiotic, indicating that the latter could be an alternative for infection control, additionally the *lactobacillus* population increased significantly, which is associated with maintaining the health of the intestinal tract and growth promotion, which was observed in this study as a 10% increase in the average daily gain of weight for chickens fed with symbiotic.

4.10 Conclusions

The definition, sources, and applications of synbiotics are subjects of continuous change and research as it showed in this chapter. Is clear that more research in the development of efficient bioreactors to produce synbiotics are needed, some good steps are taken with the implementation of membrane systems in submerged fermentation that increase yield and purification of the products in a continuous rate, deep knowledge of the metabolism and reactions involved are necessary. On the other hand, solid-state fermentation could become a cost-effective alternative due to its simplicity and reduction of waste treatments; however, the scale-up constitutes an interesting topic for improvement to become an industrial way for production.

Once produced and administrated to the host, synbiotics need to reach the desired place of action. Microencapsulation is a useful tool to achieve this goal, different wall materials and configurations are researched to assure that the microorganism and the prebiotic part could work as intended and released in a controlled manner. Additionally, nanotechnology offers a new perspective of how synbiotics could work, not only through a complementary or synergistic approach but influencing the production of certain types of metabolites by the inclusion of prebiotic nanoparticles inside of the microorganism, which can promote the production of specific substances with health benefits. According to the above, more research in this aspect and clinical studies to validate its effects, synbiotics could become an important option for personal nutrition and precision medicine.

On the other hand, some challenges as viability and interactions with host microbiome had to be addressed for synbiotic development; therefore, *in vitro*, *ex vivo*, or even *in silico* methodologies are necessary to test the synbiotic properties before its validation through *in vivo*. This approach can reduce considerably the cost and time necessary for the evaluation of promising strains of microorganism and types of prebiotics; this implies that *in vivo* testing could be overlooked at evaluating the effectiveness of synbiotic in health, food, and animal feed.

References

1 Walker, A.W. and Lawley, T.D. (2013). Therapeutic modulation of intestinal dysbiosis. *Pharmacological Research* 69 (1): 75–86. https://doi.org/10.1016/j.phrs.2012.09.008.

2 Martín, R. and Langella, P. (2019). Emerging health concepts in the probiotics field: streamlining the definitions. *Frontiers in Microbiology* 10: https://doi.org/10.3389/fmicb.2019.01047.

3 Swanson, K.S., Gleen, G.R., Hutkins, R. et al. (2020). The International Scientific Association for Probiotics and Prebiotics (ISAPP) consensus statement on the definition and scope of synbiotics. *Nature Reviews Gastroenterology & Hepatology* 17 (11): 687–701. https://doi.org/10.1038/s41575-020-0344-2.

4 Gurry, T. (2017). Synbiotic approaches to human health and well-being. *Microbial Biotechnology* 10 (5): 1070–1073. https://doi.org/10.1111/1751-7915.12789.

5 Hill, C., Guarner, F., Reid, G. et al. (2014). The International Scientific Association for Probiotics and Prebiotics consensus statement on the scope and appropriate use of the term probiotic. *Nature Reviews Gastroenterology & Hepatology* 11 (8): 506–514. https://doi.org/10.1038/nrgastro.2014.66.

6 Salminen, S., Collado, M.C., Endo, A. et al. (2021). The International Scientific Association of Probiotics and Prebiotics (ISAPP) consensus statement on the definition and scope of postbiotics. *Nature Reviews Gastroenterology & Hepatology* 18 (9): 649–667. https://doi.org/10.1038/s41575-021-00440-6.

7 Kavita, R.P., Naik, S.R., and Vakil, B.V. (2011). Probiotics, prebiotics and synbiotics – a review. *Journal of Food Science and Technology Nepal* 52 (12): 7577–7587. https://doi.org/10.1007/s13197-015-1921-1.

8 Oelschlaeger, T.A. (2010). Mechanisms of probiotic actions – a review. *International Journal of Medical Microbiology* 300 (1): 57–62. https://doi.org/10.1016/j.ijmm.2009.08.005.

9 Barboza, M., Germna, J., Lebrilla, C. et al. (2013). Prebiotic oligosaccharides. No.8.425.930. B2.

10 Dixit, Y., Wagle, A., and Vakil, B. (2016). Patents in the field of probiotics, prebiotics, synbiotics: a review. *Journal of Food: Microbiology, Safety and Hygiene* 1 (2): https://doi.org/10.4172/2476-2059.1000111.

11 Singh, S.P. and Upadhyay, S.K. (eds.) (2021). *Bioprospecting of Microorganism-Based Industrial Molecules*. John Wiley & Sons Ltd. doi:10.1002/9781119717317.

12 Johnston, B.C. (2006). Probiotics for pediatric antibiotic-associated diarrhea: a meta-analysis of randomized placebo-controlled trials. *Canadian Medical Association Journal* 175 (4): 377–383. https://doi.org/10.1503/cmaj.051603.

13 Muñoz-Atienza, E., Gómez-Salas, B., Araujo, C. et al. (2013). Antimicrobial activity, antibiotic susceptibility and virulence factors of Lactic Acid Bacteria of aquatic origin intended for use as probiotics in aquaculture. *BMC Microbiology* 13 (1): 15. https://doi.org/10.1186/1471-2180-13-15.

14 Markowiak, P. and Śliżewska, K. (2018). The role of probiotics, prebiotics and synbiotics in animal nutrition. *Gut Pathogens* 10 (1): 21. https://doi.org/10.1186/s13099-018-0250-0.

15 Anadón, A. and Castellano, V. (2013). *Probiotics and Prebiotics in Food, Nutrition and Health*, 1e, vol. 1. CRC Press https://doi.org/10.1201/b15561.

16 Patel, S., Shukla, R., and Goyal, A. (2015). Probiotics in valorization of innate immunity across various animal models. *Journal of Functional Foods* 14: 549–561. https://doi.org/10.1016/j.jff.2015.02.022.

17 Mizak, L., Gryko, R., Kwiatek, M., and Parasion, S. (2012). Probiotyki w żywieniu zwierząt. *Życie Weterynaryjne* 87 (9): 736–741.

18 Kaur, S., Kaur, R., Rani, N. et al. (2021). Sources and selection criteria of probiotics. In: *Advances in Probiotics for Sustainable Food and Medicine* (ed. G. Goel and A. Kumar), 27–43. https://doi.org/10.1007/978-981-15-6795-7_2.

19 Mathipa, M.G. and Thantsha, M.S. (2017). Probiotic engineering: towards development of robust probiotic strains with enhanced functional properties and for targeted control of enteric pathogens. *Gut Pathogens* 9 (1): 28. https://doi.org/10.1186/s13099-017-0178-9.

20 Rocha-Ramírez, L.M., Pérez-Solano, R.A., Castañón-Alonso, S.L. et al. (2017). Probiotic *Lactobacillus* strains stimulate the inflammatory response and activate human macrophages. *Journal of Immunology Research* 2017: 1–14. https://doi.org/10.1155/2017/4607491.

21 Kumar, M., Nagpal, R., Verma, V. et al. (2013). Probiotic metabolites as epigenetic targets in the prevention of colon cancer. *Nutrition Reviews* 71 (1): 23–34. https://doi.org/10.1111/j.1753-4887.2012.00542.x.

22 dos Reis, S.A., da Conceição, L.L., Siqueira, N.P. et al. (2017). Review of the mechanisms of probiotic actions in the prevention of colorectal cancer. *Nutrition Research* 37: 1–19. https://doi.org/10.1016/j.nutres.2016.11.009.

23 de Vrese, M. and Schrezenmeir, J. (2008). Probiotics, prebiotics, and synbiotics. In: *Food Biotechnology* (ed. U. Stahl, U.E. Donalies, and E. Nevoigt), 1–66. https://doi.org/10.1007/10_2008_097.

24 Gibson, G.R., Hutkins, R., Sanders, M.E. et al. (2017). Expert consensus document: the International Scientific Association for Probiotics and Prebiotics (ISAPP) consensus statement on the definition and scope of prebiotics. *Nature Reviews Gastroenterology & Hepatology* 14 (8): 491–502. https://doi.org/10.1038/nrgastro.2017.75.

25 Sharma, M., Sangwan, R.S., Khatkar, B.S., Singh, S.P.(2019). Alginate-pectin co-encapsulation of dextransucrase and dextranase for oligosaccharide production from sucrose feedstocks. *Bioprocess and Biosystems Engineering* 42 (10): 1681–1693. https://doi.org/10.1007/s00449-019-02164-z.

26 Markowiak, P. and Śliżewska, K. (2017). Effects of probiotics, prebiotics, and synbiotics on human health. *Nutrients* 9 (9): 1021. https://doi.org/10.3390/nu9091021.

27 Singh, S.P., Jadaun, J.S., Narnoliya, L.K., and Pandey, A. (2017). Prebiotic oligosaccharides: special focus on fructooligosaccharides, its biosynthesis and bioactivity. *Applied Biochemistry and Biotechnology* 183 (2): 613–635. https://doi.org/10.1007/s12010-017-2605-2.

28 Puértolas-Balint, F. and Schroeder, B.O. (2020). Does an apple a day also keep the microbes away? The interplay between diet, microbiota, and host defense peptides at the intestinal mucosal barrier. *Frontiers in Immunology* 11: https://doi.org/10.3389/fimmu.2020.01164.

29 Slavin, J. (2013). Fiber and prebiotics: mechanisms and health benefits. *Nutrients* 5 (4): 1417–1435. https://doi.org/10.3390/nu5041417.

30 Singla, V. and Chakkaravarthi, S. (2017). Applications of prebiotics in food industry: a review. *Food Science and Technology International* 23 (8): 649–667. https://doi.org/10.1177/1082013217721769.

31 Crittenden, R. and Playne, M.J. (2008). Facts and functions of prebiotics, probiotics and synbiotics. In: *Handbook of Probiotics and Prebiotics* (ed. Y.K. Lee and S. Salminen), 535–582. Hoboken, NJ; Manhattan, KS: Wiley-Interscience, Kansas State University.

32 Davani-Davari, D., Negahdaripour, M., Karimzadeh, I. et al. (2019). Prebiotics: definition, types, sources, mechanisms, and clinical applications. *Food* 8 (3): 92. https://doi.org/10.3390/foods8030092.

33 Sharma, M., Patel, S.N., Lata, K. et al. (2016). A novel approach of integrated bioprocessing of cane molasses for production of prebiotic and functional bioproducts. *Bioresource Technology* 219: 311–318. https://doi.org/10.1016/j.biortech.2016.07.131.

34 Joshi, N., Sharma, M., Singh, S.P.(2020). Characterization of a novel xylanase from an extreme temperature hot spring metagenome for xylooligosaccharide production. *Applied Microbiology and Biotechnology* 104:4889–4901. https://doi.org/10.1007/s00253-020-10562-7.

35 Jadaun, J.S., Narnoliya, L.K., Agarwal, N. et al. (2019). Catalytic biosynthesis of levan and short-chain fructooligosaccharides from sucrose-containing feedstocks by employing the levansucrase from Leuconostoc mesenteroides MTCC10508. *International Journal of Biological Macromolecules* 127: 486–495. https://doi.org/10.1016/j.ijbiomac.2019.01.070.

36 Thakur, M., Rai, A.K., Singh, S.P.(2022). An acid-tolerant and cold-active β-galactosidase potentially suitable to process milk and whey samples. *Applied Microbiology and Biotechnology* 106: 3599–3610. https://doi.org/10.1007/s00253-022-11970-7.

37 Jakubczyk, E. and Kosikowska, M. (2000). Nowa generacja mlecznych produktów fermentowanych z udziałem probiotyków i prebiotyków, produkty synbiotyczne. *Prz. Mlecz* 12: 397–400.

38 Wang, Y. (2009). Prebiotics: present and future in food science and technology. *Food Research International* 42 (1): 8–12. https://doi.org/10.1016/j.foodres.2008.09.001.

39 Macfarlane, G.T., Steed, H., and Macfarlane, S. (2008). Bacterial metabolism and health-related effects of galacto-oligosaccharides and other prebiotics. *Journal of Applied Microbiology* https://doi.org/10.1111/j.1365-2672.2007.03520.x.

40 Crittenden, R. (2012). Prebiotics and probiotics – the importance of branding. *Microbial Ecology in Health and Disease* 23: https://doi.org/10.3402/mehd.v23i0.18566.

41 Sivieri, K., Morales, M.L.V., Saad, S.M.I. et al. (2014). Prebiotic effect of fructooligosaccharide in the Simulator of the Human Intestinal Microbial Ecosystem (SHIME® Model). *Journal of Medicinal Food* 17 (8): 894–901. https://doi.org/10.1089/jmf.2013.0092.

42 Crittenden, R.G., Morris, L.F., Harvey, M.L. et al. (2001). Selection of a Bifidobacterium strain to complement resistant starch in a synbiotic yoghurt. *Journal of Applied Microbiology* 90: 266–278.

43 Gourbeyre, P., Denery, S., and Bodinier, M. (2011). Probiotics, prebiotics, and synbiotics: impact on the gut immune system and allergic reactions. *Journal of Leukocyte Biology* 89 (5): 685–695. https://doi.org/10.1189/jlb.1109753.

44 Scavuzzi, B., Henrique, F., Miglioranza, L. et al. (2014). Impact of prebiotics, probiotics and synbiotics on components of the metabolic syndrome. *Annals of Nutritional Disorders and Therapy* 1 (2): 1–13.

45 Gyawali, R., Nwamaioha, N., Fiagbor, R. et al. (2019). The role of prebiotics in disease prevention and health promotion. In: *Dietary Interventions in Gastrointestinal Diseases*, 151–167. Elsevier https://doi.org/10.1016/B978-0-12-814468-8.00012-0.

46 Kavita, R.P., Naik, S.R., and Vakil, B.V. (2015). Probiotics, prebiotics and synbiotics – a review. *Journal of Food Science and Technology* 52 (12): 7577–7587. https://doi.org/10.1007/s13197-015-1921-1.

47 Zhang, M.-M., Cheng, J.Q., Lu, Y.R. et al. (2010). Use of pre-, pro- and synbiotics in patients with acute pancreatitis: a meta-analysis. *World Journal of Gastroenterology* 16 (31): 3970. https://doi.org/10.3748/wjg.v16.i31.3970.

48 Manigandan, T., Mangaiyarkarasi, S.P., Hemalatha, R. et al. (2012). Probiotics, prebiotics and synbiotics – a review. *Biomedical and Pharmacology Journal* 5 (2): 295–304. https://doi.org/10.13005/bpj/357.

49 Krumbeck, J.A., Walter, J., and Hutkins, R.W. (2018). Synbiotics for improved human health: recent developments, challenges, and opportunities. *Annual Review of Food Science and Technology* 9 (1): 451–479. https://doi.org/10.1146/annurev-food-030117-012757.

50 Flesch, A.T., Tonial, S.T., Contu, P.D.C., and Damin, D.C. (2017). Perioperative synbiotics administration decreases postoperative infections in patients with colorectal cancer: a randomized, double-blind clinical trial. *Revista do Colégio Brasileiro de Cirurgiões* 44 (6): 567–573. https://doi.org/10.1590/0100-69912017006004.

51 Neyrinck, A.M., Rodriguez, J., Taminiau, B. et al. (2021). Improvement of gastrointestinal discomfort and inflammatory status by a synbiotic in middle-aged adults: a double-blind randomized placebo-controlled trial. *Scientific Reports* 11 (1): 2627. https://doi.org/10.1038/s41598-020-80947-1.

52 el Hadi, H., di Vincenzo, A., Vettor, R., and Rossato, M. (2019). Food ingredients involved in white-to-brown adipose tissue conversion and in calorie burning. *Frontiers in Physiology* 9: https://doi.org/10.3389/fphys.2018.01954.

53 Karamali, M., Nasiri, N., Shavazi, N.T. et al. (2018). The effects of synbiotic supplementation on pregnancy outcomes in gestational diabetes. *Probiotics and Antimicrobial Proteins* 10 (3): 496–503. https://doi.org/10.1007/s12602-017-9313-7.

54 Lalitsuradej, E., Sirilun, S., Sittiprapaporn, P. et al. (2022). The effects of synbiotics administration on stress-related parameters in Thai subjects – a preliminary study. *Food* 11 (5): 759. https://doi.org/10.3390/foods11050759.

55 Baothman, O.A., Zamzami, M.A., Taher, I. et al. (2016). The role of gut microbiota in the development of obesity and diabetes. *Lipids in Health and Disease* 15 (1): 108. https://doi.org/10.1186/s12944-016-0278-4.

56 Xiang, H., Sun-Waterhouse, D., Waterhouse, G.I.N. et al. (2019). Fermentation-enabled wellness foods: a fresh perspective. *Food Science and Human Wellness* 8 (3): 203–243. https://doi.org/10.1016/j.fshw.2019.08.003.

57 Rupchand Bansod, T. and Tah, B. (2021). Fermentation technology. *World Journal of Pharmaceutical Research* 10 (3): 260–299. https://doi.org/10.17605/OSF.IO/2JNWG.

58 Soccol, C.R., da Costa, E.S.F., Letti, L.A.J. et al. (2017). Recent developments and innovations in solid state fermentation. *Biotechnology Research and Innovation* 1: 52–71.

59 Singhania, R.R., Sukumaran, R.K., Patel, A.K. et al. (2010). Advancement and comparative profiles in the production technologies using solid-state and submerged fermentation for microbial cellulases. *Enzyme and Microbial Technology* 46 (7): 541–549. https://doi.org/10.1016/j.enzmictec.2010.03.010.

60 Kaur, P., Vohra, A., and Satyanarayana, T. (2013). Laboratory and industrial bioreactors for submerged fermentations. In: *Fermentation Processes Engineering in the Food Industry* (ed. C.R. Soccol, A. Pandey, and C. Larroche), 165–178. CRC Press.

61 Schorsch, J., Castro, C.C., Couto, L.D. et al. (2019). Optimal control for fermentative production of fructo-oligosaccharides in fed-batch bioreactor. *Journal of Process Control* 78: 124–138. https://doi.org/10.1016/j.jprocont.2019.03.004.

62 Feng, X.M., Eriksson, A.R.B., and Schnürer, J. (2005). Growth of lactic acid bacteria and *Rhizopus oligosporus* during barley tempeh fermentation. *International Journal of Food Microbiology* 104 (3): 249–256. https://doi.org/10.1016/j.ijfoodmicro.2005.03.005.

63 Martins, M., Silva, K.C.G., Ávila, P.F. et al. (2021). Xylo-oligosaccharide microparticles with synbiotic potential obtained from enzymatic hydrolysis of sugarcane straw. *Food Research International* 140: https://doi.org/10.1016/j.foodres.2020.109827.

64 Archacka, M., Celińska, E., and Białas, W. (2020). Techno-economic analysis for probiotics preparation production using optimized corn flour medium and spray-drying protective blends. *Food and Bioproducts Processing* 123: 354–366. https://doi.org/10.1016/j.fbp.2020.07.002.

65 Fang, R., Burghardt, J.P., Huang, J. et al. (2021). Purification of crude fructo-oligosaccharide preparations using probiotic bacteria for the selective fermentation of monosaccharide byproducts. *Frontiers in Microbiology* 11: 1–9. https://doi.org/10.3389/fmicb.2020.620626.

66 Vera, C., Córdova, A., Aburto, C. et al. (2016). Synthesis and purification of galacto-oligosaccharides: state of the art. *World Journal of Microbiology and Biotechnology* 32 (12): 197. https://doi.org/10.1007/s11274-016-2159-4.

67 Burghardt, J.P., Coletta, L.A., van der Bolt, R. et al. (2019). Development and characterization of an enzyme membrane reactor for fructo-oligosaccharide production. *Membranes (Basel)* 9 (11): https://doi.org/10.3390/membranes9110148.

68 Aliwarga, L., Wardani, A.K., Aryanti, P.T.P., and Wenten, I.G. (2019). Recent development of lactic acid production using membrane bioreactors. *IOP Conference Series: Materials Science and Engineering* 622 (012023): https://doi.org/10.1088/1757-899X/622/1/012023.

69 Kim, J.W., Park, B.J., Oh, T.H., and Lee, J.M. (2021). Model-based reinforcement learning and predictive control for two-stage optimal control of fed-batch bioreactor. *Computers & Chemical Engineering* 154: 107465. https://doi.org/10.1016/j.compchemeng.2021.107465.

70 Krishania, M., Sindhu, R., Binod, P. et al. (2018). Design of bioreactors in solid-state fermentation. In: *Current Developments in Biotechnology and Bioengineering*, 83–96. Elsevier https://doi.org/10.1016/b978-0-444-63990-5.00005-0.

71 Garro, M.S., Rivas, F.P., and Garro, O.A. (2020). Solid state fermentation in food processing: advances in reactor design and novel applications. In: *Innovative Food Processing Technologies: A Comprehensive Review*, 165–182. Elsevier https://doi.org/10.1016/b978-0-08-100596-5.23049-7.

72 Wu, H., Liu, H.-N., Ma, A.-M. et al. (2022). Synergetic effects of *Lactobacillus plantarum* and *Rhizopus oryzae* on physicochemical, nutritional and antioxidant properties of whole-grain oats (Avena sativa L.) during solid-state fermentation. *LWT* 154: 112687. https://doi.org/10.1016/j.lwt.2021.112687.

73 Wu, Y. and Zhang, G. (2018). Synbiotic encapsulation of probiotic *Latobacillus plantarum* by alginate -arabinoxylan composite microspheres. *LWT* 93: 135–141. https://doi.org/10.1016/j.lwt.2018.03.034.

74 Silva, K.C.G., Cezarino, E.C., Michelon, M., and Sato, A.C.K. (2018). Symbiotic microencapsulation to enhance *Lactobacillus acidophilus* survival. *LWT* 89: 503–509. https://doi.org/10.1016/j.lwt.2017.11.026.

75 Hong, L., Lee, S.M., Kim, W.S. et al. (2021). Synbiotics containing nanoprebiotics: a novel therapeutic strategy to restore gut dysbiosis. *Frontiers in Microbiology* 12: https://doi.org/10.3389/fmicb.2021.715241.

76 Fayed, B., Abood, A., El-Sayed, H.S. et al. (2018). A synbiotic multiparticulate microcapsule for enhancing inulin intestinal release and Bifidobacterium gastro-intestinal survivability. *Carbohydrate Polymers* 193: 137–143. https://doi.org/10.1016/j.carbpol.2018.03.068.

77 Handel, A.C., Lima, P.B., Tonolli, V.M. et al. (2014). Risk factors for facial melasma in women: a case–control study. *British Journal of Dermatology* 171 (3): 588–594. https://doi.org/10.1111/bjd.13059.

78 Ikino, J.K., Nunes, D.H., da Silva, V.P.M. et al. (2015). Melasma and assessment of the quality of life in Brazilian women. *Anais Brasileiros de Dermatologia* 90 (2): 196–200. https://doi.org/10.1590/abd1806-4841.20152771.

79 Piyavatin, P., Chaichalotornkul, S., Nararatwanchai, T. et al. (2021). Synbiotics supplement is effective for Melasma improvement. *Journal of Cosmetic Dermatology* 20 (9): 2841–2850. https://doi.org/10.1111/jocd.13955.

80 Kim, J., Kim, H., Jeon, S. et al. (2020). Synergistic antibacterial effects of probiotic lactic acid bacteria with *Curcuma longa* rhizome extract as synbiotic against *Cutibacterium acnes*. *Applied Sciences* 10 (24): 8955. https://doi.org/10.3390/app10248955.

81 Al-Ghazzewi, F.H. and Tester, R.F. (2010). Effect of konjac glucomannan hydrolysates and probiotics on the growth of the skin bacterium *Propionibacterium acnes in vitro*. *International Journal of Cosmetic Science* 32 (2): 139–142. https://doi.org/10.1111/j.1468-2494.2009.00555.x.

82 Mayba, J.N. and Gooderham, M.J. (2017). Review of atopic dermatitis and topical therapies. *Journal of Cutaneous Medicine and Surgery* 21 (3): 227–236. https://doi.org/10.1177/1203475416685077.

83 Noll, M., Jäger, M., Lux, L. et al. (2021). Improvement of atopic dermatitis by synbiotic baths. *Microorganisms* 9 (3): 527. https://doi.org/10.3390/microorganisms9030527.

84 Kim, Y.-H., Kang, M.S., Kim, T.H. et al. (2021). Anti-inflammatory and immune modulatory effects of Synbio-Glucan in an atopic dermatitis mouse model. *Nutrients* 13 (4): 1090. https://doi.org/10.3390/nu13041090.

85 Rai, A.K., Singh, S.P., Pandey, A. et al. (2021). *Current Developments in Biotechnology and Bioengineering; Technologies for Production of Nutraceuticals and Functional Food Products*. Elsevier. ISBN: 9780128235065. https://doi.org/10.1016/C2020-0-00498-6.

86 Agyei, D., Akanbi, T.O., and Oey, I. (2018). Enzymes for use in functional foods. In: *Enzymes in Food Biotechnology: Production, Applications, and Future Prospects*, 129–147. Elsevier https://doi.org/10.1016/B978-0-12-813280-7.00009-8.

87 García, C., Rendueles, M., and Díaz, M. (2019). Liquid-phase food fermentations with microbial consortia involving lactic acid bacteria: a review. *Food Research International* 119: 207–220. https://doi.org/10.1016/j.foodres.2019.01.043.

88 Upadhyay, S.K. and Singh, S.P. (eds.) (2021). *Bioprospecting of Plant Biodiversity for Industrial Molecules*. John Wiley & Sons Ltd. doi:10.1002/9781119718017.

89 Upadhyay, S.K. and Singh, S.P.(2023). *Plants as Bioreactors for Industrial Molecules*. John Wiley & Sons Ltd. doi:10.1002/9781119875116.

90 Zhang, N., Li, D., Zhang, X. et al. (2015). Solid-state fermentation of whole oats to yield a synbiotic food rich in lactic acid bacteria and prebiotics. *Food & Function* 6 (8): 2620–2625. https://doi.org/10.1039/C5FO00411J.

91 Ahnan-Winarno, A.D., Cordeiro, L., Winarno, F.G. et al. (2021). Tempeh: a semicentennial review on its health benefits, fermentation, safety, processing, sustainability, and affordability. *Comprehensive Reviews in Food Science and Food Safety* 20 (2): 1717–1767. https://doi.org/10.1111/1541-4337.12710.

92 Alvarado-Cóndor, P.M., Núñez-Pérez, J., Espín-Valladares, R.C., and Pais-Chanfrau, J.M. (2022). Multiple-objective optimization of lactic-fermentation parameters to obtain a functional-beverage candidate. *Electronic Journal of Biotechnology* 58: 10–13. https://doi.org/10.1016/j.ejbt.2022.04.001.

93 Feng, J.-Y., Thakur, K., Ni, Z.J. et al. (2021). Effects of okara and vitamin B2 bioenrichment on the functional properties and in vitro digestion of fermented soy milk. *Food Research International* 145: 110419. https://doi.org/10.1016/j.foodres.2021.110419.

94 Sáez-Orviz, S., Puertas, C., Marcet, I. et al. (2020). Bioactive synbiotic coatings with lactobionic acid and *Lactobacillus plantarum* CECT 9567 in the production and characterization of a new functional dairy product. *Journal of Functional Foods* 75: 104263. https://doi.org/10.1016/j.jff.2020.104263.

95 Dai, J., Yu, G., Ou, W. et al. (2021). Dietary supplementation of stachyose and *Lactobacillus casei* improves the immunity and intestinal health of turbot (*Scophthalmus maximus*. L). *Aquaculture Nutrition* 27 (S1): 48–60. https://doi.org/10.1111/anu.13401.

96 Hashem, N.M., Hosny, N.S., El-Desoky, N.I., and Shehata, M.G. (2021). Effect of nanoencapsulated alginate-synbiotic on gut microflora balance, immunity, and growth performance of growing rabbits. *Polymers (Basel)* 13 (23): 4191. https://doi.org/10.3390/polym13234191.

97 Song, D., Li, A., Wang, Y. et al. (2022). Effects of synbiotic on growth, digestibility, immune and antioxidant performance in broilers. *Animal* 16 (4): 100497. https://doi.org/10.1016/j.animal.2022.100497.

5

Microbial Asparaginase and Its Bioprocessing Significance

Susana Calderón-Toledo[1], *Amparo Iris Zavaleta*[1], *and Adalberto Pessoa-Junior*[2]

[1] *Laboratorio de Biología Molecular, Facultad de Farmacia y Bioquímica, Universidad Nacional Mayor de San Marcos, Lima, Peru*
[2] *Department of Biochemical and Pharmaceutical Technology, School of Pharmaceutical Sciences, University of São Paulo, São Paulo, Brazil*

5.1 Introduction

L-Asparaginase (E.C.3.5.1.1, L-asparaginase aminohydrolase) hydrolyzes L-asparagine to L-aspartic acid and releases ammonium in a two-step process, which is used in the treatment of leukemia. In this regard, leukemic cells, unlike normal cells, possess low levels of L-asparaginase synthetase. Consequently, serum L-asparagine deficiency leads to leukemic cells, decreased protein synthesis, and thus cell death [1]. On the other hand, in the food industry, the combination of L-asparagine and reducing sugars at high temperatures produces decarboxylated Schiff bases, which decompose to acrylamide or its precursors, which by pretreatment of foods with L-asparaginase decreases acrylamide concentrations [2].

L-Asparaginase is produced in various mammalian organs such as the liver, pancreas, brain, ovary or testis, kidneys, spleen, and lungs, as well as in fish and birds. In addition, it has been obtained from vegetables such as *Capsicum*, *Lupin*, *Pinus,* and *Phyllanthus* [3, 4] at high levels as long as L-asparagine is maintained as a nitrogen source, which is usually within the first 11 days of nitrogen source acquisition [5]. However, microorganisms are the most cost-effective source of production due to their fast growth and adaptability [6, 7]. Thus, species of the genera *Bacillus, Pseudomonas, Staphylococcus, Rouxiella, Lysinibacillus,* and *Pseudonocardia* have been isolated from soils; *Lactobacillus* from fermented food, and *Acinetobacter* from clinical samples [8–14]. In this regard, recombinant L-asparaginases with improved physicochemical characteristics that favor scale-up have been described in recent years.

Currently, on the one hand, L-asparaginase is a therapeutic agent for the treatment of acute lymphoblastic leukemia, and on the other, it decreases acrylamide concentrations in thermoprocessed food products, thus commercial L-asparaginases of high efficiency have been increased [15]. In this context, there has been an important development of innovative technologies that focus on the optimized and cost-effective production of L-asparaginase with the quality and adaptability criteria for its industrial scale-up.

Microbial Bioreactors for Industrial Molecules, First Edition.
Edited by Sudhir P. Singh and Santosh Kumar Upadhyay.

5.2 Classification of L-Asparaginase

Based on the amino acid sequences, structural homology, functionality, and sources from which they were isolated, they are divided into three families: bacterial type L-asparaginase (types I and II), plant type L-asparaginase (type III), and rhizobial type L-asparaginase. The bacterial type was obtained from *Escherichia coli* and extended for proteins of similar morphology. Type I L-asparaginase (EcAI) is cytosolic with a relatively low affinity for L-asparagine and encoded by *ansA* gene, whereas type II L-asparaginase (EcAII) is expressed in the periplasmic space [16].

EcAII is a high-molecular-mass homotetramer with a fully structured active site after dimer assembly [17]. Its enzymatic stability is related to the interactions at the interface since the hydrophobic interactions of the dimers are mainly hydrophobic, exposing polar residues on the external part of the enzyme, so it increases by hiding hydrophobic amino acids from the solvent [18]. In *E. coli*, the active sites are between the subunits in the dimers, where each one has two active sites formed between the residues of the N-domain in contact with C-terminus, which have Thr-91 as nucleophile, besides Thr-12, which is not required in the binding to the substrate, but it is necessary for catalysis [19, 20].

EcAI is a homodimer, with conserved amino acid motifs similar to EcAII, but with a different quaternary structure [21]. EcAIII is expressed as an inactive precursor composed of two heterodimers, which is autocatalytically activated by hydrolyzing a peptide bond, and two folded subunits are formed as a structural domain [22]. The distance between the catalytic threonine and arginine determines the substrate size, while the salt-bridge-like interaction of the chloride anion with Arg221 modulates the enzymatic specificity for substrates with α-carboxylate group [23].

In *Bacillus*, *ansA* and *ansB* genes are in the *ans* operon, which is expressed during vegetative growth in complex media, induced by aspartate and suppressed by nitrogen sources such as NH_4^+, while *ansR* interacts directly with aspartate and modulates the ability to affect the *ans* operon with possible repression [24]. In *E. coli*, *ansB* encodes EcAII and is regulated by the cAMP receptor protein, which is responsible for controlling the initiation of gene transcription in several catabolic pathways, and FNR, which activates genes during anaerobiosis [25]. While in *Bacillus subtilis*, *ansZ* encodes a functional L-asparaginase similar to *E. coli* asparaginase II, which is activated by TnrA during nitrogen-limited growth [26]. *Saccharomyces cerevisiae* synthesizes asparaginases I and II, where the latter is secreted according to the control of nitrogen catabolite repression and is capable of hydrolyzing exogenous asparagine [27].

5.3 Bioprocessing

5.3.1 Sources of microbial L-Asparaginase

Fungal L-asparaginases possess great relevance due to their cellular excretion and a wide diversity of members, such as ascomycetes, basidiomycetes, glomeromycetes, chytrids, microsporidia, and blastocladiomycetes [28]. Endophytes belonging to *Lasiodiplodia*, *Colletotrichum*, *Fusarium*, *Phoma*, and *Penicillium* have been described for their high levels of L-asparaginase production [29, 30]. Likewise, due to their eukaryotic nature, they are more required as they would lead to less adverse effects in their neoplastic application [31]. L-Asparaginase secreted from *Saccharomyces* is produced at the periplasmic level and expressed by nitrogen limitation [32]. *Yarrowia lipolytica* DSM3286 produces type II asparaginase with high affinity for asparagine and low specific activity to glutamine, and potential application for hematopoietic cancers [33].

Also, bacterial L-asparaginase has been one of the most studied due to its high production levels, such as those from *E. coli*, *Erwinia chrysanthemi*, *Erwinia carotovora*, *Streptomyces acrimycini*, *Pseudomonas aeruginosa*, and *B. subtilis*. Their high demand is related to their rapid production and efficiency, which lessens their manufacturing costs [34]. L-Asparaginases from gram-negative bacterial species have been further studied and categorized into types I and II, the former possessing both L-glutaminase and L-asparaginase activities, while the latter has highly specified by L-asparagine [35].

Structural and functional properties differ according to the species and the native ecosystems. From terrestrial environments, *E. coli*, *Aeromonas*, *Proteus*, *Serratia*, *Pseudomonas,* and *Bacillus* have been described [36]. Also, bacteria with high asparaginase and low glutaminase activity have been described from tropical plants, such as *Pseudomonas stutzeri* SMKLI and *Rhizobium radiobacter* SMKW2 [37], and likewise from coastal and mangrove ecosystems [38].

Microorganisms isolated from saline environments possess wide metabolic adaptability, synthesizing proteins with unique characteristics due to structural flexibility that could enhance enzymatic activity [39]. L-Asparaginase-producing microorganisms have been obtained from mangroves. However, those from forest environments have recorded higher production at the extracellular level, such as *Salinicococcus* sp. M KJ997975 [40]. Likewise, saline lakes, wetlands, salt springs, and deserts in Iran are sources of intracellular L-asparaginase-producing microorganisms with salt tolerance up to NaCl 3.5 M [41]. Also, *Bacillus*, *Enterobacter*, *Halomonas*, *Marinobacter,* and *Staphylococcus* isolated from Peruvian salterns were resistant to extreme conditions of pH, temperature, and salt concentration. However, *Bacillus* strains stood out for their higher production of L-asparaginase in liquid media [42].

Thermophilic bacteria present enzymes with high stability due to increased hydrogen bonding and decreased surface charge around the active center [43]. Thus, L-asparaginase producing *Caulobacter flavus* HS1XWS3 and HS1YWS2 with low glutaminase activity were isolated from thermal waters [44]. *Thermococcus gammatolerans* EJ3, an archae isolated from California submarine hydrothermal vents, produced stable L-asparaginase at high temperatures [45]. Consequently, the source of the L-asparaginase-producing microorganism offers different applications according to its metabolic characteristics and adaptability. In particular, extremophilic microorganisms present unique functional, physicochemical, and kinetic characteristics of better adaptability than mesophiles.

5.3.2 Upstream Bioprocessing

The increased production of L-asparaginase has been studied in microorganisms due to their growth rate and metabolic adaptability. Likewise, culture conditions modulate gene expression and can enhance its production. Table 5.1 summarizes the optimal culture conditions for the production of L-asparaginase in different microorganisms. In relation to the

Table 5.1 Optimized conditions for L-asparaginase production by different microorganisms.

Strain	Culture media (g/L)	Conditions	Activity	Reference
Acinetobacter baumannii ZAS1	1.32 Glucose, 11.50 L-asparagine, 5.30 peptone, 7.54 Na_2HPO_4	24 h, pH 7, 37 °C, 150 RPM	45.59 U/mL	[46]
Aspergillus sydowii	2 Glucose, 10 L-asparagine, 1 KH_2PO_4, 0.25 KCl, 0.52 $MgSO_4{\cdot}7\ H_2O$	5 days, pH 5, 40 °C, 100 RPM	146 U/mL	[47]
Aspergillus terreus	5 Glucose, 4.5 L-asparagine,1.5 KH_2PO_4, 0.5 K_2HPO_4, 0.75 $MgSO_4{\cdot}7\ H_2O$, 37.5 NaCl	5 days, pH 6, 35 °C	8.26 U/mg	[48]
Bacillus altitudinis BITHSP010	1.5 Lactose, 20 L-asparagine, 1.2 NaCl, 5 yeast extract	18 h, pH 7, 37 °C, 120 RPM	7.14 U/mL	[49]
Bacillus australimaris NJB19	200 Glucose, L-asparagine 24.4, $Na_2HPO_4{\cdot}2H_2O$ 6, KH_2PO_4 3, NaCl 0.5, 1 M $MgSO_4{\cdot}7\ H_2O$, 0.1 M $CaCl_2{\cdot}2H_2O$	72 h, pH 6.77, 33.5 °C	37.93 U/mL	[50]
Bacillus licheniformis PPD37	2 L-asparagine, 0.2 $MgSO_4$, 0.8 NaCl	48 h, pH 7.5, 37 °C, 120 RPM	17.09 U/mL	[51]
Bacillus sp. Asp11	3 Glucose/maltose, 6 L-asparagine, 1 $MgSO_4{\cdot}7\ H_2O$, 2 KH_2PO_4, 1 $CaCl_2{\cdot}2\ H_2O$	24 h, pH 7, 37 °C, 200 RPM	12.73 U/mg	[52]
Bacillus subtilis VUVD001	5 Glucose, 80 beef extract	6 days, pH 8.3, 49.9 °C, 200 RPM	2.88 U/mL	[53]
Cladosporium tenuissimum	2 Glucose, 10 L-asparagine, 1.52 KH_2PO_4, 0.52 $MgSO_4{\cdot}7\ H_2O$, 0.52 KCl, 0.03 $FeSO_4$, 0.03 $CuNO_3{\cdot}3\ H_2O$, 0.05 $ZnSO_4{\cdot}7\ H_2O$	72 h, pH 6.2, 37 °C, 150 RPM	2.65 U/mL	[54]
Fusarium equiseti AHMF4	10 glucose, 2 L-asparagine, 1 alanine, 1 KH_2PO_4	7 days, pH 7, 30 °C	40.78 U/mL	[55]
Lysinibacillus fusiformis B27	20 L-asparagine, 0.075 yeast extract, 0.375 $Na_3C_6H_5O_7$, 0.1 $(NH_4)_2HPO_4$, 0.00625 K_2HPO_4, 0.01 $MgSO_4{\cdot}7\ H_2O$, 0.001 $FeSO_4{\cdot}7\ H_2O$	24 h, pH 6, 35 °C, 120 RPM	8.51 U/mL	[56]
Pectobacterium carotovorum FS-4	10 Lactose, 10 NH_4NO_3, 1.52 KH_2PO_4, 0.52 KCl, 0.52 $MgSO_4{\cdot}7\ H_2O$	48 h, pH 7, 35 °C, 120 RPM	4.84 U/mL	[57]
Pseudomonas fluorescens MTCC103	1.5 Glucose,1 L-asparagine, 1.2 K_2HPO_4, 7 yeast extract, 5 tryptone	24 h, pH 7, 37 °C, 180 RPM	39.9 U/mg	[58]
Pseudomonas resinovorans IGS-131	0.2% [v/v] glycerol, 45 L-asparagine, 9 Na_2HPO_4, 1.5 KH_2PO_4, 0.25 NaCl	24 h, pH 7, 37 °C, 250 RPM	37.63 U/mL	[59]

Table 5.1 (Continued)

Strain	Culture media (g/L)	Conditions	Activity	Reference
Sarocladium strictum	10 Starch, 10 glucose, 5.05 glycerol, 2.5 yeast extract, 10 L-asparagine	5 days, pH 8.4, 28 °C, 150 RPM	1.73 U/mL	[60]
Serratia marcescens GS-7	10 Sucrose, 10 L-asparagine, 1.52 KH_2PO_4, 0.52 KCL, 0.52 $MgSO_4{\cdot}7\ H_2O$	48 h, pH 7.5, 30 °C, 120 RPM	5.22 U/mL	[57]
Streptomyces brollosae NEAE-115	15 Soybean meal + wheat bran (1 : 1), 2 glucose, 2 fructose,10 L-asparagine, 2 yeast extract, 1 KNO_3, 2 K_2HPO_4, 0.1 $MgSO_4{\cdot}7\ H_2O$, 0.1 NaCl, 0.02 $FeSO_4{\cdot}7\ H_2O$, 0.01 $CaCl_2$	7 days, pH 7, 30 °C 150 RPM	129.92 U/g substrate	[61]
Streptomyces fradiae NEAE-82	4 Glucose, 15 L-asparagine, 2 KNO_3, 0.5 $MgSO_4{\cdot}7\ H_2O$, 1 K_2HPO_4, 0.02 $FeSO_4{\cdot}7\ H_2O$, 0.2 NaCl, 0.01 $ZnSO_4$	7 days, pH 8.5, 37 °C, 100 RPM	53.57 U/mL	[62]
Streptomyces griseus NIOT-VKMA29	14.8 Starch, 15.3 L-asparagine, 20.4 yeast extract	6 days, pH 8, 35 °C	56.78 U/mL	[63]
Streptomyces labedae VSM-6	15 Starch, 10 yeast extract, 10 L-asparaginase	6 days, pH 8, 30 °C	10.17 U/mL	[64]
Streptomyces lacticiproducens 112	2.5 Lactose, 0.15 L-asparagine, 0.15 malt extract	7 days, pH 7.36, 32 °C,130 RPM	88.54 U/mL	[65]
Streptomyces olivaceus NEAE-119	4 Glucose, 10 starch, 0.1 NaCl, 10 L-asparagine, 1 KNO_3, 1 yeast extract, 1 K_2HPO_4, 0.1 $MgSO_4{\cdot}7\ H_20$	5 days, pH 5, 35 °C, 150 RPM	70.08 U/mL	[66]
Streptomyces rochei NEAE-K	2.6 Glucose, 7 L-asparagine, 3 yeast extract, 15 wheat bran-soybean meal (1 : 1), 4 fructose, 0.01 $CaCl_2$, 0.02 $FeSO_4{\cdot}7\ H_2O$, 1 KNO_3, 2 K_2HPO_4, 0.5 $MgSO_4{\cdot}7\ H_2O$, 0.1 NaCl	7 days, pH 6.7, 37 °C, 150 RPM	145.40 U/g substrate	[67]

carbon source, Moguel et al. [68] compared the effect of glycerol and sucrose on enzyme production in *Leucosporidium scottii* and obtained better performance with the former, which would be related to growth stress. However, glucose is metabolized first, and in yeast it represses the transcription of genes necessary for the consumption of other carbon sources [69].

In endophytic fungi, glucose concentration regulates L-asparaginase production; thus, Fusarium species obtain high activity with 5% and higher values would exert a repressor effect [70, 71]. In contrast, in *Streptomyces rochei* subsp. *chromatogenes* NEAE-K L-asparaginase production was optimized using 25% glucose, and in *Streptomyces olivaceus*

NEAE-119, 30% glucose was shown to exert a positive effect [66, 67]. In *Streptomyces tendae* TK-VL_333, lactose exerted minimal effect on L-asparaginase production, followed by starch, whereas the best levels (5.27 U/mL) were obtained with sucrose [72]. The comparison of substrates and levels is variable, since Gurunathan and Sahadevan [73] obtained higher L-asparaginase activity with glucose compared with sucrose, fructose, lactose, and maltose in *Aspergillus terreus*, where it was also evidenced that values higher than 0.4% inhibit production due to growth inhibition and catabolic repression.

Hamed et al. [54] evaluated in *Cladosporium tenuissimum* that the 2 : 10 carbon to nitrogen ratio improved activity using glucose and urea as the best substrates. However, *Zymomonas mobilis* ATCC 35001 evidenced that its best production is obtained with the 1 : 0.025 ratio for sucrose and yeast extract, in addition to being supplemented with L-asparagine 0.9 g/L [74]. Likewise, *Fusarium proliferatum* required two and a half times more carbon than nitrogen to reach maximum L-asparaginase production [75].

Bacillus velezensis KB-9 and *Bacillus licheniformis* KKU-KH14 also demonstrate that glucose is the source that significantly enhances L-asparaginase production, in addition to increasing their cell growth [76, 77]. *Citrobacter* sp. AS-2 obtained better activity using glucose than sucrose. Furthermore, it was evidenced that the best production is obtained in the stationary phase of growth [78]. However, high carbon concentrations generate a repressor effect due to the intracellular decrease of cAMP and of the components involved in lactate transport associated with L-asparaginase synthesis [79].

Alternative carbon sources have been used to decrease the production cost, such as tapioca starch in *Bacillus* sp. GH5, with which better activity was obtained compared to glucose and sucrose [80]. Also, coconut oil paste 6 g/L by solid fermentation induced high production in *Serratia marcescens* NCIM 2919 [81]. Furthermore, El-Naggar et al. [61] used in equal proportions soybean meal and wheat bran as higher impact sources than glucose and fructose in the production of L-asparaginase 47.67 U/g in *Streptomyces brollosae* NEAE-115.

The use of nitrogen promotes enzyme production since nitrogen availability regulates the transcription of structural genes [31]. However, organic nitrogen exerts a suppressive effect on L-asparaginase production, and the inorganic one stimulates production, such as magnesium nitrate, which was used to obtain L-asparaginase 1.09 U/mL in *Rhizopus oryzae* AM16 [82]. Likewise, *Aspergillus carneus* and *Penicillium camemberti* produced maximum levels of L-asparaginase using ammonium chloride, whereas *C. tenuissimum* required urea [54]. Similarly, L-asparagine acts as an inducer in L-asparaginase production, with better production than yeast extract, peptone, glutamine, and arginine in *P. aeruginosa* WCHPA075019 [83]. High concentrations of L-asparagine are usually applied, so *Pseudomonas resinovorans* IGS-131 needed 45 g/L to get its maximum activity; higher concentrations produce excess ammonium that can affect the enzyme [59]. Moreover, *B. licheniformis* RAM-8 required 40 g/L to produce 31.31 U/mL of L-asparaginase, while *Bacillus altitudinis* BITHSP010 decreased its activity with amounts over 20 g/L [49, 84].

Furthermore, the production medium of L-asparaginase requires ionic source, where 0.025 g/L $MgSO_4$, 0.025 g/L NaCl, and 7.54 g/L Na_2HPO_4 have been shown to increase activity in *Acinetobacter baumannii* ZAS1 as previously reported for optimized production in *E. coli* using 0.75 g/L KH_2PO_4 and 0.5 g/L NaCl [46, 85]. Also, K_2HPO_4 showed an indirectly proportional relationship with glucose and NaCl on the L-asparaginase activity of *P. aeruginosa* SN004 [86]. However, El-Naggar et al. [61] indicated that $MgSO_4$ and NaCl negatively affect production in *S. brollosae* NEAE-115, and $FeSO_4 \cdot 7\ H_2O$ and $CaCl_2$ would exert a positive effect on L-asparaginase activity. While $MgSO_4 \cdot 7\ H_2O$, K_2HPO_4, $FeSO_4 \cdot 7\ H_2O$, and $CaCl_2$ could have a positive effect on *S. rochei* subsp. *chromatogenes* NEAE-K [67].

The fermentation time has been optimized according to the metabolic characteristics of each microorganism; hence, *Streptomyces ginsengisoli* starts producing L-asparaginase after 24–32 hours and reaches its maximum level at 120 hours [87]. Likewise, L-asparaginase production has been observed at the end of the exponential phase; for example, *S. brollosae* NEAE-115 reached maximum production levels at 104 hours, before the stationary phase [88]. Production at the end of the exponential and early stationary phase in coliform fungi and bacteria is regulated by carbon and nitrogen sources [78, 89]. In contrast, recombinant L-asparaginase II production in *E. coli* BL21 (De3) takes place in the middle of the exponential phase, where the induction phase after 16–24 hours with either IPTG or lactose plays an important role [90]. *Pichia pastoris* induced recombinant L-asparaginase production using 1.0% (v/v) methanol, where no direct correlation was observed between biomass produced and specific activity when the carbon source was depleted at 8 hours [91]. IPTG concentration between 0.45 and 1 mM showed no significant difference in activity; in opposition, 10 g/L lactose increased both biomass and enzyme activity [90].

Stirring significantly influences L-asparaginase production since its application improves it by 24.6% compared to static fermentation in *B. subtilis* MK072695 [92]. Likewise, in relation to the fermentation time, recombinant *B. licheniformis* required 190 RPM in 18 hours to maximize its production, while the *E. coli* type I needed 220 RPM [39, 93]. In *S. brollosae* NEAE-115, its best production was obtained at 150 RPM despite not being a significant factor [88].

The culture media depends on the physiological requirements of each species, as well as the capacity to induce or repress the production of L-asparaginase. Culture conditions are variable and require a previous study to maximize biomass and consequently activity. In this aspect, the search for profitability is permanent, so the use of alternative media could increase production. However, genetic engineering techniques can reduce nutritional requirements and improve the specific production of L-asparaginase, which has increased the development of recombinant strains.

5.3.3 Downstream Bioprocessing

The steps of this process are the most expensive, which has led to the development of numerous techniques that improve purification and do not require high costs. Clarification by centrifugation is the most widely used recovery process due to high perfusion capacity, separation efficiency, and cell removal, then successive methods that contribute to purification are applied [94].

5.3.3.1 Protein Concentration

One of the most widely applied methods is precipitation by increasing salt concentration (salting out). Ammonium sulfate is the most widely used salt due to its effectiveness, which also depends on protein's nature [95]. The application of 70% NH_4SO_4 in *P. aeruginosa* PAO1 increased 13% specific activity, and when it increased to 90% in *Pseudomonas otitidis*

KF607097, it increased 515% [96, 97]. Likewise, its use between 40 and 95% increased 532% specific activity in *Talaromyces pinophilus* KJ372306 [98]. However, the application of high salt concentrations requires their subsequent removal by dialysis and passage through a desalting column. Other technologies, such as precipitation using cationic poly-L-lysine in low ionic strength media, have been developed and demonstrated heat tolerance, agitation, and stability of the secondary structure of L-asparaginase [99].

Organic solvents have also been used for protein concentration, such as 20–80% acetone, which improved the specific activity of L-asparaginase from *B. licheniformis* RAM-8 by 39%, and 50% ethanol increased the specific activity by 373% in *Trichoderma viride* F2 [100, 101]. Precipitated proteins are dried and reconstituted or subjected to dialysis in order to remove high percentages of organic solvents that limit subsequent purification by chromatography [101]. L-Asparaginase precipitation improves the specific activity; however, the effects it generates at the structural and functional level must be considered, in addition to the costs in subsequent treatments that contribute to its purification, so the search for innovative techniques that improve its performance and reduce the stressful effects on biomolecules is still ongoing.

5.3.3.2 L-Asparaginase Release

Intracellular L-asparaginase can be extracted using high-pressure homogenization, which is applied to intermediate and large volumes, and sonication, where the total power applied to the samples must be considered [103, 104]. Both lead to cell debris, which is critical in the optimization of the scale-up process [105]. Likewise, agitation in Triton X100 solution with potassium or sodium salts promotes release in *E. coli* ATCC 11303, where divalent and trivalent anions reported higher release compared to univalents [106]. Similarly, the use of polymers in potassium phosphate buffer accompanied by high-pressure homogenization showed better specific activity in *E. coli* compared to the two-step extraction process [103].

Also, ultrasonication in a lysis buffer with lysozyme has been combined for the release of thermostable L-asparaginase expressed in *B. subtilis* 168 [107]. Likewise, 10 mM lysozyme and 10% sodium dodecyl sulfate, and sonication have been combined in *Salmonella typhimurium* [108]. Proteins that are expressed as inclusion bodies are solubilized by high concentrations of chaotropic solvents and detergents together with reducing agents such as β mercaptoethanol, dithiothreitol, or cysteine [109].

The purification buffer minimizes the loss of activity, so the pH in accordance with the isoelectric point, stability of the buffer system, and protease inhibiting agents, among others, must be considered. The release method is related to the site of expression of L-asparaginase; however, the use of ultrasound is still the most used method in combination with detergents and enzymes.

5.3.3.3 Chromatography

There are different chromatographic techniques, such as ion exchange, hydrophobic interaction, affinity, gel filtration, among others, that have demonstrated selectivity and capacity of the chromatographic medium. Thus, the nature, ionic strength, and specificity of the interactions between the covalent bonds in the solid phase and the charged molecule of interest are considered in the separation [110]. The key factors in purification are isoelectric point, stability at different pH values, hydrophobic composition, signal peptide, affinity-Tags, and salt concentration [111].

Anion exchange chromatography is one of the most widely applied due to its high resolving power; thus, the sample is eluted in a linear salt gradient, collected and subsequently dialyzed, the purification of intracellular L-asparaginase from *Chaetomium* sp. MT484261 shows 47% specific activity [104]. Likewise, L-asparaginase from *P. otitidis* improved 76-fold specific activity, after elution with a KCl gradient and Tris-HCl buffer pH 8.6 [97].

However, due to low yields and aggregate formation, a periodic countercurrent chromatographic system has been designed for continuous folding of proteins produced in inclusion bodies in *E. coli*, thereby improving resin usage, recovery, and efficiency compared to the batch system [112].

Gel filtration removes contaminants and separates molecules according to size, which is required after ion exchange chromatography. Thus, in the purification of L-asparaginase from *P. otitidis* KF607097, Sephadex G-100 was used, which improved the specific activity by 100%, while in *B. licheniformis* RAM-8, it was over 66% [97, 100]. Sephadex G-200 contributed in improving the purification by 4.55-fold of L-asparaginase from *B. altitudinis* BITHSP010 [113].

Likewise, other methods have been used, such as separation by native gel electrophoresis, after precipitation with ammonium sulfate and dialysis, where the specific activity of L-asparaginase from *E. coli* was improved by 75% [114]. Otherwise, unique processes have been used to facilitate purification, such as adsorptive electrophoresis on Sepharose CL-6B, where extracellular recombinant L-asparaginase from *E. chrysanthemi* expressed in *E. coli* was immobilized, evidencing high specificity and affinity [115]. However, the design of recombinant proteins with Histidine tails facilitates their purification by using Nickel affinity columns, which can be complemented with anion exchange chromatography [116].

Chromatographic processes have been designed to help save reagents and time; however, the recovery percentage, sample purity, and specific activity keep anion exchange chromatography as one of the most applied techniques. It can be integrated with gel filtration or affinity techniques, depending on the protein structure and the design techniques of the recombinant proteins.

5.4 Scaled Up to Bioreactor

The scaling up of microbial processes is oriented toward profitable production for industrial applications, which requires implementation and control of the metabolic processes. Thus, among the most commonly used methods is batch fermentation, as applied in *B. licheniformis* RAM-8, which by culture at 0.75 vvm and 200 RPM produced 29.94 U/mL of L-asparaginase at 15 hours [84]. However, its limitation in scale-up has led to exponential fed culture, where glucose is added when the highest point of dissolved oxygen is reached, besides injection of compressed air and pure oxygen at 2 vvm whereby high cell density is obtained with a maximum growth rate of $0.3\,h^{-1}$. The feeding volume is performed considering the specific growth rate (maximum $0.1–0.3\,h^{-1}$), so that acetate formation, which inhibits growth and recombinant protein production, is kept low [90]. Thus, their continued feeding after reaching the end of the exponential phase is related to maintaining a constant growth rate ($0.1\,h^{-1}$), which allows a substrate supply between maintenance and threshold [117].

Likewise, the initial agitation at 600 RPM for the first 12 hours induces biomass production in less time, since higher values generate stress due to shear forces. Subsequently, due to the decrease in saturation and limitation in cell growth, it is necessary to increase it to 900 RPM, which improves growth and therefore L-asparaginase activity [118]. However, in *P. resinovorans* IGS-131, the maximum enzyme activity of 38.88 U/mL was obtained at 0.75 vvm under batch culture and 400 RPM, and in *Enterobacter aerogenes,* the maximum level of 1.2 U/mL was at 700 RPM and 1 vvm, so the amount of dissolved oxygen would be responsible for the differences in enzyme activity [59, 119].

In the production of recombinant L-asparaginase from *L. scottii*, soluble oxygen was controlled by agitation at 200–720 RPM and a flow rate of 0.22 L/min to prevent decay of air saturation by more than 30%; also L-asparaginase production (800 U/g) was stimulated by increasing the cell density [68]. Moreover, flask to bioreactor scale-up of recombinant L-asparaginase from *S. cerevisiae* in *P. pastoris* MUT showed a decrease in the maximum specific growth rate, while its production at constant pH was determinant to increase the expression of periplasmic L-asparaginase at 500 RPM and 1 vvm [91].

Scale-up requires the control of supplemented oxygen, which limits growth and thus the synthesis of L-asparagine, which occurs between the period of oxygen deprivation. Likewise, the feeding rate of a culture is related to the control of the specific growth rate, and in the case of recombinants, it is necessary to maintain low acetate production. Scale-up has been performed at the laboratory level with native and recombinant strains; however, larger scale production remains a challenge both economically and in bioprocesses.

5.5 Characterization of L-Asparaginase

Physicochemical conditions determine L-asparaginase activity, where some microorganisms produce enzymes with acidic, neutral, or alkaline tendencies. Thus, L-asparaginases I and II of *P. aeruginosa* PAO1 are alkalophilic and present better activity at pH 9.5 [96]. However, some species possess good activity under acidic conditions, as in *P. aeruginosa* SN004, which was active between 4.5 and 7.0, with optimal activity at pH 5, and retained more than 80% activity at pH 7–8 after 24 hours [86]. Its medical applicability requires enzymes with high activity at conditions close to physiological pH. Accordingly, *Staphylococcus* sp. MGM1 produced L-asparaginase of maximal activity at pH 8, which also retained more than 50% activity from pH 6.5 to 8.5 [14]. Also, *T. pinophilus* KJ372306 showed high activity at pH 8, with stability between pH 7 and 9 for 8 hours. Meanwhile, *Bacillus megaterium* MG1 retained more than 40% activity between pH 6 and 9, with maximum activity at pH 8.5, and *Bacillus* strains had optimal activity between pH 6 and 7 [98, 120, 121].

Most microorganisms produce L-asparaginase with optimal levels between 37 and 40 °C, such as *B. megaterium* MG1, which obtained it at 37 °C, and higher values led to rapid loss of activity [121]. L-Asparaginase from *P. otitidis* KF607097 exhibited maximum activity at 40 °C and significant loss after 30 minutes at 45 °C [97]. Also, *T. pinophilus* KJ372306 evidenced maximum activity at 28 °C, with high stability between 25 and 37 °C for 8 hours [98]. However, some microorganisms exhibit high tolerance to elevated temperatures, such as *T. gammatolerans* EJ3 whose L-asparaginase showed maximum activity at 85 °C and retention of over 75%

between 70 and 95 °C, while *Stenotrophomonas maltophilia* EMCC2297 retained 55% of its activity after 30 minutes at 70 °C and no significant changes at 50 °C [45, 122].

The high affinity for L-asparagine determines the effectiveness in clinical treatment and the decrease of acrylamide concentrations in food products, which is usually measured as Km, where lower values indicate higher affinity. In this regard, *Bacillus* have demonstrated high affinity for L-asparagine; thus, *B. megaterium* MG1 presented Km and apparent Vmax for L-asparagine of 0.2 mM and 1.198 mM/s; and *Bacillus* PG04 exhibited 3.3 mM and 0.024 μmol/min [120, 121]. Moreover, the maximum rate indicates the conversion of substrate to the product in a unit time under substrate saturating conditions. Thus, *Streptomyces fradiae* NEAE-82 obtained 95.08 U·mL/min, where high conversion per unit time was evidenced [123]. However, these parameters are affected according to the degree of purity and source of the enzyme, type of substrate, physicochemical conditions, and laboratory procedures. Kinetic parameters of L-asparaginase produced by different microorganisms are shown in Table 5.2.

The presence of ions or detergents in some cases has no effect on the enzymatic activity. In others, they increase or inhibit it. Among the ions that enhance L-asparaginase activity are K^+, Fe^{3+}, and Na^+, which increased it by 254, 236, and 136%, respectively [14], while Mg^{2+}, Mn^{2+}, and Na^+ increased the activity by 116, 112, and 113% in *P. aeruginosa* SN004 [86]. However, Ca^{2+}, Cu^{2+}, Co^{2+}, Hg^{2+}, and Zn^{2+} are potent inhibitors of L-asparaginase from *Staphylococcus* sp. MGM1, and Hg^{2+} is an inhibitor in *P. aeruginosa* SN004 [14, 86]. Likewise, L-asparaginase activity produced by *Streptomyces bollosae* NEAE-115 was reduced 37.55% in the presence of EDTA, while in *P. aeruginosa* SN004, Triton X-100 enhanced it by 14% [86, 126].

Table 5.2 Kinetic conditions of L-asparaginase production by microorganisms.

Strain	Size (kDa)	Vmax	Km (mM)	Reference
Aspergillus oryzae CCT 3940	115	313 U/mL	0.66	[124]
Aspergillus terreus	85	500 U/mL	31.50	[125]
Bacillus altitudinis BITHSP010	35	0.09 M/S	0.91	[49]
Bacillus licheniformis KKU-KH14	37	45.45 μmol/mL/min	49.99	[77]
Bacillus megaterium MG1	47	1.19 mM/s	0.20	[121]
Bacillus PG04	30	0.02 μmol/min	3.30	[120]
Bacillus velezensis KB-9	39.7	41.49 μmol/mL/min	0.04	[77]
Chaetomium sp. MT484261	66	38.74 μmol/min	257.10	[104]
Lasiodiplodia theobromae SCUF-TP2016	70	127.00 μM/mL/min	$9.37*10^{-3}$	[30]
Pseudomonas aeruginosa SN004	34	5.83 μmol/min	1.17	[86]
Pseudomonas fluorescens MTCC103	160	28.57 μM/mL/min	$4.51*10^{-3}$	[58]
Sarocladium strictum	-	8.19 μmol/min	9.74	[60]
Streptomyces brollosae NEAE-115	67	152.60 U/mL/min	2.14	[126]
Streptomyces lacticiproducens 112	37	57.23 U/mL	19.70	[65]
Trichoderma viride F2	57	71.3 U/mL	1.20	[101]

Immobilization techniques have shown to improve affinity for L-asparagine by decreasing Km and increasing thermal and storage stability [127, 128]. Likewise, the change in amino acid residues in *E. carotovora* and *Rhodospirillum rubrum* modulates the structural and functional properties, which alter their kinetic parameters, thermal and exposure stability to successive cycles of thawing [129, 130].

5.6 Applications of L-Asparaginase

5.6.1 Pharmaceutical Industry

L-Asparaginase has great importance in the pharmaceutical industry due to its application in the treatment of acute lymphoblastic leukemia, Hodgkin's lymphomas, and other cancers, so the World Health Organization [131] has considered it a safe medicine. Its chemotherapeutic capacity lies in decreasing serum levels of L-asparagine, an amino acid required by cancer cells to facilitate their rapid growth. These cells are dependent on serum L-asparagine concentration and exhibit minimal levels of L-asparagine synthetase, unlike normal cells [132]. Among commercial sources, *E. coli* stands out due to its ease of processing and efficacy as the first line of treatment. However, the incidence of adverse effects leads to the search for new sources and to the generation of recombinant proteins of better acceptability. Another source is *E. carotovora*, which unlike *E. coli* develops fewer cytotoxic effects with greater efficiency [133, 134].

Moreover, the main limitations are stability and the generation of hypersensitivity reactions that cause enzymatic inactivation, which has prompted the development of technologies that facilitate acceptability and effectiveness [134]. Thus, PEG-L-asparaginase has been shown to decrease allergic reactions in more than 50% of patients with acute lymphoblastic leukemia and increase serum half-life without alteration in its biological activity [135, 136]. Also, mutagenesis involves reducing L-glutaminase activity by altering amino acids around the active site that noninteract with the substrate [137]. Deficiency in resistance to proteases such as asparagine endopeptidase led to the change of residues susceptible to cleavage, which in turn compromises the stabilization of the active site and thus decreased its activity [138].

Likewise, the use of bioinformatics techniques has allowed the identification of residues involved in antigenicity, toxicity, and stability in association with mutagenesis can modulate the therapeutic response. Also, formulation techniques such as immobilization or nanoencapsulation improve their stability, permeation, and accumulation in the therapeutic area [137]. Therefore, the application of these techniques and the progress of new ones contribute to the production and formulation of L-asparaginase with desirable pharmaceutical characteristics that improve its pharmacokinetics and pharmacodynamics.

5.6.2 Food Industry

The processing of foods rich in reducing sugars and asparagine at high temperatures (>120 °C) generates the Maillard reaction and, therefore, the formation of acrylamide, which has been classified as a probable human carcinogen by the International Agency for Research on Cancer (IARC) [139]. Foods with the highest levels of acrylamide are thermo-processed products such as breads, cereals, potato chips, among others. Its exposure causes neurotoxicity, genotoxicity, decreased immunity, and tumor development, being able to cross the placental barrier [140].

Acrylamide levels have been reduced in French fries after immersion of the potatoes in the enzyme and subsequent bleaching in water, thus decreasing more than 80% of acrylamide with minimal loss in its amino acid composition [141]. Likewise, its application in bakery products has reduced up to 90% acrylamide without significant changes in appearance and 37% increase of glutamine in the dough [142]. Incubation of cereal dough for 48 hours prior to baking reduced up to 97% acrylamide, with no change in sensory quality [143]. However, prolonged pretreatment requires enzymes with thermal tolerance, such as *Thermococcus zilligii* AN1 TziAN1_1 and *Thermococcus kodakarensis* KOD1 (TkAsn) with activity at 90 °C and high stability [144].

Commercial enzymes such as Acrylaway from Novozymes A/S and PreventASe from DSM produced from *Aspergillus* by recombinant DNA technology are available on the market and have demonstrated high specificity without amino acid alteration [2] Pretreatment with L-asparaginase significantly decreases L-asparagine and thus acrylamide formation subjected to the same traditional heating conditions. Likewise, in its industrial scale-up, it was observed 43–53% removal of acrylamide in 8 tons of potatoes/hour proportional to the contact surface area [145]. Therefore, the difficulty in enzyme accessibility is a challenge for its complete removal of acrylamide in various food products. Also, thermostable enzymes are required for prolonged incubation periods or enzymes that demonstrate higher catalytic efficiency in a shorter time [146].

5.7 Conclusions

L-Asparaginase is very important both at clinical and food level; however, its production requires a meticulous control of all production and purification processes. Likewise, the selection of the strain directly influences its biochemical properties and, therefore, the search for strains with tolerance to harsh conditions and high specificity is permanent. In addition, greater knowledge of their metabolic fluxes and genetic control has facilitated the generation of recombinants with better biochemical characteristics. Scale-up has demonstrated the importance of controlling parameters that determine the production of L-asparaginase and its adaptability to laboratory scale. However, few investigations have studied its production at the industrial level and obtained high levels of catalytic activity under cost-effective conditions. Finally, the formulation of the final product can be improved through the use of technologies such as immobilization to improve its stability and recyclability, or the design of micro- or nano-encapsulations that increase the reaction rate. L-Asparaginase is in high demand, so the study of its bioprocessing is crucial to improve its profitability and quality.

References

1 Koprivnikar, J., McCloskey, J., and Faderl, S. (2017). Safety, efficacy, and clinical utility of asparaginase in the treatment of adult patients with acute lymphoblastic leukemia. *Onco. Targets. Ther.* 10: 1413–1422.

2 Xu, F., Oruna-Concha, M.J., and Elmore, J.S. (2016). The use of asparaginase to reduce acrylamide levels in cooked food. *Food Chem.* 210: 163–171.

3 Vidya, J., Sajitha, S., Ushasree, M.V. et al. (2016). Therapeutic enzymes: L-asparaginases. In: *Current Developments in Biotechnology and Bioengineering: Production, Isolation and Purification of Industrial Products* (ed. A. Pandey, S. Negi, and C. Soccol), 249–265. Elsevier B.V.

4 Singh, A., Verma, N., and Kumar, K. (2018). L-Asparaginase from *Phyllanthus emblica* (AMLA): a novel source. *Int. J. Pharm. Sci. Res.* 9 (12): 5394–5400.

5 Atkins, C.A. (2013). Mechanism of long-distance solute transport in phloem elements. In: *Symplasmic Transport in Vascular Plants* (ed. K. Sokołowska and P. Sowiński), 165–181. Springer Science.

6 Lopes, A.M., de Oliveira-Nascimento, L., Ribeiro, A. et al. (2017). Therapeutic l-asparaginase: upstream, downstream and beyond. *Crit. Rev. Biotechnol.* 37 (1): 82–99.

7 Singh, S.P. and Upadhyay, S.K. (eds.) (2021). *Bioprospecting of Microorganism-Based Industrial Molecules*. John Wiley & Sons Ltd. doi:10.1002/9781119717317.

8 Moorthy, V., Ramalingam, A., Sumantha, A., and Shankaranaya, R.T. (2010). Production, purification and characterisation of extracellular L-asparaginase from a soil isolate of *Bacillus* sp. *Afr. J. Microbiol. Res.* 4 (18): 1862–1867.

9 Kiranmayi, M.U., Poda, S., and Vijayalakshmi, M. (2014). Production and optimization of L-asparaginase by an *Actinobacterium* isolated from Nizampatnam mangrove ecosystem. *J. Environ. Biol.* 35 (5): 799–805.

10 Muslim, S.N., AL-Kadmy, I.M.S., Hussein, N.H. et al. (2016). Extraction and purification of L-asparaginase produced by *Acinetobacter baumannii* and their antibiofilm activity against some pathogenic bacteria. *Int. J. Biotechnol.* 5 (1): 7–14.

11 Gilavand, F., Marzban, A., and Kavyanifard, A. (2019). L-Asparaginase-producing *Rouxiella* species isolation, antileukemia activity evaluation, and enzyme production optimization. *Res. Mol. Med.* 6 (3): 40–59.

12 Kumar, S.N., Haseeb Nawaz, M., Shankar Mishra, S. et al. (2019). Anti-Cancer Enzyme (L-Asparaginase) production, purification and characterization from a soil isolate of *Pseudomonas* sp. *Int. J. Adv. Res.* 7 (12): 753–761.

13 Dinarvand, B., Rezaei, P.F., Akbari, N., and Branch, A. (2020). Optimization of L-asparaginase production from a *Lactobacillus* sp. isolated from traditional dairy products. *Biol. J. Microorg.* 9 (34): 71–86.

14 Ebrahimipour, G., Yaghoobi Avini, M., and Ghorbanmovahhed, M. (2020). Isolation and characterization of glutaminase-free L-asparaginase produced by *Staphylococcus* sp. MGM1. *SciMedicine J.* 2 (2): 46–55.

15 Izadpanah, F., Homaei, A., Fernandes, P., and Javadpour, S. (2018). Marine microbial L-asparaginase: biochemistry, molecular approaches and applications in tumor therapy and in food industry. *Microbiol. Res.* 208: 99–112.

16 Loch, J.I. and Jaskolski, M. (2021). Structural and biophysical aspects of L-asparaginases: a growing family with amazing diversity. *IUCrJ* 8 (4): 514–531.

17 de Moura, W.A.F., Schultz, L., Breyer, C.A. et al. (2020). Functional and structural evaluation of the antileukaemic enzyme l-asparaginase II expressed at low temperature by different *Escherichia coli* strains. *Biotechnol. Lett.* 42 (11): 2333–2344.

18 Sanches, M., Alexandre, J., Barbosa, R.G. et al. (2003). Biological crystallography structural comparison of *Escherichia coli* L-asparaginase in two monoclinic space groups. *Acta Crystallogr.* D59: 416–422.

19 Harms, E., Wehner, A., Aung, H.P., and Röhm, K.H. (1991). A catalytic role for threonine-12 of *E. coli* asparaginase II as established by site-directed mutagenesis. *FEBS Lett.* 285 (1): 55–58.

20 Swain, A.L., Jaskolski, M., Housset, D. et al. (1993). Crystal structure of *Escherichia coli* L-asparaginase, an enzyme used in cancer therapy. *Proc. Natl. Acad. Sci. U. S. A.* 90 (4): 1474–1478.

21 Jerlström, P.G., Bezjak, D.A., Jennings, M.P., and Beacham, I.R. (1989). Structure and expression in *Escherichia coli* K-12 of the L-asparaginase I-encoding ansA gene and its flanking regions. *Gene* 78 (1): 37–46.

22 Michalska, K., Hernandez-Santoyo, A., and Jaskolski, M. (2008). The mechanism of autocatalytic activation of plant-type L-asparaginases. *J. Biol. Chem.* 283 (19): 13388–13397.

23 Michalska, K., Bujacz, G., and Jaskolski, M. (2006). Crystal structure of plant asparaginase. *J. Mol. Biol.* 360 (1): 105–116.

24 Sun, D. and Setlow, P. (1993). Cloning and nucleotide sequence of the *Bacillus subtilis* ansR gene, which encodes a repressor of the ans operon coding for L-asparaginase and L- aspartase. *J. Bacteriol.* 175 (9): 2501–2506.

25 Jennings, M.P. and Beacham, I.R. (1990). Analysis of the *Escherichia coli* gene encoding L-asparaginase II, *ans*B, and its regulation by cyclic AMP receptor and FNR proteins. *J. Bacteriol.* 172 (3): 1491–1498.

26 Fisher, S.H. and Wray, L.V. (2002). *Bacillus subtilis* 168 contains two differentially regulated genes encoding L-asparaginase. *J. Bacteriol.* 184 (8): 2148–2154.

27 Dunlop, P.C., Meyer, G.M., Ban, D., and Roon, R.J. (1978). Characterization of two forms of asparaginase in *Saccharomyces cerevisiae*. *J. Biol. Chem.* 253 (4): 1297–1304.

28 da Cunha, M.C., dos Santos Aguilar, J.G., de Melo, R.R. et al. (2019). Fungal L-asparaginase: strategies for production and food applications. *Food Res. Int.* 126: 108658.

29 Chow, Y.Y. and Ting, A.S.Y. (2015). Endophytic l-asparaginase-producing fungi from plants associated with anticancer properties. *J. Adv. Res.* 6 (6): 869–876.

30 Moubasher, H.A., Balbool, B.A., Helmy, Y.A. et al. (2022). Insights into asparaginase from endophytic fungus *Lasiodiplodia theobromae*: purification, characterization and antileukemic activity. *Int. J. Environ. Res. Public Health* 19 (2): 680.

31 de Moura Sarquis, M.I., Morais Oliveira, E.M., Silva Santos, A., and da Costa, G.L. (2004). Production of L-asparaginase by filamentous fungi. *Mem. Inst. Oswaldo Cruz* 99 (5): 489–492.

32 Sotero-Martins, A., Da Silva Bon, E.P., and Carvajal, E. (2003). Asparaginase II-GFP fusion as a tool for studying the secretion of the enzyme under nitrogen starvation. *Braz. J. Microbiol.* 34 (4): 373–377.

33 Darvishi, F., Faraji, N., and Shamsi, F. (2019). Production and structural modeling of a novel asparaginase in *Yarrowia lipolytica*. *Int. J. Biol. Macromol.* 125: 955–961.

34 Ghasemian, A., Al-marzoqi, A.H., Al-abodi, H.R. et al. (2019). Bacterial l-asparaginases for cancer therapy: current knowledge and future perspectives. *J. Cell. Physiol.* 234 (11): 19271–19279.

35 Batool, T., Makky, E.A., Jalal, M., and Yusoff, M.M. (2016). A comprehensive review on l-Asparaginase and its applications. *Appl. Biochem. Biotechnol.* 178 (5): 900–923.

36 Kamble, K.D., Bidwe, P.R., Muley, V.Y. et al. (2012). Characterization of L-Asparaginase producing bacteria from water, farm and saline soil. *Biosci. Discov.* 3 (1): 116–119.

37 Sulistiyani, T.R. and Kusumawati, D.I. (2019). Diversity of endophytic bacteria as a L-asparaginase producer free of L-glutaminase. *J Kefarmas. Indones.* 9 (1): 28–39.

38 Sanjotha, G. (2016). Isolation, characterization and screening of anticancerous L-asparaginase producing microbes from costal regions. *Int. J. Res. Stud. Biosci.* 4 (8): 14–18.

39 Abdelrazek, N.A., Elkhatib, W.F., Raafat, M.M., and Aboulwafa, M.M. (2019). Experimental and bioinformatics study for production of l-asparaginase from *Bacillus licheniformis*: a promising enzyme for medical application. *AMB Exp.* 9 (1): 39.

40 Bhat, M.R., Nair, J.S., and Marar, T. (2015). Isolation and identification of L-asparginase producing *Salinicoccus* sp. M KJ997975 from soil microbial flora. *Int. J. Pharm. Sci. Res.* 6 (8): 3599–3605.

41 Barati, M., Faramarzi, M.A., Nafissi-Varcheh, N. et al. (2016). L-asparaginase activity in cell lysates and culture media of halophilic bacterial isolates. *Iran. J. Pharm. Res.* 15 (3): 435–440.

42 Montes, J., Canales, P.E., Flores-Santos, J.C. et al. (2021). Bacteria producing L-asparaginase isolated from *Peruvian saline* environments Bacterias productoras de L-asparaginasa aisladas de ambientes salinos peruanos. *Manglar* 18 (2): 193–199.

43 Li, X., Zhang, X., Xu, S. et al. (2019). Insight into the thermostability of thermophilic L-asparaginase and non-thermophilic L-asparaginase II through bioinformatics and structural analysis. *Appl. Microbiol. Biotechnol.* 103 (17): 7055–7070.

44 Setiawan, R. and Larasati, D.R. (2019). Screening of bacteria producing asparaginase free of glutaminase and urease from hot springs in west Sulawesi. *Biosaintifika J. Biol. Biol. Educ.* 11 (2): 218–225.

45 Zuo, S., Xue, D., Zhang, T. et al. (2014). Biochemical characterization of an extremely thermostable L-asparaginase from *Thermococcus gammatolerans* EJ3. *J. Mol. Catal. B: Enzym.* 109: 122–129.

46 Abhini, K.N., Rajan, A.B., Zuhara, K.F., and Sebastian, D. (2022). Response surface methodological optimization of L-asparaginase production from the medicinal plant endophyte *Acinetobacter baumannii* ZAS1. *J. Genet. Eng. Biotechnol.* 4: 22.

47 Ali, D., Ouf, S., Eweis, M., and Solieman, D.M. (2018). Optimization of L-asparaginase production from some filamentous fungi with potential pharmaceutical properties. *Egypt J. Bot.* 58 (3): 355–369.

48 Farag, A.M., Hassan, S.W., Beltagy, E.A., and El-Shenawy, M.A. (2015). Optimization of production of anti-tumor l-asparaginase by free and immobilized marine *Aspergillus terreus*. *Egypt J. Aquat. Res.* 41 (4): 295–302.

49 Prakash, P., Singh, H.R., and Jha, S.K. (2020). Production, purification and kinetic characterization of glutaminase free anti-leukemic L-asparaginase with low endotoxin level from novel soil isolate. *Prep. Biochem. Biotechnol.* 50 (3): 260–271.

50 Chakravarty, N., Priyanka, K., Singh, J., and Singh, R.P. (2021). A potential type-II L-asparaginase from marine isolate *Bacillus australimaris* NJB19: statistical optimization, in silico analysis and structural modeling. *Int. J. Biol. Macromol.* 174: 527–539.

51 Patel, P., Panseriya, H., and Dave, B. (2021). Development of process and data centric inference system for enhanced production of L-asparaginase from halotolerant *Bacillus licheniformis* PPD37. *Appl. Biochem. Biotechnol.* 194: 1659–1681.

52 Patil, R.C. and Jadhav, L.B. (2017). Screening and optimization of L-asparaginase production from *Bacillus* species. *IOSR J. Biotechnol. Biochem.* 3 (3): 32–36.

53 Erva, R.R., Venkateswarulu, T.C., and Pagala, B. (2018). Multi level statistical optimization of l-asparaginase from *Bacillus subtilis* VUVD001. *3 Biotech* 8 (1): 24.

54 Hamed, M., Osman, A., and Ates, M. (2021). Statistical optimization of L-asparaginase production by *Cladosporium tenuissimum*. *Egypt Pharm. J.* 20 (1): 51–58.

55 El-Gendy, M.M.A.A., Awad, M.F., El-Shenawy, F.S., and El-Bondkly, A.M.A. (2021). Production, purification, characterization, antioxidant and antiproliferative activities of extracellular L-asparaginase produced by *Fusarium equiseti* AHMF4. *Saudi J. Biol. Sci.* 28 (4): 2540–2548.

56 Prihanto, A.A., Yanti, I., Murtazam, M.A., and Jatmiko, Y.D. (2019). Optimization of glutaminase-free L-asparaginase production using mangrove endophytic *Lysinibacillus fusiformis* B27. *F1000Research* 8: 1938.

57 Abdel-Razik, N.E., El-Baghdady, K.Z., El-Shatoury, E.H., and Mohamed, N.G. (2019). Isolation, optimization, and antitumor activity of l-asparaginase extracted from *pectobacterium carotovorum* and *Serratia marcescens* on human breast adenocarcinoma and human hepatocellular carcinoma cancer cell lines. *Asian J. Pharm. Clin. Res.* 12 (2): 332–337.

58 Sinha, R.K., Singh, H.R., and Jha, S.K. (2015). Producton, purification and kinetic characterization of L-asparaginase from *Pseudomonas fluorescens*. *Int. J. Pharm. Pharm. Sci.* 7 (1): 135–138.

59 Mihooliya, K.N., Nandal, J., Kumari, A. et al. (2020). Studies on efficient production of a novel l-asparaginase by a newly isolated *Pseudomonas resinovorans* IGS-131 and its heterologous expression in *Escherichia coli*. *3 Biotech* 10 (4): 1–11.

60 Golbabaie, A., Nouri, H., Moghimi, H., and Khaleghian, A. (2020). L-asparaginase production and enhancement by *Sarocladium strictum*: in vitro evaluation of anti-cancerous properties. *J. Appl. Microbiol.* 129 (2): 356–366.

61 El-Naggar, N.E.A., Moawad, H., and Abdelwahed, N.A.M. (2017). Optimization of fermentation conditions for enhancing extracellular production of L-asparaginase, an anti-leukemic agent, by newly isolated *Streptomyces brollosae* NEAE-115 using solid state fermentation. *Ann. Microbiol.* 67 (1): 1–15.

62 Soliman, H.M., El-Naggar, N.E.A., and El-Ewasy, S.M. (2020). Bioprocess optimization for enhanced production of L-asparaginase via two model-based experimental designs by alkaliphilic *Streptomyces fradiae* NEAE-82. *Curr. Biotechnol.* 9 (1): 23–37.

63 Meena, B., Anburajan, L., Sathish, T. et al. (2015). L-asparaginase from *Streptomyces griseus* NIOT-VKMA29: optimization of process variables using factorial designs and molecular characterization of l-asparaginase gene. *Sci. Rep.* 5 (June): 1–12.

64 Mangamuri, U., Vijayalakshmi, M., Ganduri, V.S.R.K. et al. (2017). Extracellular L-asparaginase from *Streptomyces labedae* VSM-6: isolation, production and optimization of culture conditions using RSM. *Pharmacogn. J.* 9 (6): 932–941.

65 Arévalo-Tristancho, E., Díaz, L.E., Cortázar, J.E., and Valero, M.F. (2019). Production and characterization of L-asparaginases of *Streptomyces* isolated from the Arauca riverbank (Colombia). *Open Microbiol. J.* 13 (1): 204–215.

66 El-Naggar, N.E.A., Moawad, H., El-Shweihy, N.M., and El-Ewasy, S.M. (2015). Optimization of culture conditions for production of the anti-leukemic glutaminase free

L-asparaginase by newly isolated *Streptomyces olivaceus* NEAE-119 using response surface methodology. *Biomed. Res. Int.* 2015: 627031.

67 El-Naggar, N.E.A. and El-Shweihy, N.M. (2020). Bioprocess development for L-asparaginase production by *Streptomyces rochei*, purification and in-vitro efficacy against various human carcinoma cell lines. *Sci. Rep.* 10 (1): 1–21.

68 Moguel, I.S., Yamakawa, C.K., Pessoa, A., and Mussatto, S.I. (2020). L-asparaginase production by *Leucosporidium scottii* in a bench-scale bioreactor with co-production of lipids. *Front. Bioeng. Biotechnol.* 8: 1–11.

69 Gancedo, J.M. (1998). Yeast carbon catabolite repression. *Microbiol. Mol. Biol. Rev.* 62 (2): 334–361.

70 Thirunavukkarasu, N., Suryanarayanan, T.S., Murali, T.S. et al. (2011). L-asparaginase from marine derived fungal endophytes of seaweeds. *Mycosphere* 2 (2): 147–155.

71 Chow, Y.Y. and Ting, A.S.Y. (2017). Influence of glucose and L-asparagine concentrations on L-asparaginase production by endophytic fungi. *J. Microbiol. Biotechnol. Food Sci.* 7 (2): 186–189.

72 Kavitha, A. and Vijayalakshmi, M. (2010). Optimization and Purification of L-asparaginase Produced by *Streptomyces tendae* TK-VL_333. *Z. Naturforsch. C J. Biosci.* 65 (7–8): 528–531.

73 Gurunathan, B. and Sahadevan, R. (2011). Optimization of media components and operating conditions for exogenous production of fungal L-asparaginase. *Chiang Mai J. Sci.* 38 (2): 270–279.

74 Menegat, F.B., Baldo, C., Colabone Celligoi, M.A.P., and Buzato, J.B. (2016). Optimization of asparaginase production from *Zymomonas mobilis* by continuous fermentation. *Acta Sci. Biol. Sci.* 38 (2): 163–168.

75 Freitas, M., Souza, P., Cardoso, S. et al. (2021). Filamentous fungi producing l-asparaginase with low glutaminase activity isolated from brazilian savanna soil. *Pharmaceutics* 13 (8): 1–20.

76 Alrumman, S.A., Mostafa, Y.S., Al-izran, K.A. et al. (2019). Production and anticancer activity of an L-asparaginase from *Bacillus licheniformis* isolated from the Red Sea, Saudi Arabia. *Sci. Rep.* 9 (1): 1–14.

77 Mostafa, Y., Alrumman, S., Alamri, S. et al. (2019). Enhanced production of glutaminase-free L-asparaginase by marine *Bacillus velezensis* and cytotoxic activity against breast cancer cell lines. *Electron. J. Biotechnol.* 42: 6–15.

78 Shah, A.J., Karadi, R.V., and Parekh, P.P. (2010). Isolation, optimization and production of L-asparaginase from coliform bacteria. *Asian J. Biotechnol.* 2 (3): 169–177.

79 Garaev, M.M. and Golub, E.I. (1977). Mechanism of action of glucose on L-asparaginase synthesis by *Escherichia coli* bacteria. *Mikrobiologiia* 46 (3): 433–439.

80 Gholamian, S., Gholamian, S., Nazemi, A., and Miri Nargesi, M. (2013). Optimization of culture media for L-asparaginase production by newly isolated bacteria, *Bacillus* sp. GH5. *Microbiol (Russian Fed)* 82 (6): 856–863.

81 Ghosh, S., Murthy, S., Govindasamy, S., and Chandrasekaran, M. (2013). Optimization of L-asparaginase production by *Serratia marcescens* (NCIM 2919) under solid state fermentation using coconut oil cake. *Sustain. Chem. Process.* 1 (1): 1–8.

82 Othman, S., Mekawey, A., El-Metwally, M. et al. (2022). *Rhizopus oryzae* AM16; a new hyperactive L-asparaginase producer: semi solid-state production and anticancer activity of the partially purified protein. *Biomed. Rep.* 16 (3): 1–9.

83 Amany, B.A.E., Wesam, A.H., Zakaria, A.M., and Rabab, A.E.D. (2021). Production of chemotherapeutic agent L-asparaginase from gamma-irradiated *Pseudomonas aeruginosa* WCHPA075019. *Jordan J. Biol. Sci.* 14 (3): 403–412.

84 Mahajan, R.V., Saran, S., Kameswaran, K. et al. (2012). Efficient production of l-asparaginase from *Bacillus licheniformis* with low-glutaminase activity: optimization, scale up and acrylamide degradation studies. *Bioresour. Technol.* 125: 11–16.

85 Ghasemi, Y., Ebrahiminezhad, A., Rasoul-Amini, S. et al. (2008). An optimized medium for screening of L-asparaginase production by *Escherichia coli*. *Am. J. Biochem. Biotechnol.* 4 (4): 422–424.

86 Badoei-Dalfard, A. (2015). Purification and characterization of l-asparaginase from *Pseudomonas aeruginosa* strain SN004: production optimization by statistical methods. *Biocatal. Agric. Biotechnol.* 4 (3): 388–397.

87 Deshpande, N., Choubey, P., and Agashe, M. (2014). Studies on optimization of growth parameters for L-asparaginase production by *Streptomyces ginsengisoli*. *Sci. World J.* 2014: 1–6.

88 El-Naggar, N.E.A., Moawad, H., El-Shweihy, N.M. et al. (2019). Process development for scale-up production of a therapeutic L-asparaginase by *Streptomyces brollosae* NEAE-115 from shake flasks to bioreactor. *Sci. Rep.* 9 (1): 1–18.

89 El-Hefnawy, M.A.A., Attia, M., El-Hofy, M.E., and Ali, S.M.A. (2015). Optimization production of L-asparaginase by locally isolated filamentous fungi from. *Curr. Sci. Int.* 4 (3): 330–341.

90 Barros, T., Brumano, L., Freitas, M. et al. (2021). Development of processes for recombinant L-asparaginase II production by *Escherichia coli* Bl21 (De3): from shaker to bioreactors. *Pharmaceutics* 13 (1): 1–15.

91 Pillaca-Pullo, O., Rodrigues, D., Sánchez-Moguel, I. et al. (2021). Recombinant l-asparaginase production using *Pichia pastoris* (MUTs strain): establishment of conditions for growth and induction phases. *J. Chem. Technol. Biotechnol.* 96 (1): 283–292.

92 Ameen, F., Alshehri, W.A., Al-Enazi, N.M., and Almansob, A. (2020). L-asparaginase activity analysis, ansZ gene identification and anticancer activity of a new *Bacillus subtilis* isolated from sponges of the Red Sea. *Biosci. Biotechnol. Biochem.* 84 (12): 2576–2584.

93 Al Bahrani, M.H.A. (2016). Study the optimum parameters for production of cloned L-Asparaginase type I by *Escherichia coli*. *Int. J. Curr. Microbiol. App. Sci.* 5 (8): 479–485.

94 Tundisi, L.L., Coêlho, D.F., Zanchetta, B. et al. (2017). L-asparaginase purification. *Sep. Purif. Rev.* 46 (1): 35–43.

95 Righetti, P.G. and Boschetti, E. (2013). Detailed methodologies and protocols. In: *Low-Abundance Proteome Discovery*, 263–319. Elsevier Inc.

96 Kuwabara, T., Prihanto, A.A., Wakayama, M., and Takagi, K. (2015). Purification and characterization of *Pseudomonas aeruginosa* PAO1 asparaginase. *Procedia Environ. Sci.* 28: 72–77.

97 Husain, I., Sharma, A., Kumar, S., and Malik, F. (2016). Purification and characterization of glutaminase free asparaginase from *Pseudomonas otitidis*: induce apoptosis in human leukemia MOLT-4 cells. *Biochimie* 121: 38–51.

98 Krishnapura, P.R. and Belur, P.D. (2016). Partial purification and characterization of L-asparaginase from an endophytic *Talaromyces pinophilus* isolated from the rhizomes of *Curcuma amada*. *J. Mol. Catal. B: Enzym.* 124: 83–91.

99 Maruyama, T., Izaki, S., Kurinomaru, T. et al. (2015). Protein-poly(amino acid) precipitation stabilizes a therapeutic protein L-asparaginase against physicochemical stress. *J. Biosci. Bioeng.* 120 (6): 720–724.

100 Mahajan, R.V., Kumar, V., Rajendran, V. et al. (2014). Purification and characterization of a novel and robust L-asparaginase having low-glutaminase activity from *Bacillus licheniformis*: in vitro evaluation of anti-cancerous properties. *PLoS One* 9 (6): e99037.

101 Elshafei, A.M. and El-Ghonemy, D.H. (2021). Extracellular glutaminase-free L-asparaginase from *Trichoderma Viride* F2: purification, biochemical characterization and evaluation of its potential in mitigating acrylamide formation in starchy fried food. *J. Microbiol. Biotechnol. Food Sci.* 11 (2): 1–7.

102 Kong, R. (2005). LC/MS application in highthroughput ADME screen. In: *Handbook of Pharmaceutical Analysis by HPLC* (ed. S. Ahuja and M.W. Dong), 413–446. Elsevier Inc.

103 Zhu, J.H., Yan, X.L., Chen, H.J., and Wang, Z.H. (2007). *In situ* extraction of intracellular L-asparaginase using thermoseparating aqueous two-phase systems. *J. Chromatogr. A* 1147 (1): 127–134.

104 Arumugam, N. and Thangavelu, P. (2022). Purification and anticancer activity of glutaminase and urease free intracellular L-asparaginase from *Chaetomium* sp. *Protein Expr. Purif.* 190: 106006.

105 Middelberg, A.P.J. (2000). Microbial cell disruption by high-pressure homogenization. In: *Downstream Processing of Proteins* (ed. M.A. Desai), 11–21. Humana Press Inc.

106 Zhao, F. and Yu, J. (2001). L-asparaginase release from *Escherichia coli* cells with K_2HPO_4 and Triton X100. *Biotechnol. Prog.* 17 (3): 490–494.

107 Li, X., Zhang, X., Xu, S. et al. (2018). Simultaneous cell disruption and semi-quantitative activity assays for high-throughput screening of thermostable L-asparaginases. *Sci. Rep.* 8 (1): 1–12.

108 Kim, K., Jeong, J.H., Lim, D. et al. (2015). L-asparaginase delivered by *Salmonella typhimurium* suppresses solid tumors. *Mol. Ther. Oncolytics* 2 (March): 15007.

109 Patra, A.K., Mukhopadhyay, R., Mukhija, R. et al. (2000). Optimization of inclusion body solubilization and renaturation of recombinant human growth hormone from *Escherichia coli*. *Protein Expr. Purif.* 18 (2): 182–192.

110 Desai, M.A., Rayner, M., Burns, M., and Bermingham, D. (2000). Application of chromatography in the downstream processing of biomolecules. In: *Downstream Processing of Proteins* (ed. M.A. Desai), 73–94. Humana Press.

111 Bhat, E., Abdalla, M., and Rather, I. (2018). Key factors for successful protein purification and crystallization. *Glob. J. Biotechnol. Biomater. Sci.* 4 (1): 1–7.

112 Rajendran, V., Pushpavanam, S., and Jayaraman, G. (2022). Continuous refolding of L-asparaginase inclusion bodies using periodic counter-current chromatography. *J. Chromatogr. A* 1662: 462746.

113 Prakash, P., Chandrayan, S., Tiwari, P. et al. (2021). Development of a downstream process for purification and purity analysis of glutaminase free L-asparaginase using UPLC, DLS-ZP and DSC-TGA. *J. Taibah Univ. Sci.* 15 (1): 458–467.

114 Wei, Y., Chen, J., Jia, R. et al. (2008). Purification of *Escherichia coli* L-asparaginase mutants by a native polyacrylamide gel electrophoresis. *J. Chromatogr. Sci.* 46 (6): 556–559.

115 Karamitros, C.S. and Labrou, N.E. (2014). Extracellular expression and affinity purification of L-asparaginase from *E. chrysanthemi* in *E. coli*. *Sustain. Chem. Process.* 2 (1): 1–8.

116 Zhang, Y., Li, D., and Li, Y. (2017). Expression and purification of L-asparaginase from *Escherichia coli* and the inhibitory effects of cyclic dipeptides. *Nat. Prod. Res.* 31 (18): 2099–2106.

117 Roth, G., Nunes, J.E.S., Rosado, L.A. et al. (2013). Recombinant *Erwinia carotovora* L-asparaginase II production in *Escherichia coli* fed-batch cultures. *Braz. J. Chem. Eng.* 30 (2): 245–256.

118 Li, X., Xu, S., Zhang, X. et al. (2019). Design of a high-efficiency synthetic system for l-asparaginase production in *Bacillus subtilis*. *Eng. Life Sci.* 19 (3): 229–239.

119 Mukherjee, J. and Majumdar, S. (2000). Studies on nutritional and oxygen requirements for production of L-asparaginase by *Enterobacter aerogenes*. *Appl. Microbiol. Biotechnol.* 53: 180–184.

120 Rahimzadeh, M., Poodat, M., Javadpour, S. et al. (2016). Purification, characterization and comparison between two new L-asparaginases from PG03 and PG04. *Open Biochem. J.* 10 (1): 35–45.

121 Pal Roy, M., Das, V., and Patra, A. (2019). Isolation, purification and characterization of an extracellular L-asparaginase produced by a newly isolated *Bacillus megaterium* strain MG1 from the water bodies of Moraghat forest, Jalpaiguri, India. *J. Gen. Appl. Microbiol.* 65 (3): 137–144.

122 Abdelrazek, N.A., Elkhatib, W.F., Raafat, M.M., and Aboulwafa, M.M. (2020). Production, characterization and bioinformatics analysis of L-asparaginase from a new *Stenotrophomonas maltophilia* EMCC2297 soil isolate. *AMB Exp.* 10 (1): 71.

123 El-Naggar, N.E.A., Deraz, S.F., Soliman, H.M. et al. (2016). Purification, characterization, cytotoxicity and anticancer activities of L-asparaginase, anti-colon cancer protein, from the newly isolated alkaliphilic *Streptomyces fradiae* NEAE-82. *Sci. Rep.* 6: 1–16.

124 Gonçalves, A.B., Maia, A.C.F., Rueda, J.A., and de Figueiredo Conte Vanzela, A.P. (2016). Fungal production of the anti-leukemic enzyme l-asparaginase: from screening to medium development. *Acta Sci. Biol. Sci.* 38 (4): 387–394.

125 Hassan, S.W.M., Farag, A.M., and Beltagy, E.A. (2018). Purification, characterization and anticancer activity of L-asparaginase produced by Marine *Aspergillus terreus*. *J. Pure Appl. Microbiol.* 12 (4): 1845–1854.

126 El-Naggar, N.E.A., Deraz, S.F., El-Ewasy, S.M., and Suddek, G.M. (2018). Purification, characterization and immunogenicity assessment of glutaminase free L-asparaginase from *Streptomyces brollosae* NEAE-115. *BMC Pharmacol. Toxicol.* 19 (1): 1–15.

127 Zhang, Y.Q., Tao, M.L., De Shen, W. et al. (2004). Immobilization of L-asparaginase on the microparticles of the natural silk sericin protein and its characters. *Biomaterials* 25 (17): 3751–3759.

128 Ulu, A., Koytepe, S., and Ates, B. (2016). Synthesis and characterization of PMMA composites activated with starch for immobilization of L-asparaginase. *J. Appl. Polym. Sci.* 133 (19): 1–11.

129 Kotzia, G.A. and Labrou, N.E. (2013). Structural and functional role of Gly281 in L-asparaginase from *Erwinia carotovora*. *Protein Pept. Lett.* 20 (12): 1302–1307.

130 Pokrovskaya, M.V., Aleksandrova, S.S., Pokrovsky, V.S. et al. (2015). Identification of functional regions in the *Rhodospirillum rubrum* l-Asparaginase by site-directed mutagenesis. *Mol. Biotechnol.* 57 (3): 251–264.

131 WHO. (2015). WHO model list of essential medicines. https://list.essentialmeds.org/recommendations/872

132 Cachumba, J.J.M., Fernandes Antunes, F.A., Peres, G.F.D. et al. (2016). Current applications and different approaches for microbial L-asparaginase production. *Braz. J. Microbiol.* 47: 77–85.

133 Duval, M., Suciu, S., Ferster, A. et al. (2002). Comparison of *Escherichia coli*-asparaginase with *Erwinia*-asparaginase in the treatment of childhood lymphoid malignancies: results of a randomized European Organisation for Research and Treatment of Cancer – Children's Leukemia Group phase. *Blood* 99 (8): 2734–2739.

134 Muneer, F., Hussnain, M., Farrukh, S. et al. (2020). Microbial L-asparaginase: purification, characterization and applications. *Arch. Microbiol.* 202 (5): 967–981.

135 Narta, U.K., Kanwar, S.S., and Azmi, W. (2007). Pharmacological and clinical evaluation of l-asparaginase in the treatment of leukemia. *Crit. Rev. Oncol. Hematol.* 61: 208–221.

136 Ginn, C., Khalili, H., Lever, R., and Brocchini, S. (2014). PEGylation and its impact on the design of new protein-based medicines. *Future Med. Chem.* 6 (16): 1829–1846.

137 Brumano, L.P., da Silva, F.V.S., Costa-Silva, T.A. et al. (2019). Development of L-asparaginase biobetters: current research status and review of the desirable quality profiles. *Front. Bioeng. Biotechnol.* 6: 1–22.

138 Maggi, M., Mittelman, S.D., Parmentier, J.H. et al. (2017). A protease-resistant *Escherichia coli* asparaginase with outstanding stability and enhanced anti-leukaemic activity in vitro. *Sci. Rep.* 7 (1): 1–16.

139 IARC. (1994). IARC Monographs on the Evaluation of Carcinogenic Risks to Humans.

140 Rifai, L. and Saleh, F.A. (2020). A review on acrylamide in food: occurrence, toxicity, and mitigation strategies. *Int. J. Toxicol.* 39 (2): 93–102.

141 Onishi, Y., Prihanto, A.A., Yano, S. et al. (2015). Effective treatment for suppression of acrylamide formation in fried potato chips using L-asparaginase from *Bacillus subtilis. 3 Biotech* 5 (5): 783–789.

142 Kukurová, K., Morales, F.J., Bednáriková, A., and Ciesarová, Z. (2009). Effect of L-asparaginase on acrylamide mitigation in a fried-dough pastry model. *Mol. Nutr. Food Res.* 53 (12): 1532–1539.

143 Ciesarovát, Z., Kukurová, K., Bednáriková, A. et al. (2009). Improvement of cereal product safety by enzymatic way of acrylamide mitigation. *Czech J. Food Sci.* 27: 1–3.

144 Jia, R., Wan, X., Geng, X. et al. (2021). Microbial L-asparaginase for application in acrylamide mitigation from food: current research status and future perspectives. *Microorganisms* 9 (8): 1659.

145 Hendriksen, H.V., Budolfsen, G., and Baumann, M.J. (2013). Asparaginase for acrylamide mitigation in food. *Asp. Appl. Biol.* 116: 41–50.

146 Singh, S.P., Pandey, A., Singhania, R.R. et al. (eds.) (2020). *Biomass, Biofuels, Biochemicals: Advances in Enzyme Catalysis and Technologies.* Elsevier. ISBN 9780128198209. https://doi.org/10.1016/C2019-0-00323-8.

6

Bioreactor-Scale Strategy for Pectinase Production

Javier Ulises Hernández-Beltrán, Carlos Alberto Acosta-Saldívar, Genesis Escobedo-Morales, Nagamani Balagurusamy, and Miriam Paulina Luévanos-Escareño

Facultad de Ciencias Biológicas, Universidad Autonoma de Coahuila, Torreón, Coahuila, Mexico

6.1 Introduction

Currently, pectinases are enzymes of great importance since they have received much attention worldwide as a biological biocatalyst since they have many applications in industry, mainly in food, juice, and paper processing, among others [1–3]. It should be noted that most are produced by microorganisms, particularly fungal sources. In the industry, pectinase preparations represent 25% of global sales of enzymes for the food industry, establishing its importance in this [4, 5]. Pectinases have been extensively studied in plants, mainly in fruits. Although they are generally produced by plants, the extraction and purification of the enzyme from them may require special conditions since the enzymes obtained from plants are usually thermolabile. Therefore, pectinases of microbial origin are an alternative in the production of enzymes, because during their production they are more stable and their extraction is less complex [6, 7]. In addition, pectinases can be produced using different fermentative systems like submerged fermentation (SmF) and solid-state fermentation (SSF) [8]. Although the original source and composition of the substrate have a significant impact on pectinase production, the optimization of the fermentative process conditions and bioreactor configuration are critical factors to scaling up the process, an industry point of view, which helps to understand the phenomenology of the reaction, increase the production yield, and therefore to minimize production costs [9]. This book chapter focuses on the production of pectinase enzymes considering the optimization of process conditions in both fermentations, SmF and SSF, and as well the different bioreactor configurations and their scales to perform the fermentation process taking into account the original sources and substrates.

Microbial Bioreactors for Industrial Molecules, First Edition.
Edited by Sudhir P. Singh and Santosh Kumar Upadhyay.

6.2 Pectinase Classification and Origin Sources

Pectins are high-molecular-weight structural polysaccharides present in considerable amounts in the cell wall of vegetables and fungi together with cellulose and hemicellulose. The pectinase enzymes are responsible for hydrolyzing this polysaccharide through different mechanisms, depending on the type of enzyme involved and the specific substrate for it. Pectin consists of a complex mixture composed of a linear chain of galacturonic acid residues linked by an α-1,4-glycosidic bond. This main structure is methyl-esterified, to some extent, with rhamnose, arabinose, and galactan residues [10–12].

The pectinolytic enzyme is the common name of a family of enzymes that participate in the pectin degradation process [13–15]. Pectinases are a group composed of various related enzymes that hydrolyze the pectic substances present in plants [6] and are also very important for them since they facilitate cell elongation and growth, softening of some plant tissues, maturation, and storage [1].

Due to their role in the degradation of the cell wall, pectinases are currently considered enzymes of great industrial importance since they have improved and facilitated various processes of interest. The pectinases have various applications in industrial processes: the food industry is the one that has made the most use of these enzymes for the optimization and improvement of functional foods for man, as well as for farm animals, among many other processes such as the processing of fruits and vegetables to produce juices, obtaining antioxidant compounds and production of oligomeric or polymeric carbohydrates. In the paper industry, it is used for bleaching and reducing the number of chlorides present in the pulp, as well as reducing pollution by treating waste, specifically wastewater. In the textile industry, cleaning cotton fabrics thus improve the wettability of fabrics. Other industries where preparations of these enzymes are used are wastewater treatment, oil extraction, fermentation, and wine processing, among others [16–19]. In the same way, the important role they play in the phytopathogenicity of some fungi has been verified, since by degrading the pectins of the cell wall of the plants, the entry of this into the tissues and by infection is facilitated.

6.2.1 Pectinases

Pectinase enzymes are a very varied and widespread group in their distribution, which is why they have been classified according to their reaction mechanism and according to their reaction to different pectic substances.

Pectinases can be classified into two categories: pectinesterase or methylesterases, which remove methyl groups (EC 3.1.1.11), and depolymerases (EC 3.2.1.15), which shorten the main structure of pectic substances. Depolymerases, in turn, separate into lyases (pectin lyases EC 2.2.2.10) and hydrolases. The most studied enzymes are homogalacturonan degrading enzymes [4, 20–23]. Table 6.1 shows the classification of pectinases.

Pectinase enzymes are a very varied and widespread group in their distribution, which is why they have been classified according to their reaction mechanism and according to their reaction to different pectic substances. Pectinases can be classified into two categories: pectinesterase or methylesterases, which remove methyl groups (EC 3.1.1.11), and depolymerases (EC 3.2.1.15), which shorten the main structure of pectic substances [21];

Table 6.1 Pectinolytic enzymes classification.

Enzyme	EC number	Substrate	Mechanism of action	References
Esterases				
Pectin methyl esterase (PE)	3.1.1.11	Pectin	Random (cleaves methanol from esterified carboxyl groups to yield low methoxyl pectin and polygalacturonic acid)	[2, 24]
Pectin acetylesterase	3.1.1.6	Pectin	Random (hydrolyzes acetyl esters in HG region of pectin)	[2, 20]
Depolymerases				
Polymethylgalacturonase (PMG)		Pectin	Hydrolyze	[4]
Endo-PMG				
Exo-PMG				
Pectin lyase (PMGL) (Endo-PL)	4.2.2.10	Pectin	Trans-elimination	[4]
Polygalacturonase (PG)	3.2.1.15	Polygalacturonic acid	Hydrolyze	[6, 10]
Endopolygalacturonase (Endo-PG)	3.2.1.67 3.2.1.82			
Exopolygalacturonase 1 (Exo-PG1)				
Exopolygalacturonase (Exo-PG 2)				
Pectate lyase PGL	4.2.2.2	Pectin	Trans-elimination	[20]
Endo-PGL				
Exo-PGL	4.2.29			
Rhamnogalacturonase (RG)	3.2.1.171	Rhamnogalacturonan	Hydrolyze	[21, 25]
Rhamnogalacturonan endolyase (RGL)	4.2.2.23	Rhamnogalacturonan	Trans-elimination	[5]

(i) pectin esterase (EC 3.1.1.11); (ii) hydrolases including polygalacturonases (EC3.2.1.15); (iii) lyases or transeliminases comprising pectate lyases (EC 4.2.2.2) and pectin lyases (EC 4.2.2.10) [4]. Table 6.1 shows the classification of pectinases.

Pectinesterases (PE, EC 3.1.1.11): Commonly referred to as pectase, pectolipase, pectin methoxylase, pectin demethoxylase, and pectin methyl esterase, these enzymes catalyze the de-esterification of the methyl ester bonds in the galactan chain of pectins to release pectins acid and methanol. These enzymes are produced mainly in some plants such as bananas, tomatoes, bananas, and citrus fruits, as well as some fungi and bacteria. The optimal temperature and pH conditions are 20–60 °C and 4.0–7.0, respectively. The molecular weight of these enzymes ranges between 30 and 50 kDa. The isoelectric point is located between 4.0 and 8.0 [10, 20, 26].

Depolymerizing enzymes: they break the α-1,4-glycosidic bonds of the main chain of pectins, they can be subdivided into:

Polygalacturonases (EC 3.2.1.15): These are the enzymes that catalyze the cleavage of the polygalacturonic acid chain with the introduction of water through the oxygen bridge. The activity of these enzymes can be determined by measuring either the formation of reducing sugars or the decrease in viscosity. Polygalacturonases are of two types: Endo-polygalacturonases (Endo-PG), which randomly catalyze the hydrolysis of α-1,4-glycosidic bonds in pectic acid, and exo-polygalacturonases (Exo-PG), which catalyze the sequential hydrolysis of glycosidic bonds in pectic acid, forming oligosaccharides. Endo-polygalacturonase is extensively distributed among bacteria, fungi, and yeasts. These can be found with different molecular weights and isoelectric points ranging from 3.8 to 7.6, with optimum pH and temperatures of 2.5–6.0 and 30–50 °C. Exo-PG is produced by bacteria, fungi such as *Aspergillus*, and some plants, vegetables, and fruits such as carrots, peaches, citrus, and apples. The molecular weight of these enzymes ranges from 30 to 50 kDa and their isoelectric point is between 4.0 and 6.0 [6, 20, 27].

Polymethylgalacturonases (PGML, EC 3.2.1.67): these enzymes catalyze the breaking of the α-1,4-glycosidic bond of the main structure of pectins, preferably the region that is highly esterified, forming as a product 6-methyl-D-galactorunate [20, 28].

Lyases: These enzymes are responsible for the non-hydrolytic cleavage of pectates or pectinates by a transeliminative split mechanism of the pectic polymer. They are classified as endo-pectate transeliminase (pectate lyase, PGL, EC 4.2.2.10) and as endo-pectin transeliminase (pectin lyase, PL, EC 4.2.2.2). Pectate lyase preferentially cleaves the glycosidic bonds of polygalacturonic acid forming unsaturated products. Molecular weights vary from 30 to 50 kDa, with an optimum pH of 8.0–10.0 and a temperature of 30–40 °C. A few alkaline tolerant pectate lyases have been characterized [29]. Pectin lyase catalyzes the random breakdown of highly esterified pectin, producing methyl oligogalacturanates. The molecular mass of pectin lyases is between 30 and 40 kDa. In general, this enzyme is active in an acid pH range (4.0–7.0) although there are reports that it is also active in alkaline conditions. Its isoelectric point is at 3.5, with optimum temperature and pH of 35 °C and 4.8 [20, 28].

6.2.2 Origin Source of Production of Microbial Pectinase

Pectinolytic enzymes are produced by different organisms such as bacteria, fungi, yeasts, insects, nematodes, protozoa, and plants [6, 30–32]. Although pectinases are produced by some plants and microorganisms, however, the main source of commercial pectinases is the fungus *Aspergillus* sp., producing a complex of pectinolytic enzymes [24].

Microorganisms are currently the main source of industrial enzymes, being mostly produced by fungi and yeasts, followed by bacteria, and the rest corresponds to sources of vegetable origin [2, 33]. For commercial purposes, microbial strains are the main sources for large-scale pectinolytic enzyme production for industrial use and application due to ease of maintenance and production. Today recombinant enzymes derived from microorganisms are widely used in the recombinant industry [3, 34].

6.3 Substrates Used for Pectinase Production

The costs of microbial enzyme production depend mainly on the substrates used to induce enzyme production during fermentation [35].

In the last years, there has been a tendency to use the residues of some crops for microbial degradation and their transformation into the desired product at the industrial level. In this way, new substrates are sought as cheaper carbon sources as a requirement to efficiently produce enzymes of industrial importance [36]. Agro-industrial residues are primarily composed of complex polysaccharides to produce industrially important enzymes [37]. Microbial enzymes are produced by solid-state fermentation (SSF) and submerged (SmF) systems.

Next, some of the most used substrates for the production of pectinase enzyme are mentioned: Fruits peel wastes (pomegranate peel, orange peel, papaya peel, lemon peel, mango peel, pineapple peel, mosambi peel, banana peel), tomato pomace, coconut fiber, and date palm fruit waste (air-dried); artichokes peel; agro-residues (sugarcane bagasse, corncob, wheat bran, paddy straw, mustard oil cake, ground peel, sawdust, cottonseed oil cake, cotton straw, wheat straw, and mustard); and synthetic citrus pectin [13, 35, 38, 39]. Table 6.2 shows different microorganisms, substrates, and fermentation conditions to produce pectinases.

6.4 Fermentation Strategies

Two fundamental fermentation systems can produce enzymes like pectinases: SSF and SmF [55]. Therefore, to optimize the obtaining of pectinases in these microbial growth processes under either anaerobic or aerobic conditions, physicochemical parameters such as pH, temperature, reaction volume, and inoculum strength must be considered. In addition, particle size and moisture content are taken into account under solid-state conditions [55, 56].

6.4.1 Solid-State Fermentation

In this fermentation, substrates are placed in a solid state with low moisture content; in this way, less water is added, only in order to moisten the substrates to contribute to the growth of crops, thus harnessing of agricultural and industrial wastes like feedstocks [57, 58].

There are different types of bioreactors and modes of operation; however, for SSF, the most popular and preferred configurations are the packed bed, tray, and rotating drum bioreactors [59]. For example, [60] compared the performance in obtaining pectinases by fermentation using tray and rotary drum bioreactors. Orange pomace was used as a substrate fermented by *Aspergillus niger*. Moisture control was found to be better in the rotary drum bioreactors; however, the yields of exo- and endo-pectinase generation were 45 and 37% higher in the tray bioreactor. On the other hand, an optimization was achieved in the tray bioreactor by adding sugarcane bagasse to control the humidity to 70%, which helped to improve the production of exo- and endo-pectinases by 17 and 23%, respectively. Another study described by [61] indicates that *Aspergillus sojae* strain used at flask scale has the

Table 6.2 Microorganisms and substrates used to produce pectinolytic enzymes under different fermentation conditions.

Microorganisms	Enzyme	Substrate	Fermentation	Fermentation conditions	Reaction system	Maximum enzyme production	References
Bacteria							
Bacillus licheniformis KIBGE-IB-3	Polygalacturonase	Date fruit waste and yeast extract	Solid state	pH 7.0 37 °C Reaction time 72 h	100 mL Erlenmeyer flask	23,383 U/mL/min (enzymatic activity)	[35]
Bacillus licheniformis KIBGE IB-21	Polygalacturonase	Apple pectin, wheat bran yeast extract as a nitrogen source	Submerged	pH 7.0 37 °C Reaction time 48 h	100 mL Batch fermentation	1015 U/mg	[14]
Bacillus subtilis SAV-21	Pectinase Pectin Lyase	Combination of orange peel and coconut fiber (4 : 1)	Solid state	pH 4.0 37 °C Reaction time 18 h 60% moisture	250 mL Erlenmeyer flask Moistened 70%	3315 U/gds (pectinase) 10.5 U/gds (pectin lyase)	[38]
Bacillus subtilis BKDS1	Pectinase	Agro-waste (pineapple steam)	Submerged	pH 7.0 150 rpm Reaction time 48 h 500 rpm 40 °C 0.49 lmp aeration	100 mL Erlenmeyer flask 700 mL Fermenter	1508 U/mL	[40]

Fungi							
Aspergillus niger NRRL 322	Pectinase	Soy hull	Submerged	pH 6.0 25 °C 250 rpm 400 rpm	100 mL Shake flask 1–1.5 L Stir-tank fermenter	-	[41]
Aspergillus niger	Pectinase	Orange waste peel	Submerged	pH 5.5 30 °C Reaction time 5 days 180 rpm	250 mL Erlenmeyer flask	117.1 ± 3.4 mM/mL/min (enzyme yield)	[37]
Aspergillus niger IOC 4003	Polygalacturonase Pectin lyase	Yellow mombin pulp residue	Solid state	30 °C pH 4.5 Reaction time 240 h	250 mL Erlenmeyer flasks Moisture 40%	8.22 ± 0.63 U/g	[42]
Aspergillus niger-ATCC 1640	Pectinase	Satkara (citrus macroptera) peel	Solid state	30 °C Moisture 80% Reaction time 72 h	250 mL Erlenmeyer flask Moistened 80%	0.6045 μmol/mL	[8]
Aspergillus niger strain MCAS2	Pectinase thermostable alkaline	Defined liquid media containing pure pectin	Submerged	pH 6.0 and 30 °C. Reaction time 48 h	—	of 8.33 U/mg (specific activity	[43]

(*Continued*)

Table 6.2 (Continued)

Microorganisms	Enzyme	Substrate	Fermentation	Fermentation conditions	Reaction system	Maximum enzyme production	References
Aspergillus niger LB-02-SF	Pectinase	Wheat bran (aqueous extract) citrus pectin, glucose, yeast extract, and culture standard medium (medium M)	Submerged	pH 2.7 28 °C Reaction time135 h 300–750 rpm O_2 concentration 30% saturation	4 L Benchtop bioreactor	14 U/mL	[44]
Aspergillus niger AN07 *Aspergillus tubingensis* MP30 *Aspergillus fumigatus* M1 *Aspergillus sydowii*	Polygalacturonase	Carica papaya peel	Solid-state	Reaction time 144 h	250 mL Erlenmeyer flasks Moisture 90% 10 g	264.20, 237.31, 155.90 and 132.63 (U/gds), respectively	[45]
Aspergillus niger AUMC 4156, *Penicillium oxalicum* AUMC 4153, and *Paecilomyces variotii* AUMC 4149	Endo-polygalacturonase	Agro-industrial wastes (dried orange peel and sugar beet pulp	Submerged	pH 5, 6, and 7, respectively 30 °C 150 rpm Reaction time 5 days	100 mL Erlenmeyer flasks	4.74, 4.45, and 4.98 (mg/mL), respectively	[46]
Aspergillus sojae	Polygalacturonase	Wheat bran and sugar beet pulp	Solid state	28 °C	300 mL Culture flasks 10 g Moisture 80–120%	-	[47]

Aspergillus japonicus	Polygalacturonase (PAGj)	Passion fruit peel	Solid state	—	—	227.75 ± 13.1 U	[48]
Aspergillus aculeatus URM4953	Polygalacturonase Endo-polygalacturonase	Passion fruit peel	Submerged	pH 4.56 130 rpm 30 °C Reaction time 96 h 48 h	—	Highest activities 2.92 ± 0.12 U/mL (polygalacturonase) 6.51 ± 0.04 U/mL (endo-polygalacturonase)	[49]
Geotrichum candidum AA15	Pectinase	MSM medium with 2% of simple sugars (xylose, glucose, and galactose)	Submerged	28 °C Reaction time 72 h	Immobilized in Corncob[a]	—	[50]
Geotrichum candidum AA15	Pectinase	MSM with pectin or with pectin and yeast extract	Submerged	25 °C pH 7.2 Reaction time 48 h	Immobilized on Corncob[a]	Immobilized cells yielded 0.554 IU/mL Free cells (0.215 IU/mL).	[51]
Crypthecodinium cohnii	Polygalacturonase	Organosolv pretreatment of exhausted olive pomace (EOP)	Submerged	pH 6.5 27 °C 160 rpm Reaction time 7 days	100 mL Erlenmeyer flask	42.15 ± 0.41 U/mL	[52]
Penicillium janczewskii	Exo-polygalacturonase	Wheat bran	Solid state	70% moisture Reaction time 3–4 days	—	—	[53]
Penicillium fellutanum	Exo-polygaracturonase	Wheat bran	Solid state	pH 4.0 temperature 40°C Reaction time 96 h	40% Moisture	—	[54]

[a] These studies were carried out to produce the enzyme pectinase by immobilization of the microorganism.

reliability to be used in tray-type reactors since the experiment generated production of 298 U/g substrates without using any nutrient or inducing supplement in the substrate, so performing a process optimization could improve the enzyme activity up to 30%. Likewise, [62] evaluated the production of pectinases in onion residues in natural and onion residues dehydrated to be rehydrated by water immersion, which showed that using *Pleurotus sajor-caju* as a microorganism had a yield of 45.73% using rehydrated onion residue. However, pectinase production is stable at a temperature of 80 °C, which may not be optimal for a scalable process. The pH conditions will be given by the microorganism to be used to find the best yield of pectinase production. For example [63] used *Bacillus subtilis* as a microorganism and wheat bran as substrate searched for the best conditions for maximum enzyme activity and found that at pH 6.5 and 75% moisture content conditions obtained a maximum yield of 1272.4 U/g. It was visualized that by increasing the pH to 7.0, the yield decreased drastically, and by increasing the moisture content above 70% or decreasing it below 70%, the enzyme activity is considerably reduced. Zehra et al. [64] studied the fermentation generated by *Aspergillus fumigates* MS16 to generate pectinase and xylanase from the banana peel as substrate. It was observed that the optimum pH is 5.0; however, it was noted that increasing the pectinase production to 6.0 significantly decreased the enzyme activity from 1.5 to 0.3 U/mL. Similarly, [65] analyzed the production of pectinases by comparing two different microorganisms, *A. niger* NRRL3 and *Aspergillus oryzae* NRRL695, so that the process conditions were the same except for the incubation moisture content. It was observed that *A. niger* NRRL3 at 81–84% moisture and *A. oryzae* NRRL695 at 68–76% moisture obtained their maximum yields of 15 U/g and 13 U/g, respectively. For this reason, maintaining the pH and moisture content in their optimum range is fundamental to obtaining a better yield.

Other points to be taken into consideration to achieve an optimization in using SSF are process conditions such as temperature, inoculum concentration, and content of moisture, i.e. [66] who performed a process optimization to generate pectinases from banana peels with a moisture content of 40 and 60% (w/w) and also by varying incubation temperature conditions of 25 and 35 °C, and finally the concentration of inoculum of 1.6 × 106 spores/mL and 1.4 × 107 spores/mL. It was discovered that the incubation temperature of 35 °C was a determining parameter to achieve the maximum generation of enzymatic activity; this may be because it has been reported that the optimal temperatures for the production of some enzymes from fungi are from 25 to 37 °C [67]. On the other hand, this study indicated that the higher the moisture content, the lower the inoculum concentration required for maximum pectinase production activity, obtaining a pectinase yield of 27 U/mL, a moderately high yield. In another study, the optimization was evaluated through three process parameters: temperature, humidity, and incubation time, to determine the best conditions for maximum yield in the production of pectinase. The strain of *Rhizopus sp.* C4 was used to ferment orange peels. It was found that all three parameters studied showed quadratic effects on enzyme activity, indicating a significant interaction between those three variables for pectinase production. Thus, an optimization was achieved with a maximum yield of 11.63 IU/mL [68]. It has also been studied if pretreatment of the substrate has implications for pectinase production. Such is the case where [42] evaluated polygalacturonase production using yellow mombin pulp as substrate. They compared substrate previously washed with water with the same substrate without washing, and a maximum polygalacturonase yield of 38.22 U/g in 72 hours and 4.72 U/g in 96 hours was obtained, thus not only increasing enzyme production but also maximizing enzyme generation in less time. The substrate pretreatment suggests that the pectinase generation process can be significantly optimized.

A study done by [69] analyzed the pectin lyase production from *Schizophyllum commune* in a series of experiments with varied conditions of temperature (15–58 °C), pH (4–10), substrate concentration (3–35 g), time (1–7 days), and inoculum size (1–7 mL). They used response surface methodology to optimize the SSF, obtaining favorable results where the maximum production of pectin lyase was 480.45 U/mL with concentration of substrate of 15 g, pH value of 6.0, inoculum size of 3 mL, temperature, and incubation time of 35 °C and 24 hours, respectively. It has a 4 : 1 ratio coconut-fiber-generated higher enzyme activity than using the substrates separately. In addition, a 5.4- and 5.15-fold improvement in pectinase and pectin lyase generation was obtained by optimizing the parameters shown in Table 6.3.

This research has found that among the process conditions that have great relevance for optimizing pectinase production are substrate conditions such as moisture content and pretreatment, pH and inoculum size, and temperature. Furthermore, the most frequently used microorganism is *A. niger* due to its easy accessibility and good yields.

6.4.2 Submerged Fermentation

Liquid fermentation, also known as submerged fermentation (SmF), cultivates microorganisms in liquid substrates, such as nutrient broths or molasses to produce enzymes or other products of industrial interest. The microorganisms used in this process are bacteria or fungi, as they require a high moisture content within the medium. SmF is used to generate primary and secondary metabolites in a liquid state. The substrates used regularly are consumed quickly and need to be constantly replenished or replaced; therefore, the three main reactor types used in SmF are batch mode, fed-batch mode, and continuous mode [70, 71].

Likewise, [72] mentions that among the process parameters that most influence the optimization of pectinase production through SmF are the incubation period, temperature and volume of reaction, inoculum size, carbon and nitrogen sources, and pH, which can be related to the microorganism to be used and finally the type of substrate. They found that *A. niger* showed the best enzymatic activity among different types of fungi. Moreover, within different substrate types (banana husk, wheat bran, and rice bran), wheat bran generated a higher production of proteins, achieving optimization of this process of 15.5 U/mL in enzyme activity. In another case, [73] found that the stability of the microorganism *B. subtilis* used to generate polygalacturonase was through optimization of pH and temperature where the optimum values were pH 7.0 at 50 °C, at which there was a retention of 93% enzyme activity after seven hours. The optimum values were pH 7.0 at 50 °C, at which there was a retention of 93% enzyme activity after 7 hours, with a half-life of 21 hours. The results show that the pectinase produced can be used in the industry.

Table 6.3 Effects of different configurations of SSF bioreactions on the yield of pectinases.

Substrate	Moisture content	Microorganism	Inoculum size	pH	Temp (°C)	Particle size	Reaction system	Fermentation time	Agitation per day	Enzyme activity	References
Orange pomace	30%	*Aspergillus niger*	10^7 Spores/g of dry solid matter	4.5	30	NR	Rotating drum type. Bed volume 4.5 L	96 h	Rotating	44 U/g	[60]
Orange pomace + sugarcane bagasse	70%	*Aspergillus niger*	10^7 spores/g of dry solid matter	4.5	30	NR	Tray type. Depth of the bed 1.5 cm	96 h	Static	75 U/g	
Onion waste	70 g (dry weight) and 200 g (wet weight)	*Pleurotus sajor-caju* CCB019	10% Spawn (dry weight)	3.53	80	Between 2.36–4.75 mm	Polypropylene bags	19 days	Static	4.82 U/mL	[62]
Banana peels	65%	*Aspergillus fumigatus* MS16	10^8 Spores/mL	5.0	30	100 mesh	Erlenmeyer flask 100 mL	7 days	Static	2 U/mL	[64]
Banana peels	60%	*Aspergillus niger*	1.59×10^6 Spores/mL	NR	35	mesh of 75 μm	Erlenmeyer flask 100 mL	120 h	Static	27 U/mL	[66]
Orange peels	1 : 3.5 moisture ratios	*Rhizopus* sp. *C4*	NR	NR	30	0.8–2 mm	Wide-neck Erlenmeyer flask 250 mL	7 days	Static	11.63 IU/mL	[68]
Wheat bran	75%	*Bacillus subtilis*	2.0 mL 10% v/v	6.5	37	NR	Conical flask 250 mL	48 h	Static	1272.4 U/g	[63]

Yellow mombin pulp	40%	*Aspergillus niger* IOC 4003	10^7 Spores/g substrate	4.5	30	20 mesh screen	Erlenmeyer flask 250 mL	72 h	Static	38.22 U/g	[42]
Grape pomaces	84%	*Aspergillus niger* NRRL3	1×10^7 Spores/g	NR	30	0.25–2.38 mm	Glass flasks	72 h	Static	15 U/g	[65]
Grape pomaces	68%	*Aspergillus oryzae* NRRL695	1×10^7 Spores/g	NR	30	0.25–2.38 mm	Glass flasks	72 h	Static	13 U/g	
Wheat bran	62%	*Aspergillus sojae* ATCC 20235	10^7 Spore/g	NR	37	150–250 µm	Tray type borosilicate glass casseroles	4 days	Static	298 U/g	[61]
Mosambi peels	NR	*Schizophyllum commune*	3 mL	6.0	35	Milled into fine powder	NR	24 h	Static	480.45 U/mL	[69]
Orange peel + coconut fiber	60	*Bacillus subtilis* SAV-21	1×10^9/ mL	4.0	35	2–3 mm	Erlenmeyer flasks 250 mL	4 days 8 days	Static	3315 U/ gds pectinase 10.5 U/ gds pectin lyase	[38]

NR- No reported

Most studies have reported that *Aspergillus* spp. fungi have been shown to generate higher pectinolytic activity in the pH range of 3.0–6.0. For example, [74] have shown that *Aspergillus* spp., generating a yield of 72.3 U/mL, was obtained at pH 5.8. For example, [75] carried out the isolation of *Bacillus* sp. fungus, performing an SmF in which they found the maximum yield of 484.70 U/mg enzyme activity at pH 6 at 40 °C. In another case, [76] conducted a study using the fungus *A. niger* to ferment algal residues and generate polygalacturonase and found that the enzyme activity was stable in a temperature range from 40 to 60 °C with optimum pH of 6 where there was a maximum yield. Likewise, this microorganism can be a viable option with the best performance in both SmF and SSF because of its ease of being isolated and found in the environment, it is the best option to be used to produce pectinases.

On the other hand, if it is required to be specific with the enzyme studied, it must be analyzed separately. In a study done by [77], it was observed that using *Saccharomyces cerevisiae* to produce polygalacturonase and pectin lyase, the parameters used were the same; however, it was shown that to produce the highest yield of polygalacturonase, a pH of 4.5 is needed, contrary to pectin lyase that needs a pH 6.0 to generate the highest enzyme activity.

In another case, [78] found that adding nitrogen sources such as 0.2% sodium nitrate and carbon such as 1% fructose and the powdered wheat bran substrate in fermentation with the fermentation of *A. niger* increases the enzymatic activity of endo-polygalacturonase and endo-pectin lyase with difficulty. Therefore, adding enrichments to the substrates is feasible to optimize the production of pectinases. For example, [79] determined that the substrate concentration is also a parameter to take into account and that it affects the enzyme activity; using mango peel with concentrations of 0.2–2% in weight/volume, it was observed that between 1 and 1.5% there was the highest enzyme activity, in addition to this at a temperature of 20–50 °C, the highest pectinase activity was generated at 35 °C. Another critical parameter in optimizing pectinase production is the temperature which varies concerning the micro-organism used; for example, for fungal, the temperature that generates the maximum yield of pectinases can vary between 30 and 45 °C [64, 80, 81]. Other microorganisms can increase enzyme activity above 50 °C [73, 82, 83]. Table 6.4 shows the optimal process conditions of several studies using SmF and other related factors.

In this SmF review, it has been found that *Aspergillus fungi* are the most used because of their ease of isolation, for example, from soil or residues of fruits and vegetables, in addition to the advantages they have in the fermentation process since it has been observed that the optimal conditions are acidic pH and temperatures around 30–50 °C. These parameters are a good point since if one wants to optimize and take it to scalable processes becomes feasible.

6.5 Bioreactor-Scale Strategies

In the previous section we discussed the parameters that are involved in the SmF and SSF processes and how they affect the production of pectinases. Now we will address another fundamental aspect to carry out the scaling up of the process: the study of the different

Table 6.4 Effects of different configurations of SmF bioreactions on the yield of pectinases.

Substrate	Moisture content	Micro-organism	Inoculum size	pH	Temp. (°C)	Particle size	Reaction system	Fermentation time	Agitation per day	Enzyme activity	References
Seaweed waste	NR	*Aspergillus niger*	5×10^5 spores/g	6.0	40	NR	Erlenmeyer flask 100 mL	30 min	Shaking	4.27 mg/mL	[76]
Wheat bran	NR	*Aspergillus niger*	1 mL	6.0	30	NR	50 mL	72 h	Shaking	15.5 U/mL	[72]
Hazelnut shell	NR	*Bacillus subtilis* NRRL B-4219	10^6 CFU/mL	7.0	50	NR	Erlenmeyer flask 500 mL	7 h	Shaking	Residual activity 93%	[73]
Hankin's broth	NR	*Aspergillus* spp.	1 mL pre-fermenter inoculum	5.8	30	NR	Conical flask 50 mL	48 h	Shaking	72.3 U/mL	[74]
Banana peels	NR	*Aspergillus fumigatus* MS16	1×10^6 spores/g	5.5	45	100 mesh particle-size	Erlenmeyer flask 100 mL	7 days	Shaking	0.4 IU/mL	[64]
Fruit and vegetable peel extracts	NR	*Aspergillus niger*	NR	NR	32	NR	Erlenmeyer flask 100 mL	72 h	Shaking	60 U/mL	[80]
Pectin enriched medium	NR	*Actinomycetes*	NR	6.0	35	NR	NR	5 days	Shaking	202.7955 U/mL	[81]
Modified Kafer medium	NR	*Piriformospora indica*	NR	5.0	50	NR	Erlenmeyer flask 50 mL	12 days	Shaking	10.47 U/mL	[82]
Mineral medium with a carbon source	NR	*Calonectrla pteridis*	NR	4.0	60	20 mcsh	Erlenmeyer flask 125 mL	168 h	Shaking	0.52 mg/mL	[83]

(*Continued*)

Table 6.4 (Continued)

Substrate	Moisture content	Micro-organism	Inoculum size	pH	Temp. (°C)	Particle size	Reaction system	Fermentation time	Agitation per day	Enzyme activity	References
Pectin medium	NR	*Bacillus* sp.	NR	6.0	40	NR	Erlenmeyer flask 100 mL	48 h	Shaking	484.70 U/mg	[75]
YEPD broth	NR	*Saccharomyces cerevisiae*	1×10^4 cells/mL	4.5	30	NR	Erlenmeyer flasks 100 mL	48 h	Shaking	15.12 U/mL polygalacturonase	[77]
				6.0	30					4.15 U/mL pectin lyase	
Mango peels	NR	Pectinolytic fungi	NR	6.0	35	Powder	Erlenmeyer flask 100 mL	96 h	Shaking	0.0298 U/mL	[79]
Wheat bran with sucrose	NR	*Aspergillus niger*	10 mL of incubated broth	NR	28	Powder	Erlenmeyer flask 100 mL	16 days	Shaking	105.0 RVU Endo-polygalacturonase	[78]
										58.0 RVU Endo-pectin lyase	
Wheat bran with sodium nitrate										163 RVU Endo-polygalacturonase	
										43.66 RVU Endo-pectin lyase	

NR, no reported.

types and configurations of bioreactors. One main problem with packed bed reactors in SSF is the generation of biomass agglomerates, making nonhomogeneous flows affecting the enzyme activity yield. For example, [84] focused in a packed bed bioreactor to study the pectinases production with SSF in a pilot scale-up using *A. niger* and the substrate coppicing consisted of 3 kg of sugarcane bagasse and 27 kg of wheat bran. It was shown that agglomeration in the fermentation process of 20 hours of reaction time was minimized by stirring several times at the beginning by using a triple stirring at times of 8, 10, and 12 hours of reaction, which guarantees a uniform pectinase activity. The proposal in this research is to perform intermittent stirring during different fermentation times at different stirring regimes. On the other hand, studies conducted by [85] used *A. niger* LB-02-SF fermented on a wheat bran substrate with a minimum percentage of pectin, glucose, and saline solution inside a bench-scale rotating drum bioreactor. Fermentation presented problems inside the reactor when 30% of the total reactor volume was used; an increase in fungal growth was observed, but a considerable decrease in enzyme activity. However, maximum pectinase activity was generated when using a reaction volume of 50% with stirring intervals of 90 or 120 minutes. It has been demonstrated that at large significant volumes in SSF, agglomerations are generated so that the fermentation is not homogeneous, and the enzyme activity is drastically reduced. It is crucial to perform controlled stirring at certain intervals during SSF to produce pectinases. Likewise, [61] found that in SSF tray reactors, the interaction between substrate thickness and actual moisture is of utmost relevance, as polygalacturonase activity decreased as increased substrate thickness. A 31% improvement was observed when reducing the thickness from 14 to 11 mm at 70% humidity.

In another study, the combination of substrates with different particle sizes helps avoid compaction inside the reactor, such as [86] which used sugarcane bagasse and citrus pulp. This same research gives us a strategy to know if our process can be scalable, which first performing preliminary experiments at the flask level to obtain the best process conditions. The critical point is to perform a previous study in a column bioreactor to collect data about the pectinase activity concerning the respirometry of the substrate mixture to observe the behavior of the strain used and to determine the optimal concentration of oxygen and carbon dioxide inside the bioreactor, which the airflow inside the reactor can control. Likewise, [72] conducting preliminary studies at the flask level, using *A. sojae* fermenting powdered orange peel and apricot pomace determined that the main strategy was an airflow with dissolved oxygen tension reaching 1.7 vvm (volume of air/volume of medium per minute) and agitation at 600 rpm, since it is essential to have a greater amount of dissolved oxygen and thus the microorganism can generate greater enzyme activity. The combination of substrates mentioned above is also applied in SSF as described by [73] performing different mixtures of substrate concentrations of wheat bran, lemon peel, and orange peel using *Aspergillus giganteus*, it was observed that the ratio 66 : 17 : 17, respectively, shows the highest enzyme activity. However, a major problem with SSFs is agglomeration, which is why different aeration flow rates of 10, 15, and 20 L/min*kg of the dry substrate (kgds) were analyzed for screening, where the best pectinase production was obtained with an aeration of 20 L/min*kgds.

Another essential point to carrying out bioprocess industrialization is knowing how to identify which mode of operation the bioreactor should have since its feeding implications can improve or worsen pectinase yields, such as [87] where work was carried out to optimize pectinase production in stirred tank bioreactors by comparing a batch and semi-batch mode. The batch bioreactor was run with both pH controlled and pH uncontrolled; yields were similar at 109.63 U/mL and 99.55 U/mL at 84 hours fermentation, respectively, however; the results obtained from the experiment in a fed-batch mode of operation using a constant carbon (sucrose) feed source, a maximum yield of 450 U/mL will be maintained at 108 hours of reaction, resulting in four times more pectinase production in fed-batch than batch production. Similarly, [44] compared two SmF in batch and fed-batch operation modes, having a culture medium made up of wheat bran, glucose, ammonium sulfate, and pectin for batch mode. Furthermore, for the fed-batch mode, add wheat bran, glucose, ammonium sulfate, and later add citrus pectin after 22 hours of fermentation. In this way, the maximum production of pectinases was favored by adding citrus pectin separately since it generated 14 U/mL, 40% more production than in batch mode. This indicates that to scale an SmF process, the composition of the substrate and the mode of addition must be taken into account to have the best performance.

Within SmF, a crucial point is the agitation of bioreactors, as described by [88]. This study seeks a scale-up of a reactor from 1 to 5 L, using a lemon peel-based substrate. The strategy used was the use of logistic and Gompertz equations. This analysis showed that the number of Reynolds required for agitation in the 5 L reactor almost doubled the value of the 1 L reactor. The production of endo-pectinases and exopectinases in the 5 L reactor increased 7 and 40 times more than the enzyme activity in the 1 L reaction. This shows that having a higher degree of turbulent stirring helps to better transfer oxygen mass and nutrients that promote higher enzyme activity.

In addition, [40] has shown that the use of analytical methods such as the response surface methodology (RSM) through use of software helps the optimization of fermentation since it was possible to increase the yield production of pectinases up to 1,510,391 U/mL, as well as reducing the time of maximum production from 48 to 24 hours. In this way, the use of software also helps us validate scalable bioreactor processes to produce pectinases.

In another case, [89] conducted a study analyzing the feasibility of obtaining pectinases from tobacco wastewater scaled up to a 2 L reaction volume reactor generating a maximum enzymatic activity of endo-polygalacturonases at 347.5 U/mL, exo-polygalacturonases at 270.6 U/mL, with a semi-continuous production of 19.3 and 15.0 U/mL*h, respectively. A comparison was made between using free microorganism cells in one experiment and immobilized cells by fixing them on 45 cotton cloth matrices in three chains forming a honeycomb structure. This also helped to reduce the production time from 72 to 24 hours; this yield improvement is because the immobilization of cells improves the morphology of the fungus, and there is a greater transfer of oxygen to the internal cells of the fungus. On the other hand, the fungal mycelia adhere to the surface of the cotton fabric of the immobilization matrix and thus may help to facilitate the purification of pectinases.

Other strategies that can be taken into account to carry out a pectinase generation process at scale within the SmF is to know how to homogenize the liquid medium, as we have already seen, it can be done by mechanical agitation, but it can also be done by introducing oxygen using an airlift. For example, [90] used two bioreactors: stirred tank and airlift and compared endo-PG and exo-PG productions, obtaining an enzyme activity of 91.3 U/mL for endo-PG and 65.2 U/mL for exo-PG for the stirred tank, and 86.2 U/mL and 60.6 U/mL, respectively, for the airlift bioreactor. The airlift bioreactor has a smaller but no significant difference in production, suggesting potential viability to produce pectinases, as it would have lower set-up and operating costs compared to mechanical agitation. On the other hand, [91] compare cell immobilization in two types of stirred tank and packed bed bioreactors in continuous recirculating outflow mode. This showed a better enzyme activity in the packed bed reactor of 0.98 U/mL*h, suggesting that the cell immobilization together with the aeration of the bioreactor generates better microbial growth and higher biosynthesis disposition.

Likewise, [92] compared within an SSF the addition of pectin as a fermentation activation medium, and on the other hand with aeration within the column, the results indicate that aeration generated a production five times greater than the addition of pectin. In another study, [93] performed a comparison of pectinase production in an SmF in an aerated tank reactor with a batch and fed-batch mode of operation, and the parameters of pH, agitation, and aeration were analyzed. It was found that the reactor with fed-batch aeration and using galactose as a carbon source after 12 hours of fermentation was better than batch fermentation, generating a pectinase production of 23,300 U/L, an increase of 1.6 times better. This suggests that oxygenation in both SmF and SSF is essential to scale up to large volumes of biomass. Table 6.5 shows the scale-strategy that several studies followed in both SmF and SSF.

6.6 Conclusions

The production of pectinases has a huge market due to their multiple applications, being increased with high growth rates as well. Furthermore, it represents a sustainable and respectful tool for the environment. For this reason, a way has been sought to make its production at an industrial level more efficient. Therefore, the optimization plays a key role being the previous step for scaling up the process, and it should be carried out for every pectinase production reaction, SmF or SSF, because each of them has its optimal process conditions such as type of microorganism, substrate type and its concentration, pH, size inoculum, among others and this to achieve the maximum yield. The type of bioreactor and its operation mode can be affected by those factors, with more attention to the type of microorganism, the type of substrate, and its concentration, and consequently different bioreactor configurations are used in the scaling up in the process. For this reason, several bioreactors bring into play being the stirred tank bioreactor and packed bed bioreactors the most used in SMF and SSF being operated mainly in batch mode.

Table 6.5 Bioreactor configurations at different scales of the production process of pectinases.

Substrate	Fermentation type	Microorganism	Bioreactor system	Operation mode	Reaction volume/weight	Enzyme activity	Fermentation time	References
Wheat bran + sugar cane bagasse	SSF	*Aspergillus niger*	Packed bed bioreactor	Batch	30 kg	22 U/g	20 h	[84]
Wheat bran, citric pectin, glucose, salt solution	SSF	*Aspergillus niger* LB-02-SF	Rotating drum bioreactor	Batch	70% (2.5 kg)	178.2 U/g	96 h	[85]
Wheat bran	SSF	*Aspergillus sojae* ATCC 20235	Tray bioreactor	Batch	75 g (dry weight)	298 U/g	96 h	[61]
Sugarcane bagasse (SB) and citrus pulp (CP)	SSF	*Aspergillus oryzae* CPQBA 394–12 DRM 01	Packed bed bioreactor	Batch	26 kg SB and 7.74 kg CP, 49.45 L water	40 U/g	25 h	[86]
Mandarin peel	SmF	*Aspergillus niger* NRC1ami	Stirred tank bioreactor	Fed-batch	6 L	450 U/mL	108 h	[87]
Apricot pomace + powdered orange peel	SmF	*Aspergillus sojae* ATCC 20235	Stirred tank bioreactor	Batch	700 mL	380 U/mL	216 h	[94]
Pineapple stem	SmF	*Bacillus subtilis* BKDS1	Stirred tank bioreactor	Batch	700 mL	1510.391 U/mL	24 h	[40]
Wheat bran, glucose, ammonium sulfate + pectin	SmF	*Aspergillus niger* LB-02-SF	Stirred tank bioreactor	Fed batch	4 L	14.0 U/mL	40 h	[44]
Wheat bran, orange, and lemon peels mixture	SSF	*Aspergillus giganteus* NRRL10	static tray-type bioreactor	Batch	0.1 kg	72 kU/g*Kgds	48 h	[95]

Lemon peels	SmF	*Aspergillus flavipes* FP-500	Mechanically stirred Biostat B bioreactor	Batch	5 L	2200 U/mL Exopectinase activity	160 h	[88]
Tobacco waste	SmF	*Rhizopus oryzae* As 3.3462	Biostat B bioreactor	Semi-continuous	2 L	307.5 U/mL Endo-PG y 242.6 U/mL exo-PG	24 h	[89]
Wheat bran, citric pectin and yeast extract	SmF	*Aspergillus oryzae* IPT301	Airlift bioreactor	Batch	3.2 L	86.2 U/mL endo-PG y 60.6 U/mL exo-PG	96 h	[90]
Spent grains	SmF	*Kluyveromyces marxianus* CCT 3172	Packed bed reactor	Continuous	310 mL	0.98 U/mL*h	150 h	[91]
Wheat bran and defatted rice bran	SSF	*Aspergillus niger*	Column reactor with aeration	Batch	150 g	1800 U/g	48 h	[92]
Galactose and lemon peel	SmF	*Debaryomyces nepalensis* D3893	Aeration tank reactor	Fed batch	2.4 L	23,300 U/L	12 h	[93]

References

1 Srivastava, N., Mishra, P.K., and Upadhyay, S.N. (2020). Pectinases: significance in the digestion of pectin-rich agrowaste. In: *Industrial Enzymes for Biofuels Production*, 1e (ed. N. Srivastava, P. Mishra, and S.N. Upadhyay), 183–204. Elsevier.

2 Garg, G., Singh, A., Kaur, A. et al. (2016). Microbial pectinases: an ecofriendly tool of nature for industries. *3 Biotech* 6: 1–13.

3 Amin, F., Bhatti, H.N., and Bilal, M. (2019). Recent advances in the production strategies of microbial pectinases – a review. *Int. J. Biol. Macromol.* 122: 1017–1026. https://doi.org/10.1016/j.ijbiomac.2018.09.048.

4 Sharma, N., Rathore, M., and Sharma, M. (2013). Microbial pectinase: sources, characterization and applications. *Rev. Environ. Sci. Biotechnol.* 12: 45–60.

5 Ruiz, H.A., Rodríguez-Jasso, R.M., Hernandez-Almanza, A. et al. (2017). Pectinolytic enzymes. In: *Current Developments in Biotechnology and Bioengineering: Production, Isolation and Purification of Industrial Products*, 47–71. Elsevier B.V https://doi.org/10.1016/B978-0-444-63662-1.00003-8.

6 Nighojkar, A., Patidar, M.K., and Nighojkar, S. (2019). Pectinases: production and applications for fruit juice beverages. In: *Processing and Sustainability of Beverages* (ed. A.M. Grumezescu and A.M. Holban), 235–273. Elsevier Inc. https://doi.org/10.1016/B978-0-12-815259-1.00008-2.

7 Sharma, H. and Upadhyay, S.K. (2020). Enzymes and their production strategies. In: Biomass, *Biofuels, Biochemicals: Advances in Enzyme Catalysis and Technologies* (ed. S.P. Singh, A. Pandey, R.R. Singhania et al.), 31–48. Elsevier. https://doi.org/10.1016/B978-0-12-819820-9.00003-X.

8 Ahmed, T., Rana, M.R., Zzaman, W. et al. (2021). Optimization of substrate composition for pectinase production from Satkara (*Citrus macroptera*) peel using *Aspergillus niger*-ATCC 1640 in solid-state fermentation. *Heliyon* 7: e08133. https://doi.org/10.1016/j.heliyon.2021.e08133.

9 Tarafdar, A., Sirohi, R., Gaur, V.K. et al. (2021). Engineering interventions in enzyme production: lab to industrial scale. *Bioresour. Technol.* 326: 124771.

10 Jayani, R.S., Saxena, S., and Gupta, R. (2005). Microbial pectinolytic enzymes: a review. *Process Biochem.* 40: 2931–2944.

11 Sharma, A., Shrivastava, A., Sharma, S. et al. (2013). Microbial pectinases and their applications. In: *Biotechnology for Environmental Management and Resource Recovery*, 1e (ed. A.S. Ramesh Chander Kuhad), 107–124. New Delhi: Springer.

12 Benoit, I., Coutinho, P.M., Schols, H.A. et al. (2012). Degradation of different pectins by fungi: correlations and contrasts between the pectinolytic enzyme sets identified in genomes and the growth on pectins of different origin. *BMC Genomics* 13: 321.

13 Embaby, A.M., Masoud, A.A., Marey, H.S. et al. (2014). Raw agro-industrial orange peel waste as a low cost effective inducer for alkaline polygalacturonase production from Bacillus licheniformis SHG10. *Springerplus* 3: 327.

14 Rehman, H.U., Qader, S.A.U., and Aman, A. (2012). Polygalacturonase: production of pectin depolymerising enzyme from *Bacillus licheniformis* KIBGE IB-21. *Carbohydr. Polym.* 90: 387–391. https://doi.org/10.1016/j.carbpol.2012.05.055.

15 Dixit, S. et al. (2013). Pectin methylesterase of Datura species, purification, and characterization from *Datura stramonium* and its application. *Plant Signaling & Behavior* 8: 10. doi: 10.4161/psb.25681.

16 Moussa-Ayoub, T.E., El-Samahy, S.K., Kroh, L.W., and Rohn, S. (2011). Identification and quantification of flavonol aglycons in cactus pear (*Opuntia ficus indica*) fruit using a commercial pectinase and cellulase preparation. *Food Chem.* 124: 1177–1184. https://doi.org/10.1016/j.foodchem.2010.07.032.

17 Cui, L., Wang, P., Wang, Q., and Fan, X. (2009). The bioscouring efficiency and activity of alkaline pectinase for cotton fabric. *Fibers Polym.* 10: 476–480.

18 Jakób, A., Bryjak, J., and Polakovič, M. (2009). Selection of a method for determination of activity of pectinolytic enzymes in berry fruit materials. *Chem. Pap.* 63: 677–682.

19 Haile, S. and Ayele, A. (2022). Pectinase from microorganisms and its industrial applications. *Sci. World J.* 2022: 1881305.

20 Sharma, D.C. and Satyanarayana, T. (2012). Biotechnological potential of agro residues for economical production of thermoalkali-stable pectinase by bacillus pumilus dcsr1 by solid-state fermentation and its efficacy in the treatment of ramie fibres. *Enzyme Res.* 281384: 1–7.

21 Singh, R.S., Singh, T., and Pandey, A. (2019). Microbial enzymes-an overview. In: *Advances in Enzyme Technology* (ed. R.S. Singh, R.R. Singhania, A. Pandey, and C. Larroche), 1–40. Elsevier B.V https://doi.org/10.1016/B978-0-444-64114-4.00001-7.

22 Dixit, S., Chandrashekar, K., Upadhyay, S.K., Verma, P.C. (2023). Transcriptional plasticity and cell wall characterization in high-methanol-producing transgenic tobacco plants. *Agriculture* 13 (3): 521. https://doi.org/10.3390/agriculture13030521.

23 Dixit, S. et al. (2013). Enhanced methanol production in plants provides broad spectrum insect resistance. *PLoS One* 8 (11), e79664.

24 Favela-torres, E., Aguilar, C., Contreras-esquivel, J.C., and Viniegra-gonzalez, G. (2006). Pectinases. In: *Enzyme Technology* (ed. A. Pandey, C. Webb, C.R. Soccol, and C. Larroche), 273–296. Springer.

25 Gundala, P.B.P.C. (2017). Extremophilic Pectinases. In: *Extremophilic Enzymatic Processing of Lignocellulosic Feedstocks to Bioenergy* (ed. R.K. Sani and R.N. Krishnaraj), 155–180. Springer.

26 Tapre, A. and Jain, R. (2014). Pectinasas: Enzimas para la industria procesadora de frutas. *Rev. Int. Investig. Aliment* 21: 447–453.

27 Patil, N.P., Patil, K.P., Chaudhari, B.L., and Chincholkar, S.B. (2012). Production, purification of exo-polygalacturonase from soil isolate *Paecilomyces variotii* NFCCI 1769 and its application. *Indian J. Microbiol.* 52: 240–246.

28 Jacob, N. (2009). Pectinolytic enzymes. In: *Biotechnology for Agro-Industrial Residues Utilisation* (ed. P. Singh nee Nigam and A. Pandey), 383–396. Springer.

29 Sharma, N., Sahoo, D., Rai, A.K. et al. (2022). A highly alkaline pectate lyase from the Himalayan hot spring metagenome and its bioscouring applications. *Process Biochemistry* 115: 100–109.

30 Upadhyay, S.K. and Singh, S.P. (eds.) (2021). *Bioprospecting of Plant Biodiversity for Industrial Molecules.* John Wiley & Sons Ltd. doi:10.1002/9781119718017.

31 Upadhyay, S.K. and Singh, S.P. (2023). *Plants as Bioreactors for Industrial Molecules.* John Wiley & Sons Ltd. doi:10.1002/9781119875116.

32 Singh, S.P. and Upadhyay, S.K. (eds.) (2021). *Bioprospecting of Microorganism-Based Industrial Molecules.* John Wiley & Sons Ltd. doi:10.1002/9781119717317.

33 Gummadi, S.N. and Panda, T. (2003). Purification and biochemical properties of microbial pectinases – a review. *Process Biochem.* 38: 987–996.

34 Rebello, S., Anju, M., Aneesh, E.M. et al. (2017). Recent advancements in the production and application of microbial pectinases: an overview. *Rev. Environ. Sci. Biotechnol.* 16: 381–394.

35 Aslam, F., Ansari, A., Aman, A. et al. (2020). Production of commercially important enzymes from *Bacillus licheniformis* KIBGE-IB3 using date fruit wastes as substrate. *J. Genet. Eng. Biotechnol.* 18: 46.

36 Majumder, K., Paul, B., and Sundas, R. (2020). An analysis of exo-polygalacturonase bioprocess in submerged and solid-state fermentation by Pleurotus ostreatus using pomelo peel powder as carbon source. *J. Genet. Eng. Biotechnol.* 18: 47.

37 Ahmed, I., Zia, M.A., Hussain, M.A. et al. (2016). Bioprocessing of citrus waste peel for induced pectinase production by *Aspergillus niger*; its purification and characterization. *J. Radiat. Res. Appl. Sci.* 9: 148–154. https://doi.org/10.1016/j.jrras.2015.11.003.

38 Kaur, S.J. and Gupta, V.K. (2017). Production of pectinolytic enzymes pectinase and pectin lyase by *Bacillus subtilis* SAV-21 in solid state fermentation. *Ann. Microbiol.* 67: 333–342.

39 Wang, S., Lian, Z., Wang, L. et al. (2015). Preliminary investigations on a polygalacturonase from *Aspergillus fumigatus* in chinese pu'er tea fermentation. *Bioresour. Bioprocess.* 2: 33. https://doi.org/10.1186/s40643-015-0061-9.

40 Kavuthodi, B. and Sebastian, D. (2018). Biotechnological valorization of pineapple stem for pectinase production by *Bacillus subtilis* BKDS1: media formulation and statistical optimization for submerged fermentation. *Biocatal. Agric. Biotechnol.* 16: 715–722. https://doi.org/10.1016/j.bcab.2018.05.003.

41 Li, Q., Ray, C.S., Callow, N.V. et al. (2020). Aspergillus niger production of pectinase and α-galactosidase for enzymatic soy processing. *Enzyme Microb. Technol.* 134: 109476. https://doi.org/10.1016/j.enzmictec.2019.109476.

42 da Câmara, R.J., da Silva, A.J., de Paiva, W.K.V. et al. (2020). Yellow mombin pulp residue valorization for pectinases production by *Aspergillus niger* IOC 4003 and its application in juice clarification. *Biocatal. Agric. Biotechnol.* 30: 101876.

43 Khatri, B.P., Bhattarai, T., Shrestha, S., and Maharjan, J. (2015). Alkaline thermostable pectinase enzyme from Aspergillus niger strain MCAS2 isolated from Manaslu Conservation Area, Gorkha, Nepal. *Springerplus* 4: 488.

44 Reginatto, C., Rossi, C., Miglioranza, B.G. et al. (2017). Pectinase production by *Aspergillus niger* LB-02-SF is influenced by the culture medium composition and the addition of the enzyme inducer after biomass growth. *Process Biochem.* 58: 1–8. https://doi.org/10.1016/j.procbio.2017.04.018.

45 Patidar MK, Nighojkar S, Kumar A, Nighojkar A. Production of polygalacturonase using Carica papaya peel biowaste and its application for pomegranate juice clarification. *Environ. Sustain.* 2020;3:509–20. https://doi.org/10.1007/s42398-020-00138-6

46 Almowallad SA, Aljobair MO, Alkuraieef AN, Aljahani AH, Alsuhaibani AM, Alsayadi MM. Utilization of agro-industrial orange peel and sugar beet pulp wastes for fungal endo- polygalacturonase production. *Saudi J. Biol. Sci.* 2022;29:963–9. https://doi.org/10.1016/j.sjbs.2021.10.005

47 Heerd, D., Diercks-Horn, S., and Fernández-Lahore, M. (2014). Efficient polygalacturonase production from agricultural and agro-industrial residues by solid-state culture of *Aspergillus sojae* under optimized conditions. *Springerplus* 3: 1–14.

48 de Alencar Guimarães, N.C., Glienke, N.N., Galeano, R.M.S. et al. (2022). Polygalacturonase from *Aspergillus japonicus* (PGAj): enzyme production using low-cost carbon source, biochemical properties and application in clarification of fruit juices. *Biocatal. Agric. Biotechnol.* 39: 102233.

49 Silva J de C, de França PRL, de Melo AHF, Neves-Petersen MT, Converti A, Souza Porto T. Optimized production of Aspergillus aculeatus URM4953 polygalacturonases for pectin hydrolysis in hog plum (Spondias mombin L.) juice. *Process Biochem.* 2019;79:18–27. https://doi.org/10.1016/j.procbio.2018.12.014

50 Ahmed A, Ejaz U, Sohail M. Pectinase production from immobilized and free cells of Geotrichum candidum AA15 in galacturonic acid and sugars containing medium. *J. King Saud. Univ. – Sci.* 2020;32:952–4. https://doi.org/10.1016/j.jksus.2019.07.003

51 Ejaz U, Ahmed A, Sohail M. Statistical optimization of immobilization of yeast cells on corncob for pectinase production. *Biocatal. Agric. Biotechnol.* 2018;14:450–6. https://doi.org/10.1016/j.bcab.2018.04.011

52 Paz A, Chalima A, Topakas E. Biorefinery of exhausted olive pomace through the production of polygalacturonases and omega-3 fatty acids by *Crypthecodinium cohnii. Algal Res.* 2021;59:102470. https://doi.org/10.1016/j.algal.2021.102470

53 Amin, F., Mohsin, A., Bhatti, H.N., and Bilal, M. (2020). Production, thermodynamic characterization, and fruit juice quality improvement characteristics of an Exo-polygalacturonase from *Penicillium janczewskii. Biochim. Biophys. Acta – Proteins Proteomics* 1868: 140379. https://doi.org/10.1016/j.bbapap.2020.140379.

54 Amin, F., Arooj, T., Nazli, Z.-H. et al. (2021). Exo-polygalacturonase production from agro-waste by Penicillium fellutanum and insight into thermodynamic, kinetic, and fruit juice clarification. *Biomass Convers. Biorefinery* September: https://link.springer.com/10.1007/s13399-021-01902-2.

55 da Silva, R.R. (2018). Exploring the fermentation technology for biocatalysts production. In: *Principles and Applications of Fermentation Technology* (ed. A. Kuila and V. Sharma), 181–187. Scrivener Publishing LLC.

56 Patil, S.R. and Dayanand, A. (2006). Optimization of process for the production of fungal pectinases from deseeded sunflower head in submerged and solid-state conditions. *Bioresour. Technol.* 97: 2340–2344.

57 Pandey, A. (2003). Solid-state fermentation. *Biochem. Eng. J.* 13: 81–84. https://doi.org/10.1002/9783527807833.

58 Rudakiya, D.M. (2019). Strategies to improve solid-state fermentation technology. In: *From Cellulose to Cellulase: Strategies to Improve Biofuel Production* (ed. N. Srivastava, M. Srivastava, P.K. Mishra, et al.), 155–180. Elsevier B.V https://doi.org/10.1016/B978-0-444-64223-3.00010-2.

59 Ashok, A., Doriya, K., Rao. D.R.M., and Kumar, D.S. (2017). Design of solid state bioreactor for industrial applications: an overview to conventional bioreactors. *Biocatal. Agric. Biotechnol.* 9: 11–8. http://dx.doi.org/10.1016/j.bcab.2016.10.014.

60 Mahmoodi, M., Najafpour, G.D., and Mohammadi, M. (2019). Bioconversion of agroindustrial wastes to pectinases enzyme via solid state fermentation in trays and rotating drum bioreactors. *Biocatal. Agric. Biotechnol.* 21: 101280. https://doi.org/10.1016/j.bcab.2019.101280.

61 Demir, H. and Tari, C. (2016). Bioconversion of wheat bran for polygalacturonase production by *Aspergillus sojae* in tray type solid-state fermentation. *Int. Biodeterior. Biodegrad.* 106: 60–66. https://doi.org/10.1016/j.ibiod.2015.10.011.

62 Pereira, G.S., Cipriani, M., Wisbeck, E. et al. (2017). Onion juice waste for production of *Pleurotus sajor-caju* and pectinases. *Food Bioprod. Process.* 106: 11–18. https://doi.org/10.1016/j.fbp.2017.08.006.

63 Oumer, O.J. and Abate, D. (2018). Comparative studies of pectinase production by *Bacillus subtilis* strain Btk 27 in submerged and solid-state fermentations. *Biomed. Res. Int.* 2018: 1–10.

64 Zehra, M., Syed, M.N., and Sohail, M. (2020). Banana peels: a promising substrate for the coproduction of pectinase and xylanase from Aspergillus fumigatus MS16. *Pol. J. Microbiol.* 69: 19–26.

65 Meini, M.R., Cabezudo, I., Galetto, C.S., and Romanini, D. (2021). Production of grape pomace extracts with enhanced antioxidant and prebiotic activities through solid-state fermentation by *Aspergillus niger* and *Aspergillus oryzae*. *Food Biosci.* 42: 101168. https://doi.org/10.1016/j.fbio.2021.101168.

66 Rasit, N., Sin Sze, Y., Hassan, M.A. et al. (2021). Pectinase Production from Banana Peel Biomass via the Optimization of the Solid-state Fermentation Conditions of Aspergillus niger Strain. Pertanika. *J Sci Technol.* 30: 257–275.

67 Tripathi, M.K., Mishra, A.S., Misra, A.K. et al. (2008). Selection of white-rot basidiomycetes for bioconversion of mustard (Brassica compestris) straw under solid-state fermentation into energy substrate for rumen micro-organism. *Lett. Appl. Microbiol.* 46: 364–370.

68 Handa, S., Sharma, N., and Pathania, S. (2016). Multiple parameter optimization for maximization of pectinase production by *Rhizopus* sp. C4 under solid state fermentation. *Fermentation* 2: 10.

69 Mehmood, T., Saman, T., Irfan, M. et al. (2019). Pectinase production from *Schizophyllum commune* through central composite design using citrus waste and its immobilization for industrial exploitation. *Waste Biomass Valoriz.* 10: 2527–2536. https://doi.org/10.1007/s12649-018-0279-9.

70 Saran, S., Malaviya, A., and Chaubey, A. (2019). Introduction, scope and significance of fermentation technology. In: *High Value Fermentation Products*, 1–25. Wiley.

71 Subramaniyam, R. and Vimala, R. (2012). Solid state and submerged fermentation for the production of bioactive substances: a comparative study. *Int. J. Sci. Nat.* 3: 480–486.

72 Abdullah, R., Jafer, A., Nisar, K. et al. (2018). Process optimization for pectinase production by locally isolated fungal strain using submerged fermentation. *Biosci. J.* 34: 1025–1032.

73 Uzuner, S. and Cekmecelioglu, D. (2021). Biochemical characterization and stability of *Bacillus subtilis* polygalacturonase produced using hazelnut shells by submerged fermentation. *Biocatal. Biotransform.* 39: 269–274. https://doi.org/10.1080/10242422.2020.1871332.

74 Sudeep, K.C., Upadhyaya, J., Joshi, D.R. et al. (2020). Production, characterization, and industrial application of pectinase enzyme isolated from fungal strains. *Fermentation* 6: 59.

75 Murugan, T., Deepika, P., Kowsalya, A. et al. (2021). Production and characterization of extracellular pectinase from a newly isolated Bacillus species from fruit waste soil. *Mater. Today Proc.* 45: 2087–2090. https://doi.org/10.1016/j.matpr.2020.09.607.

76 Yaaser, Q., Almulaiky, A.A.A., Khalil, N.M. et al. (2020). Polygalacturonase by *Aspergillus Niger* using seaweed waste under submerged fermentation: production, purification and characterization. *Biomed. J. Sci. Tech. Res.* 25: 19416–19422.

77 Poondla, V., Bandikari, R., Subramanyam, R., and Reddy Obulam, V.S. (2015). Low temperature active pectinases production by *Saccharomyces cerevisiae* isolate and their characterization. *Biocatal. Agric. Biotechnol.* 4: 70–76. https://doi.org/10.1016/j.bcab.2014.09.008.

78 Begum, G. and Munjam, S. (2021). Carbon and nitrogen sources effect on pectinase synthesis by *Aspergillus niger* under submerged fermentation. *Biosci. Biotechnol. Res. Asia* 18: 185–195.

79 Mulluye, K., Kebede, A., and Bussa, N. (2021). Production and optimization of pectinase from pectinolytic fungi cultivated on mango peels and pectin subjected to submerged fermentation. *Biol. Med. Nat. Prod. Chem.* 10: 15–21.

80 Chetana, A., Kumar, V.V., Jyothi, C. et al. (2021). Production of pectinase from fruit and vegetable waste by *Aspergillus niger*. *Bull. Monum.* 22: 8–12.

81 Yugandhar, N.M., Sushma, C., Beena, C. et al. (2019). Optimizing of cultural conditions for the production of pectinolytic enzyme from marine actinomycetes by submerged fermentation along with statistical approach. *Int. J. Curr. Innov. Adv. Res.* 2: 29–43.

82 Heidarizadeh, M., Rezaei, P.F., and Shahabivand, S. (2018). Novel pectinase from *Piriformospora indica*, optimization of growth parameters and enzyme production in submerged culture condition. *Turk. J. Biochem.* 43: 289–295.

83 Ladeira Ázar, R.I.S., da Luz, M.M., Piccolo Maitan-Alfenas, G. et al. (2020). Apple juice clarification by a purified polygalacturonase from *Calonectria pteridis*. *Food Bioprod. Process.* 119: 238–245.

84 Finkler, A.T.J., Biz, A., Pitol, L.O. et al. (2017). Intermittent agitation contributes to uniformity across the bed during pectinase production by *Aspergillus niger* grown in solid-state fermentation in a pilot-scale packed-bed bioreactor. *Biochem. Eng. J.* 121: 1–12. https://doi.org/10.1016/j.bej.2017.01.011.

85 Poletto, P., Polidoro, T.A., Zeni, M., and da Silveira, M.M. (2017). Evaluation of the operating conditions for the solid-state production of pectinases by *Aspergillus niger* in a bench-scale, intermittently agitated rotating drum bioreactor. *LWT Food Sci. Technol.* 79: 92–101.

86 Biz, A., Finkler, A.T.J., Pitol, L.O. et al. (2016). Production of pectinases by solid-state fermentation of a mixture of citrus waste and sugarcane bagasse in a pilot-scale packed-bed bioreactor. *Biochem. Eng. J.* 111: 54–62. https://doi.org/10.1016/j.bej.2016.03.007.

87 El Enshasy, H.A., Elsayed, E.A., Suhaimi, N. et al. (2018). Bioprocess optimization for pectinase production using *Aspergillus niger* in a submerged cultivation system. *BMC Biotechnol.* 18: 1–13.

88 Wolf-Márquez, V.E., Martínez-Trujillo, M.A., Aguilar Osorio, G. et al. (2017). Scaling-up and ionic liquid-based extraction of pectinases from *Aspergillus flavipes* cultures. *Bioresour. Technol.* 225: 326–335. https://doi.org/10.1016/j.biortech.2016.11.067.

89 Zheng, Y.-X., Wang, Y.-L., Pan, J. et al. (2017). Semi-continuous production of high-activity pectinases by immobilized *Rhizopus oryzae* using tobacco wastewater as substrate and their utilization in the hydrolysis of pectin-containing lignocellulosic biomass at high solid content. *Bioresour. Technol.* 241: 1138–1144. https://doi.org/10.1016/j.biortech.2017.06.066.

90 Fontana, R.C., Polidoro, T.A., and da Silveira, M.M. (2009). Comparison of stirred tank and airlift bioreactors in the production of polygalacturonases by *Aspergillus oryzae*. *Bioresour. Technol.* 100: 4493–4498. https://doi.org/10.1016/j.biortech.2008.11.062.

91 Almeida, C., Brányik, T., Moradas-Ferreira, P., and Teixeira, J. (2003). Continuous production of pectinase by immobilized yeast cells on spent grains. *J. Biosci. Bioeng.* 96: 513–518.

92 Linde, G.A., Magagnin, G., Costa, J.A.V. et al. (2007). Column bioreactor use for optimization of pectinase production in solid substrate cultivation. *Braz. J. Microbiol.* 38: 557–562.

93 Gummadi, S.N. and Kumar, D.S. (2008). Batch and fed batch production of pectin lyase and .bioreactor. *Bioresour. Technol.* 99: 874–881.

94 Fratebianchi, D., Crespo, J.M., Tari, C., and Cavalitto, S. (2017). Control of agitation rate and aeration for enhanced polygalacturonase production in submerged fermentation by *Aspergillus sojae* using agro-industrial wastes. *J. Chem. Technol. Biotechnol.* 92: 305–310.

95 Ortiz, G.E., Ponce-Mora, M.C., Noseda, D.G. et al. (2017). Pectinase production by *Aspergillus giganteus* in solid-state fermentation: optimization, scale-up, biochemical characterization and its application in olive-oil extraction. *J. Ind. Microbiol. Biotechnol.* 44: 197–211.

7

Microbes as a Bio-Factory for Polyhydroxyalkanoate Biopolymer Production

Daniel Tobías-Soria[1], Julio Montañez[1], Iván Salmerón[2], Alejandro Mendez-Zavala[1], James Winterburn[3], and Lourdes Morales-Oyervides[1]

[1] *Facultad de Ciencias Quimicas, Universidad Autonoma de Coahuila, Saltillo, Coahuila, Mexico*
[2] *School of Chemical Science, Autonomous University of Chihuahua, Chihuahua, Chihuahua, Mexico*
[3] *Department of Chemical Engineering, The University of Manchester, Manchester, UK*

7.1 Introduction

Nowadays, plastics represent one of the most important materials due to their versatile applications, which substitute conventional materials like glass, wood, and metals [1]. Unsurprisingly, we use them in many applications like medicines, cars, electro-domestics, or even food packaging. Consequently, conventional plastics (petroleum-derived) have found their way into our routine livelihood [2]. About 4% of the world's oil production is converted into plastics, not considering the part used to generate the required energy for manufacturing [3]. Therefore, concerns over the quick depletion of fossil fuels have reinforced research interest in developing alternative processes to synthesize the needed plastics for daily life under innovative approaches [4].

In addition, consumers' trends and awareness about the plastic-related environmental problems due to their overgeneration, prolonged degradation, and exceeded disposal capacity have driven industry and academia to research similar but environmentally safe alternative materials. This alternative was presented as bioplastics, which can be synthesized, degraded, or consumed by microbes under specific conditions [3].

Plastic material is considered bioplastic if it is biobased, biodegradable, or features both properties [5]. Also, the term biopolymer is generally used for materials that start from renewable sources like microorganisms, plants, or trees [6, 7–9].

Bioplastics are as useful as their synthetic counterparts because their structure can be manipulated to obtain specific characteristics, e.g. higher molecular weight and low reactivity [3]. Such features result in novel functional properties that can increase the range of potential applications, such as biomedical. But more importantly, biopolymers offer environmental advantages over synthetic ones, which contribute to the efforts for pollution

Microbial Bioreactors for Industrial Molecules, First Edition.
Edited by Sudhir P. Singh and Santosh Kumar Upadhyay.

abatement, such as lower carbon footprint, lower energy requirement, and relatively low greenhouse emissions during manufacturing, litter reduction, and compostability [3, 10].

Furthermore, even when the current situation of the biopolymer industry is under development, it presents a promising future. Worldwide manufacturing of biopolymers is projected to increase to 9.6×10^6 tons by 2023, an annual growth rate of 4% [11]. Also, the global demand for bioplastics is expected to increase due to factors such as the availability of raw materials, their renewability, their technical properties and functionality, and their wide processing window, where polyhydroxyalkanoates stand out as innovative biopolymers [12, 13].

Polyhydroxyalkanoates (PHAs) are well studied within the known biopolymers and are bio-based and biodegradable. The bio-based and biodegradable concepts are not always presented together. Bio-based plastic refers to the importance of the carbon building block's origin. On the other hand, plastic is considered biodegradable when microbes use it as a food source once it breaks down [14, 15].

Some entrepreneurs' chemical companies are conscious of PHAs' potential and have begun producing and using PHA in several applications, giving way to the so-called PHA industry [16]. A list of companies has been included in a recent review of the challenges met by PHA for industrialization [17].

Now, the current volume of industrial-produced PHA has not yet exceeded 10K tons annually [18]. Even if it does not compete with the continuing production of petroleum-based plastics, it is expected to grow in the not-too-distant future considerably.

The challenges and innovations regarding the production of biopolymers such as PHA polyesters have been extensively reviewed, addressing the issue from different perspectives such as biopolymers' properties and applications [19, 20], engineering design [18], process optimization [21], scale-up and commercialization [22], sustainability [23–26], and downstream processing [27].

Therefore, this chapter presents a general overview of the microbial production of PHAs and addresses recent trends for developing a sustainable bioprocess. Special emphasis is placed on discussing the economics of the process and areas of opportunity for large-scale implementation.

7.2 Microbial Polyhydroxyalkanoates as a Novel Alternative to Substitute Petroleum-Derived Plastics

PHAs are a varied class of polymers with approximately a hundred monomers and are among the most studied bioplastic as a substitute for conventional plastics. Poly(3-hydroxybutyrate) (PHB) stands out as the most studied class of PHAs [28]. A wide range of microorganisms such as bacteria, algae, and fungi are utilized for PHAs biosynthesis. Although fungi are not directly involved in the production, they form an essential part of some production processes [7, 14].

PHAs have similar material properties to conventional petroleum-derived plastics such as polypropylene or polystyrene [4], but they can be synthesized from renewable sources. The increased interest is placed on PHAs due to their characteristics like biodegradability, biocompatibility, bioresorbable, and piezoelectricity. Also, they have been widely used in various applications such as the biomedical field due to their biocompatibility and other suitable mechanical and versatile properties. Also, in the pharmaceutical field, they are used to load drugs into the form of gels, microspheres, and nanoparticles [29, 30].

Considering the interesting characteristics of PHA bioplastics, several established companies and many start-ups have begun to synthesize and utilize PHA in a wide range of applications depending on the consumer requirements around the world, such as Biomer (Germany), Bluepha (China), PHB Industrial S.A. (Brazil), and Danimer Scientific (USA) [18]. The current increase in the understanding of PHAs biosynthesis has given rise to new approaches for producing this kind of polymer from renewable resources [31].

The life cycle of such bioplastics commonly begins with the fermentation of renewable carbon materials to synthesize PHAs, which are then manufactured for their specific application. PHAs could be degraded to CO_2 and H_2O to diminish their environmental impact [32]. Thus, the discarded PHA products can be recycled or composted as carbon feedstocks, practicing the cradle-to-cradle concept [33].

The microbial biosynthesis of PHAs has been studied for almost a hundred years. Nowadays, the study of such biopolymers has increased along with the justification for substituting petroleum-based plastics. The biosynthesis pathways have largely been studied and various attempts have been made to improve the PHA biosynthesis efficiency [34].

However, there are still aspects to improve the large-scale PHA production process, either in the downstream or upstream processing, which we will discuss in the next section.

7.3 Microbial PHAs Classification, Synthesis, and Producing Microorganisms

7.3.1 PHAs Classification

Establishing a baseline about PHAs classification, synthesis, and producer microorganisms is crucial. PHAs are polyesters that contain a characteristic ester bond, which are accumulated as a reserve of carbon and energy. PHA structure comprises 3-hydroxy fatty acid monomers [35]. Their synthesis and accumulation occur intracellularly in cells in high concentrations under limited nutrient conditions [36]. It has been studied that some adverse conditions directly influence PHA synthesis, like oxygen limitation [37].

The biosynthesized polyester compounds are stored in the cytoplasm as insoluble inclusion bodies with a typical diameter ranging from 0.2 to 0.5 μm [38]. The PHA's molecular mass is frequently reported as around 2×10^5 to 3×10^6 Da [39]. The characteristics depend on various factors such as the microbial producing species, type, and concentration of supplied carbon source, fermentation conditions (temperature, pH, DO), and also the type of fermentation (batch, fed-batch, continuous) [39]. The diversity of PHAs is due to their functional groups, molecular weight, length of chain, and monomeric units involved in the polymer [19].

Thus, the most accepted classification is based on the structure, the carbon atom's number, and the chain length of the monomers [13, 35, 40]. Table 7.1 includes PHAs classification and relevant characteristics [41, 42]. The R group displayed in the first column refers to the aliphatic alkyl side chain [41, 42]. PHAs are usually medium-chain length PHAs

Table 7.1 General structure of most commonly synthesized PHAs, their classification, and properties.

"R" group	Carbon number location	PHA polymer	PHA class	Properties
Methyl $-CH_3$	C_4	P3HB/PHB	SCL-PHAs	T_m (°C) = ~180 T_g (°C) = ~0
Ethyl $-CH_2-CH_3$	C_5	P3HV/PHV		TS (MPa) = Lower Avg. MW = Higher Crystallinity (%) = 70 Elasticity = Lower (brittle)
Propyl $(-CH_2)_2-CH_3$	C_6	PHHx	MCL-PHAs	T_m (°C) = ~60 T_g (°C) = ~ −40
Pentyl $(-CH_2)_4-CH_3$	C_8	P(3HO)/PHO		TS (MPa) = Higher Avg. MW = Lower
Heptyl $(-CH_2)_6-CH_3$	C_{10}	PHD		Crystallinity (%) = 40 Elasticity = Higher

Melting temperature (T_m); glass transition temperature (T_g); average molecular weight (Avg. MW).

(MCL-PHAs), which comprise 6–14 carbon atoms such as poly(3-hydroxyhexanoate) and poly(3-hydroxyoctanoate).

Regarding the short-chain length PHAs (SCL-PHAs), their monomers are formed by 4–5 carbon atoms, such as the poly (3-hydroxybutyrate), and the poly (4-hydroxybutyrate). The long-chain length (LCL-PHAs) consists of at least 15 carbon atoms in their monomer structure [26].

Furthermore, PHAs displayed characteristics and properties will also depend on the chain length variation. For instance, MCL-PHAs are flexible in nature, less crystalline, and have a low melting temperature (42–65 °C) [43]. On the other hand, SCL-PHAs are stiff, friable, and greatly crystalline [43].

Of the total monomers reported, most are of the type of 3-alkanoic acid, which are linear and saturated molecules composed of 3–16 carbon atoms [44]. Likewise, PHAs are divided into homopolymer PHAs or heteropolymer PHAs depending on the PHA's type of monomer [45]. Heteropolymer PHAs are when different 3-hydroxy fatty acids compose the monomer unit [45].

The homogeneous or heterogeneous characteristics of monomeric blocks also depend on the microorganism and substrate, as observed in *Pseudomonas putida* KT2440 [46]. Also, the structure and conformation of the PHAs provide them with distinct physical properties, which vary depending on their classification, as shown in Table 7.1.

7.3.2 Biosynthetic Pathways for PHAs Production

It is important to understand how the microorganism stores the PHAs, given that the PHAs granules, also called carbonosomes, are microbial carbon and energy reserves for starvation, accumulated in the presence of excess carbon source and under unfavorable growth conditions due to different limiting factors [43]. These inclusion bodies are an amorphous intracellular mixture of polymer and associated proteins [46]. They are composed of the polymer itself (on the inner side of the structure) and covered by granule-associated proteins (GAP) presenting functions of structural, biosynthesis, catabolic, and regulatory, like PHA synthases (PhaC), PHA depolymerase (PhaZ), and phasin protein (PhaF), PHA transcriptional regulators (PhaR), hydrolases, and reductases [47].

The microbial synthesis of PHAs varies depending on the used microorganism and the conditions provided for the fermentation process. These factors will define the implied metabolic pathway for the synthesis as well as the growth stage where the microorganism produces the polymer [19], for example production during the exponential phase when the conditions are favorable for growth. On the other hand, when conditions are unfavorable, like when a nutritional component is limited, production is in the stationary phase, and the excess nutrients will be stored as PHA granules until the conditions become ideal for survival [19]. However, some *Bacillus* strains can biosynthesize PHAs in both stationary and exponential phases. In this genus, eight metabolic pathways for synthesizing these polymers have been reported [32].

Generally, bacteria are known for the PHAs synthesis because they are well studied in the field, considering that some central metabolic pathways like glycolysis, TCA (tricarboxylic acid cycle), β-oxidation pathway, de novo fatty acid synthesis pathway, Calvin cycle, and serine pathway are present in the PHA synthesis [26]. They present advantages over other species like the time employed in fermentation processes.

The accumulation of PHAs in the cell is a complex and global metabolic process [46]. For didactic purposes, it can be divided into two groups of metabolic pathways: (i) the central carbon metabolism pathways, from which precursors of PHAs monomers are obtained, and (ii) the PHA synthesis pathway from these precursors [48].

It is important to consider that the cell's central carbon metabolism pathways are conserved and used to obtain cellular components independently of PHAs production. However, energy and carbon need to drive the metabolic fluxes for obtaining PHAs [48], which are produced as a reaction to certain nutrient conditions like high carbon concentration and limitations of other nutrients (nitrogen, phosphorus, and oxygen) [46]. The synthesis of PHAs in microorganisms commonly occurs via three highly reported pathways [39]; however, fourteen pathways for the biosynthesis of PHAs have been reported [47].

Pathway I is commonly reported in poly(3-hydroxybutyrate) PHB or SCL-PHAs producing organisms from sugars and is commonly found in *Azotobacter vinelandii* and *Cupriavidus necator* [49], while Pathways II and III are more for MCL-PHAs producing microbes and have been commonly reported in *Pseudomonas* spp. [20, 47]. The rest of the pathways are not common in nature or have been engineered, giving the generation of unconventional PHAs. Pathway I involves three key enzymes, the β-ketothiolase, NADPH-dependent acetoacetyl-CoA reductase, and PHA synthase (phaA, phaB and phaC, respectively). The substrates of this pathway are not structurally related to PHAs and are metabolized depending on their nature; thus, carbohydrates are catabolized by various glycolytic pathways (Embden-Meyerhof-Parnas, pentose phosphate pathway, Entner-Doudoroff pathway, xylose isomerase pathway, etc.). In comparison, the catabolism of aromatic compounds involves the oxidation of their rings and opening. Their conversion results in the generation of acetyl-CoA, generating reducing power through the synthesis of NAD(P)H [46, 48]. In this pathway, two acetyl-CoA molecules are condensed to form acetoacetyl-CoA by the enzyme β-ketothiolase; subsequently, a reduction of the obtained compound to (*R*)-3-hydroxybutyryl-CoA is catalyzed by an NADH-dependent acetoacetyl-CoA reductase, and finally, the precursor monomer is polymerized by a PHB synthase to obtain PHB [49].

Pathway II uses fatty acids by microorganisms to produce acetyl-CoA synthesized from β-oxidation of fatty acids. In this case, the substrates are structurally related to PHAs (modified fatty acids, saturated, and unsaturated) and are transported into the cell through two fatty acid transporters with different specificity depending on the length of the fatty acid chain. The presence of two different substrates structurally related to PHA metabolized by this pathway will be reflected in the monomer composition and it will depend on the ratio of the two substrates. In β-oxidation, acyl-CoA is obtained, which is oxidized to *trans*-2-enoyl-CoA with a reduction of FAD to $FADH_2$ and mediated by acyl-CoA dehydrogenases; this is the limiting step in this pathway due to the low activity of the enzymes. Subsequently, FadBA and FadB complexes are involved, allowing intermediates for PHA synthesis through the oxidation cycle by the action of the enzymatic activities, enoyl-CoA hydratase, *cis*-Δ^3-*trans*-Δ^2-enoyl-CoA isomerase, 3-hydroxyacyl-CoA dehydrogenase, and 3-hydroxyacyl-CoA epimerase. Then, 2-*trans*-enoyl-CoA is converted to (S)-3-hydroxyacyl-CoA, which is oxidized to form 3-ketoacyl-CoA with the reduction of NAD^+, and which can be converted back to acyl-CoA with the release of one molecule of acetyl-CoA per catalytic cycle. On the other hand, precursors of PHAs can be obtained through the activity of PhaJ, a stereospecific *trans*-enoyl-CoA hydratase that converts *trans*-2-enoyl-CoA into such precursors [46, 48].

Lastly, pathway III needs two key enzymes, 3-hydroxyacyl-ACP-CoA transferase (phaG) and malonyl-CoA-ACP transacylase (FabD) [19]. In the presence of substrates such as carbohydrates or alcohols, the carbon fluxes are directed toward acetyl-CoA for the biosynthesis of PHAs by *de novo*-FAS. They start with the carboxylation of acetyl-CoA to form malonyl-CoA, which is activated by *trans*-acylation mediated by an acyl-carrier protein (ACP) so that the intermediates in the cycle are of the ACP thioesters type. The relationship between this pathway and the production of PHAs is given in the conversion of (*R*)-3-hydroxyacyl-ACP into (*R*)-3-hydroxyacyl-CoA by PhaG activity; then, these monomers are polymerized into PHAs [46].

In summary, using carbon sources such as alcohols, carbohydrates, and CO_2 results in obtaining acetyl-CoA and subsequently at precursor (*R*)-3-hydroxybutyryl-CoA in P-I. From acetyl-CoA, it can be derived to TCA, where it can result in (*S*)-methylmalonyl-CoA, which is transformed into propionyl-CoA, where amino acid and propionate metabolism can converge. Subsequently, the precursor (*R*)-3-hydroxyacyl-CoA is obtained, which is a monomer of mcl-PHA synthesis as in fatty acid metabolism by β-oxidation (P-II) (this can occur from *trans*-2-enoyl-CoA, (*S*)-3-hydroxyacyl-CoA or (*R*)-3-ketoacyl-CoA), or carbohydrate or lipid metabolism by conversion of acetyl-CoA to malonyl-CoA, to the formation of (*R*)-3-ketoacyl-ACP to give (*R*)-3-hydroxyacyl-ACP by *de novo*-FAS (P-III) and generate (*R*)-3-ketoacyl-CoA which also converts from valerate metabolism; this is converted to (*R*)-3-hydroxyacyl-CoA which is also the result of butyrate metabolism and from this monomer generate PHAs [44].

Once the different pathways have metabolized the substrate, PHAs monomer intermediates are synthesized. The PHA synthases convert (*R*)-3-hydroxyacyl-CoA into a polyester chain, releasing one CoA molecule per catalytic cycle [46]. The type of PHA formed in bacteria depends on the substrate and the specificity of the PHA synthase (PhaC). The type of enzyme found in microorganisms generates PHA of different chain lengths [47].

PHAs biosynthesis reactions are regulated by the highly conserved *pha* gene cluster in *P. putida*. This cluster is composed of two operons. The first operon codes for two mcl-PHA synthases (PhaC1 and PhaC2): a depolymerase (PhaZ) and a transcriptional activator (PhaD). The second operon next to the first, but in a converse direction, codes for the phasins (PhaF and PhaI). The products of this system direct carbon fluxes toward polymer accumulation or hydrolysis. The difference between the genes coding for PHA biosynthesis in different microorganisms lies in the substrate specificity of the enzymes. The expression of the genes of the *pha* cluster is dependent on the carbon source and at the transcriptional level, it is activated by the specific regulator of the PhaD gene. The existence of a complex regulatory system that considers the PHA granules, the PhaD–PhaF complex, and the target DNA is suggested. In the same way, control of PHA metabolism by a system of regulation by catabolic repression is proposed, which may be associated at the transcriptional level in the *pha* operon and may affect the activation or repression of the components of the fatty acid oxidation pathways. On the other hand, the polymer formed can be hydrolyzed into free monomers by the PHA depolymerase enzyme, depending on metabolic conditions [46].

In conventional production systems, the nutrient limitation is used to increase PHA production. One of the strategies used is the variation in C/N ratios, which affect the production of PHA from carbohydrates. The adverse effects of these ratios appear to be on the activity of the enzyme 3-hydroxyacyl-ACP-CoA transacylase in FAS which is downregulated under conditions of excess nitrogen, tending to the production of cellular components instead of PHAs. This control is given in the catabolic repression control proteins in *P. putida* of the *pha* cluster. Depending on the cell's metabolic requirements, the synthesis of PHA precursors can be controlled; thus, under different concentrations of NAD(P)H/NAD(P)$^+$ and acetyl-CoA/CoA, β-oxidation can be regulated, tending to the biosynthesis of polymers with high proportions of such energetic components [46].

7.3.3 PHAs Producing Strains

Regarding microbial producer microorganisms, it has been reported that both prokaryotic and eukaryotic can convert biomass into biopolymers [26]. A set of specific enzymes is used depending on the microorganism and the PHA production can yield 90% of the biomass dry weight [26]. Research on PHA production focuses on various microbes like bacteria and others such as actinomycetes and algae [19]. It is worth mentioning that the microorganisms capable of producing PHA derive from any habitat, such as marine, estuarine, rhizosphere, and anthropogenic ecosystems [25].

Even though not all microorganisms can synthesize PHAs, many play an essential part in PHAs production, as shown in Table 7.2. Even when over a hundred species can biosynthesize PHAs, various factors should be considered before selecting a microorganism. For instance, the ability of the cell to grow on cheaper substrates, the growth and production rates, and the maximum possible polymer accumulation by the cell [68].

To date, a minimum of 90 different bacterial genera and extending to 300 species have been recognized as PHA producers [42], thus representing the largest group of microorganisms

Table 7.2 Microbial strains evaluation for PHA production in recent research works.

Type of microorganism	Species	Microorganism role in the PHA biosynthesis	References
Bacteria	*Paraburkholderia* sp. strain PFN29	PHA production by SmF	[50]
	Ancylobacter aquaticus ART_41	PHA production with paper industry effluents	[51]
	Schlegelella thermodepolymerons DSM 15344	PHA production with substrates rich in xylose	[52]
	Cupriavidus necátor H16	PHA production from peanut oil	[53]
	Saccharophagus degradans	PHA production through biodegradation of alginate and microalgae	[54]
	Paracoccus sp. LL1	PHA production in membrane bioreactor	[55]
	Halomonas hydrothermalis (MTCC 5445)	PHA production based on marine salt	[56]
	Bacillus tequilensis ARY86	PHA production from lactose	[57]
	Pseudomonas putida NX-1	PHA production from lignin	[58]
	Bacillus megaterium strain CAM12	PHA production with hydrolyzed sugars	[59]
	Haloferax mediterranei	PHA production in high salinity	[60]
	Kyrpidia spormannii	PHA production by thermophilic electrosynthesis	[61]
	Halogeometricum borinquense	PHA production by repeated batch process	[62]
Microalgae	*Synechococcus subsalsus*	PHA production with nitrogen reduction	[63]
	Scenedesmus obliquus *Chlorella vulgaris* *Chlorella sorokiniana*	Biomass production to use it as carbon source for PHA production	[54]
	Chlorella sorokiniana SVMIICT	PHA production through the nutrient limitation	[64]
Fungi	*Trichoderma reesei* *Aspergillus niger*	Pretreatment of corn stubble for PHA production	[65]
	Pycnoporus coccineus MScMS1	Pretreatment of sugar cane bagasse for PHA production	[66]
Yeast	*Pichia kudriavzevii* VIT-NN02	PHA production using banana peels and hydrolyzed chicken feathers as substrate	[67]

with PHA production ability. Various bacterial species have been widely studied due to their high PHA accumulation capacity, such as *Azotobacter* sp., *Bacillus* sp., *Burkholderia* sp., *Halomonas* sp., *Haloferax* sp. and *Pseudomonas* sp. [39]. Autotrophs are another option for the synthesis of PHAs since they use CO_2 and sunlight energy as raw materials [69].

In this regard, cyanobacteria can synthesize and accumulate PHB granules [19]. Although some microalgae can produce PHAs, they are generally used as raw materials to produce PHA [54]. PHAs biosynthesis could be carried out also by various fungi and yeast species, such as in the case of *Wickerhamomyces anomalus* [70]. However, the fungal production of PHAs has not been studied thoroughly [19].

Furthermore, some studies have reported using mixed microbial culture (MMC) as a promising alternative for PHA production [42].

MMC strategy relies on nature's rules of selection and competition; thus, the feast/famine regimes could be efficiently applied as a selective pressure on a microbial group of PHA producers' [25].

Other research focuses on the halophilic microorganisms as promising PHB producers because not all the microorganisms can tolerate high salt concentrations, which could meet the requirement of the sterile condition, and also that kind of microbes present the ability to use alternative waste feedstocks. BluePHA (China) used halophilic microorganisms to produce PHB; even when they reported difficulties due to the equipment damage caused by the media's salinity, the company appears to be operating at an industrial scale producing PHAs [71]. The use of extremophilic microorganisms is proposed for PHA production due to their metabolic capabilities related to their tolerance to environmental factors unconventional for most currently used microorganisms.

7.3.4 Bacteria as the Main Species for the PHA Production

Bacteria are of great importance due to their ability to synthesize various classes of polyester PHAs [72]. Both gram-positive and gram-negative bacterial strains can produce PHAs using various carbon sources like alkanes, alkenes, renewable carbon sources, wastes, volatile fatty acids (VFAs), and wastewater [12]. Gram-negative bacteria have PHA producers, for example *C. necator, Alcaligenes eutrophus, Alcaligenes latus, P. putida, Pseudomona oleovorans,* and *A. vinelandii*. In the same manner, gram-positive bacteria have some good PHA producers, including several species of *Bacillus* (*B. megaterium* and *B. cereus*) [73].

It is reported that bacterial PHA production is more cost-effective than the production from other living organisms, like plants, mostly attributed to bacterial high accumulation capacity. In this regard, some recent promising bacterial strains studied for PHA production stand out from the rest in terms of yields, such as *Aeromonas hydrophila, Bacillus* sp., *Burkholderia sacchari, Halomonas bolivinensis, Pseudomonas* sp., and *Rhodopseudomonas palustris* [39].

Nearly 150 distinct PHA analogs produced by bacterial strains with variations in their structure and properties have been reported [74]. Moreover, bacteria can produce PHA copolymers in mixed substrates [4]. Therefore, microorganisms are still being evaluated for their potential capacity to produce PHAs and novel PHAs constituents with potential applications [4]. As previously mentioned, recent research is inclining towards using extremophilic microorganisms. Examples of these are *Halomonas bluephagensis* and *Halomonas campaniensis*. Particularly *H. bluephagensis* has been engineered to redirect its metabolic fluxes toward PHA biosynthesis, reducing the specificity of PHA synthase, controlling the $NADH/NAD^+$ ratio to improve PHA accumulation, morphology engineering, etc. With this platform, a cost of USD 1.68/kg of material is possible [17].

Haloferax mediterranei is another extremophilic microorganism from marine environments that tolerates high salt concentration (halophilic); it can also degrade pollutants and produce PHA. The production of these polymers increases linearly with an increasing salt concentration in the culture, which is a protective mechanism against adverse conditions. Proteomic analysis of *H. meditarranei* culture under high salinity conditions revealed the expression of key enzymes in PHA synthesis (β-ketoacyl-ACP reductase and 3-hydroxyacyl-CoA dehydrogenase), as well as the enzymes serine-pyruvate transaminase and serine-glyoxylate transaminase, which increased the conversion of glucose to PHA [60].

7.3.5 Algae as a Feasible Alternative for PHA Production

The current bioplastic production situation related to microalgae can be addressed under two main approaches (Figure 7.1), the microalgae biomass as raw material, and microalgae as cell factories for PHA synthesis [75].

Biomass from micro and macroalgae has high carbohydrate content and does not require land cultivation in response to the problem of land overutilization. In the same way, its utilization as a PHA carbon substrate does not interfere with human nutrition and animal feeding [76]; thus, they do not compete with conventional food sources.

Algae can store energy in the form of starch; this characteristic gives them a revalorization opportunity as a low-cost and promising raw material for PHA production. Also, it has been reported that certain bacteria can utilize polysaccharides as carbon sources, thus eliminating the need for unit operations to hydrolyze the starch to glucose [77]. On the other hand, despite the promising results of using microalgal biomass as a substrate for PHA production, the pretreatments required to break it down into simpler carbohydrates (fermentable sugars) for microorganisms' utilization are still complicated [78]. For this reason, selecting a suitable microbial strain that can work with complex carbohydrates would be ideal.

It has been mentioned that macroalgae represent a sustainable alternative as a nonconventional carbon source for PHAs production by bacteria *H. mediterranei* [79]. Other studies on PHA production using macroalgae as substrate mention the brown seaweed [80] and the red algae *Gelidium amansii* [81]. *Saccharophagus degradans* was applied to

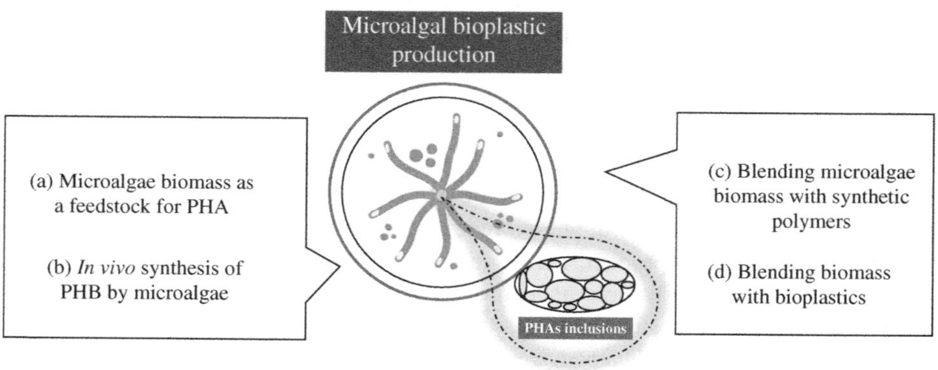

Figure 7.1 Approaches for microalgal bioplastic production.

biodegrade microalgae and alginate for the production of PHB [54]. Defatted algal biomass has also been used as the main carbon source for bacteria in PHA production [82].

Respecting microalgae as PHAs producers, the significant and increasing interest in bioplastic production from microalgae is attributed to their low nutrient requirements and mostly to their ability to utilize low-cost or inexpensive feedstock from light and CO_2 [64, 83]. Strains that can degrade polysaccharides to convert them into interesting molecules and the ones that can process non-common carbon sources such as CO_2 and methane become interesting for researchers due to their potential in the bioplastics industry. For instance, microalgae like *Chlorella minutissima, Spirulina* sp., and *Synechococcus subsalsus* fix CO_2 to synthesize PHAs [84].

Green algae have potential as a PHB-producing microbe, even when bacteria are the most studied group [85]. *Chlorella sorokiniana* SVMIICT8 was evaluated for PHB accumulation driven by nutrient constrain [64]. Different microalgae species can store PHAs, such as *Arthrospira platensis, Chlorella vulgaris, Dunaliella tertiolecta, Scenedesmus* sp., and *Synechococcus* sp., especially if growth under nutrient limitations [86].

It is essential to consider not just the type of microbial strain when attempting to produce PHA; the reactor type also influences PHA production to a large extent. For example, *Calothrix scytonemicola, Nostoc muscorum,* and *Spirulina* sp. LEB 18 performed poorly due to biofilm formation in photobioreactors [87]. The media nutrients directly influence the PHA accumulation in microalgae. Specifically, it has been observed that the culture's nitrogen and phosphorous content limitation led microalgae to divert the presence of protein in its metabolic pathway synthesizing PHAs [63].

Although PHAs are naturally present in different microalgae strains, their percentage content is relatively low compared to that reached by bacteria [88]. Despite the low content of PHAs in microalgae, converting atmospheric CO_2 into PHAs without additional carbon sources is advantageous [89].

7.4 Trends and Challenges in the PHAs Synthesis Process

Indeed, the PHA production process has been long studied and advances have been made toward understanding ways and means to improve and control their synthesis by many microorganisms. Concerning the challenges in producing PHA polyesters, it would be convenient to establish the stages into which this process is divided, as in Figure 7.2, where the challenges correspond to each stage. This section will discuss the trends and challenges associated with upstream and downstream processing.

One of the main bottlenecks is that PHAs are still more expensive to produce than their conventional counterpart, petroleum-based plastics [90]. Thus, the most recent studies address various strategies to reduce processing costs.

Moreover, the most recent trends urge processes to adhere to the goals of sustainable development of the United Nations [91]; a process must be indeed environmentally viable but also economically feasible and socially beneficial.

Thus, microbial PHAs production process must consider economic, environmental and social aspects. Regarding the economic aspects, a great effort is placed on attaining a cost-effective process by looking at both the upstream and downstream processing sections.

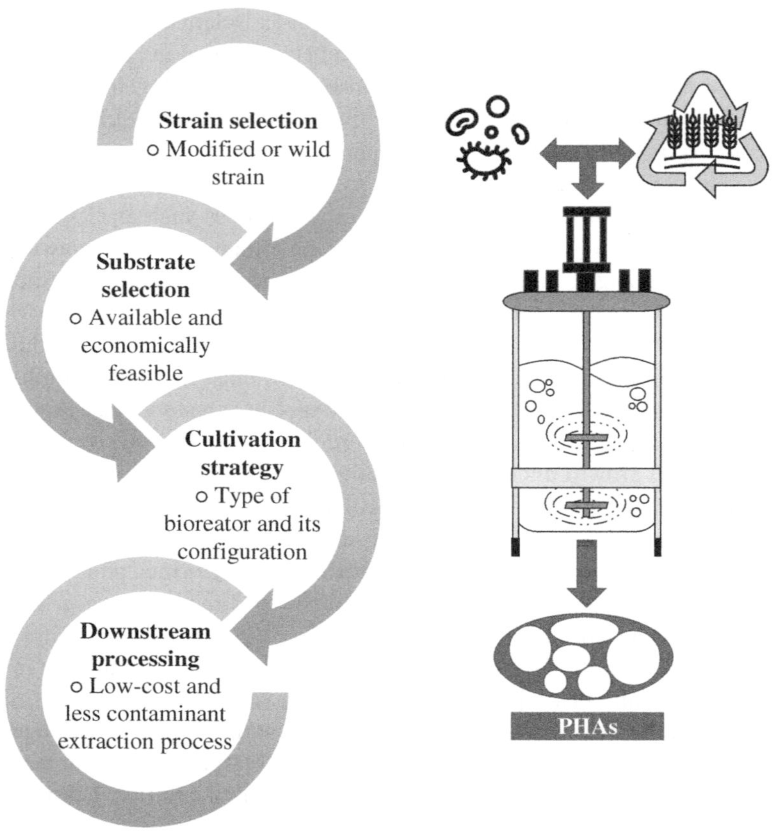

Figure 7.2 Stages and their main challenges in the production of PHAs polyesters.

Likewise, the environmental aspects cover the impact that all the raw materials and unit operations involved in the process pose the least possible risk. As the social aspect, one should be cautious about the raw materials not interfering with human nutrition and animal feed supply chains.

7.4.1 Upstream Processing Trends and Challenges

The upstream processing studies focus on the engineering aspects of the process, from making substrates available for microbial conversion and growth, cultivation strategies, process optimization, scale-up and up to metabolic engineering modification.

PHAs synthesis can be performed in solid-state fermentation (SSF) and submerged fermentation (SmF).

SSF offers the advantage of using inexpensive raw materials such as agro-food waste as support and substrate with fewer pretreatments than SmF. However, the large variety of wastes makes it challenging to establish the process conditions that will serve each type of matrix and thus the scale-up of the process [92].

Most of the available information for microbial production of PHAs is under SmF in batch, fed-batch and continuous modes [92]. The production has been studied and optimized regarding factors related to nutrient and process conditions such as carbon–naitrogen ratio, fermentation time, temperature, and pH [93]. In this regard, the feeding strategy and its operation mode should be adequate and optimized [92]. Batch cultivation is the simplest; the biomass with intracellular PHA is collected at the end of fermentation. It has some advantages, such as the low probability of contamination and short fermentation times; however, it involves some risks, such as the probability that the microorganism consumes the PHA as a carbon source [76, 94].

On the other hand, fed-batch fermentation for producing PHAs allows the ability to control the nutrient input to the system, avoiding the carbon source limitation throughout the fermentation process [95]. This fermentation type allows for higher production yields than a simple batch, and it is also feasible to scale up [96]. Similarly, a cell-recycling approach has been proposed to develop a continuous process to increase process yields and productivity [97].

Indeed, selecting a fermentation type, mode, and process conditions will depend upon the strain. Moreover, strain selection influences the properties and characteristics of the PHAs. PHA-producing microorganisms are present in almost all habitats; all adapted to diverse environmental and nutrient conditions. In this context, it is suggested to find new potential PHA-producing microorganisms in terms of extreme environmental adaptation and capable of consuming unusual substrates [25].

Further, genetic engineering represents a major contribution to the advances in optimizing the production of PHAs through the genetic modification of microbial strains [98]. For instance, the genes capable of synthesizing PHAs can be transferred to strains better characterized in terms of cultivation and ability to use low-cost substrates [99]. Alternatively, they can use unconventional substrates by sharing genes encoding to produce hydrolytic enzymes into microbial strains that already produce PHAs [100]. On the other hand, some species of bacteria do not require modification, such as *Bacillus,* due to various characteristics such as genetic stability, rapid growth, and the ability to process different carbon sources [32]. Moreover, they are recognized as safe by the US Food and Drug Administration (USFDA), giving them the capacity for biomedical applications [32].

One of the most promoted approaches is the production of PHAs via waste valorization to comply with both environmental and economic aspects of the process. Thus, several low-cost carbon sources have been studied for PHA production with promising results [23–26]. Such actions contribute to simultaneously tackling two concerns: reducing the high cost associated with fermentation substrates and the waste-associated pollution.

Generally, low-cost alternative substrates for microbial synthesis of PHAs must meet certain requirements. An obvious feature is that the material provides the required carbon or nitrogen concentration with low or noncontent of inhibitory compounds. PHAs are produced with an exceeded carbon source concentration with limitation of other nutrients, e.g. nitrogen source [29]. In addition, the selected waste should be available in sufficient quantities over time and be easy to collect and transport [24, 101]. Likewise, the waste is recommended to be stable and resistant to the sudden growth of unwanted microorganisms, or at least to mind the storage conditions.

Among the utilized waste, the use of lignocellulosic biomass, such as that provided by agro-industrial waste, has been highly used as a substrate [1]. Waste lipids have also been proposed as a feasible substrate for PHA production [101]. Some examples of waste lipids included the wastes from the biodiesel production process like glycerol [102].

Municipal wastewater has also been implemented as a cheap feedstock for PHA production, especially MMC, due to their ease of adaptation [103, 104]. Nevertheless, the use of this type of culture has the disadvantage that the production of PHAs is not uniform, i.e. it produces polymers of different types, which is not a viable option in the case of seeking uniformity in the product [25]. The potential MMC has been evaluated in various waste streams, namely municipal solid waste, sewage sludge, and cellulosic primary sludge [105].

Another alternative to produce new PHAs is residues from pyrolysis or oxidation of plastics such as polyethylene terephthalate (PET) [46]. An example is *Pseudomonas umsongensis* GO16, which can use terephthalic acid (TA) to produce PHA. It also presents genes for the metabolism of ethylene glycol (also a PET monomer), and interestingly in this strain, a high abundance of genes related to lipid β-oxidation was found [106].

The design of PHA super-producing microorganisms requires a series of conditions, such as not pathogenic, not present phage, not producing toxins, rapid growth, and use of a wide variety of substrates, as well as easy genetic manipulation. Ideally, these strains are expected to have a high cell density greater than 200 g/L with a PHA content between 90 and 95% cell dry weight [17].

Additionally, as previously mentioned, autotrophic microorganisms have also been evaluated for PHAs production. Under this theme, a breakthrough has been made in optimizing photobioreactors in conjunction with the genetic modification of cyanobacterial strains [87]. In some cases, it has been reported that the cultivation of cyanobacteria under a mixotrophic scheme increases both biomass and PHA production [107].

It is also relevant to consider that the carbon source directly affects the composition and properties of the final biopolymer [108]. In the next section, we will discuss in detail the economic aspects related to the utilization of wastes as raw materials.

7.4.2 Downstream Processing, Trends and Challenges

Downstream processing is also a crucial step in estimating the overall process economics; however, it has been lesser studied than upstream processing [27].

PHA recovery is definitively one of the main challenges concerning the environmental impact as it requires solvents and chemicals which cannot be reutilized and are hazardous, such as chloroform. For a long time, the conventional extraction methods using toxic solvents were the most extensively adopted due to the PHA recovery efficiency and purity reached. However, like the upstream, recent trends look for green, sustainable, and feasible recovery strategies by maintaining high recovery yields, using green solvents and low energy consumption [109, 110].

The downstream processing for PHA recovery comprises a series of unit operations for cell lysis, extraction, and purification [109]. Likewise, it has been proposed to pretreat the biomass prior to the cell lysis and extraction to increase recovery yields, such as pressure homogenization [111].

Overall, two routes have been proposed for PHAs recovery, dissolving the biomass to separate the PHA granules and extracting PHAs from the biomass with solvents [112].

For biomass digestion, different techniques based on the solubilization of the cell wall have been used and can be categorized into three types: chemical, enzymatic, and biological digestion [90, 93]. This approach is useful when a high PHA per biomass yield is obtained; that is it is easier to remove the minor component. It has been highly discussed about the environmental benefits of using an enzymatic process to lyse and digest the biomass, but it could significantly increase process costs due to enzymes' price [27, 109, 110]. A novel combination of hypotonic conditions and detergent (SDS) for lysis and PHA extraction was successfully proved on extremophilic microbial cells (halophilic and thermophilic) [113]. Certainly, selecting the proper digestion method will depend on the cell wall characteristics, which could be even more challenging when PHAs are produced by mixed microbial biomass. Nonionic surfactants combined with dimethyl carbonate were proposed to recover PHA from mixed microbial consortia obtaining similar yields to the obtained with chloroform [114].

Now, when it is desired to recover the biomass for further extraction, researchers have reported issues such as filter blockage and biomass losses and recent developments are the biomass harvest assisted by ultrasound [115, 116].

For PHAs recovery from the harvested cells, halogenated solvents have been the most applied, like chloroform, sodium hypochlorite multi-solvent, chloroform-methanol and sodium hypochlorite aqueous two-phase system [117]. The most significant disadvantage of halogenated solvent utilization at a large scale is the required volume, affecting the process cost and the environment. Thus, various nonhalogenated solvents, including alcohols, acids, and esters, among others, have been studied and reviewed [27, 109, 110].

Indeed, selecting a solvent requires considering the PHAs' properties and target applications. Moreover, it has been demonstrated that the extraction process also influences the properties of the extracted PHA [112]. For instance, for P(HB-coHHx) recovery, various combinations of solvents (acetone and ethyl acetate) and precipitants (1-propanol, 2-propanol, ethanol, *n*-heptane) were screened, which resulted in molecular weight variations [118]. Green solvents (ethanol, cyclohexanone, and dimethyl carbonate) were evaluated for PHBV recovery at a lab-large scale, obtaining suitable yields (80%) with dimethyl carbonate; however, a higher melt viscosity was obtained with the control ($CHCl_3$) [119]. Dimethyl carbonate has also shown promising results for PHA extraction from mixed microbial cultures [120]. The green solvent methyl levulinate showed promising results for PHB extraction with a high recovery yield (97%) and higher purity than the chloroform extracted [121].

Furthermore, another challenge to overcome is the successful solvent recycling to reduce process cost and environmental impacts since their reutilization can also affect the biopolymer recovery and purity yields and PHA properties [118, 119].

7.5 Process Economics and Perspectives Toward Industrial Implementation

The successful process development of microbial PHA production toward industrial implementation requires establishing the economic and technical feasibility early in the research stage. In addition, evaluating the process sustainability through methodologies such as life cycle assessment (LCA) and techno-economic analysis (TEA) has become imperative. LCA and TEA assess a process's environmental impact and techno-economic feasibility, respectively.

LCA studies are crucial to scale up to an industrial process. Regarding PHA production, the carbon source has been the principal factor related to environmental impact, even when the most commonly downstream methodology implies utilizing solvents [11]. Thus, most research focuses on alternative feedstocks when addressing process sustainability.

On the other hand, TEA is an approach that analyzes process economics, involving various steps from process design (unit operations), establishing mass and energy balances, estimation of capital investment and processing cost, among others. This section reviews only the techno-economic analyses that assessed the microbial PHAs production process profitability. Since TEA studies can be implemented at early development stages, they can bridge the research and development phase to the commercial stage. In other words, it allows identifying areas of opportunity to increase profitability by evaluating the impact of various scenarios or parameters such as process design, raw materials costs, unit operations time, and product yield on the economic indexes.

Some financial metrics commonly used to evaluate process profitability are the return of investment (ROI, %), payback time (years), internal rate of return (IRR, %), and net present value (NPV, $). Other important aspects are the unit production cost ($/kg) and the minimum selling price ($/kg). Different tools can be utilized for TEA evaluation, from excel to commercial software such as SuperPro Designer (Intelligen) and Aspen plus (Aspentech).

This chapter section included TEA studies from 2015 to 2022 (Table 7.3).

For comparison, we included the process simulator, microorganism, primary carbon source, selling price, production scale and/or plant capacity, economic indexes, and if the authors performed a sensitivity analysis. Included data in Table 7.3 depends on the article. Studies that do not specify utilizing a process simulator used other considerations such as using the Chemical Engineering Plant Cost Index, rules of thumb, and standard chemical engineering techniques [11, 131]. It can be observed that most of the evaluated microorganisms are bacteria but also mixed culture and photoautotrophic microorganisms.

It is interesting to note that all TEA studies regarding the production of PHAs used a low-cost substrate or waste as raw materials, including wastewater, waste stillage, slaughtering waste, glycerol, food waste, and other agro-industrial residues (see Table 7.3 for references).

The TEA studies' plant capacity varied greatly from 164 tons/year to 25,000 tons/year. Likewise, most studies included various bioreactors (number between parenthesis) working simultaneously or in staggered mode with a capacity ranging from 0.45 to 5000 m^3.

Indeed, production capacity will impact the process profitability. Regarding the selling price, software such as SuperPro designer allows the user assigning a selling price; thus, the economic indexes will vary depending on the obtained production per unit cost. Studies assigned a selling price between 4 and 12$/kg. Regarding the economic indexes, the first included number ($/kg) represents the production cost per PHB kg. It can be observed that it ranged from 1 up to nearly 50$/kg; thus, the minimum selling price must be higher than that range. Compared with petrochemical plastics, it is still difficult to compete with their commercial price (1–1.5$/kg) [23]. The lowest production cost range was achieved by

Table 7.3 Techno-economic analysis to assess the microbial PHA production process profitability.

Biopolymer	Process simulator	Microorganism	Carbon source	Production scale/ plant capacity	Selling price ($/kg)	Economic index	Sensitivity analysis	Reference
PHBV	—	*Haloferax mediterranei*	Waste stillage (rice-based ethanol industry)	240 m^3 (4) 1890 tons/year	—	2.05 $/kg Annual cost $ 3,880,868	—	[122]
PHB	AP	Bacteria	Wastewater	1500 tons/year	—	2–4 €/kg	Unit operations (cell disruption, crystallization, distillation, evaporation)	[123]
PHA	—	Bacteria	Slaughtering waste	291 m^3 (2) 10,000 tons/year	4	1.41–1.64 euros/kg Payback time 3.25–4.5 years	By-products price (biodiesel, meat and bone meal, nitrogen source)	[124]
PHA	SPD	*Cupriavidus necator* H16	Glycerol	321.3 m^3 (1) 9000 tons/year	10	5.77 $/kg ROI 25.18% IRR 21.48%	Raw materials (glycerol, surfactant) Electricity price Labor cost	[125]
PHB/PHV	—	PHA-accumulating organisms	Municipal wastewater	9000 tons/year	—	0.8 $/kg	—	[126]
PHB	AP	*Cupriavidus necator*	Citric molasses	408 m^3 (4) 10,000 tons/year	5.50	2.71 $/kg IRR 22.3% Payback time 4.3 years	Production capacity	[127]

(Continued)

Table 7.3 (Continued)

Biopolymer	Process simulator	Microorganism	Carbon source	Production scale/ plant capacity	Selling price ($/kg)	Economic index	Sensitivity analysis	Reference
PHB	AP	*Bacillus subtilis* EPAH18	Cheese whey	0.425 m^3	—	10.2–16.9 $/kg	Unit operations (evaporator, spray-dryer)	[128]
PHB	—	Photoautotrophic microorganism	Photoautotrophy	45 m^3 (4500 m^2 cultivation area)	—	24 €/kg	PHB yield Co-products	[129]
PHB	AP	Recombinant *E. coli*	Sugarcane bagasse	0.5–2.7 ton/h	11.42	2.76–9.96 $/kg IRR 24.1%	—	[130]
PHA	—	Acidogenic bacteria	Sludge (municipal wastewater)	16.3 ton/day 8 anaerobic digesters (nonspecified volume)	—	1.26–2.26 $/kg	Energy price Methane price Sludge cost PHA/COD yield	[131]
PHB	SPD	*Cupriavidus necator* DSM 545	Glycerol	40 m^3 (2) 2000 ton/year	11.42	4.88 $/kg	Plant capacity	[132]
[P(3HB)]	SPD	*Burkholderia sacchari* DSM 17165	Municipal solid waste + Plum waste juice	15 m^3 (3) 45 tons/year	4	48 $/kg Neg ROI	—	[133]
[P(3HB)]	SPD	*Burkholderia sacchari* DSM 17165	Plum waste juice	15 m^3 (3) 55 tons/year	4	28 $/kg Neg ROI	—	[133]
PHB	AP	*Ralstonia eutropha*	Carob Pod extract	5000 m^3	—	Payback time 4.8 years	—	[134]
PHA	SPD	Mixed culture microorganisms	Sewage sludge food waste	164 tons/year	2.2[a]	Neg ROI		[135]

PHB	—	Bacteria	Sunflower meal Crude glycerol	256.7 m^3 (5–16) 2500–25,000 tons/year	—	8.2–15.2 $/kg	Co-products price (protein isolate, crude phenolic extracts)	[11]
PHB	—	*Alcaligenes eutrophus*	Cellulose hydrolysate	70 m^3 1–4 tons/day	—	1.00–8.72 €/kg Payback time 3–6 years	Production capacity Glucose concentration	[136]
PHB		*Synechocystis* sp. CCALA192	Wastewater	10,000 tons/year	—	7.7 $/kg Breakeven (zero 20-year NPV)	PHB yield Raw materials Electricity	[137]
PHA	SPD	*Enterobacter aerogenes*	Food waste	2.01 tons/day	—	2.41–4.83 $/kg ROI 13.68% IRR 11.95 %	Raw materials Solid loading By-products price (chloroform, bioethanol)	[138]

[a] Indicates €/kg

processes using slaughtering waste [124] and municipal wastewater sludge [131]. Although, both TEA studies considered by-products that generated revenues.

In any case, generally, a payback time lower than five years represents an attractive process to invest in, which translates into an ROI higher than 20%. If we look at Table 7.3, the TEAs that achieved an ROI higher than 20% or a payback time lower than five years have a plant capacity of nearly 10,000 tons/year and bioreactors' approximate volume of 300–400 m^3 [124, 127, 134]. Although the high profitability indexes could be attributed to the high set selling price (10$/kg).

Regarding the sensitivity analysis, the most commonly evaluated parameters among all TEA studies were the impact of raw material cost, PHB yield or productivity, production capacity, electricity price, unit operations performance, and other by-products price.

For various biorefineries, the by-products price effect showed that it is required to include co-products that generate revenues to reduce the minimum selling price of PHB [11, 124, 138]. Moreover, to produce PHB by photoautotrophic microorganisms, it was shown that low-cost by-products such as biogas and other nutrients had little impact on the minimum selling price; thus, the authors recommended the production of high-value co-products such as pigments to increase profitability [129]. Similarly, the methane price co-produced with PHA by acidogenic bacteria had a minor effect on the minimum PHA cost [131].

Regarding the effect of raw materials cost, one would expect that since all TEAs use low-cost substrates, their price variation will not affect the minimum selling price. For instance, to produce PHA using glycerol, the authors showed that the major effect was attributed to the surfactant price even though glycerol price affected the profitability index [125]. Thus, the authors concluded on the importance of the recovery step to reduce processing costs. Similarly, the production of PHA using food waste showed that reducing solvent volume in the downstream processing aids in reducing the PHA minimum selling price [138]. Likewise, the TEA study analyzing the production of cyanobacterial PHB using wastewater showed that the cost of the nutrients had little effect on the minimum selling price [137].

As for the plant capacity, it has been shown that increasing the plant capacity positively affects the process profitability despite increasing the cost of investment. For PHB production by *C. necator* using citric molasses [127], it was shown that increasing the production capacity (2000–1000 tons/year) rose 2.5–5 times the expenditure attributed to equipment and other inputs; however, the production costs were reduced from almost 6$/kg up to 2.7$/kg. Nevertheless, in the same study, it was observed that increasing the production capacity from 8000 to 10,000 kg/year did not imply a noticeable reduction in the production costs. Similarly, producing PHB by *A. eutrophus* showed that increasing the production capacity from 1000 to 2000 kg/day significantly reduced the production cost [136]. Then, higher production capacity levels than 2000 kg/day did not considerably imply a reduction in production costs. Instead, the authors attributed a higher sensitivity to the glucose concentration utilized in the hydrolysate, which could be related to the obtained PHB/substrate yield.

Other authors did a sensitivity analysis of polymers' yield on the process profitability, either PHB/substrate or PHB/Biomass. For instance, it was shown that PHA/COD yield coefficient was the parameter with the highest effect on the production costs at small- and large-scale production using anaerobic sludge digestion [131]. Likewise, it was demonstrated that PHB/biomass yield lower than 30% significantly increases the process cost and the minimum PHB selling price for photoautotrophic microorganisms [129, 137].

Based on the above, it can be concluded that an integrated design using low-cost raw materials such as wastes aid in the process's economic feasibility.

7.6 Concluding Remarks

The acceptance, commercialization, and industrial processing for the microbial synthesis of PHAs still face significant barriers due to their noncompetitive prices compared with petrochemical plastics. This motivation and complying with sustainable development goals have driven the most recent research on PHA microbial production toward replacing the prevailing available plastics. Various strategies can be applied to improve the process performance of microbial PHA production, which will be strongly related to the selected microorganism and target application. Studies of process optimization, cultivation method, and fermentation mode are necessary to increase process yield and productivity. Certainly, novel metabolic engineering techniques are also being promoted to that end. The utilization of extremophilic microorganisms also poses a viable alternative for the process scale-up.

Undoubtedly, research has progressed toward understanding the microorganism requirements in terms of process and nutrients conditions to produce PHAs. Such comprehension has enabled one of the most widespread approaches, microbial PHA production within a biorefinery concept. Surely, waste valorization must meet specific criteria such as feedstock availability, low ecological footprint, and not interfering with human and animal feed supply chains. Likewise, the waste variability and heterogeneity may affect the PHAs' yields and properties. Regarding the recovery and purification techniques, the trend matches the same objective of the production strategies, looking to design a cost-effective, environmentally friendly, and feasible process at an industrial scale. In this regard, researchers are looking for extraction protocols and green and low-cost extractants to achieve high recovery yields, high purity, and unaltered biopolymers properties. TEAs allowed to determine that it is possible to reduce the minimum selling price of the biopolymers to start competing with petrochemical-based polymers. Aspects such as maintaining high production yields, waste valorization, by-products revenues, and plant capacity significantly reduce the processing cost and improve economic performance. Additionally, selecting the downstream processing is equally important to reduce process costs and environmental impact.

References

1 Pradhan, S., Borah, A.J., Poddar, M.K. et al. (2017). Microbial production, ultrasound-assisted extraction and characterization of biopolymer polyhydroxybutyrate (PHB) from terrestrial (*P. hysterophorus*) and aquatic (*E. crassipes*) invasive weeds. *Bioresource Technology* 242: 304–310.

2 Aaliya, B., Sunooj, K.V., and Lackner, M. (2021). Biopolymer composites: a review. *International Journal of Biobased Plastics* 3 (1): 40–84.

3 Muhammad Shamsuddin, I. (2017). Bioplastics as better alternative to petroplastics and their role in national sustainability: a review. *Advances in Bioscience and Bioengineering* 5 (4): 63.

4 Możejko-Ciesielska, J. and Kiewisz, R. (2016). Bacterial polyhydroxyalkanoates: still fabulous? *Microbiological Research* 192: 271–282.

5 Pinto, L., Bonifacio, M.A., de Giglio, E. et al. (2021). Biopolymer hybrid materials: development, characterization, and food packaging applications. *Food Packaging and Shelf Life* 28: 100676.

6 Rajesh Banu, J., Kavitha, S., Yukesh Kannah, R. et al. (2019). A review on biopolymer production via lignin valorization. *Bioresource Technology* 290: 121790.

7 Singh, S.P. and Upadhyay, S.K. (eds.) (2021). *Bioprospecting of Microorganism-Based Industrial Molecules.* John Wiley & Sons Ltd. doi:10.1002/9781119717317.

8 Upadhyay, S.K. and Singh, S.P. (eds.) (2021). *Bioprospecting of Plant Biodiversity for Industrial Molecules.* John Wiley & Sons Ltd. doi:10.1002/9781119718017.

9 Upadhyay, S.K. and Singh, S.P. (2023). *Plants as Bioreactors for Industrial Molecules.* John Wiley & Sons Ltd. doi:10.1002/9781119875116.

10 Thiruchelvi, R., Das, A., and Sikdar, E. (2020). Bioplastics as better alternative to petro plastic. *Materials Today: Proceedings* 37 (Part 2): 1634–1639.

11 Kachrimanidou, V., Ioannidou, S.M., Ladakis, D. et al. (2021). Techno-economic evaluation and life-cycle assessment of poly(3-hydroxybutyrate) production within a biorefinery concept using sunflower-based biodiesel industry by-products. *Bioresource Technology* 326: 124711.

12 Gaur, V.K., Sharma, P., Srivastava, J.K. et al. (2021). Production and application of bacterial polyhydroxyalkanoates. In: *Biomass, Biofuels, Biochemicals* (ed. P. Binod, S. Raveendran, and A. Pandey), 223–252. Elsevier.

13 Anitha, N.N.N. and Srivastava, R.K. (2021). Microbial synthesis of polyhydroxyalkanoates (PHAs) and their applications. In: *Environmental and Agricultural Microbiology: Applications for Sustainability* (ed. B.B. Mishra and S. Kumar), 151–182. Wiley.

14 Naser, A.Z., Deiab, I., and Darras, B.M. (2021). Poly(lactic acid) (PLA) and polyhydroxyalkanoates (PHAs), green alternatives to petroleum-based plastics: a review. *RSC Advances* 11 (28): 17151–17196.

15 Jadaun, J.S., Bansal, S., Sonthalia, A. (2022). Biodegradation of plastics for sustainable environment. *Bioresource Technology* 347: 126697.

16 Kourmentza, C., Pl, J., Venetsaneas, N. et al. (2017). Recent advances and challenges towards sustainable polyhydroxyalkanoate (PHA) production. *Bioengineering* 4 (55): 1–43.

17 Tan, D., Wang, Y., Tong, Y., and Chen, G.-Q. (2021). Grand challenges for industrializing polyhydroxyalkanoates (PHAs). *Trends in Biotechnology* 39 (9): 953–963.

18 Koller, M. and Mukherjee, A. (2022). A new wave of industrialization of PHA biopolyesters. *Bioengineering* 9 (2): 74.

19 Behera, S., Priyadarshanee, M., and Vandana, and Das, S. (2022). Polyhydroxyalkanoates, the bioplastics of microbial origin: properties, biochemical synthesis, and their applications. *Chemosphere* 294: 133723.

20 Boey, J.Y., Mohamad, L., Khok, Y.s. et al. (2021). A review of the applications and biodegradation of polyhydroxyalkanoates and poly (lactic acid) and its composites. *Polymers* 13 (10): 1544.

21 Lhamo, P., Behera, S.K., and Mahanty, B. (2021). Process optimization, metabolic engineering interventions and commercialization of microbial polyhydroxyalkanoates production – a state-of-the art review. *Biotechnology Journal* 16 (9): 2100136.

22 Kumar, L.R., Yadav, B., Kaur, R. et al. (2021). Process engineering and commercialization of polyhydroxyalkanoates. In: *Biomass, Biofuels, Biochemicals* (ed. P. Binod, S. Raveendran, and A. Pandey), 517–549. Elsevier.

23 Ganesh Saratale, R., Cho, S.K., Dattatraya Saratale, G. et al. (2021). A comprehensive overview and recent advances on polyhydroxyalkanoates (PHA) production using various organic waste streams. *Bioresource Technology* 325: 124685.

24 Sirohi, R., Prakash Pandey, J., Kumar Gaur, V. et al. (2020). Critical overview of biomass feedstocks as sustainable substrates for the production of polyhydroxybutyrate (PHB). *Bioresource Technology* 311: 123536.

25 Koller, M., Maršálek, L., de Sousa Dias, M.M., and Braunegg, G. (2017). Producing microbial polyhydroxyalkanoate (PHA) biopolyesters in a sustainable manner. *New Biotechnology* 37, Part A: 24–38.

26 Sathya, A.B., Sivasubramanian, V., Santhiagu, A. et al. (2018). Production of polyhydroxyalkanoates from renewable sources using bacteria. *Journal of Polymers and the Environment* 26 (9): 3995–4012.

27 Kosseva, M.R. and Rusbandi, E. (2018). Trends in the biomanufacture of polyhydroxyalkanoates with focus on downstream processing. *International Journal of Biological Macromolecules* 107, Part A: 762–778.

28 Corchado-Lopo, C., Martínez-Avila, O., Marti, E. et al. (2021). Brewer's spent grain as a no-cost substrate for polyhydroxyalkanoates production: assessment of pretreatment strategies and different bacterial strains. *New Biotechnology* 62: 60–67.

29 Miu, D.M., Eremia, M.C., and Moscovici, M. (2022). Polyhydroxyalkanoates (PHAs) as biomaterials in tissue engineering: production, isolation, characterization. *Materials* 15 (4): 1410.

30 Liu, H., Kumar, V., Jia, L. et al. (2021). Biopolymer poly-hydroxyalkanoates (PHA) production from apple industrial waste residues: a review. *Chemosphere* 284: 131427.

31 Guzik, M., Witko, T., Steinbüchel, A. et al. (2020). What has been trending in the research of polyhydroxyalkanoates? A systematic review. *Frontiers in Bioengineering and Biotechnology* 8: 959.

32 Mohapatra, S., Maity, S., Dash, H.R. et al. (2017). *Bacillus* and biopolymer: prospects and challenges. *Biochemistry and Biophysics Reports* 12: 206–213.

33 Vigneswari, S., Noor, M.S.M., Amelia, T.S.M. et al. (2021). Recent advances in the biosynthesis of polyhydroxyalkanoates from lignocellulosic feedstocks. *Life* 11 (8): 807.

34 Chen, G.Q. and Jiang, X.R. (2018). Engineering microorganisms for improving polyhydroxyalkanoate biosynthesis. *Current Opinion in Biotechnology* 53: 20–25.

35 Samrot, A., Samanvitha, S.K., Shobana, N. et al. (2021). The synthesis, characterization and applications of polyhydroxyalkanoates (PHAs) and PHA-based nanoparticles. *Polymers* 13 (19): 3302.

36 El-malek, F.A., Khairy, H., Farag, A., and Omar, S. (2020). The sustainability of microbial bioplastics, production and applications. *International Journal of Biological Macromolecules* 15 (157): 319–328.

37 Madhusoodanan, G., Hariharapura, R.C., and Somashekara, D. (2021). Dissolved oxygen as a propulsive parameter for polyhydroxyalkanoate production using *Bacillus endophyticus* cultures. *Environment, Development and Sustainability* 24 (4): 4641–4658.

38 Obruca, S., Sedlacek, P., Slaninova, E. et al. (2020). Novel unexpected functions of PHA granules. *Applied Microbiology and Biotechnology* 104 (11): 4795–4810.

39 Kumar, M., Rathour, R., Singh, R. et al. (2020). Bacterial polyhydroxyalkanoates: opportunities, challenges, and prospects. *Journal of Cleaner Production* 263: 121500.

40 Koller, M. (2020). *The Handbook of Polyhydroxyalkanoates*. CRC Press Taylor & Francis.

41 Muneer, F., Rasul, I., Azeem, F. et al. (2020). Microbial polyhydroxyalkanoates (PHAs): efficient replacement of synthetic polymers. *Journal of Polymers and the Environment* 28 (9): 2301–2323.

42 Pradhan, S., Dikshit, P.K., and Moholkar, V.S. (2020). Production, characterization, and applications of biodegradable polymer: polyhydroxyalkanoates. In: *Advances in Sustainable Polymers* (ed. V. Katiyar, A. Kumar, and N. Mulchandani), 51–94. Springer Singapore.

43 Goswami, M., Rekhi, P., Debnath, M., and Ramakrishna, S. (2021). Microbial polyhydroxyalkanoates granules: an approach targeting biopolymer for medical applications and developing bone scaffolds. *Molecules* 26 (4): 860.

44 Serafim, L.S., Xavier, A.M.R.B., and Lemos, P.C. (2018). Storage of hydrophobic polymers in bacteria. In: *Biogenesis of Fatty Acids, Lipids and Membranes* (ed. O. Geiger), 1–25. Cham: Springer.

45 Katiyar, V., Kumar, A., and Mulchandani, N. (2020). *Advances in Sustainable Polymers – Synthesis, Fabrication and Characterization*. Springer Singapore.

46 Mezzina, M.P., Tsampika Manoli, M., Prieto, M.A., and Nikel, P.I. (2021). Engineering native and synthetic pathways in *Pseudomonas putida* for the production of tailored polyhydroxyalkanoates. *Biotechnology Journal* 16 (3): 2000165.

47 Tan, D., Yin, J., and Chen, G.Q. (2017). Production of polyhydroxyalkanoates. In: *Current Developments in Biotechnology and Bioengineering* (ed. C. Larroche, M. Sanroman, G. Du, and A. Pandey), 655–692. Elsevier.

48 Alvarez-Santullano, N., Villegas, P., Mardones, M.S. et al. (2021). Genome-wide metabolic reconstruction of the synthesis of polyhydroxyalkanoates from sugars and fatty acids by *Burkholderia Sensu Lato* species. *Microorganisms* 9 (6): 1290.

49 Morales Flores, G. (2019). Biosíntesis de polihidroxialcanoatos de longitud de cadena media en *Azotobacter vinelandii OP* a partir de ácidos grasos o sacarosa para la obtención de bioplásticos. Master Thesis. Universidad Nacional Autónoma de México. Cuernavaca, México.

50 Sriyapai, T., Chuarung, T., Kimbara, K. et al. (2022). Production and optimization of polyhydroxyalkanoates (PHAs) from *Paraburkholderia sp.* PFN 29 under submerged fermentation. *Electronic Journal of Biotechnology* 56: 1–11.

51 Tyagi, P. and Sharma, A. (2021). Utilization of crude paper industry effluent for polyhydroxyalkanoate (PHA) production. *Environmental Technology and Innovation* 23: 101692.

52 Kourilova, X., Pernicova, I., Sedlar, K. et al. (2020). Production of polyhydroxyalkanoates (PHA) by a thermophilic strain of Schlegelella thermodepolymerans from xylose rich substrates. *Bioresource Technology* 315: 123885.

53 Pérez-Arauz, A.O., Aguilar-Rabiela, A.E., Vargas-Torres, A. et al. (2019). Production and characterization of biodegradable films of a novel polyhydroxyalkanoate (PHA) synthesized from peanut oil. *Food Packaging and Shelf Life* 20: 100297.

54 Hu, X., Meneses, Y.E., Stratton, J., and Huo, S. (2022). Direct processing of alginate-immobilized microalgae into polyhydroxybutyrate using marine bacterium of *Saccharophagus degradans*. *Bioresource Technology* 351: 126898.

55 Khomlaem, C., Aloui, H., Oh, W.G., and Kim, B.S. (2021). High cell density culture of *Paracoccus* sp.LL1 in membrane bioreactor for enhanced co-production of polyhydroxyalkanoates and astaxanthin. *International Journal of Biological Macromolecules* 192: 289–297.

56 Dubey, S. and Mishra, S. (2022). Natural sea salt based polyhydroxyalkanoate production by wild *Halomonas hydrothermalis* strain. *Fuel* 311: 122593.

57 Yasin, A.R., Al-Mayaly, I., and k. (2021). Biosynthesis of polyhydroxyalkanoate (PHA) by a newly isolated strain *Bacillus tequilensis* ARY86 using inexpensive carbon source. *Bioresource Technology Reports* 16: 100846.

58 Xu, Z., Xu, M., Cai, C. et al. (2021). Microbial polyhydroxyalkanoate production from lignin by *Pseudomonas putida* NX-1. *Bioresource Technology* 319: 124210.

59 Silambarasan, S., Logeswari, P., Sivaramakrishnan, R. et al. (2021). Polyhydroxybutyrate production from ultrasound-aided alkaline pretreated finger millet straw using *Bacillus megaterium* strain CAM12. *Bioresource Technology* 325: 124632.

60 Pacholak, A., Gao, Z.-L., Gong, X.-Y. et al. (2021). The metabolic pathways of polyhydroxyalkanoates and exopolysaccharides synthesized by *Haloferax mediterranei* in response to elevated salinity. *Journal of Proteomics* 232: 104065.

61 Pillot, G., Sunny, S., Comes, V., and Kerzenmacher, S. (2022). Optimization of growth and electrosynthesis of polyhydroxyalkanoates by the thermophilic bacterium *Kyrpidia spormannii*. *Bioresource Technology Reports* 17: 100949.

62 Mahansaria, R., Bhowmik, S., Dhara, A. et al. (2020). Production enhancement of poly(3-hydroxybutyrate-co-3-hydroxyvalerate) in *Halogeometricum borinquense*, characterization of the bioplastic and desalination of the bioreactor effluent. *Process Biochemistry* 94: 243–257.

63 Costa, S.S., Miranda, A.L., Andrade, B.B. et al. (2018). Influence of nitrogen on growth, biomass composition, production, and properties of polyhydroxyalkanoates (PHAs) by microalgae. *International Journal of Biological Macromolecules* 116: 552–562.

64 Kumari, P., Kiran, B.R., and Venkata Mohan, S. (2022). Polyhydroxybutyrate production by *Chlorella Sorokiniana* SVMIICT8 under nutrient-deprived mixotrophy. *Bioresource Technology* 354: 127–135.

65 Sawant, S.S., Salunke, B.K., and Kim, B.S. (2015). Degradation of corn stover by fungal cellulase cocktail for production of polyhydroxyalkanoates by moderate halophile *Paracoccus sp.* LL1. *Bioresource Technology* 194: 247–255.

66 de Souza, L., Shivakumar, S., and Das, A. (2022). Dual phase statistical optimization of biological pre-treatment of sugarcane bagasse with *Pycnoporus coccineus* MScMS1 for polyhydroxyalkanoates production. *Journal of Environmental Management* 302: 113948.

67 Ojha, N. and Das, N. (2020). Process optimization and characterization of polyhydroxyalkanoate copolymers produced by marine *Pichia kudriavzevii* VIT-NN02 using banana peels and chicken feather hydrolysate. *Biocatalysis and Agricultural Biotechnology* 27: 101616.

68 Prados, E. and Maicas, S. (2016). Bacterial production of hydroxyalkanoates (PHA). *Universal Journal of Microbiology Research* 4 (1): 23–30.

69 Costa, S.S., Miranda, A.L., de Morais, M.G. et al. (2019). Microalgae as source of polyhydroxyalkanoates (PHAs) – a review. *International Journal of Biological Macromolecules* 131: 536–547.

70 Ojha, N. and Das, N. (2018). A statistical approach to optimize the production of polyhydroxyalkanoates from *Wickerhamomyces anomalus* VIT-NN01 using response surface methodology. *International Journal of Biological Macromolecules* 107, Part B: 2157–2170.

71 Mitra, R., Xu, T., Xiang, H., and Han, J. (2020). Current developments on polyhydroxyalkanoates synthesis by using halophiles as a promising cell factory. *Microbial Cell Factories* 19 (1): 1–30.

72 Moradali, M.F. and Rehm, B.H.A. (2020). Bacterial biopolymers: from pathogenesis to advanced materials. *Nature Reviews Microbiology* 18 (4): 195–210.

73 Giraldo-Montoya, J.M., Castaño-Villa, G.J., and Rivera-Páez, F.A. (2020). Bacteria from industrial waste: potential producers of polyhydroxyalkanoates (PHAs) in Manizales, Colombia. *Environmental Monitoring and Assessment* 192 (7): 1–8.

74 Raza, Z.A., Tariq, M.R., Majeed, M.I., and Banat, I.M. (2019). Recent developments in bioreactor scale production of bacterial polyhydroxyalkanoates. *Bioprocess and Biosystems Engineering* 42 (6): 901–919.

75 Cinar, S.O., Chong, Z.K., Kucuker, M.A. et al. (2020). Bioplastic production from microalgae: a review. *International Journal of Environmental Research and Public Health* 17 (11): 1–21.

76 Khomlaem, C., Aloui, H., and Kim, B.S. (2021). Biosynthesis of polyhydroxyalkanoates from defatted chlorella biomass as an inexpensive substrate. *Applied Sciences* 11 (3): 1094.

77 Favaro, L., Basaglia, M., and Casella, S. (2019). Improving polyhydroxyalkanoate production from inexpensive carbon sources by genetic approaches: a review. *Biofuels, Bioproducts and Biorefining* 13 (1): 208–227.

78 Chandra, R., Iqbal, H.M.N., Vishal, G. et al. (2019). Algal biorefinery: a sustainable approach to valorize algal-based biomass towards multiple product recovery. *Bioresource Technology* 278: 346–359.

79 Ghosh, S., Gnaim, R., Greiserman, S. et al. (2019). Macroalgal biomass subcritical hydrolysates for the production of polyhydroxyalkanoate (PHA) by *Haloferax mediterranei*. *Bioresource Technology* 271: 166–173.

80 Azizi, N., Najafpour, G., and Younesi, H. (2017). Acid pretreatment and enzymatic saccharification of brown seaweed for polyhydroxybutyrate (PHB) production using *Cupriavidus necator*. *International Journal of Biological Macromolecules* 101: 1029–1040.

81 Sawant, S.S., Salunke, B.K., and Kim, B.S. (2018). Consolidated bioprocessing for production of polyhydroxyalkanotes from red algae *Gelidium amansii*. *International Journal of Biological Macromolecules* 109: 1012–1018.

82 Khomlaem, C., Aloui, H., Deshmukh, A.R. et al. (2020). Defatted *Chlorella* biomass as a renewable carbon source for polyhydroxyalkanoates and carotenoids co-production. *Algal Research* 51: 102068.

83 Löwe, H., Hobmeier, K., Moos, M. et al. (2017). Photoautotrophic production of polyhydroxyalkanoates in a synthetic mixed culture of *Synechococcus elongatus* cscB and *Pseudomonas putida* cscAB. *Biotechnology for Biofuels* 10 (190): 1–11.

84 Li, M. and Wilkins, M.R. (2020). Recent advances in polyhydroxyalkanoate production: feedstocks, strains and process developments. *International Journal of Biological Macromolecules* 156: 691–703.

85 Arias, D.M., Rueda, E., García-Galán, M.J. et al. (2019). Selection of cyanobacteria over green algae in a photo-sequencing batch bioreactor fed with wastewater. *Science of the Total Environment* 653: 485–495.

86 Chong, J.W.R., Yong Yew, G.Y., Shiong Khoo, K. et al. (2021). Recent advances on food waste pretreatment technology via microalgae for source of polyhydroxyalkanoates. *Journal of Environmental Management* 293: 112782.

87 Carpine, R., Olivieri, G., Hellingwerf, K.J. et al. (2020). Industrial production of poly-β-hydroxybutyrate from CO2: can cyanobacteria meet this challenge? *Processes* 8 (3): 1–23.

88 García, G., Sosa-Hernández, J.E., Rodas-Zuluaga, L.I. et al. (2021). Accumulation of PHA in the microalgae *Scenedesmus* sp. under nutrient-deficient conditions. *Polymers* 13 (1): 131.

89 Madadi, R., Maljaee, H., Serafim, L.S., and Ventura, S.P.M. (2021). Microalgae as contributors to produce biopolymers. *Marine Drugs* 19 (8): 1–27.

90 Gahlawat, G. (2019). *Polyhydroxyalkanoates Biopolymers: Production Strategies*. Chandigarh, India: Springer Nature Switzerland AG.

91 Independent Group of Scientists appointed by the United Nations Secretary-General (2019). Global Sustainable Development Report 2019: The Future is Now – Science for Achieving Sustainable Development. United Nations.

92 Koller, M. (2018). A review on established and emerging fermentation schemes for microbial production of polyhydroxyalkanoate (PHA) biopolyesters. *Fermentation* 4 (2): 30.

93 Dalton, B., Bhagabati, P., De Micco, J., and Padamati, R.B. (2022). A review on biological synthesis of the biodegradable polymers polyhydroxyalkanoates and the development of multiple applications. *Catalysts* 12 (319): 1–44.

94 Moretto, G., Russo, I., Bolzonella, D. et al. (2020). An urban biorefinery for food waste and biological sludge conversion into polyhydroxyalkanoates and biogas. *Water Research* 170: 115371.

95 Ferre-Guell, A. and Winterburn, J. (2019). Increased production of polyhydroxyalkanoates with controllable composition and consistent material properties by fed-batch fermentation. *Biochemical Engineering Journal* 141: 35–42.

96 Guzik, M.W., Duane, F., Kenny, S.T. et al. (2022). A polyhydroxyalkanoates bioprocess improvement case study based on four fed-batch feeding strategies. *Microbial Biotechnology* 15 (3): 996–1006.

97 Kacanski, M., Pucher, L., Peral, C. et al. (2022). Cell retention as a viable strategy for PHA production from diluted VFAs with *Bacillus megaterium*. *Bioengineering* 9 (3): 122.

98 García-Depraect, O., Bordel, S., Lebrero, R. et al. (2021). Inspired by nature: microbial production, degradation and valorization of biodegradable bioplastics for life-cycle-engineered products. *Biotechnology Advances* 53: 107772.

99 Kaparapu, J. (2018). Polyhydroxyalkanoate (PHA) production by genetically engineered microalgae: a review. *Journal on New Biological Reports* 7 (2): 68–73.

100 Chen, G.Q. and Jiang, X.R. (2017). Engineering bacteria for enhanced polyhydroxyalkanoates (PHA) biosynthesis. *Synthetic and Systems Biotechnology* 2 (3): 192–197.

101 Koller, M. and Braunegg, G. (2018). Advanced approaches to produce polyhydroxyalkanoate (PHA) biopolyesters in a sustainable and economic fashion. *The EuroBiotech Journal* 2 (2): 89–103.

102 Porras, M.A., Vitale, C., Villar, M.A., and Cubitto, M.A. (2017). Bioconversion of glycerol to poly(HB-co-HV) copolymer in an inexpensive medium by a *Bacillus megaterium* strain isolated from marine sediments. *Journal of Environmental Chemical Engineering* 5 (1): 1–9.

103 Shen, M.Y., Chu, C.Y., Sawatdeenarunat, C., and Bhuyar, P. (2022). Production, downstream processing, and characterization of polyhydroxyalkanoates (PHAs) boosted by pyruvate supplement using mixed microbial culture (MMC) and organic wastewater. *Biomass Conversion and Biorefinery* 1–9.

104 Vogli, L., Macrelli, S., Marazza, D. et al. (2020). Life cycle assessment and energy balance of a novel polyhydroxyalkanoates production process with mixed microbial cultures fed on pyrolytic products of wastewater treatment sludge. *Energies* 13 (11): 2706.

105 Lorini, L., Martinelli, A., Capuani, G. et al. (2021). Characterization of polyhydroxyalkanoates produced at pilot scale from different organic wastes. *Frontiers in Bioengineering and Biotechnology* 9: 628719.

106 Narancic, T., Salvador, M., Hughes, G.M. et al. (2021). Genome analysis of the metabolically versatile *Pseudomonas umsongensis* GO16: the genetic basis for PET monomer upcycling into polyhydroxyalkanoates. *Microbial Biotechnology* 14 (6): 2463–2480.

107 Afreen, R., Tyagi, S., Singh, G.P., and Singh, M. (2021). Challenges and perspectives of polyhydroxyalkanoate production from Microalgae/Cyanobacteria and bacteria as microbial factories: an assessment of hybrid biological system. *Frontiers in Bioengineering and Biotechnology* 9: 624885.

108 Gonzalez, K., Navia, R., Liu, S., and Cea, M. (2021). Biological approaches in polyhydroxyalkanoates recovery. *Current Microbiology* 78 (1): 1–10.

109 Pérez-Rivero, C., López-Gómez, J.P., and Roy, I. (2019). A sustainable approach for the downstream processing of bacterial polyhydroxyalkanoates: state-of-the-art and latest developments. *Biochemical Engineering Journal* 150: 107283.

110 Haque, M.A., Priya, A., Hathi, Z.J. et al. (2022). Advancements and current challenges in the sustainable downstream processing of bacterial polyhydroxyalkanoates. *Current Opinion in Green and Sustainable Chemistry* 36: 100631.

111 Palmieri, S., Tittarelli, F., Sabbatini, S. et al. (2021). Effects of different pre-treatments on the properties of polyhydroxyalkanoates extracted from sidestreams of a municipal wastewater treatment plant. *Science of the Total Environment* 801: 149633.

112 Pagliano, G., Galletti, P., Samorì, C. et al. (2021). Recovery of polyhydroxyalkanoates from single and mixed microbial cultures: a review. *Frontiers in Bioengineering and Biotechnology* 9: 624021.

113 Novackova, I., Kourilova, X., Mrazova, K. et al. (2022). Combination of hypotonic lysis and application of detergent for isolation of polyhydroxyalkanoates from extremophiles. *Polymers* 14 (9): 1761.

114 Colombo, B., Pereira, J., Martins, M. et al. (2020). Recovering PHA from mixed microbial biomass: using non-ionic surfactants as a pretreatment step. *Separation and Purification Technology* 253: 117521.

115 Ghosh, S., Coons, J., Yeager, C. et al. (2022). Halophyte biorefinery for polyhydroxyalkanoates production from *Ulva* sp. hydrolysate with *Haloferax mediterranei* in pneumatically agitated bioreactors and ultrasound harvesting. *Bioresource Technology* 344, Part B: 125964.

116 Martínez-Herrera, R.E., Alemán-Huerta, M.E., Almaguer-Cantú, V. et al. (2020). Efficient recovery of thermostable polyhydroxybutyrate (PHB) by a rapid and solvent-free extraction protocol assisted by ultrasound. *International Journal of Biological Macromolecules* 164: 771–782.

117 Pati, S., Maity, S., Samantaray, D.P., and Mohapatra, S. (2017). *Bacillus* and biopolymer with special reference to downstream processing. *International Journal of Current Microbiology and Applied Sciences* 6 (6): 1504–1509.

118 Bartels, M., Gutschmann, B., Widmer, T. et al. (2020). Recovery of the PHA copolymer P(HB-co-HHx) with non-halogenated solvents: influences on molecular weight and HHx-content. *Frontiers in Bioengineering and Biotechnology* 8 (944): 1–12.

119 Abbasi, M., Coats, E.R., and McDonald, A.G. (2022). Green solvent extraction and properties characterization of poly(3-hydroxybutyrate-co-3-hydroxyvalerate) biosynthesized by mixed microbial consortia fed fermented dairy manure. *Bioresource Technology Reports* 18: 101065.

120 de Souza Reis, G.A., Michels, M.H.A., Fajardo, G.L. et al. (2020). Optimization of green extraction and purification of PHA produced by mixed microbial cultures from sludge. *Water (Switzerland)* 12 (4): 1185.

121 Gnaim, R., Unis, R., Gnayem, N. et al. (2022). Turning mannitol-rich agricultural waste to poly(3-hydroxybutyrate) with *Cobetia amphilecti* fermentation and recovery with methyl levulinate as a green solvent. *Bioresource Technology* 352: 127075.

122 Bhattacharyya, A., Jana, K., Haldar, S. et al. (2015). Integration of poly-3-(hydroxybutyrate-co-hydroxyvalerate) production by *Haloferax mediterranei* through utilization of stillage from rice-based ethanol manufacture in India and its techno-economic analysis. *World Journal of Microbiology and Biotechnology* 31 (5): 717–727.

123 Fernández-Dacosta, C., Posada, J.A., Kleerebezem, R. et al. (2015). Microbial community-based polyhydroxyalkanoates (PHAs) production from wastewater: techno-economic analysis and ex-ante environmental assessment. *Bioresource Technology* 185: 368–377.

124 Shahzad, K., Narodoslawsky, M., Sagir, M. et al. (2017). Techno-economic feasibility of waste biorefinery: using slaughtering waste streams as starting material for biopolyester production. *Waste Management* 67: 73–85.

125 Leong, Y.K., Show, P.L., Lan, J.C.W. et al. (2017). Economic and environmental analysis of PHAs production process. *Clean Technologies and Environmental Policy* 19 (7): 1941–1953.

126 Liao, Q., Guo, L., Ran, Y. et al. (2018). Optimization of polyhydroxyalkanoates (PHA) synthesis with heat pretreated waste sludge. *Waste Management* 82: 15–25.

127 Pavan, F.A., Junqueira, T.L., Watanabe, M.D.B. et al. (2019). Economic analysis of polyhydroxybutyrate production by *Cupriavidus necator* using different routes for product recovery. *Biochemical Engineering Journal* 146: 97–104.

128 Peña-Jurado, E., Pérez-Vega, S., Zavala-Díaz de la Serna, F.J. et al. (2019). Production of poly (3-hydroxybutyrate) from a dairy industry wastewater using *Bacillus subtilis* EPAH18: bioprocess development and simulation. *Biochemical Engineering Journal* 151: 107324.

129 Panuschka, S., Drosg, B., Ellersdorfer, M. et al. (2019). Photoautotrophic production of poly-hydroxybutyrate – first detailed cost estimations. *Algal Research* 41: 101558.

130 Nieder-Heitmann, M., Haigh, K., and Görgens, J.F. (2019). Process design and economic evaluation of integrated, multi-product biorefineries for the co-production of bio-energy, succinic acid, and polyhydroxybutyrate (PHB) from sugarcane bagasse and trash lignocelluloses. *Biofuels, Bioproducts and Biorefining* 13 (3): 599–617.

131 Crutchik, D., Franchi, O., Caminos, L. et al. (2020). Polyhydroxyalkanoates (PHAs) production: a feasible economic option for the treatment of sewage sludge in municipal wastewater treatment plants? *Water (Switzerland)* 12 (4): 118.

132 Kumar, L.R., Kaur, R., Tyagi, R.D., and Drogui, P. (2021). Identifying economical route for crude glycerol valorization: biodiesel versus polyhydroxy-butyrate (PHB). *Bioresource Technology* 323: 124565.

133 Izaguirre, J.K., Barañano, L., Castañón, S. et al. (2021). Economic and environmental assessment of bacterial poly(3-hydroxybutyrate) production from the organic fraction of municipal solid waste. *Bioresources and Bioprocessing* 8 (39): 1–12.

134 Manikandan, N.A., Pakshirajan, K., and Pugazhenthi, G. (2021). Techno-economic assessment of a sustainable and cost-effective bioprocess for large scale production of polyhydroxybutyrate. *Chemosphere* 284: 131371.

135 Fernando-Foncillas, C. and Varrone, C. (2021). Potential of the sewage sludge valorization in Scandinavia by co-digestion with other organic wastes: a techno-economic assessment. *Journal of Cleaner Production* 324: 129239.

136 Boura, K., Dima, A., Nigam, P.S. et al. (2022). A critical review for advances on industrialization of immobilized cell bioreactors: economic evaluation on cellulose hydrolysis for PHB production. *Bioresource Technology* 349: 126757.

137 Price, S., Kuzhiumparambil, U., Pernice, M., and Ralph, P. (2022). Techno-economic analysis of cyanobacterial PHB bioplastic production. *Journal of Environmental Chemical Engineering* 10 (3): 107502.

138 Rajendran, N. and Han, J. (2022). Techno-economic analysis of food waste valorization for integrated production of polyhydroxyalkanoates and biofuels. *Bioresource Technology* 348: 126796.

8

Microbial Production of Critical Enzymes of Lignolytic Functions

M. Indira[1], S. Krupanidhi[1], K. Vidya Prabhakar[2], T. C. Venkateswarulu[1], and K. Abraham Peele[1]

[1] *Department of Biotechnology, Vignan's Foundation for Science, Technology & Research, Vadlamudi, Andhra Pradesh, India*
[2] *Department of Biotechnology, Vikrama Simhapuri University, Nellore, Andhra Pradesh, India*

8.1 Introduction

Lignin is the natural aromatic compound present in all vascular plants [1]. It accounts for 20–35% of dry weight of plants located in cell walls, between cells, and within the cells [2]. Lignolytic enzymes show a significant role in reclamation and degradation of lignocellulosic waste in the environment [3]. The lignocellulosic biomass is abundant in nature, which is the renewable source on the earth [4]. The lignocellulosic biomass consists of cellulose (40–50%), hemicellulose (25–30%), and lignin (15–20%) [5]. The cellulose is a polysaccharide material composed of β-1,4 linked D-glucose units in unbranched linear chains [4]. Hemicellulose is composed of branched polysaccharides, namely mannans, glucomannans, xyloglucans, and xylans [4]. The lignin is the complex biopolymer with low biodegradability [6]. Lignin is made up of poly phenolic organic polymers with average molecular weight of ~20,000 Da [6]. The poly phenolic compounds are 4-hydroxy, 3-methoxy cinnamyl alcohol (coniferyl alcohol), *p*-hydroxy cinnamyl alcohol (coumaryl alcohol), and 3,5-dimethoxy-4-hydroxycinnamyl alcohol (sinapyl alcohol) [7]. It is a recalcitrant compound due to the complex chemical bonding between the monomers of lignocellulosic materials [3]. Due to its complexity, lignin degradation is quite challenging compared with cellulose and hemicellulose [8]. The lignolytic enzymes are used for breakdown of lignocellulosic waste materials present in the environment [3]. Lignolytic enzymes are classified into lignin degrading auxiliary enzymes and lignin modifying enzymes (Figure 8.1) [9]. The auxiliary enzymes degrade lignin by performing a series of chemical reactions using catalytic proteins [3]. The auxiliary enzymes are glyoxal oxidase, pyranose-2-oxidase, aryl alcohol oxidases, cellobiose dehydrogenase, and glucose oxidase [9]. Lignin modifying enzymes are lignin peroxidase, versatile peroxidase, manganese peroxidase, heme-containing peroxidase, and laccases [9]. In natural environment, lignolytic enzymes play crucial role in regulating the carbon cycle by degrading the lignocellulosic waste [3].

Microbial Bioreactors for Industrial Molecules, First Edition.
Edited by Sudhir P. Singh and Santosh Kumar Upadhyay.

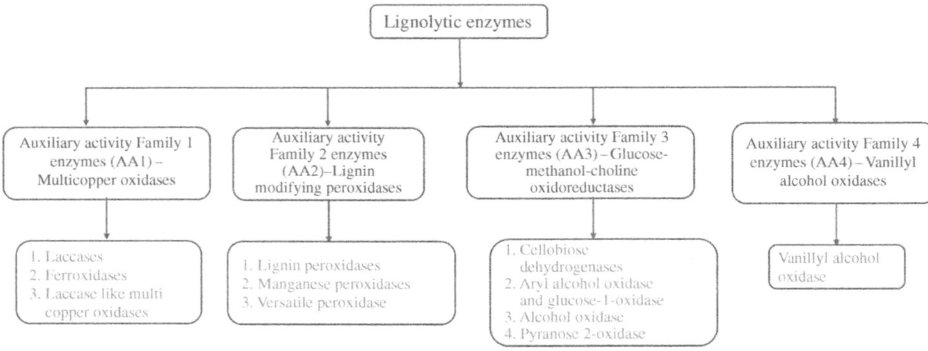

Figure 8.1 Classification of lignolytic enzymes.

In recent years, the enzymes have gained more attention in biotechnology and its allied fields (Figure 8.2) [10, 11]. The lignolytic enzymes particularly laccases are used in the preparation and stabilization of alcoholic beverages such as wine, beer, and fruit juices [12]. The laccases enhance the aroma and flavor to the aromatic beverage tea and also improves the quality of food products [13].

The lignolytic enzymes find more applications in the bioremediation of organic pollutants, degradation of dyes, polycyclic aromatic hydrocarbons, herbicides, pesticides, xenobiotics, lignocellulosic materials, and waste water treatment [3, 14, 15]. In addition to these, the enzymes are used in paper industries, coal conversion, and also polymerization in polymer industries [16]. Further the enzymes have been used in the second-generation biofuel production [17]. Bioethanol (lignocellulosic ethanol) is one of the most advanced biofuels produced from lignocellulosic material [18]. In biofuel production process, pretreatment is the primary step where the removal of lignin takes place from lignocellulosic structures by lignolytic enzymes [19]. The current pretreatment processes utilize high pressure and temperature results in undesirable compound formation [20]. To overcome this, lignolytic enzymes are used for treating the lignocellulosic biomass in pretreatment step [21]. The microorganisms are used to delignify the lignocellulosic substrates that results in the production of biofuels [22]. Among all the microorganisms, fungal organisms possess enzymatic system for effective degradation of lignin compounds [9]. The biofuel production involves delignification and detoxification [17]. In delignification, the lignolytic enzymes degrade the lignin content in several feedstocks. Later in detoxification, the lignolytic enzymes convert the toxic compounds into simpler molecules [3, 17].

8.2 Sources of Lignolytic Enzymes

Lignolytic enzymes are found in different types of organisms such as plants, insects, bacteria, actinomycetes, extremophiles, and fungi [3, 23–26]. In plants, the lignolytic enzymes are extracellular glycoproteins called as laccases [17]. They are identified in plants such as Japanese lacquer tree, tobacco, mango, mung bean, zea mays, and peach [27]. The plant laccases perform wide functions such as lignin synthesis, oxidation of Fe (II) to Fe (III), and

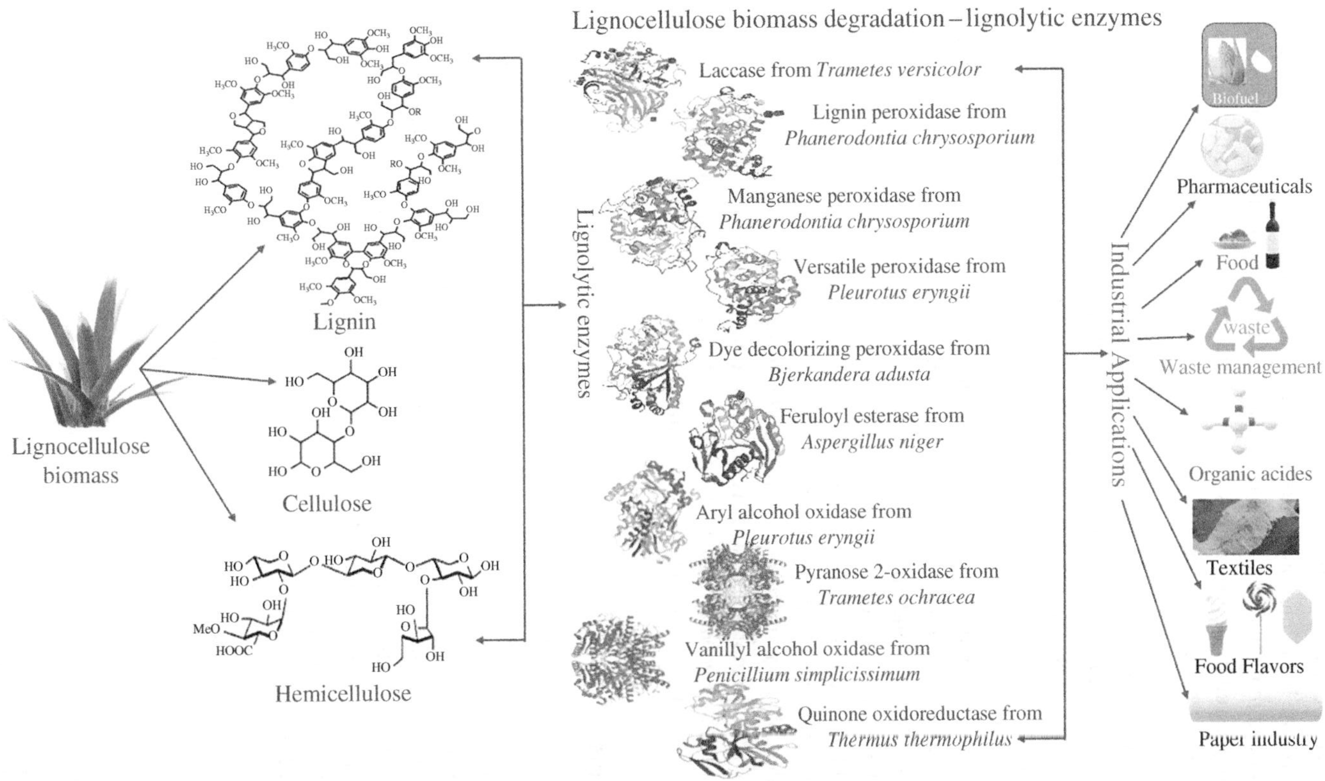

Figure 8.2 Industrial applications of lignocellulosic biomass degradation by lignolytic enzymes.

regeneration of injured tissues [28]. Further, the lignolytic activity of bacterial laccase enzyme gene *Lac51* is expressed in plant *Nicotiana benthamiana* [29]. The plant expressed the *Lac51* gene to oxidize the lignin monomers [29]. The bacterial gene coded *Lac51* is the recombinant laccase produced in plants [29]. Apart from bacterial laccase genes, plant and fungal laccase genes also expressed in plants as hosts [30]. The plant-made recombinant enzymes are used as a potential source for lignin modification [31]. The plant cells express valuable recombinant proteins such as enzymes, vaccines and biopharmaceuticals through molecular farming which is economically a viable system for production of biological products [32]. Many industrial enzymes are produced by microorganisms because they can be easily produced in culture media, inexpensive, short time, high biomass yield, strain improvement, and optimized conditions for growth [33]. Microorganisms producing lignolytic enzymes belong to three classes such as α-proteobacteria, γ-proteobacteria, and actinomycetes [17]. Strains showing lignin degrading ability are *Sphingomonas, Ensifer adhaerens* NWODO-2, (α-proteobacteria), *Pseudomonas, Raoultella ornithinolytica* OKOH-1 (γ-proteobacteria) *Rhodococcus, Nocardia,* and *Streptomyces* sp. (actinomycetes) [34, 35]. The bacterial species belongs to these classes are the major decomposers of the lignocellulosic biomass [9]. Laccase genes were identified in firmicutes, actinobacteria, and proteobacteria. The laccase genes are less expressed in bacterial strains and belong to the group bacteroides and cyanobacteria [9]. In bacteroides, phylum *Sphingobacterium* sp. HY-H is producing manganese peroxidase for degradation of lignin monomers such as *p*-coumaric acid, vanillic acid, and syringic acid [36]. Another strain *Sphingobacterium* sp. T2 produces manganese superoxide dismutase, which catalyzes the oxidative demethylation of lignin by the generation of hydroxyl radicals [37].

8.2.1 Plants

Laccases are the predominant lignolytic enzymes found in different types of plants [38]. They are widely distributed in higher plants. In plants, laccases are first identified in exudates of Japanese lacquer tree *Rhus vernicifera* [39]. Though laccase enzymes have been first identified in plants, the studies have been performed predominantly in fungal organisms [40]. For the past several years research on plant laccases have been started and identified in several plant species [28]. The plant species are *Nicotiana tabacum, Pinus taeda, Mangifera indica, Vigna radiata, Prunus persica, Pinus roxburghii, Populus trichocarpa, Acer pseudoplatanus, Lolium perenne,* and *Liriodendron tulipifera* [41]. More research has been reported in plants such as *Arabidopsis thaliana* [42], *Oryza sativa* [43], *Zea mays* [44], *Brassica napus* [45], *Saccharum officinarum* [46], and *Brachypodium distachyon* [47]. *Arabidopsis thaliana* genome encodes 17 genes for laccase enzymes, whereas the fungus *Trametes villosa* has 5 genes and insect *Nephotettix cincticeps* has 3 genes [38]. In *A. thaliana,* the laccases are found to be involved in stem lignification [48].

8.2.2 Insects

The lignolytic enzymes are produced by insects and its microflora present in different parts of the body [49]. The production of enzymes was observed in wood feeding insects such as *Anoplophora glabripennis* (Asian long-horned beetle) and *Zootermopsis angusticollis* (Pacific Damp wood termite) [49]. The insect gut consists of microbiome produce enzymes that aid in possible degradation of lignin [50]. The enzymes are also identified in salivary glands (*N. cincticeps*), midgut, fat body, epidermis, and malpighian tubes (*Manduca sexta*) [51]. The phenol oxidizing laccases are also found in cuticles (*Tribolium castaneum*) and gut (*Reticulitermes flavipes*) of the insects [52]. The insects feed on carbohydrate-rich wood consists of lignocellulosic materials for their survival [53]. A wood-feeding termite *Coptotermes formosanus* secretes both lignolytic enzymes and cellulolytic enzymes for degradation of woody biomass. In salivary glands and foregut of *C. formosanus*, *Lac1* gene was expressed and found to be a role in phenol oxidation [54]. Insects that feed on lignocellulosic biomass from agricultural crops to forest woody substrate belongs to termites, beetles, wood feeding roaches, leaf cutting ants, silver fish, leaf shredding aquatic insects, and wood wasps [53]. Gieb et al. [54] analyzed the lignin degradation by wood-feeding insects and found that degradation of lignin is due to two insects' species *A. glabripennis* (beetle) and *Z. angusticollis* (termite). They found that the depolymerization (propyl side chain oxidation) and demethylation of ring methoxyl groups were detected in both cases of beetle and termite. The wood-feeding insect *Rhynchophorus ferrugineus* is identified as potential lignin degrading organism due to its gut microbiome [55]. The lignin degradation activity is by the action of lignolytic enzymes that are produced by species of the genera *Serratia*, *Pseudomonas*, and *Enterobacter* [56, 57]. Such type of bacteria is useful for lignin compound degradation and finds wide variety of applications in biofuel, bioremediation, and bio-pulping [57].

8.2.3 Bacteria

Many bacterial organisms have lignin degradation ability and converts the lignocellulosic waste into value-added products [58]. In recent years, several of the bacterial organisms have been discovered for its ability to degrade lignocellulosic waste in the environment [59]. Several of aerobic bacteria degrade lignin content which include proteobacteria, firmicutes, and actinobacteria [60]. The bacterial species identified for lignin degradation are *Sphingobium* sp., *Novosphingobium* sp., *Rhizobium* sp., *Comamonas* sp., *Pandoraea norimbergensis*, *Pseudomonas deceptionensis*, *Pseudomonas putida*, *Burkholderia* sp., *Serratia* sp., and *Enterobacter* sp. (Proteobacteria), *Bacillus ligniphilus*, *Bacillus atrophaeus*, *Paenibacillus glucanolyticus*, *Aneurinibacillus aneurinilyticus* (Firmicutes), *Streptomyces* sp., *Rhodococcus opacus*, *Rhodococcus jostii*, *Rhodococcus pyridinivorans* (Actinobacteria) [58, 61, 62]. A recent study by Xiong et al. [60] isolated the alkali-lignin degrading potential isolates using alkali-lignin-enriched medium. The strains *Bacillus aryabhattai* BY5, *Micrococcus yunnanensis* CL32, *Acinetobacter lwoffi* LN4, and *Acinetobacter johnsonii* LN2 were evaluated for utilization of dibutyl phthalate and decomposition of this aromatic esters through dibutyl phthalate aerobic metabolic pathway. Nowadays, bacteria-based systems of lignin degradation have gained wide attention due to functional diversity of lignin degrading enzymes, diverse metabolism, and conversion of lignin waste into valuable products such as lipids, polyhydroxy alkanoates, pyruvate, and vanillin [59, 61].

8.2.4 Fungi

Generally, fungi are saprotrophic in nature and prevent the accumulation of dead and decaying matter in the environment [9]. Many basidiomycetes' fungi produce lignolytic enzymes to degrade lignin content in the plant biomass and mineralize the lignin [63]. In nature, the lignocellulosic biomass is mainly degraded by filamentous fungal organisms due to effective oxidative enzymatic system [64]. The enzymatic system in fungi is complex and capable of degrading lignin content in the biomass [65]. The enzymes produced by fungi are lignin peroxidases, manganese peroxidases, and laccases. *Phanerochaete chrysosporium*, *Myceliophthora thermophila*, *Pleurotus ostreatus*, *Trametes trogii* Berk, *Trametes versicolor*, *Pycnoporus cinnabarinus*, *Ceriporiopsis subvermispora*, *Cyathus stercoreus*, *Asperigillus niger*, *Melanocarpus albomyces*, and *Trichoderma reesei* are the lignin degrading fungal organisms [9, 63, 66]. The lignin degrading organisms are white rot fungi that convert the lignin into carbon dioxide and water [67]. *Ganoderma* species belongs to white rot fungi and are isolated from decay wood and evaluated for the production of laccases and peroxidases using PCR and cloning approach. They found the diversity of laccase encoding genes and one gene encodes the versatile peroxidase and manganese peroxidase in each strain [68]. Recombinant lignolytic enzymes are produced by using hosts for over expression of lignolytic enzyme encoding genes and the hosts are *Saccharomyces cerevisiae* and *Asperigillus flavus* var. *oryzae* [69]. CRISPR-Cas 9 technology has been used for editing of fungal genes in fungal genomes to enhance the production of lignolytic enzymes in large scale [69]. Metagenomics approaches are essential for revealing the novel lignolytic enzymes encoding genes among the fungal organisms [70].

8.2.5 Actinomycetes

Actinomycetes are one of the taxonomic groups under domain bacteria exist in the soil environment and adapted to the wide range of ecological environment [71]. They breakdown the organic materials present in the soil and lignocellulosic biomass and further recycle the materials into the environment [72]. Actinomycetes produces various enzymes that are having the ability to degrade the lignin and cellulosic content in the biomass [73]. Cellulases are enzymes that produced by these organisms have the ability to breakdown the tough compounds like lignocelluloses and converts them into simple sugar molecules [74]. In actinomycetes genera, few species are explored for the production of lignolytic enzymes, whereas several species need to be explored for its industrial and biotechnological applications [73, 75]. Streptomyces is the one genus that produces lignin degrading enzymes and decomposes the lignin [9]. The species *Streptomyces viridosporus* T7A is producing lignin degrading enzymes and resembles the fungi in filamentous form [76]. A thermophilic actinomycetes strain *Thermobifida fusca* BCRC19214 produces 4.96 U/mL of laccases after 36 hours of growth in 5 L fermenter. The laccase was found to be diphenol oxidases class and exhibited the oxidization of dye intermediates such as 2,6-dimethylphenylalanine and *p*-aminophenol [77].

8.2.6 Extremophiles

Screening of ideal organisms from extremophiles is a focused research at present due to production of lignin degrading enzymes in extreme environmental conditions [78]. Several strains from extremophiles category shown their potential for utilization of lignin due to adaptive features and structural properties of lignolytic enzymes [79]. The adaptive features are thermostability, osmotic allowance, cold adaptivity, extreme pressure, salt tolerant, and high acidic environment [80]. The extremophilic enzymes are superior compared to normal enzymes because they enable the industrial processes to be carried out in extreme conditions due to denaturation of normal enzymes [81]. Several extremophilic organisms like halotolerant bacteria *B. ligniphilus* L1 shown their potential for degradation of lignin at optimal pH 9 and produced aromatic compounds such as vanillin and vanillic acid [82]. Thermotolerant microorganisms produces thermostable enzymes that are having special interest in industrial processes compared to thermolabile enzymes [83]. The thermostable microbes find applications in composting process. The temperature determines the structure and dynamics of microbial populations and also converts the lignocellulosic structures in the composting process [84]. Lai et al. [85] isolated and screened the lignin peroxidase producing bacteria from oil palm empty fruit bunch compost and found that the strain CLMT 29 is producing 8.7673 U/mL of lignin peroxidase within 24 hours of growth in minimal salt medium supplemented with lignin. A thermotolerant strain *T. trogii* LK13 was evaluated for laccase production and found that the lignolytic enzymes are thermostable in nature [86]. A highly stable extracellular laccase producing halophilic bacteria *Aquisalibacillus elongatus* was isolated and evaluated for delignification of sugar beet pulp. The laccase production under optimal conditions was to be 4.8 U/mL [87]. In another study, Levy-Booth et al. [88] integrated the stable isotope probing technique to resolve the ligninases encoding genes using metagenomics approach and found the fourteen genomes are significantly enriched with putative ligninases. The genomes of actinomycetes, firmicutes, and γ-proteobacteria were identified. Lignin valorization using extremozymes from extremophiles have a breakthrough application in near future [89].

8.3 Lignolytic Enzymes

Lignolytic enzymes are categorized into two main groups: lignin degrading auxiliary enzymes and lignin modifying enzymes [3]. The lignin degrading auxiliary enzymes act in concert with the main degrading enzymes which belong to oxidative enzymes category [9]. The oxidative enzymes are glyoxal oxidase, pyranose-2-oxidase, and aryl alcohol oxidase [90]. The lignin modifying enzymes are laccase, manganese peroxidase, versatile peroxidase, and lignin peroxidase [3,9]. These enzymes are produced in the secondary metabolism and do not require energy to the organisms [91].

8.3.1 Lignin Peroxidase (EC 1.11.1.14)

Lignin peroxidase is firstly identified in the fungal organism *P. chrysosporium* [9]. It catalyzes the lignin in the presence of hydrogen peroxide by oxidative degradation [92]. The enzyme is produced by bacteria and fungi and many fungal organisms were studied for the lignin degradation in the plant biomass by lignin peroxidase [93]. However, several bacterial species are characterized for lignin peroxidase enzyme production by using lignin as a substrate [94]. Further, exploration of novel strains for the degradation of lignocellulosic biomass is essential due to abundance of lignocellulosic waste in the environment [95]. It is a heme protein belongs to the class oxidoreductases [3]. The enzyme is well studied in fungi than the other organisms and it uses the heme protein as an electron donor for degradation process [9]. The enzyme is nonspecifically catalyzes the lignin containing compounds and non-phenolic lignin derivatives [96]. The lignin substrate is an aromatic polymer mainly present in woody plant tissues [97]. It provides strength, rigidity to the plant cells, and resistance to microbial degradation [98]. It is degraded by two-step process: (i) depolymerization of lignin into low-molecular-weight aryl and bi aryl compounds; (ii) mineralization of the compounds by specific catabolic pathways [99]. Hirai et al. [100] characterized the lignin peroxidase from white rot fungus *Phanerochaete sordida* YK-624. The enzyme YK-LiP2 is more effective in degradation of lignin compounds. The strain *P. sordida* YK-624 oxidized veratryl alcohol at high concentrations of hydrogen peroxide (>2.5 mm). Nowadays, skin lightening agents are available to inhibit the melanin synthesis. *In vitro* studies are carried out to evaluate the melanin degradation by using lignin peroxidase enzyme that was isolated from *P. chrysosporium* NK-1 from forest soil. The enzyme showed 92% of melanin decolorization and found low cytotoxic effects [101].

8.3.2 Manganese Peroxidase (EC 1.11.1.13)

Manganese peroxidases are extracellularly produced enzymes identified in fungal and bacterial species [102]. These are glycoproteins consists of iron photo porphyrin IX prosthetic group at the center [9]. In nature, these enzymes hydrolyze the lignin and converts into smaller compounds. Due to its hydrolyzing activity, the enzymes are used in industrial applications like biofuel production, bioremediation, bio-bleaching, bio-pulping, cosmetics, dye decolorization, degradation of natural rubber, diagnostic kits, and oxidization of phenolic and non-phenolic compounds [103, 104]. They oxidize the manganese ion Mn^{2+} to Mn^{3+} and further the Mn^{3+} ion chelates takes place by organic acids (oxalic acid) [105]. The chelated Mn^{3+} acts as diffusible oxidant and degrades the phenolic moieties of the lignin compounds [106]. The peroxidases are not able to degrade lignin directly and use metal ions to attack the phenolic moiety. In *in vitro* conditions, the enzymes are capable of oxidizing and depolymerizing lignin and lignocellulose biomass [92]. Manganese peroxidase is the effective lignin degrading enzyme produced and secreted by several organisms. The fungal organisms producing these enzymes are *Agaricus bisporus*, *Penicillium chrysogenum*, *Trametes polyzona*, *Ganoderma lucidum*, and *P. chrysosporium* [102]. The bacterial species are *Bacillus cereus*, *Bacillus subtilis*, *B. aryabhattai*, *Bacillus amyloliquefaciens*, *Bacillus velezensis*, *Lactobacillus kefiri*, *Enterobacter aerogenes*, *Salmonella enterica*, *Bacillus pumilus*, *Paenibacillus* sp., and *Klebsiella pneumoniae* [102].

8.3.3 Versatile Peroxidase (EC 1.11.1.16)

Versatile peroxidases are heme-containing glycoproteins with catalytic properties of both lignin peroxidase and manganese peroxidase [107]. The enzyme was first isolated from *Bjerkandera* sp. possesses polyvalent binding sites for catalytic action [108]. The manganese ion (Mn^{2+}) binds to the binding site of versatile peroxidase and oxidizes lignin compounds similar to manganese peroxidase [107]. The enzymes catalyze various substrates including phenolic, nonphenolic compounds, aromatic alcohols, and lignin dimers. The enzyme acts with redox potential through independent redox mediators, whereas other lignolytic enzymes requires the presence of mediators [109]. The major role of versatile peroxidase is degradation of lignin present in the wood. They contain tryptophanyl environment that acts as strong oxidizing center for catalyzing large-molecular-weight substrates effectively [110]. Due to these properties the enzymes are used as novel biocatalysts for oxidizing broad spectrum of aromatic compounds, which is a significant feature for biotechnological applications [109]. The recalcitrant corn stover was efficiently degraded by versatile peroxidase isolated from *Physisporinus vitreus*. The enzyme from *P. vitreus* oxidized both phenolic and non-phenolic compounds through 5-5′ linkage and converts into monomeric compounds [111].

8.3.4 Dye Decolorizing Peroxidases (DyPs) (EC 1.11.1.19)

Dye decolorizing peroxidases are another type of peroxidases recently identified and found relatively rich in bacteria [112]. The newly discovered DyPs are heme-containing peroxidases received great attention due to its ability to degrade lignin compounds [113]. The DyPs are first isolated from fungal strain *Bjerkandera adusta* and characterized for its degradation ability of dyes [114]. The DyP enzymes contain heme b factor and belong to the family peroxidase – chlorite dismutase [115]. DyPs show dimeric ferredoxin-like fold consisting of four-stranded antiparallel β sheets surrounded by α-helices [116]. The enzymes contain a highly conserved proximal histidine and GXXDG motif [117]. The DyPs are classified into four types: class A, class B, class C, and class D [118]. Class A DyP has a Tat-dependent signal sequence mainly found in bacteria [94]. Due to this signal sequence, they perform their function outside of the cytoplasm or extracellular [94]. Class B and C type DyPs are intracellular enzymes present in bacteria [119]. Class D type DyPs are extracellular fungal representatives involved in dye decolorization [119]. DyPs act as bifunctional enzymes that catalyze both oxidative and hydrolytic reactions [120]. The DyP enzymes actively work in acidic pH and show a broad substrate profile including different classes of synthetic dyes [94]. So far, the DyP enzymes have been identified in fungi, bacteria, and archaeal genomes. DyPs encoding genes are abundant in bacterial genomes and the bacterial dye decolorizing peroxidases are emerged as a promising biocatalyst over fungal dye decolorizing peroxidases [114, 121].

8.3.5 Laccases (EC 1.10.3.2)

Laccases are first identified in *Toxicodendron vernicifluum* also called as Japanese lacquer tree [122]. Later it was identified in fungi, bacteria, and insects. Laccases are multi copper oxidase enzymes that catalyze a variety of phenolic substrates such as monophenols, polyphenols, methoxyphenols, amino phenols, and aromatic amines [40]. The enzymes consist of four copper ions and utilizes atmospheric oxygen as electron donor for oxidizing the lignin compounds and degrades the lignocellulosic waste in the environment [123]. Laccases use oxygen for lignin degradation, whereas other lignolytic enzymes use hydrogen peroxide for its catalytic activity [124]. Generally, laccases oxidize the phenolic compounds due to lower redox potential, whereas non-phenolic compounds are degraded in the presence of redox mediators [125]. The laccases attack and degrade lignin with laccase mediator system (LMS) and finds applications in decolorization of dyes, biofuels production, pulp beaching, detoxification of organic pollutants, delignification of pulp, stabilizer in wine production, transformation of antibiotics, and detoxification of waste water [3, 124]. Laccase production is highest in basidiomycetes fungi (white rot fungi) compared to deuteromycetes and ascomycetes. Lignin and lignocellulose stimulate high production of laccase [126]. Microbial laccases are potential biocatalysts for detoxification and delignification, which is a green initiative for next generation of biofuels [127].

8.3.6 Feruloyl Esterase (EC.3.1.1.73)

Feruloyl esterases (FAE) are the enzymes that degrades the lignocellulosic biomass by decoupling the lignin and plant cell wall polysaccharides [128]. The enzymes belong to a subclass of carboxylic acid esterases and found in fungi, plants, and bacteria [129]. They convert the lignocellulosic biomass into ferulic acid and hydroxy cinnamic acids [130]. Further conversion leads to formation of high-value aromatic compounds. The enzymes have biotechnological applications such as delignification of lignocellulose biomass for biofuel production, improves the digestion of forage plants by ruminants, enzymatic conversion of lignocellulose into bioethanol production, agriculture waste into ferulic acid, and subsequent conversion to high-value aromatic compounds and ferulic acid as a precursor for synthesis of flavor compounds [130]. Feruloyl esterases are classified into classes A, B, C, D, and E based on amino acid sequence similarities [131]. The enzymes are produced by *Pleurotus eryngii* (PeFaeA), *A. niger* (AnFaeA), and *Orpinomyces* sp. (OspFaeA) [130]. The FAE genes are identified in *Asperigillus* species such as *A. flavus*, *A. niger*, *Asperigillus oryzae*, *Asperigillus fumigatus*, *Asperigillus nidulans*, and *Asperigillus terreus* [132]. The genes are cloned and expressed in *Pichia pastoris* for high yield and substrate specificity [133]. The FAEs are identified in other fungal organisms such as *A. bisporus* var. *bisporus*, *Moniliophthora roreri*, *Auricularia subglabra*, *Fusarium graminearum*, and *Chaetomium globosum* [134]. The enzymes classification, family, and subfamilies are listed in the CAZy (carbohydrate-active enzymes) database [135]. Metagenomic approaches are needed to identify the novel bacterial feruloyl esterases to understand the enzyme family and applications in biofuels, pharmaceutical, food, and beverage industries [136].

8.3.7 Aryl Alcohol Oxidase (EC 1.1.3.7)

Aryl alcohol oxidases are the FAD-containing enzymes belonging to the glucose-methanol-choline (GMC) super family of proteins [137]. The enzyme activities are mainly identified in fungal species such as *P. ostreatus*, *P. eryngii*, and *P. chrysosporium* [90]. The enzymes produce hydrogen peroxide (H_2O_2) that activates the lignolytic peroxidases such as LiP, MnP, and VP [92]. These enzymes coordinate with the intracellular dehydrogenases and makes the aromatic alcohol undergo oxidation–reduction reactions and continuously supplies the H_2O_2 for the lignolytic peroxidases [138]. The enzymes catalyze broad range of substrates such as aromatic alcohols, aliphatic alcohols, and polyunsaturated alcohols [139]. The enzymes have different industrial applications in textile industry, biorefineries, fragrances, flavor synthesis, dye decolorization, and high-value compounds [137]. The enzymes catalyze secondary alcohol oxidation reactions and furan dicarboxylic acid synthesis [139]. The genes encoding aryl alcohol oxidases are expressed in *P. pastoris* and

S. cerevisiae for the development of new aryl alcohol oxidase variants with different activities toward substrates of interest [138]. The heterologous expression of these enzymes is quite challenging. Recently, progress was made in protein engineering of these enzymes for enhanced expression, activity, and selectivity [137].

8.3.8 Pyranose-2-Oxidase (EC 1.1.3.10)

Pyranose-2-oxidase belongs to a member of glucose-methanol-choline (GMC) super family of proteins [140]. The enzyme is extracellularly producing into the periplasmic space of hyphae of wood-degrading fungal organisms [141]. The enzyme is a flavin-dependent oxidoreductase that produces hydrogen peroxide which is used for other lignolytic enzymes. In coordination with lignolytic enzymes, it degrades the lignocellulosic biomass into monomeric substances [142]. The enzyme was first isolated from *Spongipellis unicolor* mycelium. The other fungal organisms producing this enzyme are *Trametes ochracea*, *P. chrysosporium*, and *Peniophora* sp. [143]. The enzyme consists of two domains such as flavin-binding domain and substrate-binding domain. The enzyme is also found in bacterial groups such as actinobacteria and proteobacteria [143]. The enzyme-encoding genes were identified in actinobacteria and firmicutes. The enzyme was classified under auxiliary activity family 3 (AA3) of the CAZy database [144].

8.3.9 Vanillyl Alcohol Oxidase (EC 1.1.3.38)

Vanillyl alcohol oxidase is a flavoenzyme first isolated from *Penicillium simplicissimum* [145]. The enzyme produces vanillin and coniferyl alcohols by converting the wide range of para-substituted phenolic compounds [146]. The value-added compounds produced due to vanillyl alcohol oxidase enzyme find applications in food, fragrance, and flavor industries [147]. The enzyme genes are identified in *P. simplicissimum*; however, little is known about its mechanism of action and its physiological role. The enzyme mainly degrades the lignin derived aromatic compounds [146]. The enzyme belongs to the auxiliary activity family 4 (AA4) of the CAZy database [148].

8.3.10 Quinone Reductase (EC 1.6.5.5)

The enzyme quinone reductase was found in brown rot fungi belonging to lignolytic enzymes category [9]. Along with other enzymes it degrades the lignocellulose biomass, particularly reduction of quinones [3]. The enzymes are produced by both bacteria and fungi. The enzyme regulates the lignin repolymerization that occurs due to lignin peroxidase, which is a promising approach for lignin valorization [149].

8.4 Microbial Production of Lignolytic Enzymes

In industrial scale, lignolytic enzymes production using microorganisms is a costly process [150]. The agricultural residues are alternative to this and to be used as inexpensive raw materials [151]. The agricultural waste materials are used due to their availability and low cost [152]. The lignocellulosic biomass content is available in large quantities across the world [4]. In fermentation process, the lignocellulosic biomass is used as a substrate for the production of lignolytic enzymes [153]. The fermentation processes are submerged fermentation (SMF) and solid-state fermentation (SSF) [154]. The former one is relatively simple due to its design, operation, and process control [154]. However, the later one is more advantageous due to its low costs, inexpensive raw materials, better oxygen circulation, less efforts in downstream processing, maintains the natural culture conditions for growth of the microorganisms, and higher yields in shorter time period [155]. The solid support materials are hard wood, soft wood, wheat straw, wheat bran, saw dust, rice straw, oat straw, grape stalks, corn leaves, corn cobs, barley bran, cane bagasse, residues from alimentary industries, agriculture, and forestry [156]. The enzyme titer is high in SSF compared with the SMF due to the fed batch mode of operation with fast oxygen penetration effect [157]. The lignolytic enzymes production is favored by the organic substrates and contains lignin that induces the lignin peroxidase production [58]. The organic substrates that contain cellulose favors the production of laccase enzymes in the fermentation process [126]. Different types of fermenters used for production of lignolytic enzymes are packed bed bioreactor, rotating drum bioreactor, tray bioreactor, immersion bioreactor, capillary membrane, stirred tank bioreactor, expanded bed, fluidized, and air lift reactors. The substrates used by various microorganisms in fermentation processes and the lignolytic enzymes production were listed in Table 8.1.

The microbial communities present in the natural environment produce biocatalysts derived from large number of microorganisms such as bacteria, fungi, and extremophiles [33]. The microorganisms are culturable and unculturable screened by metagenomic approaches [175]. The unculturable microorganisms account for 99%, which are screened for novel genes using metagenomics [176]. The metagenomics strategy reveals the microbial diversity involved in lignocellulose biomass degradation and also evaluates the potential genes encoding for novel enzymes [177]. The lignolytic consortium consists of novel genes and metabolic pathways involved in both lignin degradation and biomass valorization [70]. In recent study, the metagenomic analysis revealed the prevalence of actinobacteria, firmicutes, and proteobacteria in lignin degrading consortium, which was collected from sugarcane plantation soil [70]. In addition to gene sequencing methods, the homology-based annotation is another strategy for finding the homologous sequences based on the similarity among the microbial genes and genomes [178]. BLAST tool is extensively used for searching the homologous genes in genome annotations of lignin degrading microbial diversity [179]. Homology-based annotations cannot reveal the functional annotations, whereas the conserved domain annotations identify the conserved regions of sequences within a particular protein family [180]. This annotation strategy directly reveals the functional properties of the proteins and belongs to the members of that family [179]. The enzymes that degrade the lignin, cellulose, and hemicellulose are listed in CAZy database [181]. The lignolytic enzymes are classified into six classes based on the similarity in sequences and protein structures [9]. The CAZy web server is an expert resource for searching and analyzing the lignolytic enzymes and carbohydrate active enzymes information for the breakdown of lignocellulosic biomass [181].

Table 8.1 Fermentation processes for production of lignolytic enzymes.

Name of the microorganism	Substrate used	Fermentation process/bioreactor	Lignolytic enzymes	Enzyme activity	References
Irpex lacteus	Wheat straw	SMF	MnP	339 U/L	[158]
Pleurotus ostreatus	Potato peel waste	SSF	Lac and MnP	6708.3 ± 75 and 2503.6 ± 50 U/L, respectively	[159]
Trametes versicolor	Oak saw dust, coffee husk, and corn bran	Fixed bed solid state	Lac, MnP, endoxylanase, β-glucosidase, and cellulases	3.22 ± 0.498, 1.01 ± 0.785, 9.05 ± 2.2310, 268.6 ± 14.4714, and 9.0 ± 4.534 U/g ds, respectively	[160]
Pleurotus eryngii (DC.) Gillet (MCC58)	Apricot and pomegranate agro-industrial wastes	SSF	Lac, LiP, MnP, and AAO	1618.5 ± 25, 16.13 ± 0.8L, 570.82, and 105.99 ± 6.3 U/L, respectively	[161]
Schyzophyllum commune	Corn stover and banana stalk	SSF	LiP, MnP, and Lac	1270.40, 715.08, and 130.80 IU/mL, respectively	[162]
Penicillium sp.	Liquid PDA and solid PDA medium with textile dyes- reactive black-5, indigo, acid-blue -1 and vat brown-5	SMF and SSF	Laccase	12.6 U/mL	[163]
Marasmiellus palmivorus VE111	Glucose and casein	Stirred tank bioreactor	Lac, peroxidase, and MnP	3420, 1285, and 59 U/mL, respectively	[164]
Trametes versicolor	Wheat bran	SMF	Laccase	200 U/mL	[165]
Trametes hirsuta	Glucose and yeast extract	Air lift Bioreactor	Laccase	14704 U/L	[166]
Ganoderma lucidum	Pine apple leaves	SSF	Lip, MnP, and laccase	2885.59 ± 65.2, 889.71 ± 46.6, and 472.31 ± 41.2 IU/mL respectively.	[167]

(Continued)

Table 8.1 (Continued)

Name of the microorganism	Substrate used	Fermentation process/bioreactor	Lignolytic enzymes	Enzyme activity	References
Endomelanconiopsis sp.	Dextrose, ammonium tartarate and veratryl alcohol	SMF	LiP	345.26 ± 0.52 IU/mL	[168]
Ganoderma lucidum	Corn cobs	SSF	LiP	2807 U/mL	[169]
Burkholderia sp. SMB1	Rice bran	SMF	Cellulase, xylanase, mannanase, pectinase, and laccase	10.8, 76, 14.23, 62.18, and 24.25 U/mL, respectively	[170]
Fusarium equiseti VKF2	Saw dust	SSF	Laccase	305 U/g	[171]
Bacillus tequilensis LXM55	Mixed wood pulp	SMF	Laccase, xylanase, and mannanase	396.35, 212.95, and 153.33 IU/mL, respectively	[172]
Pseudolagarobasidium acaciicola LA1	Parthenium biomass	SSF	Laccase	34,444 U/g	[173]
Ganoderma lucidum	Wheat straw	sSF	Laccase	90,164.4 U/L	[174]

ds: dry solid; sSF: semi-solid-state fermentation; SSF: Solid state fermentation; SMF: Submerged fermentation.

8.5 Mechanism of Action of Lignolytic Enzymes

The degradation of lignocellulosic biomass is highly affected by its composition such as lignin, cellulose, and hemicellulose [5]. It is an abundant source from plants and acts as a renewable source for production of bioenergy [182]. The underutilized lignocellulosic waste is used as a source of feed stocks for biofuel production [183]. The characteristic features of lignocellulose biomass are listed in Table 8.2.

The commercial applications of lignolytic enzymes are restricted due to lack of complete information pertaining to its mechanism of action, properties, and their purification processes [3]. The fungal and bacterial species have complex mechanism of enzymes and intermediates to degrade the lignin [8]. Lignolytic enzymes have complex mechanisms in detoxification and degradation process of lignocellulosic biomass influenced by their structure, origin, type of substrate, and external conditions [3]. The plant laccases have tendency to polymerize lignin, whereas the laccase from fungi and bacteria catalyzes the lignin by oxidation process [187]. The laccase reaction occurs around four different copper ions categorized under three types of copper molecules [188]. The type I (T1 Cu) copper domain is known as substrate-reducing site [190]. The active site of the enzyme contains two histidine and one cysteine molecules [3]. The type 2 (T2 Cu) has two histidine amino acids and a water molecule [3]. The type 3 (T3 Cu) consists of two copper molecules with six histidine amino acids [190]. The laccase oxidizes a variety of substrates using oxygen and liberates hydrogen peroxide as the by-product [3]. The laccase degrades the lignin by direct oxidation and indirect oxidation. In the direct

Table 8.2 Characteristic features of lignocellulosic biomass.

Component	Characteristic features	References
1) Lignin	It is an aromatic heteropolymer. In nature, it is abundant as renewable source. It consists of cross-linked coumaryl alcohol, coniferyl alcohol, and sinapyl alcohol. The chemical composition varies from species to species. Lignin shows resistance to digestion than other naturally occurring compounds. It plays a role in carbon recycling.	[184, 185]
2) Cellulose	It is a polysaccharide consists of interlinked glucose units. The glucose molecules are linked via β (1→4) glycosidic bonds. It is degraded by the enzymes called cellulases. The degradation occurs by the hydrolysis of β-1,4 linkages.	[4, 186]
3) Hemicellulose	It is a natural polymer consists of arabinose, glucose, galactose, xylose, and mannose. It is an amorphous structure. It is located within cellulose and between the cellulose and lignin.	[4, 186]

oxidation, oxygen-mediated oxidation process occurs, whereas in the indirect oxidation process, laccase uses oxygen to oxidize mediators [3, 190]. The oxidized mediators then act as chemical oxidant for lignin degradation. The natural mediators include vanillin, acetovanillin, acetosyringine, syringaldehyde, sinapic acid, ferulic acid, and *p*-coumaric acid. The artificial mediators are 1-nitroso-napthol-3,6-disulfonic acid (NNDS), 2,2′-azinobis (3-ethylbenzthiazoline-6-sulphonate (ABTS), violuric acid, 1-hydroxy benzotriazole (1-HBT), 2,2,6,6-tetramethyl piperidine 1-oxyl (TEMPO), and promazine [190] (Figure 8.3).

Lignin peroxidase oxidizes phenol and non-phenolic compounds of lignin. The enzyme contains ferric ion inside the ferric protoporphyrin [190]. The enzyme-catalyzed reaction involves the native enzyme of ferric resting state, oxoferryl unstable intermediate compound I, and impartial oxoferryl intermediate compound II [3]. The LiP degrades various phenolic compounds in the presence of hydrogen peroxide as co-substrate and veratryl alcohol as mediator [3] (Figure 8.4).

Manganese peroxidase also uses hydrogen peroxide similar to lignin peroxidase. It requires manganese ions Mn^{2+} and in acidic medium it is oxidized by hydrogen peroxide producing Mn^{3+}. The Mn^{3+} ions oxidize the phenolic moieties in the lignin compounds [190].

$$2Mn(II) + 2H^+ + H_2O_2 \rightleftarrows 2Mn(III) + 2H_2O$$

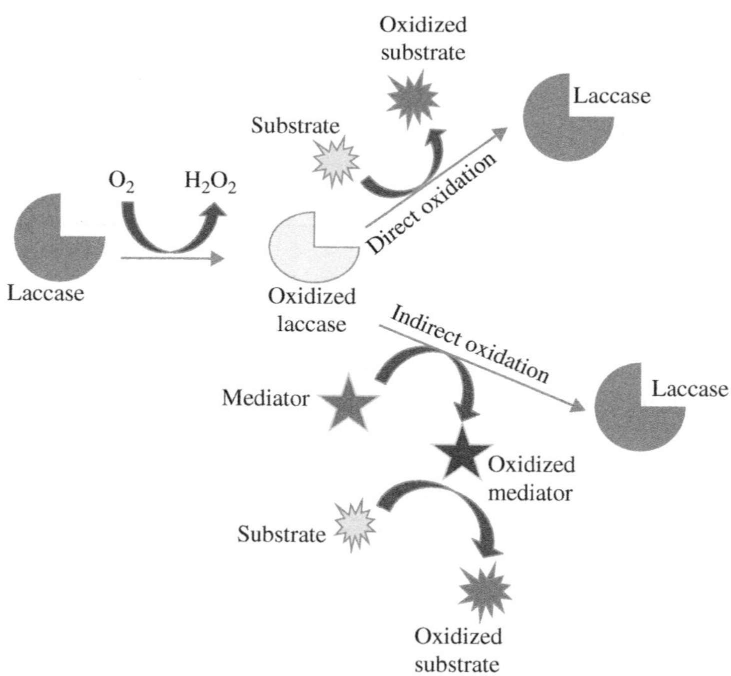

Figure 8.3 Catalytic reaction by laccase enzyme. *Source:* Kumar et al. [3] / Elsevier / CC BY-NC-ND 4.0; Agrawal et.al. [191] / Springer Nature / CC BY 4.0.

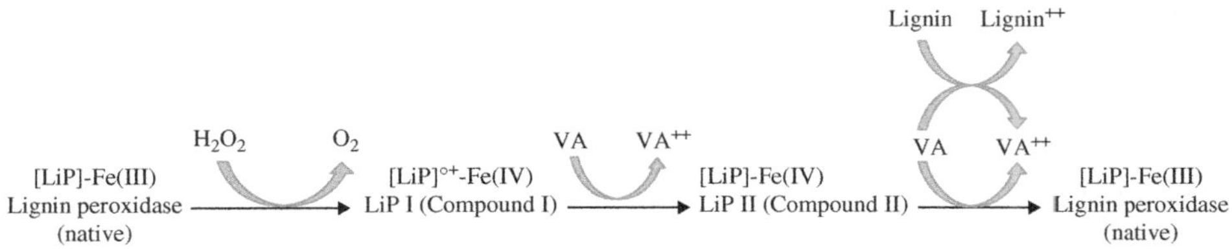

Figure 8.4 Catalytic reaction of lignin peroxidase enzyme in the presence of hydrogen peroxide and mediator veratryl alcohol. *Source:* Adapted from [92].

Versatile peroxidases do not require manganese and mediators for catalytic oxidation of phenolic and non-phenolic lignin compounds [190]. It uses hydrogen peroxide as electron acceptor. It contains heme porphyrin ring structure in central pocket and binds with the hydrogen peroxide forming the iron peroxide complex [109]. Further, the enzyme in ferric state reacts with hydrogen peroxide and forms Fe^{4+} oxo-porphyrin radical complex with heme molecule. In this state, it is called as compound I, which is electron-deficient compound and further oxidized to intermediate compound II, which is finally reduced back to ferric peroxidase the native enzyme [109, 190]. Nowadays, the lignocellulose biomass degradation is mainly focused on the key enzymes involved in lignin degradation and detoxification. The metagenomic studies are essential for identifying the novel genes involved in lignocellulose biomass degradation [70].

8.6 Conclusions

Lignocellulose biomass occupied major portion of the total biomass in the world. The lignocellulosic conversion is a challenge for waste management and several biotechnological applications have been applied for the conversion of lignocellulose waste into value-added compounds. A variety of microorganisms including fungi, bacteria, actinomycetes, extremophiles, plants, and insects have the ability to degrade the lignocellulose biomass to monomeric compounds. Currently, lignocellulose feed stocks conversion is a promising approach for the development of biofuels. However, due to various metabolic pathways for lignocellulose degradation by microorganisms limits its commercial applications. Metagenomic studies and bioprocessing strategies identify the potential microorganisms and development of engineered microbial metabolic pathways could result in the lignin degradation and biotransformation into value-added products.

Acknowledgments

The authors acknowledge the management of Vignan's Foundation for Science Technology and Research (Deemed to be University) for providing a platform to carry out the research work. Also, thanks to DST-FIST project LSI-576/2013 for assistance during the tenure of this work.

References

1 Liu, Q., Luo, L., and Zheng, L. (2018). Lignins: biosynthesis and biological functions in plants. *International Journal of Molecular Sciences* 19 (2): 335.

2 Zeng, Y., Himmel, M.E., and Ding, S.Y. (2017). Visualizing chemical functionality in plant cell walls. *Biotechnology for Biofuels* 10 (1): 1–16.

3 Kumar, A. and Chandra, R. (2020). Ligninolytic enzymes and its mechanisms for degradation of lignocellulosic waste in environment. *Heliyon* 6 (2): e03170.

4 Zoghlami, A. and Paes, G. (2019). Lignocellulosic biomass: understanding recalcitrance and predicting hydrolysis. *Frontiers in Chemistry* 7: 874.

5 Tayyab, M., Noman, A., Islam, W. et al. (2018). Bioethanol production from lignocellulosic biomass by environment-friendly pretreatment methods: a review. *Applied Ecology and Environmental Research* 16 (1): 225–249.

6 Vasquez-Garay, F., Carrillo-Varela, I., Vidal, C. et al. (2021). A review on the lignin biopolymer and its integration in the elaboration of sustainable materials. *Sustainability* 13 (5): 2697.

7 Chan, J.C., Paice, M., and Zhang, X. (2020). Enzymatic oxidation of lignin: challenges and barriers toward practical applications. *ChemCatChem* 12 (2): 401–425.

8 Iram, A., Berenjian, A., and Demirci, A. (2021). A review on the utilization of lignin as a fermentation substrate to produce lignin-modifying enzymes and other value-added products. *Molecules* 26 (10): 2960.

9 Janusz, G., Pawlik, A., Sulej, J. et al. (2017). Lignin degradation: microorganisms, enzymes involved, genomes analysis and evolution. *FEMS Microbiology Reviews* 41 (6): 941–962.

10 Chowdhary, P., More, N., Yadav, A., and Bharagava, R.N. (2019). Ligninolytic enzymes: an introduction and applications in the food industry. In: *Enzymes in Food Biotechnology* (ed. M. Kuddus), 81–195. Academic Press.

11 Singh, S.P., Pandey, A., Singhania, R.R. et al. (2020). *Biomass, Biofuels, Biochemicals: Advances in Enzyme Catalysis and Technologies.* Elsevier. ISBN 9780128198209. https://doi.org/10.1016/C2019-0-00323-8.

12 Mate, D.M. and Alcalde, M. (2017). Laccase: a multi-purpose biocatalyst at the forefront of biotechnology. *Microbial Biotechnology* 10 (6): 1457–1467.

13 Osma, J.F., Toca-Herrera, J.L., and Rodriguez-Couto, S. (2010). Uses of laccases in the food industry. *Enzyme Research* 2010 (918761): 1–8.

14 Zainith, S., Chowdhary, P., Mani, S., and Mishra, S. (2020). Microbial ligninolytic enzymes and their role in bioremediation. In: *Microorganisms for Sustainable Environment and Health* (ed. P. Chowdhary, D. Verma, A. Raj, and Y. Akhter), 179–203. Elsevier.

15 Yadav, M. and Yadav, H.S. (2015). Applications of ligninolytic enzymes to pollutants, wastewater, dyes, soil, coal, paper and polymers. *Environmental Chemistry Letters* 13 (3): 309–318.

16 Yadav, M., Singh, S.K., Yadava, S., and Singh Yadav, K.D. (2015). Ligninolytic enzymes for water depollution, coal breakdown, and paper industry. In: *CO2 Sequestration, Biofuels and Depollution* (ed. E. Lichtfouse, J. Schwarzbauer, and D. Robert), 359–386. Springer.

17 Placido, J. and Capareda, S. (2015). Ligninolytic enzymes: a biotechnological alternative for bioethanol production. *Bioresources and Bioprocessing* 2 (1): 1–12.

18 Datta, B. (2012). Socio-economic, environmental, and policy perspectives of advanced biodiesel production. In: *Advances in Biodiesel Production* (ed. R. Luque and J.A. Melero), 32–68. Woodhead Publishing.

19 Maurya, D.P., Singla, A., and Negi, S. (2015). An overview of key pretreatment processes for biological conversion of lignocellulosic biomass to bioethanol. *3 Biotech* 5 (5): 597–609.

20 Edeh, I. (2020). *Bioethanol Production: An Overview*. Bioethanol Technologies.

21 Woiciechowski, A.L., Souza Vandenberghe, L.P.D., Karp, S.G. et al. (2013). The pretreatment step in lignocellulosic biomass conversion: current systems and new biological systems. In: *Lignocellulose Conversion* (ed. V. Faraco), 39–64. Berlin, Heidelberg: Springer.

22 Adegboye, M.F., Ojuederie, O.B., Talia, P.M., and Babalola, O.O. (2021). Bioprospecting of microbial strains for biofuel production: metabolic engineering, applications, and challenges. *Biotechnology for Biofuels* 14 (1): 1–21.

23 Upadhyay, S.K. and Singh, S.P. (eds.) (2021). *Bioprospecting of Plant Biodiversity for Industrial Molecules*. John Wiley & Sons Ltd. doi:10.1002/9781119718017.

24 Upadhyay, S.K. and Singh, S.P. (eds.) (2023). *Plants as Bioreactors for Industrial Molecules*. John Wiley & Sons Ltd. doi:10.1002/9781119875116.

25 Singh, S.P. and Upadhyay, S.K. (eds.) (2021). *Bioprospecting of Microorganism-Based Industrial Molecules*. John Wiley & Sons Ltd. doi:10.1002/9781119717317.

26 Shintani, T., Upadhyay, S.K., Singh, S.P. (2021). An introduction to microbial biodiversity and bioprospection. In: *Bioprospecting of Microorganism-Based Industrial Molecules* (ed. S.P. Singh and S.K. Upadhyay). John Wiley & Sons Ltd. https://doi.org/10.1002/9781119717317.ch1.

27 Tabassum, S. and Ali, S. (2020). Characterization and advantages of lignolytic enzymes produced by solid state fermentation. *Enzyme Engineering* 9: 166.

28 Janusz, G., Pawlik, A., Swiderska-Burek, U. et al. (2020). Laccase properties, physiological functions, and evolution. *International Journal of Molecular Sciences* 21 (3): 966.

29 van Eerde, A., Várnai, A., Wang, Y. et al. (2022). Successful production and ligninolytic activity of a bacterial laccase, Lac51, made in *Nicotiana benthamiana* via transient expression. *Frontiers in Plant Science* 13: 912293.

30 Piscitelli, A., Pezzella, C., Giardina, P. et al. (2010). Heterologous laccase production and its role in industrial applications. *Bioengineered Bugs* 1 (4): 254–264.

31 Khlystov, N.A., Yoshikuni, Y., Deutsch, S., and Sattely, E.S. (2021). A plant host, *Nicotiana benthamiana*, enables the production and study of fungal lignin-degrading enzymes. *Communications Biology* 4 (1): 1–13.

32 Shanmugaraj, B., Bulaon, I., and C. J. and Phoolcharoen, W. (2020). Plant molecular farming: a viable platform for recombinant biopharmaceutical production. *Plants* 9 (7): 842.

33 Adrio, J.L. and Demain, A.L. (2014). Microbial enzymes: tools for biotechnological processes. *Biomolecules* 4 (1): 117–139.

34 Falade, A.O., Eyisi, O.A., Mabinya, L.V. et al. (2017). Peroxidase production and ligninolytic potentials of fresh water bacteria *Raoultella ornithinolytica* and *Ensifer adhaerens*. *Biotechnology Reports* 16: 12–17.

35 Furukawa, T., Bello, F.O., and Horsfall, L. (2014). Microbial enzyme systems for lignin degradation and their transcriptional regulation. *Frontiers in Biology* 9 (6): 448–471.

36 Wang, J., Liang, J., and Gao, S. (2018). Biodegradation of lignin monomers vanillic, p-coumaric, and syringic acid by the bacterial strain, Sphingobacterium sp. HY-H. *Current Microbiology* 75 (9): 1156–1164.

37 Rashid, G.M., Zhang, X., Wilkinson, R.C. et al. (2018). Sphingobacterium sp. T2 manganese superoxide dismutase catalyzes the oxidative demethylation of polymeric lignin via generation of hydroxyl radical. *ACS Chemical Biology* 13 (10): 2920–2929.

38 Wang, J., Feng, J., Jia, W. et al. (2015). Lignin engineering through laccase modification: a promising field for energy plant improvement. *Biotechnology for Biofuels* 8 (1): 1–11.

39 Upadhyay, P., Shrivastava, R., and Agrawal, P.K. (2016). Bioprospecting and biotechnological applications of fungal laccase. *3 Biotech* 6 (1): 1–12.

40 Viswanath, B., Rajesh, B., Janardhan, A. et al. (2014). Fungal laccases and their applications in bioremediation. *Enzyme Research* 2014 (163242): 1–21.

41 Kalra, K., Chauhan, R., Shavez, M., and Sachdeva, S. (2013). Isolation of laccase producing *Trichoderma* spp. and effect. *International Journal of Chem Tech Research* 5 (5): 2229–2235.

42 Turlapati, P.V., Kim, K.W., Davin, L.B., and Lewis, N.G. (2011). The laccase multigene family in *Arabidopsis thaliana*: towards addressing the mystery of their gene function (s). *Planta* 233 (3): 439–470.

43 Swetha, C. and Shivaprasad, P.V. (2019). Extraction and purification of laccases from rice stems. *Bio-Protocol* 9 (7): –e3208.

44 Caparros-Ruiz, D., Fornale, S., Civardi, L. et al. (2006). Isolation and characterization of a family of laccases in maize. *Plant Science* 171 (2): 217–225.

45 Ping, X., Wang, T., Lin, N. et al. (2019). Genome-wide identification of the LAC gene family and its expression analysis under stress in *Brassica napus*. *Molecules* 24 (10): 1985.

46 Zhang, W., Lin, J., Dong, F. et al. (2019). Genomic and allelic analyses of laccase genes in sugarcane (*Saccharum spontaneum* L.). *Tropical Plant Biology* 12 (3): 219–229.

47 Wang, Y., Bouchabke-Coussa, O., Lebris, P. et al. (2015). LACCASE5 is required for lignification of the *Brachypodium distachyon culm*. *Plant Physiology* 168 (1): 192–204.

48 Berthet, S., Thevenin, J., Baratiny, D. et al. (2012). Role of plant laccases in lignin polymerization. In: *Advances in Botanical Research* (ed. L. Jouanin and C. Lapierre), 145–172. Academic Press.

49 Geib, S.M., Filley, T.R., Hatcher, P.G. et al. (2008). Lignin degradation in wood-feeding insects. *Proceedings of the National Academy of Sciences* 105 (35): 12932–12937.

50 Jing, T.Z., Qi, F.H., and Wang, Z.Y. (2020). Most dominant roles of insect gut bacteria: digestion, detoxification, or essential nutrient provision? *Microbiome* 8 (1): 1–20.

51 Hattori, M., Tsuchihara, K., Noda, H. et al. (2010). Molecular characterization and expression of laccase genes in the salivary glands of the green rice leafhopper, *Nephotettix cincticeps* (Hemiptera: Cicadellidae). *Insect Biochemistry and Molecular Biology* 40 (4): 331–338.

52 Coy, M.R., Salem, T.Z., Denton, J.S. et al. (2010). Phenol-oxidizing laccases from the termite gut. *Insect Biochemistry and Molecular Biology* 40 (10): 723–732.

53 Sun, J. and Zhou, X.J. (2011). Utilization of lignocellulose-feeding insects for viable biofuels: an emerging and promising area of entomological science. In: *Recent Advances in Entomological Research* (ed. T. Liu and L. Kang), 434–500. Berlin, Heidelberg: Springer.

54 Geng, A., Wu, J., Xie, R.R. et al. (2018). Characterization of a laccase from a wood-feeding termite, *Coptotermes formosanus*. *Insect Science* 25 (2): 251–258.

55 Kassim, A.S.M., Ishak, N., Aripin, A.M., and Zaidel, D.N.F.A. (2016). Potential lignin degraders isolated from the gut of *Rhynchophorus Ferrugineus*. *Advance of Engineering Research* 66–72. https://doi.org/10.2991/icmmse-16.2016.22.

56 Rhoads, T.L., Mikell, A.T. Jr., and Eley, M.H. (1995). Investigation of the lignin-degrading activity of *Serratia marcescens*: biochemical screening and ultrastructural evidence. *Canadian Journal of Microbiology* 41 (7): 592–600.

57 Yang, C., Yue, F., Cui, Y. et al. (2018). Biodegradation of lignin by *Pseudomonas* sp. Q18 and the characterization of a novel bacterial DyP-type peroxidase. *Journal of Industrial Microbiology and Biotechnology* 45 (10): 913–927.

58 Lee, S., Kang, M., Bae, J.H. et al. (2019). Bacterial valorization of lignin: strains, enzymes, conversion pathways, biosensors, and perspectives. *Frontiers in Bioengineering and Biotechnology* 7: 209.

59 Georgiadou, D.N., Avramidis, P., Ioannou, E., and Hatzinikolaou, D.G. (2021). Microbial bioprospecting for lignocellulose degradation at a unique Greek environment. *Heliyon* 7 (6): e07122.

60 Xiong, Y.I., Zhao, Y., Ni, K. et al. (2020). Characterization of ligninolytic bacteria and analysis of alkali-lignin biodegradation products. *Polish Journal of Microbiology* 69 (3): 339–347.

61 Xu, Z., Lei, P., Zhai, R. et al. (2019). Recent advances in lignin valorization with bacterial cultures: microorganisms, metabolic pathways, and bio-products. *Biotechnology for Biofuels* 12 (1): 1–19.

62 Narnoliya, L.K., Agarwal, N., Patel, S.N. et al. (2019). Kinetic characterization of laccase-from Bacillus atrophaeus, and its potential in juice clarification in free and immobilized-forms. *Journal of Microbiology* 57: 900–909.

63 Suryadi, H., Judono, J.J., Putri, M.R. et al. (2022). Bio delignification of lignocellulose using ligninolytic enzymes from white-rot fungi. *Heliyon* 8: e08865.

64 de Souza, W.R. (2013). Microbial degradation of lignocellulosic biomass. In: *Sustainable Degradation of Lignocellulosic Biomass-Techniques, Applications and Commercialization* (ed. A.K. Chandel and S.S. da Silva), 207–247. InTech.

65 Del Cerro, C., Erickson, E., Dong, T. et al. (2021). Intracellular pathways for lignin catabolism in white-rot fungi. *Proceedings of the National Academy of Sciences* 118 (9): e2017381118.

66 El-Gendi, H., Saleh, A.K., Badierah, R. et al. (2021). A comprehensive insight into fungal enzymes: structure, classification, and their role in Mankind's challenges. *Journal of Fungi* 8 (1): 23.

67 Hou, L., Ji, D., Dong, W. et al. (2020). The synergistic action of electro-Fenton and white-rot fungi in the degradation of lignin. *Frontiers in Bioengineering and Biotechnology* 8: 99.

68 Torres-Farrada, G., Manzano Leon, A.M., Rineau, F. et al. (2017). Diversity of ligninolytic enzymes and their genes in strains of the genus Ganoderma: applicable for biodegradation of xenobiotic compounds. *Frontiers in Microbiology* 8: 898.

69 Asemoloye, M.D., Marchisio, M.A., Gupta, V.K., and Pecoraro, L. (2021). Genome-based engineering of ligninolytic enzymes in fungi. *Microbial Cell Factories* 20 (1) 1–18.

70 Moraes, E.C., Alvarez, T.M., Persinoti, G.F. et al. (2018). Lignolytic-consortium omics analyses reveal novel genomes and pathways involved in lignin modification and valorization. *Biotechnology for Biofuels* 11 (1): 1–16.

71 Barka, E.A., Vatsa, P., Sanchez, L. et al. (2016). Taxonomy, physiology, and natural products of Actinobacteria. *Microbiology and Molecular Biology Reviews* 80 (1): 1–43.

72 Javed, Z., Tripathi, G.D., Mishra, M., and Dashora, K. (2021). Actinomycetes – the microbial machinery for the organic-cycling, plant growth, and sustainable soil health. *Biocatalysis and Agricultural Biotechnology* 31: 101893.
73 Saini, A., Aggarwal, N.K., Sharma, A., and Yadav, A. (2015). Actinomycetes: a source of lignocellulolytic enzymes. *Enzyme Research* 2015: 279381.
74 Jayasekara, S. and Ratnayake, R. (2019). Microbial cellulases: an overview and applications. *Cellulose* 22. IntechOpen.
75 Salwan, R. and Sharma, V. (2018). The role of actinobacteria in the production of industrial enzymes. In: *New and Future Developments in Microbial Biotechnology and Bioengineering* (ed. V. Gupta), 165–177. Elsevier.
76 Zeng, J., Singh, D., Laskar, D.D., and Chen, S. (2013). Degradation of native wheat straw lignin by *Streptomyces viridosporus* T7A. *International journal of Environmental Science and Technology* 10 (1): 165–174.
77 Chen, C.Y., Huang, Y.C., Wei, C.M. et al. (2013). Properties of the newly isolated extracellular thermo-alkali-stable laccase from thermophilic actinomycetes, *Thermobifida fusca* and its application in dye intermediates oxidation. *AMB Express* 3 (1): 1–9.
78 Zhu, D., Adebisi, W.A., Ahmad, F. et al. (2020). Recent development of extremophilic bacteria and their application in biorefinery. *Frontiers in Bioengineering and Biotechnology* 8: 483.
79 Reed, C.J., Lewis, H., Trejo, E. et al. (2013). Protein adaptations in archaeal extremophiles. *Archaea* 2013: 373275.
80 Rampelotto, P.H. (2013). Extremophiles and extreme environments. *Life* 3 (3): 482–485.
81 Littlechild, J.A. (2015). Enzymes from extreme environments and their industrial applications. *Frontiers in Bioengineering and Biotechnology* 3: 161.
82 Zhu, D., Zhang, P., Xie, C. et al. (2017). Biodegradation of alkaline lignin by *Bacillus ligniniphilus* L1. *Biotechnology for Biofuels* 10 (1): 1–14.
83 Lopez, M.J., Jurado, M.M., Lopez-Gonzalez, J.A. et al. (2021). Characterization of thermophilic lignocellulolytic microorganisms in composting. *Frontiers in Microbiology* 12: 697480.
84 Antunes, L.P., Martins, L.F., Pereira, R.V. et al. (2016). Microbial community structure and dynamics in thermophilic composting viewed through metagenomics and meta transcriptomics. *Scientific Reports* 6 (1): 1–13.
85 Lai, C.M.T., Chua, H.B., Danquah, M.K., and Saptoro, A. (2017, June)). Isolation of thermophilic lignin degrading bacteria from oil-palm empty fruit bunch (EFB) compost. In: *IOP Conference Series: Materials Science and Engineering 206*, vol. 1, 012016. IOP Publishing.
86 Yan, J., Chen, Y., Niu, J. et al. (2015). Laccase produced by a thermotolerant strain of *Trametes trogii* LK13. *Brazilian Journal of Microbiology* 46: 59–65.
87 Rezaei, S., Shahverdi, A.R., and Faramarzi, M.A. (2017). Isolation, one-step affinity purification, and characterization of a polyextremotolerant laccase from the halophilic bacterium *Aquisalibacillus elongatus* and its application in the delignification of sugar beet pulp. *Bioresource Technology* 230: 67–75.
88 Levy-Booth, D.J., Navas, L.E., Fetherolf, M.M. et al. (2022). Discovery of lignin-transforming bacteria and enzymes in thermophilic environments using stable isotope probing. *The ISME Journal* 16: 1944–1956.
89 Zhu, D., Qaria, M.A., Zhu, B. et al. (2022). Extremophiles and extremozymes in lignin bioprocessing. *Renewable and Sustainable Energy Reviews* 157: 112069.

90 Hernandez-Ortega, A., Ferreira, P., and Martínez, A.T. (2012). Fungal aryl-alcohol oxidase: a peroxide-producing flavoenzyme involved in lignin degradation. *Applied Microbiology and Biotechnology* 93 (4): 1395–1410.

91 Tavares, A.P.M., Coelho, M.A.Z., Coutinho, J.A.P., and Xavier, A.M.R.B. (2005). Laccase improvement in submerged cultivation: induced production and kinetic modelling. *Journal of Chemical Technology & Biotechnology: International Research in Process, Environmental & Clean Technology* 80 (6): 669–676.

92 Falade, A.O., Nwodo, U.U., Iweriebor, B.C. et al. (2017). Lignin peroxidase functionalities and prospective applications. *Microbiology Open* 6 (1): e00394.

93 Atiwesh, G., Parrish, C.C., Banoub, J., and Le, T.A.T. (2022). Lignin degradation by microorganisms: a review. *Biotechnology Progress* 38 (2): e3226.

94 de Gonzalo, G., Colpa, D.I., Habib, M.H., and Fraaije, M.W. (2016). Bacterial enzymes involved in lignin degradation. *Journal of Biotechnology* 236: 110–119.

95 Rai, R., Agrawal, D., and Chadha, B.S. (2019). New paradigm in degradation of lignocellulosic biomass and discovery of novel microbial strains. In: *Microbial Diversity in Ecosystem Sustainability and Biotechnological Applications* (ed. T. Satyanarayana, B.N. Johri, and S.K. Das), 403–440. Springer.

96 Singh, A.K., Bilal, M., Iqbal, H.M. et al. (2021). Bioremediation of lignin derivatives and phenolics in wastewater with lignin modifying enzymes: status, opportunities and challenges. *Science of the Total Environment* 777: 145988.

97 Kai, D., Chow, L.P., and Loh, X.J. (2018). Lignin and its properties. In: *Functional Materials from Lignin: Methods and Advances* (ed. X.J. Loh, D. Kai, and Z. Li), 1–28. World Scientific Publishing.

98 Ruiz-Duenas, F.J. and Martínez, Á.T. (2009). Microbial degradation of lignin: how a bulky recalcitrant polymer is efficiently recycled in nature and how we can take advantage of this. *Microbial Biotechnology* 2 (2): 164–177.

99 Weng, C., Peng, X., and Han, Y. (2021). Depolymerization and conversion of lignin to value-added bioproducts by microbial and enzymatic catalysis. *Biotechnology for Biofuels* 14 (1): 1–22.

100 Hirai, H., Sugiura, M., Kawai, S., and Nishida, T. (2005). Characteristics of novel lignin peroxidases produced by white-rot fungus *Phanerochaete sordida* YK-624. *FEMS Microbiology Letters* 246 (1): 19–24.

101 Sadaqat, B., Khatoon, N., Malik, A.Y. et al. (2020). Enzymatic decolorization of melanin by lignin peroxidase from *Phanerochaete chrysosporium*. *Scientific Reports* 10 (1): 1–10.

102 Kumar, A. and Arora, P.K. (2022). Biotechnological applications of manganese peroxidases for sustainable management. *Frontiers in Environmental Science* 10: 365.

103 Chowdhary, P., Shukla, G., Raj, G. et al. (2019). Microbial manganese peroxidase: a ligninolytic enzyme and its ample opportunities in research. *SN Applied Sciences* 1 (1): 1–12.

104 Rahi, D.K. and Parmar, A.S. (2021). Ligninolytic peroxidases: sources and applications. *Journal of Advanced Scientific Research* 12: 69–80.

105 Martin, H. (2002). Review: lignin conversion by manganese peroxidase (MnP). *Enzyme and Microbial Technology* 30: 454–466.

106 Ayuso-Fernandez, I., Ruiz-Duenas, F.J., and Martínez, A.T. (2018). Evolutionary convergence in lignin-degrading enzymes. *Proceedings of the National Academy of Sciences* 115 (25): 6428–6433.

107 Datta, R., Kelkar, A., Baraniya, D. et al. (2017). Enzymatic degradation of lignin in soil: a review. *Sustainability* 9 (7): 1163.

108 Moreira, P.R., Duez, C., Dehareng, D. et al. (2005). Molecular characterisation of a versatile peroxidase from a Bjerkandera strain. *Journal of Biotechnology* 118 (4): 339–352.

109 Ravichandran, A. and Sridhar, M. (2017). Insights into the mechanism of lignocellulose degradation by versatile peroxidases. *Current Science* 35–42.

110 Saez-Jimenez, V., Rencoret, J., Rodríguez-Carvajal, M.A. et al. (2016). Role of surface tryptophan for peroxidase oxidation of nonphenolic lignin. *Biotechnology for Biofuels* 9 (1): 1–13.

111 Kong, W., Fu, X., Wang, L. et al. (2017). A novel and efficient fungal delignification strategy based on versatile peroxidase for lignocellulose bioconversion. *Biotechnology for Biofuels* 10 (1): 1–15.

112 Chen, C., Shrestha, R., Jia, K. et al. (2015). Characterization of dye-decolorizing peroxidase (DyP) from *Thermomonospora curvata* reveals unique catalytic properties of A-type DyPs. *Journal of Biological Chemistry* 290 (38): 23447–23463.

113 Singh, R. and Eltis, L.D. (2015). The multihued palette of dye-decolorizing peroxidases. *Archives of Biochemistry and Biophysics* 574: 56–65.

114 Fernandez-Fueyo, E., Linde, D., Almendral, D. et al. (2015). Description of the first fungal dye-decolorizing peroxidase oxidizing manganese (II). *Applied Microbiology and Biotechnology* 99 (21): 8927–8942.

115 Hofbauer, S., Schaffner, I., Furtmüller, P.G., and Obinger, C. (2014). Chlorite dismutases–a heme enzyme family for use in bioremediation and generation of molecular oxygen. *Biotechnology Journal* 9 (4): 461–473.

116 Habib, M.H. (2020). Exploring (per) oxidases as biocatalysts for the synthesis of valuable aromatic compounds. University of Groningen.

117 Li, L., Wang, T., Chen, T. et al. (2021). Revealing two important tryptophan residues with completely different roles in a dye-decolorizing peroxidase from Irpex lacteus F17. *Biotechnology for Biofuels* 14 (1): 1–21.

118 de Eugenio, L.I., Peces-Perez, R., Linde, D. et al. (2021). Characterization of a dye-decolorizing peroxidase from Irpex lacteus expressed in *Escherichia coli*: an enzyme with wide substrate specificity able to transform lignosulfonates. *Journal of Fungi* 7 (5): 325.

119 Colpa, D.I., Fraaije, M.W., and van Bloois, E. (2014). DyP-type peroxidases: a promising and versatile class of enzymes. *Journal of Industrial Microbiology and Biotechnology* 41 (1): 1–7.

120 Xu, L., Sun, J., Qaria, M.A. et al. (2021). Dye decoloring peroxidase structure, catalytic properties and applications: current advancement and futurity. *Catalysts* 11 (8): 955.

121 Chen, C. and Li, T. (2016). Bacterial dye-decolorizing peroxidases: biochemical properties and biotechnological opportunities. *Physical Sciences Reviews* 1 (9): https://doi.org/10.1515/psr-2016-0051.

122 Rodriguez-Couto, S. (2017). Microbial laccases as potential eco-friendly biocatalysts for the food processing industries. In: *Microbial Enzyme Technology in Food Applications* (ed. R.C. Ray and C.M. Rosell), 255–268. CRC Press.

123 Akram, F., Ashraf, S., Shah, F.I., and Aqeel, A. (2022). Eminent industrial and biotechnological applications of laccases from bacterial source: a current overview. *Applied Biochemistry and Biotechnology* 194: 2336–2356.

124 Christopher, L.P., Yao, B., and Ji, Y. (2014). Lignin biodegradation with laccase-mediator systems. *Frontiers in Energy Research* 2: 12.

125 Fillat, U., Ibarra, D., Eugenio, M.E. et al. (2017). Laccases as a potential tool for the efficient conversion of lignocellulosic biomass: a review. *Fermentation* 3 (2): 17.

126 Shekher, R., Sehgal, S., Kamthania, M., and Kumar, A. (2011). Laccase: microbial sources, production, purification, and potential biotechnological applications. *Enzyme Research* 2011: 217861.

127 Malhotra, M. and Suman, S.K. (2021). Laccase-mediated delignification and detoxification of lignocellulosic biomass: removing obstacles in energy generation. *Environmental Science and Pollution Research* 28 (42): 58929–58944.

128 Underlin, E.N., Frommhagen, M., Dilokpimol, A. et al. (2020). Feruloyl esterases for biorefineries: subfamily classified specificity for natural substrates. *Frontiers in Bioengineering and Biotechnology* 8: 332.

129 Xu, M., Gao, X., Chen, J. et al. (2018). The feruloyl esterase genes are required for full pathogenicity of the apple tree canker pathogen *Valsa mali*. *Molecular Plant Pathology* 19 (6): 1353–1363.

130 Oliveira, D.M., Mota, T.R., Oliva, B. et al. (2019). Feruloyl esterases: biocatalysts to overcome biomass recalcitrance and for the production of bioactive compounds. *Bioresource Technology* 278: 408–423.

131 Uraji, M., Tamura, H., Mizohata, E. et al. (2018). Loop of Streptomyces feruloyl esterase plays an important role in the enzyme's catalyzing the release of ferulic acid from biomass. *Applied and Environmental Microbiology* 84 (3): e02300–e02317.

132 Ou, S., Zhang, J., Wang, Y., and Zhang, N. (2011). Production of feruloyl esterase from *Aspergillus niger* by solid-state fermentation on different carbon sources. *Enzyme Research* 2011: 848939.

133 Dilokpimol, A., Makelä, M.R., Mansouri, S. et al. (2017). Expanding the feruloyl esterase gene family of *Aspergillus niger* by characterization of a feruloyl esterase, FaeC. *New Biotechnology* 37: 200–209.

134 Dilokpimol, A., Makela, M.R., Aguilar-Pontes, M.V. et al. (2016). Diversity of fungal feruloyl esterases: updated phylogenetic classification, properties, and industrial applications. *Biotechnology for Biofuels* 9 (1): 1–18.

135 Cantarel, B.L., Coutinho, P.M., Rancurel, C. et al. (2009). The carbohydrate-active ENZYMES database (CAZy): an expert resource for glycogenomics. *Nucleic Acids Research* 37 (1): D233–D238.

136 Mogodiniyai Kasmaei, K. and Sundh, J. (2019). Identification of novel putative bacterial feruloyl esterases from anaerobic ecosystems by use of whole-genome shotgun metagenomics and genome binning. *Frontiers in Microbiology* 10: 2673.

137 Urlacher, V.B. and Koschorreck, K. (2021). Pecularities and applications of aryl-alcohol oxidases from fungi. *Applied Microbiology and Biotechnology* 105 (10): 4111–4126.

138 Vina-Gonzalez, J. and Alcalde, M. (2020). Directed evolution of the aryl-alcohol oxidase: beyond the lab bench. *Computational and Structural Biotechnology Journal* 18: 1800–1810.

139 Serrano, A., Carro, J., and Martínez, A.T. (2020). Reaction mechanisms and applications of aryl-alcohol oxidase. *The Enzymes* 47: 167–192.

140 Sutzl, L., Foley, G., Gillam, E.M. et al. (2019). The GMC superfamily of oxidoreductases revisited: analysis and evolution of fungal GMC oxidoreductases. *Biotechnology for Biofuels* 12 (1): 1–18.

141 Hallberg, B.M., Leitner, C., Haltrich, D., and Divne, C. (2004). Crystal structure of the 270 kDa homotetrameric lignin-degrading enzyme pyranose 2-oxidase. *Journal of Molecular Biology* 341 (3): 781–796.

142 Herzog, P.L., Sutzl, L., Eisenhut, B. et al. (2019). Versatile oxidase and dehydrogenase activities of bacterial pyranose 2-oxidase facilitate redox cycling with manganese peroxidase in vitro. *Applied and Environmental Microbiology* 85 (13): e00390–e00319.

143 Abrera, A.T., Sutzl, L., and Haltrich, D. (2020). Pyranose oxidase: a versatile sugar oxidoreductase for bioelectrochemical applications. *Bioelectrochemistry* 132: 107409.

144 Sutzl, L., Laurent, C.V., Abrera, A.T. et al. (2018). Multiplicity of enzymatic functions in the CAZy AA3 family. *Applied Microbiology and Biotechnology* 102 (6): 2477–2492.

145 Ewing, T.A., Gygli, G., Fraaije, M.W., and van Berkel, W.J. (2020). Vanillyl alcohol oxidase. *The Enzymes* 47: 87–116.

146 Gygli, G., de Vries, R.P., and van Berkel, W.J. (2018). On the origin of vanillyl alcohol oxidases. *Fungal Genetics and Biology* 116: 24–32.

147 van den Heuvel, R.H., van den Berg, W.A., Rovida, S., and van Berkel, W.J. (2004). Laboratory-evolved vanillyl-alcohol oxidase produces natural vanillin. *Journal of Biological Chemistry* 279 (32): 33492–33500.

148 Gavande, P.V., Basak, A., Sen, S. et al. (2021). Functional characterization of thermotolerant microbial consortium for lignocellulolytic enzymes with central role of Firmicutes in rice straw depolymerization. *Scientific Reports* 11 (1): 1–13.

149 Majeke, B.M., Collard, F.X., Tyhoda, L., and Gorgens, J.F. (2021). The synergistic application of quinone reductase and lignin peroxidase for the deconstruction of industrial (technical) lignins and analysis of the degraded lignin products. *Bioresource Technology* 319: 124152.

150 Nigam, P.S. (2013). Microbial enzymes with special characteristics for biotechnological applications. *Biomolecules* 3 (3): 597–611.

151 Sakhuja, D., Ghai, H., Rathour, R.K. et al. (2021). Cost-effective production of biocatalysts using inexpensive plant biomass: a review. *3 Biotech* 11 (6): 1–21.

152 Bharathiraja, S., Suriya, J., Krishnan, M. et al. (2017). Production of enzymes from agricultural wastes and their potential industrial applications. *Advances in Food and Nutrition Research* 80: 125–148.

153 Shradhdha, S. and Murty, D.S. (2020). Production of lignolytic and cellulolytic enzymes by using basidiomycetes fungi in the solid-state fermentation of different agro-residues. *Research Journal of Biotechnology* 15 (9): 10–17.

154 Chukwuma, O.B., Rafatullah, M., Tajarudin, H.A., and Ismail, N. (2020). Lignocellulolytic enzymes in biotechnological and industrial processes: a review. *Sustainability* 12 (18): 7282.

155 Abu Yazid, N., Barrena, R., Komilis, D., and Sanchez, A. (2017). Solid-state fermentation as a novel paradigm for organic waste valorization: a review. *Sustainability* 9 (2): 224.

156 Kumla, J., Suwannarach, N., Sujarit, K. et al. (2020). Cultivation of mushrooms and their lignocellulolytic enzyme production through the utilization of agro-industrial waste. *Molecules* 25 (12): 2811.

157 Manan, M.A. and Webb, C. (2017). Design aspects of solid-state fermentation as applied to microbial bioprocessing. *Journal of Applied Biotechnology and Bioengineering* 4 (1): 511–532.

158 Gonzalez-Rodriguez, S., Lu-Chau, T.A., Trueba-Santiso, A. et al. (2022). Bundling the removal of emerging contaminants with the production of ligninolytic enzymes from residual streams. *Applied Microbiology and Biotechnology* 106 (3): 1299–1311.

159 Ergun, S.O. and Urek, R.O. (2017). Production of ligninolytic enzymes by solid state fermentation using *Pleurotus ostreatus*. *Annals of Agrarian Science* 15 (2): 273–277.

160 Montoya, S., Patino, A., and Sanchez, O.J. (2021). Production of lignocellulolytic enzymes and biomass of *Trametes versicolor* from agro-industrial residues in a novel fixed-bed bioreactor with natural convection and forced aeration at pilot scale. *Processes* 9 (2): 397.

161 Akpinar, M. and Urek, R.O. (2014). Extracellular ligninolytic enzymes production by *Pleurotus eryngii* on agro-industrial wastes. *Preparative Biochemistry and Biotechnology* 44 (8): 772–781.

162 Yasmeen, Q., Asgher, M., Sheikh, M.A., and Nawaz, H. (2013). Optimization of ligninolytic enzymes production through response surface methodology. *BioResources* 8 (1): 944–968.

163 Ayla, S., Golla, N., and Pallipati, S. (2018). Production of ligninolytic enzymes from *Penicillium* sp. and its efficiency to decolourise textile dyes. *The Open Biotechnology Journal* 12 (1): 112–122.

164 Schneider, W.D.H., Fontana, R.C. et al. (2018). High level production of laccases and peroxidases from the newly isolated white-rot basidiomycete *Marasmiellus palmivorus* VE111 in a stirred-tank bioreactor in response to different carbon and nitrogen sources. *Process Biochemistry* 69: 1–11.

165 Atilano-Camino, M.M., Alvarez-Valencia, L.H., Garcia-González, A., and García-Reyes, R.B. (2020). Improving laccase production from *Trametes versicolor* using lignocellulosic residues as co-substrates and evaluation of enzymes for blue wastewater biodegradation. *Journal of Environmental Management* 275: 111231.

166 Rodriguez Couto, S., Rodríguez, A., Paterson, R.R.M. et al. (2006). Laccase activity from the fungus *Trametes hirsuta* using an air-lift bioreactor. *Letters in Applied Microbiology* 42 (6): 612–616.

167 Padma, N. and Sudha, H. (2013). Optimization of lignin peroxidase, manganese peroxidase, and lac production from *Ganoderma lucidum* under solid state fermentation of pineapple leaf. *BioResources* 8 (1): 250–271.

168 Nayana, P., Aiswarya, C., Aparna, C.K., and Nambisan, P. (2020). Dataset on optimization of lignin peroxidase production by *Endomelanconiopsis* sp. under submerged fermentation using one factor at a time approach. *Data in Brief* 29: 105244.

169 Mehboob, N., Asad, M.J., Imran, M. et al. (2011). Production of lignin peroxidase by *Ganoderma leucidum* using solid state fermentation. *African Journal of Biotechnology* 10 (48): 9880–9887.

170 Beladhadi, R.V., Shankar, K., Jayalakshmi, S.K., and Sreeramulu, K. (2022). Production of cocktail of lignolytic, cellulolytic and hemicellulolytic enzymes by the novel bacterium *Burkholderia* sp SMB1 utilizing rice bran and straw: application in the saccharification of untreated agro-wastes for bioethanol production. *Waste and Biomass Valorization* 13 (3): 1565–1577.

171 Nathan, V.K., Kanthimathinathan, S.R., Rani, M.E. et al. (2018). Bio bleaching of waste paper using lignolytic enzyme from *Fusarium equiseti* VKF2: a mangrove isolate. *Cellulose* 25 (7): 4179–4192.

172 Angural, S., Kumar, A., Kumar, D. et al. (2020). Lignolytic and hemicellulolytic enzyme cocktail production from *Bacillus tequilensis* LXM 55 and its application in pulp biobleaching. *Bioprocess and Biosystems Engineering* 43 (12): 2219–2229.

173 Adak, A., Tiwari, R., Singh, S. et al. (2016). Laccase production by a novel white-rot fungus *Pseudolagarobasidium acaciicola* LA 1 through solid-state fermentation of parthenium biomass and its application in dyes decolorization. *Waste and Biomass Valorization* 7 (6): 1427–1435.

174 Gupta, A. and Jana, A.K. (2019). Production of laccase by repeated batch semi-solid fermentation using wheat straw as substrate and support for fungal growth. *Bioprocess and Biosystems Engineering* 42 (3): 499–512.

175 Handelsman, J. (2004). Metagenomics: application of genomics to uncultured microorganisms. *Microbiology and Molecular Biology Reviews* 68 (4): 669–685.

176 Indira, M., Venkateswarulu, T.C., Krupanidhi, S., and Peele, K.A. (2021). Microbial production of antimicrobial and anti cancerous biomolecules. *Bioprospecting of Microorganism-Based Industrial Molecules* 8: 147–169.

177 Le, T.T.H., Nguyen, T.B., Nguyen, H.D. et al. (2022). De novo metagenomic analysis of microbial community contributing in lignocellulose degradation in humus samples harvested from Cuc Phuong Tropical Forest in Vietnam. *Diversity* 14 (3): 220.

178 Ejigu, G.F. and Jung, J. (2020). Review on the computational genome annotation of sequences obtained by next-generation sequencing. *Biology* 9 (9): 295.

179 Gonçalves, C.C., Bruce, T., Silva, C.D.O.G. et al. (2020). Bioprospecting microbial diversity for lignin valorization: dry and wet screening methods. *Frontiers in Microbiology* 11: 1081.

180 Derbyshire, M.K., Lanczycki, C.J., Bryant, S.H., and Marchler-Bauer, A. (2012). Annotation of functional sites with the conserved domain database. *Database* 2012: bar058.

181 Levasseur, A., Drula, E., Lombard, V. et al. (2013). Expansion of the enzymatic repertoire of the CAZy database to integrate auxiliary redox enzymes. *Biotechnology for Biofuels* 6 (1): 1–14.

182 Fatma, S., Hameed, A., Noman, M. et al. (2018). Lignocellulosic biomass: a sustainable bioenergy source for the future. *Protein and Peptide Letters* 25 (2): 148–163.

183 Adewuyi, A. Underutilized lignocellulosic waste as sources of feedstock for biofuel production in developing countries. *Frontiers in Energy Research* 10: 741570.

184 Huang, J., Fu, S., and Gan, L. (2019). Structure and characteristics of lignin. *Lignin Chemistry and Applications* 2: 25–50.

185 Lu, Y., Lu, Y.C., Hu, H.Q. et al. (2017). Structural characterization of lignin and its degradation products with spectroscopic methods. *Journal of Spectroscopy* 2017: 8951658.

186 Isikgor, F.H. and Becer, C.R. (2015). Lignocellulosic biomass: a sustainable platform for the production of bio-based chemicals and polymers. *Polymer Chemistry* 6 (25): 4497–4559.

187 Agustin, M.B., de Carvalho, D.M., Lahtinen, M.H. et al. (2021). Laccase as a tool in building advanced lignin-based materials. *Chem Sus Chem* 14 (21): 4615–4635.

188 Jones, S.M. and Solomon, E.I. (2015). Electron transfer and reaction mechanism of laccases. *Cellular and Molecular Life Sciences* 72 (5): 869–883.

189 Zhu, Y., Zhang, Y., Zhan, J. et al. (2019). Axial bonds at the T1 Cu site of *Thermus thermophilus* SG 0.5 JP 17-16 laccase influence enzymatic properties. *FEBS Open Bio* 9 (5): 986–995.

190 Longe, L., Garnier, G., and Saito, K. (2016). Lignin biodegradation with fungi, bacteria and enzymes for producing chemicals and increasing process efficiency. In: *Production of Biofuels and Chemicals from Lignin*, vol. 6 (ed. Z. Fang and R.L. Smith Jr.), 147–179. Springer.

191 Agrawal, K., Chaturvedi, V., and Verma, P. (2018). Fungal laccase discovered but yet undiscovered. *Bioresources and Bioprocessing* 5 (1): 1–12.

9

Microbial Bioreactors for Biofuels

Paulo Renato Souza de Oliveira[1], Allana Katiussya Silva Pereira[1], Iara Nobre Carmona[1], and Ananias Francisco Dias Júnior[2]

[1] *Department of Forest Sciences, University of São Paulo – Luiz de Queiroz College of Agriculture, USP – ESALQ, Piracicaba, Sao Paulo, Brazil*
[2] *Department of Forestry and Wood Sciences, Federal University of Espírito Santo, UFES, Jerônimo Monteiro, Espírito Santo, Brazil*

9.1 Introduction

The search for alternatives to the continued use of fossil sources for fuel production is mandatory in order to mitigate adverse environmental and socio-economic consequences generated in recent decades. As part of the solution, investments to produce alternative fuels for the transportation sector from renewable raw materials occur [1], e.g. ethanol, bioethanol, and biodiesel. The intention is to reduce greenhouse gas emissions (GHGs), given the compatibility of these fuels with the current vehicle infrastructure. In addition, they can be minor components in fuel blends. Another alternative is gaseous fuels, such as methane or hydrogen. These fuels highlighted above aggregate microbial processes within their production chain. It uses a biochemical process that transforms a set of substrates into a product of interest through a population of living microorganisms within a closed system [2].

Bioreactors are examples of systems used in laboratory and industrial-scale processes. They are essential equipment for the production of energy inputs and bioproducts, as exemplified in Figure 9.1. It uses multidisciplinary knowledge to design a bioreactor and conduct the conversion processes involved to minimize costs and ensure that the product of interest is high quality [3].

Therefore, the successful development and operation of bioreactors at different scales is necessary to verify the feasibility of biofuel production. Thus, if it becomes possible to commercialize and structure an economy of these products, scaling up any technology must cross the boundaries of the laboratory to the proportions of the industrial sector [4]. The bioreactors used at each step vary in volume capacity and can be summarized below [4–6]. The process can start in a laboratory manner (with equipment with less than 2 L) and proceed to bench-scale or semi-pilot (capacities between 2 and 50 L). Subsequently, the pilot or

Microbial Bioreactors for Industrial Molecules, First Edition.
Edited by Sudhir P. Singh and Santosh Kumar Upadhyay.

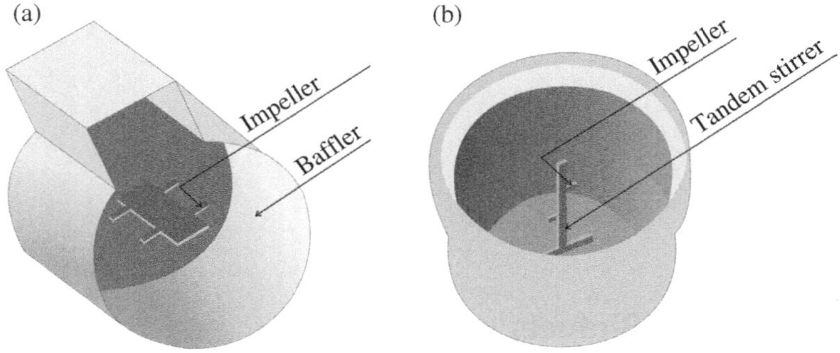

Figure 9.1 Schematic diagram from the horizontal (a) and vertical (b) mixed continuous reactors. *Source:* Brindhadevi et al. [7] / Elsevier.

demonstration scale is studied (50 L to ≤10,000 L), and finally industrial scale is reached (>10,000 L).

9.2 General Classification of Bioreactor

A bioprocess that employs microorganisms for biofuels involves the expansion of several cells, using the biomass and nutrients to form the desired products. Bioreactors are containers which require high control of environmental circumstances such as pH, dissolved oxygen, and temperature [8]. They achieve these conditions and can be classified according to the physical state of the substrate, type, and disposition of microorganism cells in the bioreactor and the feeding strategy.

9.3 Liquid-Phase Bioreactor

Cell growth and substrate metabolism occur submerged in a liquid environment.

9.3.1 Cell-Free

The cells used in the metabolism of substrates are accessible in the culture medium and are not removed from the final product until the recovery process.

9.3.1.1 Mechanically Stirred

Most industrial systems use cell-free suspension in submerged culture. In this case, the most used bioreactor, both in the laboratory and on a large scale, is the stirred tank reactor (STR), also known as a mixing tank. One of the biggest reasons to use STR is its effective mixing and lower resistance when transferring mass [9]. It is a cylindrical tank, normally made of stainless steel, carbon steel, or glass, in which the culture is stirred by the action of impellers fixed on a central axis, whose rotation is provided by a motor coupled to the axis and positioned above or below the tank. A recent study by Brindhadevi et al. [7]

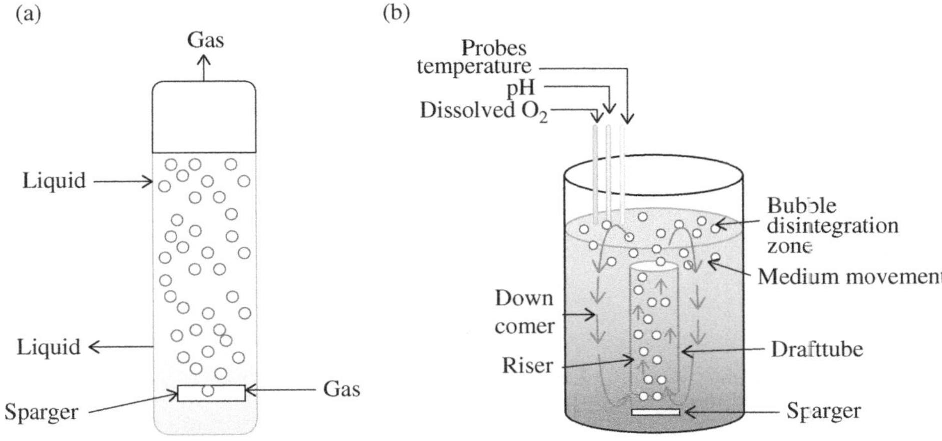

Figure 9.2 Schematic drawing of a bubble column (a) bioreactor and an airlift bioreactor (b) with riser and downcomer regions. *Source:* Sharma et al. [11] / Longdom Publishing. / CC BY 4.0.

demonstrated a method to predict the production of biogas and biohydrogen from horizontal (Figure 9.1a) and vertical (Figure 9.1b) continuous mixing reactors using the numerical method. The use of computational fluid dynamics to evaluate the efficiency of the bioreactor is important due to the uncertainty of the numerical results. Perpendicular reactors provide high productivity due to better velocity distribution in the tank. The vertical reactor also records more ethanol and acetate than the horizontal model. This proves a vertical continuous mixed tank reactor which is more promising and compelling.

9.3.1.2 Pneumatically Stirred

Pneumatically stirred bioreactors work without internal moving parts, promoting homogenization through the injection of gas and the consequent movement of the liquid. They have different applications based on displacement of oxygen from the gaseous condition to the liquid condition in enzymatic reactions, in aerobic and anaerobic cultures, where the inert gas phase is used for homogenization of the reaction medium [10]. Among the pneumatic reactors, there are bubble column types and airlift bioreactors, shown in Figure 9.2a and b, respectively. Airlift are characterized by a region sprayed with a gas called the riser region and a downcomer region, which are interconnected by the bottom region and the degassing zone located at the top of the reactor, where a total or partial detachment of the gas phase is retained in the liquid phase occurs [12].

9.3.2 Immobilized Cell

The cellular immobilization process consists of the physical confinement of cells in a particular region to preserve their catalytic activity [13]. In this way, cells that act as biocatalysts can simulate their natural growth on surfaces or within structures called supports. The ability of many microorganisms to adhere to different types of surfaces makes this process possible [14]. Immobilized cells can bring significant advantages over suspension cultures [15]. These include (i) high cell concentration, (ii) cell reuse and consequent cost reduction in cell

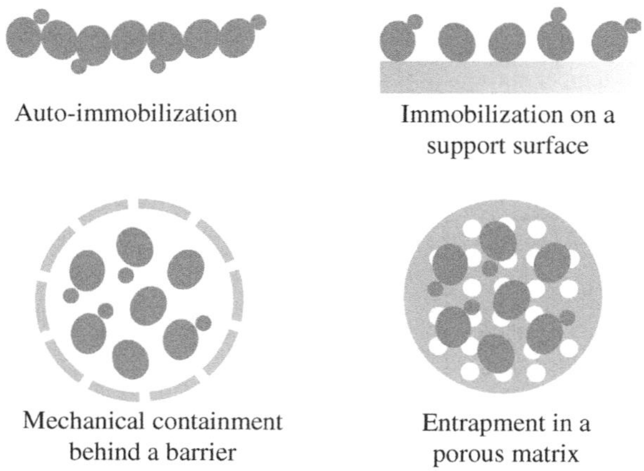

Figure 9.3 Basic methods of cell immobilization. *Source:* Moreno-García et al. [17] / Frontiers Media S.A / CC BY 4.0.

recovery and recycling processes, (iii) mitigation of problems derived from high dilution rates, (iv) increased volumetric productivity due to high cell concentration and flow rates, and (v) improved microenvironmental conditions of the cells. Furthermore, immobilization favors genetic stability in specific cases, improving biocatalysis performance [16] and protects against shear damage, an essential factor for some microorganisms (Figure 9.3).

In addition, there are essential assumptions for the support; among them are price, stability, reusability, and non-toxicity. For fermentative processes, there are still other factors that we should aim for, such as a simple immobilization process, with a high immobilization capacity and diffusion rate of the substance, and mechanical resistance of the support compatible with the continuous operation of an industrial installation [18]. Organic and soft supports are commonly used in fermentation to produce ethanol on a reduced scale. At the same time, inorganic and rigid materials can be more efficient in improving the productivity of this same process in larger volume bioreactors [19]. Recent research has also evaluated the use of 3D-printed matrices as support [20, 21].

The arrangement of the immobilized cells occurs in fixed bed support or a fluidized bed through circulation in the culture medium. Furthermore, the confinement of cells can occur between membranes, which would allow the passage of the components of the culture medium and the products and by-products of metabolism, but prevent the passage of cells. Slower systems, such as waste treatment systems, often benefit from this strategy. The advantages of membranes are more significant for sensitive organisms, as they minimize shear stresses [22].

9.4 Reactors for Solid-State Cultures

Another category of bioreactors is those for non-aqueous phase cultures, also known as solid-state fermentation (SSF). In this case, the substrate has no free water, and the cells are mixed with the substrate itself, which remains at a moisture content between 30 and

80% [22]. Recently, the major uses of SSF have been in the manufacture of enzymes, such as lipases, amylases, among others, in addition to the production of some secondary metabolites, pesticides, and biomass [23–25]. The choice of SSF as a form of cultivation is not only considered for qualitative and quantitative reasons but also favorable in economic and environmental terms. Therefore, SSF contributes to the aggregation of value of agricultural by-products as a solid medium for developing fungi [26].

The use of solid-state bioreactors gains importance for treating lignocellulosic biomass for bioethanol production, as it assists in the pretreatment through the degradation of lignin with the aid of lignin-degrading microorganisms, preliminary white rot fungi [27]. The anaerobic digestion processes, which generate gaseous biofuels, also use these bioreactors. In addition to being possible to combine it with hydrothermal carbonization, it can produce even more energy inputs [28]. Food fermentation is another example that has excellent potential for solid-state fermentation [29]. A plug-flow reactor was put into operation with positive results without the presence of an external inoculum. This research demonstrates the concept for further research on the effect of residence time, inoculum ratio, and correlations using mathematical modeling of a solid-phase fermentation system.

9.5 Bioreactor Operation Mode

Microbial bioreactors for biofuel production can operate in different ways. The inputs necessary for the development of cells may be available at the initial moment of the bioprocess, and it is called a batch or batch process (Figure 9.4a). In it, the inoculated cells also remain in the bioreactor from the beginning and, thus, can start their function as a biocatalyst. In this way, the substrate concentration gradually decreases as the cells grow and their metabolic development occurs, responsible for the effective biofuel production [22].

Alternatively, nutrient feeding can be performed during the cultivation/metabolization, called the fed-batch process (Figure 9.4b). Under these conditions, the product (broth) remains in the bioreactor for the entire course of microorganism/cell metabolism.

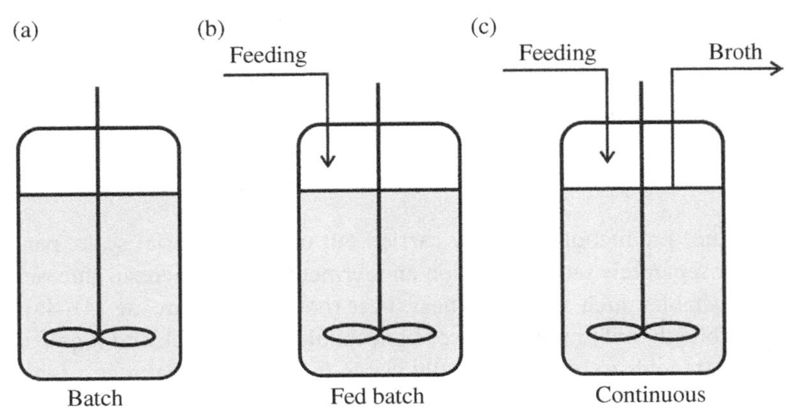

Figure 9.4 (a–c) Schematic drawing of the modes of operation of bioreactors. *Source:* Adapted from Canio et al. [30] / John Wiley & Sons.

Thus, sterilization conditions are maintained in the system until the end of the batch. In addition, the constant addition of nutrients allows the growth rate and prevents metabolic stagnation. However, batch feeding processes are labor-intensive because of the sterilization of the equipment after each batch and the accuracy of feeding nutrients to maintain particular growth characteristics [8], including biofuels. The most significant advantage is creating specific conditions for growth and generating an exponential growth curve, increasing productivity.

The last operation mode is continuous. The bioreactor receives feed continuously while residual nutrients are simultaneously removed (Figure 9.4c). The aim is to keep the volume of the bioreactor constant and help the cells to stay in a predefined growth phase that favors the production of a particular component [8]. Under these equilibrium conditions, the system reaches a steady state. However, continuous cultivation may provide contamination or mutation, resulting in a process failure. Overall, the continuous operation creates an ideal bioreactor environment for uniform biofuel production under optimal cell growth conditions [15].

9.6 Biofuels

The transport sector aggravates the widespread use of fossil fuels since it depends on raw materials and generates large amounts of polluting gases [31]. In the European Union, transport is responsible for around 21% of all greenhouse gas emissions that contribute to global warming [32]. Besides, the increase in population and globalization has been accompanied by an increase in energy and fuel consumption in critical industrial sectors, resulting in irreversible degradation of the environment and climate change [32, 33]. This scenario has emphasized the need to explore alternative fuels to replace fuels from nonrenewable sources [34, 35].

Renewables are proven technologies that are rapidly developing and have the potential for a zero carbon footprint [31], characteristics that make them suitable for tackling climate change, which is probably the most urgent challenge facing global society [36]. In this sense, biomass sources are the basis for research and development of efficient biofuels. Studies have been developed from the production of biodiesel from edible or inedible seed oils to the metabolic engineering of algae [37–40]. Below we will explore the main types of biofuel produced with the aid of bioreactors.

9.6.1 Bioethanol

First-generation bioethanol production is widely carried out on an industrial scale, performing simultaneous or separately saccharification and fermentation of sucrose, glucose, or fructose found in vegetables such as corn, wheat, beet roots, and sugarcane [41–43]. Second-generation (2G) bioethanol production uses lignocellulosic feedstock and requires high productivity and fuel yields to be economically viable for the industrial sector [44]. Lignocellulosic biomass is a promising raw material for 2G bioethanol production due to its abundance and low cost [41, 45]. Agricultural wastes come ahead of other biomasses as they do not compete with the food sector [46, 47] and become more coveted alternative sources. Third-generation (3G) bioethanol comes from algae biomass [48] and goes beyond the debate of competition between cultivated areas or food for fuel production [49, 50, 51]. In addition, 3G bioethanol can combine the use of marine biomass as a feedstock source, marine microorganisms for hydrolysis and fermentation, and the use of seawater in a much more environmentally friendly process [52]. All these different generation fuels are compatible with the world's transportation infrastructure and are a minor component of the final mix of other fuels. Recently their demand has increased [53] due to their use as a biofuel and as an active agent in asepsis and control of the new coronavirus [54].

The biochemical conversion pathway for bioethanol production can occur in steps that add up to biomass pretreatment, hydrolysis, and fermentation, with a subsequent distillation of the final product [55]. We can summarize that 1G bioethanol requires only the fermentation process of simple sugars. On the other hand, lignocellulosic and algae biomasses require prior treatment and hydrolysis of polymeric carbohydrates to have their fermentative sugars converted into 2G and 3G bioethanol, respectively [56].

The most common microorganisms used for lignocellulose hydrolysis are filamentous fungi (zygomycetes) [57]. They provide enzymes to degrade polymeric carbohydrates into less complex sugars. Different raw materials treated with filamentous fungi can result in biomass rich in protein, fat, and chitin/chitosan [43] with the potential for forming other bioproducts and bioenergy [58]. The yeasts and bacteria that stand out among the microorganisms studied for fermentation of monomeric sugars into bioethanol are *Saccharomyces cerevisiae* and *Zymomonas mobilis*, respectively [59]. We also find research with *Pichia stipites*, *Kluyveromyces fagilis*, and *Candida shehatae*, all yeasts [60]; and bacteria of the genera Aerobacter, Aeromonas, Bacillus, Bacteroides, Clostridium, Erwinia, Klebsiella, and Thermoanaerobacter [61, 62–64].

The microbial bioreactors act as essential equipment in the fermentation process of the simple sugars into these three different ethanol generations. In general, industries have experience in performing alcoholic fermentation in batches. Among the advantages provided is the knowledge acquired about this operation, besides the efficiency in ethanol production and lower risk of contamination [65]. However, the configuration the bioreactor acquires after each batch in sterilization and cleaning is usually an economic constraint. The fermentation also explores the fed-batch mode. It starts with a low substrate concentration, followed by a gradual increase of that substrate at a suitable interval to maintain the metabolic process of the cells. Then, the removal of the final product occurs at the end of each batch [66]. Finally, fermentation bioreactors can receive substrate continuously while the end product is concurrently removed [33]. This process has a higher risk of contamination than fed-batch, and the cultures change over the long term. The advantage is the economic gains since the continuous increase of substrate does not change the working volumes of the bioreactor.

For 3G bioethanol generation, there are more significant challenges when associated with algae cultivation. We can observe cultivation in open or closed systems, such as photobioreactors. In both cases, there are demands for light, CO_2, water, and organic nutrients to obtain the algae [67]. Photobioreactors are usually more expensive, but they obtain higher biomass yields and lower water costs [68]. Illuminated bubble columns, fat plate, tubular, and stirred tank bioreactors are already being explored at the lab scale. All with their own challenges and perspectives for scale-up to industrial scale [69, 70].

9.6.2 Biodiesel

Biodiesel is any biofuel equivalent to diesel derived from biological material. Biodiesel can be an excellent alternative to petroleum diesel for several reasons, as it is biodegradable, low toxicity [71, 72], lower combustion emissions resulting in lower in the release of carbon dioxide, aromatics, or other environmentally harmful chemicals [37, 73]. In addition, such characteristics make biodiesel provide improved combustion compared to petroleum-based diesel due to its high oxygen content, presenting a closed carbon cycle, which does not contribute to global warming [37].

Biodiesel is renewable feedstock fuel constituted of the monoalkyl esters of long-chain fatty acids [74]. Its production requires the transesterification and esterification of animal or vegetable fats and oils from alcohols such as methanol or ethanol. Briefly, fats containing three fatty acids and glycerol alcohol are submitted to the transesterification process under heat in acid and base catalyst [75]. The most used raw material is vegetables, divided into two categories of oils [72]. The first is edible, such as coconut oil, soybean, peanut, palm, and rapeseed. Second, nonedibles. They come from algae, Karanja, sea-mangoes, jatropha, and halophytes [72]. First-generation biodiesel is produced mainly from edible oils, about 95%. In this way, there are impacts on global food markets and food security for populations that use crops such as palm and soy for food [76]. Biodiesel production from inedible oilseeds has been extensively investigated in recent years. Residues from cooking oils, restaurant fat, and animal fats [77], like beef tallow and lard [78], have also been considered raw materials to produce second-generation (2G) biodiesel.

An alternative to the problem involving the main raw materials of biodiesel production and the limit of arable land is the use of oleaginous microorganisms [79]. Using these microorganisms is attractive because it reduces the need for arable soils and the cultivation period, promoting an increase in lipid production and presenting similarity when compared to the fatty acids of vegetable oils [80, 81]. Examples of these oleaginous microorganisms are microalgae, yeasts, and bacteria, living beings capable of accumulating high concentrations of lipids, greater than 20%, and metabolizing organic carbon sources [35, 79, 82]. Among them, microalgae attract attention for their greater photosynthetic efficiency, for producing more biomass, and for growing very quickly when compared to other energy crops [78]. *Chlorella, Neochloris, Nannochloropsis,* and *Scenedesmus* are the most common microalgae strains, which produce lipid content of 40–60% of their total cell dry weight [84–87].

Producing biodiesel is a significant obstacle to large-scale commercial use, mainly due to the high cost of feeding vegetable oils [88] and in relation to the aspects that are linked to the efficiency and sustainability of these first and 2G biodiesel raw materials [89]. On the other hand, third-generation (3G) biodiesel raw materials have microalgae as their main input and have emerged as one of the most promising alternative sources of lipids for biodiesel production. This is due to the fact that microalgae have high efficiency in photosynthesis, in the manufacture of biomass, and due to their higher growth rates [90, 91].

These algae are a large group of organisms that carry out photosynthesis produced through photobioreactors. As we saw earlier, a photobioreactor is a closed, lighted culture system designed to control biomass production. Different types of photobioreactors have been developed for algae production and have been described below. Regardless of the type of reactor, its development requires detailed knowledge of light distribution, mass transfer, shear stress, scalability, and algal cell biology [92].

1) The *vertical tubular outdoor bioreactor* has a large surface area of illumination, composed of transparent vertical tubes to allow light penetration and are subdivided into bubble column and airlift [92]. As a drawback of this bioreactor, we can cite the poor mass transfer [93, 94].
2) The *flat panel bioreactor* has a cuboidal shape with a minimal light path. Usually, manufacturing materials range from glass, plexiglass, and polycarbonate. This bioreactor has a high surface-to-volume ratio and open gas release systems [93]. We observe the stirring of the tanks pneumatically (bubble column) or mechanically.
3) The *horizontal tube bioreactor* comes in shapes ranging from a parallel set of tubes, loop, alpha, inclined tube, or horizontal. Its design provides high light conversion efficiency [92].
4) The *helical bioreactor* is a unique horizontal bioreactor because it presents a small-diameter flexible coiled transparent tube with a separate or attached degassing unit. A centrifugal pump drives the culture through a long line to the degassing team [84, 92]
5) The *stirred tank* is a conventional bioreactor, with mechanical agitation, transformed into a photobioreactor through external lighting provided by fluorescent lamps or optical fibers. We can see that the adaptation does not allow a high surface area for light contact and, consequently, affects the biomass productivity [92].

9.6.3 Butanol

It is a second-generation biofuel with attractive characteristics, including its high energy density, low volatility, and the possibility of being applied as a drop-in fuel without any changes in combustion engines and the supply infrastructure [95, 96]. In addition, butanol has been gaining interest among researchers in the automotive and internal combustion engine industries due to strict gaseous emission standards. The potential of butanol as a motor fuel was very attractive for its production [97], but it is an industrially important chemical product also in the application as essential element in the paint industry, paints, polymers, plastics, and solvents [98, 99].

The bioproduction of butanol has historically been carried out by the fermentation of acetone–butanol–ethanol, known as acetone, butanol, and ethanol (ABE) fermentation, using the *Clostridium* spp. [95, 100, 101]. In the last decades, many attempts have been made to improve the performance of ABE fermentation through solventogenic clostridia, that is through metabolic engineering [100, 102]. Research investigated the association of alterations in fermentation activities and the development of strains from the manipulation of microorganisms to obtain the levels, yield, and productivity of butanol required to meet the industry's competitiveness [101]. Research involving the co-cultivation of microorganisms has also been carried out as an alternative to overcome some challenges in the technical and economic aspects of the manufacture of biofuels. The cultivation of *Clostridium acetobutylicum* associated with *Trichoderma viride* was studied in order to produce butanol through ABE in wheat straw [103]. Another example is the co-cultivation of the fungus *S. cerevisiae* associated with *C. acetobutylicum* [104] and *Clostridium beijerinckii* [105].

In this context, clostridia are gram-positive, anaerobic bacteria, natural producers of a wide range of solvents, butanol, ethanol, acetone, isopropanol, 1,3-propanediol, 2,3-butanediol, and hexanol, of which the most important ones are acetone, butanol, and ethanol. Solventogenic clostridia such as *C. acetobutylicum* and *C. beijerinckii* can produce

ABE simultaneously at an approximately 3 : 6 : 1 ratio [100]. The primary substrates for the industrial fermentation of ABE come from crops, which are the basis of human food, such as corn and sugarcane, which may imply competition for the supply of these foods and, consequently, cause an increase in food prices [101], a problem previously reported in the section on biodiesel.

The fluidized or fixed bed bioreactors can operate in different ways in ABE production. However, the productivity of batch processes is low because bacterial growth is rapidly inhibited, and continuous fermentation has a low growth rate [106], leading to cell elimination at high dilution rates. Thus, continuous fermentation with immobilized cells can improve metabolic activity and ABE production while decreasing cell loss from the bioreactor [107].

In addition, techno-economic analysis of n-butanol production from lignocellulosic biomass hydrolysates shows the potential for industrial purposes [108]. Engineered *Clostridium tyrobutyricum* was immobilized in a fibrous-bed bioreactor, characterized by a glass column surrounded by a spiral fibrous matrix connected to a stirred tank reactor, and proved stable for long-term operation [108]. This result indicates that the fermentation of different lignocellulosic inputs can directly contribute to the manufacture of biofuels from low-cost renewable raw materials [108].

9.6.4 Biogas and Methane

The use of bioreactors in biogas production occurs through anaerobic digestion of biomass. Microorganisms metabolize different substrates, whether animal, vegetable, residual, or not. The product generated is composed mainly of methane and carbon dioxide [109]. Chemical reactions occur at different stages in biogas production. It starts with hydrolysis, acidification, acetate production, and continues to the generation of methane itself [110]. Organisms capable of producing significant amounts of methane are called methanogens. They are unique in terms of metabolism and have a high diversity [111]. We know that the microbial community acting in methanogenesis strongly depends on the substrate type. Karakashev et al. [112] indicated that *Methanosaetaceae* are usually the majority in digesters treating sludge, while digesters are operating with solid waste host more representatives of *Methanosaetaceae*.

Biogas brings environmental advantages due to its renewable character and lower emission of GHGs. The methane present in biogas contains four hydrogen atoms and only one carbon atom, so the gases emitted during combustion are mostly water vapor [113]. The general applications of biogas vary according to the system used and are summarized below [114]. Its use can generate electrical energy through internal combustion engines or gas turbine; generate heat via direct combustion in boilers; or even generate both through Otto gas engines, pilot injection gas engines, and Sterling engines. Biogas enhancement is also an application route and can produce biomethane or hydrogen. Both serve as biofuel for vehicles. In addition, biogas can supply fuel cells, which is a device that converts chemical energy directly into electricity and heat [115]. There is the feasibility of practical application with a single-chamber air-cathode microbial fuel cell integrated into an anaerobic membrane bioreactor system to improve methane production [116]. However, the number of plants based on fuel cells (most pilots) that generate electricity from biogas is incipient [114].

Recently evaluated on the pilot scale, a bioreactor design for methane production/optimization shows efficiency for recovering biomethane from organic wastes. The best performance in biomethane production was when the substrates were co-digested, pretreated, or with the presence of inoculum [117]. We can also find anaerobic bioreactors used for research on treating domestic and industrial waste in landfills, favoring the large-scale implementation of efficient systems for methane production [118].

Systems with more than one phase are standard for producing methane. In the studies by Jung et al. [119], food waste treatment occurs in a two-stage dynamic membrane bioreactor, where there was the presence of a continuously stirred tank bioreactor with a submerged immobilized cells. It was possible to observe the recovery of chemical energy from food wastes close to 80%. Another example of an integrated system is a two-stage up-flow anaerobic bioreactor with an immobilized fixed cell [120]. The researchers evaluated the efficiency and indicated a methane production rate of 15.63 L CH_4 d^{-1} in a short hydraulic retention time (one day).

9.6.5 Hydrogen

Hydrogen is not a resource per se. It is understood as an energy carrier that must be manufactured or derived from a primary energy resource [121] and can be used in land and sea transport [122]. Together with fuel cell technology, it is possible to take advantage of hydrogen in applications that include the rail sector in medium and heavy trucks. This alternative fuel promises to deal with crucial energy challenges and decarbonize potential energy systems [122]. Currently, molecular hydrogen is produced mainly by nonrenewable materials (fossil fuels), but the production of hydrogen from biomass by biological means is an emerging technology because it is sustainable and ecologically correct [123, 124]. Biohydrogen is considered a fuel with good prospects due to its powerful energy explosion and attractive ecological characteristics since the generation of biohydrogen does not use fossil fuels as raw material [3].

Recent studies have shown that among the different existing methodologies for hydrogen production, dark fermentation has gained prominence and features *Clostridium* species as protagonists [125]. It is a method by which anaerobic microorganisms use carbohydrate-rich substrates as raw material for hydrogen production [125]. In general, *Clostridium* spp., *Enterobacter* spp., *Bacillus* spp., *Escherichia coli*, and *Klebsiella* spp. are among the main producing microorganisms [126].

Biohydrogen production from microbial processes has received much attention due to its potential for clean, renewable, inexhaustible, and low-cost energy [124]. There is a diversity of research in the literature involving the production of hydrogen through residual waters and food or agricultural waste [3, 127]. Some studies are exploring the potential for biohydrogen production in horizontal and vertical continuous stirred tank reactors [3], in constant bioreactors such as continuously stirred tank reactors [128], and an up-flow anaerobic sludge blanket reactor [129].

In recent years, reactors with microbial electrolysis cells (MECs) have been investigated as a new technique to produce hydrogen from a wide variety of substrates. Electrically powered MECs enable the production of hydrogen through the conversion of a variety of organic substrates [123, 130]. MECs are capable of over 90% efficient hydrogen production [131]. However, its performance is directly linked to aspects such as the type of microorganism, type of membrane used, applied potential range, substrate composition and concentration, and MEC design [123]. Also, there are recent papers on the improvement of biohydrogen production with the aid of a packaged filter bioreactor on a laboratory scale [132]. The equipment used was made of transparent glass with a working volume of 300 mL and equipped with a magnetic stirrer. The research used sulfite-rich organic effluent from a beverage manufacturer's washing process. The results indicated that the process is stable and that the bioreactor associated with seed sludge in dark fermentation is capable of degrading the organic matter of these effluents with a short hydraulic retention time and a high rate of organic load [132].

9.6.6 Biohythane

The blend of hydrogen and methane can supply biohythane, a gaseous mixture. This product is interesting due to its environmental advantages, with lower emission of greenhouse gases (CO, CO_2, and NO_x), in addition to having high thermal efficiency [127, 133]. The hythane production chain is linked to the generation of fuels for vehicles. For its production, the independent obtainment of hydrogen and methane can come from fossil sources, which makes the process environmentally and economically unsustainable. In addition, the biological production of biohythane from renewable biomass can occur via fermentation, using bacteria such as the *Labrys portucalensis* group, bringing environmental and sustainability advantages [134].

A recent study by Chu et al. [135] used two continuously stirred anaerobic bioreactors to manufacture two-stage biohythane gaseous fuel under mesophilic conditions (Figure 9.5). This system of anaerobic bioreactors could be a very promising discovery to achieve environmental sustainability goals, in an economically and socially acceptable way, in developing

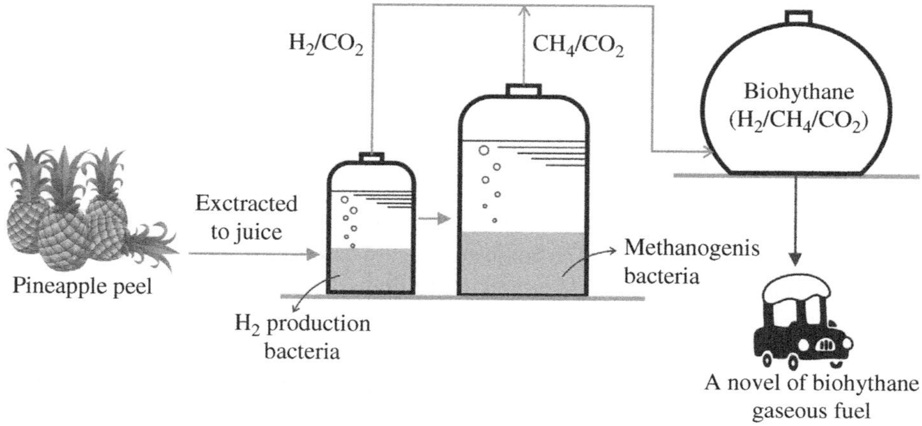

Figure 9.5 Production of biohythane from pineapple peels in continuous two-stage anaerobic reactors. *Source:* Chu et al. [135] / Elsevier.

countries. We cite other research focusing on the production of biohythane through two stages involving anaerobic digestion, using different forms of pretreatment of residual biomass for hydrolysis, such as acid-thermal pretreatment [136] and thermophilic fermentation [137].

This type of two-stage system could be a cost-effective way to generate a new gaseous automotive fuel from agro-industrial waste. However, it is still necessary to adapt the distribution system to make possible the automotive transition from nonrenewable to renewable fuel. In that way, more considerable investments in optimizing the biohythane distribution are required for vehicles [133].

9.7 Considerations and Future Perspectives

The diversity of bioreactors for biofuels is associated with the desired end product so that the microorganisms used, forms of operation, and kinetics of the bioprocess are taken into consideration. Thus, advances in bioreactor development are associated with biotechnology and process engineering. At the same time, implementing these bioreactors on a pilot and industrial scales must consider technical and economic aspects. There are successful examples of energy feedstocks, especially first-generation biofuels, produced on these scales. In addition, there is the growth potential for 2 and 3G biofuels given the need to mitigate the environmental impact of fossil fuels and to promote competition with food production.

Generally, we know that the production of biofuels helps in global energy security and is among the strategies of various governments to reduce their emissions of GHGs, without harming their production and consumption activities. Therefore, economic feasibility studies must be aligned with technical issues when establishing the best biomass conversion processes into biofuels within the regional and global reality. Integrating bioreactors with other biomass conversion processes and expanding the benefits generated is also possible. In addition, government incentives can favor the implementation of biofuel production plants and improve the infrastructure necessary for distribution and use by final consumers.

In conclusion, biofuel technology is moving ever closer to the frontier of knowledge. Technical problems are overcome, creating a cleaner, more renewable, and efficient energy base. Thus, government policies and preferences play an essential role in building the long-term future of this energy, and it is up to them to favor changes in the generation system and ensure the fundamental elements of biofuel supply.

References

1 Handler, R.M., Shonnard, D.R., Griffing, E.M. et al. (2016). Life cycle assessments of ethanol production via gas fermentation: anticipated greenhouse gas emissions for cellulosic and waste gas feedstocks. *Ind. Eng. Chem. Res.* 55 (12): 3253–3261.

2 Papagianni, M. (2017). Microbial bioprocesses. In: *Current Developments in Biotechnology and Bioengineering* (ed. C. Larroche, M. Sanroman, G. Du, and A. Pandey), 45–72. Elsevier.

3 Cooney, C.L. (1983). Bioreactors: design and operation. *Science (80-.)* 219 (4585): 728–733.

4 Ghosh, S. (2022). Assessment and update of status of pilot scale fermentative biohydrogen production with focus on candidate bioprocesses and decisive key parameters. *Int. J. Hydrog. Energy* 47 (39): 17161–17183.

5 Wynn, J.P., Hanchar, R., Kleff, S. et al. (2016). Biobased technology commercialization: the importance of lab to pilot scale-up. In: *Metabolic Engineering for Bioprocess Commercialization* (ed. S. Van Dien), 101–119. Cham: Springer International Publishing.

6 Crater, J.S. and Lievense, J.C. (2018). Scale-up of industrial microbial processes. *FEMS Microbiol. Lett.* 365 (13): 1–5.

7 Brindhadevi, K., Shanmuganathan, R., Pugazhendhi, A. et al. (2021). Biohydrogen production using horizontal and vertical continuous stirred tank reactor- a numerical optimization. *Int. J. Hydrog. Energy* 46 (20): 11305–11312.

8 Srivastava, A.K. and Gupta, S. (2011). *Fed-Batch Fermentation – Design Strategies.* Elsevier B.V.

9 Niño-Navarro, C., Chairez, I., Torres-Bustillos, L. et al. (2016). Effects of fluid dynamics on enhanced biohydrogen production in a pilot stirred tank reactor: CFD simulation and experimental studies. *Int. J. Hydrog. Energy* 41 (33): 14630–14640.

10 Kouzbour, S., Stiriba, Y., Gourich, B., and Vial, C. (2019). CFD simulation and analysis of reactive flow for dissolved manganese removal from drinking water by aeration process using an airlift reactor. *J. Water Process Eng.* 36: 101352.

11 Sharma, M., Thukral, N., Soni, N.K., and Maji, S. (2015). Microalgae as future fuel: real opportunities and challenges. *J. Thermodyn. Catal.* 6 (1): 1–11.

12 de Oliveira, D.G. (2021). Adaptação do biorreator do tipo airlift para reação enzimática empregando lipase imobilizada® na síntese de lipídeos estruturados do tipo MLM.

13 Karel, S.F., Libicki, S.B., and Robertson, C.R. (1985). The immobilization of whole cells: engineering principles. *Chem. Eng. Sci.* 40 (8): 1321–1354.

14 Kourkoutas, Y., Bekatorou, A., Banat, I.M. et al. (2004). Immobilization technologies and support materials suitable in alcohol beverages production: a review. *Food Microbiol.* 21 (4): 377–397.

15 Shuler, M.L. and Kargi, F. (2001). *Bioprocess Engineering: Basic Concepts.* New Jersey: Prentice Hall.

16 Shetty, V.K., Namitha, L., Rao, S.N., and Narayani, M. (2012). Experimental investigation and artificial neural network-based modeling of batch reduction of hexavalent chromium by immobilized cells of newly isolated strain of chromium-resistant bacteria. *Water Air Soil Pollut.* 223 (4): 1877–1893.

17 Moreno-García, J., García-Martínez, T., Mauricio, J.C., and Moreno, J. (2018). Yeast immobilization systems for alcoholic wine fermentations: actual trends and future perspectives. *Front. Microbiol.* 9: 241.

18 Zhu, Y. (2007). *Immobilized Cell Fermentation for Production of Chemicals and Fuels.* Elsevier B.V.

19 Boura, K., Dima, A., Nigam, P.S. et al. (2022). A critical review for advances on industrialization of immobilized cell bioreactors: economic evaluation on cellulose hydrolysis for PHB production. *Bioresour. Technol.* 349 (January): 126757.

20 dos Belgrano, F., Diegel, O., Pereira, N., and Hatti-Kaul, R. (2018). Cell immobilization on 3D-printed matrices: a model study on propionic acid fermentation. *Bioresour. Technol.* 249: 777–782.

21 Chacón, S.J., Matias, G., Ezeji, T.C. et al. (2021). Three-stage repeated-batch immobilized cell fermentation to produce butanol from non-detoxified sugarcane bagasse hemicellulose hydrolysates. *Bioresour. Technol.* 321: 124504.

22 Tonso, A., Badino Junior, A.C., and Schmidell, W. (2021). Tipos de biorreatores e formas de operação. In: *Biotecnologia industrial: engenharia bioquímica*, 2e, 748. Sao Paulo: Edgard Blucher.

23 De la Cruz Quiroz, R., Roussos, S., Hernández, D. et al. (2015). Challenges and opportunities of the bio-pesticides production by solid-state fermentation: filamentous fungi as a model. *Crit. Rev. Biotechnol.* 35 (3): 326–333.

24 Sarhy-Bagnon, V., Lozano, P., Saucedo-Castañeda, G., and Roussos, S. (2000). Production of 6-pentyl-α-pyrone by Trichoderma harzianum in liquid and solid state cultures. *Process Biochem.* 36 (1–2): 103–109.

25 Gutiérrez-Sánchez, G., Roussos, S., and Augur, C. (2013). Effect of caffeine concentration on biomass production, caffeine degradation, and morphology of *Aspergillus tamarii*. *Folia Microbiol. (Praha).* 58 (3): 195–200.

26 Carboué, Q., Claeys-Bruno, M., Bombarda, I. et al. (2018). Experimental design and solid state fermentation: a holistic approach to improve cultural medium for the production of fungal secondary metabolites. *Chemom. Intell. Lab. Syst.* 176: 101–107.

27 Wan, C. and Li, Y. (2012). Fungal pretreatment of lignocellulosic biomass. *Biotechnol. Adv.* 30 (6): 1447–1457.

28 Reza, M.T., Werner, M., Pohl, M., and Mumme, J. (2014). Evaluation of integrated anaerobic digestion and hydrothermal carbonization for bioenergy production. *J. Vis. Exp.* 88 (88): 51734.

29 Wang, R., Gmoser, R., Taherzadeh, M.J., and Lennartsson, P.R. (2021). Solid-state fermentation of stale bread by an edible fungus in a semi-continuous plug-flow bioreactor. *Biochem. Eng. J.* 169 (January): 107959.

30 Canio, D., Bari, D., and Patrizi, R. (2011). Latest frontiers in the biotechnologies for ethanol production from lignocellulosic biomass. In: *Biofuel Production-Recent Developments and Prospects* (ed. M.A.D. Santos). InTech.

31 García-Olivares, A., Solé, J., and Osychenko, O. (2018). Transportation in a 100% renewable energy system. *Energy Convers. Manag.* 158: 266–285.

32 Darda, S., Papalas, T., and Zabaniotou, A. (2019). Biofuels journey in Europe: currently the way to low carbon economy sustainability is still a challenge. *J. Clean. Prod.* 208: 575–588.

33 Afedzi, A.E.K., Rattanaporn, K., and Parakulsuksatid, P. (2022). Impeller selection for mixing high-solids lignocellulosic biomass in stirred tank bioreactor for ethanol production. *Bioresour. Technol. Rep.* 17: 100935.

34 Jambulingam, R., Shalma, M., and Shankar, V. (2019). Biodiesel production using lipase immobilised functionalized magnetic nanocatalyst from oleaginous fungal lipid. *J. Clean. Prod.* 215: 245–258.

35 Kumar, D., Das, T., Giri, B.S., and Verma, B. (2020). Optimization of biodiesel synthesis from nonedible oil using immobilized bio-support catalysts in jacketed packed bed bioreactor by response surface methodology. *J. Clean. Prod.* 244: 118700.

36 Sgouridis, S., Csala, D., and Bardi, U. (2016). The sower's way: quantifying the narrowing net-energy pathways to a global energy transition. *Environ. Res. Lett.* 11 (9): 094009.

37 Ahmad, A.L., Yasin, N.H.M., Derek, C.J.C., and Lim, J.K. (2011). Microalgae as a sustainable energy source for biodiesel production: a review. *Renew. Sust. Energ. Rev.* 15 (1): 584–593.

38 Dutta, K., Daverey, A., and Lin, J.G. (2014). Evolution retrospective for alternative fuels: first to fourth generation. *Renew. Energy* 69: 114–122.

39 Sims, R.E.H., Mabee, W., Saddler, J.N., and Taylor, M. (2010). An overview of second generation biofuel technologies. *Bioresour. Technol.* 101 (6): 1570–1580.

40 Singh, J. and Gu, S. (2010). Commercialization potential of microalgae for biofuels production. *Renew. Sust. Energ. Rev.* 14: 2596–2610.

41 Ferreira, J.A., Brancoli, P., Agnihotri, S. et al. (2018). A review of integration strategies of lignocelluloses and other wastes in 1st generation bioethanol processes. *Process Biochem.* 75: 173–186.

42 Ferreira, J.A. and Taherzadeh, M.J. (2020). Improving the economy of lignocellulose-based biorefineries with organosolv pretreatment. *Bioresour. Technol.* 299: 122695.

43 Devos, R.J.B. and Colla, L.M. (2022). Simultaneous saccharification and fermentation to obtain bioethanol: a bibliometric and systematic study. In: *Bioresour. Technol. Reports*, vol. 17, 100924.

44 Fu, N., Peiris, P., Markham, J., and Bavor, J. (2009). A novel co-culture process with *Zymomonas mobilis* and *Pichia stipitis* for efficient ethanol production on glucose/xylose mixtures. *Enzym. Microb. Technol.* 45 (3): 210–217.

45 Rosales-Calderon, O. and Arantes, V. (2019). A review on commercial-scale high-value products that can be produced alongside cellulosic ethanol. *Biotechnol. Biofuels* 12 (1): 1–58.

46 Lennartsson, P.R., Erlandsson, P., and Taherzadeh, M.J. (2014). Integration of the first and second generation bioethanol processes and the importance of by-products. *Bioresour. Technol.* 165 (C): 3–8.

47 Bulkan, G., Ferreira, J.A., and Taherzadeh, M.J. (2021). Retrofitting analysis of a biorefinery: integration of 1st and 2nd generation ethanol through organosolv pretreatment of oat husks and fungal cultivation. *Bioresour. Technol. Rep.* 15 (July): 100762.

48 Goh, C.S. and Lee, K.T. (2010). A visionary and conceptual macroalgae-based third-generation bioethanol (TGB) biorefinery in Sabah, Malaysia as an underlay for renewable and sustainable development. *Renew. Sust. Energ. Rev.* 14 (2): 842–848.

49 Adams, J.M., Gallagher, J.A., and Donnison, I.S. (2009). Fermentation study on *Saccharina latissima* for bioethanol production considering variable pre-treatments. *J. Appl. Phycol.* 21 (5): 569–574.

50 Singh, S.P. and Upadhyay, S.K. (eds.) (2021). *Bioprospecting of Microorganism-Based Industrial Molecules.* John Wiley & Sons Ltd. doi:10.1002/9781119717317.

51 Shintani, T., Upadhyay, S.K., Singh, S.P. (2021). An introduction to microbial biodiversity and bioprospection. In: *Bioprospecting of Microorganism-Based Industrial Molecules* (ed. S.P. Singh and S.K. Upadhyay). John Wiley & Sons Ltd. https://doi.org/10.1002/9781119717317.ch1.

52 Greetham, D., Zaky, A., Makanjuola, O., and Du, C. (2018). A brief review on bioethanol production using marine biomass, marine microorganism and seawater. *Curr. Opin. Green Sustain. Chem.* 14: 53–59.

53 International Energy Agency (2020). Global Energy Review 2020 the impacts of the Covid-19 crisis on global energy demand and CO2 emissions.

54 Kelly, S., and Weinraub, M. (2020). Ethanol makers see demand surge on hand sanitizer stockpiling. *Reuters.*

55 Balat, M. (2011). Production of bioethanol from lignocellulosic materials via the biochemical pathway: a review. *Energy Convers. Manag.* 52 (2): 858–875.

56 Melendez, J.R., Mátyás, B., Hena, S. et al. (2022). Perspectives in the production of bioethanol: a review of sustainable methods, technologies, and bioprocesses. *Renew. Sust. Energ. Rev.* 160 (January): https://doi.org/10.1016/j.rser.2022.112260.

57 Parachin, N.S., Bergdahl, B., van Niel, E.W.J., and Gorwa-Grauslund, M.F. (2011). Kinetic modelling reveals current limitations in the production of ethanol from xylose by recombinant Saccharomyces cerevisiae. *Metab. Eng.* 13 (5): 508–517.

58 Ferreira, J.A., Mahboubi, A., Lennartsson, P.R., and Taherzadeh, M.J. (2016). Waste biorefineries using filamentous ascomycetes fungi: present status and future prospects. *Bioresour. Technol.* 215: 334–345.

59 Szambelan, K., Szwengiel, A., Nowak, J. et al. (2022). Low-waste technology for the production of bioethanol from sorghum grain: comparison of *Zymomonas mobilis* and *Saccharomyces cerevisiae* in fermentation with stillage reusing. *J. Clean. Prod.* 352: 131607.

60 Sharma, B., Larroche, C., and Dussap, C.G. (2020). Comprehensive assessment of 2G bioethanol production. *Bioresour. Technol.* 313 (May): 123630.

61 Komesu, A., Oliveira, J., Neto, J.M. et al. (2020). Xylose fermentation to bioethanol production using genetic engineering microorganisms. *Genet. Metab. Eng. Improv. Biofuel Prod. Lignocellul. Biomass*: 143–154.

62 Joshi, N., Kaushal, G., Singh, S.P. (2021). Biochemical characterization of a novel thermo-halo-tolerant GH5 endoglucanase from a thermal spring metagenome. *Biotechnology and Bioengineering* 118: 1531–1544. https://doi.org/10.1002/bit.27668.

63 Joshi, N., Sharma, M., Singh, S.P. (2020). Characterization of a novel xylanase from an extreme temperature hot spring metagenome for xylooligosaccharide production. *Applied Microbiology and Biotechnology* 104: 4889–4901. https://doi.org/10.1007/s00253-020-10562-7.

64 Kaushal, G., Rai, A.K., Singh, S.P. (2021). A novel β-glucosidase from a hot-*spring* metagenome shows elevated thermal stability and tolerance to glucose and ethanol. *Enzyme and Microbial Technology* 145: 109764. doi: 10.1016/j.enzmictec.2021.109764.

65 Yu, L.P., Wu, F.Q., and Chen, G.Q. (2019). Next-generation industrial biotechnology-transforming the current industrial biotechnology into competitive processes. *Biotechnol. J.* 14 (9): 1800437.

66 Habicher, T., John, A., Scholl, N. et al. (2019). Introducing substrate limitations to overcome catabolite repression in a protease producing Bacillus licheniformis strain using membrane-based fed-batch shake flasks. *Biotechnol. Bioeng.* 116 (6): 1326–1340.

67 Sirajunnisa, A.R. and Surendhiran, D. (2016). Algae – A quintessential and positive resource of bioethanol production: a comprehensive review. *Renew. Sust. Energ. Rev.* 66: 248–267.

68 Wang, B., Lan, C.Q., and Horsman, M. (2012). Closed photobioreactors for production of microalgal biomasses. *Biotechnol. Adv.* 30 (4): 904–912.

69 Benner, P., Meier, L., Pfeffer, A. et al. (2022). Lab-scale photobioreactor systems: principles, applications, and scalability. *Bioprocess Biosyst. Eng.* 45 (5): 791–813.

70 Elkatory, M.R., Hassaan, M.A., and El Nemr, A. (2022). Algal biomass for bioethanol and biobutanol production. In: *Handbook of Algal Biofuels* (ed. M. El-Sheekh and A.E.-F. Abomohra), 251–279. Elsevier.

71 Cadenas, A. and Cabezudo, S. (1998). Biofuels as sustainable technologies: perspectives for less developed countries. *Technol. Forecast. Soc. Chang.* 58: 83–103.

72 Suzihaque, M.U.H., Alwi, H., Kalthum Ibrahim, U. et al. (2022). Biodiesel production from waste cooking oil: a brief review. *Mater. Today Proc.* 63: S490–S495.

73 Khan, S.A., Rashmi Hussain, M.Z., Prasad, S., and Banerjee, U.C. (2009). Prospects of biodiesel production from microalgae in India. *Renew. Sust. Energ. Rev.* 13 (9): 2361–2372.

74 Sytar, O. and Prasad, M.N.V. (2016). Production of biodiesel feedstock from the trace element contaminated lands in Ukraine. In: *Bioremediation and Bioeconomy* (ed. M.N.V. Prasad), 3–28. Elsevier.

75 Brito Cruz, C.H., Mendes Souza, G., and Barbosa Cortez, L.A. (2013). Biofuels for transport. In: *Futur. Energy Improv. Sustain. Clean Options our Planet* (ed. T. Letcher), 215–244. Elsevier.

76 Gui, M.M., Lee, K.T., and Bhatia, S. (2008). Feasibility of edible oil vs. non-edible oil vs. waste edible oil as biodiesel feedstock. *Energy* 33 (11): 1646–1653.

77 Ma, F. and Hanna, M.A. (1999). Biodiesel production: a review. *Bioresour. Technol.* 70 (1): 1–15.

78 Canakci, M. (2007). The potential of restaurant waste lipids as biodiesel feedstocks. *Bioresour. Technol.* 98 (1): 183–190.

79 Cho, H.U. and Park, J.M. (2018). Biodiesel production by various oleaginous microorganisms from organic wastes. *Bioresour. Technol.* 256: 502–508.

80 Beopoulos, A. and Nicaud, J.-M. (2012). Yeast: a new oil producer? *Corps Gras Lipides* 19 (1): 22–28.

81 Ratledge, C. (2004). Fatty acid biosynthesis in microorganisms being used for single cell oil production. *Biochimie* 86 (11): 807–815.

82 Chisti, Y. (2007). Biodiesel from microalgae. *Biotechnol. Adv.* 25 (3): 294–306.

83 Huang, G.H., Chen, F., Wei, D. et al. (2010). Biodiesel production by microalgal biotechnology. *Appl. Energy* 87 (1): 38–46.

84 Zhang, X. (2015). *Microalgae Removal of CO2 from Flue Gas.* London: IEA Clean Coal Centre.

85 Ma, X.N., Chen, T.P., Yang, B. et al. (2016). Lipid production from nannochloropsis. *Mar. Drugs* 14 (4): 61.

86 Altunoz, M., Montevecchi, G., Masino, F. et al. (2021). Biodiesel properties of *Neochloris oleoabundans* grown in sludge waste. *Clean. Eng. Technol.* 5: 100295.

87 Trivedi, J., Atray, N., Agrawal, D., and Ray, A. (2022). Enhanced lipid production in *Scenedesmus obliquus* via nitrogen starvation in a two-stage cultivation process andevaluation for biodiesel production. *Fuel* 316: 123418.

88 Lang, X., Dalai, A.K., Bakhshi, N.N. et al. (2001). Preparation and characterization of bio-diesels from various bio-oils. *Bioresour. Technol.* 80 (1): 53–62.

89 Patil, V., Tran, K.-Q., and Giselrød, H.R. (2008). Towards sustainable production of biofuels from microalgae. *Int. J. Mol. Sci.* 9: 1188–1195.

90 Enamala, M.K., Enamala, S., Chavali, M. et al. (2018). Production of biofuels from microalgae – a review on cultivation, harvesting, lipid extraction, and numerous applications of microalgae. *Renew. Sust. Energ. Rev.* 94: 49–68.

91 Minowa, T., Yokoyama, S., and ya, Kishimoto, M., and Okakura, T. (1995). Oil production from algal cells of *Dunaliella tertiolecta* by direct thermochemical liquefaction. *Fuel* 74 (12): 1735–1738.

92 Singh, R.N. and Sharma, S. (2012). Development of suitable photobioreactor for algae production – a review. *Renew. Sust. Energ. Rev.* 16 (4): 2347–2353.

93 Dange, P., Gawas, S., Pandit, S. et al. (2022). Trends in photobioreactor technology for microalgal biomass production along with wastewater treatment: bottlenecks and breakthroughs. *An Integr. Phycoremed. Process. Wastewater Treat.* 135–154. https://doi.org/10.1016/B978-0-12-823499-0.00001-8.

94 Narala, R.R., Garg, S., Sharma, K.K. et al. (2016). Comparison of microalgae cultivation in photobioreactor, open raceway pond, and a two-stage hybrid system. *Front. Energy Res.* 4: 29.

95 Lee, S.Y., Park, J.H., Jang, S.H. et al. (2008). Fermentative butanol production by clostridia. *Biotechnol. Bioeng.* 101 (2): 209–228.

96 Zhao, J., Lu, C., Chen, C.-C., and Yang, S.-T. (2013). Biological production of butanol and higher alcohols. In: *Bioprocessing Technologies in Biorefinery for Sustainable Production of Fuels, Chemicals, and Polymers* (ed. S.-T. Yang, H.A. El-Enshasy, and N. Thongchul), 235–262. New York: Wiley.

97 Gwalwanshi, M., Kumar, R., and Kumar Chauhan, M. (2022). A review on butanol properties, production and its application in internal combustion engines. *Mater. Today Proc.* 62: 6573–6577.

98 Jiang, Y., Liu, J., Jiang, W. et al. (2015). Current status and prospects of industrial bio-production of n-butanol in China. *Biotechnol. Adv.* 33 (7): 1493–1501.

99 Li, S., Zhou, Y., Luo, Z. et al. (2018). Dual function of ammonium acetate in acetone-butanol-ethanol fermentation by *Clostridium acetobutylicum*. *Bioresour. Technol.* 267: 319–325.

100 Cheng, C., Bao, T., and Yang, S.T. (2019). Engineering Clostridium for improved solvent production: recent progress and perspective. *Appl. Microbiol. Biotechnol.* 103 (14): 5549–5566.

101 Moon, H.G., Jang, Y.S., Cho, C. et al. (2016). One hundred years of clostridial butanol fermentation. *FEMS Microbiol. Lett.* 363 (3): https://doi.org/10.1093/femsle/fnw001.

102 Du, G., Wu, Y., Kang, W. et al. (2022). Enhanced butanol production in *Clostridium acetobutylicum* by manipulating metabolic pathway genes. *Process Biochem.* 114: 134–138.

103 Wang, Z., Cao, G., Jiang, C. et al. (2013). Butanol production from wheat straw by combining crude enzymatic hydrolysis and anaerobic fermentation using *Clostridium acetobutylicum* ATCC824. *Energy Fuel* 27 (10): 5900–5906.

104 Luo, H., Zhang, J., Wang, H. et al. (2017). Effectively enhancing acetone concentration and acetone/butanol ratio in ABE fermentation by a glucose/acetate co-substrate system incorporating with glucose limitation and *C. acetobutylicum/S. cerevisiae* co-culturing. *Biochem. Eng. J.* 118: 132–142.

105 Wu, J., Dong, L., Zhou, C. et al. (2019). Developing a coculture for enhanced butanol production by *Clostridium beijerinckii* and Saccharomyces cerevisiae. *Bioresour. Technol. Rep.* 6: 223–228.

106 Ahmed, I., Ross, R.A., Mathur, V.K., and Chesbro, W.R. (1988). Growth rate dependence of solventogenesis and solvents produced by *Clostridium beijerinckii*. *Appl. Microbiol. Biotechnol.* 28 (2): 182–187.

107 Carrié, M., Velly, H., Ben-Chaabane, F., and Gabelle, J.C. (2022). Modeling fixed bed bioreactors for isopropanol and butanol production using *Clostridium beijerinckii* DSM 6423 immobilized on polyurethane foams. *Biochem. Eng. J.* 180: 108355.

108 Li, J., Du, Y., Bao, T. et al. (2019). n-Butanol production from lignocellulosic biomass hydrolysates without detoxification by *Clostridium tyrobutyricum* Δack-adhE2 in a fibrous-bed bioreactor. *Bioresour. Technol.* 289: 121749.

109 Solarte-Toro, J.C., Chacón-Pérez, Y., and Cardona-Alzate, C.A. (2018). Evaluation of biogas and syngas as energy vectors for heat and power generation using lignocellulosic biomass as raw material. *Electron. J. Biotechnol.* 33: 52–62.

110 Nunes, N.S.P., Ansilago, M., de Oliveira, N.N. et al. (2021). Biofuel production. *Microalgae* 145–171.

111 Enzmann, F., Mayer, F., Rother, M., and Holtmann, D. (2018). Mini-review Methanogens: biochemical background and biotechnological applications. *AMB Express* 8: 1.

112 Karakashev, D., Batstone, D.J., and Angelidaki, I. (2005). Influence of environmental conditions on methanogenic compositions in anaerobic biogas reactors. *Appl. Environ. Microbiol.* 71 (1): 331–338.

113 Andriani, D., Wresta, A., Atmaja, T.D., and Saepudin, A. (2014). A review on optimization production and upgrading biogas through CO 2 removal using various techniques. *Appl. Biochem. Biotechnol.* 172 (4): 1909–1928.

114 Abanades, S., Abbaspour, H., Ahmadi, A. et al. (2022). A critical review of biogas production and usage with legislations framework across the globe. *Int. J. Environ. Sci. Technol.* 19 (4): 3377–3400.

115 Trendewicz, A.A. and Braun, R.J. (2013). Techno-economic analysis of solid oxide fuel cell-based combined heat and power systems for biogas utilization at wastewater treatment facilities. *J. Power Sources* 233: 380–393.

116 Hao, Y., Zhang, X., Du, Q. et al. (2022). A new integrated single-chamber air-cathode microbial fuel cell – anaerobic membrane bioreactor system for improving methane production and membrane fouling mitigation. *J. Membr. Sci.* 655 (February): 120591.

117 Nuhu, S.K., Gyang, J.A., and Kwarbak, J.J. (2021). Production and optimization of biomethane from chicken, food, and sewage wastes: the domestic pilot biodigester performance. *Clean. Eng. Technol.* 5 (October): 100298.

118 Srivastava, A.N. and Chakma, S. (2021). Investigating leachate decontamination and biomethane augmentation through Co-disposal of paper mill sludge with municipal solid waste in simulated anaerobic landfill bioreactors. *Bioresour. Technol.* 329: 124889.

119 Jung, J.H., Sim, Y.B., Ko, J. et al. (2022). Biohydrogen and biomethane production from food waste using a two-stage dynamic membrane bioreactor (DMBR) system. *Bioresour. Technol.* 352 (January): 127094.

120 Zainal, B.S., Danaee, M., Mohd, N.S., and Ibrahim, S. (2020). Effects of temperature and dark fermentation effluent on biomethane production in a two-stage up-flow anaerobic sludge fixed-film (UASFF) bioreactor. *Fuel* 263: 116729.

121 Farrell, A.E., Keith, D.W., and Corbett, J.J. (2003). A strategy for introducing hydrogen into transportation. *Energy Policy* 31: 1357–1367.

122 Yan, J. and Zhao, J. (2022). Willingness to pay for heavy-duty hydrogen fuel cell trucks and factors affecting the purchase choices in China. *Int. J. Hydrog. Energy* 47: 24619–24634.

123 Chandrasekhar, K., Lee, Y.J., and Lee, D.W. (2015). Biohydrogen production: strategies to improve process efficiency through microbial routes. *Int. J. Mol. Sci.* 16 (4): 8266–8293.

124 Show, K.-Y., Yan, Y., and Lee, D.-J. (2019). Bioreactor and bioprocess design for biohydrogen production. In: *Biohydrogen* (ed. A. Pandey, J.-S. Chang, P.C. Hallenbecka, and C. Larroche), 391–411. Elsevier.

125 Palomo-Briones, R., de Montoya-Rosales, J., and Razo-Flores, E. (2021). Advances towards the understanding of microbial communities in dark fermentation of enzymatic hydrolysates: diversity, structure and hydrogen production performance. *Int. J. Hydrog. Energy* 46 (54): 27459–27472.

126 Sivaramakrishnan, R., Shanmugam, S., Sekar, M. et al. (2021). Insights on biological hydrogen production routes and potential microorganisms for high hydrogen yield. *Fuel* 291: 120136.

127 Hans, M. and Kumar, S. (2019). Biohythane production in two-stage anaerobic digestion system. *Int. J. Hydrog. Energy* 44 (32): 17363–17380.

128 Stamatelatou, K., Antonopoulou, G., Tremouli, A., and Lyberatos, G. (2011). Production of gaseous biofuels and electricity from cheese whey. *Ind. Eng. Chem. Res.* 50 (2): 639–644.

129 Castelló, E., García Santos, C., Iglesias, T. et al. (2009). Feasibility of biohydrogen production from cheese whey using a UASB reactor: links between microbial community and reactor performance. *Int. J. Hydrog. Energy* 34: 5674–5682.

130 Kadier, A., Simayi, Y., Abdeshahian, P. et al. (2016). A comprehensive review of microbial electrolysis cells (MEC) reactor designs and configurations for sustainable hydrogen gas production. *Alexandria Eng. J.* 55 (1): 427–443.

131 Cheng, S. and Logan, B.E. (2007). Sustainable and efficient biohydrogen production via electrohydrogenesis. *Proc. Natl. Acad. Sci. U. S. A.* 104 (47): 18871–18873.

132 Chu, C., Zheng, J., and Bhuyar, P. (2022). Biomass and Bioenergy Enhancement of biohydrogen production by employing a packed-filter bioreactor (PFBR) utilizing sulfite-rich organic effluent obtained from a washing process of beverage manufactures. *Biomass Bioenergy* 161 (April): 106451.

133 Bolzonella, D., Battista, F., Cavinato, C. et al. (2018). Recent developments in biohythane production from household food wastes: a review. *Bioresour. Technol.* 257 (January): 311–319.

134 Liu, Z., Zhang, C., Lu, Y. et al. (2013). States and challenges for high-value biohythane production from waste biomass by dark fermentation technology. *Bioresour. Technol.* 135: 292–303.

135 Chu, C.Y., Vo, T.P., and Chen, T.H. (2020). A novel of biohythane gaseous fuel production from pineapple peel waste juice in two-stage of continuously stirred anaerobic bioreactors. *Fuel* 279 (June): 118526.

136 Lunprom, S., Phanduang, O., Salakkam, A. et al. (2019). Bio-hythane production from residual biomass of Chlorella sp. biomass through a two-stage anaerobic digestion. *Int. J. Hydrog. Energy* 44 (6): 3339–3346.

137 Mamimin, C., Kongjan, P., O-Thong, S., and Prasertsan, P. (2019). Enhancement of biohythane production from solid waste by co-digestion with palm oil mill effluent in two-stage thermophilic fermentation. *Int. J. Hydrog. Energy* 44 (32): 17224–17237.

10

Potential Microbial Bioresources for Functional Sugar Molecules

Satya Narayan Patel, Sweety Sharma, Ashish Kumar Singh, and Sudhir P. Singh

Center of Innovative and Applied Bioprocessing (DBT-CIAB), Mohali, Punjab, India

10.1 Introduction

Functional sugars refer to sugar molecules that have distinctive structural and physiological attributes but are available in significantly fewer quantities in nature; therefore, they are called as rare sugar [1]. The International Society of Rare Sugars (ISRS) defined rare sugars as "monosaccharides and their derivatives that hardly exist in nature." Functional sugars are emerging as a preferred food ingredient with a sweet taste, significantly low caloric value, and various health benefits [2]. The United States Food and Drug Administration (US-FDA) has accepted many functional sugars as generally recognized as safe (GRAS) status. The functional properties and health benefits of rare sugars offer their wide applications in the food, cosmetic, nutraceuticals, and pharmaceutical industries [3].

Further, functional sugars and their derivatives are potential candidates as antiviral and anticancer drugs [4]. The global functional sugar market was evaluated to be USD 2609.8 million in 2022, and by the end of 2029, it is projected to grow to USD 3469 million (https://www.marketwatch.com/). Functional sugars are present in a minimal amount in nature [5]. The low content of these functional sugars in the plant materials makes the extraction process very difficult. The chemical synthesis of functional sugar is possible, but the process suffers from harsh reaction conditions, by-product formation, limited yield, environmentally unfriendly, expensive starting raw materials, and safety risks [6]. Many microbial resources have been explored for genes encoding the enzymes having the potential to catalyze rare sugar biosynthesis [7, 8]. The enzymatic production of rare sugars has many advantages over chemical processes, e.g. simplicity of the process, environmentally friendliness, lower cost, and straightforward catalysis. Efficient enzyme variants have been developed by following gene mining and mutagenesis approaches for rare sugar biosynthesis.

Microbial Bioreactors for Industrial Molecules, First Edition.
Edited by Sudhir P. Singh and Santosh Kumar Upadhyay.

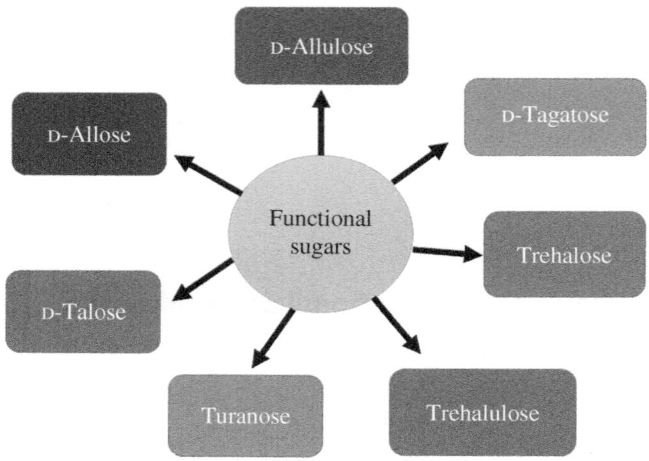

Figure 10.1 An array of high-value natural functional sugar molecules.

Moreover, immobilized enzymes or enzyme systems usually improve catalytic performance by contributing to thermal and storage stability. Whole-cell catalysis is a fermentation-based approach to manufacturing functional sugars [3]. Using whole-cell reactions for synthesizing D-allulose is advantageous over enzyme-based reactions in many ways; enzyme purification steps are reduced, the cellular microenvironment provides suitable conditions for cofactor regeneration, enzymes remain protected from harsh reaction conditions, etc. [9]. This chapter discusses the biosynthesis and health attributes of functional sugars, e.g. D-allulose, D-tagatose, Trehalose, Turanose, Trehalulose, D-allose, and D-talose (Figure 10.1 and Table 10.1).

10.2 D-Allulose

D-Allulose (also known as D-psicose) is a functional sugar, which rarely occurs in nature. D-Allulose is a C-3 epimer of D-fructose. It has ultralow calories (0.4 kcal/g) with a sweetness of about 70% compared to table sugar (i.e. sucrose) [10]. Its presence has been reported in traces in a few plant species, such as wheat [11], *Itea* plants [12], and cane molasses [13]. In addition, a few high sugar-containing foods like candy and sauces may undergo an isomerization reaction during heating, resulting in the synthesis of a trace amount of D-allulose [14]. Recently, several researchers have reported D-allulose to have numerous physiological benefits such as improvement in serum lipid metabolism, suppression of blood glucose elevation, antiobesity, antioxidant properties, decrease in body fat accumulation, rheological properties, and neuroprotective activity [15–19]. Therefore, there is an increasing preference for D-allulose in the food industry mainly due to being a sweet sugar with low-calorie, health benefactions, pleasant taste via Maillard reaction, and good gelling properties [20–22]. The Food and Drug Administration (FDA, US) has accorded it a safe food ingredient (FDA, GRAS Notice 400, 498, 624, 647, and 693) and exempted it from the total added sugars count in the "Supplement Facts" nutrition information [23]. In 2020, the

Table 10.1 Summary of the common properties and physiological application of the functional sugars.

Functional sugar	Properties and physiological application
D-Allulose	Nearly zero calorie sweetener, low glycemic, decreases the body fat accumulation, improves serum lipid metabolism, suppression of blood glucose elevation, anti-obesity, antioxidant, rheological properties, and neuro-protective activity
D-Tagatose	Low-calorie sweetener, low glycemic, non-cariogenicity, antidiabetic and obesity effect, reducing cholesterol, preventing colon cancer, and prebiotic action
Trehalose	Stabilizer for the proteins and cell membrane, protection of the organism against drought, anti-stress agent in hyperthermia, osmotic stress, reduced cariogenic activity, enhancing the shelf life of food item, reduces the progression of insulin resistant, antioxidant property, cryopreservation of blood stem cells and sperm cells without hampering cell viability
Turanose	Nearly zero calorie sweetener, being a sugar of non-cariogenic nature, turanose consumption can help avoid the dental plaque to erode the enamel, low glycemic effect, and suppressive effect of lipid accumulation in adipose tissues
Trehalulose	Considered as s bioactive compound and non-carcinogenic sweetener, low glycemic index with nearly 70% of the sweetness of sucrose. It also shows several beneficial properties like antioxidant, antidiabetic, and high solubility in water, attributed to its potential food and pharmacological applications
D-Allose	Nearly low-calorie sugar, antioxidant, antitumor, antihypertension, protection from ischemic perfusion injury, suppression of blood glucose elevation, etc.
D-Talose	D-Talose and its derivatives, specifically its glycoconjugate derivatives, are well known for antimicrobial (such as Caminoside A, Telbivudine, and Emtricitabine) and anticancer activities. The adenine derivative of L-talose, i.e. L-talofuranosyladenine, has shown promising activity against inhibiting leukemia. Moreover, the O2 and O3 methylated forms of D-talose have been demonstrated to be submillimolar inhibitors of galactose-binding galectin-8 and galectin-4 proteins that are responsible for cancer and inflammation.

global market size of D-allulose was US$34 million, which could reach US$59 million by 2027 [13]. Thus, D-allulose has influenced substantial notice from researchers worldwide and has an intense future in the food, nutraceutical, and pharmaceutical companies [24].

The isolation and extraction of D-allulose from natural sources are challenging and economically unviable because of its very low concentration. Although chemical synthesis of D-allulose from D-fructose is possible, the process is accompanied by the release of unwanted byproducts. The complex reaction steps and chemical pollution make this process non-environmentally friendly [24, 25]. Therefore, the enzymatic biosynthesis of D-allulose from D-fructose is a promising strategy in view of industrial viability. Ketose 3-epimerase (KE) is the enzyme that reversibly catalyzes the epimerization of D-fructose at the C-3 position into D-allulose. Four types of KEs have been demonstrated from various microbial sources: D-tagatose 3-epimerase, L-ribulose 3-epimerase, D-fructose 3-epimerase, and D-allulose 3-epimerase. The four KE types are named as per their maximum substrate specificity toward the substrates, e.g. D-tagatose, L-ribulose, D-fructose, and D-allulose substrate. However, the four types of KE have more or less similar attributes to catalytic transformation of D-fructose into D-allulose [26–28]. Till date, more than 26 KEs have been characterized for the production of D-allulose from D-fructose from different microorganisms *viz.*, *Pseudomonas cichorii* [29], *Caballeronia fortuita* [28], *Sinorhizobium* sp. [30], *Christensenella minuta* [31], *Agrobacterium tumefaciens* [32], *Clostridium cellulolyticum* H10 [33], *Clostridium bolteae* [34], *Ruminococcus* sp. [35], *Clostridium* sp.[36], *Clostridium scindens* [37], *Desmospora* sp. 8437 [38], *Treponema primitia* ZAS-1 [39], *Dorea* sp. CAG317 [40], *Arthrobacter globiformis* [41], *Flavonifractor plautii* [42], *Paenibacillus senegalensis* [43], *Agrobacterium* sp. ATCC 31749 [44], *Rhodopirellula baltica* [45], *Staphylococcus aureus* [46], DaeM [47], *Bacillus* sp. (DaeB) [26], *Halanaerobium congolense* [48], *Pirellula* sp. [49], *Thermoclostridium caenicola* [50], *Rhodobacter sphaeroides*, and *Mesorhizobium loti* [51] (Table 10.2).

Table 10.2 Microbial ketose 3-epimerase for D-allulose production with biochemical properties.

Organism	Optimum temperatue (°C)	Optimum pH	Half-life (min)	Conversion (%)	Reference
Pseudomonas cichorii	60	7.5	NR	20 : 80 (30 °C)	[29]
Caballeronia fortuita	65	7.5	63 (60 °C) 307 (55 °C)	28.1 : 71.9	[28]
Christensenella minuta	50	6.0	40 (50 °C)	30 : 70	[31]
Agrobacterium tumefaciens	50	8.0	8.9 (55 °C) 64 (50 °C)	32 : 68 (30 °C)	[32]
Clostridium cellulolyticum	55	8.0	408 (60 °C)	32 : 68	[33]
Ruminococcus sp.	60	7.5–8.0	~96 (60 °C)	28 : 72	[35]
Clostridium scindens	60	7.5	50 (60 °C) 108 (50 °C)	28 : 72	[37]
Clostridium sp.	65	8.0	15 (60 °C)	28 : 72 (65 °C)	[36]
Desmospora sp.	60	7.5	120 (50 °C)	30 : 70	[38]
Clostridium bolteae	55	7.0	156 (55 °C)	32 : 68 (60 °C)	[34]
Dorea sp.	70	6.0	30 (60 °C)	30 : 70 (70 °C)	[40]
Treponema primitia	70	8.0	30 (50 °C)	28 : 72	[39]
Flavonifractor plautii	65	7.0	40 (65 °C)	31 : 69	[42]
Arthrobacter globiformis	70	7.0–8.0	NA	27 : 73 (70 °C)	[41]
Agrobacterium sp.	55–60	7.5–8.0	267 (55 °C) 28.2 (60 °C) 3.8 (65 °C)	30 : 70	[44]
Paenibacillus senegalensis	55	8.0	140 (60 °C)	30 : 70	[43]
Staphylococcus aureus	70	7.5	120 (70 °C)	38.9 : 61.1	[46]
Rhodopirellula baltica	60	8.0	52 (60 °C)	28.6 : 61.4	[45]

Table 10.2 (Continued)

Organism	Optimum temperatue (°C)	Optimum pH	Half-life (min)	Conversion (%)	Reference
DaeM (from metagenome)	80	7.0	9900 (60 °C) 3240 (70 °C) 49 (80 °C)	31 : 69	[47]
Bacillus sp. (DaeB)	55	8.0	36,000 (50 °C) 1320 (55 °C)	28.5 : 71.5	[26]
Halanaerobium congolense	70	8.0	66 (70 °C)	35 : 65	[48]
Pirellula sp.	60	7.5	360 (60 °C)	30 : 70	[49]
Thermoclostridium caenicola	65	7.5	816 (50 °C)	28 : 72	[50]
Rhodobacter sphaeroides	40	9.0	60 (50 °C)	23 : 77 (40 °C)	[52]

NA, not available.

KEs have been demonstrated to show activity at temperature and pH ranges of 50–80 °C and 7–9, respectively [25, 47]. D-Allulose 3-epimerase can be employed for conversion of fructose present in plant biomass or agro-industrial by-products into D-allulose [53–55]. A higher reaction temperature is advisable not only to avoid contamination issues but also to decrease the viscosity and increase the substrate's solubility, which favors the reaction kinetics. Most of the KEs are reported to achieve conversion yield of 30–33% from D-fructose to D-allulose [32, 47]. Recently, D-allulose 3-epimerase has been expressed in food-safe microorganisms such as *Bacillus subtilis* [47, 56], *Saccharomyces cerevisiae* [57], and *Corynebacterium glutamicum* [2, 43]. Further, multienzyme cascade systems have also been developed to convert high-calorie sugars, like sucrose, D-glucose, and D-fructose derived from fruit juices or other sources, into D-allulose [49]. The microbial cells were engineered with D-allulose-3-epimerase expression constructs and employed as a microbial cell factory to produce D-allulose utilizing D-fructose. These whole-cell systems produce D-allulose with a conversion in the range of 30–32% [42, 47]. The use of a cell factory for D-allulose biosynthesis avoids cell disruption, enzyme extraction, and purification steps. However, D-allulose may be considered a GMO-derived product, as it is directly recovered from the recombinant cell. Using purified enzyme or immobilized enzyme, followed by making the enzymatic product, D-allulose, free from any enzyme contamination, has a greater chance for consumption acceptability.

10.3 D-Tagatose

D-Tagatose is a naturally occurring ketohexose, an isomer of D-galactose and C-4 epimer of D-fructose. It is counted as a rare sugar due to its limited existence in nature [55]. It is naturally found mainly as a constituent of gum exudate from a tropical tree (*Sterculia setigera*) and some lichens. It has nearly 92% sweetness but only 38% caloric value of sucrose [59, 60]. Since it received the GRAS status from US-FDA regulation, D-tagatose became a favorable functional sweetener in the food as well as pharmaceutical industries. It has been found useful as a sweet ingredient in soft drinks, cereals, yogurt, diabetes-specific food, meat product, cough syrups, oral disinfectants, candies, and confectionery, for enhancing flavors and lowering calories [61, 62]. D-Tagatose has gained global awareness due to its ability to reduce cholesterol levels, treat type II obesity, prevent colon cancer, have prebiotic activity, and so on [63].

D-Tagatose can be synthesized by biological and chemical routes, where researchers produce D-tagatose using the substrate, D-galactose, via a chemical process using potassium aluminate, alkaline earth, or soluble alkali metal salt. However, chemical methods have a few limitations, such as the formation of by-products, difficult purification steps, and environmental pollution [64]. On the other hand, biological methods offer an eco-friendly approach in which D-galactose is enzymatically converted into D-tagatose by the biocatalyst, L-arabinose isomerase (EC 5.3.1.4; L-AI) [65]. Many attempts have been made for the discovery of novel L-AIs with a high-temperature tolerance and catalytic efficiency from several natural and engineered microorganisms (Table 10.3). Till date, more than 30 L-AIs have been specified from different microorganisms, such as *Escherichia coli* [66], *Pediococcus pentosaceus* [67], *Bacillus coagulans* NL01 [62], *Shigella flexneri* [68], *Bacillus stearothermophilus* IAM1101 [69], *Acidothermus cellulolyticus*

Table 10.3 Microbial L-arabinose isomerase for D-tagatose biosynthesis with its biochemical properties.

Microbial sources	Metal ions	Temperature optima (°C)	pH optima	Reference
Escherichia coli	Mn^{2+}	30	8	[66]
Pediococcus Pentosaceus PC-5	Co^{2+}, Mn^{2+}	50	6	[67]
Bacillus coagulans NL01	Co^{2+}, Mn^{2+}	60	7.5	[62]
Shigella flexneri	Co^{2+}, Mn^{2+}	40	8	[68]
Bacillus stearothermophilus IAM1101	Mn^{2+}	65	7.5	[69]
Acidothermus cellulolyticus ATCC 43068	Co^{2+}, Mn^{2+}	75	7.5	[70]
Anoxybacillus flavithermus	Ni^{2+}	95	9.5–10.5	[71]
Alicyclobacillus hesperidum URH17-3-68	Co^{2+}	70	7	[72]
Thermoanaerobacterium saccharolyticum NTOU1	Co^{2+}, Mn^{2+}	70	7–7.5	[73]
Thermotoga neapolitana	Co^{2+}, Mn^{2+}	85	7	[74]
Thermotoga maritime	Co^{2+}, Mn^{2+}	90	7–7.5	[72]

NA, not available.

ATCC 43068 [70], *Anoxybacillus flavithermus* [71], *Alicyclobacillus hesperidum* URH17-3-68 [72], *Thermoanaerobacterium saccharolyticum* NTOU1[73], and *Thermotoga neapolitana* [74] for D-tagatose formation. Most of the L-AIs enzymes are active at higher temperatures, optima ranging from 60 to 95 °C and neutral pH. Moreover, all the reported L-AIs require divalent metal ions, namely Co^{2+} and Mn^{2+}, for stability and activity [75]. However, for the production of D-tagatose at the industrial level, the enzyme, L-AI, needs to function at acidic pH for reduced by-product formation. Also, enzymatic reactions should be conducted without additional metal ions, mainly Co^{2+} ions, because it is unacceptable for food applications [75]. At elevated temperatures, the reversible reaction between D-galactose and D-tagatose is shifted toward D-tagatose, so biocatalysts with activity at higher temperatures are favored. Thus, the study of L-arabinose isomerases from thermophilic and hyper-thermophilic microbes is of special interest for the biocatalytic production of D-tagatose [76].

The higher bioconversion of D-tagatose from D-galactose has been demonstrated by employing L-arabinose isomerase from *Lactobacillus fermentum* expressed at the surface of *B. subtilis* cell, achieving about 75% conversion in 24 hours at 70 °C [77]. L-AI from *T. neapolitana* produced D-tagatose having a conversion yield close to 68% at 80 °C [74]. Several generally regarded as safe (GRAS) microorganisms, such as *C. glutamicum* KCTC 13032, *B. subtilis* 168, *Lactococcus lactis*, and *Lactobacillus delbrueckii*, have been used for D-tagatose production, with the conversion yield in the range of 24–75%. Several types of the matrix have been examined for the immobilization of L-AIs *viz.*, alginate, calcium alginate, glutaraldehyde, polyethyleneimine, chitopearl beads, and agarose, experiencing improvement toward thermal stability and recyclability [58, 78]. Nevertheless, research gaps still exist, e.g. the biocatalytic affinity of L-arabinose isomerase for D-galactose is comparatively low as compared to its natural substrate L-arabinose. The combination of D-tagatose and D-galactose can be purified by applying simulated moving bed chromatography (SSMB) (Patent No. CN103992362A) and by employing *S. cerevisiae* that can consume D-galactose from the reaction mixture, obtaining about 95% pure D-tagatose [79]. However, the downstream processing and purification of the reaction product is a challenge due to the analogous property of D-tagatose and D-galactose.

10.4 Trehalose

Trehalose or α-D-glucopyranosyl-(1→1)-α-D-glucopyranoside is a naturally occurring non-reducing di-saccharide. It is made up of two glucose units linked with an α,α-1,1-glycosidic bond [80]. Naturally, it is available in a wide number of organisms, e.g. insects, nematodes, yeasts, mycobacteria, and in plants. It is also known as mushroom sugar or mycose because edible mushrooms carry a high content of trehalose [81]. However, it has been found in many vertebrates but is not ascertained in mammalian cells [82]. Recent studies suggested many biological applications of trehalose, such as a stabilizer for the proteins and cell membrane, protecting the organism during drought conditions by acting as a water replacement or vitrifying agent, anti-stress agent in hyperthermia, hypothermia, and osmotic stress. It is a part of the cell wall in *Corynebacteria* and *Mycobacteria*. It is a crucial energy source in insects, utilized during flight [80, 82].

Trehalose was accorded GRAS status by US-FDA in 2000 (GRN No. 000045) [83]. Its calorific value is close to sucrose, nearly 4 kcal/g [84]. However, it shows 45% sweetness compared to sucrose [85] and reduced cariogenic activity [86]. It is very stable in a broad range of pHs because of its low-energy glycosidic bond in contrast to other disaccharides [82]. With these characteristics, trehalose is a favored candidate for flavor improvement in the food industry [83]. The self-life of food items can also be enhanced with trehalose, and it also prevents starch aging [87].

Furthermore, it decreases the freezing point of food products [83, 88]. Trehalose was found to be an essential additive that could reduce the progression of insulin resistance [89] and maintain the antioxidant property and polyphenol content of food products [90]. Due to its high water retention capacity, it reduces water loss in pharmaceutical, food, and cosmetic products [91]. Without hindering cell viability, trehalose is also applicable for the cryopreservation of blood stem cells and sperm cells [92]. Furthermore, it is a favorable molecule for preserving numerous types of tissues and organs for transplantation [93]. In the cosmetic sector, it is used as a moisture-retaining agent, enhances storage constancy, and as a suppressor of foul odor in lotion and creams [83].

Initially, it was chemically synthesized via a reaction involving ethylene oxide addition using 2,3,4,5-tetra-*O*-acetyl-D-glucose and 3, 4, 6-tri-*O*-acetyl-1,2-anhydro-D-glucose as substrates. However, the chemical production of trehalose has remained very expensive [80, 94]. The biosynthesis of trehalose is feasible by applying a microbial enzymatic system. The enzymatic synthesis of trehalose is favored due to its straightforward reaction and low expenses. Three enzymatic pathways are involved in the biosynthesis of trehalose: (i) maltooligosyl trehalose synthase and maltooligosyl trehalose trehalohydrolase, (ii) trehalose-6-phosphate synthase and trehalose-6-phosphate phosphatase, and (iii) trehalose synthase [95]. The biosynthesis of trehalose via trehalose synthase is straightforward as well as relatively economical. Trehalose synthase reversibly synthesizes trehalose from maltose [96]. Trehalose synthase belongs to glycosyl hydrolase 13 family of enzymes [97]. To date, more than 22 trehalose synthases (from bacterial species) have been biochemically optimized for trehalose biosynthesis (Table 10.4). Many microbial strains demonstrated >80% yield of trehalose, such as *Pimelobacter* sp. (81.8%), *Pseudomonas* sp. P8005 (>80%), *Thermus thermophiles* (80%), *Thermus aquaticus* (80.7%), and *Thermus caldophilus* (86%) (Table 10.4). The immobilization techniques were used for enhanced stability, improved activity, separation from the reaction, and enzyme reusability. The immobilization of trehalose synthase has been demonstrated on the carrier of eupergit C250L and silanized magnetic ferrous-ferric oxide. Immobilization of trehalose synthase leads to improvement in thermal activity, stability, and pH stability of the enzyme [82, 113]. In addition, trehalose has also been synthesized using maltose as a substrate via the whole cell *Pseudomonas monteilii* TBRC 1196, harboring trehalose synthase enzyme [114].

10.5 Turanose

Turanose (α-D-glucopyranosyl-1,3-D-fructofuranose), a reducing disaccharide, is a sugar made up of glucose and fructose units with α-1,3-glycosidic linkages. It is an isomer of sucrose. It naturally exists in honey as a rare sugar. Turanose displays 50%

Table 10.4 Biochemical property of trehalose synthase characterize from different microbial sources.

Organism	Optimum pH	Optimum temperature (°C)	Trehalose yield (%)	Reference
Corynebacterium glutamicum (ATCC 13032)	7	35	69% (25 °C) in 9 h	[98]
Thermobifida fusca (DSM 43792)	6.5	25	55–65% (25 °C)	[99]
Thermomonospora curvata (DSM 43183)	6.5	35	70% (60 °C) in 24 h	[97]
Rhodococcus opacus (ATCC 41021)	7	25	67% (25 °C)	[100]
Thermus antranikianii	7	60	76.8% (40 °C) in 8 h; 62.6% (60 °C) in 2 h	[101]
Deinococcus geothermalis (DSMZ 11300)	7.6	40	60.4% (40 °C) in 24 h	[102]
Thermobaculum terrenum	7.5	45	70% (45 °C) in 10 h	[95]
Meiothermus ruber	6.5	50	64% (20 °C), 61% (30 °C), 56% (40 °C), 47% (50 °C) in 24 h	[103]
Pimelobacter sp. R48	7.5	20	81.8% at 5 °C in 48 h	[104]
Pseudomonas sp. P8005	7.2	37	>80% (10–40 °C, 30 min) 10% (50 °C, 30 min)	[105]
Thermus thermophilus (ATCC 33923)	6.5	65	80% (65 °C) in 48 h	[106]
Arthrobacter aurescens	6.5	35	60% (35 °C) in 8 h	[107]
Picrophilus torridus (DSM 9790)	6	45	71 (20 °C), 68 (30 °C), 61 (45 °C), 50 (60 °C) in 72 h	[107]
Mycobacterium smegmatis	7.2	37	42–45% (37 °C) in 6 h	[109]
Enterobacter hormaechei	6	37	48% (40 °C) in 30 min	[110]
Pseudomonas stutzeri CJ 38	8.5	35	72% (35 °C) in 19 h	[111]
Thermus aquaticus (ATCC 33923)	6.5	65	80.7% (40 °C) in 48 h	[104]
Thermus caldophilus	6.3	40	86% (40 °C)	[112]
Hot spring metagenome (TreM)	7	45	66% (30 °C), 63% (45 °C) in 3 h; 70% (20 °C), 74% (5 °C) in 12 h; 52% (45 °C) in 15 min	[80]

sweetness compared to sucrose, and it is also known to hydrolyze more slowly than sucrose and maltose [115, 116]. Being a sugar of non-cariogenic nature, turanose consumption can help to avoid dental plaque eroding the enamel. It is considered a functional sweetener due to the potentially low glycemic effect and suppression of lipid accumulation in the adipose tissues [117]. Furthermore, turanose's ability to reduce inflammation in macrophages showed that it might have a role in preventing metabolic disorders. [118]. Apart from health beneficiary applications, it can be favored as a functional component in food-based companies. The aforementioned application of turanose made it a promising candidate in the category of non-fermentable and low-calorie next-generation sweetener. Turanose has also shown potential for identifying Pompe's disease of the kidneys, where it functions as an inhibitor of the enzyme glucosidase [119].

The chemical production of turanose is a tedious process and demands a high energy consumption [115, 116]. It can be enzymatically produced from the mixture of fructose and α-cyclodextrin using a dual-enzyme system, including the glucoamylase and cyclomaltodextrin glucanotransferase from *B. stearothermophilus* [120]. Amylosucrase, a glycoside hydrolases family 13 protein, offers a one-step enzymatic reaction for turanose biosynthesis via the transglycosylation of sucrose [121].

Amylosucrase works as glucansucrase, which performs the transglycosylation reaction without extra energy [122]. Interestingly, in the reaction where amylosucrase is catalyzed via transglycosylation, the energy required to form a new α-1,4 glycosidic bond is fulfilled by breaking glucosidic linkage (α1-β2) of the sucrose substrate [123]. Utilizing sucrose, amylosucrase catalyzes the isomerization, hydrolysis, and polymerization reaction. The significant enzymatic activity of amylosucrase is a catalytic-polymerization reaction resulting in the biosynthesis of glucan chain-like amylose by joining glucose moieties with α-1,4 linkages and, concurrently, losing the fructose moieties. Otherwise, an isomerization reaction may occur, depending on the substrate's concentration [124]. Amylosucrase, which hydrolyzes sucrose into D-fructose and D-glucose molecules at higher concentrations, may accelerate the isomerization reaction by transglycosylating D-glucosyl molecules into free D-fructose molecules. The synthesis of sucrose isomer, turanose, utilizes fructose as an acceptor molecule during the transglycosylation of the D-glucosyl moiety of sucrose [125]. In turanose, D-glucose and D-fructose molecules are linked via α-1,3 linkage.

The first amylosucrase was characterized from *Neisseria perflava* [126]. Afterward, more than 12 microbial amylosucrase were recognized and characterized from different bacterial species, e.g. *Arthrobacter chlorophenolicus* [127], *Alteromonas macleodii* [128], *Methylomicrobium alcaliphilum* [129], *Bifidobacterium thermophilum* [130], *Cellulomonas carbonis* [131], *Deinococcus geothermalis* [132], *Deinococcus radiodurans* [133], *Deinococcus radiopugnans* [134], *Methylobacillus flagellatus* [135], *Neisseria subflava* [125], *Neisseria polysaccharea* [136], *Synechococcus* sp., [137], *Truepera radiovictrix* [138], and from thermal aquatic habitat metagenome, As_{met} (Table 10.5). Among these the most investigated amylosucrases are from the *N. polysaccharea*. Most of the characterized amylosucrases display the highest activity between 30 and 50 °C temperature and 7–9 pH range [115]. The highest turanose yield of about 74% was reported from *N. polysaccharea* amylosucrase [136] (Table 10.5).

Table 10.5 Comparison of biochemical property of amylosucrase characterize from difference sources.

Name of organism	Optimum temperature (°C)	Optimum pH	Thermal stability (min)	Turanose yield	Reference
Arthrobactor chlorophenicolicus	NR	8	104 (40 °C), 3 (45 °C)	23%, 1M sucrose	[127]
Alteromonas macleodii	45	8	708 (40 °C), 30 (45 °C)	NR	[128]
Bifidobacterium thermophilum	50	7	23104 h (45 °C), 577 h (50 °C), 70.7 h (55 °C)	43.3% (1 M sucrose + 1 M fructose) at 48 h, 35 °C	[130]
Cellulomonas carbonis	40	7	16620 (40 °C), 44 (45 °C)	NR	[131]
Deinococcus geothermalis	45	8	1686 (50 °C), 408 (55 °C)	22%	[132]
Deinococcus radiodurans	50	8	900 (30 °C)	15%, 100 mM sucrose, 24 h	[133]
Deinococcus radiopugnans	40–45	8	25260 (40 °C), 198 (45 °C)	NR	[134]
Methylobacillus flagellatus	45	8.5	20 (50 °C), 4 (55 °C)	15%, 48 h	[135]
Neisseria subflava	45	8	23104 (40 °C), 546 (45 °C)	29% (1 M sucrose) in 24 h	[125]
Synechococcus sp.	30	6.5–7	NR	NR	[137]
Truepera radiovictrix	45	7.5	2400 (55 °C), 384 (60 °C), 73 (65 °C)	NR	[138]
Neisseria polysaccharea	37	8	1260 (30 °C), 25 (45 °C)	73.7% (2 M sucrose + 0.75 M fructose) in 80 h, 35 °C	[136]
As_{met}.	60	9	60 (60 °C), 390 (55 °C)	47% (1.5 M sucrose + 0.5 M fructose)	[115]

10.6 Trehalulose

Trehalulose, a ketose analogue of trehalose, is a disaccharide made up of glucose and fructose residues linked with an α-1,1 glycosidic bond [80]. This disaccharide is chemically close to isomaltulose and has been considered a bioactive compound and non-carcinogenic sweetener [139]. It is a rare sugar as it occurs in minute quantities in nature, primarily reported in high amounts from stingless bees, around 13–44 g/100 g of honey [140, 141]. It can be enzymatically synthesized from sucrose. However, it exhibits a low glycemic index with nearly 70% of the sweetness of sucrose [142]. It also shows several beneficial properties like antioxidant, antidiabetic, and high solubility in water. Therefore, it has a future in food and pharmacological applications [141]. Trehalulose may be synthesized chemically via solid-phase extraction followed by hydrophilic interaction chromatography [143]. However, chemical synthesis suffers from a lack of specificity and by-product formation [142].

Various enzymes from different microbial sources have been reported to produce trehalulose from sucrose. Some of these enzymes are sucrose isomerase, isomaltulose synthase, trehalose synthase, and amylosucrase. The enzyme sucrose isomerase isolated from *Erwinia rhapontici* was first reported for the biosynthesis of trehalulose. Later, many microbial sources were applied for sucrose isomerases with different selectivity. Some produce trehalulose as a by-product of isomaltulose, and some biosynthesize only trehalulose [144]. Some microbial sources of sucrose isomerase enzymes are *E. rhapontici* NX-5, *Ervinia* sp. D12, *Klebsiella* sp., *Pseudomonas mesoacidophila* MX-45, *Serratia plymuthica* ATCC15928, and *E. rhapontici* NCPPB 1578, *E. rhapontici* DSM 4484, *E. rhapontici* NX-5, *P. mesoacidophila* MX-45, *Enterobacter* sp. FMB-1, *Protaminobacter rubrum* CBS 57477, *Pantoea dispersa* UQ68J, *Klebsiella pneumoniae* NK33-98-8, *Klebsiella planticola* UQ14S, and *Klebsiella* sp. [145]. Apart from this, the enzyme has also been reported in *Bemisia argentifolii*, *E. rhapontici*, and whiteflies which naturally harbor this enzyme and produce trehalulose-rich honey [146].

Trehalose synthase from *T. aquaticus* ATCC 33923 and *Thermus thermophilus* HB-8 also show trehalulose synthesizing capacity, which later can synthesize trehalulose selectively, i.e. without any by-product [142]. Amylosucrase catalyzes transglycosylation of D-glucosyl moiety onto D-fructose molecule and biosynthesizing sucrose isomers, i.e. turanose or trehalulose. Amylosucrase enzyme was first discovered from *N. perflava* and later characterized from multiple microbial sources such as *N. subflava*, *N. polysaccharea*, *B. thermophilum*, *T. radiovictrix*, *M. alcaliphilum*, *C. carbonis*, *M. flagellates*, *Synchococcus* sp., *D. geothermalis*, *A. macleodii*, *D. radiodurans*, *A. chlorophenolicus*, and *D. radiopugnans* [80] and very recently discovered from *Deinococcus deserti* [142]. Therefore, there is a need for genetic engineering of microbial strains to develop a system that can selectively and efficiently produce trehalulose.

10.7 D-Allose

D-Allose is a monosaccharide naturally present in trace amounts. It is an aldohexose and a C-3 epimer of D-glucose. It is a bioactive compound and a potential low-calorie sugar (approximately 80% sweet to sucrose and gives an energy value close to zero) [147]. A trace amount of D-allose has been reported in human cord blood, whereas Indian seaweed *Halodule pinifolia* contains about 3.7% D-allose [148, 149]. It has been extracted from the African shrub *Protea rubropilosa*, *Veronica filiformis*, *Mentzelia*, and potato leaves. Several medicinal herbs, such as *Tamarindus indica*, *Acalypha hispida* leaves, and *Crataeva nurvala* have also been reported to contain an ultralow amount of free D-allose [150, 151]. Toxicity tests performed in rats suggested nontoxicity of D-allose [152]. Several studies revealed great pharmacological potentials of D-allose, such as antioxidant [153], antitumor [153], antihypertension [154], protection from ischemic perfusion injury [155], and suppression of blood glucose elevation [149]. The role of rare sugars has also been reported in plant metabolisms, such as triggering molecules of rice defense against reactive oxygen species [156], the regulator of plant immunity in tomatoes and other plants [157, 158], etc.

Due to the above-mentioned applications of D-allose, it is emerging as an important molecule in food and pharmaceutical industries and agriculture. Hence, there is a need for mass production of this promising molecule to get more insights into its applications. However, the extraction and isolation of D-allose from natural sources is a cumbersome process, thus impractical to perform. There are both chemical and biological methods of D-allose synthesis. The biological approach is preferred over the chemical method due to several drawbacks associated with chemical synthesis. D-Allose can be produced biologically using the Izumoring approach through enzymatic and microbiological processes [159]. D-Allulose is bioconverted into D-allose by the enzyme L-rhamnose isomerase. The D-allulose is also a rare sugar and an expensive substrate that can be obtained from D-fructose by employing D-allulose 3-epimerase enzyme. D-Allulose 3-epimerase converts D-fructose into D-allulose, which can further be used for D-allose production via L-rhamnose isomerase catalysis. Many enzymes are known to transform D-allulose into D-allose, like L-rhamnose isomerase, D-galactose-6-phosphate isomerase, D-ribose-5-phosphate isomerase, and glucose-6-phosphate isomerase. However, D-allose mass production is done using L-rhamnose isomerase [160]. D-Allulose to D-allose isomerization is reversibly catalyzed by L-rhamnose isomerase. This enzyme has been characterized from *T. saccharolyticum*, *Thermotoga maritima*, *Caldicellulosiruptor saccharolyticus*, *Bacillus halodurans*, *M. loti*, *Dictyoglomus turgidum*, and *B. subtilis* [161] (Table 10.6).

It has been demonstrated that the temperature range for working D-allose-producing enzymes is 65–85 °C, and the pH range is 7–9 [161] (Table 10.6). The higher reaction

Table 10.6 Microbial L-rhamnose isomerase for D-allose production with biochemical properties.

Organism name	Optimum temperature (°C)	Optimum pH	Half-life (h)	Conversion (%)	Reference
Escherichia coli	60	7.6	0.1 (50 °C)	NA	[162]
Pseudomonas stutzeri	60	9	0.1 (50 °C)	25	[163]
Bacillus pallidus Y25	65	7	1 (60 °C)	35	[164]
Thermoanaerobacterium saccharolyticum	75	7	2 (70 °C)	34	[165]
Thermotoga maritima	85	8	773 (75 °C)	NA	[166]
Mesorhizobium loti	60	9	>1 (50 °C)	NA	[167]
Caldicellulosiruptor saccharolyticus ATCC 43494	90	7	6 (80 °C)	33	[168]
Bacillus halodurans	70	7	0.42 (70 °C)	NA	[169]
Dictyoglomus turgidum	75	8	71.3 (65 °C)	NA	[170]
Bacillus subtilis 168	70	8.5	10 (60 °C)	37	[171]
Thermobacillus composti	65	7.5	10 (60 °C)	23.34	[172]
Caldicellulosiruptor saccharolyticus OB47	90	7	1 (90 °C)	25	[173]
Clostridium stercorarium	75	7	22.8 (65 °C)	33	[161]

temperature is favored at the industrial level to prevent contamination, reduce viscosity, boost substrate solubility, and favor reaction kinetics. By utilizing *B. subtilis* L-rhamnose isomerase, the highest conversion of D-allulose into D-allose is achieved, which is close to 38%. Microbial genome engineering and enzyme engineering approaches should be explored for a higher conversion yield of this high-value sugar.

10.8 D-Talose

D-Talose is an aldohexose sugar a C-2 epimer of galactose. It is present in some plants and bacteria in traces. Among the rare sugars, D-talose is very costly due to its scarcity in nature. D-Talose and its derivative, specifically glycoconjugate derivatives, are well known for antimicrobial (such as Caminoside A, Telbivudine, and Emtricitabine) and anticancer activities. The adenine derivative of L-talose, i.e. L-talofuranosyladenine, has shown promising activity against inhibiting leukemia [76, 174–177]. Moreover, O_2 and O_3 methylated forms of D-talose have been demonstrated to be submillimolar inhibitors of galactose-binding galectin-8 and galectin-4 proteins that are responsible for cancer and inflammation [176]. It also acts as a maker for the O-antigen [177]. Therefore, developing a process to produce this expensive sugar in bulk for use in other applications is crucial. Due to the aforementioned beneficial applications, including as a raw ingredient to create value-added products and as a food additive to increase lifespan and promote health, it has immense application in the food and pharmaceutical industries.

D-Talose production by chemical route is complicated or requires lots of solvents with more than five consecutive chemical steps [176]. Therefore, researchers are exploring the biological route for synthesizing D-talose for high efficiency, stereoselectivity, and environmental friendliness. However, there are only a few reports on the enzymatic synthesis of D-talose. The enzyme L-ribose isomerase (L-RI) catalyzes the conversion of D-tagatose to D-talose. A few L-RIs have been identified from *Acinetobacter* sp. [178], *Actinotalea fermentans* ATCC 43,279 [179], *Cellulomonas parahominis* MB426 [180], *Geodermatophilus obscurus* [181], and *Mycetocola miduiensis* [182]. Researchers have also attempted immobilization of L-RI in cobalt metal-based micro-composite construct and carbon-made nanomaterials, such as MWCNT and GO*x*, for higher stability and reusability. Immobilized L-RI has been estimated to produce 12–14% D-talose from the D-tagatose substrate [174, 175]. These processes need to be streamlined for the continuous production of rare sugar molecules.

10.9 Conclusions

The industrial-scale production and physiological attribute study of functional sugars have emerged today's hot topic owing to its incredible demand and application in several industrial fields. However, the limited availability of functional sugars in nature has restricted its application. Nevertheless, its chemical synthesis requires several reaction steps that are cumbersome, very expensive, non-environmentally friendly, and result in lower yields. Mainly, D-allulose, D-tagatose, and trehalose have excellent potential to occupy the market of functional sugar biomolecules. D-Allose has potential in the pharmaceutical industry.

The enzymatic synthesis of functional sugar offers superiority over chemical synthesis because of its high selectivity, environment-friendly, mild reaction conditions, and economic values. Also, with the surge in demands for more sustainable and bio-renewable products, the biological way of functional sugar production has become an eco-friendly approach for functional sugar biosynthesis. The biotransformation efficiency (higher turnover number, catalytic efficiency, thermal stability with acidic pH active enzymes) can be further improved by employing new genetic tools and technology to create highly efficient enzymes. The enzyme's yield, immobilization, constant production of functional sugar molecules, and purification are critical challenges in the production and downstream processing of the functional ingredient.

Acknowledgment

The Department of Biotechnology (DBT), Govt. of India, is acknowledged for all kinds of support. SS and AKS acknowledge UGC-JRF and ICMR fellowships, respectively.

References

1 Bilal, M., Xu, S., Iqbal, H.M.N., and Cheng, H. (2020). Yarrowia lipolytica as an emerging biotechnological chassis for functional sugars biosynthesis. *Crit. Rev. Food Sci. Nutr.* 61 (4): 535–552.

2 Park, Y.-C., Oh, E.J., Jo, J.-H. et al. (2016). Recent advances in biological production of sugar alcohols. *Curr. Opin. Biotechnol.* 37: 105–113.

3 Fickers, P., Cheng, H., and Lin, C.S.K. (2020). Sugar alcohols and organic acids synthesis in yarrowia lipolytica: where are we? *Microorganisms* 8 (4): 574.

4 Beerens, K., Desmet, T., and Soetaert, W. (2012). Enzymes for the biocatalytic production of rare sugars. *J. Ind. Microbiol. Biotechnol.* 39 (6): 823–834.

5 Abbasi, A.R., Liu, J., Wang, Z. et al. (2021). Recent advances in producing sugar alcohols and functional sugars by engineering *Yarrowia lipolytica*. *Front. Bioeng. Biotechnol.* 9: 648382.

6 Huang, J., Chen, Z., Zhang, W. et al. (2018). d-Lyxose isomerase and its application for functional sugar production. *Appl. Microbiol. Biotechnol.* 102 (5): 2051–2062.

7 Singh, S.P. and Upadhyay, S.K. (eds.) (2021). *Bioprospecting of Microorganism-Based Industrial Molecules*. John Wiley & Sons Ltd. doi:10.1002/9781119717317.

8 Shintani, T., Upadhyay, S.K., Singh, S.P. (2021). An introduction to microbial biodiversity and bioprospection. In: *Bioprospecting of Microorganism-Based Industrial Molecules* (ed. S.P. Singh and S.K. Upadhyay). John Wiley & Sons Ltd. https://doi.org/10.1002/9781119717317.ch1.

9 Wu, S. and Li, Z. (2018). Whole-cell cascade biotransformations for one-pot multistep organic synthesis. *ChemCatChem* 10 (10): 2164–2178.

10 Maeng, H.J., Yoon, J.H., Chun, K.H. et al. (2019). Metabolic stability of d-allulose in biorelevant media and hepatocytes: comparison with fructose and erythritol. *Foods* 8 (10): 448.

11 Miller, B.S. and Swain, T. (1960). Chromatographic analyses of the free amino-acids, organic acids and sugars in wheat plant extracts. *J. Sci. Food Agric.* 11 (6): 344–348.

12 Hough, L. and Stacey, B.E. (1966). Variation in the allitol content of Itea plants during photosynthesis. *Phytochemistry* 5 (1): 171–175.

13 Hu, M., Li, M., Jiang, B., and Zhang, T. (2021). Bioproduction of d-allulose: properties, applications, purification, and future perspectives. *Compr. Rev. Food Sci. Food Saf.* 20 (6): 6012–6026.

14 Oshima, H., Kimura, I., and Izumori, K. (2006). Psicose contents in various food products and its origin. *Food Sci. Technol. Res.* 12 (2): 137–143.

15 Chung, M.Y., Oh, D.K., and Lee, K.W. (2012). Hypoglycemic health benefits of d -Psicose. *J. Agric. Food Chem.* 60 (4): 863–869.

16 Han, Y., Park, H., Choi, B.R. et al. (2020). Alteration of microbiome profile by d-allulose in amelioration of high-fat-diet-induced obesity in mice. *Nutrients* 12 (2): 352.

17 Suna, S., Yamaguchi, F., Kimura, S. et al. (2007). Preventive effect of d-psicose, one of rare ketohexoses, on di-(2-ethylhexyl) phthalate (DEHP)-induced testicular injury in rat. *Toxicol. Lett.* 173 (2): 107–117.

18 Mooradian, A.D. (2019). In search for an alternative to sugar to reduce obesity. *Int. J. Vitam. Nutr. Res.* 89 (3–4).

19 Patel, S.N., Sharma, M., Lata, K. et al. (2016). Improved operational stability of d-psicose 3-epimerase by a novel protein engineering strategy, and d-psicose production from fruit and vegetable residues. *Bioresour. Technol.* 216: 121–127.

20 Lee, P., Oh, H., Kim, S.Y., and Kim, Y.S. (2020). Effects of d-allulose as a sucrose substitute on the physicochemical, textural, and sensorial properties of pound cakes. *J. Food Process. Preserv.* 44 (6): e14472.

21 Matsuo, T., Suzuki, H., Hashiguchi, M., and Izumori, K. (2002). d-Psicose is a rare sugar that provides no energy to growing rats. *J. Nutr. Sci. Vitaminol. (Tokyo)* 48 (1): 77–80.

22 Sun, Y., Hayakawa, S., and Izumori, K. (2004). Modification of ovalbumin with a rare ketohexose through the maillard reaction: effect on protein structure and gel properties. *J. Agric. Food Chem.* 52 (5): 1293–1299.

23 FDA (2019). FDA allows the low-calorie sweetener allulose to be excluded from total and added sugars counts on Nutrition and Supplement Facts labels when used as an ingredient. In: FDA In Brief [Internet], p. 1. https://www.fda.gov/news-events/fda-brief/fda-brief-fda-allows-low-calorie-sweetener-allulose-be-excluded-total-and-added-sugars-counts.

24 Emmadi, M. and Kulkarni, S.S. (2014). Recent advances in synthesis of bacterial rare sugar building blocks and thessssir applications. *Nat. Prod. Rep.* 31 (7): 870–879.

25 Patel, S.N., Singh, V., Sharma, M. et al. (2018). Development of a thermo-stable and recyclable magnetic nanobiocatalyst for bioprocessing of fruit processing residues and d-allulose synthesis. *Bioresour. Technol.* 247: 633–639.

26 Patel, S.N., Kaushal, G., and Singh, S.P. (2021). d-Allulose 3-epimerase of *Bacillus* sp. origin manifests profuse heat-stability and noteworthy potential of d-fructose epimerization. *Microb. Cell Factories* 20 (1): 1–16.

27 Jiang, S., Xiao, W., Zhu, X. et al. (2020). Review on D-allulose: in vivo metabolism, catalytic mechanism, engineering strain construction, bio-production technology. *Front. Bioeng. Biotechnol.* 8: 26.

28 Li, S., Chen, Z., Zhang, W. et al. (2019). Characterization of a d-tagatose 3-epimerase from *Caballeronia fortuita* and its application in rare sugar production. *Int. J. Biol. Macromol.* 138: 536–545.

29 Ishida, Y., Kamiya, T., Itoh, H. et al. (1997). Cloning and characterization of the D-tagatose 3-epimerase gene from *Pseudomonas cichorii* ST-24. *J. Ferment. Bioeng.* 83 (6): 529–534.

30 Zhu, Z., Li, C., Liu, X. et al. (2019). Biochemical characterization and biocatalytic application of a novel d-tagatose 3-epimerase from: *Sinorhizobium* sp. *RSC Adv.* 9 (6): 2919–2927.

31 Wang, Y., Ravikumar, Y., Zhang, G. et al. (2020). Biocatalytic synthesis of d-allulose using novel D-tagatose 3-epimerase from *Christensenella minuta*. *Front. Chem.* 8: 622325.

32 Kim, H.J., Hyun, E.K., Kim, Y.S. et al. (2006). Characterization of an *Agrobacterium tumefaciens* D-psicose 3-epimerase that converts D-fructose to D-psicose. *Appl. Environ. Microbiol.* 72 (2): 981–985.

33 Mu, W., Chu, F., Xing, Q. et al. (2011). Cloning, expression, and characterization of a d-psicose 3-epimerase from *Clostridium cellulolyticum* H10. *J. Agric. Food Chem.* 59 (14): 7785–7792.

34 Jia, M., Mu, W., Chu, F. et al. (2014). A d-psicose 3-epimerase with neutral pH optimum from *Clostridium bolteae* for d-psicose production: cloning, expression, purification, and characterization. *Appl. Microbiol. Biotechnol.* 98 (2): 717–725.

35 Zhu, Y., Men, Y., Bai, W. et al. (2012). Overexpression of D-psicose 3-epimerase from Ruminococcus sp. in *Escherichia coli* and its potential application in D-psicose production. *Biotechnol. Lett.* 34 (10): 1901–1906.

36 Mu, W., Zhang, W., Fang, D. et al. (2013). Characterization of a d-psicose-producing enzyme, d-psicose 3-epimerase, from *Clostridium* sp. *Biotechnol. Lett.* 35 (9): 1481–1486.

37 Zhang, W., Fang, D., Xing, Q. et al. (2013). Characterization of a novel metal-dependent D-psicose 3-epimerase from *Clostridium scindens* 35704. *PLoS One* 8 (4): e62987.

38 Zhang, W., Fang, D., Zhang, T. et al. (2013). Characterization of a metal-dependent D-psicose 3-epimerase from a novel strain, *Desmospora* sp. 8437. *J. Agric. Food Chem.* 61 (47): 11468–11476.

39 Zhang, W., Zhang, T., Jiang, B., and Mu, W. (2016). Biochemical characterization of a d-psicose 3-epimerase from *Treponema primitia* ZAS-1 and its application on enzymatic production of d-psicose. *J. Sci. Food Agric.* 96 (1): 49–56.

40 Zhang, W., Li, H., Zhang, T. et al. (2015). Characterization of a D-psicose 3-epimerase from *Dorea* sp. CAG317 with an acidic pH optimum and a high specific activity. *J. Mol. Catal. B Enzym.* 120: 68–74.

41 Yoshihara, A., Kozakai, T., Shintani, T. et al. (2017). Purification and characterization of D-allulose 3-epimerase derived from *Arthrobacter globiformis* M30, a GRAS microorganism. *J. Biosci. Bioeng.* 123 (2): 170–176.

42 Park, C.S., Kim, T., Hong, S.H. et al. (2016). D-allulose production from D-fructose by permeabilized recombinant cells of *Corynebacterium glutamicum* cells expressing D-allulose 3-epimerase *Flavonifractor plautii*. *PLoS One* 11 (7): e0160044.

43 Yang, J., Tian, C., Zhang, T. et al. (2019). Development of food-grade expression system for d-allulose 3-epimerase preparation with tandem isoenzyme genes in *Corynebacterium glutamicum* and its application in conversion of cane molasses to D-allulose. *Biotechnol. Bioeng.* 116 (4): 745–756.

44 Tseng, W.C., Chen, C.N., Hsu, C.T. et al. (2018). Characterization of a recombinant D-allulose 3-epimerase from *Agrobacterium* sp. ATCC 31749 and identification of an important interfacial residue. *Int. J. Biol. Macromol.* 112: 767–774.

45 Mao, S., Cheng, X., Zhu, Z. et al. (2020). Engineering a thermostable version of d-allulose 3-epimerase from *Rhodopirellula baltica* via site-directed mutagenesis based on B-factors analysis. *Enzym. Microb. Technol.* 132: 10944.

46 Zhu, Z., Gao, D., Li, C. et al. (2019). Redesign of a novel d-allulose 3-epimerase from *Staphylococcus aureus* for thermostability and efficient biocatalytic production of d-allulose. *Microb. Cell Factories* 18 (1): 1–10.

47 Patel, S.N., Kaushal, G., and Singh, S.P. (2020). A novel d-allulose 3-epimerase gene from the metagenome of a thermal aquatic habitat and d-Allulose production by *Bacillus subtilis* whole-cell catalysis. *Appl. Environ. Microbiol.* 86 (5): e02605–e02619.

48 Zhu, Z., Li, L., Zhang, W. et al. (2021). Improving the enzyme property of d-allulose 3-epimerase from a thermophilic organism of *Halanaerobium congolense* through rational design. *Enzym. Microb. Technol.* 149: 109850.

49 Li, C., Li, L., Feng, Z. et al. (2021). Two-step biosynthesis of d-allulose via a multienzyme cascade for the bioconversion of fruit juices. *Food Chem.* 357: 129746.

50 Chen, J., Chen, D., Ke, M. et al. (2021). Characterization of a Recombinant d-allulose 3-epimerase from *Thermoclostridium caenicola* with potential application in d-allulose production. *Mol. Biotechnol.* 63 (6): 534–543.

51 Uechi, K., Takata, G., Fukai, Y. et al. (2013). Gene cloning and characterization of l-ribulose 3-epimerase from *Mesorhizobium loti* and its application to rare sugar production. *Biosci. Biotechnol. Biochem.* 77 (3): 511–515.

52 Zhang, L., Mu, W., Jiang, B., and Zhang, T. (2009). Characterization of d-tagatose-3-epimerase from *Rhodobacter sphaeroides* that converts d-fructose into d-psicose. *Biotechnol. Lett.* 31 (6): 857–862.

53 Sharma, M., Patel, S.N., Lata, K. et al. (2016). A novel approach of integrated bioprocessing of cane molasses for production of prebiotic and functional bioproducts. *Bioresour. Technol.* 219: 311–318. https://doi.org/10.1016/j.biortech.2016.07.131.

54 Sharma, M., Patel, S.N., Sangwan, R.S., Singh, S.P. (2017). Biotransformation of banana pseudo-stem extract into a functional juice containing value-added biomolecules of potential health benefits. *Indian J. Exp. Biol.* 55: 453–462.

55 Lata, K., Sharma, M., Patel, S.N. et al. (2018). An integrated bio-process for production of functional biomolecules utilizing raw and by-products from dairy and sugarcane industries. *Bioprocess. Biosyst. Eng.* 41 (8): 1121–1131. https://doi.org/10.1007/s00449-018-1941-0.

56 He, W., Mu, W., Jiang, B. et al. (2016). Food-grade expression of d-psicose 3-epimerase with tandem repeat genes in *Bacillus subtilis*. *J. Agric. Food Chem.* 64 (28): 5701–5707.

57 Li, Z., Li, Y., Duan, S. et al. (2015). Bioconversion of d-glucose to d-psicose with immobilized d-xylose isomerase and d-psicose 3-epimerase on *Saccharomyces cerevisiae* spores. *J. Ind. Microbiol. Biotechnol.* 42 (8): 1117–1128.

58 Roy, S., Chikkerur, J., Roy, S.C. et al. (2018). Tagatose as a potential nutraceutical: production, properties, biological roles, and applications. *J. Food Sci.* 83 (11): 2699–2709.

59 Donner, T.W., Wilber, J.F., and Ostrowski, D. (1999). d-tagatose, a novel hexose: acute effects on carbohydrate tolerance in subjects with and without type 2 diabetes. *Diabetes Obes. Metab.* 1 (5): 285–291.

60 Guo, Q., An, Y., Yun, J. et al. (2018). Enhanced d-tagatose production by spore surface-displayed l-arabinose isomerase from isolated Lactobacillus brevis PC16 and biotransformation. *Bioresour. Technol.* 247: 940–946.

61 Kim, P. (2004). Current studies on biological tagatose production using l-arabinose isomerase: a review and future perspective. *Appl. Microbiol. Biotechnol.* 65 (3): 243–249.

62 Mei, W., Wang, L., Zang, Y. et al. (2016). Characterization of an L-arabinose isomerase from *Bacillus coagulans* NL01 and its application for d-tagatose production. *BMC Biotechnol.* 16 (1): 1–11.

63 Zhang, G., An, Y., Parvez, A. et al. (2020). Exploring a highly d-galactose specific l-arabinose isomerase from *Bifidobacterium adolescentis* for d-tagatose production. *Front. Bioeng. Biotechnol.* 8: 377.

64 Oh, D.-K. (2007). Tagatose: properties, applications, and biotechnological processes. *Appl. Microbiol. Biotechnol.* 76 (1): 1–8.

65 Zhang, Y., Fan, Y., Hu, H. et al. (2017). d-Tagatose production by *Lactococcus lactis* NZ9000 cells harboring lactobacillus plantarum l-arabinose isomerase. *Indian J. Pharm. Educ. Res.* 51: 288–294.

66 Yoon, S.H., Kim, P., and Oh, D.K. (2003). Properties of l-arabinose isomerase from *Escherichia coli* as biocatalyst for tagatose production. *World J. Microbiol. Biotechnol.* 19 (1).

67 Men, Y., Zhu, Y., Zhang, L. et al. (2014). Enzymatic conversion of d-galactose to d-tagatose: cloning, overexpression and characterization of l-arabinose isomerase from *Pediococcus pentosaceus* PC-5. *Microbiol. Res.* 169 (2–3): 171–178.

68 Patel, M.J., Akhani, R.C., Patel, A.T. et al. (2017). A single and two step isomerization process for d-tagatose and l-ribose bioproduction using l-arabinose isomerase and d-lyxose isomerase. *Enzym. Microb. Technol.* 97: 27–33.

69 Cheng, L., Mu, W., and Jiang, B. (2010). Thermostable l-arabinose isomerase from *Bacillus stearothermophilus* IAM 11001 for D-tagatose production: gene cloning, purification and characterisation. *J. Sci. Food Agric.* 90 (8): 1327–1333.

70 Cheng, L., Mu, W., Zhang, T., and Jiang, B. (2010). An l-arabinose isomerase from *Acidothermus cellulolytics* ATCC 43068: cloning, expression, purification, and characterization. *Appl. Microbiol. Biotechnol.* 86 (4): 1089–1097.

71 Li, Y., Zhu, Y., Liu, A., and Sun, Y. (2011). Identification and characterization of a novel l-arabinose isomerase from *Anoxybacillus flavithermus* useful in d-tagatose production. *Extremophiles* 15 (3): 441–450.

72 Fan, C., Liu, K., Zhang, T. et al. (2014). Biochemical characterization of a thermostable l-arabinose isomerase from a thermoacidophilic bacterium, *Alicyclobacillus hesperidum* URH17-3-68. *J. Mol. Catal. B Enzym.* 102: 120–126.

73 Hung, X.G., Tseng, W.C., Liu, S.M. et al. (2014). Characterization of a thermophilic L-arabinose isomerase from *Thermoanaerobacterium saccharolyticum* NTOU1. *Biochem. Eng. J.* 83: 121–128.

74 Kim, B.C., Lee, Y.H., Lee, H.S. et al. (2002). Cloning, expression and characterization of l-arabinose isomerase from *Thermotoga neapolitana*: bioconversion of D-galactose to d-tagatose using the enzyme. *FEMS Microbiol. Lett.* 212 (1): 121–126.

75 Wanarska, M. and Kur, J. (2012). A method for the production of d-tagatose using a recombinant Pichia pastoris strain secreting β-D-galactosidase from *Arthrobacter chlorophenolicus* and a recombinant l-arabinose isomerase from Arthrobacter sp. 22c. *Microb. Cell Factories* 11: 1–15.

76 Li, Z., Gao, Y., Nakanishi, H. et al. (2013). Biosynthesis of rare hexoses using microorganisms and related enzymes. *Beilstein J. Org. Chem.* 9: 2434–2445.

77 Liu, Y., Li, S., Xu, H. et al. (2014). Efficient production of d-tagatose using a food-grade surface display system. *J. Agric. Food Chem.* 62 (28): 6756–6762.

78 Xu, Z., Li, S., Fu, F. et al. (2012). Production of D-tagatose, a functional sweetener, utilizing alginate immobilized *Lactobacillus fermentum* CGMCC2921 cells. *Appl. Biochem. Biotechnol.* 166 (4): 961–973.

79 Liang, M., Chen, M., Liu, X. et al. (2012). Bioconversion of d-galactose to d-tagatose: continuous packed bed reaction with an immobilized thermostable l-arabinose isomerase and efficient purification by selective microbial degradation. *Appl. Microbiol. Biotechnol.* 93 (4): 1469–1474.

80 Agarwal, N. and Singh, S.P. (2021). A novel trehalose synthase for the production of trehalose and trehalulose. *Microbiol. Spectr.* 9 (3): e01333–e01321.

81 Elbein, A.D., Pan, Y.T., Pastuszak, I., and Carroll, D. (2003). New insights on trehalose: a multifunctional molecule. *Glycobiology* 13 (4): 17R–27R.

82 Cai, X., Seitl, I., Mu, W. et al. (2018). Biotechnical production of trehalose through the trehalose synthase pathway: current status and future prospects. *Appl. Microbiol. Biotechnol.* 102 (7): 2965–2976.

83 Ohtake, S. and Wang, Y.J. (2011). Trehalose: current use and future applications. *J. Pharm. Sci.* 100 (6): 2020–2053.

84 Wal, P., Saxena Pal, R., and Wal, A. (2019). A Review on the sugar alternates. *Int. J. Pharm. Sci. Res.* 10 (4): 1595–1604.

85 Schiraldi, C., Di Lernia, I., and De Rosa, M. (2002). Trehalose production: exploiting novel approaches. *Trends Biotechnol.* 20 (10): 420–425.

86 Neta, T., Takada, K., and Hirasawa, M. (2000). Low-cariogenicity of trehalose as a substrate. *J. Dent.* 28 (8): 571–576.

87 Liang, J., Wang, S., and Ludescher, R.D. (2015). Effect of additives on physicochemical properties in amorphous starch matrices. *Food Chem.* 171: 298–305.

88 Newman, Y.M., Ring, S.G., and Colaco, C. (1993). The role of trehalose and other carbohydrates in biopreservation. *Biotechnol. Genet. Eng. Rev.* 11 (1): 263–294.

89 Arai, C., Arai, N., Mizote, A. et al. (2010). Trehalose prevents adipocyte hypertrophy and mitigates insulin resistance. *Nutr. Res.* 30 (12): 840–848.

90 Kopjar, M., Hribar, J., Simčič, M. et al. (2013). Effect of trehalose addition on volatiles responsible for strawberry aroma. *Nat. Prod. Commun.* 8 (12): 1934578X1300801229.

91 Yu, H., Yang, S., Yuan, C. et al. (2018). Application of biopolymers for improving the glass transition temperature of hairtail fish meat. *J. Sci. Food Agric.* 98 (4): 1437–1443.

92 Martinetti, D., Colarossi, C., Buccheri, S. et al. (2017). Effect of trehalose on cryopreservation of pure peripheral blood stem cells. *Biomed. Rep.* 6 (3): 314–318.

93 Ikeda, M., Bando, T., Yamada, T. et al. (2015). Clinical application of ET-Kyoto solution for lung transplantation. *Surg. Today* 45 (4): 439–443.

94 Lemieux, R.U. and Bauer, H.F. (1954). A chemical synthesis of d-trehalose. *Can. J. Chem.* 32 (4): 340–344.

95 Wang, J., Ren, X., Wang, R. et al. (2017). Structural characteristics and function of a new kind of thermostable trehalose synthase from *Thermobaculum terrenum*. *J. Agric. Food Chem.* 65 (35): 7726–7735.

96 Li, Y., Sun, X., Feng, Y., and Yuan, Q. (2015). Cloning, expression and activity optimization of trehalose synthase from *Thermus thermophilus* HB27. *Chem. Eng. Sci.* 135: 323–329.

97 Liang, J., Huang, R., Huang, Y. et al. (2013). Cloning, expression, properties, and functional amino acid residues of new trehalose synthase from *Thermomonospora curvata* DSM 43183. *J. Mol. Catal. B Enzym.* 90: 26–32.

98 Kim, T.K., Jang, J.H., Cho, H.Y. et al. (2010). Gene cloning and characterization of a trehalose synthase from *Corynebacterium glutamicum* ATCC13032. *Food Sci. Biotechnol.* 19 (2): 565–569.

99 Wei, Y.T., Zhu, Q.X., Luo, Z.F. et al. (2004). Cloning, expression and identification of a new trehalose synthase gene from *Thermobifida fusca* genome. *Acta Biochim. Biophys. Sin. Shanghai* 36 (7): 477–484.

100 Yan, J., Qiao, Y., Hu, J., and Ding, H. (2013). Cloning, expression and characterization of a trehalose synthase gene from *Rhodococcus opacus*. *Protein J.* 32 (3): 223–229.

101 Lin, Y.F., Su, P.C., and Chen, P.T. (2020). Production and characterization of a recombinant thermophilic trehalose synthase from *Thermus antranikianii*. *J. Biosci. Bioeng.* 129 (4): 418–422.

102 Filipkowski, P. (2012). Expression of *Deinococcus geothermalis* trehalose synthase gene in *Escherichia coli* and its enzymatic properties. *Afr. J. Biotechnol.* 11 (67): 13131–13139.

103 Zhu, Y., Wei, D., Zhang, J. et al. (2010). Overexpression and characterization of a thermostable trehalose synthase from *Meiothermus ruber*. *Extremophiles* 14 (1): 1–8.

104 Nishimoto, T., Nakada, T., Chaen, H. et al. (1996). Purification and characterization of a thermostable trehalose synthase from *Thermus aquaticus*. *Biosci. Biotechnol. Biochem.* 60 (5): 835–839.

105 Gao, Y., Xi, Y., Lu, X.L. et al. (2013). Cloning, expression and functional characterization of a novel trehalose synthase from marine *Pseudomonas* sp. P8005. *World J. Microbiol. Biotechnol.* 29 (11).

106 Wang, J.H., Tsai, M.Y., Chen, J.J. et al. (2007). Role of the C-terminal domain of *Thermus thermophilus* trehalose synthase in the thermophilicity, thermostability, and efficient production of trehalose. *J. Agric. Food Chem.* 55 (9): 3435–3443.

107 Xiuli, W., Hongbiao, D., Ming, Y., and Yu, Q. (2009). Gene cloning, expression, and characterization of a novel trehalose synthase from *Arthrobacter aurescens*. *Appl. Microbiol. Biotechnol.* 83 (3): 477–482.

108 Chen, Y.S., Lee, G.C., and Shaw, J.F. (2006). Gene cloning, expression, and biochemical characterization of a recombinant trehalose synthase from *Picrophilus torridus* in *Escherichia coli*. *J. Agric. Food Chem.* 54 (19): 7098–7104.

109 Pan, Y.T., Koroth Edavana, V., Jourdian, W.J. et al. (2004). Trehalose synthase of *Mycobacterium smegmatis*: purification, cloning, expression, and properties of the enzyme. *Eur. J. Biochem.* 271 (21): 4259–4269.

110 Yue, M., Wu, X., Gong, W.N., and Ding, H.B. (2009). Molecular cloning and expression of a novel trehalose synthase gene from *Enterobacter hormaechei*. *Microb. Cell Factories* 8: 1–7.

111 Lee, J.H., Lee, K.H., Kim, C.G. et al. (2005). Cloning and expression of a trehalose synthase from *Pseudomonas stutzeri* CJ38 in *Escherichia coli* for the production of trehalose. *Appl. Microbiol. Biotechnol.* 68 (2): 213–219.

112 Koh, S., Shin, H.J., Kim, J.S. et al. (1998). Trehalose synthesis from maltose by a thermostable trehalose synthase from *Thermus caldophilus*. *Biotechnol. Lett.* 20 (8): 757–761.

113 Zheng, J., Chen, Y., Yang, L. et al. (2014). Preparation of cross-linked enzyme aggregates of trehalose synthase via co-aggregation with polyethyleneimine. *Appl. Biochem. Biotechnol.* 174 (6): 2067–2078.

114 Trakarnpaiboon, S. and Champreda, V. (2022). Integrated whole-cell biocatalysis for trehalose production from maltose using permeabilized *Pseudomonas monteilii* cells and bioremoval of byproduct. *J. Microbiol. Biotechnol.* 32 (8): 1054–1063.

115 Agarwal, N., Narnoliya, L.K., and Singh, S.P. (2019). Characterization of a novel amylosucrase gene from the metagenome of a thermal aquatic habitat, and its use in turanose production from sucrose biomass. *Enzym. Microb. Technol.* 131: 109372.

116 Park, M.O., Chandrasekaran, M., and Yoo, S.H. (2019). Production and characterization of low-calorie turanose and digestion-resistant starch by an amylosucrase from *Neisseria subflava*. *Food Chem.* 300: 125225.

117 Wang, S., Suh, J.H., Zheng, X. et al. (2017). Identification and quantification of potential anti-inflammatory hydroxycinnamic acid amides from wolfberry. *J. Agric. Food Chem.* 65 (2): 364–372.

118 Chung, J.-Y., Kim, Y.-S., Kim, Y., and Yoo, S.-H. (2017). Regulation of Inflammation by Sucrose Isomer, Turanose, in Raw 264.7 Cells. *J. Cancer Prev.* 22 (3): 195.

119 Broadhead, D.M. and Butterworth, J. (1978). Pompe's disease: diagnosis in kidney and leucocytes using 4-methylumbelliferyl-α-D-glucopyranoside. *Clin. Genet.* 13 (6): 504–510.

120 Shibuya, T., Mandai, T., Kubota, M. et al. (2004). Production of turanose by cyclomaltodextrin glucanotransferase from *Bacillus stearothermophilus*. *J. Appl. Glycosci.* 51 (3): 223–227.

121 De Montalk, G.P., Remaud-Simeon, M., Willemot, R.M. et al. (1999). Sequence analysis of the gene encoding amylosucrase from *Neisseria polysaccharea* and characterization of the recombinant enzyme. *J. Bacteriol.* 181 (2): 375–338.

122 Van Der Veen, B.A., Potocki-Véronèse, G., Albenne, C. et al. (2004). Combinatorial engineering to enhance amylosucrase performance: construction, selection, and screening of variant libraries for increased activity. *FEBS Lett.* 560 (1–3): 91–97.

123 Zhang, H., Zhou, X., He, J. et al. (2017). Impact of amylosucrase modification on the structural and physicochemical properties of native and acid-thinned waxy corn starch. *Food Chem.* 220: 413–419.

124 Wang, R., Bae, J.S., Kim, J.H. et al. (2012). Development of an efficient bioprocess for turanose production by sucrose isomerisation reaction of amylosucrase. *Food Chem.* 132 (2): 773–779.

125 Park, M.-O., Chandrasekaran, M., and Yoo, S.-H. (2018). Expression, purification, and characterization of a novel amylosucrase from *Neisseria subflava*. *Int. J. Biol. Macromol.* 109: 160–166.

126 Hehre, E.J. and Hamilton, D.M. (1946). Bacterial synthesis of an amylopectin-like polysaccharide from sucrose. *J. Biol. Chem.* 166 (2): 777.

127 Seo, D.H., Jung, J.H., Choi, H.C. et al. (2012). Functional expression of amylosucrase, a glucan-synthesizing enzyme, from *Arthrobacter chlorophenolicus* A6. *J. Microbiol. Biotechnol.* 22 (9): 1253–1257.

128 Ha, S.-J., Seo, D.H., Jung, J.H. et al. (2009). Molecular cloning and functional expression of a new amylosucrase from *Alteromonas macleodii*. *Biosci. Biotechnol. Biochem.* 73 (7): 1505–1512.

129 But, S.Y., Khmelenina, V.N., Reshetnikov, A.S. et al. (2015). Sucrose metabolism in halotolerant methanotroph *Methylomicrobium alcaliphilum* 20Z. *Arch. Microbiol.* 197 (3): 471–480.

130 Choi, S.W., Lee, J.A., and Yoo, S.H. (2019). Sucrose-based biosynthetic process for chain-length-defined α-glucan and functional sweetener by *Bifidobacterium amylosucrase*. *Carbohydr. Polym.* 205: 581–588.

131 Wang, Y., Xu, W., Bai, Y. et al. (2017). Identification of an $α$-(1, 4)-glucan-synthesizing amylosucrase from *Cellulomonas carboniz* T26. *J. Agric. Food Chem.* 65 (10): 2110–2119.

132 Guérin, F., Barbe, S., Pizzut-Serin, S. et al. (2012). Structural investigation of the thermostability and product specificity of amylosucrase from the bacterium *Deinococcus geothermalis*. *J. Biol. Chem.* 287 (9): 6642–6654.

133 Pizzut-Serin, S., Potocki-Véronèse, G., Van Der Veen, B.A. et al. (2005). Characterisation of a novel amylosucrase from *Deinococcus radiodurans*. *FEBS Lett.* 579 (6): 1405–1410.

134 Kim, M.D., Seo, D.H., Jung, J.H. et al. (2014). Molecular cloning and expression of amylosucrase from highly radiation-resistant *Deinococcus radiopugnans*. *Food Sci. Biotechnol.* 23 (6): 2007–2012.

135 Jeong, J.W., Seo, D.H., Jung, J.H. et al. (2014). Biosynthesis of glucosyl glycerol, a compatible solute, using intermolecular transglycosylation activity of amylosucrase from *Methylobacillus flagellatus* KT. *Appl. Biochem. Biotechnol.* 173 (4): 904–917.

136 Park, M.O., Lee, B.H., Lim, E. et al. (2016). Enzymatic process for high-yield turanose production and its potential property as an adipogenesis regulator. *J. Agric. Food Chem.* 64 (23): 4758–4764.

137 Perez-Cenci, M. and Salerno, G.L. (2014). Functional characterization of *Synechococcus amylosucrase* and fructokinase encoding genes discovers two novel actors on the stage of cyanobacterial sucrose metabolism. *Plant Sci.* 224: 95–102.

138 Zhu, X., Tian, Y., Xu, W. et al. (2018). Biochemical characterization of a highly thermostable amylosucrase from *Truepera radiovictrix* DSM 17093. *Int. J. Biol. Macromol.* 116: 744–752.

139 Côté, G.L. (2007). Flavorings and other value-added products from sucrose. In: *Novel Enzyme Technology for Food Applications*, 243–269. Elsevier.

140 Hungerford, N.L., Zhang, J., Smith, T.J. et al. (2021). Feeding sugars to stingless bees: identifying the origin of trehalulose-rich honey composition. *J. Agric. Food Chem.* 69 (35): 10292–10300.

141 Fletcher, M.T., Hungerford, N.L., Webber, D. et al. (2020). Stingless bee honey, a novel source of trehalulose: a biologically active disaccharide with health benefits. *Sci. Rep.* 10 (1): 1–8.

142 Bae, J., Jun, S.J., Chang, P.S., and Yoo, S.H. (2022). A unique biochemical reaction pathway towards trehalulose synthesis by an amylosucrase isolated from *Deinococcus deserti*. *New Biotechnol.* 70: 1–8.

143 Fu, D., Zhang, X., Zhang, H. et al. (2021). Simple and efficient preparation of high-purity trehalulose from the waste syrup of isomaltulose production using solid-phase extraction followed by hydrophilic interaction chromatography. *J. Sep. Sci.* 44 (12): 2334–2342.

144 Cheetham, P.S. (1984). The extraction and mechanism of a novel isomaltulose-synthesizing enzyme from *Erwinia rhapontici*. *Biochem. J.* 220 (1).

145 Mu, W., Li, W., Wang, X. et al. (2014). Current studies on sucrose isomerase and biological isomaltulose production using sucrose isomerase. *Appl. Microbiol. Biotechnol.* 98 (15): 213–220.

146 Salvucci, M.E. (2003). Distinct sucrose isomerases catalyze trehalulose synthesis in whiteflies, *Bemisia argentifolii*, and *Erwinia rhapontici*. *Comp. Biochem. Physiol. B Biochem. Mol. Biol.* 135 (2): 385–395.

147 Iida, T. and Okuma, K. (2013). Properties of three rare sugars d-psicose, d-allose, d-tagatose and their applications. *Oleoscience* 13 (9): 435–440.

148 Hashimoto, F., Nishiumi, S., Miyake, O. et al. (2013). Metabolomics analysis of umbilical cord blood clarifies changes in saccharides associated with delivery method. *Early Hum. Dev.* 89 (5): 315–320.

149 Kannan, R.R.R., Arumugam, R., and Anantharaman, P. (2012). Chemical composition and antibacterial activity of Indian seagrasses against urinary tract pathogens. *Food Chem.* 135 (4): 2470–2473.

150 Shintani, H., Shintani, T., Sato, M., and Sato, M. (2020). d-allose, a trace component in human serum, and its pharmaceutical applicability. *Int. J. Appl. Biol. Pharm. Technol.* 11: 200–213.

151 Chen, Z., Chen, J., Zhang, W. et al. (2018). Recent research on the physiological functions, applications, and biotechnological production of d-allose. *Appl. Microbiol. Biotechnol.* 102 (10): 4269–4278.

152 Iga, Y., Nakamichi, K., Shirai, Y., and Matsuo, T. (2010). Acute and sub-chronic toxicity of d-allose in rats. *Biosci. Biotechnol. Biochem.* 74 (7): 1476–1478.

153 Ishihara, Y., Katayama, K., Sakabe, M. et al. (2011). Antioxidant properties of rare sugar d-allose: effects on mitochondrial reactive oxygen species production in Neuro2A cells. *J. Biosci. Bioeng.* 112 (6): 638–642.

154 Kimura, S., Zhang, G.X., Nishiyama, A. et al. (2005). d-allose, an all-cis aldo-hexose, suppresses development of salt-induced hypertension in Dahl rats. *J. Hypertens.* 23 (10): 1887–1894.

155 Shinohara, N., Nakamura, T., Abe, Y. et al. (2016). d-Allose attenuates overexpression of inflammatory cytokines after cerebral ischemia/reperfusion injury in Gerbil. *J. Stroke Cerebrovasc. Dis.* 25 (9): 2184–2188.

156 Kano, A., Fukumoto, T., Ohtani, K. et al. (2013). The rare sugar d-allose acts as a triggering molecule of rice defence via ROS generation. *J. Exp. Bot.* 64 (16): 4939–4951.

157 Zhang, H., Jiang, M., and Song, F. (2020). d-allose is a critical regulator of inducible plant immunity in tomato. *Physiol. Mol. Plant Pathol.* 111: 101507.

158 Mijailovic, N., Nesler, A., Perazzolli, M. et al. (2021). Rare sugars: recent advances and their potential role in sustainable crop protection. *Molecules* 26 (6): 1720.

159 Izumori, K. (2006). Izumoring: a strategy for bioproduction of all hexoses. *J. Biotechnol.* 124 (4): 717–722.

160 Bhuiyan, S.H., Itami, Y., Rokui, Y. et al. (1998). d-allose production from d-psicose using immobilized l-rhamnose isomerase. *J. Ferment. Bioeng.* 85 (5): 539–541.

161 Seo, M.J., Choi, J.H., Kang, S.H. et al. (2018). Characterization of l-rhamnose isomerase from *Clostridium stercorarium* and its application to the production of d-allose from d-allulose (d-psicose). *Biotechnol. Lett.* 40 (2): 325–334.

162 Korndörfer, I.P., Fessner, W.D., and Matthews, B.W. (2000). The structure of rhamnose isomerase from *Escherichia coli* and its relation with xylose isomerase illustrates a change between inter and intra-subunit complementation during evolution. *J. Mol. Biol.* 300 (4): 917–933.

163 Leang, K., Takada, G., Fukai, Y. et al. (2004). Novel reactions of l-rhamnose isomerase from *Pseudomonas stutzeri* and its relation with d-xylose isomerase via substrate specificity. *Biochim. Biophys. Acta (BBA)-General Subj.* 1674 (1): 68–77.

164 Poonperm, W., Takata, G., Okada, H. et al. (2007). Cloning, sequencing, overexpression and characterization of l-rhamnose isomerase from *Bacillus pallidus* Y25 for rare sugar production. *Appl. Microbiol. Biotechnol.* 76 (6): 1297–1307.

165 Lin, C.-J., Tseng, W.-C., Lin, T.-H. et al. (2010). Characterization of a thermophilic l-rhamnose isomerase from *Thermoanaerobacterium saccharolyticum* NTOU1. *J. Agric. Food Chem.* 58 (19): 10431–10436.

166 Park, C.S., Yeom, S.J., Lim, Y.R. et al. (2010). Characterization of a recombinant thermostable l-rhamnose isomerase from *Thermotoga maritima* ATCC 43589 and its application in the production of l-lyxose and l-mannose. *Biotechnol. Lett.* 32 (12): 1947–1953.

167 Takata, G., Uechi, K., Taniguchi, E. et al. (2011). Characterization of *Mesorhizobium loti* l-rhamnose isomerase and its application to l-talose production. *Biosci. Biotechnol. Biochem.* 75 (5): 1006–1009.

168 Lin, C.J., Tseng, W.C., and Fang, T.Y. (2011). Characterization of a thermophilic l-rhamnose isomerase from *Caldicellulosiruptor saccharolyticus* ATCC 43494. *J. Agric. Food Chem.* 59 (16): 8702–8708.

169 Prabhu, P., Doan, T.T.N., Jeya, M. et al. (2011). Cloning and characterization of a rhamnose isomerase from *Bacillus halodurans*. *Appl. Microbiol. Biotechnol.* 89 (3): 635–644.

170 Kim, Y.S., Shin, K.C., Lim, Y.R., and Oh, D.K. (2013). Characterization of a recombinant l-rhamnose isomerase from *Dictyoglomus turgidum* and its application for l-rhamnulose production. *Biotechnol. Lett.* 35 (2): 259–264.

171 Bai, W., Shen, J., Zhu, Y. et al. (2015). Characteristics and kinetic properties of l-rhamnose isomerase from *Bacillus subtilis* by isothermal titration calorimetry for the production of d-allose. *Food Sci. Technol. Res.* 21 (1): 13–22.

172 Xu, W., Zhang, W., Tian, Y. et al. (2017). Characterization of a novel thermostable l-rhamnose isomerase from *Thermobacillus composti* KWC4 and its application for production of d-allose. *Process Biochem.* 53: 153–161.

173 Chen, Z., Xu, W., Zhang, W. et al. (2018). Characterization of a thermostable recombinant l-rhamnose isomerase from *Caldicellulosiruptor obsidiansis* OB47 and its application for the production of l-fructose and l-rhamnulose. *J. Sci. Food Agric.* 98 (6): 2184–2193.

174 Singh, A., Rai, S.K., and Yadav, S.K. (2022). Metal-based micro-composite of L-arabinose isomerase and l-ribose isomerase for the sustainable synthesis of l-ribose and d-talose. *Colloids Surf. B: Biointerfaces* 217: 112637.

175 Singh, A., Rai, S.K., Manisha, M., and Yadav, S.K. (2021). Immobilized L-ribose isomerase for the sustained synthesis of a rare sugar d-talose. *Mol. Catal.* 511: 111723.

176 Van Overtveldt, S., Gevaert, O., Cherlet, M. et al. (2018). Converting galactose into the rare sugar talose with cellobiose 2-epimerase as biocatalyst. *Molecules* 23 (10): 2519.

177 Xiao, H., Wang, G., Wang, P., and Li, Y. (2010). Convenient synthesis of d-talose from d-galactose. *Chin. J. Chem.* 28 (7): 1229–1232.

178 Mizanur, R.M., Takata, G., and Izumori, K. (2001). Cloning and characterization of a novel gene encoding L-ribose isomerase from *Acinetobacter* sp. Strain DL-28 in *Escherichia coli*. *Biochim. Biophys. Acta Gene Struct. Expr.* 1521 (1–3): 141–145.

179 Tseng, W.C., Wu, T.J., Chang, Y.J. et al. (2017). Overexpression and characterization of a recombinant l-ribose isomerase from *Actinotalea fermentans* ATCC 43279. *J. Biotechnol.* 259: 168–174.

180 Morimoto, K., Terami, Y., Maeda, Y.I. et al. (2013). Cloning and characterization of the l-ribose isomerase gene from *Cellulomonas parahominis* MB426. *J. Biosci. Bioeng.* 115 (4): 377–381.

181 Hung, X.G., Yu, M.Y., Chen, Y.C., and Fang, T.Y. (2015). Characterization of a recombinant l-ribose isomerase from *Geodermatophilus obscurus* DSM 43160 and application of this enzyme to the production of l-ribose from l-arabinose. *J. Mar. Sci. Technol.* 23 (4): 20.

182 Mahmood, S. et al. (2020). Characterization of a recombinant l-ribose isomerase from *Mycetocola miduiensis* and its application for the production of l-ribulose. *Enzym. Microb. Technol.* 135: 109510.

11

Microbial Production of Bioactive Peptides

Adriano Gennari[1,2], *Fernanda Leonhardt*[1], *Graziela Barbosa Paludo*[1,2], *Daniel Neutzling Lehn*[1], *Gaby Renard*[3], *Giandra Volpato*[4], *and Claucia Fernanda Volken de Souza*[1,2]

[1] *Food Biotechnology Laboratory, University of Vale do Taquari – Univates, Lajeado, Rio Grande do Sul, Brazil*
[2] *Biotechnology Graduate Program, University of Vale do Taquari – Univates, Lajeado, Rio Grande do Sul, Brazil*
[3] *Quatro G Pesquisa & Desenvolvimento Ltda, TECNOPUC, Porto Alegre, Rio Grande do Sul, Brazil*
[4] *Federal Institute of Education, Science and Technology of Rio Grande do Sul, Porto Alegre, Rio Grande do Sul, Brazil*

11.1 Introduction

Peptides are protein fragments consisting of short amino acid sequences, usually from 2 to 20 residues, derived from animal, plant, and microbial sources. These peptides are called bioactive when they have one or more beneficial effects on human or animal health, such as antioxidants, antimicrobials, antihypertensive, antidiabetic, immunomodulatory, antitumor, opioid, and antithrombotic effects [1–3]. The methods to obtain bioactive peptides are (i) conventional enzymatic hydrolysis, (ii) microbial fermentation (natural or induced), (iii) combination of (i) and (ii), (iv) in vivo gastrointestinal digestion, (v) chemical hydrolysis, and (vi) chemical synthesis [4].

For the microbial production of bioactive peptides, different protein sources such as dairy, meat, fish and shellfish, vegetables, cereals, pseudocereals, microalgae, and residues from food processing, as well as complex culture media, can be employed in the fermentation process [2, 5]. The generation of these bioactive peptides involves yeast, filamentous fungi, or bacteria, whose enzymes hydrolyze protein(s) releasing the microbial bioactive peptides. Employing different cultivation conditions and microorganisms, it is possible to generate hydrolysates containing peptides with different amino acid residues, according to the enzymatic specificities of microbial proteases, and thus produce various bioactive peptides. In contrast, this bioactivity diversity usually does not occur in conventional hydrolysis processes employing specific commercial enzymes [6–8].

Among the microorganisms used in the production of bioactive peptides, yeast stands out, mainly due to their hydrolytic activity. Yeast strains express aminopeptidases and carboxypeptidases, generating a variety of biologically active peptides [9]. The fermentation of proteins from different sources with proteolytic starter cultures is another method of

Microbial Bioreactors for Industrial Molecules, First Edition.
Edited by Sudhir P. Singh and Santosh Kumar Upadhyay.

producing these biologically active peptides. The proteolytic activity of lactic acid bacteria, for example, involves a variety of proteases with different enzymatic specificities. Lactic acid bacteria possess cell membrane-associated proteinases and intracellular peptidases, such as endopeptidases, aminopeptidases, tripeptidases, and dipeptidases, for possible application in the production of bioactive peptides [10]. Fermentation employing lactic acid bacteria is considered a cost-effective and scalable way to obtain bioactive peptides. However, once the growth conditions that promote the release of these peptides are determined, modifications such as denaturation or degradation are undesirable, as they can affect the bioactivity of these biomolecules [11].

The use of bioactive peptides, mainly in the pharmaceutical and food industries, arouses great interest due to their unique characteristics. Nevertheless, its production on an industrial scale is still a challenge, especially concerning cost, yield, purity, and environmental sustainability. Bioactive peptides are obtained mainly by chemical synthesis, which leads to large consumption of solvents and as a consequence the accentuated generation of residues [12, 13]. Studies and technological advances in metabolic engineering of the enzymatic system of microorganisms and in the recombinant obtaining of active peptides are important strategies to overcome these challenges and enable the large-scale bioproduction of these microbial peptides [4, 14–17].

In this context, this chapter describes the most recent investigations on microbial generation of biologically active peptides, highlighting the identification and biological activity of peptides, the microorganisms used, and the characteristics of the culture processes. Finally, some examples of the production of recombinant peptides in microbial expression systems and the main challenges related to the concentration and purification of microbial biologically active peptides are presented.

11.2 Microbial Production of Peptides with Antioxidant Activity

Oxidation is a vital process in aerobic organisms, which also occurs in the lipid portion of foods, resulting in the formation of free radicals [18]. Synthetic or natural antioxidant compounds can prevent the effects of these radicals and reactive oxygen species on the human body and food products. Considering the health risks of synthetic antioxidants, it is necessary to identify alternative and safe natural sources of these compounds [19, 20]. Among synthetic antioxidants, microbial peptides with antioxidant activity stand out.

This biological activity of peptides is influenced by the processing conditions of protein raw material; the type of protein; the extent of hydrolysis; the proteolytic enzyme; and the structure, molecular weight, and concentration of the peptide. In addition, hydrolysis conditions, such as enzyme/substrate ratio, pH, reaction time, and temperature, also influence the antioxidant activity of the bioactive peptide [18, 21].

Several studies have shown that protein hydrolysate fractions with less than 3 kDa and peptides with less than 10 amino acid residues have higher antioxidant activity [21]. This occurs because smaller peptides are more accessible to oxidant molecules and have higher bioavailability [22]. Moreover, bioactive peptides with higher antioxidant potential consist of specific amino acids, such as histidine – which has peroxyl radical trapping and

Table 11.1 Bioprocess conditions to produce microbial peptides with antioxidant activity using different protein sources.

Source of proteins	Microorganism	Fermentation conditions	Reference
Goat milk	*Lactobacillus plantarum*	37 °C for 36 h	[23]
Bovine milk	*Lactobacillus delbrueckii*	37 °C for 96 h	[22]
Bovine milk	*Kluyveromyces marxianus* *Lactobacillus bulgaricus* and *Streptococcus thermophilus*	35 °C for 6 h	[24]
Colostrum	*Candida lipolytica* and kefir grain	30 °C for 72 h	[9]
Colostrum	*Candida lipolytica* and kefir grain	30 °C for 48 h	[25]
Bitter beans	*Lactobacillus fermentum*	37 °C for 8 days	[26]
Defatted wheat	*Bacillus subtilis*	37 °C for 3 days	[27]
Ground sorghum	*Saccharomyces cerevisiae*	37 °C for 72 h	[28]
Soft chhurpi	*Kluyveromyces marxianus* and *S. cerevisiae*	30 °C for 48 h	[29]

chelation power – and hydrophobic amino acids – which increase peptides' accessibility to target molecules [19]. Table 11.1 presents the bioprocess conditions to produce microbial peptides with antioxidant activity, using different protein sources for production of these compounds.

The in vitro antioxidant activity of bioactive peptides of microbial origin can be evaluated through different methods. The most commonly used method for this determination is 2,2-diphenyl-1-picryl-hydrazyl-hydrate (DPPH) free radical reduction and the evaluation of color decrease when an antioxidant is added to the compound 2,2-azino-bis(3-ethylbenz thiazoline-6-sulfonic acid) (ABTS) (blue-green chromophore) [19, 22]. In addition to these, several other methodologies are less frequently employed for microbial fermentation products, such as oxygen radical absorbance capacity (ORAC) assay, phosphomolybdenum method, β-carotene bleaching assay, ferric reducing antioxidant power (FRAP assay), and the cupric ions (Cu^{2+}) reducing antioxidant power (CUPRAC assay) [30].

11.3 Microbial Production of Peptides with Antimicrobial Activity

The antimicrobial activity of bioactive peptides of microbial origin depends on several structural characteristics, such as net charge and hydrophobic and hydrophilic properties. A relevant aspect for bioactive peptides to exhibit antimicrobial activity is their electrostatic interaction with negatively charged components of microbial membranes, such as lipopolysaccharide, lipoteichoic acid, and mannoproteins of bacterial membranes, and chitin and β-1,3-glucan chains of fungal membranes [31]. Thus, the higher the net positive charge of the bioactive peptide, the greater is its electrostatic interaction with the cell membrane, and consequently the greater is its antimicrobial activity [32]. The positive charge of the

Table 11.2 Production of bioactive peptides with antimicrobial activity using different protein sources and microorganisms.

Source of proteins	Microorganism	Peptides sequence	Inhibitory activity	Reference
Sheep milk	*Lactobacillus fermentum*	FAWPQYLK	>13% with *E. coli*, *Bacillus cereus*, *Salmonella typhimurium*, and *Enterococcus faecalis*	[35]
Camel milk	*Lactobacillus plantarum*	HPPGSGLL, ELLPDMPLNQ, RGLVPL	>30% with *E. coli* and *Staphylococcus aureus*	[32]
Camel milk	*L. plantarum*	32 identified peptides	5–20% with *Staphylococcus faecalis*, *Shigella dysenteriae*, *S. aureus*, and *E. coli*	[36]
Cheese whey and milk permeate	*Lactobacillus acidophilus* and *Bifidobacterium lactis*	—	MIC[a] 8–64 μg/mL with *S. aureus*, *Listeria monocytogenes*, *E. coli*, *Pseudomonas aeruginosa*, and *Salmonella enterica*	[37]
Fresh paneer whey	*L. plantarum* and *Kluyveromyces lactis*	—	17–30 mm of diameter with *Salmonella typhi*, *Salmonella abony*, *Shigella dysentariae*, *E. coli*, *B. cereus*, *L. monocytogenes*, *S. aureus* and *E. faecalis*	[38]
Corn steep liquor	*Brevibacillus laterosporus*	—	MIC[a] 1–62.5 μg/mL with *L. monocytogenes*, *S. aureus*, *B. cereus*, *Bacillus subtilis*, *E. coli*, *Salmonella* sp., *Penicillium citrinum*, *Aspergillus brasiliensis*	[39]

[a] MIC, minimum inhibitory concentration.

bioactive peptide is directly proportional to the number of amino acids arginine, lysine, and histidine present in its peptide chain [33]. Furthermore, the hydrophobicity of the peptide chain positively affects the antimicrobial activity of the bioactive peptide. Shang et al. [34] found that 40–60% of amino acid residues are hydrophobic in peptides with antimicrobial activity. Table 11.2 presents different sources of proteins and microorganisms for the production of peptides with antimicrobial activity.

11.4 Microbial Production of Peptides with Antihypertensive Activity

Bioactive peptides have several positive effects on health, among them the antihypertensive activity and inhibition of the angiotensin-converting enzyme (ACE). This protein converts angiotensin I into vasoconstrictor angiotensin II, resulting in inactivation of the vasodilator bradykinin. Thus, compounds with ACE-inhibiting effects can be used to control blood pressure in patients with hypertensive symptoms. Synthetic ACE inhibitors are widely used to treat these symptoms and heart disease [40]. However, these inhibitors may have different collateral effects, such as diarrhea, allergy, taste disturbances, and skin rashes [18]. Thus, there has been an increasing search for natural ACE inhibitors, such as bioactive peptides. Even though these peptides are less efficient compared to synthetic inhibitors, they have high tissue affinities and can be eliminated more slowly from tissues, increasing the antihypertensive effect [41].

The main characteristics of ACE-inhibiting bioactive peptides are directly related to the quantity and sequence of the amino acid residues. Crystallographic analyses have shown that long-chain peptides cannot bind to ACE-active sites, and thus, short-chain peptides (2–12 amino acids) have higher ACE-inhibitory activity. However, regardless of the size of the peptide chain, peptides that have hydrophobic (proline) and aliphatic (isoleucine and leucine) amino acids at the N-terminus may have higher ACE-inhibitory activities [42].

Biologically active peptides with ACE-inhibitory activity can be obtained using enzymes of various sources (microbial, animal, and plant). Among the commercial microbial enzymes, Alcalase, Thermolysin, Flavourzyme, Proteinase K, and Corolase PP stand out [18]. Connolly et al. [43] hydrolyzed brewers' spent grain protein using commercial enzymes Flavourzyme, Alcalase, and Corolase PP to produce bioactive peptides. The authors verified that the peptides present in the extracts permeated through a 5 kDa ultrafiltration membrane showed in vitro ACE-inhibitory activity. Contreras et al. [44] used the enzymes Thermolysin and Corolase PP in the optimization of the hydrolysis process of whey protein concentrate enriched with β-lactoglobulin. In the resulting hydrolysate, 19 peptides derived from β-lactoglobulin were identified, and among them two showed amino acid composition with potential use as ACE inhibitor. Karami et al. [45] also optimized the protein hydrolysis process of an agro-industrial byproduct (wheat germ protein hydrolysate) using Proteinase K to obtain antihypertensive peptides. Two of the released peptides showed antioxidant activity (MDATALHYENQK and SGGSYADELVSTAK) and five had an inhibitory effect on ACE activity (VALTGDNGHSDHVVHF, VDSLLTAAK, MDATALHYENQK, IGGIGTVPVGR, and SGGSYADELVSTAK).

ACE-inhibiting microbial bioactive peptides can also be obtained by lactic acid bacteria through the proteolysis of matrices from different origins: animal, such as milk and meat; marine organisms; and plants. Among them, bioactive peptides hydrolyzed from milk proteins stand out, since caseins are a promising source of bifunctional biologically active peptides, and can be applied in the initial treatment of hypertension symptoms [13].

Wu et al. [46] isolated and purified peptides from the fermentation of milk casein by *Lactobacillus delbrueckii*. The authors isolated a pentapeptide (LPYPY) with ACE-inhibitory activity, whose activity was kept after the process of simulated pepsin gastrointestinal digestion (pH 1.3). Furthermore, these authors found that peptides with molecular weights below 3 kDa had the highest ACE-inhibitory activities.

Parmar et al. [47] employed *Lactobacillus casei* and *Lactobacillus fermentum* to obtain biologically active peptides with ACE-inhibitory activity from goat milk. The product of milk fermentation by *L. fermentum* was ultrafiltered on a 10 kDa membrane, and the highest concentration of microbial peptides was identified in the retentate. In the fermentation product by *L. casei*, the peptide sequence AFPEHK was identified, with evident ACE-inhibitory activity among the biopeptides present in the permeate.

The application of *Lactobacillus* to obtain bifunctional products, probiotics, and products enriched with bioactive peptides has also been highlighted. Rutella et al. [48] added *L. casei* to the traditional process of obtaining yogurt (fermentation of cow's milk with *Streptococcus thermophilus* and *L. delbrueckii* subsp. *bulgaricus*) and evidenced the release of peptides with antihypertensive and antioxidant activities, both in the fermentation process and during the cold storage time. In this study, Rutella et al. [48] identified two ACE-inhibiting peptides, isoleucine-proline-proline (IPP) and valine-proline-proline (VPP). Anayotova, Pashova-Baltova, and Dimitrov [49] evaluated the obtainment of bioactive peptides with ACE-inhibitory activity in milk fermented by *Lactobacillus helveticus*, *S. thermophilus,* and *Lactobacillus bulgaricus*. The ACE-inhibitory effect was confirmed and the peptide sequences IPP, VPP, and ALPM were identified, demonstrating that the *L. helveticus* added to the yogurt production process has promising potential in the generation of biologically active peptides with antihypertensive activity.

Besides milk, dairy coproducts, such as cheese whey, can be used as a protein source to obtain microbial bioactive peptides by lactic acid bacteria. Daliri et al. [41] evaluated the proteolytic capacity of 34 lactic acid bacteria to produce biologically active peptides with ACE-inhibitory activity, using cheese whey as protein source. Seven hydrolysates showed ACE-inhibitory activity, whose highest values were obtained using *Pediococcus acidilactici*, and the peptides showed molecular weight below 7 kDa.

11.5 Microbial Production of Peptides with Antidiabetic Activity

Diabetes mellitus is a chronic disorder of glucose metabolism with severe clinical consequences, such as nephropathy, retinopathy, and stroke [50]. Type 2 diabetes is a metabolic disorder that results in increased blood glucose due to decreased insulin secretion by pancreatic cells and deficiency or resistance to insulin action or both. Thus, it is necessary to prevent the disease through the development of natural antidiabetic products without adverse collateral effects. Inhibition of intestinal α-glucosidase is a strategy to control hyperglycemia by delaying carbohydrate digestion and thus reducing glucose absorption. Microbial bioactive peptides can be used to regulate the metabolism of this monosaccharide [51].

Ofosu et al. [28] assessed the effects of heat treatment of sorghum in fermentation with lactic acid bacteria to obtain bioactive peptides. After the pressurized cooking and fermentation process using the microorganism *P. acidilactici*, a significant increase in the inhibition of α-glucosidase was evidenced. Lactic acid bacteria such as *L. casei*, *Lactobacillus plantarum*, and *Lactobacillus brevis* were also identified by Mushtaq et al. [52] as promising in releasing bioactive peptides with antidiabetic activity in the aqueous extract of Kalari cheese.

Lermen et al. [53] obtained a soy protein hydrolysate using a partially purified *Bacillus* sp. protease produced in a culture medium with salts, peptone, and chicken feathers. The enzyme was characterized as a serine protease and could hydrolyze isolated soy protein releasing bioactive peptides with antidiabetic effects in vitro.

11.6 Microbial Production of Peptides with Immunomodulatory Activities

Microbial bioactive peptides with immunomodulatory activities can be obtained from traditional fermented products and by alternative sources of proteins, in which these peptides are produced from controlled bioprocesses. In fermented food products, the lactic acid bacteria are the main ones responsible for producing peptides with immunomodulatory activities, helping in the prevention of chronic diseases through modulation of the immune system and anti-inflammatory action [54, 55].

An example is the maturation process of Cheddar-type cheeses made with bubaline and bovine milks, in which water-soluble peptides with in vitro anti-inflammatory property were identified [7]. In addition, Dharmisthaben et al. [56] in the process of fermentation of camel milk with *L. plantarum* verified peptides with anti-inflammatory activity by employing suitable temperature and inoculation conditions. On other hand, Luti et al. [57], in addition to the acid-lactic bacteria, employed yeast in bread production and found that the peptides generated showed in vitro anti-inflammatory activity in both yeast and baked bakery products.

Bioactive peptides may also exhibit immunomodulatory action due to their potential interactions with human cell membranes [58]. In this context, Izquierdo-González et al. [59], when investigating bioactive peptide precursors formed during kefir production, observed an increase in the immunomodulatory activity of the product after 24 hours of fermentation. Furthermore, Ebner et al. [60] identified 236 peptides derived from kefir fermentation, where two of them (YQEPVLGPVRGPFPIIV and LYQEPVLGPVRGPFPIIV) showed immunomodulatory properties. Immunomodulatory activity has also been observed in fermented products of plant origin, as reported by Zhao et al. [61] who observed the formation of peptide as secondary metabolites in fig fermentation, and attributed this biological activity to the original plant active compounds and secondary metabolites of microorganisms.

Peptides with immunomodulatory and anti-inflammatory activities can also be consumed as supplements through incorporation into food and drug formulations [6, 62]. A challenge for the employment of active bioactive peptides as supplements is the reduction of their bitter taste, which negatively influences consumer perception. For this, one alternative is the encapsulation of these bioactive peptides, which also allows for increased shelf life and controlled release of the product [63].

11.7 Microbial Production of Peptides with Antitumoral Activity

The antitumoral activity of bioactive peptides released by microbial production is related to the inhibition effect of peptide extracts during fermentation on cancer cells proliferation [64]. Bioactive peptides may induce the apoptosis of cancer cells. This mechanism is a programmed cell death regulated by genes and cannot be stopped after its start. The in vitro cytotoxicity assay of microbial bioactive peptides with anticancer activity is performed using human cancer cell lines and cisplatin as a positive control [65]. The cytotoxicity of peptides is determined by the half-maximal inhibitory concentration (IC_{50}), which is an informative measure of a drug's efficacy [66]. Table 11.3 shows peptides obtained via microbial fermentation with antitumoral activity.

Table 11.3 Peptides obtained via microbial fermentation with antitumoral, opioid, or antithrombotic activity.

Source of proteins	Microorganism	Fermentation conditions	Reference
	Antitumoral activity		
Cheonggukjang	*Bacillus subtilis*	37 °C for 48 h	[67]
Skim milk	*Lactobacillus helveticus*	37 °C for 24 h	[64]
MRS medium	*Lactococcus lactis*	—	[68]
—	*L. lactis*	—	[69]
Nutrient broth	*B. subtilis*	32 °C for 48 h, 160 rpm	[70]
Nutrient broth	*B. subtilis*	37 °C for 4 h	[71]
GYM broth	*Streptomyces* sp.	28 °C for 168 h, 95 rpm, pH 7.8	[72]
Milk	*Lactobacillus acidophilus*	37 °C for 24 h	[73]
Soymilk	*Schleiferilactobacillus harbinensis*	37 °C for 24 h	[74]
Goat milk	*Lactobacillus plantarum*	37 °C for 24 h	[75]
Cow milk	*L. plantarum* and *Lactobacillus casei*	37 °C for 24 h	[76]
Peptone from soya	*Aspergillus* spp.	—	[77]
Potato dextrose broth	*Acremonium persicinum*	28 °C for 7 days, 200 rpm	[78]
Beef extract-peptone medium agar	*Brevibacillus* sp.	25–30 °C for 36–48 h, 180–200 rpm	[79]
Medium broth containing beef extract	*Micrococcus luteus*	37 °C for 1 day, 150 rpm	[80]
	Opioid activity		
Milk	*L. helveticus*	37 °C for 12 h	[81]
Bovine milk	Kefir microorganisms	23 °C for 24 h, 800 rpm	[82]
Bovine milk	*L. lactis*, *Leuconostoc* spp., *Streptococcus thermophilus*, *Lactobacillus* spp., kefir yeast, and kefir grain microflora	25 °C until the pH decreased to 4.8	[60]
Cheese	*L. lactis* subsp. *lactis* and *L. lactis* subsp. *cremoris*	25 °C for 24 h	[83]
Casein	*L. helveticus*	37 °C for 7 h	[84]
	Antithrombotic activity		
Milk	*Lactobacillus casei* Shirota	37 °C for 42 h	[85]

Table 11.3 (Continued)

Source of proteins	Microorganism	Fermentation conditions	Reference
Milk	*Lactobacillus casei* Shirota and *Lactobacillus johnsonii* LA1	37 °C until the pH decreased to 4.5	[86]
Milk	*Lactobacillus casei* Shirota and *S. thermophilus*	—	[87]
Milk	*Lactobacillus casei* Shirota	37, 39.5, and 42 °C for 12 and 20 h, pH 6.0, 6.2, and 6.5	[88]
Amaranth	*L. casei* Shirota and *S. thermophilus*	37 °C for 36 h	[89]
Skim milk	*L. lactis, Leuconostoc* spp., *S. thermophilus, Lactobacillus* spp., kefir yeast, and kefir grain microflora	25 °C until the pH decreased to 4.8	[60]
Milk	*Lactobacillus delbrueckii* subsp. *bulgaricus* and *S. thermophilus*	37 °C for 12 h	[90]
Milk	*L. helveticus*	37 °C for 12 h	[55]
Milk	*L. lactis*	30 °C for 24 and 48 h	[91]
Milk	*Lactobacillus gasseri* H10, *L. gasseri* H11, *Lactobacillus fermentum* H4, and *L. fermentum* H9	37 °C for 12, 24, 36, and 48 h	[92]
Skim milk	*Lactobacillus paracasei* B-4564	42 °C for 16 h and 25 °C for 48 h	[93]
Goat milk kefir	Microbiota from kefir grains	25 °C for 12, 24, and 36 h	[59]
Bovine milk	Microbiota from kefir grains	23 °C for 24 h, 800 rpm	[82]

Zheng et al. [74] observed antiproliferative activity on breast cancer (MCF-7) and liver cancer (HepG2) cells of soy milk fermented by *Schleiferilactobacillus harbinensis* M1 due to the presence of hydrolyzed antitumoral peptides. Gholamhosseinpour et al. [73] observed antitumoral activity of bioactive peptides against colon adenocarcinoma (Caco-2) cells using *Lactobacillus acidophilus* PTCC1643 to release secondary metabolites from the fermentation process of cow milk. The antitumoral activity of these bioactive peptides against Caco-2 cells increased over the fermentation time due to the bacteria proteolytic system activity.

In another study, Hashemi and Gholamhosseinpour [75] used *L. plantarum* LP3 and LU5 as a source of microbial peptides derived from goat milk. The produced peptides were tested regarding their capacity to inhibit Caco-2 cells and human primary colon cell (T4056) lines. Concurrently, the effects of ongoing ultrasound treatment were investigated. The use of ultrasound at 60% amplitude increased the antitumoral effects of peptides compared with the control and ultrasound at 30% amplitude samples. Praveesh et al. [76] employed a co-culture of *L. plantarum* and *L. casei* to ferment the cow milk and observed the release of bioactive peptides with antitumor activity against human cervical carcinoma (HeLa) cells.

In addition to fermented foods, complex culture media have been employed to produce bioactive peptides with antitumoral activity. Ebrahimi et al. [80] demonstrated that

Micrococcus luteus protease was capable of releasing bioactive peptides with antitumor activity from a culture medium containing beef extract, saline solution, and thiamine. The cytotoxicity of the produced peptide was determined using MCF-7 cell lines in vitro. The peptide was found to have a 4.5 kDa molecular mass, and its IC_{50} value was 59.5 μg/mL. Using four *L. helveticus* strains (in monoculture), Elfahri et al. [64] provided in vitro evidence of the antitumoral activity of peptides released by microbial production, using reconstituted skim milk as medium. In this regard, this study evaluated the inhibition effect of peptide extracts on colon cancer (HT-29) and healthy T4056 cells.

Xiong et al. [65] reported that *Chromobacterium violaceum* No. 968 was capable of releasing bioactive peptides while fermented in broth medium. Romipeptides A and B were released, and they were tested against human cancer cell lines including acute promyelocytic leukemia (HL60), colonic carcinoma (SW620), and human lung carcinoma (A549). The two new romipeptides (A and B) presented cytotoxic activity with the IC_{50} (mol/L) value of 42.5 and 12.5 against SW620, 21.8 and 6.7 against HL-60, and 40.6 and 5.7 against A549, respectively.

Besides, marine-derived microorganisms have been applied to produce antitumoral bioactivity peptides employing complex culture media. Zheng et al. [79] demonstrated that *Brevibacillus* sp. S-1 was capable of releasing antitumoral peptides from fermentation on beef extract-peptone agar. This microbial production released a novel cytotoxic peptide (SBP) that exhibited cytotoxicity against human lung carcinoma (A549), human glioma (U251), human hepatocellular carcinoma (BEL-7402), human colon carcinoma (RKO), and MCF-7 cells. In contrast, the peptide did not exhibit substantial cytotoxicity against human normal fibroblast lung (HFL1) cells.

Ebada et al. [77] used two strains of *Aspergillus* in a co-culture to release antiproliferative peptides from peptone from soy. Three main peptide compounds were sterigmatocystin, 5-methoxysterigmatocystin, and psychrophilin E, and they were tested for their antiproliferative effects against human tumor cell lines: HCT116 (colon), K562 (leukemia), A2780CisR (cisplatin-resistant mutant), and A2780 (ovary). These three peptides exhibited selective antiproliferative activities particularly against colon cell lines with IC_{50} values (mM) of 10.3 (sterigmatocystin), 4.4 (5-methoxysterigmatocystin), and 28.5 (psychrophilin E).

Chen et al. [78] performed a microbial production with *Acremonium persicinum* SCSIO115 in potato dextrose medium. Three peptides, cordyheptapeptides C–E, were isolated from the fermentation. The cytotoxicity of the peptides was tested using MCF-7, human lung cancer (NCI-H460), and human glioblastoma (SF-268) cell lines. Cordyheptapeptide E demonstrated anticancer activity against the cell lines, with the respective IC_{50} values of 2.7, 4.5, and 3.2 μM. Cordyheptapeptide C was found to exhibit cytotoxicity against MCF-7 ($IC_{50} = 3.0$ μM) and SF-268 ($IC_{50} = 3.7$ μM) as well and a weaker cytotoxicity against NCI-H460. Although the cordyheptapeptides E and C presented good results, the peptide D exhibited no activity against the cell lines.

Lipopeptides stood out by their antitumoral activity. Zhao et al. [70] performed the production of lipopeptides with biological activity with *Bacillus subtilis* using nutrient broth. The released bacterial lipopeptides inhibited leukemia (K562) cells significantly and induced apoptosis of these cells. In Zhao et al. [70], only 7 out of more than 30 lipopeptide fractions were found to have antitumor effects related to K562 cells; these seven peptides showed similar molecular weight and sequence to those of iturin. In a more recent study, Zhao et al. [71] used *B. subtilis* strains in nutrient broth to release iturin A, a lipopeptide with multiple bioactivities, which contains a fatty acid chain and cyclic peptide. In the in vitro analysis, Zhao et al. [71] found that iturin A could enter HepG2 cells and inhibit their growth by inducing apoptosis, autophagy, and paraptosis.

Another example of lipopeptide is the surfactin, a cyclic lipopeptide with the GLLVDLL sequence. Surfactin was released by *B. subtilis* CSY191 from a fermented soybean paste [67]. This hydrolyzed lipopeptide showed antiproliferative effects against MCF-7 cells with IC_{50} value of 10 μg/mL. Aftab and Sajid [72] investigated the use of *Streptomyces* sp. SSA13 in a glucose yeast minerals salts (GYM) broth as a source to produce antitumor lipopeptides. The released cyclic lipopeptide (iturin A6) showed bioactivity against the tumor cell lines HepG2, HeLa, and MCF-7 with the respective IC_{50} values of 8.9, 1.73, and 6.44 μg/mL. Lu et al. [94] performed an experimental study on the capacity of *Bacillus amyloliquefaciens* X030 in producing peptides from fermentation in a Lennox broth (LB). A cyclic lipopeptide was isolated from the bacterial broth, bacillomycin Lb (known as bacillopeptin B). This secondary metabolite exhibited antitumor activity against several cancerous cells.

Among cyclic peptides, one of the most studied is nisin. This is a polycyclic peptide produced by *Lactococcus lactis* bacterial strains during fermentation. A study performed by Avand et al. [95] determined that the optimum culture condition for producing nisin with *L. lactis* is in the De Man, Rogosa, and Sharpe (MRS) medium supplemented with tryptone. After the cytotoxicity assay, Avand et al. [95] proved that nisin exhibits a high activity (IC_{50}: 5 μM) against MCF-7 cells. In addition, Ahmadi et al. [69] determined that nisin exhibits cytotoxic impacts on epithelial-like colon cancer (SW480) cells and induces apoptosis through an intrinsic pathway by modifying the expression levels of Bcl-2 and Bax genes. Goudarzi et al. [68] investigated the effects of nisin on K562 cells. They proved that nisin attacks these cells through mitochondrial pathway by modifying the expression of the aforesaid genes.

11.8 Microbial Production of Peptides with Opioid Activity

The opioid activities of bioactive peptides released by microbial production have been obtained mainly by the hydrolysis of casein, present in milk and dairy products, due to the action of microbial enzymes. Table 11.3 shows peptides obtained via microbial fermentation with opioid activity.

Skrzypczak et al. [55] demonstrated that *L. helveticus* was capable of releasing opioid peptides from bovine milk, especially from casein. Four peptides were found to have opiate-like activities in the study; they were matched in the BIOPEP (Bioactive Peptides) database. Two peptides were hydrolyzed from αS1-casein; one presented the sequence TTMPLW, with 747.3 Da and peptide identity f(209–214), and the other presented 1266.6 Da, YLGYLEQLLR sequence, and f(106–115) identity. The other two peptides were released from κ-casein. The f(74–83) peptide showed 1274.5 Da and the following sequence: NQFLPYPYYA. The last one, f(76–86), presented the sequence FLPYPYYAKPA with 1328.0 Da.

Other studies revealed peptides generated from casein. Ebner et al. [60] used commercial kefir starter culture (*L. lactis*, *Leuconostoc* spp., *S. thermophilus*, *Lactobacillus* spp., kefir yeast, and kefir grain microflora) to ferment bovine milk. One opioid peptide, VYPFPGPIPN, hydrolyzed from β-casein was found in the kefir peptide profile, confirming the starter culture hydrolyzing potential. Fan et al. [84] fermented casein with *L. helveticus* using the MASCOT database. One of the peptides identified showed an opioid potential, so it was hydrolyzed from β-casein presenting the sequence YPFPGPIHNSLPQ with opioid agonist activity.

Dallas et al. [82] conducted an experiment with kefir microorganisms to ferment bovine milk and release peptides with biological activities. All peptides had their sequence aligned with a peptide database. Dallas et al. [82] matched five opioid peptides: β-casomorphin-7 (YPFPGPI) with an agonist opioid activity, casoxin-A (YPSYGLN) presenting an antagonist action, β-neocasomorphin-6 (YPVEPF), pro-8-β-casomorphin-13 (YPFPGPIPNSLPQ), and pro-8-β-casomorphin 9 (YPFPGPIPN).

Galli et al. [83] used *L. lactis* subsp. *lactis* and *L. lactis* subsp. *cremoris* to produce peptides during the ripening of Camembert cheese. One of the peptides, the β-CN f(57–72), hydrolyzed from β-casein, showed opioid activity. This shows the capacity of the studied microorganisms to generate bioactive peptides with morphine-like activity.

11.9 Microbial Production of Peptides with Antithrombotic Activity

The antithrombotic activity of bioactive peptides released by microbial fermentation is related to the inhibition of fibrin cross-linking, responsible for the clotting activity [85]. This bioactivity is determined by the inhibition of fibrinogen-fibrin conversion [86]. According to Oh et al. [92], milk is known as the main source of antithrombotic peptides. Milk-derived compounds have shown potential preventive cardiovascular effects, such as reduction of cholesterol uptake, fibrinolytic activity, and inhibition of thrombin and 3-hidroxi-3-methyl-glutaril-CoA reductase (HMGR). Table 11.3 shows the peptides obtained via microbial fermentation with antithrombotic activity.

Studies have employed species of *Lactobacillus* to produce peptides with antithrombotic activity. El-Fattah et al. [93] reported that the microorganism *Lactobacillus paracasei* B-4564 is capable of releasing bioactive peptides related to cardiovascular diseases with different inhibition rates of thrombin in a skim milk matrix. Overall, the inhibition rate of thrombin increased significantly due to the decrease in fermentation temperature and increase in fermentation time. Rojas-Ronquillo et al. [85] demonstrated that *L. casei* Shirota produced biologically active peptides with antithrombotic activity from bovine milk, specifically from the β-casein protein. The most active peptide found had the following amino acid sequence, YQEPVLGPVRGPFPIIV, with 1.88 kDa, showing a thrombin inhibition efficiency rate of 4.6%. Besides the antithrombotic activity, this peptide shows bioactivity against ACEs. Guzmán-Rodríguez et al. [88] also employed *L. casei* Shirota to generate antithrombotic bioactive peptides. They used milk as culture medium to hydrolyze casein under different fermentation conditions, showing the highest antithrombotic activity of 79.1%.

Pérez-Escalante et al. [86] used strains of *L. casei* Shirota and *Lactobacillus johnsonii* LA1 separately for microbial production of peptides with fermentation of skim milk powder and lactose. The microorganisms released antithrombotic peptides with low molecular weight from bovine casein. Additionally, Oh et al. [92] used *L. fermentum* H9, *L. fermentum* H4, *Lactobacillus gasseri* H10, and *L. gasseri* H11 to release antithrombotic peptides from milk proteins using fermented Maillard reaction products. *L. gasseri* H11 was responsible for the greatest activity (inhibition rate of 41.78 ± 2.22%).

Due to the fact that κ-casein was considered to be the main source of short peptides released by microbial fermentation, Skrzypczak et al. [81] used it as a matrix of short-chain bioactive peptides. The microorganism selected for microbial production was *L. helveticus*. The study proved that κ-casein was susceptible to the activity of proteases produced by bacterial cultures and was capable of producing a peptide with antithrombotic activity: HPHLSFMAIPPK, with peptide identity f(121–132) and mass of 1373.7 Da.

Besides *Lactobacillus*, other microorganisms can be used to obtain peptides with antithrombotic activity. El-Fattah et al. [90] performed casein hydrolysis using *S. thermophilus* and *L. delbrueckii* subsp. *bulgaricus* using milk as culture medium. The antithrombotic activity increased during fermentation, reaching 52.88% after 14 days. Domínguez-González et al. [87] used *L. casei* Shirota and *S. thermophilus* concurrently to ferment milk. As a result, six antithrombotic peptides hydrolyzed from casein were released. Although fermentation was carried out with both bacteria, Domínguez-González et al. [87] assumed that the antithrombotic bioactivity should come from *L. casei* Shirota considering the results of the study performed by Rojas-Ronquillo et al. [85], who reported that no thrombin inhibition activity was produced by casein with *S. thermophilus*. In contrast, Ayala-Niño et al. [89] used the same microorganisms to hydrolyze amaranth proteins and conduct fermentation in a monoculture and with the two bacteria together. The highest antithrombotic activity was reached with the combined culture.

Kefir grains have also been highlighted to generate peptides with antithrombotic activity. Izquierdo-González et al. [59] aimed to release antithrombotic peptides from goat milk kefir using the kefir grains microbiota for microbial production. The antithrombotic peptide TAQVTSTEV, hydrolyzed from κ-casein, was identified and matched in sequence entries in the BIOPEP database. Dallas et al. [82] reported the peptide known as casoplatein (MAIPPKKNQDK), capable of producing thrombin inhibitors, hydrolyzed from a milk matrix, using a microbiota from kefir grains for microbial production. A trial performed by Ebner et al. [60] found the same peptide sequence from β-casein while performing fermentation with *Lactobacillus* spp., *Leuconostoc* spp., kefir yeast, *S. thermophilus*, kefir grain microflora, and *L. lactis*.

Additionally, Rendon-Rosales et al. [91], using strains of *L. lactis*, provided in vitro evidence of the antithrombotic activity of peptide fractions hydrolyzed from casein, using bovine milk, when exposed to simulated gastrointestinal digestion. The authors observed that microbial peptide biological activities were maintained after the simulation. In this study, the inhibition of thrombin activity was evaluated as a mechanism to prevent clots.

11.10 Production of Recombinant Peptides in Microbial Expression Systems

Recombinant DNA techniques have enabled the production of different molecules of interest by means of genetic manipulation. Biotechnological processes employing these techniques favor productivity in obtaining microbial bioactive peptides with less environmental impact [4, 14].

Recombinant expression systems involve different types of microbial cells, such as bacteria, filamentous fungi, and yeast. Among the main microorganisms of recombinant enzyme

Table 11.4 Bioactive recombinant peptides obtained in microbial expression systems.

Host microorganism	Identified bioactive peptides	Biological activities	Reference
Bacillus subtilis	Enniatin	Antibacterial, insecticidal, antifungal, herbicidal, anthelmintic, and antitumoral	[97]
Cordyceps militaris	Magainin II-cecropin B	Antibacterial and immunomodulatory	[98]
Escherichia coli	Lasioglossin LL III	Antioxidant and antimicrobial	[12]
Pichia pastoris	Recombinant Abaecin	Antimicrobial	[99]
	Recombinant PaDef	Antimicrobial	[100]
	β-Defensin 3	Antimicrobial	[101]
	Bovine Lactoferrampin and Bovine Lactoferricin	Antimicrobial	[102]
Saccharomyces cerevisiae	Pediocin PA-1	Antimicrobial	[103]

expression are *Escherichia coli, Lactobacillus* spp., *Aspergillus* spp., *Pichia pastoris,* and *Saccharomyces cerevisiae* [96]. Within this approach concerning the production of bioactive peptides, *E. coli* and *P. pastoris* stand out. They are used to obtain peptides with antimicrobial, antioxidant, immunomodulatory, and antitumor bioactivities. Table 11.4 presents microorganisms used to obtain recombinant peptides with different bioactivities.

Zhang et al. [98] expressed via the medicinal fungus *Cordyceps militaris* the hybrid antimicrobial peptide magainin II-cecropin B from *Xenopus laevis*. The peptide presented antibacterial and immunomodulatory activity in mice infected with *E. coli* ATCC25922, with potential for use as antibiotics or food additive for bovine feed.

Among the host microorganisms, *P. pastoris* stands out in the bioproduction of peptides with antimicrobial activity. Luiz et al. [99] expressed abaecine (broad-spectrum proline-rich antibacterial peptide from *Apis mellifera*) in *P. pastoris*. The gene was synthesized, cloned into the pUC57 vector, and subcloned into the pPIC9 expression vector, producing a 5200 Da peptide, which significantly inhibited *E. coli* growth after 24 hours of treatment. Another peptide with antimicrobial activity was cloned by Meng et al. [100]. The authors expressed in *P. pastoris*, transformed with the expression vector pPICZαA, the peptide produced by Mexican avocado (*Persea americana* var. *drymifolia defensin* – PaDef) tagged with 6×His at the N-terminal. This expression system allowed the production of 32.8 mg/L of recombinant PaDef, with 95.7% purity, indicating that this bioactive peptide is a promising antibiotic against pathogenic bacteria. Also, in *P. pastoris,* the mzfDB3 gene encoding the *zebrafish* β-defensin 3 peptide was expressed by employing codon optimization, tagged with 6×His at the 3′ end [101]. *Pichia pastoris* X-33 was transformed with the pPICZαA vector. The 5.9 kDa β-defensin 3 peptide showed antibacterial activity against Gram-negative (*Salmonella lignieres*, *Vibrio parahaemolyticus*, *E. coli*, and *Pseudomonas aeruginosa*) and Gram-positive (*Staphylococcus aureus*, *Listeria monocytogenes*, and *Bacillus cereus*) bacteria.

The recombinant DNA tool has also been used to produce hybrid peptides with high bioactivity, stability, and half-life. Wanmakok et al. [15] produced hybrid peptides

(L-31 and P-113) with antimicrobial activity. The sequence was cloned into the pTXB-1 vector and the peptides expressed in *E. coli* BL21, presenting antimicrobial activity against Gram-negative bacteria, and can be applied for microbial infection treatment.

11.11 Purification and Identification of Microbial Bioactive Peptides

The optimization of the purification of bioactive peptides of microbial origin is a relevant step for the development of scalable bioprocesses. The stability of the biological activity of these compounds is a challenge in large-scale production to obtain an active, safe, and standardized product [58]. The fractionation and enrichment methods should allow the recovery of these peptides with a high yield. Advances in bioseparation technologies employed in the isolation and purification of biomolecules can offer an economical and scalable production platform for microbial bioactive peptides [18].

There are several methods for separation, isolation, identification, and analysis of bioactive peptides (Figure 11.1), from detection to quantitative determination by mass

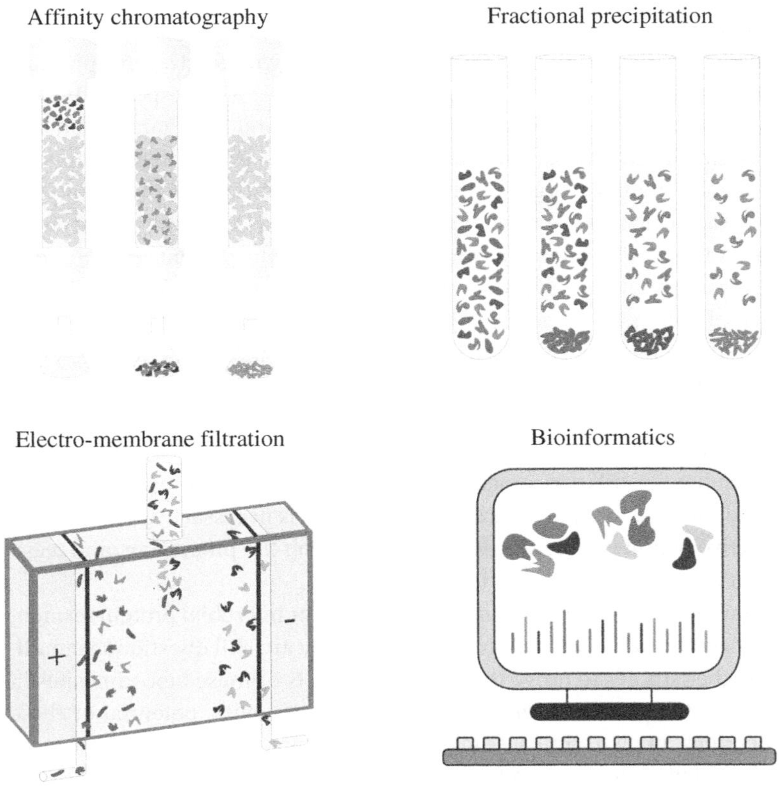

Figure 11.1 Overview of methods for isolation by affinity chromatography and fractional precipitation, concentration by electro-membrane filtration, and interaction analysis using bioinformatics tools of bioactive peptides.

spectrometry [27]. First, proteins must be separated from the other components of the fermentation matrix, and subsequently the bioactive peptides of interest are isolated. Aqueous extraction is the most commonly used technique for the separation of these biocomposites due to their solubility and stability [104].

Fractional precipitation and chromatographic methods are employed for the separation, purification, and identification of biologically active peptides resulting in products with high yield and purity. Affinity chromatography techniques are the most robust for downstream processing of these biocomposites considering selectivity and recovery parameters [104, 105]. In addition to these methods, electro-membrane filtration can be used efficiently for charged molecules such as biologically active peptides, since this technique combines electrophoresis and membrane filtration [106].

Besides, bioinformatics assists in the analysis of microbial bioactive peptides. In silico techniques help predict the nature of peptides and their bioactivities before, during, and after fermentation processes [104, 107–111].

11.12 Conclusions and Perspectives

The cultivation process of microorganisms employing different protein sources is a promising biotechnological tool for the generation of bioactive peptides with food and/or pharmaceutical grade. The bioactivity of peptides depends on protein origin, specificity, and selectivity of the hydrolytic enzymes, the amino acid sequence, and the characteristics of the uptake mechanism by the organism. Considering the microbial production of peptides, in besides to these factors, the microorganism used and the conditions of the cultivation process also influence the bioactivity. In this context, aspects such as separation, concentration, and purification of bioactive microbial peptides for food and pharmaceutical areas should be improved aiming at industrial production. Thus, the production of recombinant bioactive peptides employing microorganisms generally recognized as safe (GRAS) will play a relevant role in the competitive scenario of biotechnology industries, aiming yield, purity, environmental sustainability, and cost-effectiveness in their large-scale production processes.

Another challenge to be overcome is the scale-up of microbial production of bioactive peptides, since most studies are still performed at laboratory scale. Therefore, collaboration between academia and the food and pharmaceutical industries is necessary to promote the development of bioactive peptides with efficacy, stability during the processing, and feasible production cost for a promising commercialization.

Although studies have shown that biopeptides generated after microbial protein fermentation may present bioactivity, their effects on the human body are still questionable, indicating the need for further studies to prove the beneficial effects of these biocomposites in the body. In the current state of knowledge to identify the therapeutic potential of these biologically active peptides, in vivo clinical studies in humans are needed to clarify factors such as dosage, mode of action, toxicity, and adverse effects in the body.

References

1 Pavlicevic, M., Marmiroli, N., and Maestri, E. (2022). Immunomodulatory peptides – a promising source for novel functional food production and drug discovery. *Peptides* 148: 170696.

2 Chai, K.F., Voo, A.Y.H., and Chen, W.N. (2020). Bioactive peptides from food fermentation: a comprehensive review of their sources, bioactivities, applications, and future development. *Comprehensive Reviews in Food Science and Food Safety* 19 (6): 3825–3885.

3 Chakrabarti, S., Guha, S., and Majumder, K. (2018). Food-derived bioactive peptides in human health: challenges and opportunities. *Nutrients* 10 (11): 1738.

4 Akbarian, M., Khani, A., Eghbalpour, S., and Uversky, V.N. (2022). Bioactive peptides: synthesis, sources, applications, and proposed mechanisms of action. *International Journal of Molecular Sciences* 23 (3): 1445.

5 Chaudhary, A., Bhalla, S., Patiyal, S. et al. (2021). FermFooDb: a database of bioactive peptides derived from fermented foods. *Heliyon* 7 (4): e06668.

6 Hajfathalian, M., Ghelichi, S., García-Moreno, P.J. et al. (2018). Peptides: production, bioactivity, functionality, and applications. *Critical Reviews in Food Science and Nutrition* 58 (18): 3097–3129.

7 Rafiq, S., Huma, N., Rakariyatham, K. et al. (2018). Anti-inflammatory and anticancer activities of water-soluble peptide extracts of buffalo and cow milk Cheddar cheeses. *International Journal of Dairy Technology* 71 (2): 432–438.

8 Zhao, Y., Hu, Y., and Luo, H.-Y. (2015). Optimization of conditions for plastein reaction and its application in improving the flavor of protein hydrolyzates from *Yellowfin Tuna*. *Journal of Food Engineering and Technology* 4: 1–8.

9 Gaspar-Pintiliescu, A., Oancea, A., Cotarlet, M. et al. (2020). Angiotensin-converting enzyme inhibition, antioxidant activity and cytotoxicity of bioactive peptides from fermented bovine colostrum. *International Journal of Dairy Technology* 73 (1): 108–116.

10 Soleymanzadeh, N., Mirdamadi, S., Kianirad, M., and Mirdamadi Mirdamadi, S. (2016). Antioxidant activity of camel and bovine milk fermented by lactic acid bacteria isolated from traditional fermented camel milk (Chal). *Dairy Science and Technology* 96 (4): 443–457.

11 Baptista, D.P., Galli, B.D., Cavalheiro, F.G. et al. (2018). *Lactobacillus helveticus* LH-B02 favours the release of bioactive peptide during Prato cheese ripening. *International Dairy Journal* 87: 75–83.

12 Tanhaeian, A., Habibi Najafi, M.B., Rahnama, P., and Azghandi, M. (2020). Production of a recombinant peptide (Lasioglossin LL III) and assessment of antibacterial and antioxidant activity. *International Journal of Peptide Research and Therapeutics* 26 (2): 1021–1029.

13 Sánchez, A. and Vázquez, A. (2017). Bioactive peptides: a review. *Food Quality and Safety* 1 (1): 29–46.

14 Romero-Luna, H.E., Hernández-Mendoza, A., González-Córdova, A.F., and Peredo-Lovillo, A. (2022). Bioactive peptides produced by engineered probiotics and other food-grade bacteria: a review. *Food Chemistry: X* 13: 100196.

15 Wanmakok, M., Orrapin, S., Intorasoot, A., and Intorasoot, S. (2018). Expression in *Escherichia coli* of novel recombinant hybrid antimicrobial peptide AL32-P113 with enhanced antimicrobial activity in vitro. *Gene* 671: 1–9.

16 Singh, S.P., Upadhyay, S.K. (eds.) (2021). *Bioprospecting of Microorganism-Based Industrial Molecules*. Wiley.

17 Shintani, T., Upadhyay, S.K., and Singh, S.P. (2021). An introduction to microbial biodiversity and bioprospection. In: *Bioprospecting of Microorganism-Based Industrial Molecules* (ed. S.P. Singh and S.K. Upadhyay). Wiley. https://doi.org/10.1002/9781119717317.ch1.

18 Peighambardoust, S.H., Karami, Z., Pateiro, M., and Lorenzo, J.M. (2021). A review on health-promoting, biological, and functional aspects of bioactive peptides in food applications. *Biomolecules* 11 (5): 631.

19 Lorenzo, J.M., Munekata, P.E.S., Gómez, B. et al. (2018). Bioactive peptides as natural antioxidants in food products – a review. *Trends in Food Science & Technology* 79: 136–147.

20 Hettiarachchy, N.S., Glenn, K.C., Gnanasambandam, R., and Johnson, M.G. (1996). Natural antioxidant extract from Fenugreek (*Trigonella foenumgraecum*) for ground beef patties. *Journal of Food Science* 61 (3): 516–519.

21 Amorim, F.G., Coitinho, L.B., Dias, A.T. et al. (2019). Identification of new bioactive peptides from Kefir milk through proteopeptidomics: bioprospection of antihypertensive molecules. *Food Chemistry* 282: 109–119.

22 Wu, N., Zhao, Y., Wang, Y., and Shuang, Q. (2022). Effects of ultra-high pressure treatment on angiotensin-converting enzyme (ACE) inhibitory activity, antioxidant activity, and physicochemical properties of milk fermented with *Lactobacillus delbrueckii* QS306. *Journal of Dairy Science* 105 (3): 1837–1847.

23 Chen, L., Hui, Y., Gao, T. et al. (2021). Function and characterization of novel antioxidant peptides by fermentation with a wild *Lactobacillus plantarum* 60. *LWT - Food Science and Technology* 135: 110162.

24 Zhang, D.D., Liu, J.L., Jiang, T.M. et al. (2017). Influence of *Kluyveromyces marxianus* on proteins, peptides, and amino acids in *Lactobacillus*-fermented milk. *Food Science and Biotechnology* 26 (3): 739.

25 Cotârleţ, M., Vasile, A.M., Cantaragiu, A.M. et al. (2019). Colostrum-derived bioactive peptides obtained by fermentation with kefir grains enriched with selected yeasts. *Food Technology* 43 (1): 54–68.

26 Muhialdin, B.J., Abdul Rani, N.F., and Meor Hussin, A.S. (2020). Identification of antioxidant and antibacterial activities for the bioactive peptides generated from bitter beans (*Parkia speciosa*) via boiling and fermentation processes. *LWT - Food Science and Technology* 131: 109776.

27 Liu, F., Chen, Z., Shao, J. et al. (2017). Effect of fermentation on the peptide content, phenolics and antioxidant activity of defatted wheat germ. *Food Bioscience* 20: 141–148.

28 Ofosu, F.K., Elahi, F., Daliri, E.B.M. et al. (2022). Impact of thermal treatment and fermentation by lactic acid bacteria on sorghum metabolite changes, their antioxidant and antidiabetic activities. *Food Bioscience* 45: 101502.

29 Rai, A.K., Kumari, R., Sanjukta, S., and Sahoo, D. (2016). Production of bioactive protein hydrolysate using the yeasts isolated from soft chhurpi. *Bioresource Technology* 219: 239–245.

30 Navarro-Orcajada, S., Conesa, I., Matencio, A. et al. (2022). The use of cyclodextrins as solubility enhancers in the ORAC method may cause interference in the measurement of antioxidant activity. *Talanta* 243: 123336.

31 Farkas, A., Pap, B., Kondorosi, É., and Maróti, G. (2018). Antimicrobial activity of NCR plant peptides strongly depends on the test assays. *Frontiers in Microbiology* 9: 2600.

32 Algboory, H.L. and Muhialdin, B.J. (2021). Novel peptides contribute to the antimicrobial activity of camel milk fermented with *Lactobacillus plantarum* IS10. *Food Control* 126: 108057.

33 Sridhar, K. and Charles, A.L. (2019). In vitro antioxidant activity of Kyoho grape extracts in DPPH and ABTS assays: estimation methods for EC50 using advanced statistical programs. *Food Chemistry* 275: 41–49.

34 Shang, D., Zhang, Q., Dong, W. et al. (2016). The effects of LPS on the activity of Trp-containing antimicrobial peptides against Gram-negative bacteria and endotoxin neutralization. *Acta Biomaterialia* 33: 153–165.

35 Ashokbhai, J.K., Basaiawmoit, B., Das, S. et al. (2022). Antioxidative, antimicrobial and anti-inflammatory activities and release of ultra-filtered antioxidative and antimicrobial peptides during fermentation of sheep milk: *in-vitro*, *in-silico* and molecular interaction studies. *Food Bioscience* 47: 101666.

36 Muhialdin, B.J. and Algboory, H.L. (2018). Identification of low molecular weight antimicrobial peptides from Iraqi camel milk fermented with *Lactobacillus plantarum*. *PharmaNutrition* 6 (2): 69–73.

37 Amiri, S., Rezaei Mokarram, R., Sowti Khiabani, M. et al. (2022). Characterization of antimicrobial peptides produced by *Lactobacillus acidophilus* LA-5 and *Bifidobacterium lactis* BB-12 and their inhibitory effect against foodborne pathogens. *LWT - Food Science and Technology* 153: 112449.

38 Kadyan, S., Rashmi, H.M., Pradhan, D. et al. (2021). Effect of lactic acid bacteria and yeast fermentation on antimicrobial, antioxidative and metabolomic profile of naturally carbonated probiotic whey drink. *LWT - Food Science and Technology* 142: 111059.

39 Ning, Y., Han, P., Ma, J. et al. (2021). Characterization of brevilaterins, multiple antimicrobial peptides simultaneously produced by *Brevibacillus laterosporus* S62-9, and their application in real food system. *Food Bioscience* 42: 101091.

40 Ahmad, I., Yanuar, A., Mulia, K., and Mun'Im, A. (2017). Review of angiotensin-converting enzyme inhibitory assay: rapid method in drug discovery of herbal plants. *Pharmacognosy Reviews* 11 (21): 1.

41 Daliri, E.B.M., Lee, B.H., Park, B.J. et al. (2018). Antihypertensive peptides from whey proteins fermented by lactic acid bacteria. *Food Science and Biotechnology* 27 (6): 1781.

42 Lee, S.Y. and Hur, S.J. (2017). Antihypertensive peptides from animal products, marine organisms, and plants. *Food Chemistry* 228: 506–517.

43 Connolly, A., O'Keeffe, M.B., Piggott, C.O. et al. (2015). Generation and identification of angiotensin converting enzyme (ACE) inhibitory peptides from a brewers' spent grain protein isolate. *Food Chemistry* 176: 64–71.

44 del Contreras, M., Hernández-Ledesma, B., Amigo, L. et al. (2011). Production of antioxidant hydrolyzates from a whey protein concentrate with thermolysin: optimization by response surface methodology. *LWT - Food Science and Technology* 44 (1): 9–15.

45 Karami, Z., Peighambardoust, S.H., Hesari, J. et al. (2019). Identification and synthesis of multifunctional peptides from wheat germ hydrolysate fractions obtained by proteinase K digestion. *Journal of Food Biochemistry* 43 (4): e12800.

46 Wu, N., Xu, W., Liu, K. et al. (2019). Angiotensin-converting enzyme inhibitory peptides from *Lactobacillus delbrueckii* QS306 fermented milk. *Journal of Dairy Science* 102 (7): 5913–5921.

47 Parmar, H., Hati, S., and Sakure, A. (2018). In vitro and in silico analysis of novel ACE-inhibitory bioactive peptides derived from fermented goat milk. *International Journal of Peptide Research and Therapeutics* 24 (3): 441–453.

48 Rutella, G.S., Tagliazucchi, D., and Solieri, L. (2016). Survival and bioactivities of selected probiotic lactobacilli in yogurt fermentation and cold storage: new insights for developing a bi-functional dairy food. *Food Microbiology* 60: 54–61.

49 Panayotova, T., Pashova-Baltova, K., and Dimitrov, Z. (2018). Production of ACE-inhibitory peptides in milk fermented with selected lactic acid bacteria. *Journal of BioScience and Biotechnology* 7 (1): 31–37.

50 Alihosseini, N., Moahboob, S.A., Farrin, N. et al. (2017). Effect of probiotic fermented milk (kefir) on serum level of insulin and homocysteine in type 2 diabetes patients. *Acta Endocrinologica (Bucharest)* 13 (4): 431.

51 Vieira, C.P., Rosario, A.I.L.S., Lelis, C.A. et al. (2021). Bioactive compounds from Kefir and their potential benefits on health: a systematic review and meta-analysis. *Oxidative Medicine and Cellular Longevity* 2021: 9081738.

52 Mushtaq, M., Gani, A., and Masoodi, F.A. (2019). Himalayan cheese (Kalari/Kradi) fermented with different probiotic strains: in vitro investigation of nutraceutical properties. *LWT - Food Science and Technology* 104: 53–60.

53 Lermen, A.M., Clerici, N.J., and Daroit, D.J. (2020). Biochemical properties of a partially purified protease from *Bacillus* sp. CL18 and its use to obtain bioactive soy protein hydrolysates. *Applied Biochemistry and Biotechnology* 192 (2): 643–664.

54 Aguilar-Toalá, J.E., Santiago-López, L., Peres, C.M. et al. (2017). Assessment of multifunctional activity of bioactive peptides derived from fermented milk by specific *Lactobacillus plantarum* strains. *Journal of Dairy Science* 100 (1): 65–75.

55 Skrzypczak, K., Gustaw, W., Szwajgier, D. et al. (2017). κ-Casein as a source of short-chain bioactive peptides generated by *Lactobacillus helveticus*. *Journal of Food Science and Technology* 54 (11): 3679–3688.

56 Dharmisthaben, P., Basaiawmoit, B., Sakure, A. et al. (2021). Exploring potentials of antioxidative, anti-inflammatory activities and production of bioactive peptides in lactic fermented camel milk. *Food Bioscience* 44: 101404.

57 Luti, S., Mazzoli, L., Ramazzotti, M. et al. (2020). Antioxidant and anti-inflammatory properties of sourdoughs containing selected Lactobacilli strains are retained in breads. *Food Chemistry* 322: 126710.

58 Venegas-Ortega, M.G., Flores-Gallegos, A.C., Martínez-Hernández, J.L. et al. (2019). Production of bioactive peptides from lactic acid bacteria: a sustainable approach for healthier foods. *Comprehensive Reviews in Food Science and Food Safety* 18 (4): 1039–1051.

59 Izquierdo-González, J.J., Amil-Ruiz, F., Zazzu, S. et al. (2019). Proteomic analysis of goat milk kefir: profiling the fermentation-time dependent protein digestion and identification of potential peptides with biological activity. *Food Chemistry* 295: 456–465.

60 Ebner, J., Aşçi Arslan, A., Fedorova, M. et al. (2015). Peptide profiling of bovine kefir reveals 236 unique peptides released from caseins during its production by starter culture or kefir grains. *Journal of Proteomics* 117: 41–57.

61 Zhao, J., Gong, L., Wu, L. et al. (2020). Immunomodulatory effects of fermented fig (*Ficus carica* L.) fruit extracts on cyclophosphamide-treated mice. *Journal of Functional Foods* 75: 104219.

62 Giroux, H.J., Robitaille, G., and Britten, M. (2016). Controlled release of casein-derived peptides in the gastrointestinal environment by encapsulation in water-in-oil-in-water double emulsions. *LWT - Food Science and Technology* 69: 225–232.

63 Nirmal, N.P., Santivarangkna, C., Rajput, M.S. et al. (2022). Valorization of fish byproducts: sources to end-product applications of bioactive protein hydrolysate. *Comprehensive Reviews in Food Science and Food Safety* 21 (2): 1803–1842.

64 Elfahri, K.R., Vasiljevic, T., Yeager, T., and Donkor, O.N. (2016). Anti-colon cancer and antioxidant activities of bovine skim milk fermented by selected *Lactobacillus helveticus* strains. *Journal of Dairy Science* 99 (1): 31–40.

65 Xiong, L., Chen, C.F., Min, T.L., and Hu, H.F. (2019). Romipeptides A and B, two new romidepsin derivatives isolated from *Chromobacterium violaceum* No.968 and their antitumor activities in vitro. *Chinese Journal of Natural Medicines* 17 (2): 155–160.

66 Aykul, S. and Martinez-Hackert, E. (2016). Determination of half-maximal inhibitory concentration using biosensor-based protein interaction analysis. *Analytical Biochemistry* 508: 97–103.

67 Lee, J.H., Nam, S.H., Seo, W.T. et al. (2012). The production of surfactin during the fermentation of cheonggukjang by potential probiotic *Bacillus subtilis* CSY191 and the resultant growth suppression of MCF-7 human breast cancer cells. *Food Chemistry* 131 (4): 1347–1354.

68 Goudarzi, F., Esmaeilzadeh, M., and Yaghoubi, H. (2021). The mechanisms of anticancer activity of Nisin peptide on myelogenous leukemia cell line (K562) as a new treatment: inducing apoptosis by changing in the expression of Bax and Bcl-2 Genes. *International Journal of Peptide Research and Therapeutics* 27 (4): 2661–2670.

69 Ahmadi, S., Ghollasi, M., and Hosseini, H.M. (2017). The apoptotic impact of nisin as a potent bacteriocin on the colon cancer cells. *Microbial Pathogenesis* 111: 193–197.

70 Zhao, H., Yan, L., Xu, X. et al. (2018). Potential of *Bacillus subtilis* lipopeptides in anti-cancer I: induction of apoptosis and paraptosis and inhibition of autophagy in K562 cells. *AMB Express* 8 (1): 1–16.

71 Zhao, H., Yan, L., Guo, L. et al. (2021). Effects of *Bacillus subtilis* iturin A on HepG2 cells in vitro and vivo. *AMB Express* 11 (1): 1–12.

72 Aftab, U. and Sajid, I. (2017). Antitumor peptides from Streptomyces sp. SSA 13, isolated from Arabian Sea. *International Journal of Peptide Research and Therapeutics* 23 (2): 199–211.

73 Gholamhosseinpour, A., Hashemi, S.M.B., Raoufi Jahromi, L., and Sourki, A.H. (2020). Conventional heating, ultrasound and microwave treatments of milk: fermentation efficiency and biological activities. *International Dairy Journal* 110: 104809.

74 Zheng, Y., Li, L., Jin, Z. et al. (2021). Characterization of fermented soymilk by *Schleiferilactobacillus harbinensis* M1, based on the whole-genome sequence and corresponding phenotypes. *LWT - Food Science and Technology* 144: 111237.

75 Hashemi, S.M.B. and Gholamhosseinpour, A. (2020). Effect of ultrasonication treatment and fermentation by probiotic *Lactobacillus plantarum* strains on goat milk bioactivities. *International Journal of Food Science & Technology* 55 (6): 2642–2649.

76 Praveesh, B.V., Angayarkanni, J., and Palaniswamy, M. (2011). Antihypertensive and anticancer effect of cow milk fermented by *Lactobacillus plantarum* and *Lactobacillus casei*. *International Journal of Pharmacy and Pharmaceutical Sciences* 3 (5): 452–456.

77 Ebada, S.S., Fischer, T., Hamacher, A. et al. (2014). Psychrophilin E, a new cyclotripeptide, from co-fermentation of two marine alga-derived fungi of the genus *Aspergillus*. *Natural Product Research* 28 (11): 776–781.

78 Chen, Z., Song, Y., Chen, Y. et al. (2012). Cyclic heptapeptides, cordyheptapeptides C-E, from the marine-derived fungus *Acremonium persicinum* SCSIO 115 and their cytotoxic activities. *Journal of Natural Products* 75 (6): 1215–1219.

79 Zheng, L., Yi, Y., Liu, J. et al. (2014). Isolation and characterization of marine *Brevibacillus* sp. S-1 collected from south China Sea and a novel antitumor peptide produced by the strain. *PLoS One* 9 (11): e111270.

80 Ebrahimi, L., Nawaz, K.A.A., Kiran, K.G. et al. (2016). Production and purification of angiotensin-converting enzyme inhibitor by selected bacterial strain for cancer therapy. *International Journal of Pharmacy and Pharmaceutical Sciences* 8 (2): 80.

81 Skrzypczak, K.W., Gustaw, W.Z., Jabłonska-Ryś, E.D. et al. (2017). Antioxidative properties of milk protein preparations fermented by Polish strains of *Lactobacillus helveticus*. *Acta Scientiarum Polonorum. Technologia Alimentaria* 16 (2): 199–207.

82 Dallas, D.C., Citerne, F., Tian, T. et al. (2016). Peptidomic analysis reveals proteolytic activity of kefir microorganisms on bovine milk proteins. *Food Chemistry* 197 (Pt A): 273–284.

83 Galli, B.D., Baptista, D.P., Cavalheiro, F.G. et al. (2019). Peptide profile of Camembert-type cheese: effect of heat treatment and adjunct culture *Lactobacillus rhamnosus* GG. *Food Research International* 123: 393–402.

84 Fan, M., Guo, T., Li, W. et al. (2019). Isolation and identification of novel casein-derived bioactive peptides and potential functions in fermented casein with *Lactobacillus helveticus*. *Food Science and Human Wellness* 8 (2): 156–176.

85 Rojas-Ronquillo, R., Cruz-Guerrero, A., Flores-Nájera, A. et al. (2012). Antithrombotic and angiotensin-converting enzyme inhibitory properties of peptides released from bovine casein by *Lactobacillus casei* Shirota. *International Dairy Journal* 26 (2): 147–154.

86 Pérez-Escalante, E., Jaimez-Ordaz, J., Castañeda-Ovando, A. et al. (2018). Antithrombotic activity of milk protein hydrolysates by lactic acid bacteria isolated from commercial fermented milks. *Brazilian Archives of Biology and Technology* 61: 18180132.

87 Domínguez-González, K.N., Cruz-Guerrero, A., González-Márquez, H. et al. (2014). Antihypertensive and antithrombotic activities of a commercial fermented milk product made with *Lactobacillus casei* Shirota and *Streptococcus thermophilus*. *International Journal of Dairy Technology* 67 (3): 358–364.

88 Guzmán-Rodríguez, F., Gómez-Ruizy, L., Rodríguez-Serrano, G. et al. (2019). Iron binding and antithrombotic peptides released during the fermentation of milk by *Lactobacillus casei* Shirota. *Revista Mexicana de Ingeniera Quimica* 18 (3): 1161–1166.

89 Ayala-Ninõ, A., Rodríguez-Serrano, G.M., Jiménez-Alvarado, R. et al. (2019). Bioactivity of peptides released during lactic fermentation of amaranth proteins with potential cardiovascular protective effect: an in vitro study. *Journal of Medicinal Food* 22 (10): 976–981.

90 Abd El-Fattah, A., Sakr, S., El-Dieb, S., and Elkashef, H. (2018). Developing functional yogurt rich in bioactive peptides and gamma-aminobutyric acid related to cardiovascular health. *LWT - Food Science and Technology* 98: 390–397.

91 Rendon-Rosales, M.Á., Torres-Llanez, M.J., González-Córdova, A.F. et al. (2019). In vitro antithrombotic and hypocholesterolemic activities of milk fermented with specific strains of *Lactococcus lactis*. *Nutrients* 11 (9): 2150.

92 Oh, N.S., Kwon, H.S., Lee, H.A. et al. (2014). Preventive effect of fermented Maillard reaction products from milk proteins in cardiovascular health. *Journal of Dairy Science* 97 (6): 3300–3313.

93 Abd El-Fattah, A., Sakr, S., El-Dieb, S.M., and Elkashef, H. (2017). Biological activities of lactobacilli relevant to cardiovascular health in skim milk. *Food Science and Biotechnology* 26 (6): 1613.

94 Lu, J.Y., Zhou, K., Huang, W.T. et al. (2019). A comprehensive genomic and growth proteomic analysis of antitumor lipopeptide bacillomycin Lb biosynthesis in *Bacillus amyloliquefaciens* X030. *Applied Microbiology and Biotechnology* 103 (18): 7647–7662.

95 Avand, A., Akbari, V., and Shafizadegan, S. (2018). In vitro cytotoxic activity of a *Lactococcus lactis* antimicrobial peptide against breast cancer cells. *Iranian Journal of Biotechnology* 16 (3): e1867.

96 Baghban, R., Farajnia, S., Rajabibazl, M. et al. (2019). Yeast expression systems: overview and recent advances. *Molecular Biotechnology* 61: 365–384.

97 Zobel, S., Kumpfmüller, J., Süssmuth, R.D., and Schweder, T. (2015). *Bacillus subtilis* as heterologous host for the secretory production of the non-ribosomal cyclodepsipeptide enniatin. *Applied Microbiology and Biotechnology* 99 (2): 681–691.

98 Zhang, M., Shan, Y., Gao, H. et al. (2018). Expression of a recombinant hybrid antimicrobial peptide magainin II-cecropin B in the mycelium of the medicinal fungus *Cordyceps militaris* and its validation in mice. *Microbial Cell Factories* 17 (1): 1–14.

99 Luiz, D.P., Almeida, J.F., Goulart, L.R. et al. (2017). Heterologous expression of abaecin peptide from *Apis mellifera* in *Pichia pastoris*. *Microbial Cell Factories* 16 (1): 1–7.

100 Meng, D.M., Zhao, J.F., Ling, X. et al. (2017). Recombinant expression, purification and antimicrobial activity of a novel antimicrobial peptide PaDef in *Pichia pastoris*. *Protein Expression and Purification* 130: 90–99.

101 Tao, Y., Song, C.F., and Li, W. (2017). Expression of the Zebrafish β-defensin 3 mature peptide in *Pichia pastoris* and its purification and antibacterial activity 1. *Applied Biochemistry and Microbiology* 53 (6): 661–668.

102 Tang, X.S., Tang, Z.R., Wang, S.P. et al. (2012). Expression, purification, and antibacterial activity of bovine lactoferrampin-lactoferricin in *Pichia pastoris*. *Applied Biochemistry and Biotechnology* 166 (3): 640–651.

103 Nguyen, T.P.A., Nguyen, T.T.M., Nguyen, N.H. et al. (2020). Application of yeast surface display system in expression of recombinant pediocin PA-1 in *Saccharomyces cerevisiae*. *Folia Microbiologica* 65 (6): 955–961.

104 Franca-Oliveira, G., Fornari, T., and Hernández-Ledesma, B. (2021). A review on the extraction and processing of natural source-derived proteins through eco-innovative approaches. *Processes* 9 (9): 1626.

105 Liu, S., Li, Z., Yu, B. et al. (2020). Recent advances on protein separation and purification methods. *Advances in Colloid and Interface Science* 284: 102254.

106 Esfandi, R., Walters, M.E., and Tsopmo, A. (2019). Antioxidant properties and potential mechanisms of hydrolyzed proteins and peptides from cereals. *Heliyon* 5 (4): e01538.

107 Abedin, M.M., Chourasia, R., Phukon, L.C. et al. (2022). Characterization of ACE inhibitory and antioxidant peptides in yak and cow milk hard chhurpi cheese of the Sikkim Himalayan region. *Food Chemistry: X* 13: 100231.

108 Padhi, S., Chaurasia, R., Kumari, M., et al. (2022). Production and characterization of bioactive peptides from rice beans using *Bacillus subtilis*. *Bioresource Technology* 351: 126932.

109 Samurailatpam, S., Padhi, S., Sarkar, P. et al. (2021). Production, characterization and molecular docking of antioxidant peptides from peptidome of kinema fermented with proteolytic Bacillus spp. *Food Research International* 141: 110161.

110 Padhi, S., Samurailatpam, S., Chourasia, R. et al. (2021). A multifunctional peptide from Bacillus fermented soybean for effective inhibition of SARS-CoV-2 S1 receptor binding domain and modulation of Toll like receptor 4: a molecular docking study. *Frontiers in Molecular Biosciences* 8: 636647.

111 Chourasia, R., Padhi, S., Phukon, L.C. et al. (2020). A potential peptide from soy cheese produced using Lactobacillus delbrueckii WS4 for effective inhibition of SARS-CoV-2 main protease and S1 glycoprotein. *Frontiers in Molecular Biosciences* 7: 601753.

12

Trends in Microbial Sources of Oils, Fats, and Fatty Acids for Industrial Use

Alaa Kareem Niamah[1], Deepak Kumar Verma[2], Shayma Thyab Gddoa Al-Sahlany[1], Soubhagya Tripathy[2], Smita Singh[3], Nihir Shah[4], Ami R. Patel[4], Mamta Thakur[5], Gemilang Lara Utama[6,7], Mónica L. Chávez-González[8], and Cristobal Noe Aguilar[8]

[1] *Department of Food Science, College of Agriculture, University of Basrah, Basra City, Iraq*
[2] *Agricultural and Food Engineering Department, Indian Institute of Technology Kharagpur, Kharagpur, West Bengal, India*
[3] *Department of Allied Health Sciences, Chitkara School of Health Sciences, Chitkara University, Rajpura, Punjab, India*
[4] *Division of Dairy Microbiology, Mansinhbhai Institute of Dairy and Food Technology-MIDFT, Dudhsagar Dairy Campus, Mehsana, Gujarat, India*
[5] *Department of Food Technology, School of Sciences, ITM University, Gwalior, Madhya Pradesh, India*
[6] *Faculty of Agro-Industrial Technology, Universitas Padjadjaran, Sumedang, Indonesia*
[7] *Center for Environment and Sustainability Science, Universitas Padjadjaran, Bandung, Indonesia*
[8] *Bioprocesses and Bioproducts Research Group, Food Research Department, School of Chemistry, Autonomous University of Coahuila, Saltillo, Coahuila, Mexico*

12.1 Introduction

The vast majority of oils and fats found around the globe come from either plants or animals. Triacylglycerols, which are more commonly known as triglycerides, may be found in these in practically every instance. The cost of producing oils and fats from microorganisms is substantially higher than the cost of acquiring the same oils and fats from plants, which results in microorganisms contributing a much smaller proportion of the total. Animal fats, which are often produced as byproducts or primary products of the meat and dairy industries, were historically quite affordable. This is because animal fats are normally produced as a byproduct of other products. As a result, there is an urgent need to produce high-value oils and lipids to compensate for the high production costs so unless we are to rely on large-scale fermentation technology to grow bacteria in sufficient quantities to deliver practical and usable amounts of their triglycerides [1].

The production of oils and fats from microorganisms as an alternative to supplies obtained from agriculture and animals is an idea that has been proposed in several scientific discussions. However, in order for microbial oils to be produced on a commercial scale, they will eventually have to compete with traditional lipid products [2, 3, 4]. Given the continuously falling prices of plant oils and animal fats, it is self-evident that there is zero

Microbial Bioreactors for Industrial Molecules, First Edition.
Edited by Sudhir P. Singh and Santosh Kumar Upadhyay.

possibility that microbial oils comparable to these could ever be produced in an economically viable manner. In light of this, if microorganisms are to be regarded as potential providers of single-cell oils (SCOs), then these oils will either need to be exceedingly specialized, which is currently expensive to get from animal and agricultural sources, or not often produced by animal and agricultural sources. There is an astonishing variety of fatty acid structures to be discovered in the world of microbes; yet, many minor fatty acids are difficult to get in sufficient quantities, despite the fact that they may have potential applications. The increasing need for lipids with a high economic value has been an impetus behind the production of polyunsaturated fatty acids (PUFAs) [5]. These acids are used for a variety of functions, including nutrition and health.

After the development of gas-liquid chromatography (GLC) analysis in the 1960s, there was no scientific basis for presuming that the fatty acids that are found in the majority of microorganisms would be harmful in any way to people or other animals that might ingest them. This is because there was no evidence to support such a hypothesis. Instead, it became clear very soon that these fatty acids were the same as those that are present in microorganisms of a higher level. In addition, research showed that triacylglycerols and storage techniques in the oils produced by microbes were identical to those found in plants and animals [6].

The term "oil-bearing" is what is meant to be understood by the word "oleaginous," which refers to microorganisms that store oils within their cells. It was initially proposed that the term should only be used to describe any microbe that accumulates more than 20% of its biomass as storage lipid, as levels somewhat lower than this may simply be caused by the characteristics of the organism's development. However, it was ultimately decided that the term should be used to describe any microbe that accumulates more than 20% of its biomass as storage (Table 12.1). In hindsight, it has been shown that a minimum lipid content of 20% serves as a reasonable empirical criterion for establishing oleaginicity [20]. Recent research has focused on determining whether or not several bacteria that store hydrophobic lipids might serve as possible sources of SCOs, particularly those that are utilized in the production of biodiesel. Because the maximal levels of lipid accumulation that are possible for various species and even different strains of the same species can vary substantially [22], the amount of lipid that can be accumulated depends on the genetic make-up of the microorganisms. The term SCOs is presently being used extensively to describe oils that are being examined for usage in the biodiesel industry to make methyl fatty acid esters in addition to oils that are meant for both human and animal use. Because there is presently no commercially accessible technology that is developed particularly to produce SCOs for the biofuel sector, this chapter will not address these concerns because they are not relevant to the topic at hand.

As a result, the primary objective of this chapter is to provide knowledge of the most recent tendencies and technical breakthroughs in microbial sources of oils, fats, and fatty acids that have been carried out by a variety of groups of researchers and scientists. In addition, this chapter delves into the topic of several types of microorganisms, such as algae, bacteria, fungi, and yeasts, which have been linked to diverse sources of oils, fats, and fatty acids. In addition to this, microbial oils, fats, and fatty acids have also received a lot of attention for the industrial uses that they have in the food and health industries. At the end of the chapter, possibilities for study in the field of microbial oils, fats, and fatty acids as

Table 12.1 Oil content of some oleaginous microorganisms.

Microorganisms		Oil content (%) in cell on a dry weight basis	References
Microalgae	*Botryococcus braunii*	10–40	[7, 8]
	Chlorella vulgaris	25–63	[8, 10]
	Crypthecodinium cohnii	20–50	[11]
	Cylindrotheca fusiformis	4.68	[12]
	Nitzschia spp.	28–37	[13]
	Phaeodactylum tricornutum	20–30	[14]
	Schizochytrium spp.	50–77	[15]
	Tetraselmis suecia	15–23	[15]
Bacteria	*Acinetobacter calcoaceticus*	2–25	[16]
	Rhodococcus spp.	4–85	[17]
	Bacillus spp.	57	[17]
	Pseudomonas spp.	38–92	[18]
	Serratia spp.	64	[17]
Yeasts	*Aureobasidium melanogenum*	32–66	[19]
	Rhodosporidium spp.	48–70	[19]
	Yarrowia lipolytica	40–50	[19]
	Rhodosporidium spp.	62–74	[19]
	Rhodosporidiobolus fluvialis	55–64	[19]
Molds	*Aspergillus oryzae*	57	[20]
	Mucor spp.	28–32	[21]
	Mortierella isabellina	29	[21]
	Phanerochaete chrysosporium	>40	[21]

well as their potential future have been highlighted. These opportunities are aimed at researchers, scientists, and industry professionals.

12.2 Microbial Sources

The oleochemicals sector of the economy places a significant emphasis on the use of oils and fats derived from either plant or animal sources. The debate regarding whether it is more important to prioritize food or fuel has, on the other hand, brought to light concerns regarding the long-term viability and security of food supplies. As a result, traditional feedstocks have had to be replaced with oils made from inedible crops, used cooking oil, and animal tallow. The necessity to refine oils acquired from waste and non-edible products before they can be utilized in industrial processing boosts the cost of the entire production process [15]. This requirement places a limitation on the use of such resources at the current time.

Because the oil that is generated in microbial reactors has a fatty acid profile that is equivalent to that of plant oils, it is capable of taking the place of traditional sources in industries that are focused on the production of oleochemicals. The accumulation of lipids can reach as high as 70% of the dry biomass depending on the strain of various oleaginous microorganisms (bacteria, microalgae, yeast, and fungus) as well as the conditions of their cultivation. Oleaginous microorganisms can be used to produce biodiesel or as dietary supplements, depending on how their fatty acids are produced [16, 19, 21].

Researchers have spent the past three to four decades investigating the role that oleaginous bacteria can play in the conversion of feed material into triacylglycerols. This conversion involves the use of materials like glucose. Because of this, researchers and scientists now have a reasonable knowledge of how this is accomplished on a biochemical level [1, 2, 5]. When microorganisms move from the equitable phase of growth into the lipid consolidation stage, they are no longer required to produce large amounts of adenosine triphosphate (ATP). ATP is the metabolically available energy that is used by cells to continue synthesizing new cells and cellular components. This is the crucial step in the procedure. In the mitochondria of the cell, the tricarboxylic acid (TCA) cycle, which is also known as the Krebs cycle, and the process of oxidative phosphorylation, in which the reduced cofactors involved in the enzyme reactions are converted to their oxidized counterparts, work together to produce ATP from adenosine diphosphate (ADP). One of the most significant mechanisms is called isocitrate dehydrogenase (ICDH) catalysis (Figure 12.1).

12.2.1 Microalgal Sources

Microalgae have been used by humans and animals alike for their nutritional value for a very long time; nevertheless, it was not until recently that their cultivation and harvesting on an industrial scale began to considerably increase. The biology of microalgae has been researched more thoroughly than it has been in the previous 50 years owing to the

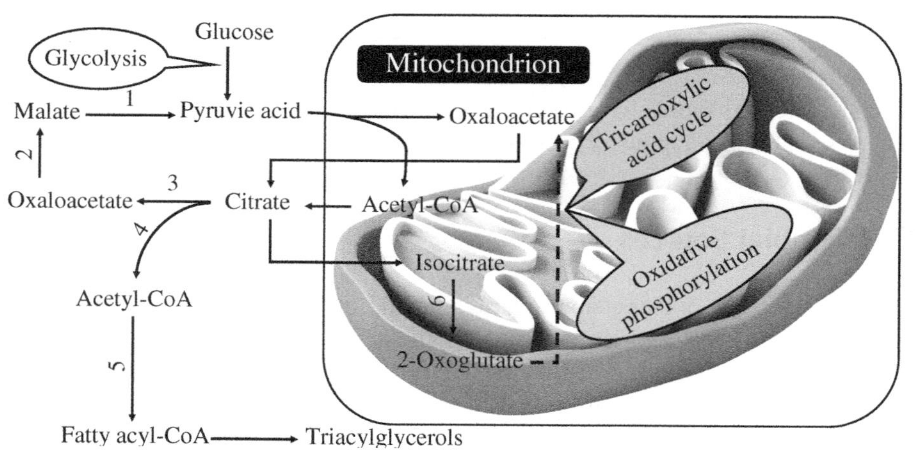

Figure 12.1 The metabolism ways of oleaginous microorganisms to fatty acid production and lipid accumulation in cells (1: malic enzyme; 2: malate dehydrogenase; 3: citrate lyase; 4: adenosine triphosphate + co-enzyme A; 5: fatty acid synthase; 6: isocitrate dehydrogenase).

incorporation of genetics and molecular biology methodologies into research programs in the 1970s. This is the longest period of time during which such research has been conducted. In recent times, the generation of energy and resources through the use of microalgae has garnered a significant amount of thought [23]. In contrast to terrestrial crops, which take a whole growing season to mature and only have an oil content that can reach a maximum of around 5% of their dry weight, microalgae develop quickly and have a high oil content [23]. The construction of large-scale photobioreactors that are capable of functioning under certain ideal circumstances while also posing a minimum danger of contamination is essential for the future of microalgal biotechnology. Although closed systems are able to eliminate the bulk of the problems that are caused by open-air systems, there is still a need for the development of closed culturing systems that are both more inexpensive and more effective [24].

There are three processes that must be carefully considered in order to successfully produce microalgal lipids that may be used for a number of purposes. These include (i) maximizing the extraction of the lipids that have accumulated in the microalgal cells, (ii) optimizing the culture conditions to achieve maximum lipid production, and (iii) determining the fatty acid profile that is best suited for biodiesel production and achieving maximum lipid production. However, environmental factors and the conditions of the production process can influence the type and quantity of lipid that is produced by microalgae. Because this might pose a risk to the process as a whole, it is imperative that rigorous monitoring takes place [25]. During the procedure for producing biodiesel, these microalgal lipids go through a process called "transesterification," in which they interact with alcohol in the presence of a catalyst. In the course of the process, glycerin, which is a structural component of triglycerides, is removed. As a result, around 10% of the oil's initial weight is retained (Figure 12.2).

In order to comprehend the single-cell dynamics that occur during the process of lipid production in microalgae and to conduct an analysis of a well-known model experiment, a novel single-cell analytic approach was utilized. Sandmann and coworkers asserted that the findings of these investigations can be helpful for further study into the method by which microalgal lipids are accumulated [26]. The assessment methodologies for enhancing the overall economic viability of the lipid for trade revealed that a microalgal bioprocess utilizing a microalgal strain with desirable features is essential for reducing the expenses associated with the production of biodiesel. The manufacture of biodiesel from algae is now the primary focus of research that attempts to address both the issue of energy sustainability and the issue of environmental sustainability. Tan and coworkers emphasized that it is of the utmost significance to conduct research into the practicability of using this

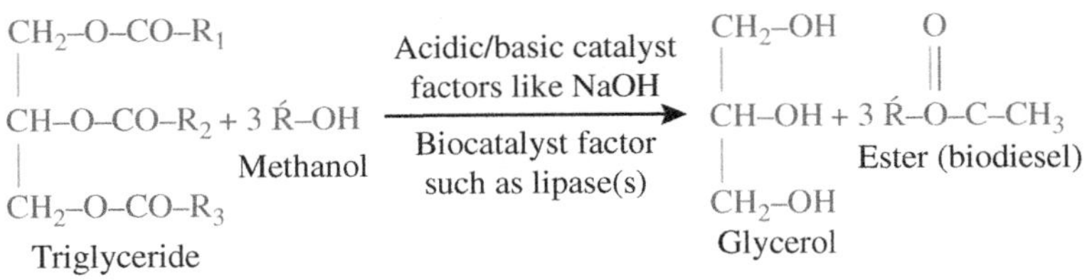

Figure 12.2 Catalyzed biochemical process for transesterification of a triglyceride with methanol.

innovative feedstock for the production of valuable byproducts and renewable fuels on an industrial scale [27].

12.2.2 Bacterial Sources

Although bacteria have a fast rate of cell proliferation, in comparison to fungi and algae, they acquire less amount of lipids. Under standard culture conditions, bacterial lipids are produced in the form of minute droplets inside the bacterial cytoplasm at high cell growth rates. Certain strains of bacteria are able to collect oil when exposed to a certain growth condition. Polyhydroxyalkanoic acids are the most common form of neutral lipid found in the majority of bacterial species. These acids are also utilized as intracellular carbon and energy storage compounds. According to the findings of Koreti and coworkers on triacylglycerol, the accumulation of lipids is more likely to take place during the stationary growth phase of a bacterial culture. The composition and concentrations of microbial lipids change in different ways depending on the metabolic processes involved in their formation. The processes involved in lipid synthesis are susceptible to influence from a variety of attributes, such as culture conditions, carbon sources, nitrogen sources, temperature, pH, and the availability of nutrients [18].

A variety of bacterial species, including *Rhodococcus*, *Mycobacterium*, *Streptomyces*, *Nocardia*, and *Acinetobacter*, are responsible for the production of the largest quantities of triacylglycerols. Gram-positive bacteria such as *Rhodococcus opacus* and *Arthrobacter* spp. have the potential to store fatty acids making up as much as 87% of their cellular dry weight. These bacteria also have significant biomass. *R. opacus*, an oleaginous bacterial strain that is amenable to pilot-scale fermentation and may be optimized, has been the subject of the most study. This strain can collect triacylglycerol at a rate of up to 86% of its cellular dry weight. According to reports, *Clostridium* is capable of producing and storing hydrocarbons with a molecular weight range that extends from C11 to C35 [28]. The majority of these hydrocarbons are classified as upper-range *n*-alkanes and medium-chain *n*-alkanes, which fall within the range of C18 to C27 [28]. The halotolerant microorganisms include the bacterium *Vibrio furnissii*, which produces internal and extracellular hydrocarbons with molecular weights between C15 and C24 and characteristics similar to kerosene and light oil [29]. These hydrocarbons are produced by the bacterium *V. furnissii*. In contrast to Gram-positive bacteria, Gram-negative bacterial strains have not been subjected to nearly as much research about the accumulation of triacylglycerol. The cultivation of these bacterial strains on various carbon sources results in an increase in certain numbers (*n*-alkanes or olive oil). A Gram-negative strain of Aeromonas can have fatty acids comprising as much as 12% of its cellular dry weight [30]. Of these fatty acids, 30% are eicosatetraenoic acid, which is a kind of PUFA.

For the purpose of treating wastewater, oleaginous Gram-negative bacterial strains of the genus *Nitratireductor* have been developed. These bacteria utilize short-chain organic acids as their primary source of carbon. This particular bacterial strain is also capable of producing combinations of fatty acids such as triacylglycerol, squalene, and the methyl ester of 2-butenoic acid. These are just a few examples. Genetic engineering is the most cutting-edge technique now available for the purpose of optimizing and selecting carbon sources for the synthesis of lipids and fatty acids [31, 32].

12.2.3 Fungal and Yeast Sources

Fungi are classified as oleaginous if they have the ability to accumulate lipids at a rate that is higher than 20% of their dry cell weight. In point of fact, depending on the dry weight of their cells, many oleaginous yeasts are capable of accumulating more than 40% lipids. Triacylglycerols and sterol esters are the principal types of lipids that are produced by oleaginous yeasts. These lipid classes are found deposited in the lipid particles that are found in the cells of oleaginous yeast. The filamentous fungus *Mortierella alpina* has the potential to accumulate considerable amounts of the triacylglycerol that contains PUFAs with a C20 chain length. In point of fact, *M. alpina* is capable of producing up to 20 g/L of culture broth worth of triacylglycerol, and between 30 and 70% of the total fatty acid is arachidonic acid, which is an essential PUFA for cellular signaling and structural purposes. The positive effects that functional PUFAs have on one's health have sparked an increased interest in the search for plentiful sources of these molecules, particularly fungal strains that exhibit a better synthesis of specific PUFAs. Because of mutant screening and targeted gene editing in *M. alpina*, the functions of various enzymes involved in the biosynthesis of PUFAs have been elucidated. This has also led to the development of lines with increased PUFA production [33].

An oleaginous filamentous fungus known as *Mucor circinelloides* grew to notoriety due to its great efficiency in producing and accumulating lipids, including a substantial amount of γ-linolenic acid. Mycelium obtained from *M. circinelloides* has, in recent times, garnered a great deal of attention due to the fact that it has been suggested as a convenient source of raw materials for the synthesis of biodiesel by means of lipid transformation. Metabolic engineering has been shown to be essential for a considerable increase in the yields of lipids produced by *M. circinelloides*, despite the fact that this organism naturally accumulates lipids. Fazili and coworkers found that *M. circinelloides* has the capacity for both the modification of already-existing biosynthetic pathways as well as the creation of new biosynthetic pathways [34].

As noted by Xue and coworkers, the ascomycetous yeasts and the basidiomycetous yeasts group are both included in the oleaginous yeasts (Figure 12.3) that have been successfully isolated and described to this point [19]. There have been confirmations that certain strains of *Yarrowia lipolytica* are oleaginous yeasts. For example, *Y. lipolytica* ACA-DC 50109 has the ability to store a significant quantity of lipids (44–54% of the yeast cells' dry weight is composed of fat). Although the same transformant was able to obtain 50.6% of their cell dry weight as lipid from an extract of *Jerusalem artichoke* tubers in their cells, and the cell dry weight was 14.6 g/L within 78 hours of fermentation, the cell dry weight was much higher. Due to the overexpression of an inulinase gene in *Y. lipolytica* ACA-DC 50109, the transformant was capable of obtaining 48.3% of their cell dry weight as lipid from inulin within their cells within 78 hours, and their cell dry weight was 13.3 g/L [35]. According to Kreger-van Rij, N. J. W., the wild-type strain of *Y. lipolytica* is incapable of absorbing inulin, sucrose, lactose, soluble starch, D-xylose, cellobiose, or maltose and can only utilize glucose and glycerol for the formation of its cells [36].

The yeast known as *Cryptococcus albidus* is classified as an oleaginous yeast. In the continuous cultures with a single stage, initial research focused on determining how dilution rates affected the culture. To attain a high cell density at a constant dilution rate of 0.36/h

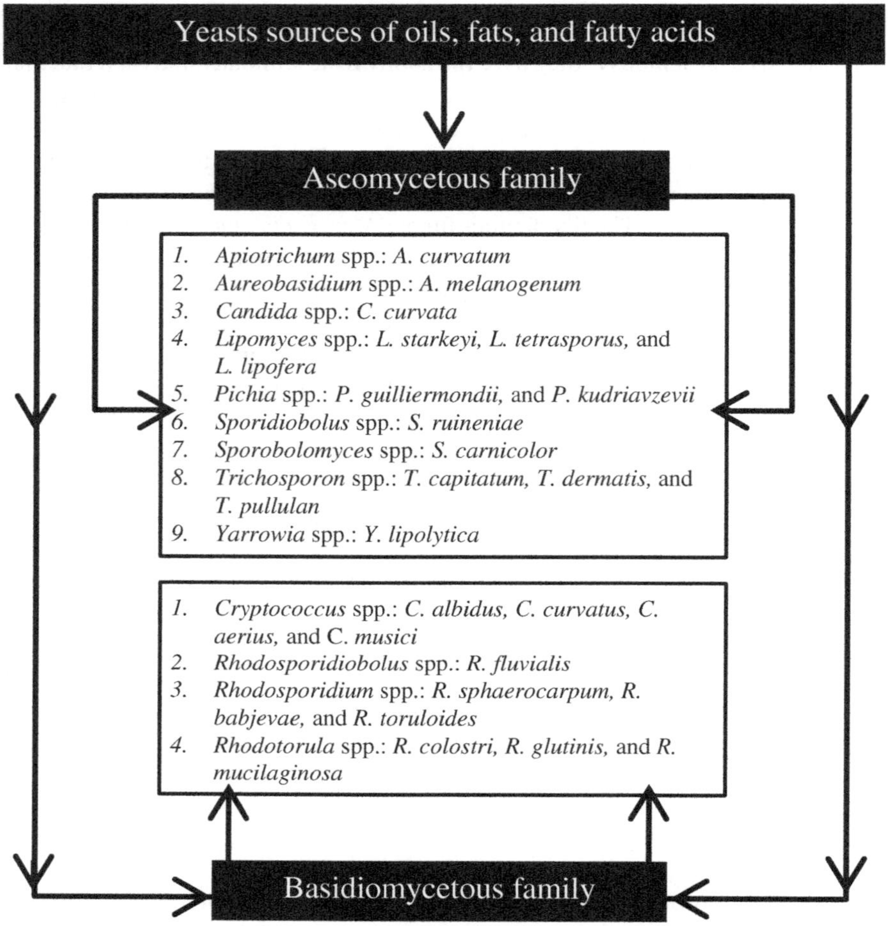

Figure 12.3 Yeasts are sources of oils, fats, and fatty acids, and they come from a variety of species belonging to the ascomycetous and basidiomycetous families. *Source:* Adapted from [19]/Taylor and Francis.

with varying bleeding ratios, a single-stage continuous culture was carried out utilizing a membrane cell recycling system. This culture was carried out in order to use. The highest bleeding ratio, which was 0.4, resulted in the highest amount of lipid production, which was 0.69 g/L/h. A two-stage continuous culture was carried out in order to get a higher lipid production and content. This was accomplished by modifying the C/N ratio during two separate periods of the culture. In the end, researchers got a lipid yield of 0.32 g/g and a lipid content that was 56.4% [37].

Arachidonic acid is one of the fatty acids, and because of its unique biological properties, it has a broad variety of applications in pharmacology, in food and agricultural industries, and in the care of infants. It has been proven that the production of arachidonic acid is extremely sensitive to pH values in an acidic environment and that it is completely inhibited at a pH of 3. The range of 20–22 °C was considered to be the optimal temperature for the production of arachidonic acid. *M. alpina* NRRL-A-10995 was continually cultured in a medium that included glycerol, and the development of the organism was limited by the addition of nitrogen and the selection of optimal conditions (pH 6.0 and 20 °C). This resulted in the active synthesis of arachidonic acid, which accounted for 25.2% of the lipids and 3.1% of the biomass [38]. The product yield from the consumed glycerol was 1.6% by mass and 3.4% by energy.

12.3 Application in Food and Health

Isolating the microbial lipid that is rich in PUFA allows for the creation of pure oil or stable emulsions that can then be added to a wide range of foods. Alternately, a wide variety of agricultural products (such as cereals) and byproducts (such as orange peels, apple or pear pomace, and sweet sorghum) can be augmented with PUFAs by solid or semi-solid-state fermentation with PUFA-producing microbes and then utilized directly as food and/or feed supplements [39].

In view of the continually growing human population as well as the limited number of natural PUFA sources, a significant number of studies have been carried out on the production of PUFA. According to Bellou and coworkers, the bulk of the commercially available microbial oils includes large quantities of PUFAs. This includes the oils that are generated from the yeast *Y. lipolytica*, the fungus *M. circinelloides*, and *M. alpina* [40]. Microorganisms were grown in the distillery effluent in prior research so that they could make biodiesel from the lipids that they accumulated [41]. *Metschnikowia pulcherrima* was utilized to inoculate the raw waste, and it was then cultured under different circumstances regarding pH, temperature, and the duration of the culture period. The raw wastewater had a total dissolved solids concentration of 46.9 g/L, and its COD concentration was 86 g/L. An analysis was done to determine the ideal growth conditions for a C/N ratio of 11.4%, which is already obtainable. It was demonstrated that the optimal conditions for growth in culture were a pH of 6.2, a temperature of 300 °C, and a period of 120 hours [41]. After the procedure of lipid extraction was finished, the extracted lipids were put to use in the production of biodiesel.

Infectious diseases need more than one treatment option in order to be effectively treated because many bacterial species are growing increasingly resistant to antibiotics [42]. According to the World Health Organization (WHO) and the United States Centers for Disease Control and Prevention (US CDC), it is possible that infections such as those caused by *Neisseria gonorrhoeae* will no longer be treatable in a few short years. One area of research interest is looking at the possibility of employing antimicrobial fatty acids and the derivatives of these acids in the therapeutic prevention or treatment of bacterial infections. Numerous research has been conducted to study the efficacy of monoglycerides, fatty acids, and derivatives of both of these in the fight against a wide variety of bacterial species [43].

The pathological implications of certain fatty acids, such as omega(ω)-3 PUFAs, which influence human health conditions and slow the progression of some illnesses, make certain fatty acids particularly important to people. Their anti-arrhythmic, anti-inflammatory, anti-thrombotic, anti-atherosclerotic, anti-fibrotic, and endothelial relaxing capabilities are known as being advantageous in the prevention of a number of illnesses [44]. This is especially true in the case of CVDs. According to a number of studies, the administration of dietary supplements rich in ω-3 PUFAs to young children diagnosed with attention deficit hyperactivity disorder or any other kind of comparable developmental abnormality causes a number of positive effects. Low levels of erythrocyte docosahexaenoic acid and eicosapentaenoic acid have been associated with a variety of major issues, including bipolar disorder, schizophrenia, and other psychiatric conditions. ω-3 PUFAs have favorable benefits on any inflammatory region of the body, including those associated with rheumatoid arthritis, asthma, lupus erythematosus, diabetes, migraine, nephritis, and psoriasis. These effects may be seen in the body's physiological functioning. In addition, they protect against or treat atherosclerosis, hypertriglyceridemia, and hypertension [45]. In a different area of research, it was demonstrated that oleic, linoleic, capric, lauric, and myristic acids are antiviral agents that are effective against vesicular herpes simplex type1, visna, and the stomatitis virus. Thormar and coworkers revealed that the monoglycerides monocaprylin, monocaprin, and monolaurin made these viruses inactive [46].

12.4 Opportunities and Prospective Future

Since the first time microbial oils were made available to the general public in 1985, the significance and value of microbial oils have been steadily increasing in the specialized market for high-value nutraceuticals. However, there is a very small proportion of people who loathe taking fish oil capsules, and the primary reason for this aversion is that the capsules cause them to have "fishy burps." In addition, certain religious groups, individuals who follow a vegetarian or vegan diet, and vegans do not desire to ingest these oils. Therefore, the only thing that can fulfill these requirements is the use of microbial oils. Therefore, *Crypthecodinium cohnii* and several species of *Schizochytrium* or *Thraustochytrium* serve as the basis for the many primary sources that are now available. For the production of biodiesel from wet oleaginous microorganisms with a water content of more than 90% (weight basis), a novel direct saponification-esterification of fatty acids microbial conversion approach has been developed [47]. This method involves the esterification of fatty acids rather than their saponification. This approach is extremely simple, and it has the potential to supplant more conventional methods of feedstock drying and lipid extraction in order to boost biodiesel production at an affordable cost. In addition, the findings of the trial and the suggested kinetic model indicated that the system operates primarily through the conversion of lipids produced from microalgae cells into soap during the saponification stage and biodiesel during the subsequent esterification step. This was indicated by the fact that the system produced soap from the lipid during the saponification stage. The production of biodiesel from waste oil is rather confined since there is a shortage of waste oil, despite the fact that the process is successful for independent small-scale producers. Large-scale commercial producers typically make use of the oil that is extracted from seeds such as corn, soybeans, rapeseed, and palm, among others. Unfortunately, the discussion around whether biodiesel ought to be categorized as food or feed has resulted in the discovery that this particular resource, when utilized on a commercial basis, carries a higher price tag. The higher yield of bacterial biomass that is produced when waste materials are utilized as the carbon source is one potential solution that could be used to bring the cost of the raw materials that are used in the production of biodiesel down to an affordable level. Following this step, the biomass may be transformed into fatty acids. The efficient production of several groups of fatty acids and the derivatives of these acids was made possible by the use of substrates such as alkanoic acids and alkanes. Nevertheless, additional challenges are presented by their high toxicity, low miscibility, and rapidly increasing prices on the market [48]. The utilization of ultrasound as a cutting-edge method in the production of fatty acids and biodiesel from oleaginous bacteria has just come to light. Ultrasonic treatment, in general, has the effect of enhancing the mass-transfer capabilities of a material, which, in turn, leads to an improved reaction rate, a quicker response time, and maybe even cheaper production costs [49]. However, further research is required, particularly in the area of the techno-economic feasibility of the proposed solution. The major impediment to the synthesis of lipids derived from the associated carbon source is produced from microorganisms, which account for up to 85% of the entire production expenses, hence making the manufacturing process expensive. Therefore, the cost would be reduced if inexpensive carbon or nitrogen sources were used, such as hydrolyzed plant biomass, molasses, crude glycerol from the biodiesel industry, whey from the cheese industry, or sludge from wastewater treatment facilities. These are all examples of carbon or nitrogen sources.

12.5 Conclusion

Because of the growing population density and the shift in lifestyle attitudes, there is a greater amount of pressure being placed on the manufacturing market. This effort is necessary in order to satisfy the demands and wishes of society. The production and consumption patterns that have been built in recent times mainly rely on fossil fuels, which have a severe impact on both the natural resources and the environment. The effective production of biological materials is a burgeoning sector that shows signs of further expansion and provides a diverse variety of options for businesses to expand their operations. The production of biofuels from bacterial lipids, which is more applicable to real-world situations and suitable for usage in production environments, is progressively becoming the primary focus of study. The associated carbon source, which accounts for more than half of the production costs, poses the greatest challenge in the whole process of creating lipid-derived fuels from microorganisms. This presents the greatest challenge since it accounts for more than half of the production costs. As a consequence of this, the production of lipids and biodiesel from bacteria using a variety of waste materials as carbon sources, the application of contemporary biotechnological techniques, and the improvement of transesterification processes will make it possible to produce biodiesel at a price that is more affordable.

References

1 Ratledge, C. and Lippmeier, C. (2017). Microbial production of fatty acids. In: *Fatty Acids*, 237–278. AOCS Press.

2 Koritala, S., Hesseltine, C.W., Pryde, E.H., and Mounts, T.L. (1987). Biochemical modification of fats by microorganisms: a preliminary survey. *Journal of the American Oil Chemists Society 64* (4): 509–513.

3 Singh, S.P. and Upadhyay, S.K. (eds.) (2021). *Bioprospecting of Microorganism-Based Industrial Molecules*. John Wiley & Sons Ltd. doi:10.1002/9781119717317.

4 Shintani, T., Upadhyay, S.K., Singh, S.P. (2021). An introduction to microbial biodiversity and bioprospection. In: *Bioprospecting of Microorganism-Based Industrial Molecules* (ed. S.P. Singh and S.K. Upadhyay). John Wiley & Sons Ltd. https://doi.org/10.1002/9781119717317.ch1.

5 Rattray, J.B. (1984). Biotechnology and the fats and oils industry – an overview. *Journal of the American Oil Chemists' Society 61* (11): 1701–1712.

6 Ackman, R.G. (1972). The analysis of fatty acids and related materials by gas-liquid chromatography. *Progress in the Chemistry of Fats and Other Lipids 12*: 165–284.

7 Li, Y., Moore, R.B., Qin, J.G. et al. (2013). Extractable liquid, its energy and hydrocarbon content in the green alga *Botryococcus braunii*. *Biomass and Bioenergy 52*: 103–112.

8 Shiho, M., Kawachi, M., Horioka, K. et al. (2012). Business evaluation of a green microalgae *Botryococcus braunii* oil production system. *Procedia Environmental Sciences 15*: 90–109.

9 San Cha, T., Chee, J.Y., Loh, S.H., and Jusoh, M. (2018). Oil production and fatty acid composition of *Chlorella vulgaris* cultured in nutrient-enriched solid-agar-based medium. *Bioresource Technology Reports 3*: 218–223.

10 Teh, K.Y., Loh, S.H., Aziz, A. et al. (2021). Lipid accumulation patterns and role of different fatty acid types towards mitigating salinity fluctuations in *Chlorella vulgaris*. *Scientific Reports 11* (1): 1–12.

11 Moniz, P., Andrade, G., Reis, A., and da Silva, T.L. (2022). *Crypthecodinium Cohnii* lipid fractionation for the simultaneous DHA and biodiesel production. *Chemical Engineering Transactions 93*: 253–258.

12 Gevorgiz, R.G., Gontcharov, A.A., Zheleznova, S.N. et al. (2022). Biotechnological potential of a new strain of *Cylindrotheca fusiformis* producing fatty acids and fucoxanthin. *Bioresource Technology Reports* 18: 101098.

13 Sahin, M.S., Khazi, M.I., Demirel, Z., and Dalay, M.C. (2019). Variation in growth, fucoxanthin, fatty acids profile and lipid content of marine diatoms *Nitzschia* sp. and *Nanofrustulum shiloi* in response to nitrogen and iron. *Biocatalysis and Agricultural Biotechnology 17*: 390–398.

14 Valenzuela, J., Mazurie, A., Carlson, R.P. et al. (2012). Potential role of multiple carbon fixation pathways during lipid accumulation in *Phaeodactylum tricornutum*. *Biotechnology for Biofuels 5* (1): 1–17.

15 Kumar, M.S. and Buddolla, V. (2019). Future prospects of biodiesel production by microalgae: a short review. *Recent Developments in Applied Microbiology and Biochemistry* 161–166. https://doi.org/10.1016/B978-0-12-816328-3.00012-X.

16 Kumar, M., Rathour, R., Gupta, J. et al. (2020). Bacterial production of fatty acid and biodiesel: opportunity and challenges. In: *Refining Biomass Residues for Sustainable Energy and Bioproducts* (ed. R.P. Kumar, E. Gnansounou, J.K. Raman, and G. Baskar), 21–49. Academic Press.

17 Kumar, S., Gupta, N., and Pakshirajan, K. (2015). Simultaneous lipid production and dairy wastewater treatment using *Rhodococcus opacus* in a batch bioreactor for potential biodiesel application. *Journal of Environmental Chemical Engineering 3* (3): 1630–1636.

18 Koreti, D., Kosre, A., Jadhav, S.K., and Chandrawanshi, N.K. (2022). A comprehensive review on oleaginous bacteria: an alternative source for biodiesel production. *Bioresources and Bioprocessing* 9 (1): 1–19.

19 Xue, S.J., Chi, Z., Zhang, Y. et al. (2018). Fatty acids from oleaginous yeasts and yeast-like fungi and their potential applications. *Critical Reviews in Biotechnology 38* (7): 1049–1060.

20 Meng, X., Yang, J., Xu, X. et al. (2009). Biodiesel production from oleaginous microorganisms. *Renewable Energy 34* (1): 1–5.

21 Chintagunta, A.D., Zuccaro, G., Kumar, M. et al. (2021). Biodiesel production from lignocellulosic biomass using oleaginous microbes: prospects for integrated biofuel production. *Frontiers in Microbiology 12: 658284.*

22 Patel, A., Karageorgou, D., Rova, E. et al. (2020). An overview of potential oleaginous microorganisms and their role in biodiesel and omega-3 fatty acid-based industries. *Microorganisms 8* (3): 434.

23 Ahmad, A.L., Yasin, N.M., Derek, C.J.C., and Lim, J.K. (2011). Microalgae as a sustainable energy source for biodiesel production: a review. *Renewable and Sustainable Energy Reviews 15* (1): 584–593.

24 Gupta, P.L., Lee, S.M., and Choi, H.J. (2015). A mini review: photobioreactors for large scale algal cultivation. *World Journal of Microbiology and Biotechnology 31* (9): 1409–1417.

25 Hena, S., Fatihah, N., Tabassum, S., and Ismail, N. (2015). Three stage cultivation process of facultative strain of *Chlorella sorokiniana* for treating dairy farm effluent and lipid enhancement. *Water Research 80*: 346–356.

26 Sandmann, M., Schafberg, M., Lippold, M., and Rohn, S. (2018). Analysis of population structures of the microalga *Acutodesmus obliquus* during lipid production using multi-dimensional single-cell analysis. *Scientific Reports 8* (1): 1–9.

27 Tan, X.B., Lam, M.K., Uemura, Y. et al. (2018). Cultivation of microalgae for biodiesel production: a review on upstream and downstream processing. *Chinese Journal of Chemical Engineering 26* (1): 17–30.

28 Alvarez, H.M., Hernández, M.A., Lanfranconi, M.P. et al. (2021). Rhodococcus as biofactories for microbial oil production. *Molecules 26* (16): 4871.

29 Park, M.O., Tanabe, M., Hirata, K., and Miyamoto, K. (2001). Isolation and characterization of a bacterium that produces hydrocarbons extracellularly which are equivalent to light oil. *Applied Microbiology and Biotechnology 56* (3): 448–452.

30 Niamah, A.K. (2021). Detected of aero gene in *Aeromonas hydrophila* isolates from shrimp and peeled shrimp samples in local markets. *Journal of Microbiology, Biotechnology and Food Sciences 2021*: 634–639.

31 Singh, S., Verma, D.K., Thakur, M. et al. (2021). Supercritical fluid extraction (SCFE) as green extraction technology for high-value metabolites of algae, its potential trends in food and human health. *Food Research International 150*: 110746. https://doi.org/10.1016/j.foodres.2021.110746.

32 Verma, D.K., Al-Sahlany, S.T.G., Niamah, A.K. et al. (2022). Recent trends in microbial flavour compounds: a review on chemistry, synthesis mechanism and their application in food. *Saudi Journal of Biological Sciences 29* (3): 1565–1576.

33 Kikukawa, H., Sakuradani, E., Ando, A. et al. (2018). Arachidonic acid production by the oleaginous fungus *Mortierella alpina* 1S-4: a review. *Journal of Advanced Research 11*: 15–22.

34 Fazili, A.B.A., Shah, A.M., Zan, X. et al. (2022). *Mucor circinelloides*: a model organism for oleaginous fungi and its potential applications in bioactive lipid production. *Microbial Cell Factories 21* (1): 1–19.

35 Dobrowolski, A., Drzymała, K., Rzechonek, D.A. et al. (2019). Lipid production from waste materials in seawater-based medium by the yeast *Yarrowia lipolytica*. *Frontiers in Microbiology 10*: 547.

36 Kreger-van Rij, N.J.W. (ed.) (2013). *The Yeasts: A Taxonomic Study*. Elsevier.

37 Fu, R., Fei, Q., Shang, L. et al. (2018). Enhanced microbial lipid production by *Cryptococcus albidus* in the high-cell-density continuous cultivation with membrane cell recycling and two-stage nutrient limitation. *Journal of Industrial Microbiology and Biotechnology 45* (12): 1045–1051.

38 Mironov, A.A., Nemashkalov, V.A., Stepanova, N.N. et al. (2018). The effect of pH and temperature on arachidonic acid production by glycerol-grown *Mortierella alpina* NRRL-A-10995. *Fermentation 4* (1): 17.

39 Economou, C.N., Makri, A., Aggelis, G. et al. (2010). Semi-solid state fermentation of sweet sorghum for the biotechnological production of single cell oil. *Bioresource Technology 101* (4): 1385–1388.

40 Bellou, S., Triantaphyllidou, I.E., Aggeli, D. et al. (2016). Microbial oils as food additives: recent approaches for improving microbial oil production and its polyunsaturated fatty acid content. *Current Opinion in Biotechnology 37*: 24–35.

41 Anbarasan, T., Jayanthi, S., and Ragina, Y. (2018). Investigation on synthesis of biodiesel from distillery spent wash using oleaginous yeast *Metschnikowia pulcherrima*. *Materials Today: Proceedings 5* (11): 23293–23301.

42 Sharma, N., Kumari, N., Thakur, M. et al. (2022). Molecular dissemination of emerging antibiotic, biocide, and metal co-resistomes in the Himalayan hot springs. *Journal of Environmental Management* 307: 114569.

43 Churchward, C.P., Alany, R.G., and Snyder, L.A. (2018). Alternative antimicrobials: the properties of fatty acids and monoglycerides. *Critical Reviews in Microbiology 44* (5): 561–570.

44 Saini, R.K., Prasad, P., Sreedhar, R.V. et al. (2021). Omega-3 polyunsaturated fatty acids (PUFAs): emerging plant and microbial sources, oxidative stability, bioavailability, and health benefits – a review. *Antioxidants 10* (10): 1627.

45 Kannan, N., Rao, A.S., and Nair, A. (2021). Microbial production of omega-3 fatty acids: an overview. *Journal of Applied Microbiology 131* (5): 2114–2130.

46 Thormar, H., Isaacs, C.E., Kim, K., and Brown, H.R. (1994). Inactivation of visna virus and other enveloped viruses by free fatty acids and monoglycerides. *Annals of the New York Academy of Sciences 724* (1): 465–471.

47 Fang, Y.R., Yeh, Y., and Liu, H.S. (2018). A novel strategy of biodiesel production from wet microalgae by direct saponification-esterification conversion (DSEC). *Journal of the Taiwan Institute of Chemical Engineers 83*: 23–31.

48 Al Rayaan, M. and Alshayqi, I.A. (2020). A review on oleaginous microorganisms for biological wastewater treatment: current and future prospect. *Journal of Environmental Treatment Techniques 9*: 280–288.

49 Sivaramakrishnan, R. and Incharoensakdi, A. (2018). Microalgae as feedstock for biodiesel production under ultrasound treatment – a review. *Bioresource Technology 250*: 877–887.

13

Microbial Bioreactors for Secondary Metabolite Production

Luis V. Rodríguez-Durán[1], Mariela R. Michel[2], Alejandra Pichardo[3], and Pedro Aguilar-Zárate[2]

[1] *Biochemical Engineering Department, UAM-Mante, Universidad Autónoma de Tamaulipas, Ciudad Mante, Tamaulipas, Mexico*
[2] *Engineering Department, Tecnológico Nacional de México/I. T. de Ciudad Valles, Ciudad Valles, San Luis Potosí, Mexico*
[3] *Department of Biotechnology, Universidad Autonoma Metropolitana-Unidad Iztapalapa, Colonia Vicentina, Mexico City, Mexico*

13.1 Introduction

Microbial secondary metabolites include several compounds such as antibiotics, growth hormones, and pigments, among others. They are not involved directly in microbial growth but have applications in pharmaceuticals, food, cosmetics, and biocontrol [1, 2]. Secondary metabolites are produced by bacteria and fungi, and their production is influenced by culture conditions [3] affecting microbial physiology, metabolism, and stress responses [4–6].

The composition of culture media mainly the carbon-to-nitrogen (C/N) ratio, salinity, and metal ion can regulate the degree and pattern of secondary metabolites expression by genes. As well the culture conditions such as adequate temperature, pH, oxygen concentration, and cultivation status are necessary for the growth of the microbes and the correct biochemistry reactions allowing the production of secondary metabolites [7]. The control or variation of the above-mentioned factors may affect the production of microbial secondary metabolites or change the chemical diversity of the compounds.

The type of cultivation affects directly the microbe's metabolic process. The secondary metabolites are produced generally during the late growth phase of the microorganisms and are repressed in the logarithmic phase and depressed in the stationary phase [8, 9]. The production of secondary metabolites has been carried out mainly by submerged and solid-state cultures. The submerged culture has the characteristic that the parameters can be monitored on-line and can be automated. It allows the scaling up of the processes. However, the solid-state culture imitates the natural environment for microorganisms and imitates the performance in their natural habitat [1].

The bioreactor is a basic requirement for the fermentation process. The bioreactor is a device or a system that allows a biologically active environment. They have been tools for scientists in the biotechnology field for obtaining particular microbial products, such as

Microbial Bioreactors for Industrial Molecules, First Edition.
Edited by Sudhir P. Singh and Santosh Kumar Upadhyay.

secondary metabolites [10]. The bioreactor design must assure homogeneity in the system and optimal conditions for the microbial growth and obtention of products. Also, the design must consider problems such as oxygen transfer, which depends on the complexity of the matrix and is the factor that affects the design and control strategies [11].

According to the above described, the use of the correct bioreactor under the correct conditions and the adequate nutrients will allow the microbes to produce (or not) the metabolites. Hence, the present chapter describes the commonly used bioreactors, their design, and their impact on the production of microbial secondary metabolites.

13.2 Design of Bioreactors

Bioreactors are core elements for biological reactions. Into the bioreactors, the microbial processes occur. The bioreactors must have the conditions and satisfy the requirements (chemical and physical) of microbes [12]. In submerged fermentation and solid-state culture, secondary metabolites are produced generally by batch and fed-batch even at an industrial scale. For submerged fermentation, the control parameters are medium composition, pH, temperature, agitation, and aeration rate. In solid-state culture, the parameters to control are similar to the parameters for submerged fermentation but is necessary to control initial moisture, particle size, and medium concentration [13].

The bioreactors exist in numerous designs and configurations according to their applications. The integrated control systems allow working from lab scale to industrial bioreactors [14]. It is important to take into account different considerations for the design of the bioreactors.

The above-depicted information about the bioreactors highlights the purpose of the design efforts. The bioreactors provide the conditions where diverse cell types can grow and produce a variety of biologicals. Hence, this is derived in an assortment of bioreactors systems with different design solutions.

The challenges start when cells start growing need nutrients and growth factors, also generating the necessity of mass and energy transfer. The bioreactor must provide the nutrients and the correct heat removal to the microbes [15]. The cells are variable and sensible to microenvironmental conditions that affect microbial growth. Gradients of chemicals, mainly nutrients, are generated into the bioreactor when lacks a good design or has agitation problems. It causes growth and heat gradients, and changes in metabolic pathways causing variation in the production of metabolites.

The purity of the strains inoculated is important since the bioreactor must assure an environment free of external microorganisms with the correct temperature and moisture to carry out the bioprocess. Hence, sterilization is another challenge in bioreactor design since it depends on the shape and construction material.

The lab-scale bioreactors are generally constructed on glass, while the industrial-scale bioreactors are constructed on stainless steel [16]. Nonetheless, there are other materials such as polyethylene (bags) and acrylic used for the elaboration of bioreactors. The material for the construction of the bioreactor must be chemically inert and not allow the trespassing of external elements to the media. The construction material for the bioreactor will be determined by the operating scale of the process, the process itself, and the economic considerations and requirements of the operator [17, 18]. Also, it is important that bioreactors provide functional characteristics such as transparency, mechanical force, wear resistance, chemical stability, and be easy to clean [19].

Based on the above premises, several designs of bioreactors have emerged. The commonly used bioreactor for submerged fermentation is the stirred tank. Despite its versatile design, operability, and manufacturability, mechanical agitation could be its main disadvantage depending on the used microorganism. Since mechanical agitation causes shear stress to the cells changing the metabolic pathways or causing the death of the microbes. This problem was solved by designing the bubble column and air-lift bioreactors. They can be used in aerobic and anaerobic fermentations [20]. Both types of bioreactors have the characteristic of substituting the mechanical impeller with rising bubbles. But in the design, it must be taken into account the possible low volumetric oxygen transfer overall for the fast-growing microorganisms and/or for high-density culture media. Similar advantages and disadvantages are found in the fixed bed, fluidized bed, and biofilm bioreactors. The main challenge in the design of these bioreactors is the integration of the external pumping, the metabolic heat control, and the efficient supply of oxygen and CO_2 removal.

The design of solid-state bioreactors can be as simple as a polyethylene bag or as complex as the instrumented rotating/stirred drum bioreactors. There are several types of bioreactors with different geometries for developing solid-state culture of microbes. It includes bioreactors without aeration such as tray bioreactors and bags, bioreactors with continuous agitation as the rotating/stirred drum, and bioreactors with forced aeration such as packed bed bioreactors, among others. However, it is important to consider that the design of the bioreactor must fit the requirements of the microbes. For example, some fungal strains are sensitive to mechanical agitation. In that case, the bioreactor must be without rotating or stirred agitation. Instead, a tray bioreactor or a forced aeration column must be used. The main challenges in the design of solid-state bioreactors are the control of metabolic heat, the control or monitoring of growth, and the control of factors such as pH, moisture, and oxygen supply [21, 22]. The control of the temperature of the process in solid-state culture has been solved by using an incubator (for tray bioreactors and bags). Other geometries, such as packed bed bioreactors, are easy to introduce into a fish tank with tempered water for heat transfer. In this latter case also the forced aeration acts as metabolic heat removal [23, 24]. The stirred drum and rotating drum generally include a heat jacket [25]. The pH and moisture content are difficult factors for controlling. Many of the literature reports indicate the adjustment of the pH and humidity content at the very beginning of the fermentation process [23, 26–30]. An alternative for controlling the moisture content is by the supply of sterile humid air at the air inlet port [23]. The heterogeneity of the solid-state culture makes difficult the measurement of microbial growth. The CO_2 production and O_2 consumption rates have been used to estimate biomass production [21]. Hence, it is important to consider the implementation of gaseous O_2 and CO_2 sensors for microbial growth monitoring. Generally, they are installed at the gas exit of the bioreactor [24]. Sterilization is a process that can be handled in different ways depending on the bioreactor type, dimensions, building material, substrate, and solid support. Some bioreactors can be autoclaved, but large bioreactors only can be sanitized using chemical agents, steam, or hot water. However, the most often procedure is the separate sterilization of substrate/support and the bioreactor [25].

The microbial secondary metabolites have been produced mostly in the above-mentioned bioreactors. Nevertheless, still there are challenges in bioreactors design, for example the scale-up, the integration of all features in a single bioreactor, or the development of a single bioreactor capable to handle most of the fermentation processes or that accommodates most microbes. So, investigations are still needed.

13.3 Types of Bioreactors for Secondary Metabolite Production

This section discusses the information related to the main reported bioreactors used for the production of microbial secondary metabolites. Also, there are examples of the microorganisms and the produced metabolites.

13.3.1 Stirred Tank Bioreactor (STB)

STB is the most widely used reactor for industrial applications. It consists of a glass or stainless-steel vessel equipped with a motor and a shaft with one or more impellers attached to it (Figure 13.1). The height: diameter ratio of the vessel varies from 2 : 1 to 6 : 1 and the working volume is usually 75% of the total volume. Baffles (4–12) are used to prevent the formation of a central vortex and to improve mass and heat transfer. The most common type of impeller used is the four-blade disc turbine, but, depending on the characteristics of the culture medium, other types of impeller can be used. Gases are supplied from the bottom of the reactor through a sparger. A multiple-orifice ring sparger is generally used for efficient mass transfer [31].

STB is highly versatile as it can operate in batch, fed-batch, and continuous modes. It provides good mixing and high mass and heat transfer. This reactor is ideal for the aerobic cultivation of microorganisms, where high oxygen transfer rates are required to maintain the high growth rates. On the other hand, the high shear stress can be detrimental to the growth of eukaryotic cells [32].

STB has been used for the production of secondary metabolites including antibiotics, pigments, biosurfactants, biopolymers, phenolic compounds, alkaloids, and terpenes,

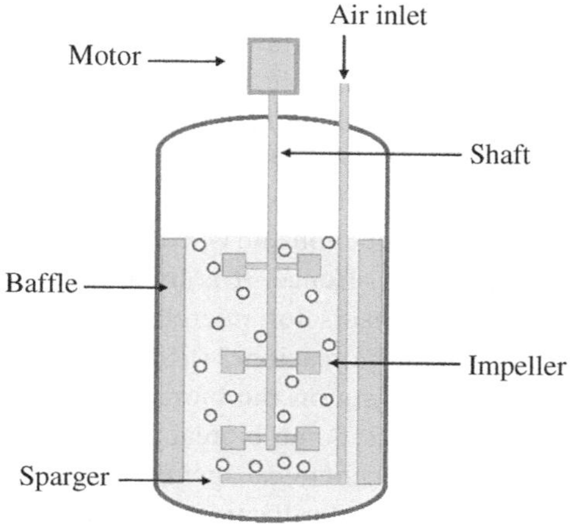

Figure 13.1 Main components of stirred tank bioreactor (STB).

among others (Table 13.1). These compounds are obtained by microbial culture or by plant cell culture.

Penicillin was one of the first secondary metabolites produced at an industrial scale by submerged culture. The high demand for this antibiotic prompted the development of modern STBs in the 1940s [54]. In STB the filamentous fungi grow in the form of spherical pellets. This form of growth has advantages over the cultivation in static bioreactors. For example, in static culture, fungal growth increases viscosity and hinders oxygen transfer. The formation of pellets facilitates the mixing of the culture medium and increases the oxygen transfer rate [55]. Penicillin, like other secondary metabolites, is produced at the end of the exponential growth phase. For this reason, the industrial production of penicillin is carried out mainly by fed-batch cultivation. In the first stage, the fungus grows at a high rate. In the second stage, the carbon source is fed at a low rate to maintain penicillin production [56]. A similar approach has been followed for the production of other antibiotics, such as streptomycin [51] and cephalosporin [47].

Table 13.1 Summary of representative secondary metabolites produced in STB.

Product	Organism	Mode of operation	Reference
Mycophenolic acid	*Penicillium brevicompactum*	Batch, fed batch, and continuous	[33]
Sclerotiorin	*Penicillium sclerotiorum*	Batch	[34]
Compactin	*Penicillium solitum*	Batch	[35]
Prodigiosin	*Serratia* sp.	Batch	[36]
Resveratrol	*Vitis labrusca*	Fed-batch	[37]
Surfactin	*Bacillus subtilis*	Batch	[38]
Polyhydroxybutyrate	*Cupriavidus necator*	Fed-batch	[39]
Rubromycin	*Streptomyces* sp.	Batch	[40]
Sanguinarine	*Papaver somniferum*	Batch	[41]
Lavendamycin	*Streptomyces flocculus*	Batch	[42]
Anthraquinones	*Rubia tinctorum*	Batch	[43]
Red and orange pigments	*Monascus purpureus*	Batch	[44]
Ginsenoside	*Panax quinquefolium*	Batch and fed-batch	[45]
Gibberellic acid	*Gibberella fujikuroi*	Batch	[46]
Cephalosporin	*Acremonium chrysogenum*	Fed batch	[47]
Lovastatin and geodin	*Aspergillus terreus*	Batch and Fed batch	[48]
Azadirachtin	*Azadirachta indica*	Batch	[49]
Guggulsterone	*Commiphora wightii*	Batch	[50]
Streptomicyn	*Streptomyces griseus*	Fed batch	[51]
Sophorolipids	*Candida bombicola*	Fed batch	[52]
Penicillin	*Penicillium chrysogenum*	Fed batch	[53]

Filamentous fungi have been used for the production of natural pigments since ancient times. For example, in Asia, fungi of the genus *Monascus* are used to make a traditional fermented food known as fermented red rice, or red koji. During the preparation of this food, a series of red, orange, and yellow pigments are produced. These pigments are a mixture of azaphilones, which are secondary metabolites with important biological activities [57]. Although the commercial production of *Monascus* pigments is carried out mainly by solid-state fermentation, several authors have studied the production of these metabolites in STB [44, 58]. The production of pigments in STB allows for better control of fermentation parameters and facilitates the scaling up and the downstream processing.

Biosurfactants are amphiphilic molecules of biological origin produced by plants and microorganisms. For example, *Bacillus subtilis* produces biosurfactant lipopeptides such as surfactin, iturin, and fengycin. Currently, the production costs of these metabolites are very high, so their production at the industrial level has not been established [59]. However, some attempts have been made to scale-up surfactin production to stirred-tank bioreactors at benchtop [60] and pilot scale [61]. Sophorolipids are biosurfactant glycolipids produced mainly by yeasts of the genus *Starmerella* [62]. Sophorolipids are produced in the presence of a hydrophilic and a hydrophobic carbon source. *Starmerella bombicola* fed-batch culture in STB can reach concentrations of more than 400 g/L of sophorolipids. First, the microorganism grows in a medium composed of glucose and corn oil; when the glucose is depleted, corn oil is added discontinuously [52].

Plants produce a wide variety of secondary metabolites, such as terpenes, phenolic compounds, and alkaloids. These compounds are traditionally produced by field cultivation. However, this approach has some problems such as low yields, variations due to environmental factors, the need for large cultivation areas, and intensive labor [63]. Plant cell culture in bioreactors is an alternative that has some advantages over traditional field cultivation, for example control of environmental conditions, higher and reproducible yields, simpler extraction and recovery processes, and ease of scaling [64]. STB has been used for the production of guggulsterone by *Commiphora wightii* cells [50], azadirachtin by *Azadirachta indica* cells [49], ginsenoside by *Panax quinquefolium* cells [45], anthraquinones by *Rubia tinctorum* cells [43], sanguinarine by *Papaver somniferum* cells, and resveratrol by *Vitis labrusca* cells [37], among others.

13.3.2 Bubble Column

Bubble column bioreactors have no mechanical agitation. Instead, they have pneumatic agitation, which means a gas (generally air) is injected into the bottom of the bioreactor (Figure 13.2). The gas enters the bioreactor via a diffuser. The bubbles travel through the liquid media and along the cylinder (that is the common shape of this type of bioreactors) until reaching the air exit port. The cylinder generally has a diameter-height ratio of 6 : 1. Temperature keep constant by using a heat jacket or coolant coil installed outside of the vessel.

Despite the simple design of the bioreactor, it is important to consider the rheological properties of the fluid. Since the oxygen transfer and agitation are affected by these properties [65]. Branco et al. [66] mentioned the production of xylitol by immobilized yeast cells is influenced by a high aeration rate improving the oxygenation of the system and the

Figure 13.2 Bubble column bioreactor.

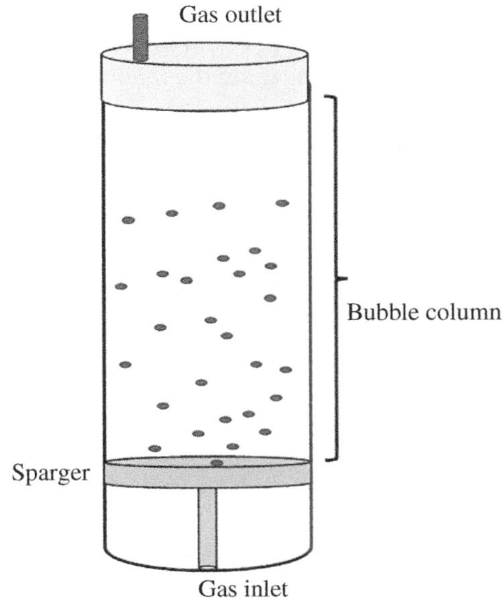

homogeneity of the high concentration of cells. This latter point is really important mainly when filamentous fungi are involved since the morphology of the microorganisms changes the rheology of the culture media. For example, in the biosynthesis of antibiotics using a bubble column is necessary to supply high oxygen transfer. For that reason, the high superficial gas velocity is necessary for increasing the gas holdup [67]. This parameter is important when gases are the main entrance product into the bioreactor. For microbial ethanol production using CO, CO_2, and H_2, solubilization of the gases is important. Here appears another parameter as thermodynamic, necessary for mass transfer and microbial growth [68, 69]. Kheradmandnia et al. [70] reported that by increasing the temperature of the process the K_La, maximum growth rate, and oxygen uptake rate are increased in a bioprocess for the culture of *Escherichia coli*. The composition of the gases is also important to induce microbes to produce secondary metabolites. Rahnama et al. [71] evaluated the methane to air ratio and nitrogen content in a bioprocess for the production of poly-3-hydroxybutyrate (PHB) by *Methylocystis hirsuta*. Methane to air ratio of 1 : 1 and 50% of nitrogen provided the highest accumulation of PHB. For these bioprocesses, the presence of oxygen is important for microbial growth. The increase in methane concentration (10–50%) did not affect the biosynthesis of PHB [72].

Most of the applications of bubble column bioreactors are for biological wastewater treatment. The commonly used microorganisms for this purpose are *Chlorella* species. They are used for the removal of contaminants such as organic matter, ammonium, and phosphorous and for increasing the chemical and biochemical oxygen demand. However, the *Chlorella* cells produced valued secondary metabolites such as pigments and lipids [73–75]. The production of metabolites such as chlorophyll and carotenoids are affected besides the shear rate by the presence or absence of light [74].

The synthesis of bio-oils has been carried out by using refinery wastewater. For this purpose, strains such as *Rhodococcus opacus* are used. The strain besides removing chemical oxygen demand produces high amounts of lipids transformed *a posteriori* into bio-oil [76]. The production of lipids and microbial growth is also affected by the bubble size. The gas transfer and the shear are the main factors affected. Hoseinkhani et al. [77] reported high production of docosahexaenoic acid (DHA) by *Crypthecodinium cohnii* by using a bigger bubble diameter despite the low biomass production. So, the oxygen limitation derived from the big bubble diameter induces the microorganism to survive producing DHA.

Very recent and interesting use of the bubble column bioreactors is for the treatment of the e-waste. Nili et al. [78] developed a bioleaching process to extract copper and nickel from waste mobile phones using a pure culture of *Penicillium simplicissimum* in a bubble column bioreactor and molasses as a carbon source. They recovered 96.94% and 71.51% of Cu and Ni, respectively.

13.3.3 Air-Lift

The air-lift bioreactors have received attention due to their relatively simple construction and the less shear damage to microbial cells that are sensitive to stirred tanks. The air-lift bioreactors consider also gas–liquid–solid pneumatic contacting devices. The main characteristic of these bioreactors is the fluid circulation in a defined cyclic pattern through channels built specifically for this purpose [20, 79]. The bioreactor is composed generally of the following elements: the main vessel, the draft tube, gas sparger, and gas outlet (Figure 13.3). The presence of the draft tube generates the formation of the gas raiser and downcomer zones. The zones are generated because of the different densities of gas and liquid. The use of air-lift bioreactors in the production of microbial secondary metabolites is due to the advantages of the above-mentioned characteristics.

Similar to the bubble column bioreactor, the aeration rate is a factor that affects the microbial bioprocesses. Hence, it is important to consider the rheology of the culture media for an efficient mass transfer, the dissolution of oxygen, and the homogeneity of the system [80]. Godó et al. [81] evaluated the mixing time during the production of citric acid by *Aspergillus niger*. They discussed that the changes in viscosity of culture media due to fungal growth

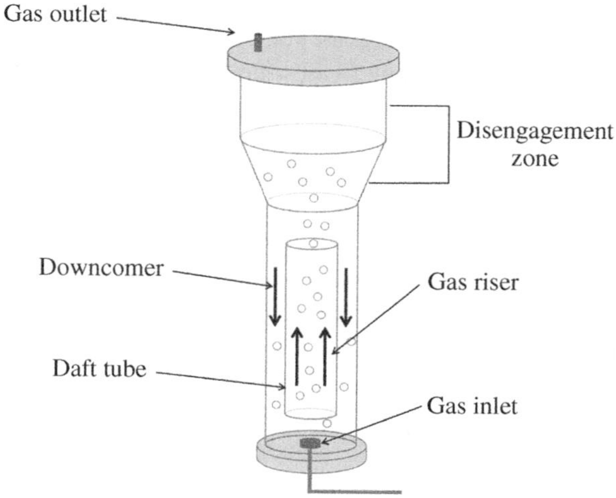

Figure 13.3 Air-lift bioreactor.

changes the mixing efficiency. Despite keeping the same gas flow rate, the velocity of liquid in the downcomer at the end of fermentation was four times lower than at the beginning.

The gas transfer into the air-lift bioreactor is important in the bioprocess for the conversion of methane into methanol. The solubilization of methane through a hydraulic transfer chamber was an important factor in the efficient bioconversion to methanol by methanogenic bacteria [82]. When no devices are available for the gas transfer, high aeration rates are used for dissolving the gases. Manowattana et al. [83] applied 6 vvm (volume of air/volume of culture media/min) to maintain 60% of dissolved oxygen in a bioprocess for the production of lipids, carotene, and carotenoids by the red yeast *Sporidiobolus pararoseus* KM281507. In all the cases mentioned previously, the sparger is a very important accessory for adequate shear and efficient gas transfer [84, 85]. Hence, there exist wide and very innovative gas spargers for air-lift bioreactors.

The air-lift bioreactors are very versatile devices that allow to researchers innovate and improve bioprocesses. Matsumoto and Furuta [86] designed a bioreactor for the production of lactic acid by *Rhizopus oryzae*. The same air-lift bioreactor was designed with an extraction section/phase for the *in situ* extraction of the lactic acid. The use of moderately elevated pressure in an air-lift bioreactor is a very interesting trend in bioreactor design. Pressures between 5 and 10 bar have been explored for the solubilization of gases such as CO_2, CH_4, CO, H_2, and O_2. However, microbial growth rate and metabolite formation need to be improved [87]. Żywicka et al. [88] designed a novel magnetically assisted external loop air-lift bioreactor for the production of bacterial cellulose by *Komagataeibacter xylinus*. It was equipped with a rotating magnetic field generator that induces the microorganism to a stable metabolic activity.

13.3.4 Biofilm Bioreactor

Biofilm bioreactors are used with microorganisms capable to get attached to a surface and adhering within the reactor. They have mostly been used for the treatment of wastewater, and the organisms present in the biofilm absorb and break down toxic substances in the water [89]. The rotating disc contractors and membrane bioreactors are the most common examples where biofilm production occurs on the bioreactor surface [90]. In most biofilm bioreactors, the microorganisms are attached to organic or inorganic support materials, generate a biofilm, then catalyze reactions within the bioreactor, and finally are removed or inactivated once the bioreaction is completed [91].

The biofilm bioreactors are classified according to the flow pattern into the groups: fixed-bed and expanded-bed bioreactors. The fixed-bed bioreactors are designs where the microbial biofilm is formed in a static media. Basically, the biofilm bioreactors are designed from a stirred tank, fixed-bed, rotating disc, fluidized bed, air-lift, or membrane biofilm. Combination and modification of these basic reactors systems broaden the range of biofilm bioreactors. An important part of the biofilm bioreactors is the biofilm support (Figure 13.4). They can be inorganic and organic materials [92].

The biofilm formation is regulated by different genetic and environmental factors such as nutrient availability and hydrodynamics. Hydrodynamics (provided by the bioreactor) greatly influence the mass transfer mechanisms and also create stresses that create direct action on the biostructure and the production of the metabolites (Table 13.2).

Figure 13.4 Stainless steel biofilm support with the mycelium of *Beauveria bassiana*.

Table 13.2 Recent applications of biofilm bioreactors for the production of secondary metabolites.

Product	Organism	Material support/type of bioreactor	Productivity (g/L/h)	Reference
Bioenergy				
Ethanol	*Zymomonas mobilis* and *Saccharomyces cerevisiae*	Packed-bed reactor with PCS ring	5.0–124.0	[93] [94]
Butanol	*Clostridium acetobutylicum*	Packed-bed reactor with Tygon® rings	4.4	[95]
Organic acids				
Acetic acid	*Acetobacter aceti* M7	Multistage shallow flow biofilm reactor	4.3	[96]
Citric acid	*Aspergillus niger*	Polyurethane foam, fluidized-bed reactor	0.13	[97]
	Aspergillus niger	Polyurethane foam particles	0.11	[98]
	Aspergillus niger	Rotating disc reactor	0.9	[99]
Fumaric acid	*Rhizopus oryzae*	Rotating disc reactor	3.78	[100]
Lactic acid	*Lactobacillus amylophilus*, *Lacticaseibacillus casei*, *Lactobacillus delbrueckii*	Packed-bed reactor with PCS	13–60 g/L	[101]
	Lacticaseibacillus casei	Packed-bed reactor with PCS	7.6 g/L	[102]

Table 13.2 (Continued)

Product	Organism	Material support/type of bioreactor	Productivity (g/L/h)	Reference
	Lacticaseibacillus casei	Grid-like orientation PCS biofilm reactor	9.0	[103]
Succinic acid	*Actinobacillus succinogenes*	Packed-bed reactor with PCS	2.08	[104]
Antimicrobials				
Cephalosporin C	*Cephalosporium acremonium*	Air-lift biofilm reactor	0.2–1.3 g/L	[105]
Nisin	*Lactococcus lactis*	PCS tubes attached on the agitator shaft	4314 U/mL	[106]
Oosporein	*Beauveria bassiana*	Bubbled bottle bioreactor with stainless steel fiber	183 mg/L	[107]

Source: Cheng et al. [106] / Modified with permission from Springer Nature.

The performance of a biofilm reactor depends on the biofilm characteristics such as biofilm density and biofilm thickness since the overall reaction critically depends on these parameters. Into biofilm bioreactors occurs the most complex microbial metabolic processes since liquid and solid (biofilm) fermentations are carried out. The availability of oxygen in the gaseous phase is necessary even when there is oxygen saturation in the liquid phase.

In addition, the biomass attachment to the support confers some advantages, such as the low viscosity of the culture broth and easy recovery of the products [109]. However, the attachment of biomass to the solid support represents a disadvantage mainly to fungi because the growth measurement is complicated. The adhesion of biomass to solid support can be estimated by measuring the CO_2 production [23, 24]. The gaseous CO_2 monitoring of the whole fermentation process (liquid and solid phases) can indicate the physiological stage of the microorganism in the biofilm bioreactors [107].

13.3.5 Solid-State Fermentation (SSF) Bioreactors

Solid-state fermentation (SSF) is a biological process that occurs in the absence or near absence of free water, which increases the efficiency of fermentation. SSF leads to concentrated products, reduces catabolic repression, and facilitates the growth of microorganisms in hydrophobic substrates. In addition, SSF allows the use of agro-industrial residues to obtain high-value-added products [110].

One of the main drawbacks of SSF is the difficulty in measuring and controlling process variables, such as temperature, pH, the concentration of substrates/products, and microbial growth. Furthermore, the scale-up of SSF reactors is generally more complex than in submerged culture reactors. The main challenge for SSF scaling is metabolic heat dissipation. This is because in SSF the inter-particle spaces are filled with air, and the thermal conductivity of air is much lower than that of water. Also, in most cases, in SSF there is no mixing or mixing is less efficient than in submerged fermentation [111].

SSF bioreactors are classified into four types: static bioreactors without forced aeration (e.g. tray bioreactor), forcefully-aerated bioreactors but without mixing (e.g. packed-bed bioreactor), bioreactors with mixing but without forced aeration (e.g. rotating-drum and stirred-drum bioreactors), and bioreactors with mixing and forced aeration (e.g. fluidized-bed, rocking-drum, and stirred-aerated bioreactors) [112]. In this section, the main bioreactors used for the SFF are reviewed.

13.3.6 Tray Bioreactor

Tray reactors are characterized by a thin layer of solid substrate spread on a horizontal tray with a bed height of 1–4 cm. The trays are placed inside a closed chamber (Figure 13.5). This type of bioreactor allows good aeration and dissipation of metabolic heat. In this type of bioreactors, the humidity, temperature, and composition of the gaseous atmosphere can be monitored and controlled. The investment cost for this type of reactor is relatively low [113].

These types of reactors are used for the production of enzymes and other products, including some secondary metabolites. For example, fungi of the genus *Monascus* produce orange, red, and yellow pigments during their growth on starch as a substrate. These pigments are a mixture of at least six secondary compounds: ankaflavin, monascin, monascorubrin, rubropunctatin, monascorubramine, and rubropunctamine [114]. Monascus is cultivated on a solid medium in Asian countries to produce a red colorant named "Anka,", or "red mold rice," which is used as a food ingredient. The classical method consists of steaming grains of rice, spreading them on large trays and inoculating them with a strain of *Monascus* sp. The trays are incubated for 20 days in a room with controlled temperature and aeration. [115]. Zhang et al. [116] conducted a comparative study between submerged fermentation and solid-state fermentation with *Monascus purpureus* AS3.531 and they found that the pigment yield was 1.2 times higher in submerged fermentation compared to SSF. However, citrinin production was 100 times higher in submerged fermentation than in solid fermentation. This study shows how SSF is a competitive system with submerged fermentation and decreases the synthesis of citrinin (a nephrotoxic mycotoxin).

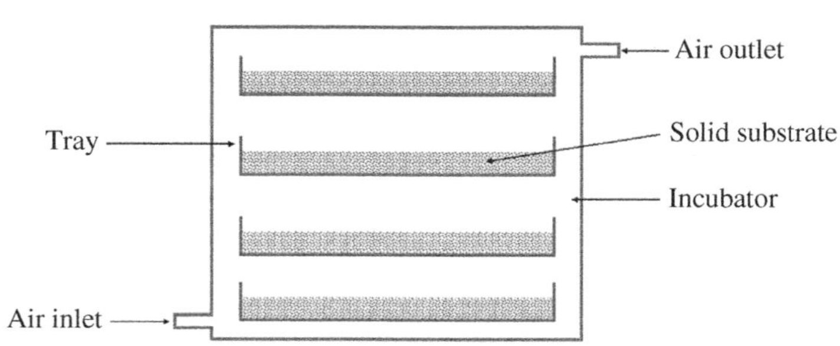

Figure 13.5 Tray bioreactor used for solid-state fermentation.

Another of the secondary metabolites produced by solid fermentation is lovastatin, a hypocholesterolemic agent that inhibits HMG-CoA reductase [117]. Conventional production of lovastatin is carried out by submerged culture. However, SSF has become an interesting alternative for the industrial production of this drug. Baños et al. [118] conducted a comparative study on the production of lovastatin in SSF and submerged fermentation by *Aspergillus terreus*. Lovastatin yields were 30 times higher in SSF than those in submerged fermentation and lovastatin biomass was almost 15 times more productive. In 2001, a patent for a semi-automated tray technology for SSF was granted to the Indian biopharmaceutical company Biocon. This bioreactor, known as PlaFractor, allows the sterilization of the solid support, the inoculation under aseptic conditions, and extraction of the product in the same device [119]. Currently, Biocon produces lovastatin, cultivating *A. terreus* on wheat bran SSF, probably using the PlaFractor technology [120].

13.3.7 Packed Bed Bioreactor

Packed bed bioreactors are static reactors of cylindrical geometry. Oxygen is supplied by forced aeration (Figure 13.6). The air stream passes through a humidifier before entering the reactor to regulate the temperature and humidity of the air. The bioreactor temperature can be controlled by immersion in a thermostatic bath or the use of water jackets [113]. This type of bioreactor allows the measurement of exhaust gases.

Respirometry is a technique based on the measurement of O_2 consumption and of CO_2 production. Respirometry is used to estimate microbial growth and metabolic heat generation in SSF. These measurements can be used for process optimization and as scale-up criteria [121].

The most important challenges for the use of this type of reactor on a large scale are the accumulation of metabolic heat, the formation of temperature gradients, the drying of the solid bed, and the channeling [122].

An example of the application of a packed bed bioreactor where the gases were monitored is the work carried out by Ranjbar and Hejazi [123]. The authors used packed glass columns to study the kinetic parameters of growth and production of secondary metabolites by

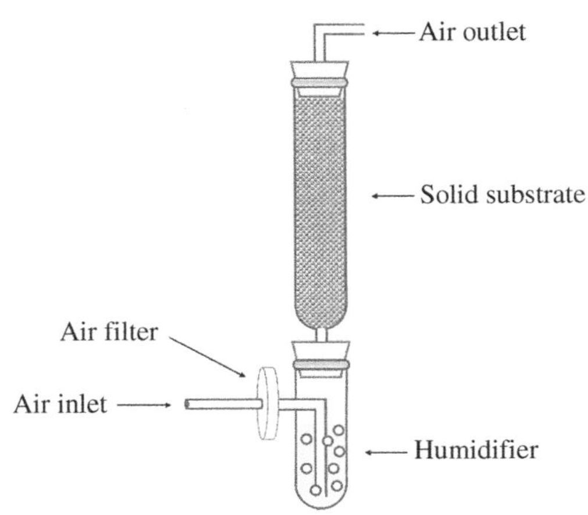

Figure 13.6 Packed bed bioreactor used for solid-state fermentation.

Pseudomonas aeruginosa. The kinetic study was carried out in small reactors (12 cm × 2.5 cm; height × internal diameter) with 10 g of initial dry substrate, each column was connected to a flow of humidified air, and the outlet gases were monitored by sensors to quantify the CO_2 production and oxygen consumption. They evaluated the effect of bed temperature and initial moisture content on kinetic parameters and rhamnolipid production. The results were used to model the kinetics and transport phenomena in the column reactor. The model generated was successfully validated in a larger scale jacketed glass column bioreactor (3 L, 60 cm height × 6 cm internal diameter).

13.3.8 Stirred and Rotating Drum Bioreactor

Stirred and rotating drum bioreactors are two types of reactors with mixing but without forced aeration. These bioreactors are typically horizontally lying cylindrical drums partially filled with a solid substrate. Oxygen is supplied by an air current blowing through the headspace of the reactor (Figure 13.7). In a rotating drum reactor, mixing is conducted by rotating the entire bioreactor around its central axis. In stirred drum bioreactor, mixing is provided by mechanical devices attached to a shaft such as paddles or scrapers [112].

An interesting example of the application of this type of reactor for the production of secondary metabolites in SSF is the production of biosurfactant rhamnolipids by *P. aeruginosa* from soybean meal. Dabaghi et al. [124] studied the effect of the air-flow rate, initial moisture content, incubation temperature, and rotation time on the production of rhamnolipid biosurfactant by *P. aeruginosa* in a laboratory-scale rotating drum bioreactor. They found that the aeration of the bed and intermittent rotation of the baffled drum enhanced rhamnolipid production.

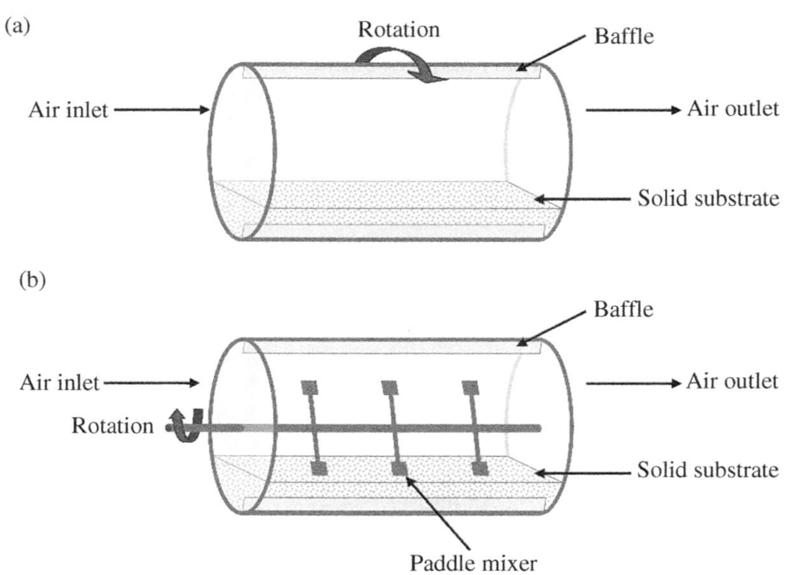

Figure 13.7 Stirred (a) and rotating (b) drum bioreactors used for solid-state fermentation.

13.4 Conclusion

Bioreactors are an important part of the biosynthesis of microbial secondary metabolites since they are the environment for microbes' growth. Understanding the physical, chemical, and mechanical requirements of microorganisms is an important topic. This could help operators and/or researchers to select/design the adequate bioreactor configuration for producing the desired secondary metabolites. The high demands of microbial secondary metabolites worldwide promote the necessity of improving or developing bioreactors. However, most of the literature reports used laboratory-scale bioreactors. Despite the good productivity yields and the very interesting bioreactor developments, the real challenge is to scale up the bioreactor designs and to keep or enhance the production of microbial secondary metabolites. For that reason, research works are still necessary for understanding how the microorganism pattern is and how can be induced to produce secondary metabolites. Also, the partnering between researchers and companies will promote the rapid dissemination and application of knowledge.

Acknowledgment

The chapter is part of the projects 6691.18-P and 10394.21-P funded by Tecnológico Nacional de México and the project SEP-CONACYT A1-S-29456 Identificación de esterasas fúngicas capaces de catalizar la síntesis de derivados bioactivos del ácido cafeico.

References

1 Kumar, V., Ahluwalia, V., Saran, S. et al. (2021). Recent developments on solid-state fermentation for production of microbial secondary metabolites: challenges and solutions. *Bioresour. Technol.* 323: 124566.

2 Yadav, A.N., Kour, D., Rana, K.L. et al. (2019). Metabolic engineering to synthetic biology of secondary metabolites production. In: *New and Future Developments in Microbial Biotechnology and Bioengineering: Microbial Secondary Metabolites Biochemistry and Applications* (ed. G. Vijai Kumar and A. Pandey), 279–320. Elsevier.

3 Singh, B.P., Rateb, M.E., Rodriguez-Couto, S. et al. (2019). Editorial: Microbial secondary metabolites: recent developments and technological challenges. *Front. Microbiol.* 10 (Apr): 914.

4 Perry, E.K., Meirelles, L.A., and Newman, D.K. (2021). From the soil to the clinic: the impact of microbial secondary metabolites on antibiotic tolerance and resistance. *Nat. Rev. Microbiol.* 20 (3): 129–142.

5 Singh, S.P. and Upadhyay, S.K. (eds.) (2021). *Bioprospecting of Microorganism-Based Industrial Molecules*. John Wiley & Sons Ltd. doi:10.1002/9781119717317.

6 Shintani, T., Upadhyay, S.K., Singh, S.P. (2021). An introduction to microbial biodiversity and bioprospection. In: *Bioprospecting of Microorganism-Based Industrial Molecules* (ed. S.P. Singh and S.K. Upadhyay). John Wiley & Sons Ltd. https://doi.org/10.1002/9781119717317.ch1.

7 Pan, R., Bai, X., Chen, J. et al. (2019). Exploring structural diversity of microbe secondary metabolites using OSMAC strategy: a literature review. *Front. Microbiol.* 10: 294.

8 Thirumurugan, D., Cholarajan, A., Raja, S.S., and Vijayakumar, R. (2018). An introductory chapter: secondary metabolites. In: *Secondary Metabolites – Sources and Applications* (ed. R. Vijayakumar and S.S.S. Raja), 3–21. IntechOpen.

9 Gonçalves, S. and Romano, A. (2018). Production of plant secondary metabolites by using biotechnological tools. In: *Secondary Metabolites – Sources and Applications* (ed. R. Vijayakumar and S.S.S. Raja). INTECH.

10 Singhal, A., Kumar, M., Bhattacharya, M. et al. (2018). Pretreatment of *Leucaena leucocephala* wood by acidified glycerol: optimization, severity index and correlation analysis. *Bioresour. Technol.* 265: 214–223.

11 Ruíz-Leza, H.A., Rodríguez-Jasso, R.M., Rodríguez-Herrera, R. et al. (2007). Bioreactors design for solid state fermentation. *Rev. Mex. Ing. Qum.* 6 (1): 33–40.

12 Pino, M.S., Rodríguez-Jasso, R.M., Michelin, M. et al. (2018). Bioreactor design for enzymatic hydrolysis of biomass under the biorefinery concept. *Chem. Eng. J.* 347: 119–136.

13 Barrios-Gonzalez, J., Fernandez, F.J., and Tomasini, A. (2003). Microbial secondary metabolites production and strain improvement. *Indian J. Biotechnol.* 2 (3): 322–333.

14 Leathers, R.R., Smith, M.A.L., and Aitken-Christie, J. (1995). Automation of the bioreactor process for mass propagation and secondary metabolism. In: *Automation and Environmental Control in Plant Tissue Culture* (ed. J. Aitken-Christie, T. Kozai, and M.A.L. Smith), 187–214. Dordrecht: Springer.

15 Mandenius, C.-F. (2016). Challenges for bioreactor design and operation. In: *Bioreactors: Design, Operation and Novel Applications* (ed. C.F. Mandenius), 1–34. Wiley.

16 McNeil, B. and Harvey, L.M. (2008). *Practical Fermentation Technology*. Chichester, UK: Wiley.

17 Ramchandra Gaikwad, V. (2018). Designing and manufacturing of fermenter for Idli. M.Sc. Thesis. Lovely Professional University, Phagwara, India.

18 Stanbury, P.F., Whitaker, A., and Hall, S.J. (2013). *Principles of Fermentation Technology*. Oxford, UK: Elsevier.

19 Georgiev, M.I. (2014). Design of bioreactors for plant cell and organ cultures. In: *Production of Biomass and Bioactive Compounds Using Bioreactor Technology*, vol. 9789401792, 3–15. Netherlands, Dordrecht: Springer.

20 Aguilar-Zarate, P., Cruz-Hernandez, M.A., Montañez, J.C. et al. (2014). Enhancement of tannase production by *Lactobacillus plantarum* CIR1: validation in gas-lift bioreactor. *Bioprocess Biosyst. Eng.* 37 (11): 2305–2316.

21 Viniegra-González, G. (1997). Solid state fermentation: definition, characteristics, limitations and monitoring. In: *Advances in Solid State Fermentation* (ed. S. Roussos, B.K. Lonsane, M. Raimbault, and G. Viniegra-Gonzalez), 5–22. Dordrecht: Springer.

22 Viniegra-González, G. and Favela-Torres, E. (2006). Why solid-state fermentation seems to be resistant to catabolite repression? *Food Technol. Biotechnol.* 40 (3): 397–406.

23 Aguilar-Zárate, P., Wong-Paz, J.E., Rodríguez-Duran, L.V. et al. (2018). On-line monitoring of *Aspergillus niger* GH1 growth in a bioprocess for the production of ellagic acid and ellagitannase by solid-state fermentation. *Bioresour. Technol.* 247: 412–418.

24 Saucedo-Castañeda, G., Trejo-Hernández, M.R., Lonsane, B.K. et al. (1994). On-line automated monitoring and control systems for CO2 and O2 in aerobic and anaerobic solid-state fermentations. *Process Biochem.* 29 (1): 13–24.

25 Arora, S., Rani, R., and Ghosh, S. (2018). Bioreactors in solid state fermentation technology: design, applications and engineering aspects. *J. Biotechnol.* 269: 16–34.

26 Andrade-Damián, M.F., Muñiz-Márquez, D.B., Wong-Paz, J.E. et al. (2019). Exploratory study of pigment extraction from *Curcuma longa* L. By solid-state fermentation using five fungal strains. *Mex. J. Biotechnol.* 4 (3): 1–11.

27 de la Cruz-Quiroz, R., Carrillo-Nieves, D., Aguilar-Zárate, P. et al. (2018). Utilization of lignocellulose-based orange peel waste for induced sporulation of *Trichoderma asperellum* via box-behnken matrix design. *Bioresources* 13 (2): https://doi.org/10.15376/biores.13.2.3971-3985.

28 Robledo, A., Aguilera-Carbo, A., Rodriguez, R. et al. (2008). Ellagic acid production by *Aspergillus niger* in solid state fermentation of pomegranate residues. *J. Ind. Microbiol. Biotechnol.* 35 (6): 507–513.

29 Saldaña-Mendoza, S.A., Ascacio-Valdés, J.A., Palacios-Ponce, A.S. et al. (2020). Use of wastes from the tea and coffee industries for the production of cellulases using fungi isolated from the Western Ghats of India. *Syst. Microbiol. Biomanuf.* 1 (1): 33–41.

30 Ascacio-Valdés, J.A., Aguilera-Carbó, A.F., Buenrostro, J.J. et al. (2016). The complete biodegradation pathway of ellagitannins by *Aspergillus niger* in solid-state fermentation. *J. Basic Microbiol.* 56 (4): 329–336.

31 Najafpour, G.D. (2015). *Biochemical Engineering and Biotechnology*. Amsterdam: Elsevier.

32 Liu, S. (2020). *Bioprocess Engineering*. Elsevier.

33 Anand, S. and Srivastava, P. (2022). Comparative study for the production of mycophenolic acid using *Penicillium brevicompactum* in batch, bed-batch and continuous fermentation process. *Biointerface Res. Appl. Chem.* 12 (1): 366–376.

34 Amache, R., Yerramalli, S., Giovanni, S., and Keshavarz, T. (2019). Quorum sensing involvement in response surface methodology for optimisation of sclerotiorin production by *Penicillium sclerotiorum* in shaken flasks and bioreactors. *Ann. Microbiol.* 69 (13): 1415–1423.

35 Boruta, T., Przerywacz, P., Ryngajllo, M., and Bizukojc, M. (2018). Bioprocess-related, morphological and bioinformatic perspectives on the biosynthesis of secondary metabolites produced by *Penicillium solitum*. *Process Biochem.* 68: 12–21.

36 Granada, S.D., Ramirez-Restrepo, S., Lopez-Lujan, L. et al. (2018). Screening of a biological control bacterium to fight avocado diseases: from agroecosystem to bioreactor. *Biocatal. Agric. Biotechnol.* 14: 109–115.

37 Chastang, T., Pozzobon, V., Taidi, B. et al. (2018). Resveratrol production by grapevine cells in fed-batch bioreactor: experiments and modelling. *Biochem. Eng. J.* 131: 9–16.

38 Mubarak, M.Q.E., Jufri, S.H.M., Smsns, Z. et al. (2017). Kinetics of surfactin production by *Bacillus subtilis* in a 5 L stirred-tank bioreactor. *Sains Malays.* 46 (9): 1541–1548.

39 Ienczak, J.L., Schmidt, M., Quines, L.K. et al. (2016). Poly(3-hydroxybutyrate) production in repeated fed-batch with cell recycle using a medium with low carbon source concentration. *Appl. Biochem. Biotechnol.* 178 (2): 408–417.

40 Boumehira, A.Z., Malek, R.A., Othman, N.Z. et al. (2016). Bioprocess development for beta- and gamma-rubromycin production: a human telomerase inhibitors, by *Streptomyces* sp ADR1. *J. Sci. Ind. Res.* 75 (10): 609–614.

41 Verma, P., Khan, S.A., Mathur, A.K. et al. (2014). Improved sanguinarine production via biotic and abiotic elicitations and precursor feeding in cell suspensions of latex-less variety of *Papaver somniferum* with their gene expression studies and upscaling in bioreactor. *Protoplasma* 251 (6): 1359–1371.

42 Xia, X., Lin, S.J., Xia, X.X. et al. (2014). Significance of agitation-induced shear stress on mycelium morphology and lavendamycin production by engineered *Streptomyces flocculus*. *Appl. Microbiol. Biotechnol.* 98 (10): 4399–4407.

43 Busto, V.D., Calabro-Lopez, A., Rodriguez-Talou, J. et al. (2013). Anthraquinones production in *Rubia tinctorum* cell suspension cultures: down scale of shear effects. *Biochem. Eng. J.* 77: 119–128.

44 Vendruscolo, F., Luise Müller, B., Esteves Moritz, D. et al. (2013). Thermal stability of natural pigments produced by *Monascus ruber* in submerged fermentation. *Biocatal. Agric. Biotechnol.* 2 (3): 278–284.

45 Wang, J., Gao, W.Y., Zhang, J. et al. (2012). Production of ginsenoside and polysaccharide by two-stage cultivation of *Panax quinquefolium* L. cells. *In vitro Cell Dev. Biol.* 48 (1): 107–112.

46 Rios-Iribe, E.Y., Flores-Cotera, L.B., Chavira, M.M.G. et al. (2011). Inductive effect produced by a mixture of carbon source in the production of gibberellic acid by *Gibberella fujikuroi*. *World J. Microbiol. Biotechnol.* 27 (6): 1499–1505.

47 Shin, H.Y., Lee, J.Y., Choi, H.S. et al. (2011). Production of cephalosporin C using crude glycerol in fed-batch culture of *Acremonium chrysogenum* M35. *J. Microbiol.* 49 (5): 753–758.

48 Bizukojc, M. and Ledakowicz, S. (2008). Biosynthesis of lovastatin and (+)-geodin by *Aspergillus terreus* in batch and fed-batch culture in the stirred tank bioreactor. *Biochem. Eng. J.* 42 (3): 198–207.

49 Prakash, G. and Srivastava, A.K. (2008). Statistical elicitor optimization studies for the enhancement of azadirachtin production in bioreactor *Azadirachta indica* cell cultivation. *Biochem. Eng. J.* 40 (2): 218–226.

50 Mathur, M. and Ramawat, K.G. (2008). Improved guggulsterone production from sugars, precursors, and morphactin in cell cultures of *Commiphora wightii* grown in shake flasks and a bioreactor. *Plant Biotechnol. Rep.* 2 (2): 133–136.

51 Meanwell, R.J.L. and Shama, G. (2008). Production of streptomycin from chitin using *Streptomyces griseus*. *Bioresour. Technol.* 99 (13): 5634–5639.

52 Pekin, G., Vardar-Sukan, F., and Kosaric, N. (2005). Production of sophorolipids from *Candida bombicola* ATCC 22214 using Turkish corn oil and honey. *Eng. Life Sci.* 5 (4): 357–362.

53 Jüsten, P., Paul, G.C., Nienow, A.W., and Thomas, C.R. (1998). Dependence of *Penicillium chrysogenum* growth, morphology, vacuolation, and productivity in fed-batch fermentations on impeller type and agitation intensity. *Biotechnol. Bioeng.* 59 (6): 762–775.

54 Bud, R. (2011). Innovators, deep fermentation and antibiotics: promoting applied science before and after the Second World War. *Dynamis* 31: 323–341.

55 Veiter, L., Rajamanickam, V., and Herwig, C. (2018). The filamentous fungal pellet – relationship between morphology and productivity. *Appl. Microbiol. Biotechnol.* 102 (7): 2997–3006.

56 Papagianni, M. (2017). Microbial bioprocesses. In: *Current Developments in Biotechnology and Bioengineering* (ed. C. Larroche, M.Á. Sanromán, G. Du, and A. Pandey), 45–72. Elsevier.

57 Feng, Y., Shao, Y., and Chen, F. (2012). Monascus pigments. *Appl. Microbiol. Biotechnol.* 96 (6): 1421–1440.

58 Kongruang, S. (2011). Growth kinetics of biopigment production by Thai isolated *Monascus purpureus* in a stirred tank bioreactor. *J. Ind. Microbiol. Biotechnol.* 38 (1): 93–99.

59 Henkel, M., Geissler, M., Weggenmann, F., and Hausmann, R. (2017). Production of microbial biosurfactants: status quo of rhamnolipid and surfactin towards large-scale production. *Biotechnol. J.* 12 (7): 1600561.

60 Amani, H., Mehrnia, M.R., Sarrafzadeh, M.H. et al. (2010). Scale up and application of biosurfactant from *Bacillus subtilis* in enhanced oil recovery. *Appl. Biochem. Biotechnol.* 162 (2): 510–523.

61 Hu, J., Luo, J., Zhu, Z. et al. (2021). Multi-scale biosurfactant production by *Bacillus subtilis* using tuna fish waste as substrate. *Catalysts* 11 (4): 456.

62 Qazi, M.A., Wang, Q., and Dai, Z. (2022). Sophorolipids bioproduction in the yeast *Starmerella bombicola*: current trends and perspectives. *Bioresour. Technol.* 346: 126593.

63 Murthy, H.N., Dandin, V.S., Zhong, J.-J., and Paek, K.-Y. (2014). Strategies for enhanced production of plant secondary metabolites from cell and organ cultures. In: *Production of Biomass and Bioactive Compounds Using Bioreactor Technology* (ed. K.-Y. Paek, H.N. Murthy, and J.-J. Zhong), 471–508. Netherlands, Dordrecht: Springer.

64 Sharma, S. and Shahzad, A. (2013). Bioreactors: a rapid approach for secondary metabolite production. In: *Recent Trends in Biotechnology and Therapeutic Applications of Medicinal Plants* (ed. M. Shahid, A. Shahzad, A. Malik, and A. Sahai), 25–49. Netherlands, Dordrecht: Springer.

65 Shu, S., Vidal, D., Bertrand, F., and Chaouki, J. (2019). Multiscale multiphase phenomena in bubble column reactors: a review. *Renew. Energy* 141: 613–631.

66 Branco, R.F., Santos, J.C., Murakami, L.Y. et al. (2007). Xylitol production in a bubble column bioreactor: influence of the aeration rate and immobilized system concentration. *Process Biochem.* 42 (2): 258–262.

67 Gavrilescu, M. and Roman, R.V. (1994). Oxygen mass transfer and gas holdup in a bubble column bioreactor with biosynthesis liquids. *Acta Biotechnol.* 14 (1): 27–36.

68 Almeida Benalcázar, E., Noorman, H., Maciel Filho, R., and Posada, J.A. (2020). Modeling ethanol production through gas fermentation: a biothermodynamics and mass transfer-based hybrid model for microbial growth in a large-scale bubble column bioreactor. *Biotechnol. Biofuels* 13 (1): 1–19.

69 Sonego, J.L.S., Lemos, D.A., Pinto, C.E.M. et al. (2016). Extractive fed-batch ethanol fermentation with CO2 stripping in a bubble column bioreactor: experiment and modeling. *Energy Fuel* 30 (1): 748–757.

70 Kheradmandnia, S., Hashemi-Najafabadi, S., Shojaosadati, S.A. et al. (2015). Development of parallel miniature bubble column bioreactors for fermentation process. *J. Chem. Technol. Biotechnol.* 90 (6): 1051–1061.

71 Rahnama, F., Vasheghani-Farahani, E., Yazdian, F., and Shojaosadati, S.A. (2012). PHB production by *Methylocystis hirsuta* from natural gas in a bubble column and a vertical loop bioreactor. *Biochem. Eng. J.* 65: 51–56.

72 Karthikeyan, O.P., Chidambarampadmavathy, K., Nadarajan, S. et al. (2015). Effect of CH4/O2 ratio on fatty acid profile and polyhydroxybutyrate content in a heterotrophic–methanotrophic consortium. *Chemosphere* 141: 235–242.

73 Xue, C., Gao, K., Qian, P. et al. (2021). Cultivation of *Chlorella sorokiniana* in a bubble-column bioreactor coupled with cooking cocoon wastewater treatment: effects of initial cell density and aeration rate. *Water Sci. Technol.* 83 (11): 2615–2628.

74 Benavente-Valdés, J.R., Méndez-Zavala, A., Morales-Oyervides, L. et al. (2017). Effects of shear rate, photoautotrophy and photoheterotrophy on production of biomass and pigments by *Chlorella vulgaris*. *J. Chem. Technol. Biotechnol.* 92 (9): 2453–2459.

75 Benavente-Valdés, J.R., Aguilar, C., Contreras-Esquivel, J.C. et al. (2016). Strategies to enhance the production of photosynthetic pigments and lipids in chlorophycae species. *Biotechnol. Rep.* 10: 117–125.

76 Paul, T., Sinharoy, A., Pakshirajan, K., and Pugazhenthi, G. (2020). Lipid-rich bacterial biomass production using refinery wastewater in a bubble column bioreactor for bio-oil conversion by hydrothermal liquefaction. *J. Water Process Eng.* 37: 101462.

77 Hoseinkhani, N., Jalili, H., Ansari, S., and Amrane, A. (2019). Impact of bubble size on docosahexaenoic acid production by *Crypthecodinium cohnii* in bubble column bioreactor. *Biomass Convers. Biorefinery* 11 (4): 1137–1144.

78 Nili, S., Arshadi, M., and Yaghmaei, S. (2022). Fungal bioleaching of e-waste utilizing molasses as the carbon source in a bubble column bioreactor. *J. Environ. Manag.* 307: 114524.

79 Chisti, M.Y. and Moo-Young, M. (1987). Airlift reactors: characteristics, applications and design considerations. *Chem. Eng. Commun.* 60 (1–6): 195–242.

80 Roukas, T. and Mantzouridou, F. (2001). Effect of the aeration rate on pullulan production and fermentation broth rheological properties in an airlift reactor. *J. Chem. Technol. Biotechnol.* 76 (4): 371–376.

81 Godó, Š., Klein, J., Báleš, V., and Annus, J. (1999). Mixing time in airlift reactors during citric acid fermentation. *Bioprocess Eng.* 21 (3): 245–248.

82 Ghaz-Jahanian, M.A., Khoshfetrat, A.B., Hosseinian Rostami, M., and Haghighi, M. (2018). An innovative bioprocess for methane conversion to methanol using an efficient methane transfer chamber coupled with an airlift bioreactor. *Chem. Eng. Res. Des.* 134: 80–89.

83 Manowattana, A., Techapun, C., Watanabe, M., and Chaiyaso, T. (2018). Bioconversion of biodiesel-derived crude glycerol into lipids and carotenoids by an oleaginous red yeast *Sporidiobolus pararoseus* KM281507 in an airlift bioreactor. *J. Biosci. Bioeng.* 125 (1): 59–66.

84 Esperança, M.N., Mendes, C.E., Rodriguez, G.Y. et al. (2020). Sparger design as key parameter to define shear conditions in pneumatic bioreactors. *Biochem. Eng. J.* 157: 107529.

85 Wang, Z., Guo, H., Zhou, T. et al. (2022). Influence of sparger type on mass transfer in a pilot-scale internal loop airlift reactor. *Process* 10 (2): 429.

86 Matsumoto, M. and Furuta, H. (2018). In situ extractive fermentation of lactic acid by *Rhizopus oryzae* in an Air-lift Bioreactor. *Chem. Biochem. Eng. Q.* 32 (2): 275–280.

87 Van Hecke, W., Bockrath, R., and De Wever, H. (2019). Effects of moderately elevated pressure on gas fermentation processes. *Bioresour. Technol.* 293: 122129.

88 Żywicka, A., Ciecholewska-juśko, D., Drozd, R. et al. (2021). Preparation of *Komagataeibacter xylinus* inoculum for bacterial cellulose biosynthesis using magnetically assisted external-loop airlift bioreactor. *Polymers* 13 (22): 3950.

89 Jabiba, P., Naga Vignesh, S., and Hariharan, S. (2020). Working principle of typical bioreactors. In: *Bioreactors: Sustainable Design and Industrial Applications in Mitigation of GHG Emissions* (ed. L. Singh, A. Yousuf, and D.M. Mahapatra), 145–173. Elsevier.

90 Taşkan, B., Hasar, H., and Lee, C.H. (2020). Effective biofilm control in a membrane biofilm reactor using a quenching bacterium (*Rhodococcus* sp. BH4). *Biotechnol. Bioeng.* 117 (4): 1012–1023.

91 Kretschmer, M., Hayta, E.N., Ertelt, M.J. et al. (2022). A rotating bioreactor for the production of biofilms at the solid–air interface. *Biotechnol. Bioeng.* 119 (3): 895–906.

92 Muffler, K., Lakatos, M., Schlegel, C. et al. (2014). Application of biofilm bioreactors in white biotechnology. *Adv. Biochem. Eng. Biotechnol.* 146: 123–161.

93 Demirci, A. and Pometto, A.L. (1995). Repeated-batch fermentation in biofilm reactors with plastic-composite supports for lactic acid production. *Appl. Microbiol. Biotechnol.* 43 (4): 585–589.

94 Kunduru, M.R. and Pometto, A.L. (1996). Evaluation of plastic composite-supports for enhanced ethanol production in biofilm reactors. *J. Ind. Microbiol.* 16 (4): 241–248.

95 Napoli, F., Olivieri, G., Russo, M.E. et al. (2010). Butanol production by *Clostridium acetobutylicum* in a continuous packed bed reactor. *J. Ind. Microbiol. Biotechnol.* 37 (6): 603–608.

96 Park, Y.S. and Toda, K. (1992). Multi-stage biofilm reactor for acetic acid production at high concentration. *Biotechnol. Lett.* 14 (7): 609–612.

97 Sanromán, A., Feijoo, G., Lema, M., and J. (1996). Immobilization of *Aspergillus niger* and *Phanerochaete chrysosporium* on polyurethane foam. In: *Progress in Biotechnology*, vol. 11, 132–135. Elsevier.

98 Ricciardi, A., Parente, E., Volpe, E., and Clementi, F. (1997). Citric acid production from glucose by *Aspergillus niger* immobilized in polyurethane foam. *Ann. Microbiol. Enzimol.* 47 (1): 63–76.

99 Wang, J. (2000). Production of citric acid by immobilized *Aspergillus niger* using a rotating biological contactor (RBC). *Bioresour. Technol.* 75 (3): 245–247.

100 Cao, N., Du, J., Gong, C.S., and Tsao, G.T. (1996). Simultaneous production and recovery of fumaric acid from immobilized *Rhizopus oryzae* with a rotary biofilm contactor and an adsorption column. *Appl. Environ. Microbiol.* 62 (8): 2926–2931.

101 Demirei, A., Pometto Iil, A.L., and Johnson, K.E. (1993). Evaluation of biofilm reactor solid support for mixed-culture lactic acid production. *Appl. Microbiol. Biotechnol.* 38: 728–733.

102 Ho, K.-L.G., Pometto, A.L. III, and Hinz, P.N. (1997). Optimization of L-(+)-lactic acid production by ring and disc plastic composite supports through repeated-batch biofilm fermentation. *Appl. Environ. Microbiol.* 63 (7): 2533–2542.

103 Cotton, J.C., Pometto, A.L. III, and Gvozdenovic-Jeremic, J. (2001). Continuous lactic acid fermentation using a plastic composite support biofilm reactor. *Appl. Microbiol. Biotechnol.* 57 (5): 626–630.

104 Urbance, S.E., Pometto, A.L., DiSpirito, A.A., and Demirci, A. (2003). Medium evaluation and plastic composite support ingredient selection for biofilm formation and succinic acid production by *Actinobacillus succinogenes*. *Food Biotechnol.* 17 (1): 53–65.

105 Srivastava, P. and Kundu, S. (1999). Studies on cephalosporin-C production in an air lift reactor using different growth modes of *Cephalosporium acremonium*. *Process Biochem.* 34 (4): 329–333.

106 Pongtharangkul, T. and Demirci, A. (2006). Effects of pH profiles on nisin production in biofilm reactor. *Appl. Microbiol. Biotechnol.* 71 (6): 804–811.

107 Lara-Juache, H.R., Ávila-Hernández, J.G., Rodríguez-Durán, L.V. et al. (2021). Characterization of a biofilm bioreactor designed for the single-step production of aerial conidia and oosporein by *Beauveria bassiana* PQ2. *J. Fungi* 7 (8): 582.

108 Cheng, K.C., Demirci, A., and Catchmark, J.M. (2010). Advances in biofilm reactors for production of value-added products. *Appl. Microbiol. Biotechnol.* 87 (2): 445–456.

109 Zune, Q., Delepierre, A., Gofflot, S. et al. (2015). A fungal biofilm reactor based on metal structured packing improves the quality of a Gla::GFP fusion protein produced by *Aspergillus oryzae*. *Appl. Microbiol. Biotechnol.* 99 (15): 6241–6254.

110 Chilakamarry, C.R., Mimi Sakinah, A.M., Zularisam, A.W. et al. (2022). Advances in solid-state fermentation for bioconversion of agricultural wastes to value-added products: opportunities and challenges. *Bioresour. Technol.* 343: 126065.

111 Mitchell, D.A., von Meien, O.F., Luz, L.F.L., and Berovič, M. (2006). The scale-up challenge for SSF bioreactors. In: *Solid-State Fermentation Bioreactors: Fundamentals of Design and Operation* (ed. D.A. Mitchell, M. Berovič, and N. Krieger), 57–64. Berlin Heidelberg, Berlin, Heidelberg: Springer.

112 Ge, X., Vasco-Correa, J., and Li, Y. (2017). Solid-state fermentation bioreactors and fundamentals. In: *Current Developments in Biotechnology and Bioengineering* (ed. C. Larroche, M.Á. Sanromán, G. Du, and A. Pandey), 381–402. Elsevier.

113 Méndez-González, F., Loera-Corral, O., Saucedo-Castañeda, G., and Favela-Torres, E. (2018). Bioreactors for the production of biological control agents produced by solid-state fermentation. In: *Current Developments in Biotechnology and Bioengineering* (ed. A. Pandey, C. Larroche, and C.R. Soccol), 109–121. Elsevier.

114 Chaudhary, V., Katyal, P., Poonia, A.K. et al. (2021). Natural pigment from *Monascus*: the production and therapeutic significance. *J. Appl. Microbiol.* 133: 18–38.

115 Dufossé, L., Galaup, P., Yaron, A. et al. (2005). Microorganisms and microalgae as sources of pigments for food use: a scientific oddity or an industrial reality? *Trends Food Sci. Technol.* 16 (9): 389–406.

116 Zhang, L., Li, Z., Dai, B. et al. (2013). Effect of submerged and solid-state fermentation on pigment and citrinin production by *Monascus purpureus*. *Acta Biol. Hung.* 64 (3): 385–394.

117 Ábrego-Gacía, A., Poggi-Varaldo, H.M., Robles-González, V. et al. (2021). Lovastatin as a supplement to mitigate rumen methanogenesis: an overview. *J. Anim. Sci. Biotechnol.* 12 (1): 123.

118 Baños, J.G., Tomasini, A., Szakács, G., and Barrios-González, J. (2009). High lovastatin production by *Aspergillus terreus* in solid-state fermentation on polyurethane foam: an artificial inert support. *J. Biosci. Bioeng.* 108 (2): 105–110.

119 Mazumdar-Shaw, K. and Suryanarayan, S. (2003). Commercialization of a novel fermentation concept. In: *Biotechnology in India II* (ed. T.K. Ghose, P. Ghosh, S. Chand, et al.), 29–42. Berlin Heidelberg: Springer.

120 Barrios-González, J. and Miranda, R.U. (2010). Biotechnological production and applications of statins. *Appl. Microbiol. Biotechnol.* 85 (4): 869–883.

121 Torres-Mancera, M.T., Figueroa-Montero, A., Favela-Torres, E. et al. (2018). Online monitoring of solid-state fermentation using respirometry. In: *Current Developments in Biotechnology and Bioengineering* (ed. A. Pandey, C. Larroche, and C.R. Soccol), 97–108. Elsevier.

122 Finkler, A.T.J., de Lima Luz, L.F., Krieger, N. et al. (2021). A model-based strategy for scaling-up traditional packed-bed bioreactors for solid-state fermentation based on measurement of O2 uptake rates. *Biochem. Eng. J.* 166: 107854.

123 Ranjbar, S. and Hejazi, P. (2019). Modeling and validating *Pseudomonas aeruginosa* kinetic parameters based on simultaneous effect of bed temperature and moisture content using lignocellulosic substrate in packed-bed bioreactor. *Food Bioprod. Process.* 117: 51–63.

124 Dabaghi, S., Ataei, S.A., and Taheri, A. (2021). Performance analysis of a laboratory scale rotating drum bioreactor for production of rhamnolipid in solid-state fermentation using an agro-industrial residue. *Biomass Convers. Biorefinery* 1–8.

14

Microbial Cell Factories for Nitrilase Production and Its Applications

Neerja Thakur[1], Vinay Kumar[2], and Shashi Kant Bhatia[3]

[1] *Department of Biotechnology and Microbiology, RKMV, Shimla, Himachal Pradesh, India*
[2] *Department of Physiology and Cell Biology, The Ohio State University Wexner Medical Center, Columbus, OH, USA*
[3] *Department of Biological Engineering, College of Engineering, Konkuk University, Seoul, South Korea*

14.1 Introduction

Enzymes are gaining importance in various industries due to numerous catalytic properties such as specificity, catalytic efficiency, selectivity, and mild reaction conditions [1, 2]. These useful properties make them commercially useful tools for the production of various industrially important products [3, 4]. Enzyme-based processes are environment friendly generating less or negligible waste compounds and utilizing mild reaction conditions when compared to catalysts-mediated reactions [5, 6]. Enzymes can be purified, well characterized, and prepared for large-scale production by applying advanced techniques and bioprocesses [7, 8]. Thereby, enzyme industries are raising continuously due to the making of value-added chemical compounds and products [9, 10]. Among various commercially important enzymes, nitrilase enzymes are having a role in the production of various commodity chemicals [11, 12]. Nitrilase is one of the valuable biocatalysts that are employed for the transformation of nitriles into their corresponding carboxylic acids [13]. Nitrilases belong to the superfamily of thiol enzymes having a conserved catalytic triad (glutamate-lysine-cysteine) at their active site and are also referred to CN-hydrolases due to their capability to hydrolyze non-peptide carbon-nitrogen bonds [14]. Nitrilases are classified based on substrate catalyzed by them, e.g. aliphatic- [15], aromatic- [16, 17], heterocyclic [18], and arylaceto-nitrilases [19]. This substrate specificity of nitrilases is exploited to transform nitriles into valuable chemicals like acrylic acid, benzoic acid, *p*-hydroxybenzoic acid, nicotinic acid, glycolic acid, and mandelic acid [19–23]. Nitrilases can be isolated from different sources such as microorganisms, plants, animals, algae, yeast, nematodes, and archaea [24–27]. These enzymes were discovered initially in certain plants and then isolated from several microorganisms such as *Acinetobacter*, *Alcaligenes*, *Arthrobacter*, *Bacillus*, *Nocardia*, *Pseudomonas*, *Rhodococcus*, and *Rhodobacter* [27–29]. The selection of a particular source depends upon the use of the product, enzyme efficiency, manufacturing

Microbial Bioreactors for Industrial Molecules, First Edition.
Edited by Sudhir P. Singh and Santosh Kumar Upadhyay.

process, etc. Microbes are preferred candidates for the production of commercially important enzymes due to their ease of handling, manipulation, and cultivation under controlled conditions [30]. Keeping in view the commercial application of nitriles and products (acid and amides) produced by nitrilase-mediated reactions, this chapter was planned to provide an overview of the importance of nitrilase.

14.2 Nitrilase Categorization, Sources, Metabolism, and Production Process

14.2.1 Nitrilase Categorization

Nitrilases are broadly classified as aliphatic, aromatic, and arylaceto-nitrilases based on their substrates [31]. The aromatic nitrilases prefer aromatic and heteroaromatic substrates with reduced activity against aliphatic nitriles. Aliphatic nitrilases are active against a wide variety of aliphatic nitriles and some also exhibited activity toward aromatic and heterocyclic nitriles [32]. Arylacetonitrilases are specific to arylacetonitrile compounds with a preference for phenyl acetonitrile and (*R*, *S*)-mandelonitrile as substrates while very less or negligible activity was observed for aromatic and heteroaromatic nitriles [26, 33].

14.2.2 Nitrilase Sources

Nitrile hydrolyzing enzymes are distributed in plants, animals, and microbes [15, 33–35]. Thimann and Mahadevan [35] reported indoleacetonitrile (IAN) to indole acetic acid (IAA) degrading enzyme from barley leaves. In the same year, the first bacterial nitrilase from *Pseudomonas* sp. was reported by Hook and Robinson [36] that utilized naturally occurring nitrile ricinine.

The majority of aromatic nitrilase has been reported from the genera *Rhodococcus*, *Nocardia*, *Pseudomonas* (bacteria); *Aspergillus*, *Fusarium*, *Gibberella*, *Penicillium* (fungi); and *Exophiala*, *Cryptococcus* (yeast) [26, 37–39]. Aliphatic nitrilases are active against aliphatic nitriles and have been reported from genera *Rhodococcus*, *Acidovorax*, *Pseudomonas*, *Alcaligenes*, *Acidovorax*, *Comamonas*, *Pyrococcus*, *Synechocystis* as per earlier reports by [32, 37]. Arylacetonitrilases have been reported from several bacteria belonging to the genera *Alcaligenes*, *Pseudomonas*, *Burkholderia*, *Bradyrhizobium*, *Halomonas*, *Labrenzia*, etc. [20, 22, 28, 29, 40–42].

14.2.3 Nitrilase in the Metabolism of Nitriles

Nitriles (R—C≡N) are organic compounds synthesized naturally by a diverse set of plants and animals from terrestrial and marine habitats [43]. Naturally, these are present in the form of cyanoglycosides, cyanolipids, and phenylacetonitriles, which are the defensive metabolites in plants and able to release toxic hydrogen cyanide (HCN), ketones, and aldehydes [44]. They are commonly found in higher plants, especially almonds, *Brassica* crops (cabbage, cauliflower), and microorganisms such as fungi, bacteria, sponges, and algae [43, 45–47]. Nitrilases have versatile functions in plant and microbe's metabolism and play a crucial role in cyanide detoxification and nitrogen recycling [21, 48]. A chemical warfare among plants and microbes enforces co-evolution. Plants can synthesize a wide variety of cyanogenic glucosides (CNglcs), glucocynolates, and phenolics in their metabolic pathways [48, 49]. These secondary metabolites perform various functions such as defense and signaling. Microbes generally that live in close association with plants possess nitrilase superfamily enzymes to metabolize nitrile and amides of plant origin [50].

14.2.4 Isolation and Screening of Nitrilase-Producing Microorganisms

There is an increased interest in the isolation, identification, and screening of nitrile-hydrolyzing biocatalysts to find an alternative and ecofriendly way to replace the chemical synthetic reactions [51]. Nitrile-converting biocatalysts are useful due to their chemo-, regio-, or enantioselectivity activity that is not easy to attain in chemical reactions [52]. These capabilities of nitrilases are enhancing the interest of several researchers in the isolation of nitriles hydrolyzing bacteria, archaea, yeasts, and fungi [31]. A variety of microorganisms harboring nitrilase production capability have been isolated. Shen et al. [53] used acrylonitrile as a nitrogen source to isolate *Arthrobacter nitroguajacolicus* from soil samples that can hydrolyze acrylonitrile to acrylic acid. In another study, researchers used glucose and acetonitrile for the isolation of an acid-tolerant nitrilase-producing black yeast *Exophiala oligosperma* [54]. Nitrilase activity detection is generally based on the estimation of ammonia, which is produced equimolarly with acids from the hydrolysis of nitriles. The use of enrichment method using nitriles as the sole carbon and nitrogen source is a conventional method for screening for nitrilase-producing strains. This method helps in the survival of microbes harboring nitrile degrading activity and able to grow on nitriles. These conventional enrichment methods are time consuming. Researchers have developed a high-throughput screening method that involves the use of fluorogenic and chromogenic substrates or pH indicator reagents for the screening of new biocatalysts. Santoshkumar et al. [55] used pH indicators (phenol red, bromothymol blue, and phenolphthalein) to screen microbes able to degrade aliphatic nitrile. Banerjee et al. [56] proposed a fluorometric method involving *o*-phthaldialdehyde-2-mercaptoethanol reagent that forms a fluorochrome with the nitrilase reaction solution. The use of metagenomics technology has also been reported in nitrilase screening. DeSantis et al. [57] used metagenomics approach to screen environmental DNA mandelonitrile-hydrolyzing nitrilases. Bayer et al. [58] created four metagenomic libraries to search nitriles hydrolyzing enzymes. Genome mining and metagenomics methods are advantageous over the conventional methods as they have greatly shortened the working time. The rational genome mining method along with functional analysis was used for various nitrilase screening [59, 60].

14.2.5 Cultivation of Nitrilase-Producing Microbes

The nitrilase-producing microorganisms are generally cultivated in a medium supplemented with nitriles (inducer). Different nitriles, amides, and acids have been used for the induction of the nitrilase enzyme (Table 14.1). Constitutive expression of nitrilases has also been reported in some bacterial species [69, 70]. Aliphatic nitriles are potential inducers for their ability to induce various nitrilases, which are evident from the literature [31].

Table 14.1 Culture conditions for nitrilase production by various microorganisms.

Microorganism	Inducer	Incubation conditions	Reference
Alcaligenes faecalis CCTCC M 208168	*n*-Butyronitrile	30 °C, pH 7.5, 24 h	[61]
Rhodobacter sphaeroides LHS-305	Acetonitrile	30 °C for 24 h and 160 rpm	[27]
Alcaligenes faecalis MTCC 10757	Various nitriles like benzonitrile/e-caprolactum/acetonitrile/3-cyanopyridine	24 h at 35 °C and 200 rpm	[62]
Alcaligenes faecalis ZJUTB10	*n*-Butyronitrile	30 °C at 120 rpm for 24 h	[63]
Gibberella intermedia CA3-1	Caprolactam	30 °C on a shaker (120 rpm) for 48 h	[64]
Gordonia terrae MTCC 8139	Isobutyronitrile	30 °C at 150 rpm for 42 h	[17]
Isoptericola variabilis RGT01	Acetonitrile	30 °C, 200 rpm, 48 h	[65]
Rhodococcus rhodochrous BX2	—	20 h at 30 °C and 180 rpm	[66]
Alcaligenes sp. MTCC 10675	Isobutyronitrile	30 °C, pH 7.0, 24 h	[12]
Alcaligenes faecalis MTCC 12629	Isobutyronitrile	30 °C, pH 7.0, 21 h	[67]
Rhodococcus rhodochrous ATCC-BAA870	Dimethylformamide	30 °C, 72 h	[68]
Halomonas sp. IIIMB2797	Glutaronitrile	37 °C, 160 rpm	[40]

The induction of nitrilase varies with the capability of various nitriles, species, and strains, which further depends on transcription regulators [31]. In some cases, it has been observed that substrate specificity of nitrilase differs when induced with different inducers, which shows the presence of more than one nitrilase enzyme [62, 65]. In many cases, the nitrile inhibiting the growth of a microorganism was hydrolyzed by the nitrilase-induced cells of the same organism [67]. A benzonitrile hydrolyzing nitrilase was induced in *Nocardia* sp. 11216 and *Fusarium solani* by adding benzonitrile 0.005% inducer with in mineral salt medium containing yeast extract as a carbon source [71, 72]. Many aliphatic nitriles have been reported to act as both C and N sources for the growth and inducer for the production of nitrilase, i.e. acetonitrile for *Geotrichum* sp. JR1, isobutyronitrile for *Nocardia globerula* NHB2, and *Alcaligenes* sp. MTCC 10674 [73, 74]. Glycerol, glucose, sorbitol, ammonium acetate, sodium succinate, sucrose, starch, and sodium citrate have been added to the production medium to serve as a carbon source to support the growth of various microorganisms. The growth medium of various microorganisms also comprises inorganic nitrogen source: ammonium acetate, sodium nitrate, ammonium sulfate [61], and complex organic nitrogen sources (peptone, yeast extract, malt extract, tryptone) [17, 27, 67]. Some microbial strains *Rhodococcus rhodochrous* K22, *R. rhodochrous* J1, *N. globerula* NHB2, and *Alcaligenes* sp. MTCC 10674 exhibited slower growth in the minimal salt medium [73, 74].

14.2.6 Nitrilase Production in Bioreactor

Nitrilase-mediated biocatalysis has proved to be an attractive tool for an industrial scale chemical production (Figure 14.1). It is a green catalyst to synthesize commercially important compounds due to clean reactions and other catalytic activities [38]. To meet the industrial demands, scale-up of process at a large scale is required and fermentation is the preferred method [75]. Nitrilase production was carried out in a stirred tank bioreactor from *Streptomyces* sp. MTCC 7546 and reported 10.26 g/L cell biomass [75]. Nitrilase from *A. faecalis* was overexpressed in *Escherichia coli* and performed fermentation in a 7 L lab scale bioreactor [76]. A fed-batch method was performed using DO-stat feeding approach and under induction strategy and process results in a yield of nitrilase (247 kU/L).

14.2.6.1 Factors Affecting Nitrilase Production in a Bioreactor

Aeration is one of the key fermentation parameters required for the growth of microbes in the bioreactor. It is provided in the reactor to prevent the depletion of oxygen in the broth [77]. Stirring is provided in stirred tank reactors, especially to provide mixing and proper aeration in the culture broth. The effect of aeration has also been studied in reactors

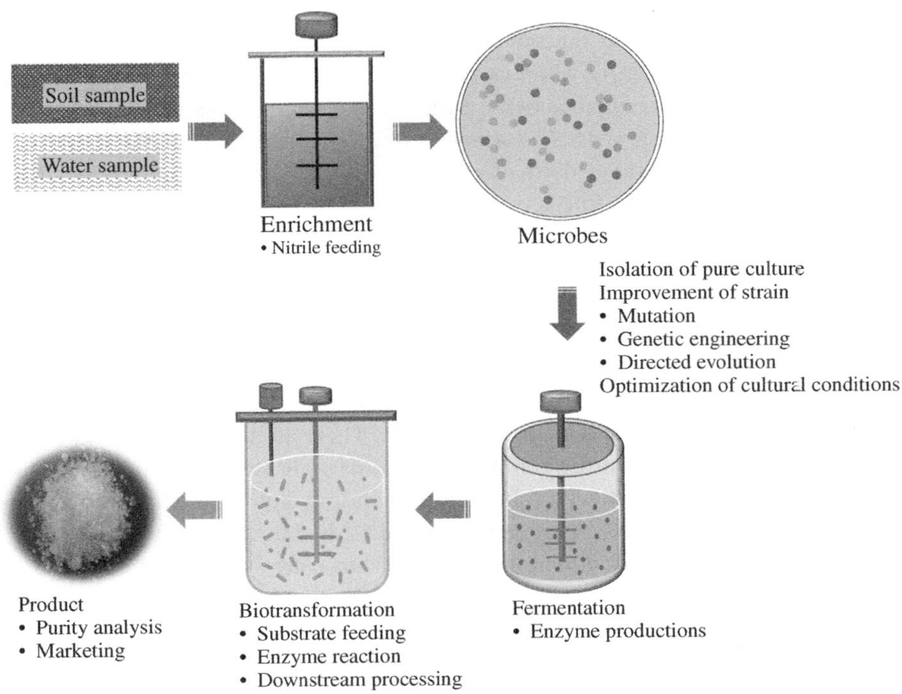

Figure 14.1 The overview of various steps of nitrilase production and biotransformation.

in various studies as it also affects product formation. Aeration rate also affects the growth and nitrilase production of microbes in the stirred tank bioreactors whether the reaction is made in batch or fed-batch mode. It has been seen that the growth of *E. coli* cells was increased and enantioselective nitrilase activity was decreased with an increase in aeration rate at higher values [78]. A minor boost in the aeration rate did not affect the growth of cells and nitrilase activity in the case of thermostable mutant nitrilase of recombinant *E. coli* [77].

Dissolved oxygen concentration in the surrounding of microorganisms is very crucial to obtain the maximum growth of microorganisms and heterologous protein expression. The optimum value of aeration and agitation is required to retain the oxygen for appropriate growth and activity of the enzyme produced by the microorganism [77]. Impellers are the equipment installed in the bioreactor to agitate the culture broth. The speed of the impeller affects the growth of microorganisms and nitrilase production, thereby it is varied to check the optimum value of agitation in the bioreactor. The agitation rate above the optimum value decreases the nitrilase activity due to shear stress, whereas the lower agitation affects the cell mass concentration adversely. Reduced agitation leads to lower biomass and nitrilase due to insufficient mixing and less availability of oxygen to the microbial culture. A similar trend was observed for *E. coli* expressing nitrilase where the agitation rate was varied in the bioreactor from 200 to 500 rpm at 37 °C. Above and below this value resulted in decreased biomass and nitrilase activity. A similar effect of agitation on nitrilase activity and specific growth rate of microorganisms was observed with the reduction and enhancement of agitation values from the optimum value in the previous studies on nitrilase transformation in stirred tank reactor [77]. Temperature and pH are other important parameters that have a remarkable effect on the microbial productivity in the bioreactor. Both the parameters are required to control while carrying out the reaction in the reactor as they have a drastic effect on the enzyme activity. Different microorganisms require optimum temperature and pH for their growth. The temperature has a direct effect on the stability of enzymes and pH also affects the activity of enzymes, thereby reducing productivity. The bioreactor studies pertaining to nitrilase conversions involve the investigation of such parameters for high yield and nitrilase activity [77, 79].

14.3 Nitrilase in the Biotransformation of Nitriles

Enzymes capable of nitrile and amide degradation are continuously evolving as nitriles and amide compounds are widespread and occur naturally. There are two pathways that have been reported for hydrolysis of nitrile compounds where single-step reaction is mediated by nitrilase and nitriles are directly hydrolyzed into acids and ammonia while in two-step reaction, nitriles are first hydrolyzed into amide by nitrile hydratase and then amides are hydrolyzed by amidase into acids and ammonia [21]. Nitrile are synthesized and used as solvents, precursors in pharmaceutical, plastic industry, herbicides, and other industrially important pharmaceutical precursors [80, 81]. There are various nitrile compounds that are aromatic, aliphatic, heterocyclic, and aryl in nature that are catalyzed by nitrilase to respective products (Table 14.2). Biotransformation of nitriles into acids can be performed using whole resting cells, free enzymes, or immobilized cells, which depends upon the

Table 14.2 Bioprocess development using the whole cell and immobilized nitrilase enzyme from various microorganisms.

Nitrile	Nitrilase source	Product	Biocatalyst form/ Matrix	Reaction	Applications	Reference
Glycolonitrile (GLN)	*Alcaligenes* sp. ECU0401	Glycolic acid	Whole cells	100 mL	Polymers synthesis and pharmaceuticals	[82]
3-Cyanopyridine	*Escherichia coli* JM109 harboring nitrilase gene from *Alcaligenes faecalis MTCC 126*	Nicotinic acid	Immobilized/ sodium alginate	250 mL	Food additives and pharmaceutical intermediates	[73]
	Nocardia globerula NHB-2		Whole cells	1 L, fed batch		[83]
4-Cyanopyridine	*Nocardia globerula* NHB-2	Isonicotinic acid	Whole cells	1 L, fed batch	Antituberculosis drugs	[18]
o-Chloromandelonitrile	*Labrenzia aggregate*	(*R*)-*o*-chloromandelic	Whole cells	250 mL	Synthesis of clopidogrel (cardiovascular drug)	[42]
	Recombinant *E. coli* M15 harboring nitrilase from *Burkholderia cenocepacia* J2315	(*R*)-*o*-Chloromandelic acid	Recombinant cells	2 L		[22]
	Burkholderia cenocepacia J2315	(*R*)-*o*-Chloromandelic acid	Whole cell enzyme	250 mL		[84]
	Recombinant *E. coli* cells expressing nitrilase of *Alcaligenes faecalis* ZJUTB10	(*R*)-*o*-Chloromandelic acid	Recombinant cells	100 mL		[85]
Isobutyronitrile	*Alcaligenes* sp. MTCC 10674	Isobutyric acid	Whole cells	40 mL, fed batch	Pharmaceutical intermediates and polymer synthesis	[59]

(*Continued*)

Table 14.2 (Continued)

Nitrile	Nitrilase source	Product	Biocatalyst form/ Matrix	Reaction	Applications	Reference
Mandelonitrile	Recombinant *Escherichia coli* harboring nitrilase of *Alcaligenes* sp. ECU0401	(*R*)-(−)-mandelic acid	Whole cells	2 L	Semi-synthetic penicillins, cephalosporins, and cosmetics	[86]
	Alcaligenes sp. MTCC 10675	(*R*)-(−)-mandelic acid	Whole cells	1 L, fed batch		[12]
	Alcaligenes faecalis ECU0401	(*R*)-(−)-mandelic acid	Immobilized enzyme	100 mL		[87]
	Escherichia coli BL21(DE3)/pET-Nit	(*R*)-(−)-mandelic acid	Recombinant cells Whole cell	20,000 L		[88]
4-Hydroxybenzonitrile	*Gordonia terrae*	4-Hydroxybenzoic acid	Whole cells	500 L, fed batch	Cosmetics, fragrance, food, preservatives, and polymer synthesis	[17]
	Alcaligenes faecalis MTCC 12629	4-Hydroxyphenylacetic acid	Whole-cell enzyme	500 mL		[67]
Phenylacetonitrile	*Alcaligenes* sp. MTCC 10675	Phenylacetic acid	Whole-cell enzyme	1 L	Used as an adjunct to treat acute hyperammonemia	[89]
	Recombinant *Escherichia coli* M15 harboring nitrilase from *Burkholderia cenocepacia* J2315	Phenylacetic acid	Recombinant cells	300 mL		[20]
	Alcaligenes faecalis MTCC 12629	4-Aminophenylacetic acid	Whole-cell enzyme	500 mL		[28]

product and their applications. Biotransformation using the whole cell is economical as it provides a natural environment to the enzyme inside [90]. Some of the very important commercial compounds and their reactor synthesis are discussed below.

14.3.1 Aliphatic Acids

14.3.1.1 Acrylic Acid

Acrylic acid is used in textiles, surface coatings, and many other applications [91]. Acrylic acid can be produced by biotransformation of acrylonitrile using nitrilase-mediated hydrolysis reactions [53, 92, 93]. Use of the whole cell of *Alcaligenes* sp. having nitrilase activity has been reported for the synthesis of acrylic acid (115 g/L) [93]. Acrylic acid (390 g/L) production using nitrilase activity of *R. rhodochrous* J1 was reported [94]. An accumulation of 414.5 g/L acrylic acid in 10 hours of reaction using a mutant of *R. rhodochrous* J1 as biocatalyst has been reported by Luo et al. [94].

14.3.1.2 Glycolic Acid

Glycolic acid is α-hydroxycarboxylic acid having wide applications in medicine and pharmaceuticals. Many researchers have explored the enzymatic transformation of glyconitrile to glycolic acid [82, 95, 96]. An engineered nitrilase of *Acidovorax facilis* 72W with improved catalytic efficiency has been used for the biotransformation of glycolonitrile into glycolic acid [97]. *Acidovorax facilis* mutant F168V has exhibited very high productivity of 1010 g/g dcw after 55 cycles of reactions. He et al. [82] immobilized the whole cell of *Alcaligenes* sp. having nitrilase using carrageenan cross-linked with glutaraldehyde/PEI and used for biotransformation (18.0 g/L/h glycolic acids).

14.3.2 Aromatic Acids

14.3.2.1 Nicotinic Acid

Nicotinic acid is an important vitamin being synthesized through chemical routes involving high-energy inputs. Nicotinic acid (Vitamin B_3) or 3-pyridine carboxylic acid is an important vitamin and its deficiency causes pellagra. There are reports on the biological synthesis of nicotinic acid through nitrilase-mediated hydrolysis of 3-cyanopyridine. Nitrilase-mediated synthesis of nicotinic acid was first performed by using free cells of *R. rhodochrous* Jl at 50 mL scale in fed-batch system and the process resulted in 172 g/L nicotinic acids [98]. A column reactor packed with calcium-alginate-entrapped *Nocardia rhodochrous* LL100-21 cells were employed for the synthesis of nicotinic acid by Vaughan et al. [99]. The column was maintained at 30 °C through a circulatory water jacket and 3-cyanopyridine (0.3 M) substrate was pumped upward at 14 mL/h through the column. Through this method, 96 g of nicotinic acid was produced in 150 hours of reaction [99]. A similar column bioreactor was used for the 3-cyanopyridine conversion to nicotinic acid by taking thermophilic nitrilase of *Bacillus pallidus* Dac 521 [100]. The glass column was packed with calcium-alginate-immobilized biocatalyst and operated at 50 °C. The substrate 3-cyanopyridine (0.1 M) was pumped upward to the column and a total of 3.12 g of 3-cyanopyridine conversion to the product was achieved in 100 h at a rate of 104 mg/g cells/h [100]. Fed-batch reaction of 1 L scale was carried out to produce nicotinic acid using free cells of *N. globerula* NHB-2 with the productivity of 3.21 g/g dcw [101]. A continuous 3-cyanopyridine conversion to nicotinic acid using membrane bioreactor and resting cells of *Microbacterium imperiale* CBS 498-74 have been also reported [102].

14.3.2.2 Isonicotinic Acid

Isonicotinic acid (pyridine-4-carboxylic acid) is widely used for the production of isoniazid antituberculastic drug. It has various applications in the synthesis of inabenfide, terefenadine, an antihistamine, nialamide, and antidepressant [18]. Isobutyronitrile-induced nitrilase of *N. globerula* NHB-2 has been used in fed-batch reaction at 1 L scale for isonicotinic acid production from 4-CP (21.1 g/h/mg) [18].

14.3.2.3 Benzoic Acid

Benzoic acid is used in homeopathic formulations and the production of glycol benzoate. *Nocardia globerula* NHB-2 nitrilase was utilized for benzonitrile hydrolysis into benzoic acid in fed-batch mode and reported 108 and 84 g/L benzoic acid production using free and agar confined cells [103]. *p*-Hydroxybenzoic acid has applications as antioxidant, esters synthesis, food preservatives, flavors, polymers, cosmetics, and pharmaceutical products [17, 104]. Free cells of *Gordonia terrae* possessing nitrilase activity were used for the conversion of *p*-hydroxybenzonitrile to *p*-hydroxybenzoic acid and 98.7% conversion has been reported by Kumar and Bhalla [17]. Biotransformation of 1-cyanocyclohexaneacetonitrile to 1-cyanocyclohexaneacetic acid was performed in a 2 L stirred reactor with a productivity of 244.5 g/L/d [105]. Table 14.2 presents the nitrilase-mediated synthesis of some important aromatic acids.

14.3.3 Arylacetic Acids

Nitrilases, which preferably utilizes aryl as substrates, are termed arylacetonitrilases and they specifically catalyze arylacetonitriles to corresponding acid and ammonia. The preferable substrates for bacterial arylacetonitrilases are phenylacetonitrile, mandelonitrile, and their substitutive compounds such as methoxy or hydroxyphenylacetonitriles and chloromandelonitrile. Nitrilase-mediated bioreactor studies have been reported for the synthesis of mandelic acid, phenylacetic acid, *p*-methoxyphenylacetic acid, and *o*-chloromandelic acid [20, 28, 42, 67, 104]. Arylacetonitrilase-mediated synthesis of some important arylacetic acid is discussed in the successive sections and summarized in Table 14.2.

14.3.3.1 Mandelic Acid

Many researchers have reported the enantioselective nitrilase-based reactions for the synthesis of (*R*)-(−)-mandelic acid from racemic mandelonitrile [12, 106–108]. A number of bacteria have been reported for arylacetonitrilase production and biotransformation of mandelonitrile to mandelic acid, i.e. *Pseudomonas putida* MTCC 5110 (0.39 g/g dcw), *A. faecalis* ECU0401 (3.8 g/g dcw), *Alcaligenes* sp. MTCC 10675 (3.9 g/g dcw). Zhang et al., used engineered *E. coli* cells overexpressing nitrilase gene of *Alcaligenes* sp. overexpressed and reported 4.5 g/L/h *R*(−) mandelic acids with 99% enantiomeric excess (ee) [86]. Fed-batch reaction at 1 L scale using free cells of *Alcaligenes* sp. MTCC 10675 and 10 mM substrate per feed leads to the production of 130 mM of mandelic acid with 0.78 g/L/h productivity [12]. Zhang et al. [70] used immobilized cells in a 2 L stirred reactor with toluene–water biphasic system to relieve substrate inhibition and reported 13.8 g/g dcw *R*-(−)-mandelic acid with 98.0% ee. Engineered *E. coli* M15/BCJ2315 cells overexpressing nitrilase from *Burkholderia cenocepacia* J2315 immobilized in magnetized chitosan nanoparticles have been used for hydrolysis of mandelonitrile and the process resulted in 37.3 g/L/h *R*-(−)-mandelic acid with 95% ee [107].

Enantiopure (*R*)-*o*-chloromandelic acid is used as a precursor for the synthesis of Clopidogrel®, a platelet aggregation inhibitor [42, 109, 110]. (*R*)-*o*-chloromandelic acid production from the hydrolysis of *o*-chloromandelonitrile using *Labrenzia aggregata* nitrilase in toluene–water (1 : 9, v/v) biphasic system has been reported with 154.4 g/L/day productivity and 96.3% ee [42].

14.3.3.2 Phenylacetic Acid

Phenylacetic acid is widely used in medicine, pesticide, and perfume industries and as a precursor penicillin G production [111]. Other applications include making phenobarbital, primidone, pesticide, and fungicide (3-chlorophenylacetic acid and rodenticide) and antimicrobial activity [89]. The arylacetonitrile hydrolyzing activity of nitrilase from *A. faecalis* MTCC 12629 was explored for conversion of 4-hydroxyphenlyacetonitrile to 4-hydroxyphenylacetic acid and fed-batch reaction at 500 mL scale 4.13 g/g dcw/h product [67]. *p*-Methoxyphenylacetic acid is another aryl acid used as an intermediate for the synthesis of formoterol fumarate (anti-asthmatic drug), venlafaxine (antidepressant agent), and a flavoring agent [112]. Biotransformation of *p*-methoxyphenylacetonitrile to *p*-methoxyphenylacetic acid was carried out by using resting cells of *Bacillus subtilis* ZJB-063 [113].

14.4 Conclusion

Nitrilase enzyme holds great potential in the industrial production of useful compounds in biotransformation reactions and extensively studied enzyme in this field. This enzyme is an important tool over chemical catalysts for the synthesis of useful chemicals and contributes to green chemistry. Nitrilase can be used as a whole cell, purified, or immobilized form in the bioreactors to catalyze the reactions. However, the bioprocess reactions suffer from substrate/product inhibitions that affect the productivity of enzymes at a large scale. This problem can be overcome by overcoming product inhibition using *in situ* product removal or enzyme engineering to improve substrate and product tolerance. The most preferable reaction modes in a reactor are batch and fed-batch reactions to obtain a high yield. Development of improved methods, processes, and employing different strategies based on these enzymes will further improve the prospects of these enzymes for wider use in the industrial synthesis of several compounds that have not been commercialized to date.

References

1 Martinkova, L., Rucka, L., Nesvera, J., and Patek, M. (2017). Recent advances and challenges in the heterologous production of microbial nitrilases for biocatalytic applications. *World Journal of Microbiology and Biotechnology* 33: 8.

2 Singh, S.P., Pandey, A., Singhania, R.R. et al. (eds.) (2020). *Biomass, Biofuels, Biochemicals: Advances in Enzyme Catalysis and Technologies*. Elsevier. ISBN 9780128198209. https://doi.org/10.1016/C2019-0-00323-8.

3 Han, Y.-H., Choi, T.-R., Park, Y.-L. et al. (2020). Enhancement of pipecolic acid production by the expression of multiple lysine cyclodeaminase in the *Escherichia coli* whole-cell system. *Enzyme and Microbial Technology* 140: 109643.

4 Kim, J.-H., Bhatia, S.K., Yoo, D. et al. (2015). Lipase-catalyzed production of 6-O-cinnamoyl-sorbitol from D-sorbitol and cinnamic acid esters. *Applied Biochemistry and Biotechnology* 176: 244–252.

5 Park, J.Y., Park, Y.-L., Choi, T.-R. et al. (2020). Production of γ-aminobutyric acid from monosodium glutamate using *Escherichia coli* whole-cell biocatalysis with glutamate decarboxylase from *Lactobacillus brevis* KCTC 3498. *Korean Journal of Chemical Engineering* 37: 2225–2231.

6 Ham, S., Bhatia, S.K., Gurav, R. et al. (2022). Gamma aminobutyric acid (GABA) production in *Escherichia coli* with pyridoxal kinase (pdxY) based regeneration system. *Enzyme and Microbial Technology* 155: 109994.

7 Hong, Y.-G., Moon, Y.-M., Choi, T.-R. et al. (2019). Enhanced production of glutaric acid by NADH oxidase and GabD-reinforced bioconversion from l-lysine. *Biotechnology and Bioengineering* 116: 333–341.

8 Sharma, H. and Upadhyay, S.K. (2020). Enzymes and their production strategies. In: *Biomass, Biofuels, Biochemicals: Advances in Enzyme Catalysis and Technologies* (ed. S.P. Singh, A. Pandey, R.R. Singhania et al.) 31–48. Elsevier. https://doi.org/10.1016/B978-0-12-819820-9.00003-X.

9 Choi, T.-R., Jeon, J.-M., Bhatia, S.K. et al. (2020). Production of low molecular weight P(3HB-co-3HV) by butyrateacetoacetate CoA-transferase (cftAB) in *Escherichia coli*. *Biotechnology and Bioprocess Engineering* 25: 279–286.

10 Mehta, P.K., Bhatia, S.K., Bhatia, R.K., and Bhalla, T.C. (2013). Purification and characterization of a novel thermo-active amidase from *Geobacillus subterraneus* RL-2a. *Extremophiles* 17: 637–648.

11 Mehta, P.K., Bhatia, S.K., Bhatia, R.K., and Bhalla, T.C. (2015). Thermostable amidase catalyzed production of isonicotinic acid from isonicotinamide. *Process Biochemistry* 50: 1400–1404.

12 Bhatia, S.K., Mehta, P.K., Bhatia, R.K., and Bhalla, T.C. (2014). Optimization of arylacetonitrilase production from *Alcaligenes* sp. MTCC 10675 and its application in mandelic acid synthesis. *Applied Microbiology and Biotechnology* 98: 83–94.

13 Chen, Z., Chen, H., Ni, Z. et al. (2015). Expression and characterization of a novel nitrilase from hyperthermophilic bacterium *Thermotoga maritima* MSB8. *Journal of Microbiology and Biotechnology* 25: 1660–1669.

14 Rodriguez, J.R. (2014). Understanding nitrile-degrading enzymes: classification, biocatalytic nature and current applications. *Revista Latinoamericana de Biotecnologia Ambientaly Algal* 4: 8–25.

15 Mukram, I., Ramesh, M., Monisha, T.R. et al. (2016). Biodegradation of butyronitrile and demonstration of its mineralization by *Rhodococcus* sp. MTB5. *3 Biotech* 6: 1–7.

16 Zhu, X.Y., Gong, J.S., Li, H. et al. (2014). Bench-scale biosynthesis of isonicotinic acid from 4-cyanopyridine by *Pseudomonas putida*. *Chemical Papers* 68: 739–744.

17 Kumar, V. and Bhalla, T.C. (2013). Transformation of *p*-hydroxybenzonitrile to *p*-hydroxybenzoic acid using nitrilase activity of *Gordonia terrae*. *Biocatalysis and Biotransformation* 31: 42–48.

18 Sharma, N.N., Sharma, M., and Bhalla, T.C. (2012). *Nocardia globerula* NHB-2 nitrilase catalyzed biotransformation of 4-cyanopyridine to isonicotinic acid. *AMB Express* 2: 25–32.

19 Martinkova, L. and Kren, V. (2018). Biocatalytic production of mandelic acid and analogues: a review and comparison with chemical processes. *Applied Microbiology and Biotechnology* 102: 3893–3900.

20 Fan, H., Chen, L., Sun, H. et al. (2017). A novel nitrilase from *Ralstonia eutropha* H16 and its application to nicotinic acid production. *Bioprocess and Biosystems Engineering* 40: 1271–1281.

21 Bhalla, T.C., Kumar, V., Kumar, V. et al. (2018). Nitrile metabolizing enzymes in biocatalysis and biotransformation. *Applied Biochemistry and Biotechnology* 185: 925–946.

22 Wang, H., Sun, H., Gao, W., and Wei, D. (2013). Efficient production of (R)-c-Chloromandelic acid by recombinant *Escherichia coli* cells harboring nitrilase from *Burkholderia cenocepacia* J2315. *Organic Process Research and Development* 18: 767–773.

23 Muller, E., Sosedov, O., Alexander, J. et al. (2021). Synthesis of (R)-mandelic acid and (R)-mandelic acid amide by recombinant *E. coli* strains expressing a (R)-specific oxynitrilase and an arylacetonitrilase. *Biotechnology Letters* 43: 287–296.

24 Beck, A., Divakar, P.K., Zhang, N. et al. (2015). Evidence of ancient horizontal gene transfer between fungi and the terrestrial alga *Trebouxia*. *Organisms, Diversity and Evolution* 15: 235–248.

25 Dennett, G.V. and Blamey, J.M. (2016). A new thermophilic nitrilase from an antarctic hyperthermophilic microorganism. *Frontiers in Bioengineering and Biotechnology* 4: 1–9.

26 Martinkova, L. (2019). Nitrile metabolism in fungi: a review of its key enzymes nitrilases with focus on their biotechnological impact. *Fungal Biology Reviews* 33: 149–157.

27 Yang, C., Wang, X., and Wei, D. (2011). A new nitrilase-producing strain named *Rhodobacter sphaeroides* LHS-305: biocatalytic characterization and substrate specificity. *Applied Biochemistry and Biotechnology* 165: 1556–1567.

28 Thakur, N., Sharma, N., Thakur, S. et al. (2019). Bioprocess development for the synthesis of 4-aminophenylacetic acid using nitrilase activity of whole cells of *Alcaligenes faecalis* MTCC 12629. *Catalysis Letters* 149: 2854–2863.

29 Brunner, S., Eppinger, E., Fischer, S. et al. (2018). Conversion of aliphatic nitriles by the arylacetonitrilase from *Pseudomonas fluorescens* EBC191. *World Journal of Microbiology and Biotechnology* 34: 91.

30 Bhatia, S.K., Vivek, N., Kumar, V. et al. (2021). Molecular biology interventions for activity improvement and production of industrial enzymes. *Bioresource Technology* 324: 124596.

31 Chhiba-Govindjee, V.P., van der Westhuyzen, C.W., Bode, M.L., and Brady, D. (2019). Bacterial nitrilases and their regulation. *Applied Microbiology and Biotechnology* 103: 4679–4692.

32 Ramteke, P.W., Maurice, N.G., Joseph, B., and Wadher, B.J. (2013). Nitrile-converting enzymes: an eco-friendly tool for industrial biocatalysis. *Biotechnology and Applied Biochemistry* 60: 459–481.

33 Dooley-Cullinane, T.M., O'Reilly, C., Weiner, D.P.A.B. et al. (2018). The use of clade-specific PCR assays to identify novel nitrilase genes from environmental isolates. *MicrobiologyOpen* 2018: e700.

34 Gunther, J., Irmisch, S., Lackus, N.D. et al. (2018). The nitrilase PtNIT1 catabolizes herbivore-induced nitriles in *Populus triochocarpa*. *BMC Plant Biology* 18: 251.

35 Thimann, K.V. and Mahadevan, S. (1958). Enzymatic hydrolysis of indoleacetonitrile. *Nature* 181: 1466–1467.

36 Hook, R.H. and Robinson, W.G. (1964). Ricinine nitrilase II: purification and properties. *Journal of Biological Chemistry* 239: 4263–4267.

37 Agarwal, A., Nigam, V.K., and Vidyarthi, A.S. (2012). Nitrilases-an attractive nitrile degrading biocatalyst. *International Journal of Pharma and Bioscience* 3: 232–246.

38 Gong, J.S., Lu, Z.M., Li, H. et al. (2012). Nitrilases in nitrile biocatalysis: recent progress and forthcoming research. *Microbial Cell Factories* 11: 142–148.

39 Vesela, A.B., Rucka, L., Kaplan, O. et al. (2016). Bringing nitrilase sequences from databases to life: the search for novel substrate specificities with a focus on dinitriles. *Applied Microbiology and Biotechnology* 100: 2193–2202.

40 Singh, R.V., Sharma, H., Koul, A., and Babu, V. (2018). Exploring a broad spectrum nitrilase from moderately halophilic bacterium *Halomonas* sp. IIIMB2797 isolated from saline lake. *Journal of Basic Microbiology* 58: 876–874.

41 Radisch, R., Chmatal, M., Rucka, L. et al. (2018). Overproduction and characterization of the first enzyme of a new aldoxime dehydratase family in *Bradyrhizobium* sp. *International Journal of Biological Macromolecules* 115: 746–753.

42 Zhang, C.S., Zhang, Z.J., Li, C.X. et al. (2012). Efficient production of (*R*)-*o*-chloromandelic acid by deracemization of *o*-chloromandelonitrile with a new nitrilase mined from *Labrenzia aggregata*. *Applied Microbiology and Biotechnology* 95: 91–99.

43 Egelkamp, R., Schneider, D., Robert, H., and Daniel, R. (2017). Nitrile-degrading bacteria isolated from compost. *Frontiers in Environmental Science* 5: 1–8.

44 Zagrobelny, M. and Moller, B.L. (2011). Cyanogenic glucosides in the biological warfare between plants and insects: the Burnet moth-birds foot trefoil model system. *Phytochemistry* 72: 1585–1592.

45 Upadhyay, S.K. and Singh, S.P. (eds.) (2021). *Bioprospecting of Plant Biodiversity for Industrial Molecules*. John Wiley & Sons Ltd. doi:10.1002/9781119718017.

46 Upadhyay, S.K. and Singh, S.P. (2023). *Plants as Bioreactors for Industrial Molecules*. John Wiley & Sons Ltd. doi:10.1002/9781119875116.

47 Singh, S.P. and Upadhyay, S.K. (eds.) (2021). *Bioprospecting of Microorganism-Based Industrial Molecules*. John Wiley & Sons Ltd. doi:10.1002/9781119717317.

48 Moller, B.L. (2010). Functional diversifications of cynogenic glucosides. *Current Opinion in Plant Biology* 13: 338–347.

49 Mandal, S.M., Chakraborty, D., and Dey, S. (2010). Phenolic acids act as signaling molecules in plant-microbe symbioses. *Plant Signaling & Behavior* 5: 359–368.

50 Howden, A.J.M. and Preston, G.M. (2009). Nitrilase enzymes and their role in plant-microbe interactions. *Microbial Biotechnology* 2: 441–451.

51 Gong, J.S., Shi, J.S., Lu, Z.M. et al. (2015). Nitrile converting enzymes as a tool to improve biocatalysis in organic synthesis: recent insight and promises. *Critical Reviews in Biotechnology* 23: 1–13.

52 Vergne-Vaxelaire, C., Bordier, F.Z., Fossey, A. et al. (2013). Nitrilase activity screening on structurally diverse substrates: providing biocatalytic tools for organic synthesis. *Advanced Synthesis and Catalysis* 355: 1763–1779.

53 Shen, M., Zheng, Y.G., and Shen, Y.C. (2009). Isolation and characterization of a novel *Arthrobacter nitroguajacolicus* ZJUTB06-99, capable of converting acrylonitrile to acrylic acid. *Process Biochemistry* 44: 781–785.

54 Rustler, S. and Stolz, A. (2007). Isolation and characterization of a nitrile hydrolyzing acidotolerant black yeast-*Exophiala oligosperma* R1. *Applied Microbiology and Biotechnology* 75: 899–908.

55 Santoshkumar, M., Nayak, A., Anjaneya, O., and Karegoudar, T. (2010). A plate method for screening of bacteria capable of degrading aliphatic nitriles. *Journal of Industrial Microbiology and Biotechnology* 37: 111–115.

56 Banerjee, A., Sharma, R., and Banerjee, U.C. (2003). A rapid and sensitive fluorometric assay method for the determination of nitrilase activity. *Biotechnology and Applied Biochemistry* 37: 289–293.

57 DeSantis, G., Wong, K., Farwell, B. et al. (2003). Creation of a productive, highly enantioselective nitrilase through gene site saturation mutagenesis. *Journal of the American Chemical Society* 125: 11476–11477.

58 Bayer, S., Birkemeyer, C., and Ballschmiter, M. (2011). A nitrilase from a metagenomic library acts regioselectively on aliphatic dinitriles. *Applied Microbiology and Biotechnology* 89: 91–98.

59 Zhu, D., Mukherjee, C., Biehl, E.R., and Hua, L. (2007). Discovery of a mandelonitrile hydrolase from *Bradyrhizobium japonicum* USDA110 by rational genome mining. *Journal of Biotechnology* 129: 645–650.

60 Zhu, D., Mukherjee, C., Yang, Y. et al. (2008). A new nitrilase from *Bradyrhizobium japonicum* USDA 110 Gene cloning, biochemical characterization and substrate specificity. *Journal of Biotechnology* 133: 327–333.

61 Xue, Y.P., Liu, Z.Q., Xu, M. et al. (2010). Enhanced biotransformation of (*R*,*S*)-mandelonitrile to (*R*)-(−)-mandelic acid with *in situ* production removal by addition of resin. *Biochemical Engineering Journal* 53: 143–149.

62 Nageshwar, Y.V.D., Sheelu, G., Shambhu, R.R. et al. (2011). Optimization of nitrilase production from *Alcaligenes faecalis* MTCC 10757 (IICT-A3): effect of inducers on substrate specificity. *Bioprocess and Biosystems Engineering* 34: 515–523.

63 Xue, Y.P., Xu, S.Z., Liu, Z.Q. et al. (2011). Enantioselective biocatalytic hydrolysis of (R,S) -mandelonitrile for production of (R)-(−)-mandelic acid by a newly isolated mutant strain. *Journal of Industrial Microbiology and Biotechnology* 38: 337–345.

64 Wu, Y., Gong, J.S., Lu, Z.M. et al. (2013). Isolation and characterization of *Gibberella intermedia* CA3-1, a novel and versatile nitrilase-producing fungus. *Journal of Basic Microbiology* 53: 1–8.

65 Kaur, G., Soni, P., Tewari, R., and Sharma, R. (2014). Isolation and characterization of a nitrile-hydrolysing bacterium *Isoptericola variabilis* RGT01. *Indian Journal of Microbiology* 54: 232–238.

66 Fang, S., An, X., Liu, H. et al. (2015). Enzymatic degradation of aliphatic nitriles by *Rhodococcus rhodochrous* BX2, a versatile nitrile degrading bacterium. *Bioresource Technology* 185: 28–34.

67 Thakur, N., Kumar, V., Thakur, S. et al. (2018). Biotransformation of 4-hydroxyphenylacetonitrile to 4-hydroxyphenylacetic acid using whole cell arylacetonitrilase of *Alcaligenes faecalis* MTCC 12629. *Process Biochemistry* 73: 117–123.

68 Chhiba-Govindjee, V.P., Mathiba, K., van der Westhuyzen, C.W. et al. (2018). Dimethylformamide is a novel nitrilase inducer in *Rhodococcus rhodochrous*. *Applied Microbiology and Biotechnology* 102: 10055–10065.

69 Kamal, A., Kumar, M.S., Kumar, C.G., and Shaik, T.B. (2011). Bioconversion of acrylonitrile to acrylic acid by *Rhodococcus ruber* strain AKSH-84. *Journal of Microbiology and Biotechnology* 21: 37–42.

70 Zhang, Z.J., Xy, J.H., He, Y.C. et al. (2011). Cloning and biochemical properties of a highly thermostable and enantioselective nitrilase from *Alcaligenes* sp. ECU0401 and its potential for (R)-(-)-mandelic acid production. *Bioprocess and Biosystems Engineering* 34: 315–322.

71 Harper, D.B. (1977). Fungal degradation of aromatic nitriles: enzymology of C-N cleavage by *Fusarium solani*. *Biochemistry Journal* 167: 685–692.

72 Harper, D.B. (1977). Microbial metabolism of aromatic nitriles: enzymology of C-N cleavage by *Nocardia* sp. (*Rhodochrous* group) NCIB 11216. *Biochemistry Journal* 165: 309–319.

73 Sharma, N.N., Monica, S., and Bhalla, T.C. (2011). An improved nitrilase-mediated bioprocess for synthesis of nicotinic acid from 3-cyanopyridine with hyperinduced *Nocardia globerula* NHB-2. *Journal of Industrial Microbiology and Biotechnology* 38: 1235–1243.

74 Bhatia, S.K., Mehta, P.K., Bhatia, R.K., and Bhalla, T.C. (2012). An isobutyronitrile induced bienzymatic system of *Alcaligenes* sp. MTCC 10674 and its application in the synthesis of α-hydroxyisobutyric acid. *Bioprocess and Biosystems Engineering* 36: 613–625.

75 Nigam, V., Khandelwal, A.K., Agarwal, A., and Vidyarthi., Ambarish. (2012). Production of a thermostable nitrilase in a lab scale stirred tank bioreactor. *International Journal of Bio-Science and Bio-Technology* 4: 1–90.

76 Jain, D., Meena, V.S., Kaushik, S. et al. (2012). Production of nitrilase by a recombinant *Escherichia coli* in a laboratory scale bioreactor. *Fermentation Technology* 1: 1.

77 Kiran, D., Mahesh, D., Manoj, J. et al. (2017). Production of thermostable mutant nitrilase by recombinant *Escherichia Coli*. *Advances in Biotechnology & Microbiology* 7: 555713.

78 Banerjee, A., Dubey, S., Kaul, P. et al. (2009). Enantioselective nitrilase from *Pseudomonas putida*: cloning, heterologous expression, and bioreactor studies. *Molecular Biotechnology* 41: 35–41.

79 Sohoni, S.V., Kundalia, P.H., Shetty, A.G. et al. (2018). A plug-and-play system for enzyme production at commercially viable levels in fed-batch cultures of *Escherichia coli* BL21 (DE3). *bioRxiV* https://doi.org/10.1101/263582.

80 Gong, J.S., Shi, J.S., Lu, Z.M. et al. (2017). Nitrile converting enzymes as a tool to improve biocatalysis in organic synthesis: recent insights and promises. *Critical Reviews in Biotechnology* 37: 69–81.

81 Chen, Z., Wang, H., Yang, L. et al. (2021). Significantly enhancing the stereoselectivity of a regioselective nitrilase for the production of (S)-3-cyano-5-methylhexanoic acid using an MM/PBSA method. *Chemical Communications* 57: 931.

82 He, Y.C., Xu, J.H., Su, J.H., and Zhou, L. (2009). Bioproduction of glycolic acid from glycolonitrile with a new bacterial isolate of *Alcaligenes* sp. ECU0401. *Applied Biochemistry and Biotechnology* 160: 1428–1440.

83 Pai, O., Banoth, L., Ghosh, S. et al. (2014). Biotransformation of 3-cyanopyridine to nicotinic acid by free and immobilized cells of recombinant *Escherichia coli*. *Process Biochemistry* 49: 655–659.

84 Wang, H., Fan, H., and Sun, H. (2015). Process development for the production of (*R*)-(−)-mandelic acid by recombinant *Escherichia coli* cells harboring nitrilase from *Burkholderia cenocepacia* J2315. *Organic Process Research and Development* 19: 2012–2016.

85 Han, R., Chen, C., Wang, F. et al. (2016). Whole-cell biotransformation with recombinant nitrilase for enantioselective production of (R)-o-chloromadelic acid. *International Journal of Applied Microbiology and Biotechnology Research* 4: 75–89.

86 Zhang, Z.J., Xu, J.H., He, Y.C. et al. (2010). Efficient production of (*R*)-(−)-mandelic acid with highly substrate/product tolerant and enantioselective nitrilase of recombinant *Alcaligenes* sp. *Process Biochemistry* 45: 887–891.

87 He, Y.C., Zhang, Z.J., Xu, J.H., and Liu, Y.Y. (2010). Biocatalytic synthesis of (R)-(-)-mandelic acid from racemic mandelonitrile by cetyltrimethylammonium bromide-permeabilized cells of *Alcaligenes faecalis* ECU0401. *Journal of Industrial Microbiology and Biotechnology* 37: 741–750.

88 Zhang, X.-H., Wang, C.-Y., Cai, X. et al. (2020). Upscale production of (R)-mandelic acid with a stereospecific nitrilase in an aqueous system. *Bioprocess and Biosystems Engineering* 43 (7): 1299–1307.

89 Bhatia, S.K., Kumar, D., Bhatia, R.K., and Bhalla, T.C. (2013). Bench scale production of phenylacetic acid using *Alcaligenes* sp. MTCC 10675. *International Journal of Universal Pharmacy and Bio Sciences* 2: 16–25.

90 Gong, J.S., Lu, Z.M., Heng, L. et al. (2012). Metagenomic technology and genome mining: emerging areas for exploring novel nitrilases. *Applied Microbiology and Biotechnology* 97: 6603–6611.

91 Thakur, N., Kumar, V., Sharma, N.K. et al. (2016). Aliphatic amidase of *Rhodococcus rhodochrous* PA-34: purification, characterization and application in synthesis of acrylic acid. *Protein and Peptide Letters* 23: 152–158.

92 Nagasawa, T., Nakamura, T., and Yamada, H. (1990). ε-Caprolactam, a new powerful inducer for the formation of *Rhodococcus rhodochrous* J1 nitrilase. *Archives of Microbiology* 155: 13–17.

93 Zabaznaya, E., Kozulin, S., and Voronin, S. (1998). Selection of strains transforming acrylonitrile and acrylamide into acrylic acid. *Applied Biochemistry and Microbiology* 34: 341–345.

94 Luo, H., Wang, T., Yu, H. et al. (2006). Expression and catalyzing process of the nitrilase in *Rhodococcus rhodochrous* tg1-A6. *Modern Chemistry and Industry* 26: 109–111.

95 Panova, A., Mersinger, L.J., Liu, Q. et al. (2007). Chemoenzymatic synthesis of glycolic acid. *Advanced Synthesis and Catalysis* 349: 1462–1474.

96 Wu, S., Fogiel, A.J., Petrillo, K.L. et al. (2006). Protein engineering of *Acidovorax facilis* 72W nitrilase for bioprocess development. *Biotechnology and Bioengineering* 97: 689–693.

97 Wu, S., Fogiel, A.J., Petrillo, K.L. et al. (2007). Protein engineering of nitrilase for chemoenzymatic production of glycolic acid. *Biotechnology and Bioengineering* 99: 717–720.

98 Mathew, C.D., Nagasawa, T., Kobayashi, M., and Yamada, H. (1988). Nitrilase-catalyzed production of nicotinic acid from 3-cyanopyridine in *Rhodococcus rhodochrous* J1. *Applied and Environmental Microbiology* 54: 1030–1032.

99 Vaughan, P.A., Knowles, C.J., and Cheetham, P.S.J. (1989). Conversion of 3-cyanopyridine to nicotinic acid by *Nocardia rhodochrous* LL100-21. *Enzyme and Microbial Technology* 11: 815–823.

100 Almatawah, Q.A. and Cowan, D.A. (1999). Thermostable nitrilase catalysed production of nicotinic acid from 3-cyanopyridine. *Enzyme and Microbial Technology* 25: 718–724.

101 Sharma, N.N., Sharma, M., Kumar, H., and Bhalla, T.C. (2006). *Nocardia globerula* NHB-2: bench scale production of nicotinic acid. *Process Biochemistry* 41: 2078–2081.

102 Cantarella, L., Gallifuoco, A., Malandra, A. et al. (2011). High-yield continuous production of nicotinic acid via nitrile hydratase–amidase cascade reactions using cascade CSMRs. *Enzyme and Microbial Technology* 48: 345–350.

103 Raj, J., Singh, N., Prasad, S. et al. (2007). Bioconversion of benzonitrile to benzoic acid using free and agar entrapped cells of *Nocardia globerula* NHB-2. *Acta Microbiologica et Immunologica Hungerica* 54: 79–88.

104 Bhalla, T.C., Kumar, V., and Bhatia, S.K. (2013). Hydroxy acids: production and applications. In: *Advances in Industrial Biotechnology* (ed. R.S. Singh, A. Pandey, and C. Larroche), 56–76. IK International Publishing House Pvt Ltd.

105 Zou, S.P., Huang, J.W., Xue, Y.P., and Zheng, Y.G. (2018). Highly efficient production of 1-cyanocyclohexaneacetic acid by cross-linked cell aggregates (CLCAs) of recombinant *E. coli* harboring nitrilase gene. *Process Biochemistry* 65: 93–99.

106 Xue, Y., Xu, M., Chen, H. et al. (2013). A novel integrated bioprocess for efficient production of (R)-(−)-mandelic acid with immobilized *Alcaligenes faecalis* ZJUTB10. *Organic Research Development* 17: 213–220.

107 Ni, K., Wang, H., Zhao, L. et al. (2013). Efficient production of (R)-(-)-mandelic acid in biphasic system by immobilized recombinant *E. coli*. *Journal of Biotechnology* 167: 433–440.

108 Vesela, A.B., Krenkova, A., and Martinkova, L. (2015). Exploring the potential of fungal arylacetonitrilases in mandelic acid synthesis. *Molecular Biotechnology* 57: 466–474.

109 Osprian, I., Fechter, M.H., and Griengl, H. (2003). Biocatalytic hydrolysis of cyanohydrins: an efficient approach to enantiopure α-hydroxy carboxylic acids. *Journal of Molecular Catalysis B: Enzymatic* 24–25: 89–98.

110 Ema, T., Ide, S., Okita, N., and Sakai, T. (2008). Highly efficient chemoenzymatic synthesis of methyl (R)-o-chloromandelate, a key intermediate for clopidogrel, via asymmetric reduction with recombinant *Escherichia coli*. *Advanced Synthesis and Catalysis* 350: 2039–2044.

111 Zhu, Y.J., Zhou, H.T., Hu, Y.H. et al. (2011). Antityrosinase and antimicrobial activities of 2-phenylethanol, 2-phenylacetaldehyde and 2-phenylacetic acid. *Food Chemistry* 124: 298–302.

112 Yardley, J.P., Husbands, G.E., Stack, G. et al. (1990). 2-Phenyl-2-(1-hydroxycycloalkyl) ethylamine derivatives: synthesis and antidepressant activity. *Journal of Medicinal Chemistry* 33: 2899–2905.

113 Chen, J., Zheng, Y.G., and Shen, Y.C. (2008). Biotransformation of *p*-methoxyphenylacetonitrile into *p*-methoxyphenylacetic acid by resting cells of *Bacillus subtilis*. *Biotechnology and Applied Biochemistry* 50: 147–153.

15

Chemistry and Sources of Lactase Enzyme with an Emphasis on Microbial Biotransformation in Milk

Alaa Kareem Niamah[1], Shayma Thyab Gddoa Al-Sahlany[1], Deepak Kumar Verma[2], Smita Singh[3], Soubhagya Tripathy[2], Deepika Baranwal[4], Nihir Shah[5], Ami R. Patel[5], Mamta Thakur[6], Gemilang Lara Utama[7,8], Mónica L. Chávez-González[9], and Cristobal Noe Aguilar[9]

[1] *Department of Food Science, College of Agriculture, University of Basrah, Basra City, Iraq*
[2] *Agricultural and Food Engineering Department, Indian Institute of Technology Kharagpur, Kharagpur, West Bengal, India*
[3] *Department of Allied Health Sciences, Chitkara School of Health Sciences, Chitkara University, Rajpura, Punjab, India*
[4] *Department of Home Science, Arya Mahila PG College, Banaras Hindu University, Varanasi, Uttar Pradesh, India*
[5] *Division of Dairy Microbiology, Mansinhbhai Institute of Dairy and Food Technology-MIDFT, Dudhsagar Dairy Campus, Mehsana, Gujarat, India*
[6] *Department of Food Technology, School of Sciences, ITM University, Gwalior, Madhya Pradesh, India*
[7] *Faculty of Agro-Industrial Technology, Universitas Padjadjaran, Sumedang, Indonesia*
[8] *Center for Environment and Sustainability Science, Universitas Padjadjaran, Bandung, Indonesia*
[9] *Bioprocesses and Bioproducts Research Group, Food Research Department, School of Chemistry. Autonomous University of Coahuila, Saltillo, Coahuila, Mexico*

15.1 Introduction

Enzymes are referred to as the "powerhouses" of the metabolic system because they catalyze all of the chemical reactions that are necessary for life and make it possible for these reactions to take place far more quickly than they would be able to without enzymes [1]. Glycosidases are enzymes that hydrolyze glycosides into oligosaccharides, polysaccharides, and glycoconjugates in a manner that is efficient and inexpensive. Lactase is an enzyme that is found in higher plants, animals, and microbes. It is a member of the β-glycosidases enzyme family and may be found in all three. Lactose in milk is digested by enzymes known as β-glycosidases, which results in lactose-free milk that is sweeter than ordinary milk and is suited for persons who are unable to digest lactose due to a lactose intolerance [2]. β-Galactosidase is an enzyme that breaks down lactose and is utilized in the food industry to make dairy products easier to digest, sweeter, more soluble, and have a better flavor. β-Galactosidase is put to use in the food processing industry for a variety of purposes, including the production of hydrolyzed milk products, whey, and galactooligosaccharides [3]. As a consequence of this, this enzyme is a functional protein that can now be produced through the utilization of recombinant technology [4, 5].

Microbial Bioreactors for Industrial Molecules, First Edition.
Edited by Sudhir P. Singh and Santosh Kumar Upadhyay.

Lactase is an enzyme that is generated by a broad range of animals. It is located along the brush boundary of the small intestine in animals, including humans and other mammals. Lactase is required to digest milk because it breaks down lactose, the sugar responsible for milk's distinctive sweetness. A person who consumes dairy products but lacks sufficient lactase production may develop lactose intolerance. Lactase is a dietary supplement added to milk in order to generate lactose-free dairy products. Lactase is the digestive enzyme responsible for breaking down lactose, a sugar present in milk and other dairy products [6].

Lactase, sometimes referred to as lactase-phlorizin hydrolase or LPH, is the glycoside hydrolase that is responsible for the hydrolysis of lactose into the monomers of glucose and galactose. The enzyme family known as β-galactosidase includes lactase as one of its own members. The majority of lactase is found on the brush border membranes of differentiated enterocytes that line the villi in the small intestine. The *lct* gene, which is located on chromosome 2 in humans, is responsible for encoding lactase [7].

Both the dimer and the tetramer forms of the lactase enzyme have been shown to have biological activity, with the predominant form being determined by factors such as pH and temperature during experimentation. The strength of the interaction between two different subunits is significantly impacted by the parameters of the experiment. Within the range of their optimal pH, multimeric enzymes are stable and active; nevertheless, more extreme pH values in both directions (the acidic and alkaline zones) may cause them to become inactive [8].

Lactase preparations that have been purified have been produced by *Escherichia coli*, *Kluyveromyces* species, and fungi. Only analytical chemists have made use of the first enzyme, which comes from *E. coli* and breaks down lactose. It is possible to get economically valuable enzymes from the yeast *Kluyveromyces lactis* and the fungi *Aspergillus niger*, which are both used in the dairy and other food industries. *Kluyveromyces lactis* lactase is the commercial preparation that is utilized the most frequently at the present time [8, 9].

In this chapter, we explore the origins of β-galactosidase as well as its structure, recombinant synthesis, and the important modifications that were made to the enzyme in order to improve its performance.

15.2 Lactase Enzyme

The hydrolysis of β-galactosides is catalyzed by the enzyme known as lactase. Other names for this enzyme include β-galactosidase or β-D-galactosidegalactohydrolase (EC 3.2.1.23). Lactose (milk sugar) is a naturally occurring substrate of lactase that may be present in cow's milk in concentrations of up to 5%. Lactose is a disaccharide composed of two sugar molecules: β-galactose and glucose (Figure 15.1). Human lactose intolerance is due to a lack of lactase activity. Due to an inability to hydrolyze the disaccharide and absorb its components into circulation, lactose accumulates in the digestive tracts of affected individuals who ingest milk products. Lactose, if undigested, can produce flatulence, diarrhea, and abdominal discomfort [11, 12].

The well-known biocatalyst known as β-galactosidase is responsible for catalyzing both the hydrolytic and transgalactosylation reactions. It is conceivable for it to take part in the

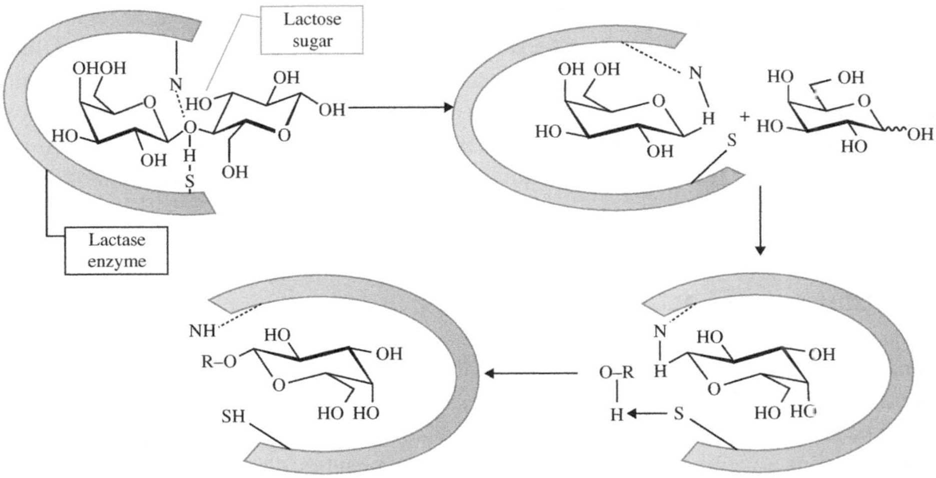

Figure 15.1 A mechanism for the breakdown of the sugar lactose by the enzyme lactase.

formation of prebiotic galactooligosaccharide (GOS) or lactulose in certain circumstances due to its activity as an associative transglycosylase, which makes this participation feasible. β-Galactosidase is an enzyme that performs two distinct enzymatic activities: on the one hand, it cleaves or separates the β-glycosidic link between galactose and its organic residues; on the other hand, it also cleaves cellobiose, calories, collaterals, and cellulose. Both of these activities are necessary for the digestion of galactose. On the other side, it catalyzes the transgalactosylation reaction, which is the process that converts lactose to allolactose [13].

Lactases are categorized as neutral or acidic based on their optimal pH for action. The food and pharmaceutical industries utilized β-galactosidase extensively [14]. Due to lactose maldigestion or intolerance caused by lactose maldigestion in a significant portion of the world's population, lactose's nutritional value has decreased. Lactose is a hygroscopic sugar that has a powerful capacity to take on the flavors and odors of its surroundings. This leads to a variety of problems with frozen foods, including the crystallization of milk products, an improvement in sandy or gritty texture, and the creation of layers [15]. Therefore, β-galactosidase may have certain advantages. Lactose hydrolysis by enzymes may help in the digestion of lactose-rich foods. Lactose intolerant people and the food industry both need the lactase enzyme because it is used to prevent lactose crystallization, increase the solubility of milk products, and address the issue of whey consumption and disposal, which may cause environmental or health concerns [16]. Lactose crystallization is a problem for lactose intolerant people and the food industry.

β-Galactosidase, which is the inducer of the lac operon, converts lactose to allolactose through a process known as transgalactosylation [17]. As a consequence of this, a virtuous cycle of positive feedback can be produced. The affinity of the *lac* operon is decreased as a result of the binding of allolactose to the *lac* repressor. When the *lac* operon is turned on, it causes the synthesis of the enzyme β-galactosidase. In the field of molecular research, the *LacZ* gene is frequently utilized as a reporter marker in order to analyze the expression of genes located in the lac operon [18]. The enzyme β-galactosidase uses lactose as its natural substrate. However, it is just selective for the galactose residue; thus, it may convert other substrates as well. The enzyme β-galactosidase has the ability to convert several different aglycones, such as X-gal, oNPG, and pNPG. The oNPG and pNPG are the substrates that are employed for enzyme assays the majority of the time. The screening method known as blue/white screening, which more commonly goes by the name α-complementation, makes use of X-gal [19].

15.3 Sources of Lactase

The enzyme is found in a diverse selection of natural environments across the world. Many different kinds of species, including plants, animals, and microbes, are capable of producing the enzyme known as β-galactosidase. It is possible to find it in many different plants, such as almonds (*Prunus amygdalus*, syn. *Prunus dulcis*), peaches (*Prunus persica*), apricots (*Prunus armeniaca*), apples (*Malus domestica*), wild rose (*Rosa canina*) tips, alfalfa (*Medicago sativa*), coffee (*Coffea arabica*), and soybean (*Glycine max*) seeds [20, 21]. It was discovered that the enzyme is present in animal organs such as the intestines, the brain, the placenta, and the testicles of dogs, rabbits, snails, calves, sheep, goats, and rats. In addition, lactase was found in the saliva of humans, primates, and farm animals, as well as in the tissues of rats and mice, as well as in the plasma, serum, and urine of dogs [3].

15.3.1 Plants

Plant tissues have an abundance of lactase. These enzymes have been associated with several biological procedures, including development of plant, fruit ripening, and lactose breakdown. Using molecular techniques, lactase's role in the development of fruit and their ripening was also examined. β-Galactosidase (lactase)/exo-galactanase activity was observed in ripening tomato (*Lycopersicon esculentum* Mill.) fruit, and a family of seven tomato β-galactosidase (TBG)-cDNAs were identified [22]. In addition, softening-related lactase enzyme has also been identified from *L. esculentum*, *M. domestica*, muskmelon (*Cucumis melo*), avocado (*Persea americana*), kiwi (*Actinidia deliciosa*), *C. arabica*, mango (*Mangifera indica*), and Japanese pear (*Pyrus pyrifolia*) plant fruits [23]. Moreover, Hussien and Doosh [24] assessed the enzyme activity of *L. esculentum*-isolated β-galactosidase by its ability to hydrolyze the substrate 2-nitrophenyl-β-D-galactopyranoside. The isoelectric point of the enzyme was 4.4. Using the phenol–sulfuric acid method, the enzyme's carbohydrate content was confirmed to be 19.5%. The Km and Vmax of the enzyme are 3.65 mM and 0.18 mol/min, respectively. The *P. dulcis* extract β-galactosidase was isolated *via.* ammonium sulfate ($[NH_4]_2SO_4$) precipitation. The optimal pH and temperature for the partially purified β-galactosidase were 5.5 °C and 50 °C, respectively. Heat, pH, Ca^{2+} ions, Mg^{2+} ions, and D-galactose all demonstrated significant effects on the stability of the enzyme. *Prunus dulcis* β-galactosidase retained approximately 89% of its activity after being stored at 4 °C for two months. This enzyme was utilized in a stirring batch process to hydrolyze lactose in milk and whey, and it was observed that the rate of lactose hydrolysis rose steadily over time [25]. A study performed recently demonstrated that these two phytohormones might influence rice seed germination via β-galactosidase (lactase) activity. The increased transcriptional expression of *OsBAGL1*, *OsBAGL4*, *OsBAGL8*, and *OsBAGL11* during seed germination compared to other comparable genes suggests that these four genes are involved in the germination process [26].

15.3.2 Bacteria

Microbial β-galactosidase has a number of advantages over other sources of the enzyme that are already available on the market. These advantages include how simple it is to work with, how quickly it can reproduce, how productive it can be, how active and stable it is, and how simple fermentation can be. For lactose hydrolysis, β-galactosidase, which is derived from bacterial sources, has been used because of its many advantages, including its high activity level, its simplicity in terms of fermentation, and its stability as an enzyme [3]. Lactic acid bacteria (LAB), which include streptococci, lactococci, and lactobacilli, have recently become the focus of research in the scientific community for the following three reasons [27–29, 30]: (i) People who have trouble digesting lactose can safely consume fermented dairy products; (ii) LAB are Generally regarded as safe (GRAS); as a result, enzymes generated from them can be used without extensive purification; and (iii) certain strains exhibit probiotic activity, such as enhanced lactose digestion.

There have been reports of the production of lactase (β-galactosidase) by *Bacillus* spp. (*Bacillus aryabhattai*), *Pseudoalteromonas* spp., *Bifidobacterium longum*, *Alicyclobacillus acidocaldarius*, *Lactobacillus leichmannii*, *Lactobacillus acidophilus*, *Streptococcus thermophilus*, *Enterobacter* spp., *Aspergillus oryzae* (*Aspergillus alliaceus*, *Aspergillus lacticoffeatus*, *A. niger*), *Rhizomucor* spp., *Talaromyces thermophilus,* and *Teratosphaeria acidotherma* (Tables 15.1 and 15.2). The pH range of 6.5–7.5 is maintained due to the presence of bacterial β-galactosidase. They perform at their best at optimal temperature ranging from 50 to 60 °C. Enzymes produced by bacteria can have molecular weights that range anywhere from 20,000 to 50,000 Da. Thermophilic bacteria are also responsible for the production of a temperature-stable form of β-galactosidase. Recently, metagenomic resource generated from environmental niches has been shown as a potent source for β-galactosidase with acid and cold activity of the enzyme [52]. This β-galactosidase variant showed catalytic activity at acidic pH, a catalytic property useful in the processing of acidic whey samples.

The highest levels of β-galactosidase activity were discovered in *Bifidobacterium infantis* strain CCRC 14633, and *B. longum* strain CCRC 15708, respectively [32]. *Bifidobacterium* spp. and *Lactobacillus* spp. are the species of bacteria that are most frequently utilized in the production of probiotics due to the potential health benefits they offer [29]. The microbiota in the colon has chosen *Bifidobacterium* to serve as a model organism for the study of lactose fermentation [53, 54]. In an earlier investigation, Zárate and Chaia [55] used *Propionibacterium acidipropionici* and found that the highest level of β-galactosidase activity was observed in a solution that consisted only of lactate. This was the case when they tested the bacteria. They investigated how the addition of lactose and lactate as primary and secondary sources of energy affected the development of *P. acidipropionici* Q4 as well as the activity of the β-galactosidase enzyme. There was a large rise in intracellular pyruvate when this strain used lactate as a secondary energy source. This was followed by lactate ingesting and an increase in particular β-galactosidase activity; however, lactose

Table 15.1 Synthesis of β-galactosidase by bacteria during submerged fermentation.

Bacteria species	Location	Substrate	Yield of enzyme	Reference
Pseudoalteromonas spp.	Extracellular	Lactose	1 IU/mL	[31]
Bifidobacterium longum B6	Extracellular	Fermentation media contain 4% lactose	18.6 IU/mL	[32]
Alicyclobacillus acidocaldarius subsp. *rittmannii*	Intracellular	Lactose	0.6 IU/mg protein	[33]
Bacillus sp. MSP7	Extracellular	Lactose	65 IU/mg protein	[34]
Lactobacillus acidophilus ATCC 4356	Extracellular	Modified MRS	1.01 IU/mL	[35]
Streptococcus thermophilus	Extracellular	Whey	11 IU/mL	[36]
Enterobacter sp. 3TP2A	Extracellular	Lactose	76.5 IU/mg protein	[37]
Lactobacillus leichmannii 313	Extracellular	MRS	4.5 IU/mg protein	[38]
Bacillus aryabhattai GEL-09	Extracellular	Milk	8 IU/mL	[39]

Table 15.2 The characteristics of the lactase (β-galactosidases) that is produced by fungus.

Molds source of lactase	Optimal temperature (°C)	Optimal pH	Molecular weight (kDa)	Reference
Aspergillus oryzae	55	3.5–8	—	[40]
Rhizomucor spp.	60	4.5	—	[41]
Talaromyces thermophilus	50	5.5–6.0	50	[42]
Guehomyces pullulans	50	4	—	[43]
Aspergillus alliaceus	45	7.2	—	[44]
Teratosphaeria acidotherma	70	2.5–4.0	140	[45]
Aspergillus lacticoffeatus	50–60	3.5–4.5	—	[46]
Aspergillus niger	50	5	76	[47]
Aspergillus terreus	40	6	42	[48]
Kluyveromyces lactis	30	6.6	135	[13]
Aspergillus terreus	60	6	—	[49]
Thielaviopsis ethacetica	60	7	50	[50]
Cladosporium tenuissimum	35–55	3.0–4.5	—	[51]

intake was very modest. After the addition of lactose as a second source of energy, the processing of lactic acid stopped, the amount of pyruvate contained inside the cell decreased, and the activity of β-galactosidase rapidly reverted to a level that was comparable to that of glucose [27, 55].

Ultrasound treatment was one of the many methods that were tried in an effort to boost the amount of lactase enzyme that was produced by bacteria. The effect of treatment with ultrasonic waves of low frequency on the β-galactosidase enzyme that is produced by probiotic bacteria. The use of ultrasound induced all strains of probiotic bacteria to rupture led to an increase in the extracellular release of the β-galactosidase enzyme. When the time of exposure was extended, there was a corresponding rise in enzymatic activity. *Lactobacillus reuteri* (*Limosilactobacillus reuteri*) had 2.9 unit/mL of β-galactosidase after being subjected to ultrasonic treatment for a period of 20 minutes [56]. In all fermented camel milk prepared with mixed cells of *L. acidophilus*, *Lactobacillus delbrueckii* subsp. *bulgaricus*, and *S. thermophilus*, the highest activity of β-galactosidase was found after four hours of fermentation. The values for this activity were 1.970.12, 1.770.06, and 1.700.01, respectively. According to the findings of this study, an appropriate rupture-cell technique has the potential to simultaneously boost the rate of microbial growth in fermented camel milk [57].

People who have trouble digesting lactose can still eat fermented dairy products without suffering many of the unpleasant side effects that are often associated with lactose consumption, and the probiotic activity of LAB will assist with the digestion of lactose. The strains of *Bifidobacterium* spp. and *Lactobacillus* spp. that are most often utilized as probiotics in food and food systems are the two kinds of bacteria mentioned here. This is because there is speculation that some strains of bacteria may provide certain health advantages. The bacterium *Bifidobacterium* has been chosen to serve as a model organism for the research project that investigates the process of lactose fermentation carried out by bacteria found in the colon.

15.3.3 Yeasts

Due to the fact that its native habitat is the environment of dairy production, the yeast *K. lactis* is a vital component in the commercial production of lactase (β-galactosidase). This is because it is an enzyme that breaks down lactose. The production of β-galactosidase by yeast seems to be of interest because this enzyme is utilized in the food sector to produce reduced lactose milk, which is a remarkable commercial product that is consumed by a considerable number of lactose-intolerant individuals [58]. It is possible for *K. marxianus* to produce homologous enzymes such as β-galactosidase as well as heterologous proteins, and it has the capacity to thrive on a wide variety of substrates, including lactose as the only source of carbon and energy. Other substrates include glucose, xylose, and mannose. Due to the high lactose-hydrolyzing activity of this yeast lactase, it is employed in the commercial production of low-lactose milk for persons who are lactose intolerant [59]. β-Galactosidase, which is produced by a type of psychrophilic yeast known as *Guehomyces pullulans*, has been put to use in the food sector to hydrolyze whey and milk. In addition, the temperature of 30 °C, pH of 6.0, and enzyme concentration of 3% (v/v) were the optimum operational conditions for maximizing lactose hydrolysis and optimizing enzyme activity for produced β-galactosidase from *Saccharomyces fragilis*. The optimum operational conditions for permeabilized cells were temperature of 44 °C, pH 7.0, and enzyme concentration of 4% (v/v), respectively [60]. Lactase (β-D-galactosidase) is produced by *Candida pseudotropicalis* when it is cultivated in deproteinized whey. At a temperature of 37 °C, the lactase enzyme can hydrolyze 50% and 100% of the lactose in whey and milk in four and five hours, respectively. The lyophilized enzyme retained 95% of its original activity even after being stored at 20 °C for three months [61]. The most important properties of yeast β-galactosidase are stated in Table 15.2.

15.3.4 Molds

The Food and Drug Administration (FDA) has determined that certain species of *Aspergillus* are "Generally regarded as safe" (GRAS). *Aspergillus oryzae* is responsible for the production of extracellular β-galactosidase, which finds use in the commercial sector. The research was conducted on purified β-galactosidase from *A. oryzae* at optimum pH and temperatures of 5 and 50 °C [62]. An enzyme derived from *A. oryzae* was shown to be more appropriate for the use of whey when compared to lactose hydrolysis. Fungi are capable of breaking down lactose through two primary pathways: (i) the extracellular hydrolysis and subsequent absorption of monomers, and (ii) the uptake of disaccharides. In place of lactose hydrolysis, the enzyme β-galactosidase, which is derived from the fungus *A. niger*, is typically utilized in the process of removing galactose residues from plant-derived oligosaccharides and polysaccharides [63]. This has been shown by different expression rates on different carbon sources, with arabinose, pectin, and xylose having the highest expression of the *lacA* gene, which codes for the enzyme that is being produced [64]. In a fermentation process that takes place in the solid state and uses wheat bran as the solid substrate, *Trichoderma* spp. generates the enzyme β-galactosidase. The optimal conditions for enzyme activity were found to be 55 °C and a pH of 45. According to the research of De Jesus and Guimares from 2021, the catalytic activity remained stable for up to 180 minutes when incubated at temperatures between 35 and 45 °C and for up to 24 hours when the pH was acidic or alkaline [65]. Purification of fungal β-galactosidase has been shown to be possible through the application of a variety of chromatographic techniques, including DEAE-cellulose chromatography, ammonium sulfate fractionation, and DEAE-Sephadex column chromatography in a number of different studies. The characteristics of β-galactosidase found in molds are described in Table 15.2.

15.4 Microbial Biotransformation of Lactase Enzyme

15.4.1 Improvement of Microbial Strains

In order to get β-galactosidase to the market, researchers have tried a variety of different approaches. The use of genetic engineering technologies to produce recombinant β-galactosidase and metagenomic research to uncover sources of β-galactosidase with better kinetic and catalytic characteristics have both been described in the published research [66, 67]. Utilizing recombinant enzymes may result in a variety of possible benefits that have been indicated in the Figure 15.2. Utilizing recombinant DNA technology for the overexpression of lactase with interesting features from microbial sources that are already known for highly resourceful heterologous protein synthesis enables the cost-effective use of lactase in industrial processes and their potential range of applications to be substantially broadened. Recombinant DNA technology also allows for the production of lactase with intriguing properties from microbial sources that are already recognized for producing heterologous proteins. Because of this, recombinant DNA technology enables the overexpression of lactase with fascinating features from microbial sources that are already recognized for producing extremely useful heterologous proteins. This is because recombinant DNA technology allows for the overexpression of lactase [68]. Protein engineering

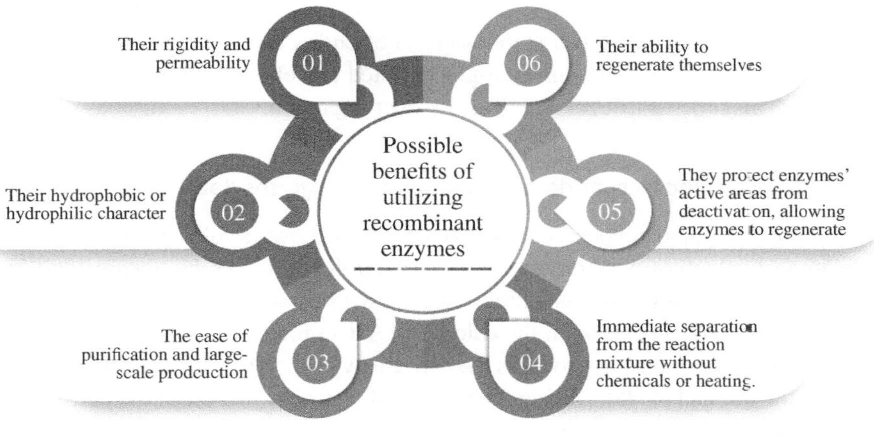

Figure 15.2 Graphical representation of a number of potential advantages that might result from the utilization of recombinant enzymes. *Source:* Adapted from de Andrade et al. [67].

techniques that are technologically advanced allow for the construction of unique properties into a specific lactase, such as reduced product inhibition, increased product yields, or secretion signals. These properties can be engineered into the lactase while genetic engineering techniques are used [66, 69].

It has been established that β-galactosidase derived from bacterial as well as nonbacterial sources are capable of being created using recombinant technology. Those from filamentous fungi like *A. niger*, *Penicillium expansum*, *Pichia pastoris*, and others, as well as those from the yeast *K. lactis*, are examples of nonbacterial β-galactosidase that are overexpressed in recombinant yeast hosts. According to the findings of Bury et al. [70], increasing the content of lactose and yeast extract by 0.2–0.8% greatly increased the activity of recombinant β-galactosidase. In addition to this, the activity of the recombinant β-galactosidase was purified to a significant degree. In addition to this, they demonstrated that lactose was entirely digested in a timeframe of less than 40 hours when recombinant β-galactosidase was produced using cheese whey permeate. This leads one to believe that the recombinant system is capable of performing both the biosynthesis of β-galactosidase and the bioremediation of cheese whey at the same time. The researchers discovered a 21-fold increase in β-galactosidase synthesis in a 10-L bioreactor under optimum circumstances when compared to fermentation in Erlenmeyer flasks. These results were obtained by comparing the two processes [71].

In the most recent decades, a lot of work has been put into developing efficient heterologous expression techniques that may be used for the production of β-galactosidase. Some enzymes exhibit dual catalytic activity of glucosidase and galactosidase [72]. One of the expression hosts is *E. coli*, while others include yeast, *Lactiplantibacillus plantarum*, *Lactococcus lactis*. Galactosidases that have been recombinantly produced in *E. coli*, *Lactobacillus*, and *Lactococcus* are often found in the cytoplasm, which makes purification challenging and costly. In addition to the low secretion efficiency of β-galactosidase in yeast strains, other problems include plasmid instability, the presence of toxic methyl alcohol, and the presence of antibiotic resistance markers on the host genome. In addition to these issues, yeast strains often have a low capacity for efficiently secreting β-galactosidase [4]. *Escherichia coli* is the bacteria of choice in the pharmaceutical industry for the production of recombinant non-glycosylated proteins utilizing recombinant DNA technology. This process utilizes *E. coli* as the host organism.

15.4.2 Galactooligosaccharide Synthesis and Transglycosylation

Prebiotics include galactooligosaccharides (GOS), also known as oligogalactosyllactose, oligogalactose, oligolactose, or transgalactooligosaccharides (TOS). Prebiotics are nondigestible components of food that have a favorable effect on the host by encouraging the growth and/or activity of beneficial bacteria in the colon [73]. This process is known as probiotic activity [29]. GOS can be found in products that are easily available commercially, such as food for babies and adults [74, 75]. During the process of lactose hydrolysis, GOSs are simultaneously produced because of the transglycosylation activity of β-galactosidase. The origin of the enzyme determines which of these processes takes priority, and it is possible for both processes to occur simultaneously. *Aspergillus oryzae* and *Bacillus circulans* β-galactosidase have strong transgalactosylation activity, whereas *Kluyveromyces* β-galactosidases have high hydrolytic activity but moderate transgalactosylation activity. The origin of the enzyme has an effect on both the affinity of β-galactosidase for the source (lactose or lactulose) and acceptor (lactose, lactulose, or fructose) of transgalactosylated galactose, which can be either lactose, lactulose, or fructose. In addition to hydrolyzing the lactose saccharide link, they also catalyze processes that result in galactooligosaccharides that are prebiotic in nature. Dextransucrase can catalyse the transglucosylation of sucrose and lactose into 2-α-D-glucopyranosyl-lactose (4-galactosyl-kojibiose), a prebiotic molecule [76].

Figure 15.3 depicts the enzyme β-galactosidase in operation throughout the manufacturing process of GOS. In order to produce oligosaccharides from mono- and disaccharides, either transglycosylation or polysaccharide hydrolysis must first take place. There are three stages involved in the production of oligosaccharide [77], and they are as follows: (i) the release of glucose residue, which, in turn, leaves galactosyl residue, which can then be processed further by the complex enzyme galactosyl; (ii) the transfer of this complex to an acceptor such as saccharides or water molecules, both of which contain a hydroxyl group. When there is a low concentration of lactose in a solution, the water molecule functions as an acceptor and creates galactose. This occurs because galactose is a more stable sugar than

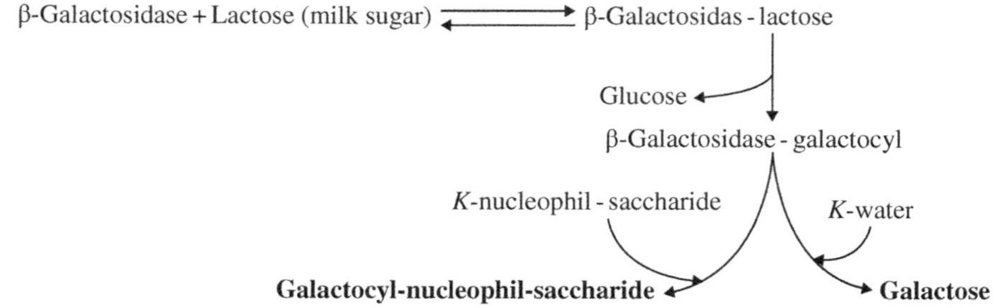

Figure 15.3 A schematic representation of the reaction that takes place when β-galactosidase produces galactooligosaccharides, where *K* denotes a reaction constant.

lactose. (iii) If there is a high concentration of lactose in the solution, the lactose acts as an acceptor, binding the complex and leading to the formation of galactosyloligosaccharides. There were 5.06% lactose, 8.76% monosaccharides, and 13.43% GOS detected in this GOS syrup. The transgalactosylation and lactose conversion rates in this GOS syrup were at 25.2% and 83%, respectively, with a maximum GOS production of 40.6% [78]. Oligosaccharides offer a number of beneficial features for one's health, including anti-carcinogenic capabilities and the ability to reduce blood cholesterol levels (as a result of oligosaccharides binding with cholesterol in the small intestine) and the ability to improve liver function. Because of this, there has been a significant increase in the general population's desire for GOS and low-cost oligosaccharides. GOS are now being utilized in a wide range of items, some of which include cosmetics, low-calorie sweeteners, soft drinks, cereals, powdered milk, and infant food [79].

This article published by Zerva et al. [80] describes the heterologous production of a novel β-galactosidase from the fungus *Thermothielavioides terrestris* in *P. pastoris*. It was determined that the enzyme, TtbGal1, had the highest level of activity at a temperature of 60 °C and a pH of 4. TtbGal1 is thermostable, maintaining almost all of its activity for a whole day at a temperature of 50 °C [80].

Overexpression of the glycosyl hydrolase B-gal42 in *E. coli* allowed for its use in the synthesis of GOS from lactose or milk whey. B-gal42 was isolated from the *Pantoea anthophila* strain that was found in Tejuino. Because of their superior stability, crude enzyme extracts that are devoid of cells were utilized in the manufacturing processes of GOS. In reactions with 400 g/L of lactose, HPAEC-PAD found that a GOS yield of 40% (w/w), which is equivalent to an 86% conversion rate, was optimal. This enzyme showed a strong predilection for producing GOS with galactosyl links (1–6), and it also produced GOS with galactosyl links (1–3). Both milk whey and pure lactose produced the same product profile and yielded 38% of GOS when tested at a concentration of 300 g/L for synthesis [81]. This was determined by comparing the two substrates in a GOS production experiment.

The presence of GOS in human milk is associated with an increase in the number of bifidobacteria in the small intestine of a breastfed newborn. As a result of their bifidogenic effect, these GOS reduce the number of potentially dangerous bacteria. As a direct consequence of this, companies that produce infant food are increasingly including GOS in their milk- and cereal-based products [82]. Oligosaccharides offer a number of beneficial features for one's health, including anticarcinogenic capabilities, and the ability to reduce blood cholesterol levels (as a result of oligosaccharides binding with cholesterol in the small intestine), and the ability to improve liver function. As a direct consequence of this, there has been a significant surge in demand for GOS and low-cost oligosaccharides. GOS are now being utilized in a wide range of items, some of which include cosmetics, low-calorie sweeteners, soft drinks, cereals, powdered milk, and infant food [83, 84].

15.4.3 Lactose Intolerance

Glucose and galactose are the two monosaccharides that combine to form the disaccharide known as lactose, which is also commonly referred to as milk sugar. These two monosaccharides constitute milk's primary and most important sources of carbohydrates. Lactase, also known as β-galactosidase, is an enzyme that plays a role in the process of absorbing lactose. This enzyme is situated on the brush boundary of the small intestine. In lactose intolerance, the body is unable to hydrolyze lactose, a kind of sugar that may be found in dairy products such as milk, curd, butter, cheese, and yoghurt. This disease is known as lactose intolerance. Because there is less available β-galactosidase, there are greater levels of lactose that have not been digested. It has been determined that lactose intolerance, which is also referred to as lactose malabsorption, is a significant health problem that affects more than 70% of the globe [85]. Individuals who are lactose intolerant may already be at a disadvantage when it comes to getting enough calcium in their diets because there are few lactose-free foods that are also high in calcium. In this review, we offer data from research conducted on both humans and animals about the effects of lactose and lactase deficiency on the body's ability to absorb calcium and maintain healthy bones. According to the findings of the research, neither consuming lactose in the diet nor having a lactase deficiency has a significant impact on the amount of calcium that individuals absorb [86]. As a consequence of this, roughly 60% of the population has a diminished capacity to digest lactose as a result of low levels of β-galactosidase enzyme activity. Live bacteria or yeasts are known as probiotics, and they are used to help restore a healthy balance to the gut flora that is found in the digestive tract. Studies have shown that probiotics give a number of health advantages, some of which include enhanced gut health, increased immune system responses, and decreased blood cholesterol levels. According to an ever-expanding body of research, the presence of particular probiotic bacteria in fermented and unfermented milk products has been shown to assist in the reduction of clinical symptoms associated with lactose intolerance. These symptoms include abdominal pain, bloating, gas, and diarrhea [29, 87].

A precise diagnosis is necessary before beginning any kind of treatment, and several methods have been tried out in an effort to achieve this objective. Included in this category are genetic tests, hydrogen breath tests (also known as HBT), fast lactase tests, and lactose tolerance tests. The HBT method is the one that is utilized the most frequently as a result of its noninvasive nature, cheap cost, high sensitivity and specificity, and ease of application. In clinical practice, other methodologies are often used for HBT integration testing. Additionally, there are several therapy options available. A suitable kind of intervention is a change in dietary styles, such as the consumption of lactose-free meals that have nutritional values comparable to those of dairy products. Other feasible choices include choosing milk that has specific forms of β-caseins, consuming exogenous enzymes, and taking probiotics and/or prebiotics [88].

15.5 Conclusion

Lactases (β-galactosidase) research is becoming increasingly popular, both for the purpose of locating new sources of the enzyme that have the potential to generate high enzyme titers for large-scale manufacturing and for the purpose of identifying β-galactosidase enzymes that have distinctive properties. When cold-active and thermophilic enzymes are required for lactose breakdown in milk or whey, large-scale synthesis of β-galactosidase may frequently be achieved through the utilization of recombinant enzyme expression systems in combination with a variety of genetic engineering approaches. The large-scale production of lactose breakdown in milk or whey, where cold-active and thermophilic enzymes are utilized, is often accomplished through the use of recombinant enzyme expression systems in conjunction with various genetic engineering. Investigations are now being carried out to discover new cold-active sources of β-galactosidase that may be utilized in the process of removing lactose from milk in a chilled environment. In the processing of dairy products, which involves the simultaneous breakdown of lactose and the application of heat, thermostable enzymes are also utilized. In addition to the hydrolysis of lactose, more research is required to locate microbial sources of β-galactosidase that have enhanced transglycosylation characteristics. In order to produce microbial sources that are capable of galactooligosaccharide synthesis in a more efficient manner, considerable use of genetic engineering technologies will be used.

References

1 Holley, A.K., Bakthavatchalu, V., Velez-Roman, J.M., and St Clair, D.K. (2011). Manganese superoxide dismutase: guardian of the powerhouse. *International Journal of Molecular Sciences* 12 (10): 7114–7162.

2 Divakar, S. (2013). Glycosidases. In: *Enzymatic Transformation*, 5–21. India: Springer.

3 Saqib, S., Akram, A., Halim, S.A., and Tassaduq, R. (2017). Sources of β-galactosidase and its applications in food industry. *3 Biotech* 7 (1): 1–7.

4 Movahedpour, A., Ahmadi, N., Ghalamfarsa, F. et al. (2022). β-Galactosidase: from its source and applications to its recombinant form. *Biotechnology and Applied Biochemistry* 69 (2): 612–628.

5 Sharma, H. and Upadhyay, S.K. (2020). Enzymes and their production strategies. In: *Biomass, Biofuels, Biochemicals: Advances in Enzyme Catalysis and Technologies* (ed. S.P. Singh, A. Pandey, R.R. Singhania et al.) 31–48. Elsevier. https://doi.org/10.1016/B978-0-12-819820-9.00003-X.

6 Kretchmer, N. (1972). Lactose and lactase. *Scientific American* 227 (4): 70–79.

7 Marten, L.M., Wanes, D., Santer, R., and Naim, H.Y. (2020). Molecular and cellular analysis of intestinal lactase-phlorizin hydrolase gene variants unravel a heterogeneous pathogenic pattern of congenital lactase deficiency. *The FASEB Journal* 34 (S1): 1-1.

8 Souza, C.J., Garcia-Rojas, E.E., and Favaro-Trindade, C.S. (2018). Lactase (β-galactosidase) immobilization by complex formation: impact of biopolymers on enzyme activity. *Food Hydrocolloids* 83: 88–96.

9 Leksmono, C.S., Manzoni, C., Tomkins, J.E. et al. (2018). Measuring lactase enzymatic activity in the teaching lab. *JoVE (Journal of Visualized Experiments)* 138: e54377.

10 Nolan, V., Collin, A., Rodriguez, C., and Perillo, M.A. (2020). Effect of polyethylene glycol-induced molecular crowding on the enzymatic activity and thermal stability of β-galactosidase from *Kluyveromyces lactis*. *Journal of Agricultural and Food Chemistry* 68 (33): 8875–8882.

11 Verma, D.K., Mahato, D.K., Billoria, S. et al. (2017). Microbial spoilage in milk products, potential solution, food safety and health issues. In: *Microorganisms in Sustainable Agriculture, Food and the Environment* (ed. D.K. Verma and P.P. Srivastav). Volume-1, as part of book series on *Innovations in Agricultural Microbiology*, 171–196. Palm Bay, FL: Apple Academic Press.

12 Vaessen, E.M., den Besten, H.M., Esveld, E.D., and Schutyser, M.A. (2019). Accumulation of intracellular trehalose and lactose in *Lactobacillus plantarum* WCFS1 during pulsed electric field treatment and subsequent freeze and spray drying. *LWT* 115: 108478.

13 de Freitas, M.D.F.M., Hortêncio, L.C., de Albuquerque, T.L. et al. (2020). Simultaneous hydrolysis of cheese whey and lactulose production catalyzed by β-galactosidase from *Kluyveromyces lactis* NRRL Y1564. *Bioprocess and Biosystems Engineering* 43 (4): 711–722.

14 O'Connell, S. and Walsh, G. (2010). A novel acid-stable, acid-active β-galactosidase potentially suited to the alleviation of lactose intolerance. *Applied Microbiology and Biotechnology* 86 (2): 517–524.

15 Fox, P.F. (2009). Lactose: chemistry and properties. In: *Advanced Dairy Chemistry* (ed. P.F. Fox and P.L.H. McSweeney), 1–15. New York: Springer.

16 Nivetha, A. and Mohanasrinivasan, V. (2017). Mini review on role of β-galactosidase in lactose intolerance. In: *IOP Conference Series: Materials Science and Engineering* (Vol. 263, No. 2, p. 022046). IOP Publishing.

17 Velazco, S., Kambo, D., Yu, K. et al. (2021). Modeling gene expression: lac operon. In: *2021 43rd Annual International Conference of the IEEE Engineering in Medicine & Biology Society (EMBC)*, 1086–1091. IEEE.

18 Hamed, A.A., Khedr, M., and Abdelraof, M. (2020). Activation of LacZ gene in *Escherichia coli* DH5α via α-complementation mechanism for β-galactosidase production and its biochemical characterizations. *Journal, Genetic Engineering & Biotechnology* 18 (1): 1–14.

19 Sedzro, D.M., Bellah, S.M.F., Akbar, H., and Billah, S.M.S. (2018). Structure, function, application and modification strategy of β–galactosidase. *Journal of Multidisciplinary Research and Reviews* 1: 10–16.

20 Upadhyay, S.K. and Singh, S.P. (eds.) (2021). *Bioprospecting of Plant Biodiversity for Industrial Molecules*. John Wiley & Sons Ltd. doi:10.1002/9781119718017.

21 Upadhyay, S.K. and Singh, S.P. (eds.) (2023). *Plants as Bioreactors for Industrial Molecules*. John Wiley & Sons Ltd. doi:10.1002/9781119875116.

22 Smith, D.L. and Gross, K.C. (2000). A family of at least seven β-galactosidase genes is expressed during tomato fruit development. *Plant Physiology* 123 (3): 1173–1184.

23 Seddigh, S. and Darabi, M. (2014). Comprehensive analysis of beta-galactosidase protein in plants based on *Arabidopsis thaliana*. *Turkish Journal of Biology* 38 (1): 140–150.

24 Hussien, S.A. and Doosh, K. (2021, May). Extraction, purification and characterization of β-galactosidase from tomato (*Lycopersicom esculentum*). In: *IOP Conference Series: Earth and Environmental Science* (Vol. 761, No. 1, p. 012123). IOP Publishing https://doi.org/10.1088/1755-1315/761/1/012123.

25 Pal, A., Lobo, M., and Khanum, F. (2013). Extraction, purification and thermodynamic characterization of almond (*Amygdalus communis*) β-galactosidase for the preparation of delactosed milk. *Food Technology and Biotechnology* 51 (1): 53–61.

26 Zhang, Q., Peng, Y., Li, X. et al. (2021). β-Galactosidase is involved in rice seed germination. *Seed Science and Technology* 49 (3): 261–274.

27 Patel, A., Shah, N., and Verma, D.K. (2017). Lactic acid bacteria (LAB) bacteriocins: an ecological and sustainable biopreservative approach to improve the safety and shelf-life of foods. In: *Microorganisms in Sustainable Agriculture, Food and the Environment* (ed. D.K. Verma and P.P. Srivastav). Volume-1, as part of book series on *Innovations in Agricultural Microbiology*, 197–258. Palm Bay, FL: Apple Academic Press.

28 Xavier, J.R., Ramana, K.V., and Sharma, R.K. (2018). β-Galactosidase: biotechnological applications in food processing. *Journal of Food Biochemistry* 42 (5): E12564.

29 Verma, D.K., Patel, A., Prajapati, J.B. et al. (2020). Starter culture and probiotic bacteria in dairy food products. In: *Microbiology for Food and Health Technological Developments and Advances* (ed. D.K. Verma, A.R. Patel, P.P. Srivastav, et al.), 3–49. Palm Bay, FL: Apple Academic Press.

30 Singh, S.P. and Upadhyay, S.K. (eds.) (2021). *Bioprospecting of Microorganism-Based Industrial Molecules*. John Wiley & Sons Ltd. doi:10.1002/9781119717317.

31 Fernandes, S., Geueke, B., Delgado, O. et al. (2002). β-Galactosidase from a cold-adapted bacterium: purification, characterization and application for lactose hydrolysis. *Applied Microbiology and Biotechnology* 58 (3): 313–321.

32 Hsu, C.A., Yu, R.C., and Chou, C.C. (2005). Production of β-galactosidase by Bifidobacteria as influenced by various culture conditions. *International Journal of Food Microbiology* 104 (2): 197–206.

33 Gul-Guven, R., Guven, K., Poli, A., and Nicolaus, B. (2007). Purification and some properties of a β-galactosidase from the thermoacidophilic *Alicyclobacillus acidocaldarius* subsp. rittmannii isolated from Antarctica. *Enzyme and Microbial Technology* 40 (6): 1570–1577.

34 Zavaleta, A.I., Ávila, J., Chávez-Hidalgo, E.L., and Izaguirre, V. (2013). Partial characterization of the β-galactosidase gen from *Bacillus* sp. MSP7 isolated from Pilluana Saltern, San Martin-Peru. *Ciencia e Investigación* 16 (1): 18–23.

35 Carević, M., Vukašinović-Sekulić, M., Grbavčić, S. et al. (2015). Optimization of β-galactosidase production from lactic acid bacteria. *Hemijska Industrija* 69 (3): 305–312.

36 Geiger, B., Nguyen, H.M., Wenig, S. et al. (2016). From by-product to valuable components: efficient enzymatic conversion of lactose in whey using β-galactosidase from *Streptococcus thermophilus*. *Biochemical Engineering Journal* 116: 45–53.

37 Shaikhan, B.A., Güven, K., Bekler, F.M. et al. (2020). A highly inducible β-galactosidase from *Enterobacter* sp. *Journal of the Serbian Chemical Society* 85 (5): 609–622.

38 Deng, Y., Xu, M., Ji, D., and Agyei, D. (2020). Optimization of β-galactosidase production by batch cultures of *Lactobacillus leichmannii* 313 (ATCC 7830™). *Fermentation* 6 (1): 27.

39 Luan, S. and Duan, X. (2022). A novel thermal-activated β-galactosidase from *Bacillus aryabhattai* GEL-09 for lactose hydrolysis in milk. *Food* 11 (3): 372.

40 Park, Y.K., De Santi, M.S.S., and Pastore, G.M. (1979). Production and characterization of β-galactosidase from *Aspergillus oryzae*. *Journal of Food Science* 44 (1): 100–103.

41 Shaikh, S.A., Khire, J.M., and Khan, M.I. (1997). Production of β-galactosidase from thermophilic fungus Rhizomucor sp. *Journal of Industrial Microbiology and Biotechnology* 19 (4): 239–245.

42 Nakkharat, P. and Haltrich, D. (2006). Purification and characterisation of an intracellular enzyme with β-glucosidase and β-galactosidase activity from the thermophilic fungus *Talaromyces thermophilus* CBS 236.58. *Journal of Biotechnology* 123 (3): 304–313.

43 Song, C., Chi, Z., Li, J., and Wang, X. (2010). β-Galactosidase production by the psychrotolerant yeast *Guehomyces pullulans* 17-1 isolated from sea sediment in Antarctica and lactose hydrolysis. *Bioprocess and Biosystems Engineering* 33 (9): 1025–1031.

44 Sen, S., Ray, L., and Chattopadhyay, P. (2012). Production, purification, immobilization, and characterization of a thermostable β-galactosidase from *Aspergillus alliaceus*. *Applied Biochemistry and Biotechnology* 167 (7): 1938–1953.

45 Isobe, K., Yamashita, M., Chiba, S. et al. (2013). Characterization of new β-galactosidase from acidophilic fungus, *Teratosphaeria acidotherma* AIU BGA-1. *Journal of Bioscience and Bioengineering* 116 (3): 293–297.

46 Cardoso, B.B., Silvério, S.C., Abrunhosa, L. et al. (2017). β-Galactosidase from *Aspergillus lacticoffeatus*: a promising biocatalyst for the synthesis of novel prebiotics. *International Journal of Food Microbiology* 257: 67–74.

47 Martarello, R.D.A., Cunha, L., Cardoso, S.L. et al. (2019). Optimization and partial purification of beta-galactosidase production by *Aspergillus niger* isolated from Brazilian soils using soybean residue. *AMB Express* 9 (1): 1–13.

48 Vidya, B., Palaniswamy, M., Angayarkanni, J. et al. (2020). Purification and characterization of β-galactosidase from newly isolated *Aspergillus terreus* (KUBCF1306) and evaluating its efficacy on breast cancer cell line (MCF-7). *Bioorganic Chemistry* 94: 103442.

49 Abd El-Salam, B.A., Ibrahim, O.A., and Amer, A.E. (2020). Efficient enzymatic conversion of lactose in milk using fungal β-galactosidase. *Biocatalysis and Agricultural Biotechnology* 29: 101813.

50 Dissanayaka, D.M.S., De Silva, S.N.T., Attanayaka, D.P.S.T.G., and Vimukthi, E.M.M. (2021). Production and characterization of β-galactosidase from the fungus *Thielaviopsis ethacetica* Went. *Journal of the National Science Foundation of Sri Lanka* 49 (4): 563–571.

51 Paulo, A.J., Wanderley, M.C.D.A., de Oliveira, R.J.V. et al. (2021). Production and partial purification by PEG/citrate ATPS of a β-galactosidase from the new promising isolate *Cladosporium tenuissimum* URM 7803. *Preparative Biochemistry & Biotechnology* 51 (3): 289–299.

52 Thakur, M., Kumar Rai, A., and Singh, S.P. (2022). An acid-tolerant and cold-active β-galactosidase potentially suitable to process milk and whey samples. *Applied Microbiology and Biotechnology* 106: 3599–3610.

53 Xu, X., Fan, X., Fan, C. et al. (2019). Production optimization of an active-galactosidase of *Bifidobacterium animalis* in heterologous expression systems. *BioMed Research International* 2019: 8010635.

54 Kolev, P., Rocha-Mendoza, D., Ruiz-Ramírez, S. et al. (2022). Screening and characterization of β-galactosidase activity in lactic acid bacteria for the valorization of acid whey. *JDS Communications* 3 (1): 1–6.

55 Zárate, G. and Chaia, A.P. (2012). Influence of lactose and lactate on growth and β-galactosidase activity of potential probiotic *Propionibacterium acidipropionici*. *Anaerobe* 18 (1): 25–30.

56 Niamah, A.K. (2019). Ultrasound treatment (low frequency) effects on probiotic bacteria growth in fermented milk. *Future of Food: Journal on Food, Agriculture and Society* 103: https://doi.org/10.17170/kobra-20190709592.

57 Ibrahim, A.H. (2018). Enhancement of β-galactosidase activity of lactic acid bacteria in fermented camel milk. *Emirates Journal of Food and Agriculture* 30: 256–267.

58 Dagbagli, S. and Goksungur, Y. (2008). Optimization of b-galactosidase production using *Kluyveromyces lactis* NRRL Y-8279 by response surface methodology. *Electronic Journal of Biotechnology* 11 (4): 11–12.

59 de Albuquerque, T.L., de Sousa, M., de Silva, N.C.G. et al. (2021). β-Galactosidase from *Kluyveromyces lactis*: characterization, production, immobilization and applications – a review. *International Journal of Biological Macromolecules* 191: 881–898.

60 Morioka, L.R.I., Viana, C.D.S., Alves, E.D.P. et al. (2019). Concentrated beta-galactosidase and cell permeabilization from *Saccharomyces fragilis* IZ 275 for beta-galactosidase activity in the hydrolysis of lactose. *Food Science and Technology* 39: 524–530.

61 de Bales, S.A. and Castillo, F.J. (1979). Production of lactase by *Candida pseudotropicalis* grown in whey. *Applied and Environmental Microbiology* 37 (6): 1201–1205.

62 Palumbo, M.S., Smith, P.W., Strange, E.D. et al. (1995). Stability of β-galactosidase from *Aspergillus oryzae* and *Khyveromyces lactis* in dry milk powders. *Journal of Food Science* 60 (1): 117–119.

63 Kazemi, S., Khayati, G., and Faezi-Ghasemi, M. (2016). β-Galactosidase production by *Aspergillus niger* ATCC 9142 using inexpensive substrates in solid-state fermentation: optimization by orthogonal arrays design. *Iranian Biomedical Journal* 20 (5): 287.

64 van den Bogaard, P.T., Hols, P., Kuipers, O.P. et al. (2004). Sugar utilisation and conservation of the gal-lac gene cluster in *Streptococcus thermophilus. Systematic and Applied Microbiology* 27 (1): 10–17.

65 De Jesus, L.F.M.C. and Guimarães, L.H.S. (2021). Production of β-galactosidase by *Trichoderma* sp. through solid-state fermentation targeting the recovery of galactooligosaccharides from whey cheese. *Journal of Applied Microbiology* 130 (3): 865–877.

66 Verma, D.K., Kimmy, G., Singhal, P. et al. (2018). Genetically modified organisms (GMOs) produced enzymes: multifarious applications in food manufacturing industries. In: *Bioprocess Technology in Food and Health: Potential Applications and Emerging Scope* (ed. D.K. Verma, A.R. Patel, and P.P. Srivastav), as part of book series on *Innovation in Agricultural Microbiology*, 77–120. Palm Bay, FL: Apple Academic Press.

67 de Andrade, B.C., Migliavacca, V.F., Okano, F.Y. et al. (2019). Production of recombinant β-galactosidase in bioreactors by fed-batch culture using DO-stat and linear control. *Biocatalysis and Biotransformation* 37 (1): 3–9.

68 Ren, Z.Y., Liu, G.L., Chi, Z. et al. (2017). Overexpression of both the lactase gene and its transcriptional activator gene greatly enhances lactase production by *Kluyveromyces marxianus. Process Biochemistry* 61: 38–46.

69 Zhou, Y., Lu, Z., Wang, X. et al. (2018). Genetic engineering modification and fermentation optimization for extracellular production of recombinant proteins using *Escherichia coli. Applied Microbiology and Biotechnology* 102 (4): 1545–1556.

70 Bury, D.E.A.N., Geciova, J.A.N.A., and Jelen, P.A.V.E.L. (2001). Effect of yeast extract supplementation on beta-galactosidase activity of *Lactobacillus delbrueckii* subsp. bulgaricus 11842 grown in whey. *Czech Journal of Food Sciences* 19 (5): 166–170.

71 Oliveira, C., Guimarães, P.M., and Domingues, L. (2011). Recombinant microbial systems for improved β-galactosidase production and biotechnological applications. *Biotechnology Advances* 29 (6): 600–609.

72 Kaushal, G., Rai, A.K., Singh, S.P. (2021). A novel β-glucosidase from a hot-spring metagenome shows elevated thermal stability and tolerance to glucose and ethanol. *Enzyme and Microbial Technology* 145: 109764. doi: 10.1016/j.enzmictec.2021.109764.

73 Singh, S.P., Jadaun, J.S., Narnoliya, L.K., Pandey, A. (2017). Prebiotic oligosaccharides: special focus on fructooligosaccharides, its biosynthesis and bioactivity. *Applied Biochemistry and Biotechnology* 183 (2): 613–635. https://doi.org/10.1007/s12010-017-2605-2.

74 Park, A.R. and Oh, D.K. (2010). Galacto-oligosaccharide production using microbial β-galactosidase: current state and perspectives. *Applied Microbiology and Biotechnology* 85 (5): 1279–1286.

75 Mano, M.C.R., Paulino, B.N., and Pastore, G.M. (2019). Whey permeate as the raw material in galacto-oligosaccharide synthesis using commercial enzymes. *Food Research International* 124: 78–85.

76 Lata, K., Sharma, M., Patel, S.N. et al. (2018). An integrated bio-process for production of functional biomolecules utilizing raw and by-products from dairy and sugarcane industries. *Bioprocess and Biosystems Engineering* 41 (8): 1121–1131. https://doi.org/10.1007/s00449-018-1941-0.

77 Hassan, N., Geiger, B., Gandini, R. et al. (2016). Engineering a thermostable Halothermothrix orenii β-glucosidase for improved galacto-oligosaccharide synthesis. *Applied Microbiology and Biotechnology* 100 (8): 3533–3543.

78 Jung, S.J., Houde, R., Baurhoo, B. et al. (2008). Effects of galacto-oligosaccharides and a *Bifidobacteria lactis*-based probiotic strain on the growth performance and fecal microflora of broiler chickens. *Poultry Science* 87 (9): 1694–1699.

79 Martins, G.N., Ureta, M.M., Tymczyszyn, E.E. et al. (2019). Technological aspects of the production of fructo and galacto-oligosaccharides. Enzymatic synthesis and hydrolysis. *Frontiers in Nutrition* 6: 78.

80 Zerva, A., Limnaios, A., Kritikou, A.S. et al. (2021). A novel thermophile β-galactosidase from *Thermothielavioides terrestris* producing galactooligosaccharides from acid whey. *New Biotechnology* 63: 45–53.

81 Yañez-Ñeco, C.V., Cervantes, F.V., Amaya-Delgado, L. et al. (2021). Synthesis of β (1→ 3) and β (1→ 6) galactooligosaccharides from lactose and whey using a recombinant β-galactosidase from *Pantoea anthophila*. *Electronic Journal of Biotechnology* 49: 14–21.

82 Thomson, P., Medina, D.A., and Garrido, D. (2018). Human milk oligosaccharides and infant gut bifidobacteria: molecular strategies for their utilization. *Food Microbiology* 75: 37–46.

83 Nijman, R.M., Liu, Y., Bunyatratchata, A. et al. (2018). Characterization and quantification of oligosaccharides in human milk and infant formula. *Journal of Agricultural and Food Chemistry* 66 (26): 6851–6859.

84 Khassaf, W.H., Niamah, A.K., and Al-Manhel, A.J. (2019). study of the optimal conditions of levan production from a local isolate of *Bacillus subtilis* subsp. subtilis w36. *Basrah Journal of Agricultural Sciences* 32 (2): 213–222.

85 Szilagyi, A. and Ishayek, N. (2018). Lactose intolerance, dairy avoidance, and treatment options. *Nutrients* 10 (12): 1994.

86 Hodges, J.K., Cao, S., Cladis, D.P., and Weaver, C.M. (2019). Lactose intolerance and bone health: the challenge of ensuring adequate calcium intake. *Nutrients* 11 (4): 718.

87 Oak, S.J. and Jha, R. (2019). The effects of probiotics in lactose intolerance: a systematic review. *Critical Reviews in Food Science and Nutrition* 59 (11): 1675–1683.

88 Catanzaro, R., Sciuto, M., and Marotta, F. (2021). Lactose intolerance: an update on its pathogenesis, diagnosis, and treatment. *Nutrition Research* 89: 23–34.

16

Microbial Biogas Production: Challenges and Opportunities

Diana B. Muñiz-Márquez[1], *Christian Iván Cano-Gómez*[2], *Jorge Enrique Wong-Paz*[1], *Victor Emmanuel Balderas-Hernández*[3], *and Fabiola Veana*[2]

[1] *Facultad de Estudios Profesionales Zona Huasteca, Universidad Autónoma de San Luis Potosí, Ciudad Valles, San Luis Potosí, Mexico*
[2] *Tecnológico Nacional de México/IT de Ciudad Valles, Ciudad Valles, San Luis Potosí, Mexico*
[3] *División de Biología Molecular, Instituto Potosino de Investigación Científica y Tecnológica, A. C. San Luis Potosí, San Luis Potosí, Mexico*

16.1 Introduction

Food loss and waste (FLW) represent a great amount of squandered food in the world. In 2019, about 158 million-tons (17% of total food) accessible to consumers ended up as trash garbage cans from households, retail business, restaurants, and food services, according to Food Waste Index Report 2021 [1]. In agreement with Sustainable Development Goals (SDGs13.2), for 2030 it is necessary to reduce by 50% the per-capita global food waste at retail and consumer levels. In addition, cut down the food losses of forward production, supply chains, and postharvest losses [2]. Generation of food wastes is a problem that needs an urgent solution, because during food production natural resources are involved, such as water, land, and energy, which are also wasted. Around 8–10% of global emissions of "greenhouse gases" involved the not consumed food, and to 25% of the total freshwater used by agriculture, each year [1,3]. These situations causes ecosystem degradation and loss of biodiversity [3]. Circular bioeconomy is an excellent alternative to reduce FLW, since food wastes have been used as sources of bioactive compounds, such as phenols, carotenoids, anthocyanins, peptides, fatty acids, fibers, and enzymes. These compounds can be recovered for the introduction of new products into the market, such as beverage and food fermentation [4–6]. Additionally, these waste are excellent for usage as feedstock during biogas production, including food waste related (29.1%), sludge related (22.8%), manure related (20.3%), agricultural and horticulture waste related (15.2%), industrial bioethanol waste (6.3%) and others (6.3%) [7]. In contrast, an alternative to reduce these waste generation problems is the usage of food wastes in biogas production too, which is a promising strategy to decrease the wastes and use them as energy sources [8].

Microbial Bioreactors for Industrial Molecules, First Edition.
Edited by Sudhir P. Singh and Santosh Kumar Upadhyay.

The composition of municipal solid wastes (MSW) is mainly of organic nature with a high energy content; however, its recovery is inefficient, costly, and limited to the conversion of biodegradable fractions into methane (CH_4) within landfills and thermal conversion in incinerators [9]. Besides, biogas is an excellent alternative as an economic energy carrier that is applicable to move vehicles designed to burn gaseous fuel, where biogas can be purified as 95% of CH_4 [10]. As well as evaluation of its efficiency in microturbines and micro humid air turbine. As well as other systems, such as solid oxide fuel cells (SOFC) and its hybrid systems [11].

Biogas is an interesting product obtained during anaerobic digestion (AD) from biodegradable organic materials or solid residues [12, 13]. Biogas composition is composed from major to minor components: CH_4 (50–70%), carbon dioxide (30–50%), water vapors (5–10%), nitrogen (0–3%), oxygen (0–1%), an others compounds such as hydrogen sulfide (0–10,000 ppm), siloxanes, ammonia, and hydrocarbons [14]. Essentially, in biogas production, an AD is developed following the next steps: hydrolysis, acidogenesis, acetogenesis, and methanogenesis [15, 16]. In each stage microorganisms are protagonists and responsible for the biochemical reactions occurred, including bacteria of *Clostridium* genus (*Clostridium ultunense* and *Clostridium bornimense*), *Herbinix hemicellulosilytica*, *Peptoniphilus* sp., and thermophilic bacterium *Ruminiclostridium cellulosi* [17], fungi (*Aspergillus nidulans, Rhizomucor miehei, Gilbertella persicari,* and *Trichoderma reesei*) [18], archaeal (*Methanoculleus bourgensis, Methanothermobacter* sp., *Methanothermobacter defluvii, Methanosacarcina mazeis*), among others [19], and even macro- and microalgae (*Ulva lactuca, Chlorella minutissima, Chlorella pyrenoidosa,* and others) [20, 21, 22]. At the end of AD, a digestate is obtained and can be used as fertilizer, soil amendment, and livestock bedding, which also contributes to circular bioeconomy [23].

New trends focused on emerging bioelectrochemical technologies, such as power-to-gas AD (P2G-AD), AD microbial electrosynthesis (AD-MES), and microbial electrolysis cell AD (MEC-AD), which are considered in the design of future "cascading circular biosystems" with the objective to produce sustainable advanced biofuels [24]. In this sense, biogas implementation in developing countries is an opportunity with potential, where infrastructure, capital, and policy are focused for its success, from household or domestic implementation (small-scale) to large scale [25]. As an example, India has worked with government programs focused on biogas production and establishing goals in production of energy and in the promotion of social and environment conditions with this energetic production [25].

This chapter deals with the generalities in biogas production, such as substrates, microorganisms, and enzymes involved in the process, as well as challenges and opportunities in this topic.

16.2 Generalities of Biogas Production: the Process and Its Yields

The biogas production associated with the utilization of organic wastes is an interesting area to be explored, since circular bioeconomy is focused in reducing FWL. In AD, around 50–70% of biogas is produced, being CH_4 or biomethane the major compound after biogas purification [26–28]. However, many factors have influence over biogas yield, such as organic loading rate, biomass pretreatments, temperature, co-digestion, reactor design, among others [29].

A first step for biogas production is the "pretreatment" of substrates to create the right scenario for feedstock degradation (enzyme-cellulose or enzyme-hemicellulose bonding), including the maximum degradation of carbohydrates, a decreased release of growth inhibitors and, the reduction of the environmental impact [29, 30]. The biomass pretreatments include addition of sodium hydroxide, sulfuric acid, sodium carbonate, and aqueous ammonia [31], which might affect the microbial diversity and their growth during the AD process.

In a second step, acidogenesis microorganisms produce short-chain fatty acid (SCFA), CO_2, and hydrogen. Next step is acetogenesis where SCFA are converted to acetic acid by acetogenic microoganisms; in this step, the acetic acid formed originates 70% of the CH_4 produced during AD in last step that is realized by methanogenic microorganisms. The formed CH_4 is from substrates with one or two carbon atoms covalently bonded, such as acetate, H_2, CO_2, formate, methanol, and some methylamines [32].

Having clarified the process, it is important to talk about the yields of the biogas produced. Since a 15-year-old agricultural biogas plant in Poland reported different CH_4 yields, using cheese production waste and post floating settlements from slaughterhouses, generating yields of 610.2 and 680 m^3/tone dry organic mass, respectively, while minor yields were obtained using cow slurry and chicken waste as substrate (~230 m^3/tone dry organic mass) [33]. In this sense, some of the previously reported biogas yields obtained using different types of substrates are presented in Table 16.1.

Table 16.1 Biogas production yields obtained using different types of wastes as substrates and conditions.

		Biogas composition			
Wastes	**Conditions**	**Biogas generated/yield**	**Methane**	**Other gases**	**References**
Cow dung	32–36 °C for 21 days	280	67.9%	CO_2: 27.2% CO: 4.7% H_2S: 0.1%	[26]
Wood chips Corn stover Dust mills Leachate Manure Oxidation lagoon water Rumen	Mesophilic conditions (35–40 °C) for 22 days	(25.5 L)	55.3%	NM	[27]

(Continued)

Table 16.1 (Continued)

		Biogas composition			
Wastes	**Conditions**	**Biogas generated/yield**	**Methane**	**Other gases**	**References**
Shavings from the thickness adjustment operation of the tannery process Sludge from the wastewater treatment plant of the tannery Nutrient solution (yeast extract, peptone, K_2HPO_4, KH_2PO_4, and water)	35 °C for 150 days	26.1 mL of biogas/gVSS of waste	52%	NM	[28]
Ulva lactuca *Chlorella minutissima* Sludge	35 and 55 °C; particle sizes (0.2 and 0.75 mm) for 30 days	342.59 cm^3/g VS	166.16 cm^3/g VS	NM	[20]

NM, not mentioned.

16.3 Feedstocks Used in Biogas Production and Their Characteristics

Biogas is a biofuel product obtained from chemical reactions in natural environments or specific devices through the biodegradation and fermentation of organic matter by the activity of specialized microorganisms, absence of oxygen, and other factors [34]. This kind of gas can be produced from different materials, such as food waste, lignocellulosic materials, animal manure, and environment samples, such as sewage sludge, wastewater, sewage, among others. The latter has recently been of interest for biogas production; however, it is necessary to subject them to a pretreatment, either physical, thermal, chemical, or thermochemical, to facilitate their degradation prior to AD. In this sense, lignocellulosic materials are constituted in three fractions which are cellulose between 40 and 50%, hemicellulose between 25 and 35%, and lignin between 15 and 20%, the quantity and quality will depend on the type of plant and its origin [35]. This material can be found as agricultural residues, forestry residues, and domestic food waste. Crops of wheat, rice, corn, and sugarcane after being harvested produce large amount of crop residues such as wheat straw, rice straw, husk, stalks, and bagasse, which can be effectively used as a source of biomass energy; this lignocellulosic biomass is part of the production of second generation of biofuels. For example, wheat straw is economical and most available renewable biomasses worldwide with a production of 500 Mt/year, so it is a potential biomass substrate for biogas production [35, 36]. Another available agricultural waste with a high potential is sugarcane bagasse through its pretreatment and AD to obtain biogas [38]; similarly, acid-pretreated rice straw improves AD for biogas production [39].

On the other hand, the chicken, pig, and cow manure are commonly used for the biogas production, with chicken manure having a higher potential than pig and cow manure, being high content of organic matter and nutrients (potassium, nitrogen, and phosphorus). However, during microbial degradation in biogas production, organic nitrogen is transformed into ammonia, which is a capable inhibitor during microbiological transformation in the AD. A total ammonia nitrogen concentration above 3 g/L causes a decrease in biogas production and should be avoided. In order to stop such inhibition, several techniques have been implemented, including a co-digestion with a high carbon content in substrates, temperature and pH adjustment, dilution of the substrate, the use of materials with adsorption capacity (activated carbon, bentonite, and zeolite), trace elements, use of microflora adaptation and bio-augmentation, pickling, struvite precipitation, membrane processes, ultrasound and microwave irradiation, and biological process such as anaerobic ammonium ion oxidation (Anammox) [40, 41].

On the other hand, co-digestion has been used to enhance biogas production, i.e. utilizing the mixture of two substrates such as straw and manure. For example, Sumardiono et al. [42] used the mixture of corn stalk and cow manure, through a physical pretreatment (smaller particle size), biological, and chemical with the addition of NaOH placed in a biodigester in a 1 : 1 ratio for fermentation obtaining a yield of 215.77 L/kg of total biogas production. Sánchez-Sánchez et al. [43] made a mixture of sheep manure (20% w/w), cheese serum (80% w/w), and porous materials (almond shells, walnut shells, kenaf fiber, and charcoal) crushed them and placed them in a bioreactor with a fixed bed. The result was an increase of more than 27% in biogas production compared to biomethanization without porous materials and a 50% decrease in chemical oxygen demand. Ihoeghian et al. [39], studied the biogas production and process stability. The co-digestion was realized with rumen cattle and food waste, where the 50 : 50 ratio of substrates was the optimal treatment and the one with the highest yield with a production of 320.52 mL/g added, plus the co-digestion characteristics of volatile fatty acids, pH, and ammonia nitrogen in the AD.

16.4 Microbial Biodiversity in Biogas Production

16.4.1 Generalities

The microbial communities are dynamic during biogas production. Studies about biogas production have revealed the increase around 50% of *Bacteroidetes* phyla at 150 days and reduction of *Firmicutes*. At class level, *Bacteroidia* and *Clostridia* were increased and decreased, respectively. These results were observed when treated wastewater from tannery (sludge) or tannery process (shavings) were used [28]. In other studies, microorganisms present in a full-scale anaerobic digester using on food waste were monitored through qPCR during 18 months. The presence of substrates or inhibitors impact under the methanogenic population in digesters. *Methanosaetaceae* was dominant in the digester, suggesting that the acetic acid methanogenic as principal methane production pathway. However, a decrease to 69% from *Methanosaetaceae* was presented with an accumulation of volatile fatty-acids, but later a recovery of microorganisms caused the volatile fatty-acids consumption. Inhibition by ammonia was observed and changes in acetate utilization at hydrogen was reported. In addition, oligo elements and alkalinity with high concentrations of propionate stimulated the microbial growth. These results are interesting for stability and optimization during the process [44].

Multiple studies have been development to describe the microbial communities that participate in the different steps of AD during biogas production, where the microbial dynamics is direct relationship with the feedstock used for biogas production. Whole genome of mesophilic microorganisms found in AD, such as *M. bourgensis, C. bornimense, C. ultunense, Peptoniphilus* sp. and *H. hemicellulosilytica*, and thermophilic *R. cellulosi* have been sequenced and characterized for different steps of AD [17].

Also, during the biogas production step, microbial communities have been monitored. The Bacteria domain highlights the phylum *Bacteroidetes, Chloroflexi,* and *Firmicutes.* In the latter, the *Clostridiales* order was found with the highest abundance of 28% [7]. Specifically, in the first stage (hydrolysis), *Trichococcus* was predominant; in the second stage (acidogenesis), *Aminobacterium* was present; and in the third stage (methanogenesis), *Levilinea* was the most abundant [7]. According to Amoozegar et al. [45], bacterial and archaeal strains have been reported with activity for CH_4 production, such as the family *Fusobacteriaceae* and the genus *Methanosaeta*, respectively.

Another study about the monitoring of microbial communities during biogas production using cow dung (CD) and fruit and vegetable waste mix (FVWM) with individual mono-digestion and co-digestion (CO) of both substrates. The phylogenetic analysis revealed that the phylas *Bacteroidetes, Firmicutes* (55% in FVWM), *Proteobacteria,* and *Actinobacteria* (0.2–15.91% in CO) were dominant in the three treatments. In addition, the class *Bacteroidia* and *Clostridia* were abundant in CD (67%) and CO (59.5%). At genus level, *Syntrophomonas* predominated in co-digestion. Therefore, at genus level, the relative abundance of archaeal communities in all samples were *Methanobrevibacter, Methanosarcina,* and *Methanobacterium*. Particularly, *Methanobrevibacter* was present in great quantities in FVWM and CD, and *Methanosarcina* in CD too [12].

16.4.2 Anaerobic Fungi in Biogas Production

Biogas is mostly a fuel from anaerobic fermentation of organic and agricultural wastes or sewage sludge. Particularly, substrates such as municipal solid wastes, animal wastes, or other by-products derived from industrial processes such as pulp, paper, forestry, agriculture, and food production are used for biogas production with anaerobic fungi [46]. The phylum *Neocallimastigomycota* are anaerobic fungi present in the digestive tract of the herbivorous that decompose a portion of the ingested forage. Dollhofer et al. [47] studied the hydrolysis of various lignocellulolytic materials (grass silage, maize silage, hay, straw, molasses, and soja) to increase biogas production with *Neocallimastix frontalis* and the preprocessing of the lignocellulosic substrates was suitable for future biotechnological applications. Aydin et al. [48] studied biogas production with microalgae *Haematococcus pluvialis* using anaerobic rumen fungi such as *N. frontalis*, *Piromyces* sp., *Orpinomyces* sp., and *Anaeromyces* sp. and obtained an increase in the CH_4 production up to 41% with a 91% more algae biomass degradation. According to Li et al. [49], the anaerobic fungi genera that have been identified as a biogas producer have the following morphology: monocentric, filamentous, and uniflagellate, such as *Agriosomyces, Aklioshbomyces, Buwchfawromyces, Capellomyces, Joblinomyces, Khoyollomyces, Liebetanzomyces, Pecoramyces, Piromyces, Oontomyces,* and *Tahromyces*; monocentric, filamentous, and polyflagellate, such as *Feramyces, Ghazallomyces,* and *Neocallimastix;* monocentric, bulbous and uniflagellate, such as *Caecomyces;* polycentric, filamentous, and uniflagellate, such as *Anaeromyces;* polycentric, filamentous, and polyflagellate, such as *Orpinomyces;* and polycentric, bulbous, and uniflagellate, such as *Cyllamyces*. These microorganisms have been isolated from sheep rumen contents, horse caecum, Holstein steer rumen, Holstein steer rumen, cow rumen, cow feces, buffalo feces, Indian camel stomach, sheep feces, Barbary sheep, goats' rumen samples, Mouflon sheep feces, white-tailed deer feces, Boer goat feces, Axis deer feces, goat feces, Grevy's zebra feces, and Nilgiri Tahr feces, respectively. Yildirim et al. [50] studied biogas production in anaerobic digesters with filamentous fungi rumen, such as *Anaeromyces* sp., *Orpinomyces* sp., *N. frontalis,* and *Piromyces* sp. and their results showed an increase by 60% of CH_4 production using animal manure [18], also tested filamentous fungi for the pretreatment of lignocellulosic substrates for enhancing biogas production (*A. nidulans, G. persicaria, R. miehei* and *T. reesei*), and they observed that the addition of these microorganisms confirmed to be an excellent β-glucosidase and endo-(1,4)-β-D-glucanase sources for lignocellulosic materials pretreatment. The obtained CH_4 can be used for rural cooking, vehicular fuel, or power generation; therefore, the researches continues to focus on the biogas production and in the establishment and optimization process to enhance the production of CH_4 from agricultural wastes with fungal or bacterial strains (Figure 16.1) [51, 52].

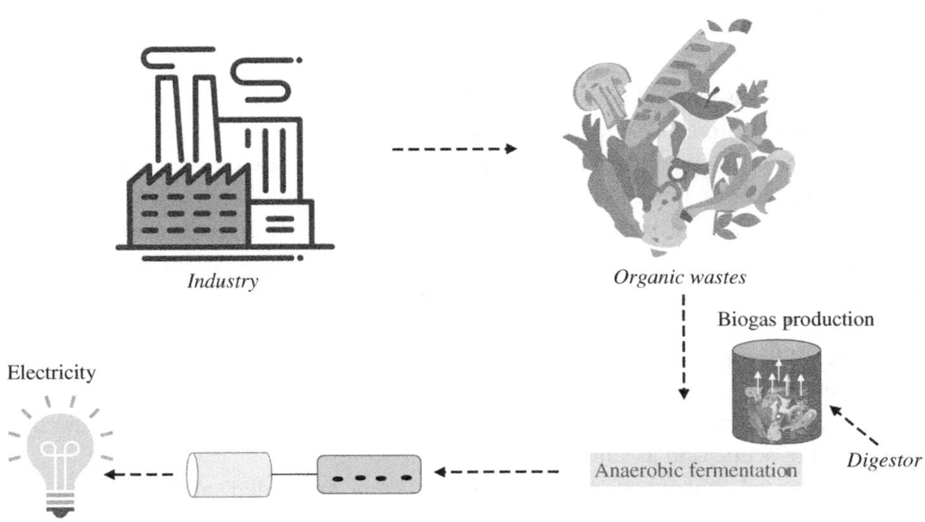

Figure 16.1 Biogas production with organic wastes for electricity generation.

16.4.3 Anaerobic Bacteria in Biogas Production

In oxygen-free conditions, the organic materials are broken down by different microorganisms that are classified into hydrolyzing/fermenting bacteria, obligate hydrogen-producing acetogenic and methanogenic bacteria. On the other hand, information on biogas-producing fungi is scarce [46]. The biogas production can be carried out in a digester or in a natural way in anaerobic environment in the forestomach of animals. Here, a wide variety and great amount of microorganisms are involved, such as *Clostridium, Fibrobacter, Ruminobacter, Ruminococcus*, and methanogens (*Methanobrevibacter, Methanobacterium, Methanomicrobium*) and are found in this habitat [18]. Methanogenic archaea are important organisms in the production of biogas in anaerobic conditions with yields of 50–75% of CH_4. Chen et al. [53] evaluated the *Methanobacterium congolense* for the CH_4 production; however, the cellular biomass and CH_4 production were affected at low CO_2 concentrations; therefore, the inorganic carbon, which is availability, with low CO levels might not greatly reduce methanogenic activities control also other parameters such as pH. Gopinath et al. [54], mentioned that CH_4 production is a complex process classified into four stages: (i) hydrolysis, (ii) acidogenesis, (iii) acetogenesis, and (iv) methanogenesis. The first microorganisms' group (hydrolytic bacteria) is composed of strict anaerobes (*Bacteriocides, Bifidobacteria,* and *Clostridia*) and other facultative anaerobes (*Enterobacteriaceae* and *Streptococci*). The second microorganism group corresponds to acidogenic bacteria and the third microorganism group is acetogenic bacteria (homoacetogenic bacteria) such as *Acetobacterium woodii* and *Clostridium acetium.* In the last degradation process, two classes of methanogenic bacteria synthesize CH_4 from acetate/hydrogen and carbon dioxide, which are strict bacteria that want a minor redox potential for reproduction. Only few classes are capable to downgrade acetate into CH_4 and carbon dioxide such as *M. mazei, Methanosarcina barkeri,* and *Methanotrix soehngenii.*

16.4.4 Methanogenic Archaeal and Algae in Biogas Production

The microbia involved in biogas production is diverse, fungi and bacteria, as well as archaea. In the biogas plant "Luchki" (AltEnergo L.L.C., Belgorod oblast, Russia), archaea population was monitored. Particularly, the plant has four tanks and operates for biogas production at 39 °C using a mix of swine manure, sugar beet pulp, silage, meat waste, and other organic substrates. In tanks 3 and 1, the major percentages of archaea, with values of 7.9–8.3%, was observed, respectively. In addition, the microbial diversity was variable at different temperatures (9, 15, 21, 35, 45, and 55 °C) during biogas production. The most dominant archaeal group was at phylum level *Euryarchaeota* in all temperatures and *Crenarchaeota* in minor dominance [55]. Therefore, other archaea have been detected during AD when oil palm was used as substrate, *Methanothermobacter* sp., *M. defluvii,* and *M. mazei* were found, these microorganisms were detected by PCR-DGGE [19]. The thermophilic microorganisms of genus *Methanosacarcina* grow at 50–60 °C and tolerate high concentration of acetic acid as substrate [19].

Reports suggested that when anaerobic rumen fungi are added, an increase in CH_4 production is observed, until 60%. The anaerobic rumen fungi were compound by *Piromyces* sp., *Anaeromyces* sp., *Orpinomyces* sp., and *N. frontalis* and mixed in equal ratios.

The biodigester with inoculum at different percentages (v/v) of animal manure as inoculum: 0% (R0), 5% (R1), 15% (R2), 20% (R3). The presence of archaeal communities was composed as follows: *Methanosarcinales* (digester R0), *Methanoasaeta* (digester R1), *Methanobacteriales* (digester R2), and 3 *Methanomicrobiales* (digester R3) with dominances of *Methanosarcinales*, *Methanoasaeta*, *Methanobacteriales*, and 3 *Methanomicrobiales*, respectively. Particularly, *Methanobacterium kanagiense* was detected in digester R2, which registered the higher biogas production [50].

Archaeal communities of 10 household biogas digesters (YL1 to YL10) with pig, sheep and cow manures, human feces, and food wastes have been described. The most dominant archaea were *Methanomicrobiales* (13.5–81.34% of abundance), specifically *Methanocorpusculum* was present in all digesters. Particularly, the major abundances were observed as follows: *Methanogenium* in digesters YL4 (100% pig manure) and YL6 (80% pig manure and 20% human feces); *Methanosarcina* and *Methanosaeta* in digester YL1 (60% sheep manure and 40% pig manure) and YL9 (100% vegetable wastes) [56]. The high amount of biogas/CH_4 observed in day 7 (sample P2) was associated with *Methanobacteriaceae* family in a digester and usage of different wastes as float and activated sludges, pig and poultry bloods, mainly [57].

In the other hand, microalgae have been used during more than 60 years as feedstock for biogas production and other high-value products. The microalgae are considered renewable and sustainable biomass and are important to climate change mitigation due to its facility for sequestering atmospheric carbondioxide [58, 59]. The *U. lactuca* macroalgae and *C. minutissima* microalgae have been used for optimal biogas production conditions, where the major yield was observed with macroalgae *U. lactuca* [20]. Other strains of alga have been reported, such as *Chlamydomonas reinhardtii*, *Chollera vulgaris*, *Nannochloropsis* sp. [23].

16.5 The Role of the Enzymes in Biogas Production

As previously described, production of biogas can arise from a wide variety of biodegradable feedstocks, many of which are a mixture of complex polymers such as lipids, carbohydrates, or proteins [60–62]. These polymeric carbon skeletons cannot be directly fermented into biogas production; they require their preliminary breakdown. In order to obtain simpler and more assimilable monomers, specialized microorganisms can secrete saccharolytic enzymes such as amylases, cellulases, xylanases, as well as lipases and proteases [63]. Enzymatic liberation of sugar monomers from complex carbohydrate polymers is denominated saccharification [64]. Plant-based polysaccharides include mostly cellulose, hemicellulose, gums, xylans, or starch, and from their enzymatic depolymerization glucose, xylose, fructose, arabinose, galactose, or some of its dimers can be obtained, conditional on the nature of the raw material [65, 66]. As the complexity of the polymers in the feedstock increases the more difficult for a single microorganism to produce all the required hydrolytic enzymes. Thus, utilization of microbial consortiums, including bacteria, fungi, and archaea, to produce a large range of hydrolytic enzymes is an advantageous strategy for the efficient degradation of complex substrates [67, 68]. In this sense, AD of organic materials is the basis to produce biogas in which waste is used for energy generation. *Bacteroides* and *Firmicutes* are the main phyla that contributes with the hydrolytic bacteria found in anaerobic digesters [69, 70].

Thus, to fully and optimally degrade the organic substrates availability of diverse hydrolytic bacteria is a key requisite. However, even in the presence of a wide microbial diversity, organic substrates will require long residence times in the anaerobic digester to depolymerize most of the components [13, 29]. Also, organic material can be recalcitrant to degradation, i.e. cellulose and hemicellulose, requiring physicochemical pretreatments such as acid or alkali hydrolysis, organosolv processes, or steam explosion, making the process economically and environmentally inefficient [71, 72]. Also, these pretreatment processes are responsible for the generation of inhibitor compounds that are toxic for the fermentative microorganisms, such as acetic and formic acids, furfural, and hydroxymethylfurfural [73, 74]. An alternative to the requirement of pretreatment to improve the digestibility of recalcitrant materials is the biological treatment using specialized hydrolytic microorganisms [75–77]. In nature, fungi efficiently degrade plant biomass by secreting multiple hydrolytic enzymes, also known as carbohydrate active enzymes (CASy) [78, 79]. Hydrolysis of β-1→4 glycosidic bonds in cellulose requires the action of endoglucanases (EC 3.2.1.4) that release oligosaccharides from amorphous regions of cellulose, then exoglucanases (EC 3.2.1.176; EC 3.2.1.91) act gradually on the ends of the oligosaccharides releasing cellobiose disaccharides. The latter is then degraded by β-glucosidases (EC 3.2.1.21) [46]. For the lignocellulose residues, CASy family members found in some filamentous fungi include mannanases (GH26), pectinases (GH28, GH78, GH93, PL1, CE8, and CE12), xyloglucanases (GH29 and gh74), amylases (GH31), inulinases (GH32), cellulases (GH45 and AA9), and xylanases (GH115 and CE15). As well as members of the family's lipase (abH03 and abH23), cutinase (abH36), and protease (S09 and A01) [80–82]. Also, wood decay fungi, especially white root fungi, can delignify the plant material by the combined action of the lignin peroxidase (EC 1.11.1.14), catalase (EC 1.10.3.2), and/or by the action of manganese peroxidase (EC 1.11.1.13), oxidizing or cleaving the phenolic and non-phenolic aromatic lignin rings [83, 84]. Addition of fungal strains *Cephalotrichum stemonitis* MUT 6326, *Coprinopsis cinerea* MUT 6385, and *Cyclocybe aegerita* MUT 5639 in the anaerobic digester of solid fractions from agricultural wastes improved the hemicellulose, lignin, and cellulose digestibility. This was also accompanied by an increment in the biogas and CH_4 yields by around two times, in contrast with the production obtained from the untreated material (no fungal addition) [85]. Similar results, reporting increments ranging from 20 to 300% in the biogas yields obtained by the fungal pretreatment of the biomass substrate, have been described elsewhere [46, 86–89]. A major drawback in the biological pretreatment of biological substrates is the slow growth rates of some of the microbial sources for hydrolytic enzymes. A strategy to improve the hydrolysis rate of biomass substrates independent of the microbial growth rates is the *in situ* addition of the specialized hydrolytic enzymes (Figure 16.2). For plant-based materials, the most common strategy is the direct addition of commercial and already available cellulases and other polysaccharases as a main treatment [90, 91] or posterior to an alkaline hydrolysis –pretreatment [92, 93]. Enzymatic pretreatment of different types of biomasses has been used to produce biogas, such as algae treated with cellulase, chitinase, and protease [94], pulp and paper biosludge treated with glucosidases (EC 3.2.1) and proteases (EC 3.4) [90], poultry waste and feathers digested with keratinases [95], wastewater plant treatment primary sludge digested with lipases and proteases isolated from different wild-type bacteria [96]. In all these reports, the enzymatic pretreatment improved the corresponding biogas production. These results indicate that under appropriate conditions,

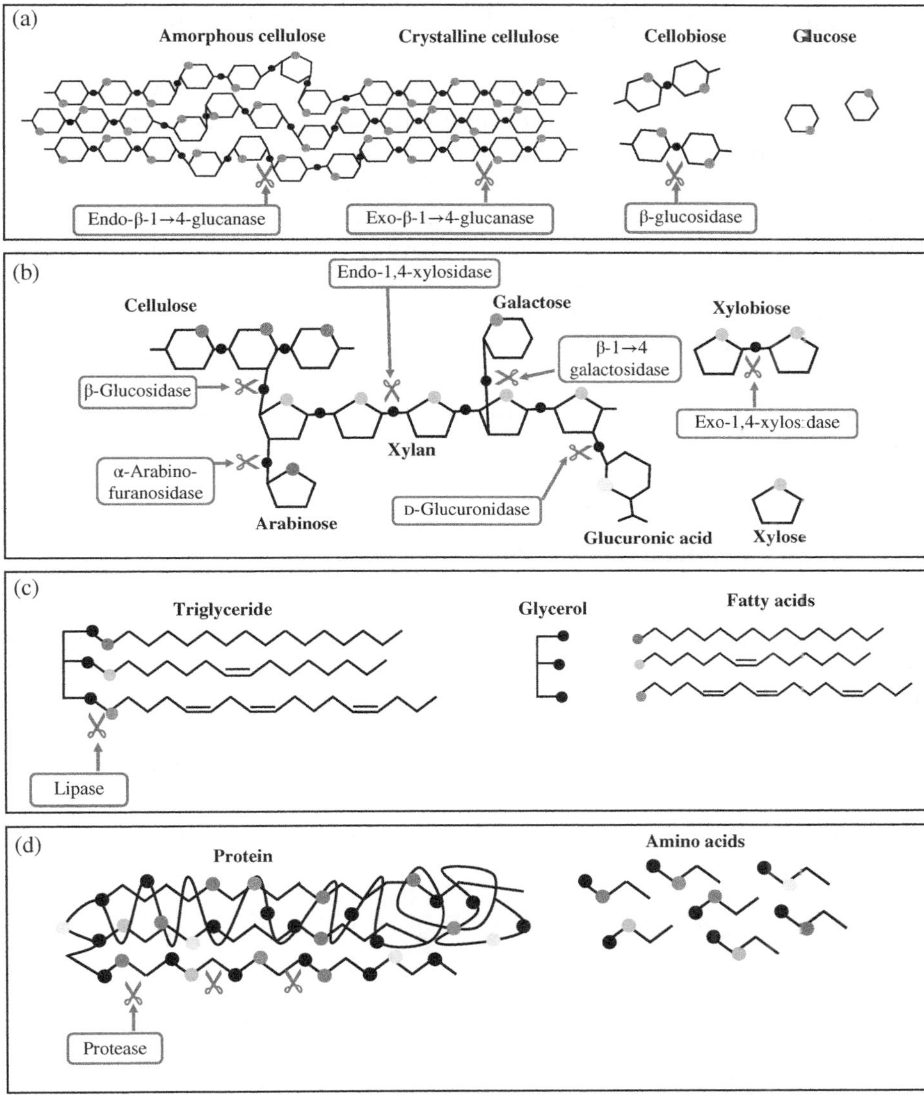

Figure 16.2 Enzymes used in the depolymerization of complex substrates for biogas production. Enzymes used for the degradation of (a) cellulose, (b) hemicellulose, (c) proteins, and (d) lipidic-based materials. Enzymes are indicated with scissors.

recalcitrant organic materials such as fats, lignocellulose, or even hard protein-based fibers can be digested and serve as substrate for biogas and other biofuels and value-added metabolites (Figure 16.2). Optimization of enzyme concentration, type of enzymes to be added, operational conditions, material pretreatment requirements, among other parameters is strongly suggested to enhance the hydrolysis rate, which is fundamental to improve the production yield and the production rate of biogas formation, and also will have a positive impact on the process economics [29, 97].

16.6 Challenges and Opportunities in Biogas Production

In the face of global concern about climate change and efforts to reduce the carbon footprint, biomass conversion technologies are an effective way to decrease carbon dioxide emissions, cutdown fossil fuel consumption, and gradually replace them with sources of renewable energy, which are now a necessary element of the energy supply system. The development of renewable resources around the world and their scale of utilization have continuously expanded, and the operation costs have reduced; therefore, the progress of renewable resources has become important for a global climate change management and national energy transformation in various countries [98]. Biogas production is considered a cleaner and renewable energy source that is the most imminent "biorefinery" solution to global energy problems. AD is an appropriate complex biological process that requires accelerated efforts to determine the most important factors and optimal conditions for its stabilization and to generate higher yields and productivity for new high-value products [99].

16.6.1 Challenges for Biogas Production

The implementation of biogas-based plants and the utilization of biofuels imply facing some challenges, which are mainly related to production, social, economic, and political issues (Figure 16.3).

The challenges of biogas production are of diverse nature; for instance, the first decision to make is the selection of the raw material to be used and how it will provide the adequate carbon monomers; this can be solved with efficient pretreatment strategies. Other challenges involve the high cost of biogas production, as well as the production technology, utility requirements, and biofuel properties [100]. Some promising approaches reveal the

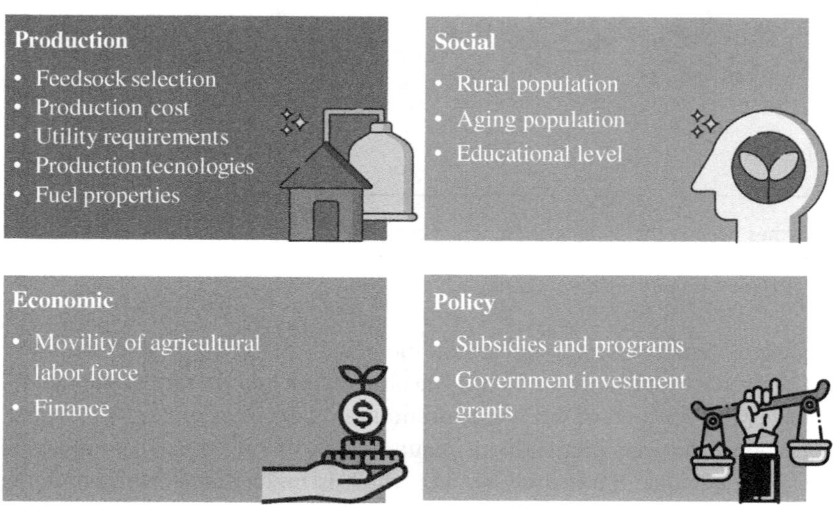

Figure 16.3 Challenges involved in the microbial biogas production.

microbial communities involved in AD, including the application of traditional and more advanced methods, such as molecular techniques and meta-omics approaches, which includes PCR, real-time PCR, RT-qPCR, PCR-DGGE, FISH, RFLP, mainly, as well as a correct inspection of the process, real-time control, and the application of mathematical models to characterize the performance of the microbial communities that participate during AD. The latter has been reached from an omics perspective approach, since it provides information to understand the complexity of microbial communities (structure, function, activities, and interactions) to reveal their biodiversity and organization in different anaerobic digesters, although not all species have been identified yet [99].

It is known that the high level of production, processing, and capital investment costs for biogas production has led to a negative energy balance [101, 102]. Currently, it is estimated that a household biodigester (50 kg) costs approximately $1500/day, being a large investment for low-income people who would not be able to afford it. In addition, there are no successful models of biogas plants that are adopted in factories, which in turn do not capture or use CH_4. There are no regulations or restrictions in this regard [101].

Other factors influencing biogas production (decrease in the use of rural digesters) are social and economic factors. Particularly, a drastic decrease in the rural population, the mobility of the agricultural labor force to the city, and the aging of the population, as well as the educational level or the lack of information regarding maintenance and technical support in the use of rural digesters. This has resulted in the failure of the establishment of biogas projects due to poor management [98, 103].

There is a lack of incentives for electricity generation from biogas, which conditions further research on biogas, digestate, and other end products that support biogas production and circular bioeconomy of food waste for urban and rural residents. Government investment grants, industry investment, and capital investment are needed to provide financial support for the construction of large-scale biogas plants, as the area of renewable energy has not been given the attention it deserves [25, 98, 103]. However, in developing countries, biogas technology continues to advance from small to large scale because of the issues of sustainability of financing, policy issues, technical services, awareness raising, and education. These areas are key factors for a correct implementation to maximize the use of biogas. Financial support is also crucial to facilitate the installation of biogas plants up to large-scale level for energy and electricity generation, and transportation [25, 101].

Environmental policy is very deficient and its application in renewable energy is not well regulated since waste managers deposit their waste in unregulated landfills or burn it in open. There is a lack of government commitment and lack of follow-up on biogas programs that become key challenges limiting the progress of biogas deployment. In addition, corruption is a complex challenge, among others in the biogas value chain (substrate supply, biogas production, distribution, and use). In addition, policy coherence and coordination are necessary [100]. Policies should be reviewed according to the priorities and characteristics of each sector for improvement, and a system of biogas standards should be established at the national level [104].

In countries such as China, India, and Nepal, the government provides financial and technical support for biogas programs. In fact, when the government decreased subsidies and programs, new biogas plants also decreased significantly [103]. Biogas utilization is important, but it requires capital investment and management by large companies, which may be easier to use private capital. For small and medium farms, financing is necessary [105]. At the level of rural communities, microfinance institutions can be chosen to enable them to dispense biogas technology in their homes. In addition, collaboration and cooperation between institutions should be promoted to improve the structure of biogas [103, 104, 106]. There is a need for comprehensive policies for biogas deployment and reduction of the rate of return on investment [107]. Ideally, multisectoral policies should drive the adoption of more sustainable technologies, such as carbon trading mechanisms and an operation to improve biogas carbon emission reduction schemes, and carbon trading pilot projects can be developed [25, 98, 100]. Cost-effective transformation of power system infrastructure and fuel consumption mode is required, as well as the combination of different renewable generation technologies. Thus, biogas plants are an option to improve system integration of intermittent renewables [108]. The strong modification of technologies, social behavior, economic aspirations, and government policies cause a vital analysis of these factors to successfully generate energy from waste [109]. In summary, the lack of adequate infrastructure, sufficient capital, and convenient policies have hindered the successful application of biogas.

16.6.2 Opportunities for Biogas Production

Concerns about the exhaustion of fossil fuels has conducted an increased research activity on renewable energy development, such as biogas production from waste for a sustainable energy generation. The fulfillment of future energy demands is a big challenge considering the growing greenhouse gas (GHG) emissions and the socioeconomic stability [101, 102]. The use of food waste for AD instead of its landfill accumulation can counteract climate change by avoiding food losses or wastage, as CH_4 production contributes globally to 90% of total GHG, which is mainly generated in landfills. It has been evaluated that the use of biogas as a fuel boosted with more than 90% CH_4 can lead to a decrease in GHG emissions of 60–80% compared to conventional fossil fuels [23]. Biogas is a suitable alternative with a huge potential and a conceivable outcome; it has an implied potential in producing clean energy, improving waste management, reducing workload, and building employment opportunities for local communities. The conversion of "waste to bioenergy" offers to decrease our dependence on fossil fuels. For example, biohydrogen and biogas production using agricultural residues [23, 103]. Technologies have been developed to promote the biogas process, such as the use of biomass ash as a high buffering additive not only increases the efficiency during biogas production but also significantly reduces the usage of commercial alkaline reagents and the constant pH adjustment in anaerobic digesters [110].

New trends in energy allow us to produce advanced and sustainable biofuels, and understanding endogenous enzyme activities (distribution and relative activity in AD) can lead to improved biogas production from high solids-organic matter. In addition, three emerging bioelectrochemical technologies, power-to-gas DA (P2G-DA), microbial electrolysis cell DA (MEC-DA), and DA microbial electrosynthesis (AD-MES), have been evaluated, three future circular cascade bioelectrochemical systems with different configurations in terms of potential to reduce GHG and augment CH_4 production [24, 111].

The reasons for biofuel production lies in energy security issues and environmental concerns and the need to produce clean energy and the future potential of bioenergy. But among the different biofuels, biogas stands out as a key player in the circular bioeconomy, reducing food losses or waste, as well as being a clean, environmentally friendly and versatile fuel [23]. AD not only helps to generate biogas as a source of bioenergy, it is a conversion process from which versatile uses of the products, CH_4 and digestate, are possible; the latter can be used as organic field supplements, compost, and feedstock for biochar synthesis, it is still rich in macronutrients and micronutrients and, when applied to land, improves attributes of the soil (physical, chemical, and biological) and increases crop productivity [23, 103, 112]. The use of biogas and its derivatives afford a clean energy source to farmers, care for the ecological environment and living conditions, and improves the quality of agricultural products, contributing to the circular bioeconomy [23].

The building of biogas plants at large scale in regions where agriculture is the main activity and crop straw/livestock manure are abundant as feedstocks for biogas production can give gas and heat to farmers and rural people. In addition, electricity generation from biogas can be utilized in the power grid or by enterprises, and biogas residues can be returned to fields, composted, or used in other ways [98]. To decrease installation costs and reduce operation and care of digesters, several strategies have been implemented, such as polyethylene film tubes to reduce cost in digesters, which are built using easily available materials (polyvinyl chloride and plastic bags). Finding low-cost alternatives makes it affordable for developing countries [103].

In general, the development of domestic biogas digesters (small and medium size levels) of low maintenance for agricultural regions could concede the biogas usage in households and farms to accommodate the clean fuel needs, where biogas residues and sludge obtained can be used as fertilizer to generate green and organic agricultural products [98]. Finally, biomass energy is expected to contribute greatly to future sustainability since it is a renewable and sustainable energy system, becoming an important global energy source driving a green civilization, a low-carbon economy, the evolution of sustainable rural villages, and the response to climate change [23, 98].

References

1 United Nations Environment Programm (2021). *Food Waste Index Report 2021*, 100 p. Nairobi: UNEP.

2 Koester, U. and Galaktionova, E. (2021). FAO food loss index methodology and policy implications. *Stud. Agric. Econ.* 123 (1): 1–7.

3 Food and Agriculture Organization (2021). Código de conducta voluntario para la reducción de las pérdidas y el desperdicio de alimentos. In: 42° período de sesiones 2021, 40.

4 Ricci, A., Cirlini, M., Guido, A. et al. (2019). From byproduct to resource: fermented apple pomace as beer flavoring. *Foods* 8 (8): 309.

5 Prandi, B., Faccini, A., Lambertini, F. et al. (2019). Food wastes from agrifood industry as possible sources of proteins: a detailed molecular view on the composition of the nitrogen fraction, amino acid profile and racemisation degree of 39 food waste streams. *Food Chem.* 286: 567–575.

6 Del Rio Osorio, L.L., Flórez-López, E., and Grande-Tovar, C.D. (2021). The potential of selected agri-food loss and waste to contribute to a circular economy: applications in the food, cosmetic and pharmaceutical industries. *Molecules* 26 (2): 515.

7 Zhang, L., Loh, K.C., Lim, J.W., and Zhang, J. (2019). Bioinformatics analysis of metagenomics data of biogas-producing microbial communities in anaerobic digesters: a review. *Renew. Sust. Energ. Rev.* 100: 110–126. https://doi.org/10.1016/j.rser.2018.10.021.

8 Hublin, A., Schneider, D.R., and Džodan, J. (2014). Utilization of biogas produced by anaerobic digestion of agro-industrial waste: energy, economic and environmental effects. *Waste Manag. Res.* 32 (7): 626–633.

9 Reinhart, D.R., Podder, A., and Bolyard, S.C. (2022). A brief history of energy recovery from municipal solid waste. In: *Energy from Waste Production and Storage*, 1e (ed. R.K. Gupta and T.A. Nguyen), 51–68. CRC Press.

10 Ciuła, J., Gaska, K., Iljuczonek, Ł. et al. (2019). Energy efficiency economics of conversion of biogas from the fermentation of sewage sludge to biomethane as a fuel for automotive vehicles. *Archit. Civ. Eng. Environ.* 2: 131–140.

11 Golmakani, A., Ali Nabavi, S., Wadi, B., and Manovic, V. (2022 Jun). Advances, challenges, and perspectives of biogas cleaning, upgrading, and utilisation. *Fuel* 1 (317): 123085. https://doi.org/10.1016/j.fuel.2021.123085.

12 Mukhuba, M., Roopnarain, A., Moeletsi, M.E., and Adeleke, R. (2020). Metagenomic insights into the microbial community and biogas production pattern during anaerobic digestion of cow dung and mixed food waste. *J. Chem. Technol. Biotechnol.* 95 (1): 151–162.

13 Shi, X.S., Dong, J.J., Yu, J.H. et al. (2017). Effect of hydraulic retention time on anaerobic digestion of wheat straw in the semicontinuous continuous stirred-tank reactors. *Biomed. Res. Int.* 2457805. https://doi.org/10.1155/2017/2457805.

14 Naveed, M., Anwar Muneeba, S., Saif, H. et al. (2021). Microbial and biotechnological advancement in biogas production. In: *Environmental Microbiology and Biotechnology* (ed. A. Singh, S. Srivastava, D. Rathore, and P. Deepak), 31–64. Singapore: Springer Nature Singapore Pte Ltd.

15 Sivamani, S., Saikat, B., Naveen Prasad, B.S. et al. (2021). A comprehensive review on microbial technology for biogas production. In: *Bioenergy Research* (ed. M. Srivastava, N. Srivastava, and R. Singh), 53–78.

16 Czekała, W. (2022). Biogas as a sustainable and renewable energy source. *Energy Environ. Sustain.* 201–214.

17 Jouzani, G.S. and Sharafi, R. (2018). Metadata of the chapter that will be visualized in SpringerLink. In: *Biogas, Biofuel and Biorefinery Technologies*, vol. 6 (ed. B. Tabatabaei and H. Ghanavati), 419–436. Springer International Publishing AG.

18 Szűcs, C., Kovács, E., Bagi, Z. et al. (2021). Enhancing biogas production from agroindustrial waste pre-treated with filamentous fungi. *Biol. Futur.* 72 (3): 341–346. https://doi.org/10.1007/s42977-021-00083-3.

19 Singkhala, A., Mamimin, C., Reungsang, A., and O-Thong S. (2021). Enhancement of thermophilic biogas production from palm oil mill effluent by pH adjustment and effluent recycling. *Processes* 9 (5): 878.

20 Koçer, A.T. and Özçimen, D. (2018). Investigation of the biogas production potential from algal wastes. *Waste Manag. Res.* 36 (11): 1100–1105.

21 Kumari, P., Varma, A.K., Shankar, R. et al. (2021). Phycoremediation of wastewater by *Chlorella pyrenoidosa* and utilization of its biomass for biogas production. *J. Environ. Chem. Eng.* 9 (1): 104974. https://doi.org/10.1016/j.jece.2020.104974.

22 Singh, S.P. and Upadhyay, S.K. (eds.) (2021). *Bioprospecting of Microorganism-Based Industrial Molecules*. John Wiley & Sons Ltd. doi:10.1002/9781119717317.

23 Mishra, A., Kumar, M., Bolan, N.S. et al. (2021). Multidimensional approaches of biogas production and up-gradation: opportunities and challenges. *Bioresour. Technol.* 338 (May): 125514. https://doi.org/10.1016/j.biortech.2021.125514.

24 Ning, X., Lin, R., O'shea, R. et al. (2021). Emerging bioelectrochemical technologies for biogas production and upgrading in cascading circular bioenergy systems. *iScience* 24 (9): 102998. https://doi.org/10.1016/j.isci.

25 Patinvoh, R.J. and Taherzadeh, M.J. (2019). Challenges of biogas implementation in developing countries. *Curr. Opin. Environ. Sci. Health* 12: 30–37. https://doi.org/10.1016/j.coesh.2019.09.006.

26 Adebayo, A., Adeleke, S., Tiekuro, E. et al. (2019). The production of biogas from cow dung. *J. Energy Res. Rev.* 2 (2): 1–5.

27 Senés-Guerrero, C., Colón-Contreras, F.A., Reynoso-Lobo, J.F. et al. (2019). Biogas-producing microbial composition of an anaerobic digester and associated bovine residues. *Microbiologyopen* 8 (9): 1–13.

28 Agustini, C.B., da Costa, M., and Gutterres, M. (2020). Biogas from tannery solid waste anaerobic digestion is driven by the association of the bacterial order bacteroidales and archaeal family methanosaetaceae. *Appl. Biochem. Biotechnol.* 192 (2): 482–493.

29 Olatunji, K.O., Ahmed, N.A., and Ogunkunle, O. (2021). Optimization of biogas yield from lignocellulosic materials with different pretreatment methods: a review. *Biotechnol. Biofuels* 14: 1–34.

30 Westerholm, M. and Schnürer, A. (2019). Microbial responses to different operating practices for biogas production systems. In: *Anaerobic Digestion* (ed. J.R. Banu and R.Y. Kannah), 1–36. IntechOpen.

31 Mirmohamadsadeghi, S., Karimi, K., Azarbaijani, R. et al. (2021). Pretreatment of lignocelluloses for enhanced biogas production: a review on influencing mechanisms and the importance of microbial diversity. *Renew. Sust. Energ. Rev.* 135: 110173. https://doi.org/10.1016/j.rser.2020.110173.

32 Reyes Aguilera, E.A. (2018). Generación de biogás mediante el proceso de digestión anaerobia, a partir del aprovechamiento de sustratos orgánicos. *Revista Científica de FAREM-Estelí* 24: 60–81.

33 Igliński, B., Piechota, G., Iwański, P. et al. (2020). 15 Years of the polish agricultural biogas plants: their history, current status, biogas potential and perspectives. *Clean Techn. Environ. Policy* 22 (2): 281–307. https://doi.org/10.1007/s10098-020-01812-3.

34 Ghofrani-Isfahani, P., Tsapekos, P., Peprah, M. et al. (2022). Ex-situ biogas upgrading in thermophilic trickle bed reactors packed with micro-porous packing materials. *Chemosphere* 296: 133987. https://doi.org/10.1016/j.chemosphere.2022.133987.

35 Khan, M.U., Usman, M., Ashraf, M.A. et al. (2022). A review of recent advancements in pretreatment techniques of lignocellulosic materials for biogas production: opportunities and limitations. *Chem. Eng. J. Adv.* 10: 100263. https://doi.org/10.1016/j.ceja.2022.100263.

36 Rani, P., Bansal, M., and Pathak, V.V. (2022). Experimental and kinetic studies for improvement of biogas production from KOH pretreated wheat straw. *Curr. Res. Green Sustain. Chem.* 5: 100283. https://doi.org/10.1016/j.crgsc.2022.100283.

37 Kaur, M. (2022). Effect of particle size on enhancement of biogas production from crop residue. *Mater. Today: Proc.* 57: 1950–1954.

38 Agarwal, N.K., Kumar, M., Ghosh, P. et al. (2022). Anaerobic digestion of sugarcane bagasse for biogas production and digestate valorization. *Chemosphere* 295: 133893. https://doi.org/10.1016/j.chemosphere.2022.133893.

39 Ihoeghian, N.A., Amenaghawon, A.N., Ajieh, M.U. et al. (2022). Anaerobic co-digestion of cattle rumen content and food waste for biogas production: establishment of co-digestion ratios and kinetic studies. *Bioresour. Technol. Rep.* 18: 101033. https://doi.org/10.1016/j.biteb.2022.101033.

40 Yılmaz, Ş. and Şahan, T. (2020). Utilization of pumice for improving biogas production from poultry manure by anaerobic digestion: a modeling and process optimization study using response surface methodology. *Biomass Bioenergy* 138: 105601. https://doi.org/10.1016/j.biombioe.2020.105601.

41 Fakkaew, K. and Polprasert, C. (2021). Air stripping pre-treatment process to enhance biogas production in anaerobic digestion of chicken manure wastewater. *Bioresour. Technol. Rep.* 14: 100647. https://doi.org/10.1016/j.biteb.2021.100647.

42 Sumardiono, S., Hawali Abdul Matin, H., Ivan Hartono, I. et al. (2022). Biogas production from corn stalk as agricultural waste containing high cellulose material by anaerobic process. *Mater. Today: Proc.* 63: S477–S483. https://doi.org/10.1016/j.matpr.2022.04.135.

43 Sánchez-Sánchez, C., González-González, A., Cuadros-Salcedo, F., and Cuadros-Blázquez, F. (2018). Using low-cost porous materials to increase biogas production: a case study in Extremadura (Spain). *J. Clean. Prod.* 198: 1165–1172.

44 Williams, J., Williams, H., Dinsdale, R. et al. (2013). Monitoring methanogenic population dynamics in a full-scale anaerobic digester to facilitate operational management. *Bioresour. Technol.* 140: 234–242. https://doi.org/10.1016/j.biortech.2013.04.089.

45 Amoozegar, M.A., Safarpour, A., Noghabi, K.A. et al. (2019). Halophiles and their vast potential in biofuel production. *Front. Microbiol.* 10: 1895.

46 Dollhofer, V., Podmirseg, S.M., Callaghan, T.M. et al. (2015). Anaerobic fungi and their potential for biogas production. *Adv. Biochem. Eng. Biotechnol.* 151: 41–61.

47 Dollhofer, V., Dandikas, V., Dorn-In, S. et al. (2018). Accelerated biogas production from lignocellulosic biomass after pre-treatment with *Neocallimastix frontalis*. *Bioresour. Technol.* 264: 219–227. https://doi.org/10.1016/j.biortech.2018.05.068.

48 Aydin, S., Yıldırım, E., Ince, O., and Ince, B. (2017). Rumen anaerobic fungi create new opportunities for enhanced methane production from microalgae biomass. *Algal Res.* 23: 150–160.

49 Li, Y., Meng, Z., Xu, Y. et al. (2021). Interactions between anaerobic fungi and methanogens in the rumen and their biotechnological potential in biogas production from lignocellulosic materials. *Microorganisms* 9 (1): 1–17.

50 Yıldırım, E., Ince, O., Aydin, S., and Ince, B. (2017). Improvement of biogas potential of anaerobic digesters using rumen fungi. *Renew. Energy* 109: 346–353. https://doi.org/10.1016/j.renene.2017.03.021.

51 Nasir, I.M., Ghazi, T.I.M., and Omar, R. (2012). Production of biogas from solid organic wastes through anaerobic digestion: a review. *Appl. Microbiol. Biotechnol.* 95 (2): 321–329.

52 Baky, A.H., Khan, M.N.H., Kader, F., and Chowdhury, H.A. (2014). Production of biogas by anaerobic digestion of food waste and process simulation. *ASME 2014 8th International*

Conference on Energy Sustainability collocated with the ASME 2014 12th International Conference on Fuel Cell Science, Engineering and Technology 2 (3): 79–83.

53 Chen, X., Ottosen, L.D.M., and Kofoed, M.V.W. (2019). How low can you go: methane production of *Methanobacterium congolense* at low CO2 concentrations. *Front. Bioeng. Biotechnol.* 7: 34.

54 Gopinath, L.R., Christy, P.M., Mahesh, K. et al. (2014). Identification and evaluation of effective bacterial consortia for efficient biogas production. *IOSR J. Environ. Sci., Toxicol. Food Technol.* 8 (3): 80–86.

55 Tian, G., Yang, B., Dong, M. et al. (2018). The effect of temperature on the microbial communities of peak biogas production in batch biogas reactors. *Renew. Energy* 123: 15–25. https://doi.org/10.1016/j.renene.2018.01.119.

56 Han, R., Liu, L., Meng, Y. et al. (2021). Archaeal and bacterial community structures of rural household biogas digesters with different raw materials in Qinghai Plateau. *Biotechnol. Lett.* 43 (7): 1337–1348.

57 Granada, C.E., Hasan, C., Marder, M. et al. (2018). Biogas from slaughterhouse wastewater anaerobic digestion is driven by the archaeal family Methanobacteriaceae and bacterial families Porphyromonadaceae and Tissierellaceae. *Renew. Energy* 118: 840–846. https://doi.org/10.1016/j.renene.2017.11.077.

58 Zabed, H.M., Akter, S., Yun, J. et al. (2020). Biogas from microalgae: technologies, challenges and opportunities. *Renew. Sust. Energ. Rev.* 117: 109503. https://doi.org/10.1016/j.rser.2019.109503.

59 Yukesh Kannah, R., Kavitha, S., Parthiba Karthikeyan, O. et al. (2021). A review on anaerobic digestion of energy and cost effective microalgae pretreatment for biogas production. *Bioresour. Technol.* 332 (March): 125055. https://doi.org/10.1016/j.biortech.2021.125055.

60 Fantozzi, F. and Buratti, C. (2009). Biogas production from different substrates in an experimental Continuously Stirred Tank Reactor anaerobic digester. *Bioresour. Technol.* 100 (23): 5783–5789.

61 Kazimierowicz, J. (2015). The effect of substrate on the amount and composition of biogas in agricultural biogas plant. *Infrastrukt i Ekol Teren Wiej* III (2): 809–818.

62 Nwokolo, N., Mukumba, P., Obileke, K., and Enebe, M. (2020). Waste to energy: a focus on the impact of substrate type in biogas production. *Processes* 8: 1224.

63 Zverlov, V.V., Köck, D.E., and Schwarz, W.H. (2015). The role of cellulose-hydrolyzing bacteria in the production of biogas from plant biomass. In: *Microorganisms in Biorefineries* (ed. B. Kamm), 335–361.

64 Ramachandran, P., Joshi, J.B., Kasirajan, L. et al. (2022). Enzymatic saccharification technologies for biofuel production: challenges and prospects. In: *Microbial Biotechnology for Renewable* (ed. J.K. Saini and R.K. Sani), 297–320. Clean Energy Production Technologies.

65 Artzi, L., Bayer, E.A., and Moraïs, S. (2016). Cellulosomes: bacterial nanomachines for dismantling plant polysaccharides. *Nat. Rev. Microbiol.* 15 (2): 83–95.

66 Bayer, E.A., Belaich, J.P., Shoham, Y., and Lamed, R. (2004). The cellulosomes: multienzyme machines for degradation of plant cell wall polysaccharides. *Annu. Rev. Microbiol.* 58: 521–554.

67 Blair, E.M., Dickson, K.L., and O'Malley, M.A. (2021). Microbial communities and their enzymes facilitate degradation of recalcitrant polymers in anaerobic digestion. *Curr. Opin. Microbiol.* 64: 100–108.

68 Brethauer, S. and Studer, M.H. (2014). Consolidated bioprocessing of lignocellulose by a microbial consortium. *Energy Environ. Sci.* 7 (4): 1446–1453.

69 Sun, L., Liu, T., Müller, B., and Schnürer, A. (2016). The microbial community structure in industrial biogas plants influences the degradation rate of straw and cellulose in batch tests. *Biotechnol. Biofuels* 9: 128.

70 Weiss, B., Souza, A.C.O., Constancio, M.T.L. et al. (2021). Unraveling a lignocellulose-decomposing bacterial consortium from soil associated with dry sugarcane straw by genomic-centered metagenomics. *Microorganisms* 9 (5): 995.

71 Brodeur, G., Yau, E., Badal, K. et al. (2011). Chemical and physicochemical pretreatment of lignocellulosic biomass: a review. *Enzyme Res.* 787532. https://doi.org/10.4061/2011/787532.

72 Taylor, M.J., Alabdrabalameer, H.A., and Skoulou, V. (2019). Choosing physical, physicochemical and chemical methods of pre-treating lignocellulosic wastes to repurpose into solid fuels. *Sustainability* 11: 3604.

73 Jönsson, L.J. and Martín, C. (2016). Pretreatment of lignocellulose: formation of inhibitory by-products and strategies for minimizing their effects. *Bioresour. Technol.* 199: 103–112.

74 Pienkos, P.T. and Zhang, M. (2009). Role of pretreatment and conditioning processes on toxicity of lignocellulosic biomass hydrolysates. *Cellulose* 16 (4): 743–762.

75 Ali, S.S., Abomohra, A.E.F., and Sun, J. (2017). Effective bio-pretreatment of sawdust waste with a novel microbial consortium for enhanced biomethanation. *Bioresour. Technol.* 238: 425–432.

76 Mishra, S., Singh, P.K., Dash, S., and Pattnaik, R. (2018). Microbial pretreatment of lignocellulosic biomass for enhanced biomethanation and waste management. *3 Biotech* 8 (11): 1–12.

77 Shrestha, S., Fonoll, X., Khanal, S.K., and Raskin, L. (2017). Biological strategies for enhanced hydrolysis of lignocellulosic biomass during anaerobic digestion: current status and future perspectives. *Bioresour. Technol.* 245: 1245–1257.

78 Lange, L., Pilgaard, B., Herbst, F.A. et al. (2019). Origin of fungal biomass degrading enzymes: evolution, diversity and function of enzymes of early lineage fungi. *Fungal Biol. Rev.* 33: 82–97.

79 Mäkelä, M.R., Donofrio, N., and De Vries, R.P. (2014). Plant biomass degradation by fungi. *Fungal Genet. Biol.* 72: 2–9.

80 Barrett, K., Jensen, K., Meyer, A.S. et al. (2020). Fungal secretome profile categorization of CAZymes by function and family corresponds to fungal phylogeny and taxonomy: example Aspergillus and Penicillium. *Sci. Rep.* 10 (1): 1–12.

81 Benoit, I., Culleton, H., Zhou, M. et al. (2015). Closely related fungi employ diverse enzymatic strategies to degrade plant biomass. *Biotechnol. Biofuels* 8 (1): 1–14.

82 Nagel, J.H., Wingfield, M.J., and Slippers, B. (2021). Increased abundance of secreted hydrolytic enzymes and secondary metabolite gene clusters define the genomes of latent plant pathogens in the Botryosphaeriaceae. *BMC Genomics* 22 (1): 1–24.

83 Castoldi, R., Bracht, A., de Morais, G.R. et al. (2014). Biological pretreatment of *Eucalyptus grandis* sawdust with white-rot fungi: study of degradation patterns and saccharification kinetics. *Chem. Eng. J.* 258: 240–246.

84 Wan, C. and Li, Y. (2012). Fungal pretreatment of lignocellulosic biomass. *Biotechnol. Adv.* 30 (6): 1447–1457.

85 Zanellati, A., Spina, F., Rollé, L. et al. (2020). Fungal pretreatments on non-sterile solid digestate to enhance methane yield and the sustainability of anaerobic digestion. *Sustainability.* 12 (20): 8549.

86 Ge, X., Matsumoto, T., Keith, L., and Li, Y. (2014). Fungal pretreatment of albizia chips for enhanced biogas production by solid-state anaerobic digestion. *Energy Fuel* 29 (1): 200–204.

87 Mackul'ak, T., Prousek, J., Švorc, L., and Drtil, M. (2012). Increase of biogas production from pretreated hay and leaves using wood-rotting fungi. *Chem. Pap.* 66 (7): 649–653.

88 Müller, H.W. and Trösch, W. (1986). Screening of white-rot fungi for biological pretreatment of wheat straw for biogas production. *Appl. Microbiol. Biotechnol.* 24 (2): 180–185.

89 Rouches, E., Zhou, S., Steyer, J.P., and Carrere, H. (2016). White-Rot fungi pretreatment of lignocellulosic biomass for anaerobic digestion: impact of glucose supplementation. *Process Biochem.* 51 (11): 1784–1792.

90 Bonilla, S., Choolaei, Z., Meyer, T. et al. (2018). Evaluating the effect of enzymatic pretreatment on the anaerobic digestibility of pulp and paper biosludge. *Biotechnol. Rep.* 17: 77–85.

91 Schroyen, M., Vervaeren, H., Van Hulle, S.W.H., and Raes, K. (2014). Impact of enzymatic pretreatment on corn stover degradation and biogas production. *Bioresour. Technol.* 173: 59–66.

92 Michalska, K., Bizukojć, M., and Ledakowicz, S. (2015). Pretreatment of energy crops with sodium hydroxide and cellulolytic enzymes to increase biogas production. *Biomass Bioenergy* 80: 213–221.

93 Rollini, M., Sambusiti, C., Musatti, A. et al. (2014). Comparative performance of enzymatic and combined alkaline-enzymatic pretreatments on methane production from ensiled sorghum forage. *Bioprocess Biosyst. Eng.* 37 (12): 2587–2595.

94 Avila, R., Carrero, E., Vicent, T., and Blánquez, P. (2021). Integration of enzymatic pretreatment and sludge co-digestion in biogas production from microalgae. *Waste Manag.* 124: 254–263.

95 Brandelli, A., Sala, L., and Kalil, S.J. (2015). Microbial enzymes for bioconversion of poultry waste into added-value products. *Food Res. Int.* 73: 3–12.

96 Tongco, J.V., Kim, S., Oh, B.R. et al. (2020). Enhancement of hydrolysis and biogas production of primary sludge by use of mixtures of protease and lipase. *Biotechnol. Bioprocess Eng.* 25 (1): 132–140.

97 Gopal, L.C., Govindarajan, M., Kavipriya, M.R. et al. (2021). Optimization strategies for improved biogas production by recycling of waste through response surface methodology and artificial neural network: sustainable energy perspective research. *J. King Saud Univ. Sci.* 33: 101241. https://doi.org/10.1016/j.jksus.2020.101241.

98 Lu, J. and Gao, X. (2021). Biogas: potential, challenges, and perspectives in a changing China. *Biomass Bioenergy* 150 (January): 106127. https://doi.org/10.1016/j.biombioe.2021.106127.

99 Harirchi, S., Wainaina, S., Sar, T. et al. (2022). Microbiological insights into anaerobic digestion for biogas, hydrogen or volatile fatty acids (VFAs): a review. *Bioengineered* 13 (3): 6521–6557. https://doi.org/10.1080/21655979.2022.2035986.

100 Kanda, W., Zanatta, H., Magnusson, T. et al. (2022). Policy coherence in a fragmented context: the case of biogas systems in Brazil. *Energy Res. Soc. Sci.* 102454. https://doi.org/10.1016/j.erss.2021.102454.

101 Chin, M.J., Poh, P.E., Tey, B.T. et al. (2013). Biogas from palm oil mill effluent (POME): opportunities and challenges from Malaysia's perspective. *Renew. Sust. Energ. Rev.* 26: 717–726.

102 Kumar, M., Sun, Y., Rathour, R. et al. (2020). Algae as potential feedstock for the production of biofuels and value-added products: opportunities and challenges. *Sci. Total Environ.* 716: 137116. https://doi.org/10.1016/j.scitotenv.2020.137116.

103 Surendra, K.C., Takara, D., Hashimoto, A.G., and Khanal, S.K. (2014). Biogas as a sustainable energy source for developing countries: opportunities and challenges. *Renew. Sust. Energ. Rev.* 31: 846–859. https://doi.org/10.1016/j.rser.2013.12.015.

104 Giwa, A.S., Ali, N., Ahmad, I. et al. (2020). Prospects of China's biogas: fundamentals, challenges and considerations. *Energy Rep.* 6 (189): 2973–2987.

105 Villarroel-Schneider, J., Höglund-Isaksson, L., Mainali, B. et al. (2022). Energy self-sufficiency and greenhouse gas emission reductions in Latin American dairy farms through massive implementation of biogas-based solutions. *Energy Convers. Manag.* 261: 115670. https://doi.org/10.1016/j.enconman.2022.115670.

106 Namugenyi, I., Coenen, L., and Scholderer, J. (2022). Realising the transition to bioenergy: integrating entrepreneurial business models into the biogas socio-technical system in Uganda. *J. Clean. Prod.* 333: 130135. https://doi.org/10.1016/j.jclepro.2021.130135.

107 Huang, X., Wang, S., Shi, Z. et al. (2022). Challenges and strategies for biogas production in the circular agricultural waste utilization model: a case study in rural China. *Energy* 241: 122889. https://doi.org/10.1016/j.energy.2021.122889.

108 Lauer, M. and Thrän, D. (2017). Biogas plants and surplus generation: cost driver or reducer in the future German electricity system? *Energy Policy* 109 (April): 324–336. https://doi.org/10.1016/j.enpol.2017.07.016.

109 Glivin, G., Kalaiselvan, N., Mariappan, V. et al. (2021). Conversion of biowaste to biogas: a review of current status on techno-economic challenges, policies, technologies and mitigation to environmental impacts. *Fuel* 302: 121153. https://doi.org/10.1016/j.fuel.2021.121153.

110 Alavi-Borazjani, S.A., Capela, I., and Tarelho, L.A.C. (2020). Valorization of biomass ash in biogas technology: opportunities and challenges. *Energy Rep.* 6: 472–476. https://doi.org/10.1016/j.egyr.2019.09.010.

111 Parawira, W. (2012). Enzyme research and applications in biotechnological intensification of biogas production. *Crit. Rev. Biotechnol.* 32 (2): 172–186.

112 Bedoić, R., Špehar, A., Puljko, J. et al. (2020). Opportunities and challenges: experimental and kinetic analysis of anaerobic co-digestion of food waste and rendering industry streams for biogas production. *Renew. Sust. Energ. Rev.* 130: 109951. https://doi.org/10.1016/j.rser.2020.109951.

17

Molecular Farming and Anticancer Vaccine

Current Opportunities and Openings

Yashwant Kumar Ratre[1], *Arundhati Mehta*[1], *Sapnita Shinde*[1], *Vibha Sinha*[1], *Vivek Kumar Soni*[1], *Subash Chandra Sonkar*[2], *Dhananjay Shukla*[1], *and Naveen Kumar Vishvakarma*[1]

[1] *Department of Biotechnology, Guru Ghasidas Vishwavidyalaya, Bilaspur, Chhattisgarh, India*
[2] *Multidisciplinary Research Unit, Maulana Azad Medical College and Associated Hospitals, University of Delhi, New Delhi, India*

17.1 Introduction

Cancer is one of the most devastating community health concerns of the twenty-first century. The prevalence and mortality rate is significantly increasing worldwide [1]. According to International Agency for Research on Cancer (IARC), cancer alone is responsible for more than 19.3 million new cases and 10 million deaths globally in 2020 [1]. As per an estimate, expectedly cancer cases may raise up to 28.4 million by 2040 [1]. The high rate of increasing cancer burden, aggressiveness, and drug resistance is the major obstacle to the better management of cancer. Until recently, conventional therapeutic options including chemotherapy, radiotherapy, surgery, targeted therapy, hormonal therapy, and immunotherapy are the commonly used interventions for all types of cancer. However, treatment potency differs according to clinical conditions. Moreover, the currently available therapeutic approaches are effective in improving the overall survival of patients. Existing therapies are mainly designed to disrupt the dysregulated pathways and mechanisms underlying hallmarks of cancer. In recent trends, combination therapies have gained attention among cancer researchers to synergistically provide benefits to patients via improving drug efficiency. However, conventional therapeutic approaches are less specific to particular tumors and have uninvited side effects, multi-organ toxicities, and off-target effects [2]. Therefore, more prominent and promising advanced solutions are needed.

Over the last few decades, cancer is introducing new challenges to the public health community. Thus, the revolution in cancer management strategies is in demand. Vaccination is rising as one of the most effective and significant ways to prevent and control diseases. It is very popular against healthcare-associated infections, multidrug resistance microbes, and minimizing antimicrobial use. Vaccine mitigates or controls infections in all age groups, particularly in the older population. The remarkable success in designing and

Microbial Bioreactors for Industrial Molecules, First Edition.
Edited by Sudhir P. Singh and Santosh Kumar Upadhyay.

developing immunization against communicable diseases (i.e. smallpox, polio, tetanus, measles, diphtheria, and rabies) was a boon for modern science [2,3]. This milestone success can guide future vaccine production and development for diseases of noncommunicable nature like cancer that have not conventionally been addressed by vaccination as standard therapy. However, several challenges remain to exist in vaccine manufacturing and development for malignant disorders. These concerns need to be addressed to achieve optimal utilization of the prophylactic benefits of vaccination against cancer. Consequently, it becomes imperative to highlight the significance of vaccines from classical to modern times followed by discussing exciting opportunities in vaccine biology to produce various therapeutically important molecules, proteins, and antigens using bioengineering of microbes to treat cancer in the future. Such large-scale production strategies for pharmaceutical molecules is called "molecular pharming" also known as "molecular farming." A focused discussion on research progress in the cancer vaccine, current microbial-based clinical trials, and their clinical efficiency will help the scientific audience to design a future vaccine.

17.2 Vaccines and the Possibility in Noncommunicable Diseases

The vaccine is one of the finest and greatest healthcare inventions all the time in history to provide ensured protection against various communicable diseases. A vaccine is a biological inducer designed as an inactivated or attenuated pathogen or a component of a pathogen such as protein, DNA, or RNA to stimulate immune response via adapting defensive mechanism against a given disease [4]. Vaccination is a great initiative for all countries worldwide to deliver benefits, especially to pregnant women, infants, children, older individuals, and those who have significant susceptibility, and high risk of contracting infections. To date, a significant number of causative agents have been reported against which vaccines are approved and many microbial agents are in pipeline for the development and production of vaccines [5–7]. The committee of the World Health Organization (WHO) estimates that vaccines prevent two to three million human deaths every year. The toll of lives saved can rise to six million if all children receive the recommended vaccines and specified times. Vaccination exceptionally contributed to mitigating mortality of children aged below 5 years from 93 deaths per 1000 live births in 1990 to 39 deaths per 1000 live births in 2018 [7]. These statistics indicate the power of vaccination in preventing the contract of infection, and/or developing pathological consequences afterward in a large fraction of the population.

In the past century, transmissible diseases such as tuberculosis, diphtheria, smallpox, pertussis, measles, influenza, and typhoid fever were the leading causes of death. They had been associated with high morbidity in affected individuals. According to a study conducted in the United States, since 1924 more than 40 million cases of diphtheria, 35 million cases of measles, and 103 million cases of infants and childhood diseases were protected by the vaccine [8]. Today life expectancy has increased, however, in the last few decades, the incidence and prevalence of noncommunicable diseases have dramatically increased. Such noncommunicable diseases include cancer, hypertension, diabetes, stroke, Alzheimer's, and cardiovascular disease [9]. These diseases are becoming a leading cause of death worldwide [9].

The area of vaccine biology has received much more attention since the innovation of the small-pox vaccine by British physician Edward Jenner in 1978. In the early 1980s, the development of a vaccine was tentatively introduced to fight against pathogenic microbes. In modern days, vaccines are used to improve and accelerate the defensive ability of the body to fight severe human infections and diseases. Conventionally, the vaccine is ideally administered to combat various viral diseases such as Tdap, HPV, and Meningitis. Even studies dictated that three times administration of HPV vaccine has able to protect (90%) human against HPV infection for up to five years [10]. Biomedical researchers are now exploring preventive measures through vaccination strategies to avert the menace associated with cancer and other noncommunicable diseases.

Despite the great success in vaccine biology, there are still several challenges that remain to unfold in vaccine development and administration. The most common adverse effect reported is hypersensitivity, neurological manifestations, autoimmunity, etc. A study also reported that a more number of vaccines administered are associated with the incidence of adverse effects [11]. Therefore, scientists need to validate, upgrade, and innovate key strategies to overcome the risk associated with vaccine administration followed by enhancing the safety measures to establish vaccines at the global level against current deadly noncommunicable diseases like cancer.

17.3 Vaccine Production

The invention of vaccination has revolutionized the whole world more than any other invention or discovery yet did [12]. Progressively, it has become a key to improving life expectancy, overall survival, and health outcomes. Currently, the development of highly advanced, cost-effective, rapid, and specific techniques for the production of vaccines is highly in demand. Even the current COVID-19 pandemic [13] has witnessed the importance of the vaccine for public health. The journey of vaccine manufacturing is a multistep time-consuming process. It takes around 7–10 years for a vaccine to be available for public use. Therefore, update in current vaccine development strategies and techniques are further warranted. Although, there have been several "eras of vaccine" with aided scientific knowledge and technological advancement that continually improve the vaccine journey from a trial-and-error approach to reverse vaccinology [14]. Although vaccine designing and validation strategies have undergone multifold improvisation, large-scale production warrants microbial farming utilizing bioreactor-based culture of genetically manipulated microbes for non-whole-cell vaccines. Nevertheless, microbes play a vital role in the production of many medically important molecules including antigenic components as vaccine candidates. Various etiologic agents have been used as a vector for the cloning, expression, and purification of antigenic vaccines. The level of success may differ according to the suitability of chosen expression platform. The current advancement in microbiology, molecular biology, and immunology allows the expression of antigenic peptides in both eukaryotes (mammalian cells, yeast, and cells of insects, plants, and animals) and prokaryotes (*Escherichia coli*, *Bacillus subtilis*). As the molecular farming approach utilizes the production of molecules in plants or microbes after the insertion of gene encoding peptides of interest [15, 16]. The microbial hosts have several merits as farmlands are required for growing plant hosts. Biotic and abiotic stress-induced damages are frequent due to absolute control of growing conditions for genetically modified plants in molecular farming [17]. However, in a bioreactor, microbes carrying genes of interest grow in optimally regulated conditions. The bacterial-based expression system is one of the most common approaches used for the efficient production of vaccines. Although microbes and microbial bioreactors are apt for large production of vaccines, mammalian or insect cell culture has been preferred where post-translational modifications (e.g. glycosylation) are necessary [18, 19]. Moreover, there are various challenges associated with vaccine production, which need to be addressed including process development, production facilities, equipment varieties, time length, product portfolio management, and life cycle management [20]. Therefore, currently, strategies are being optimized for emphasizing the significance of a highly robust, rapid, and stable production process to ensure the efficiency and long life cycle of a vaccine. Over the decades, production can be done in the traditional bioreactor (stainless steel fermenters), single-use system, or mixed approach as per necessities and the production scale [21]. Currently, from the COVID-19 pandemic, the use of the next-generation vaccine platform is prospering. Therefore, establishing novel technical approaches, which can shorten the time cycle and elicit rapid responses against given diseases, is expected to improve the producibility and efficacy.

17.3.1 Cancer Vaccine

A vaccine is a foremost weapon to prevent disease, which is even more efficient and promising than curative efforts through treatment and the use of therapeutic drugs [22]. Historically, a vaccine is specially introduced to eradicate infectious diseases. However, the global burden of NCDs is quite large. In the present scenario, the burden of NCDs has significantly increased over communicable diseases. Surprisingly, the death ratio is also observed to increase in the case of NCDs such as cancer and cardiovascular disease as compared to CDs [23]. In the recent past, cancer has emerged as one of the leading causes of death globally. In the past few decades, scientists have observed that conventional therapeutic interventions are not quite effective to eradicate cancer from the root. Moreover, it also has very adverse events during and after treatment. Likewise, tumor recurrence and multidrug resistance (MDR) circumvent a major obstacle to interrupting the treatment strategies in cancer. Therefore, new therapies are warranted to be introduced to enormously increase the survival outcome of cancer patients by providing the best remedial solutions.

Over the last few years, vaccine biology has earned much more attention for developing cancer vaccines. Cancer vaccines are specially designed as an active immunotherapeutic regimen, particularly in an established disease state in which cancer expresses all functional phenotypes or hallmarks [24]. The principles of cancer vaccine formulation have been ideally structured to trigger or increase immune fitness via adopting various strategies such as immune checkpoint inhibitors (ICIs) and engineered T-cell-based therapies against cancer [25–27]. The journey of therapeutic cancer vaccine development is still challenging and tough trying. However, the clinical and preclinical cancer researchers along with whole biomedical researchers are excited after the great success of vaccines as therapeutics against hepatitis B virus (HBV) and HPV to avert liver and cervical cancer, respectively [28, 29]. For the first time, Hoover et al. developed a tumor cell/lysate-based cancer vaccine to treat colorectal cancer [30]. Over time, researchers developed cancer vaccines by using various sources of tumor antigens such as purified or synthesized tumor cell surface molecules like peptides, proteins, tumor cells, or their lysates of allogeneic or autologous tumor cells. In the early 1990s, melanoma-associated antigen 1 (MAA1) was discovered as a tumor antigen, which opened new opportunities for the scientific community to use tumor antigen in cancer vaccines [31]. To date, only two therapeutic vaccines have been clinically approved, namely sipuleucel-T, a first dendritic-cell (DC)-based vaccine to treat prostate cancer, and Bacillus Calmette–Guerin (BCG) for treating early-stage bladder cancer [32]. These successes in vaccines against cancer provide strength to the idea of tumor-cell-based cancer vaccines for future use [32]. Tumor antigens include tumor-associated antigens (TAAs) and tumor-specific antigens (TSAs). These antigens play an essential role in tumorigenesis and cancer progression. These are molecule of interest and key for the development of vaccines against neoplastic disorders. These vaccines augment the antitumor immune response in the host. Conventionally, the vaccine could stimulate both cell- and humoral-mediated immunity against tumor growth [33]. Today, a number of cancer vaccines are in the pre-clinical and clinical phases [34]. Antigen-specific vaccination has a focus of attention due to its ability to modulate not only the course of pathogenic acute and chronic illness but also graft rejection, autoimmunity, and cancer [34–36]. The period from 1990 to 2010 was proven very effective to discover and develop more TAAs to use in combination to achieve the best possible immunogenicity and clinical outcomes [37]. The first clinical trial either tested was a peptide-based vaccine and vaccines formulated from tumor cells/lysates containing various TAAs including the breast cancer antigen human epidermal growth factor receptor 2 (HER2), tumor antigen mucin 1 (MUC1) [38], and melanoma-associated antigen 3 (MAGEA3), etc. [39]. The discovery of antigen-presenting cells (APCs) including dendritic cells (DCs) has opened an existing window for the delivery of TAAs. After that more than hundreds of DC-based vaccines were designed and produced and are currently under clinical trials against different malignancies. However, several biological models were also used for the high-yield production and development of vaccines such as bacterial and viral vectors, virus-like particles, and nucleic-acid (DNA, RNA)-based vaccines [40–43].

Unfortunately, due to poor immunogenicity of tumor antigen and undesirable safety, very few cancer vaccines have been approved in the clinical trial. Thus, understanding the link between the structure and function of cancer vaccines is crucial to increase their opportunities to trigger the immune system for prolonging survival and quality of life in cancer patients. In addition, improvements in vaccine delivery techniques including proactive adjuvant and novel antigen expression systems like microbes have profoundly upgraded and accelerated antigen-based immunity in cancer patients.

17.4 Types of Cancer Vaccine

The fundamental aim of vaccination is to deliver the best and most affordable remedial to overcome the disease burden. A vaccine provokes the immune system by eliciting possible immune responses to destroy foreign particles like antigens via producing defensive molecules including antibodies, cytotoxic cells, and memory cells. Thus, identifying and discovering complex mechanisms adapted by the cancer cells to modulate immune system might be a gold standard. This is approach to revolutionize the invention of key tools including competent immune cells (both lymphocyte and leukocytes), peptides, and proteins including antibodies, nucleic acids, and other molecular moieties for immunotherapy against malignancies. In the coming future, a cancer vaccine might be a choice of therapy to prevent malignant disorders. Presently, various types of cancer vaccines has been emerging as a fruitful means to fight cancer. Few common types of cancer vaccines are recombinant live vector vaccines (viral/bacterial), nucleic acid (DNA, RNA) vaccines, protein/peptide vaccines, viral-like particle (VLP) vaccines, whole-cell vaccines (DC or tumor immune cell-based), edible vaccines, and combined approaches (e.g. prime-boost vaccination) [44].

A cell-based vaccine is one of the classical approaches evaluated using a tumor antigen-based method. Different kinds of biological components such as whole cells/cell fragments are used as a key source of tumor antigen (TA) to elicit an immune response. DC vaccine is a form of the cell-based vaccine. Personalized neoantigens cancer vaccine relies on DCs and has resulted in very effective anti-tumor efficiency in clinical settings. This method has been implemented for many cancers including colon [45], prostate [46], lung [47], melanoma [48], and renal cell carcinoma [49]. In many varieties of malignancies, heterogenic populations of tumor cells are bioengineered to gain immune functions including production immuno-stimulants such as interleukins and colony-stimulating factors. GVAX is a cancer cell-based vaccine, which is engineered to produce GM-CSF [50]. These types of strategies are used after radiation therapy to stop the uncontrollable growth of cancer cells [51]. However, obtaining a high yield of cells is sometimes difficult, limiting their further application in cancer vaccine development and production [52, 53]. Protein/peptide-based therapeutic cancer vaccines were mainly produced by incorporating 20–30 amino acid peptides from specific TA coding sequences to boost immunity. This boosted immunity is against key antigenic determinants identified to be expressed on malignant cells and are used in the preparation of anticancer vaccine. These artificial antigenic peptides are administered and are taken up by APCs to complex with human leukocyte antigen (HLA) molecules on their cell surface. HLA-antigenic peptide complexes are then recognized by T cells to induce a cancer-specific immune response. Peptide-based cancer vaccine could offer a range of benefits such as cost-effectiveness, convenient manufacturing, and production, low risk of carcinogenic potential, reduced contamination risk, and high chemical stability. After many decades of hard work, scientists completed the sequencing of the whole human genome. The human genome project has opened diverse windows to understand the genetic material and its associated opportunities broadly. Nucleic acid (DNA, RNA) vaccine is emerging as a promising platform for the development of genetic vaccines because they can induce MHC I-mediated $CD8^{+}T$ immune responses against multiple epitopes to trigger humoral and cell-mediated immunity. Several preclinical studies reveal the importance and future of DNA vaccines [53]. For instance, currently, a DNA-based vaccine VGX3100 is in phase 3 clinical trials [54]. Unlike DNA vaccine, RNA vaccine is not integrated into the genome, thereby preventing malignancies. Collectively, findings indicate that the nucleic acid vaccine might be a suitable strategy for the development of a personalized neoantigen cancer vaccine.

17.5 Microbial Production of Anticancer Vaccine: Challenges and Opportunities

The trend of using the microbial platform as vehicles to deliver recombinant antigens has gained much more attention for the mass production and development of a vaccine for various NCDs including cancer. Over the last two decades, the evolution of tools of genetic manipulation has increased the opportunity to produce therapeutic molecules by incorporating the concept of microbiology, immunology, and molecular biology. The implementation of new strategies has enabled the construction of recombinant microorganisms with the potential to express heterogeneous proteins in different components of the cell to enhance their immunogenic ability for the production of vaccines against pathogenic bacteria, viruses, parasites, and other deadly diseases like cancer (Figure 17.1).

Currently, the development of an efficient, affordable, and reproducible microbial system is needed for the mass production of immuno-protective molecules such as monoclonal antibodies (mAbs), TAs, and DCs to defeat cancer cells. Additionally, safety concerns are one of the biggest hurdles in achieving the regulatory standard and quality control in vaccine formulation. Recently, microbe-based cancer immunotherapy has developed as an effective approach for accelerating immune functioning. To date, various classes of microbes inclusding bacteria [55–57], yeast and fungus [58], and oncolytic viruses [59] have been implemented to sensitize adaptive and innate immunity to deliver the best antitumor immune response. For instance, a genetically engineered attenuated *Salmonella typhimurium* can be used to induce infiltration of immune cells and pro-inflammatory cytokine production in the tumor microenvironment (TME) to augment immunity against cancer [60]. Presently, many yeast and bacteria-based vaccines have been clinically tested and are in the trial phase (Tables 17.1 and 17.2) [61].

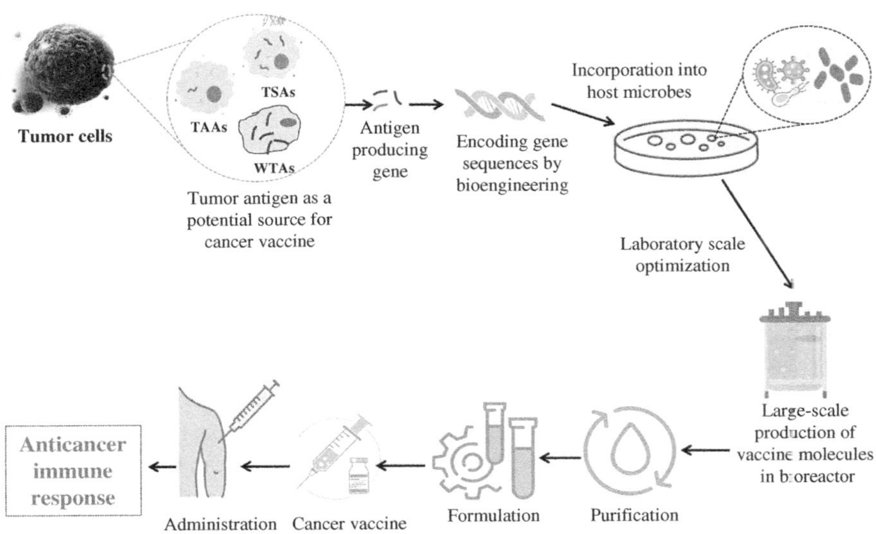

Figure 17.1 Illustration of bioengineering of tumor antigen molecules for anticancer vaccine development. *Source:* TILT Biotherapeutics LLC.

Table 17.1 Yeast-based pre-clinical and clinical studies for cancer vaccine development.

Microbial species	Tumor antigen	Strategies	Cancer	Status	Reference
Saccharomyces cerevisiae	Brachyury (GI-6301)	Whole recombinant yeast	Advanced malignant solid tumors	Phase-I	[62]
	Brachyury (GI-6301)		Epithelial mesenchyma		[63]
	CEA		Carcinoma		[64]
	Human MART-1 (hMART-IT) derived granulocyte		Melanoma	Pre-clinical	[65]
	KRAS		Lung adenocarcinoma	Phase-II	[66]
	BCR-ABLT^{315I}		Leukemias	Pre-clinical	[67]
	Cancer testis antigen NY-ESO-1	Yeast surface display	Melanoma	Phase-I	[68]
	Derived GM-CSF	Purified protein	Melanoma	Phase-III	[69]
	Single-chain chimeric peptide composed of hCGβ &oLHα	Purified protein	—	Pre-clinical	[70]
Pichia pastoris	HPV16 L1 antigen	Whole recombinant yeast	Papilloma virus associated cancer		[71]

17.5.1 Yeast-Based Cancer Vaccine (YBCV)

Yeast or unicellular fungus occupied an honored place in the field of biotechnology, commonly used in bioreactor and particularly play a translational role in food industries. Yeasts emerge as an ideal choice for the routine expression of therapeutically important biomolecules such as protein. Despite its nonpathogenic nature (*Saccharomyces cerevisiae, Pichia pastoris*), yeast carries some key salient features like efficient heterologous gene expression ability. This makes them suitable for the expression of various heterogeneous proteins for clinical as well as veterinary purposes [72]. The component of the yeast cell wall including β-glucan and chitin [73] does not exist in the mammalian system but possess a strong signal to stimulate a multiepitope immune response [74]. Recently, yeast-derived β-glucan was reported to activate DCs and macrophages leading to the activation of T cells and enhanced antitumor immune response [75]. To explore yeast as an expression system, it is very crucial to know that yeast-based vaccine approaches can trigger antitumor immune protective molecules such as CD81 CTLs cells to recognize and kill tumor cells. The current efforts should include constructing yeasts capable of expressing an adequate and defensive level of tumor-specific or TAAs. However, many studies have incorporated these strategies to enhance immune response which has been nicely reviewed earlier [76].

Table 17.2 Recent and ongoing clinical studies using bacteria for cancer vaccine development.

Trail ID	Status	Microbial agent	Cancer type	Mode of administration	Participant	Phase
NCT02302170[a]	Completed	*Helicobacter pylori* vaccine	*H. pylori*-associated cancer	Oral	4464	III
NCT01838200[a]	Terminated	Bacillus Calmette-Guerin	Metastatic Melanoma	Subcutaneous	5	I
NCT02371447[a]	Active	Recombinant Bacillus Calmette–Guérin (VPM1002BC)	Bladder cancer	Intravenous	39	I/II
NCT02243371[a]	Completed	*Listeria monocytogenes*-expressing mesothelin (CRS-207)	Previously treated metastatic adenocarcinoma of the pancreas	Intravenous	93	II
NCT04025307[a]	Completed	*Bifidobacterium longum* expressing IL-12 (bacTRL-IL-12)	Advanced and treatment-refractory solid tumors	Intravenous	38	I
NCT03762291[a]	Recruiting	Salmonella CVD908ssb strain producing Survivin (TXSVN)	Multiple myeloma	Oral	18	I
NCT03847519[a]	Recruiting	*Listeria monocytogenes* engineered to express 22 tumor antigens commonly found in non–small cell lung cancer (NSCLC; i.e. 11 hotspot mutations and 11 tumor-associated antigens (ADXS-503 or A503)	Lung cancer, non-small cell Metastatic squamous cell carcinoma Metastatic non–squamous cell carcinoma	Intravenous	74	I/II
NCT02002182[a]	Active	*B. longum* expressing cytosine deaminase (APS001F)	Head and neck cancer squamous cell carcinoma of the head and neck HPV positive oropharyngeal squamous cell carcinoma	Intravenous	15	II
NCT02325557[a]	Unknown	*Listeria monocytogenes* secreting an antigen-adjuvant fusion protein tLLO-HPV-16 E7 (ADXS11-001)	Prostate cancer	Intravenous	51	I/II
NCT03750071[a]	Recruiting	Attenuated *Salmonella typhimurium* encoding murine vascular endothelial growth factor receptor 2 (VEGFR-2) (VXM01)	Recurrent and progressive glioblastoma	Oral	30	I/II

[a] https://clinicaltrials.gov/ For more details, readers can visit mentioned URL link.

Heery and colleagues demonstrated that *S. cerevisiae* expressing brachyury (embryonic transcription factor) was able to activate human T-cells [62]. In one more finding, *S. cerevisiae*-derived micro-particles conjugated with ova albumin are recognized by DCs leading to inducing an immune response against malignant tumors [77]. Recently, the yeast-CEA (GI-6207) vaccine is under Phase I clinical trial. Yeast-CEA is an engineered moiety generated using heat-killed *S. cerevisiae* to express recombinant carcinoembryonic antigen (CEA) protein [64]. This vaccine is effective against metastatic CEA-expressing carcinoma in adults [64]. Jiang et al. demonstrated that a budding yeast-derived rabbit anti-hCGβ-oHLα IgG was able to *in vitro* inhibit the growth of human chorionic gonadotropin (hCG) expressing colorectal cancer cell lines and also able to neutralize its bioactivity [66]. In another finding, yeast cells were used to express and purify melanoma protein for the use possibly as a prophylactic vaccine to provide defense against melanoma cancer, despite enough amount of knowledge about epitopes recognized by MHC class molecules. Further, this expressed protein is also able to protect mice from developing tumors [78]. Currently, two antiviral vaccines, HBV and HPV, were developed using *S. cerevisiae* as a model system. HPV is potentially associated with cervical cancer [79, 80]. Another species of yeast called *P. pastoris* can hold great value for the production of vaccine antigens and immunotherapeutics. *Pichia pastoris* was introduced in the 1960s as a food additive and later its large-scale production was achieved via a fermentation approach [81]. Now, *P. pastoris* has been established as a tightly regulated expression system and has been studied to cultivate recombinant antigens for human vaccines [82, 83]. This accumulated evidence established the yeast as a strong microbial model for the expression, purification, and production of tumor antigens. It can be possible that extending the application of recombinant yeast may be transformed the era of vaccine development with high specificity against the tumor.

17.5.2 Bacteria-Based Cancer Vaccine (BBCV)

Historically, bacteria are a well-practiced model for the cultivation of industrial and recombinant therapeutic molecules. Of note, the strategy of using bacteria as vehicles to deliver recombinant antigens has continually evolved for the development of new recombinant vaccines and immunotherapeutic molecules. The motility is a very critical feature of bacteria that allows them to move away from the vasculature and deeply penetrate hypoxic regions of the tumor [84] and proliferate within tumor cells [85]. However, as a delivery and expression system bacteria can offer a spectrum of advantages such as they carry well-identified virulence mutations, having the ability to regulate *in vivo* expression of antigens and their number and amount, they can provide multiple delivery routes and also can regulate both innate and adaptive immune response. Additionally, the growing number of knowledge helped in unfolding the key immunological events of bacterial physiology linked with host–pathogen interactions to use attenuated bacteria as conventional vaccine vectors [86]. All these features play various crucial roles in designing and developing any vaccines using bacteria.

Almost a century back, William Coley transformed medical science by establishing bacterial products as immunotherapy agents; later on, he prepared a vaccine using live and attenuated bacteria (*Streptococcus pyrogenes*, *Seretia marcescenes*). This preparation, known as Cooley's toxin, has served as a gold standard alternative to defeat carcinoma, lymphoma, sarcoma, melanoma, and myeloma [87–89]. Previously, various bacteria have been used as a vector including *E. coli*, *Listeria monocytogenes*, *Yersinia*, *Salmonella*, and *Shigella* [90]. Many findings support the use of bacterial vectors specifically *L. monocytogenes*, and *P. aeruginosa* as an expression system for the development of cancer vaccines [91, 92]. Nowadays, bacterial cells have been genetically modified to carry and express TAAs, recombinant protein, deliver genes, or transport anticancer molecules [93]. Thus, cultivating bioengineered bacteria as a high-throughput system can boost the expression, production, and purification of antitumor molecules used for the generation of cancer vaccines. In addition, advancements in large-scale fermentation tools and highly efficient bioengineering techniques provide a convenient platform to produce cost-effective and affordable vaccines using bacterial cells.

Bacterial vectors can be used to supply and propagate molecules that overcome or kill tumor cells by protecting them from self-antigen and heterologous antigens. In one study, it was noted that an attenuated *S. typhimurium* vector has been used to elicit an immune response in tumor-bearing mice and some cases of humans against the malignant tumor. These outcomes highlight this vector as a great model for vaccine development [94]. Although bacterial type, species, mode of antigen delivery to APCs, and route of administration remain to understand fully to match the safety concerns of vectors against patients and the environment [42].

To date, various bacterial strains have been genetically modified for adding some cancer-protective properties to enhance immune fitness. Some bacteria-based vaccines are currently in the clinical trial phases [56]. A genetically manipulated bacteria *S. typhimurium* has been developed by editing the cyp/crp gene (which encodes a protein associated with cyclic AMP regulation). This is used to express interleukin-2 (IL-2) to possibly treat liver cancer in the preclinical study [95, 96]. In another study, transformed *S. typhimurium* has been used as a vehicle to orally administer immunotherapeutic molecules expressed in eukaryotic vectors (IL-2, mIL-12, human interleukin-12 [hIL-12], human granulocyte/macrophage colony-stimulating factor [hGM-CSF], mGM-CSF, and green fluorescent protein [GFP]). These transformants increased the number of cytotoxic T cells and facilitated the prolonging survival in mice with lung tumors [86]. In addition, *Bifidobacterium adolescentis* has also been considered a putative gene delivery vector after its success at the preclinical level. This *B. adolescentis* strain has been found effective to inhibit the growth of various cancer cells (liver cancer, breast cancer) and is also able to induce an immune response [98]. Most recently, Cheng et al. developed a bacteria-derived genetically modified outer membrane vesicle (OMV)-based vaccine platform via plug and display technique for tumor antigen display to elicit antitumor immunity using a preclinical tumor model [99].

17.6 Conclusion

In the current century, cancer or malignant disease is growing as one of the leading causes of death [100]. With time, the toll of concerns associated with standard and conventional cancer treatment is rising due to drug resistance, tumor dormancy, and other relevant consequences discovered with the advancement of science. Vaccination, an old approach devised against communicable diseases, is being implemented against malignant disorders. Vaccines against cancer were used first as a therapeutic setup; however, preventive effects will be of immense advantage to prevent mortality and morbidity. A variety of vaccines against different malignant disorders are tested or currently under pre-clinical or clinical evaluation. One of the concerns about the requirement of a large number of cancer vaccines (either peptides of cancer origin, their nucleic acid, etc.). Conventional approaches for commercial production of cancer vaccines or their components utilize the cell-culture-based method. Nowadays, the advancement of genetic manipulation can easily assist the production of vaccines against cancer through microbial bioreactors. The insertion of a gene-encoding antigenic peptides in the genome of a microbial host can be exploited for commercial production of vaccines even against cancers after laboratory-level optimization. Vaccines are the cornerstone for the management of pathogenic diseases and provide the surest means of defusing pandemics and epidemics.

The delivery of antitumor antigen has special attention for successful vaccine delivery. Of note, various forms of vaccine delivery have been reported including peptide, nucleic acid (mRNA/DNA) vaccines, or loaded on DCs *ex vivo*. Microbial-based culture platforms can offer large-scale production and purification of antigens for loading on DCs to promote specific immune responses and boost T-cells. These approaches have increased the opportunity for personalized vaccine development for future use. However, the field of cancer vaccine is not fully matured and needs further development to enhance the quality of epitopes and neoepitopes to maximize their efficiency. However, the microbial system for expression of tumor antigen candidates with suitability for vaccines will be available for any development. The previously optimized process for upstream and downstream processing of products in microbial bioreactors offers new opportunities. In such a way, molecular farming is not only serving as a platform for the production of pharmaceutically active molecules but also aids in cancer vaccine production. Thus, incorporating the diversified mechanisms of microbes for cancer vaccine development and production can transform the era of cancer therapy. Additionally, the administration of the traditional approach with the vaccine in combination might be also a good idea for the management of cancer.

References

1 Sung, H., Ferlay, J., Siegel, R.L. et al. (2021). Global cancer statistics 2020: GLOBOCAN estimates of incidence and mortality worldwide for 36 cancers in 185 countries. *CA: a Cancer Journal for Clinicians* 71 (3): 209–249. https://doi.org/10.3322/caac.21660.

2 Strassburg, M.A. (1982). The global eradication of smallpox. *American Journal of Infection Control* 10 (2): 53–59. https://doi.org/10.1016/0196-6553(82)90003-7.

3 Robinson, H.L. and Amara, R.R. (2005). T cell vaccines for microbial infections. *Nature Medicine* 11 (4 Suppl): S25–S32. https://doi.org/10.1038/nm1212.

4 Plotkin, S.A., Orenstein, W., and Offit, P.A. (2013). *Vaccines*, 6e, 1141–1196. Philadelphia, PA: Saunders.

5 World Health Organization (2012). *Global Immunization Data 2011*. Geneva: WHO. www.who.int/hpvcentre/Global_Immunization_Data.pdf (accessed 30 May 2012).

6 World Health Organization (2020). *Child Mortality and Causes of Death*. WHO.

7 Fenner, F., Henderson, D.A., Arita, I. et al. (1988). *Smallpox and Its Eradication*, 369–371. Geneva: World Health Organization.

8 van Panhuis, W.G., Grefenstette, J., Jung, S.Y. et al. (2013). Contagious diseases in the United States from 1888 to the present. *The New England Journal of Medicine* 369 (22): 2152–2158. https://doi.org/10.1056/NEJMms1215400.

9 Murphy, S.L., Xu, J., and Kochanek, K.D. (2013). Deaths: final data for 2010. National vital statistics reports: from the Centers for Disease Control and Prevention, National Center for Health Statistics. *National Vital Statistics System* 61 (4): 1–117.

10 Cutts, F.T., Franceschi, S., Goldie, S. et al. (2007). Human papillomavirus and HPV vaccines: a review. *Bulletin of the World Health Organization* 85 (9): 719–726. https://doi.org/10.2471/blt.06.038414.

11 Greydanus, D.E., Leonov, A., Elisa, A., and Azmeh, R. (2019). Should rare immunologic, neurologic, and other adverse events be indications to withhold vaccination? *Translational Pediatrics* 8 (5): 419–427. https://doi.org/10.21037/tp.2019.06.01.

12 Ulmer, J.B., Valley, U., and Rappuoli, R. (2006). Vaccine manufacturing: challenges and solutions. *Nature Biotechnology* 24 (11): 1377–1383. https://doi.org/10.1038/nbt1261.

13 Ratre, Y.K., Kahar, N., Bhaskar, L. et al. (2021). Molecular mechanism, diagnosis, and potential treatment for novel coronavirus (COVID-19): a current literature review and perspective. *3 Biotech* 11 (2): 94. https://doi.org/10.1007/s13205-021-02657-3.

14 Delany, I., Rappuoli, R., and De Gregorio, E. (2014). Vaccines for the 21st century. *EMBO Molecular Medicine* 6 (6): 708–720. https://doi.org/10.1002/emmm.201403876.

15 Shintani, T., Upadhyay, S.K., Singh, S.P. (2021). An introduction to microbial biodiversity and bioprospection. In: *Bioprospecting of Microorganism-Based Industrial Molecules* (ed. S.P. Singh and S.K. Upadhyay). John Wiley & Sons Ltd. https://doi.org/10.1002/9781119717317.ch1.

16 Upadhyay, S.K. and Singh, S.P. (eds.) (2023). *Plants as Bioreactors for Industrial Molecules*. John Wiley & Sons Ltd. doi:10.1002/9781119875116.

17 Upadhyay, S.K. (ed.) (2021). *Genome Engineering for Crop Improvement*. John Wiley & Sons Ltd. doi:10.1002/9781119672425.

18 Clark, T.G. and Cassidy-Hanley, D. (2005). Recombinant subunit vaccines: potentials and constraints. *Developments in Biologicals* 121: 153–163.

19 Singh, S.P. and Upadhyay, S.K. (eds.) (2021). *Bioprospecting of Microorganism-Based Industrial Molecules*. John Wiley & Sons Ltd. doi:10.1002/9781119717317.

20 Plotkin, S., Robinson, J.M., Cunningham, G. et al. (2017). The complexity and cost of vaccine manufacturing – an overview. *Vaccine* 35 (33): 4064–4071. https://doi.org/10.1016/j.vaccine.2017.06.003.

21 Healthcare, G.E. (2015). Process economy and production capacity using single-use versus stainless steel fermentation equipment. *Brochure* 29143348 A: 1–12.

22 Wilson, C.B. and Marcuse, E.K. (2001). Vaccine safety – vaccine benefits: science and the public's perception. *Nature reviews. Immunology* 1 (2): 160–165. https://doi.org/10.1038/35100585.

23 Mohan, P., Mohan, S.B., and Dutta, M. (2019). Communicable or noncommunicable diseases? Building strong primary health care systems to address double burden of disease in India. *Journal of Family Medicine and Primary Care* 8 (2): 326–329. https://doi.org/10.4103/jfmpc.jfmpc_67_19.

24 Hanahan, D. and Weinberg, R.A. (2011). Hallmarks of cancer: the next generation. *Cell* 144 (5): 646–674. https://doi.org/10.1016/j.cell.2011.02.013.

25 Dalgleish, A.G. and Whelan, M.A. (2006). Cancer vaccines as a therapeutic modality: the long trek. *Cancer Immunology, Immunotherapy* 55 (8): 1025–1032. https://doi.org/10.1007/s00262-006-0128-8.

26 Yang, Y. (2015). Cancer immunotherapy: harnessing the immune system to battle cancer. *The Journal of Clinical Investigation* 125 (9): 3335–3337. https://doi.org/10.1172/JCI83871.

27 Melero, I., Gaudernack, G., Gerritsen, W. et al. (2014). Therapeutic vaccines for cancer: an overview of clinical trials. *Nature Reviews. Clinical Oncology* 11 (9): 509–524. https://doi.org/10.1038/nrclinonc.2014.111.

28 Kim, B.K., Han, K.H., and Ahn, S.H. (2011). Prevention of hepatocellular carcinoma in patients with chronic hepatitis B virus infection. *Oncology* 81 (Suppl 1): 41–49. https://doi.org/10.1159/000333258.

29 Roden, R. and Stern, P.L. (2018). Opportunities and challenges for human papillomavirus vaccination in cancer. *Nature Reviews Cancer* 18 (4): 240–254. https://doi.org/10.1038/nrc.2018.13.

30 Hoover, H.C. Jr., Surdyke, M.G., Dangel, R.B. et al. (1985). Prospectively randomized trial of adjuvant active-specific immunotherapy for human colorectal cancer. *Cancer* 55 (6): 1236–1243. https://doi.org/10.1002/1097-0142(19850315)55:6<1236::aid-cncr2820550616>3.0.co;2-#.

31 van der Bruggen, P., Traversari, C., Chomez, P. et al. (1991). A gene encoding an antigen recognized by cytolytic T lymphocytes on a human melanoma. *Science (New York, N.Y.)* 254 (5038): 1643–1647. https://doi.org/10.1126/science.1840703.

32 Gardner, T.A., Elzey, B.D., and Hahn, N.M. (2012). Sipuleucel-T (Provenge) autologous vaccine approved for treatment of men with asymptomatic or minimally symptomatic castrate-resistant metastatic prostate cancer. *Human Vaccines & Immunotherapeutics* 8 (4): 534–539. https://doi.org/10.4161/hv.19795.

33 Miao, L., Zhang, Y., and Huang, L. (2021). mRNA vaccine for cancer immunotherapy. *Molecular Cancer* 20 (1): 41. https://doi.org/10.1186/s12943-021-01335-5.

34 Saxena, M., van der Burg, S.H., Melief, C., and Bhardwaj, N. (2021). Therapeutic cancer vaccines. *Nature Reviews Cancer* 21 (6): 360–378. https://doi.org/10.1038/s41568-021-00346-0.

35 Parks, R.J. and Gussoni, E. (2018). Building immune tolerance through DNA vaccination. *Proceedings of the National Academy of Sciences of the United States of America* 115 (39): 9652–9654. https://doi.org/10.1073/pnas.1813461115.

36 Samy, K.P. and Brennan, T.V. (2018). Dendritic cell therapy in transplantation, phenotype governs destination and function. *Transplantation* 102 (10): 1593–1594. https://doi.org/10.1097/TP.0000000000002238.

37 Finn, O.J. (2017). Human tumor antigens yesterday, today, and tomorrow. *Cancer Immunology Research* 5 (5): 347–354. https://doi.org/10.1158/2326-6066.CIR-17-0112.

38 Goydos, J.S., Elder, E., Whiteside, T.L. et al. (1996). A phase I trial of a synthetic mucin peptide vaccine. Induction of specific immune reactivity in patients with adenocarcinoma. *The Journal of Surgical Research* 63 (1): 298–304. https://doi.org/10.1006/jsre.1996.0264.

39 Marchand, M., van Baren, N., Weynants, P. et al. (1999). Tumor regressions observed in patients with metastatic melanoma treated with an antigenic peptide encoded by gene

MAGE-3 and presented by HLA-A1. *International Journal of Cancer* 80 (2): 219–230. https://doi.org/10.1002/(sici)1097-0215(19990118)80:2<219::aid-ijc10>3.0.co;2-s.

40 Palucka, K. and Banchereau, J. (2013). Dendritic-cell-based therapeutic cancer vaccines. *Immunity* 39 (1): 38–48. https://doi.org/10.1016/j.immuni.2013.07.004.

41 Larocca, C. and Schlom, J. (2011). Viral vector-based therapeutic cancer vaccines. *Cancer Journal (Sudbury, Mass.)* 17 (5): 359–371. https://doi.org/10.1097/PPO.0b013e3182325e63.

42 Toussaint, B., Chauchet, X., Wang, Y. et al. (2013). Live-attenuated bacteria as a cancer vaccine vector. *Expert Review of Vaccines* 12 (10): 1139–1154. https://doi.org/10.1586/14760584.2013.836914.

43 Pejawar-Gaddy, S., Rajawat, Y., Hilioti, Z. et al. (2010). Generation of a tumor vaccine candidate based on conjugation of a MUC1 peptide to polyionic papillomavirus virus-like particles. *Cancer Immunology, Immunotherapy* 59 (11): 1685–1696. https://doi.org/10.1007/s00262-010-0895-0.

44 Park, J.W., Lagniton, P., Liu, Y., and Xu, R.H. (2021). mRNA vaccines for COVID-19: what, why and how. *International Journal of Biological Sciences* 17 (6): 1446–1460. https://doi.org/10.7150/ijbs.59233.

45 de Weger, V.A., Turksma, A.W., Voorham, Q.J. et al. (2012). Clinical effects of adjuvant active specific immunotherapy differ between patients with microsatellite-stable and microsatellite-instable colon cancer. *Clinical Cancer Research: An Official Journal of the American Association for Cancer Research* 18 (3): 882–889. https://doi.org/10.1158/1078-0432.CCR-11-1716.

46 Tani, K., Azuma, M., Nakazaki, Y. et al. (2004). Phase I study of autologous tumor vaccines transduced with the GM-CSF gene in four patients with stage IV renal cell cancer in Japan: clinical and immunological findings. *Molecular Therapy* 10 (4): 799–816. https://doi.org/10.1016/j.ymthe.2004.07.001.

47 Rüttinger, D., van den Engel, N.K., Winter, H. et al. (2007). Adjuvant therapeutic vaccination in patients with non-small cell lung cancer made lymphopenic and reconstituted with autologous PBMC: first clinical experience and evidence of an immune response. *Journal of Translational Medicine* 5: 43. https://doi.org/10.1186/1479-5876-5-43.

48 Méndez, R., Ruiz-Cabello, F., Rodríguez, T. et al. (2007). Identification of different tumor escape mechanisms in several metastases from a melanoma patient undergoing immunotherapy. *Cancer Immunology, Immunotherapy* 56 (1): 88–94. https://doi.org/10.1007/s00262-006-0166-2.

49 Fishman, M., Hunter, T.B., Soliman, H. et al. (2008). Phase II trial of B7-1 (CD-86) transduced, cultured autologous tumor cell vaccine plus subcutaneous interleukin-2 for treatment of stage IV renal cell carcinoma. *Journal of Immunotherapy (Hagerstown, Md.: 1997)* 31 (1): 72–80. https://doi.org/10.1097/CJI.0b013e31815ba792.

50 Salgia, R., Lynch, T., Skarin, A. et al. (2003). Vaccination with irradiated autologous tumor cells engineered to secrete granulocyte-macrophage colony-stimulating factor augments antitumor immunity in some patients with metastatic non-small-cell lung carcinoma. *Journal of Clinical Oncology: Official Journal of the American Society of Clinical Oncology* 21 (4): 624–630. https://doi.org/10.1200/JCO.2003.03.091.

51 Nemunaitis, J., Jahan, T., Ross, H. et al. (2006). Phase 1/2 trial of autologous tumor mixed with an allogeneic GVAX vaccine in advanced-stage non-small-cell lung cancer. *Cancer Gene Therapy* 13 (6): 555–562. https://doi.org/10.1038/sj.cgt.7700922.

52 Berger, M., Kreutz, F.T., Horst, J.L. et al. (2007). Phase I study with an autologous tumor cell vaccine for locally advanced or metastatic prostate cancer. *The Journal of Pharmacy and Pharmaceutical Sciences* 10 (2): 144–152.

53 Ferraro, B., Morrow, M.P., Hutnick, N.A. et al. (2011). Clinical applications of DNA vaccines: current progress. *Clinical Infectious Diseases: An Official Publication of the Infectious Diseases Society of America* 53 (3): 296–302. https://doi.org/10.1093/cid/cir334.

54 Trimble, C.L., Morrow, M.P., Kraynyak, K.A. et al. (2015). Safety, efficacy, and immunogenicity of VGX-3100, a therapeutic synthetic DNA vaccine targeting human papillomavirus 16 and 18 E6 and E7 proteins for cervical intraepithelial neoplasia 2/3: a randomised, double-blind, placebo-controlled phase 2b trial. *Lancet (London, England)* 386 (10008): 2078–2088. https://doi.org/10.1016/S0140-6736(15)00239-1.

55 Leventhal, D.S., Sokolovska, A., Li, N. et al. (2020). Immunotherapy with engineered bacteria by targeting the STING pathway for anti-tumor immunity. *Nature Communications* 11 (1): 2739. https://doi.org/10.1038/s41467-020-16602-0.

56 Sedighi, M., Zahedi Bialvaei, A., Hamblin, M.R. et al. (2019). Therapeutic bacteria to combat cancer; current advances, challenges, and opportunities. *Cancer Medicine* 8 (6): 3167–3181. https://doi.org/10.1002/cam4.2148.

57 Yi, X., Zhou, H., Chao, Y. et al. (2020). Bacteria-triggered tumor-specific thrombosis to enable potent photothermal immunotherapy of cancer. *Science Advances* 6 (33): eaba3546. https://doi.org/10.1126/sciadv.aba3546.

58 Limon, J.J., Skalski, J.H., and Underhill, D.M. (2017). Commensal fungi in health and disease. *Cell Host & Microbe* 22 (2): 156–165. https://doi.org/10.1016/j.chom.2017.07.002.

59 Park, A.K., Fong, Y., Kim, S.I. et al. (2020). Effective combination immunotherapy using oncolytic viruses to deliver CAR targets to solid tumors. *Science Translational Medicine* 12 (559): eaaz1863. https://doi.org/10.1126/scitranslmed.aaz1863.

60 Zheng, J.H., Nguyen, V.H., Jiang, S.N. et al. (2017). Two-step enhanced cancer immunotherapy with engineered *Salmonella typhimurium* secreting heterologous flagellin. *Science Translational Medicine* 9 (376): eaak9537. https://doi.org/10.1126/scitranslmed.aak9537.

61 Kumar, R. and Kumar, P. (2019). Yeast-based vaccines: new perspective in vaccine development and application. *FEMS Yeast Research* 19 (2): foz007. https://doi.org/10.1093/femsyr/foz007.

62 Heery, C.R., Singh, B.H., Rauckhorst, M. et al. (2015). Phase I trial of a yeast-based therapeutic cancer vaccine (GI-6301) targeting the transcription factor brachyury. *Cancer Immunology Research* 3 (11): 1248–1256. https://doi.org/10.1158/2326-6066.CIR-15-0119.

63 Hamilton, D.H., Litzinger, M.T., Jales, A. et al. (2013). Immunological targeting of tumor cells undergoing an epithelial-mesenchymal transition via a recombinant brachyury-yeast vaccine. *Oncotarget* 4 (10): 1777–1790. https://doi.org/10.18632/oncotarget.1295.

64 Bilusic, M., Heery, C.R., Arlen, P.M. et al. (2014). Phase I trial of a recombinant yeast-CEA vaccine (GI-6207) in adults with metastatic CEA-expressing carcinoma. *Cancer Immunology, Immunotherapy* 63 (3): 225–234. https://doi.org/10.1007/s00262-013-1505-8.

65 Tanaka, A., Jensen, J.D., Prado, R. et al. (2011). Whole recombinant yeast vaccine induces antitumor immunity and improves survival in a genetically engineered mouse model of melanoma. *Gene Therapy* 18 (8): 827–834. https://doi.org/10.1038/gt.2011.28.

66 Chaft, J.E., Litvak, A., Arcila, M.E. et al. (2014). Phase II study of the GI-4000 KRAS vaccine after curative therapy in patients with stage I-III lung adenocarcinoma harboring a KRAS G12C, G12D, or G12V mutation. *Clinical Lung Cancer* 15 (6): 405–410. https://doi.org/10.1016/j.cllc.2014.06.002.

67 Bui, M.R., Hodson, V., King, T. et al. (2010). Mutation-specific control of BCR-ABL T315I positive leukemia with a recombinant yeast-based therapeutic vaccine in a murine model. *Vaccine* 28 (37): 6028–6035. https://doi.org/10.1016/j.vaccine.2010.06.085.

68 Mischo, A., Bubel, N., Cebon, J.S. et al. (2011). Recombinant NY-ESO-1 protein with ISCOMATRIX adjuvant induces broad antibody responses in humans, a RAYS-based analysis. *International Journal of Oncology* 39 (1): 287–294. https://doi.org/10.3892/ijo.2011.1032.

69 Lawson, D.H., Lee, S., Zhao, F. et al. (2015). Randomized, placebo-controlled, phase III trial of yeast-derived granulocyte-macrophage colony-stimulating factor (GM-CSF) versus peptide vaccination versus GM-CSF plus peptide vaccination versus placebo in patients with no evidence of disease after complete surgical resection of locally advanced and/or stage IV melanoma: a trial of the eastern cooperative oncology group-American College of Radiology Imaging Network Cancer Research Group (E4697). *Journal of Clinical Oncology: Official Journal of the American Society of Clinical Oncology* 33 (34): 4066–4076. https://doi.org/10.1200/JCO.2015.62.0500.

70 Jiang, C., Jiang, Y., Huang, Z. et al. (2010). Evaluation of the immunogenicity of a single chain chimeric peptide composed of hCGβ and oLHα for inhibition of the growth of hCGβ-expressing cancer cells. *Cancer Immunology, Immunotherapy* 59 (12): 1771–1779. https://doi.org/10.1007/s00262-010-0902-5.

71 Bolhassani, A., Muller, M., Roohvand, F. et al. (2014). Whole recombinant *Pichia pastoris* expressing HPV16 L1 antigen is superior in inducing protection against tumor growth as compared to killed transgenic Leishmania. *Human Vaccines & Immunotherapeutics* 10 (12): 3499–3508. https://doi.org/10.4161/21645515.2014.979606.

72 van Ooyen, A.J., Dekker, P., Huang, M. et al. (2006). Heterologous protein production in the yeast *Kluyveromyces lactis*. *FEMS Yeast Research* 6 (3): 381–392. https://doi.org/10.1111/j.1567-1364.2006.00049.x.

73 Ruiz-Herrera, J. and Ortiz-Castellanos, L. (2019). Cell wall glucans of fungi. A review. *Cell Surface (Amsterdam, Netherlands)* 5: 100022. https://doi.org/10.1016/j.tcsw.2019.100022.

74 Kalafati, L., Kourtzelis, I., Schulte-Schrepping, J. et al. (2020). Innate immune training of Granulopoiesis promotes anti-tumor activity. *Cell* 183 (3): 771–785.e12. https://doi.org/10.1016/j.cell.2020.09.058.

75 Geller, A., Shrestha, R., and Yan, J. (2019). Yeast-derived β-glucan in cancer: novel uses of a traditional therapeutic. *International Journal of Molecular Sciences* 20 (15): 3618. https://doi.org/10.3390/ijms20153618.

76 Ardiani, A., Higgins, J.P., and Hodge, J.W. (2010). Vaccines based on whole recombinant *Saccharomyces cerevisiae* cells. *FEMS Yeast Research* 10 (8): 1060–1069. https://doi.org/10.1111/j.1567-1364.2010.00665.x.

77 Pan, Y., Li, X., Kang, T. et al. (2015). Efficient delivery of antigen to DCs using yeast-derived microparticles. *Scientific Reports* 5: 10687. https://doi.org/10.1038/srep10687.

78 Riemann, H., Takao, J., Shellman, Y.G. et al. (2007). Generation of a prophylactic melanoma vaccine using whole recombinant yeast expressing MART-1. *Experimental Dermatology* 16 (10): 814–822. https://doi.org/10.1111/j.1600-0625.2007.00599.x.

79 Moreira, E.D. Jr., Block, S.L., Ferris, D. et al. (2016). Safety profile of the 9-valent HPV vaccine: a combined analysis of 7 phase III clinical trials. *Pediatrics* 138 (2): e20154387. https://doi.org/10.1542/peds.2015-4387.

80 Ratre, Y.K., Jain, V., Amle, D. et al. (2019). Association of TP53 gene codon 72 polymorphism with the incidence of cervical cancer in Chhattisgarh. *Indian Journal of Experimental Biology* 57: 580–585.

81 Gasser, B., Prielhofer, R., Marx, H. et al. (2013). *Pichia pastoris*: protein production host and model organism for biomedical research. *Future Microbiology* 8 (2): 191–208. https://doi.org/10.2217/fmb.12.133.

82 Bill, R.M. (2015). Recombinant protein subunit vaccine synthesis in microbes: a role for yeast? *The Journal of Pharmacy and Pharmacology* 67 (3): 319–328. https://doi.org/10.1111/jphp.12353.

83 Curti, E., Kwityn, C., Zhan, B. et al. (2013). Expression at a 20L scale and purification of the extracellular domain of the Schistosomamansoni TSP-2 recombinant protein: a vaccine candidate for human intestinal schistosomiasis. *Human Vaccines & Immunotherapeutics* 9 (11): 2342–2350. https://doi.org/10.4161/hv.25787.

84 Forbes, N.S. (2010). Engineering the perfect (bacterial) cancer therapy. *Nature Reviews Cancer* 10 (11): 785–794. https://doi.org/10.1038/nrc2934.

85 Wood, L.M., Guirnalda, P.D., Seavey, M.M., and Paterson, Y. (2008). Cancer immunotherapy using *Listeria monocytogenes* and listerial virulence factors. *Immunologic Research* 42 (1–3): 233–245. https://doi.org/10.1007/s12026-008-8087-0.

86 Lin, I.Y., Van, T.T., and Smooker, P.M. (2015). Live-attenuated bacterial vectors: tools for vaccine and therapeutic agent delivery. *Vaccine* 3 (4): 940–972. https://doi.org/10.3390/vaccines3040940.

87 Coley, W.B. (1991). The treatment of malignant tumors by repeated inoculations of erysipelas. With a report of ten original cases. 1893. *Clinical Orthopaedics and Related Research* 262: 3–11.

88 Richardson, M.A., Ramirez, T., Russell, N.C., and Moye, L.A. (1999). Coley toxins immunotherapy: a retrospective review. *Alternative Therapies in Health and Medicine* 5 (3): 42–47.

89 McCarthy, E.F. (2006). The toxins of William B. Coley and the treatment of bone and soft-tissue sarcomas. *The Iowa Orthopaedic Journal* 26: 154–158.

90 Kaimala, S., Al-Sbiei, A., Cabral-Marques, O. et al. (2018). Attenuated bacteria as immunotherapeutic tools for cancer treatment. *Frontiers in Oncology* 8: 136. https://doi.org/10.3389/fonc.2018.00136.

91 Shahabi, V., Maciag, P.C., Rivera, S., and Wallecha, A. (2010). Live, attenuated strains of Listeria and Salmonella as vaccine vectors in cancer treatment. *Bioengineered Bugs* 1 (4): 235–243. https://doi.org/10.4161/bbug.1.4.11243.

92 Le Gouëllec, A., Chauchet, X., Polack, B. et al. (2012). Bacterial vectors for active immunotherapy reach clinical and industrial stages. *Human Vaccines & Immunotherapeutics* 8 (10): 1454–1458. https://doi.org/10.4161/hv.21429.

93 Gardlik, R. and Fruehauf, J.H. (2010). Bacterial vectors and delivery systems in cancer therapy. *IDrugs: the Investigational Drugs Journal* 13 (10): 701–706.

94 Carleton, H.A. (2010). Pathogenic bacteria as vaccine vectors: teaching old bugs new tricks. *The Yale Journal of Biology and Medicine* 83 (4): 217–222.

95 Saltzman, D.A., Heise, C.P., Hasz, D.E. et al. (1996). Attenuated *Salmonella typhimurium* containing interleukin-2 decreases MC-38 hepatic metastases: a novel anti-tumor agent. *Cancer Biotherapy & Radiopharmaceuticals* 11 (2): 145–153. https://doi.org/10.1089/cbr.1996.11.145.

96 Saltzman, D.A., Katsanis, E., Heise, C.P. et al. (1997). Patterns of hepatic and splenic colonization by an attenuated strain of *Salmonella typhimurium* containing the gene for human interleukin-2: a novel anti-tumor agent. *Cancer Biotherapy & Radiopharmaceuticals* 12 (1): 37–45. https://doi.org/10.1089/cbr.1997.12.37.

97 Yuhua, L., Kunyuan, G., Hui, C. et al. (2001). Oral cytokine gene therapy against murine tumor using attenuated *Salmonella typhimurium*. *International Journal of Cancer* 94 (3): 438–443. https://doi.org/10.1002/ijc.1489.

98 Reddy, B.S. and Rivenson, A. (1993). Inhibitory effect of *Bifidobacterium longum* on colon, mammary, and liver carcinogenesis induced by 2-amino-3-methyli-midazo[4][5-f]quinoline, a food mutagen. *Cancer Research* 53: 3914–3918.

99 Cheng, K., Zhao, R., Li, Y. et al. (2021). Bioengineered bacteria-derived outer membrane vesicles as a versatile antigen display platform for tumor vaccination via Plug-and-Display technology. *Nature Communications* 12 (1): 2041. https://doi.org/10.1038/s41467-021-22308-8.

100 Ferlay, J., Ervik, M., Lam, F. et al. (2020). *Global Cancer Observatory: Cancer Today*. Lyon: International Agency for Research on Cancer. https://gco.iarc.fr/today (accessed February 2021).

18

Microbial Bioreactors at Different Scales for the Alginate Production by *Azotobacter vinelandii*

Belén Ponce[1], Viviana Urtuvia[1], Tania Castillo[2], Daniel Segura[3], Carlos Peña[2], and Alvaro Díaz-Barrera[1]

[1] *Escuela de Ingeniería Bioquímica, Pontificia Universidad Católica del Valparaíso, Valparaíso, Chile*
[2] *Departamento de Ingeniería Celular y Biocatálisis, Instituto de Biotecnología, Universidad Nacional Autónoma de México, Cuernavaca, Morelos, Mexico*
[3] *Departamento de Microbiología Molecular, Instituto de Biotecnología, Universidad Nacional Autónoma de México, Cuernavaca, Morelos, Mexico*

18.1 Introduction

Alginates are polysaccharides composed of β,D-mannuronic acid (M) and its C-5 epimer, α, L-guluronic acid (G) [1]. Alginates are extracted from seaweed and are mainly applied in the food and pharmaceutical industries. These polymers can also be produced by bacteria of the genera *Pseudomonas* and *Azotobacter* as a component of the extracellular matrix [1, 2]. However, bacterial alginates are acetylated and typically have a higher molecular weight than algal polymers. It is important to mention that the acetylation and the molecular weight determine the viscosifying capacity and, in general, the rheological properties of the polymer solutions. This has an important impact on the specific applications of alginate [3]. Alginates mainly are used as a viscosifying, gelling, thickening agent, or as a source of dietary fiber in the food industry [4, 5, 6]. In the pharmaceutical industry, they have potential application as an agent for the controlled release of drugs and the design of matrices, allowing cell anchoring and proliferation in tissue engineering [7].

Azotobacter vinelandii is a Gram-negative, strictly aerobic bacterium capable of fixing atmospheric nitrogen. This bacterium produces two polymers of industrial interest, polyhydroxy butyrate (PHB) and alginate. The alginate produced by *A. vinelandii* has molecular weights higher than 1000 kDa and presents acetylation in positions 2 and/or 3 of the mannuronic acid residues [8]. This acetylation causes an expansion of the molecular chain of the polymer [9], which, together with the high molecular weight, gives the alginate a higher viscosifying capacity than that obtained with algal alginate [3].

In recent decades, the study of the production of alginates using microbial sources has been very extensive, highlighting the use of *A. vinelandii* for the large-scale production of these polymers. In the present review, the following topics are addressed: research

Microbial Bioreactors for Industrial Molecules, First Edition.
Edited by Sudhir P. Singh and Santosh Kumar Upadhyay.

concerning alginate biosynthesis and genetic regulation, bacterial production on a bioreactor scale, different cultivation modalities for the production of alginate, as well as the influence of cultivation parameters on the quantity and quality of the alginate and finally the main studies on the scaling-up of alginate production are discussed.

18.2 Bacterial Alginate

18.2.1 Compositions and Structures

Alginates are formed by monomers of α-L-guluronic (G) and β-D-mannuronic (M) acids, and these monomers are linked by glycosidic bonds (1–4) [10, 11]. The mannuronic and guluronic acids, in the alginates, can be grouped as MMMM, GGGG, and MGMG blocks. In addition, the carbons 2 and 3 of the mannuronic residues, the bacterial alginates can present acetyl groups. These polymers are polysaccharides, which are hydrophilic, biodegradable, edible, and biocompatible [11], and show high water absorption and crosslinking capability, as well as chemical versatility [11, 12]. Due to the presence of carboxylic groups, this polymer is characterized by its polyanionic nature, and in the presence of cations, alginates can form gels mainly by the interaction of the G residues with the cations, forming structures called egg boxes. On the other hand, alginates can increase the viscosity of the solutions. The gel-forming capability of alginates is mainly determined by the G/M ratio and GGGG blocks, whereas the presence of acetyl groups increases the swelling and viscosifying capabilities of alginates [8, 13], in addition, the viscosity of the solutions of alginate is determined by the mean molecular weight of this polymer [3, 14].

These polymers are naturally produced by brown seaweed such as *Laminaria, Macrocystis,* and *Sargasso* and species of the bacterial genus *Azotobacter* and *Pseudomonas* [11]. The algal alginates show high variability in their composition and can be contaminated with heavy metals, in contrast, the chemical composition of the alginates produced in cultivations of *A. vinelandii* can be modulated and controlled by the growth conditions [15].

18.2.2 Applications

Alginates have been widely recognized because of their applications as viscosifying, thickening, and gel-forming agents in the food, textile, and cosmetic industries. However, in the past decade, alginate applications have been diversified through ecofriendly packaging and functional food, bioremediation, and agricultural purposes, as well as, for pharmaceutical and biomedical approaches, some examples of the most novel applications are summarized in Table 18.1.

18.3 Alginate Biosynthesis and Genetic Regulation

The biosynthesis and the regulation of the expression of the alginate biosynthetic genes have been studied both in *Pseudomonas* and *Azotobacter* species. Because of the interest in the use of a nonpathogenic bacterium for alginate production, here we will focus on what is known in *A. vinelandii*. We present only a summary because this theme has been reviewed by Hay et al. [24], Urtuvia et al. [1], and more recently by Núñez *et al.* [25].

Table 18.1 Novel applications of alginate.

Area of application	Application	Description	Reference
Bioremediation and strategies for agriculture	Sodium alginate-based aerogel with antimicrobial activity for oil absorption	Sodium alginate (apparent viscosity 20 mPa s). Combined with graphene and ZIF-8 nanoparticles and methyltrimethoxysilane	[16]
	Amphiphilic calcium alginate hydrogels for soil aggregation and pesticide (acetamiprid) retention	Alginate showed an apparent viscosity <100 mPa s. Alginate hydrogels reduce the presence of pesticides in soil and improve the quality of soils	[17]
	Hydrogels as adsorbents of antibiotics in water treatment	Sodium alginate hydrogels combined with graphene oxide modified κ-carrageenan, the moieties of carrageenan increase swelling capability and viscosity	[18]
Ecofriendly packaging	Biofilm made with alginate-carboxymethylcellulose-gluten-onion waste extract for food packaging	The mixture of alginate-carboxymethylcellulose-gluten have improved properties such as water vapor barrier	[19]
	Biofilm of calcium alginate – nanoparticles (NP) of Fe_2TiO_5 for food packaging with antibacterial activity	Addition of NP improved preservation of food and can be used as carriers of other active compounds, such as peptides and oils	[20]
	A composite made with sodium alginate and citrus pectin and pterostilbene with antioxidant properties	Composite of alginate and citrus pectin showed a good tensile strength and elongation at break, and addition of pterostilbene increased the moisture resistance	[21]
Biomedical	Hydrogel bioink loaded with stem cell for complex 3D tissue engineering	A bioink based on oxidized and methacrylated alginate showed shear-thinning behavior; these were mechanically stable, printable, and cytocompatible	[22]
	Sponges of alginate/chitosan/carbon dots as wound dressings	These sponges had an improved water vapor transmission rate, porosity, water absorption rate, and hydrophilicity able to absorb the moisture of the blood with hemocompatibility and without cytotoxicity	[23]
	Nanogels in aerosol for drug (quercetin) delivery in treatment of acute lung injury	Sodium alginate in an oil phase was combined with quercetin in calcium ethanolic phase; nanogels were lyophilized for their application. Nanogels of less than 100 nm were obtained increasing the quercetin bioavailability. In rats, inhalation of this aerosol reduced pulmonary inflammation and prevented pulmonary fibrosis	[16]

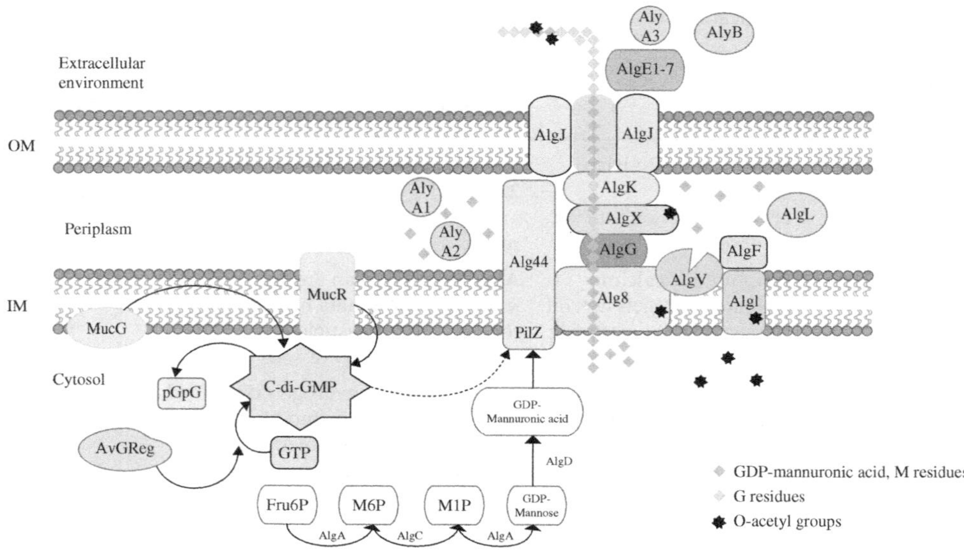

Figure 18.1 Schematic representation of the alginate biosynthesis in *A. vinelandii*. *Source:* Adapted from Hay et al. [26], Ertesvåg [27], and Ahumada-Manuel et al. [28].

Alginates are polymerized first as polymannuronic acid (Figure 18.1). The activated monomer required for its synthesis is guanosine diphosphate (GDP) mannuronic acid. The production of this monomer starts with fructose-6-phosphate, which can be generated from gluconeogenesis [29]. The reaction converting fructose-6-phosphate to GDP-mannuronic acid is achieved through four enzymatic reactions: the production of mannose-6-phosphate from fructose-6-phosphate catalyzed by the activity phosphomannose isomerase of the bifunctional enzyme AlgA; later, the phosphomannomutase AlgC isomerizes mannose-6-phosphate to mannose-1-phosphate, the mannose-1-phosphate is converted to GDP-mannose by the GDP-mannose pyrophorylase activity of AlgA (using GTP), and finally, the GDP-mannose is oxidized to GDP mannuronic acid by the GDP-mannose dehydrogenase named AlgD. These reactions occur in the cytosol [26]. The subsequent polymerization needs the activity of two inner membrane-bound enzymes: Alg8 and Alg44. These are transmembrane proteins, Alg8 is presumably a glycosyltransferase (polymerase), and Alg44 probably plays an indirect role in the polymerization. It may also play a role in bridging the polymerase to the periplasmic multiprotein scaffold formed by AlgG, AlgX, AlgK, AlgV, AlgI, AlgF, and AlgL. This complex translocates alginate to the outer membrane and also modifies the polymer by acetylating some mannuronic acid residues (AlgI, AlgF, AlgV, and AlgX) and epimerizing some mannuronate residues to guluronate (AlgG). The alginate lyase AlgL degrades alginates that fail to be secreted out of the periplasm. Finally, alginates cross the outer membrane through the porin protein AlgJ [1, 24, 25, 29].

In *A. vinelandii,* contrary to what occurs in *Pseudomonas spp.*, additional modification of the alginates occurs outside the cell. Six additional mannuronan C-5 epimerases (AlgE1-6), four alginate lyases (AlyA1-3 and AlyB), and a bifunctional epimerase-lyase (AlgE7) participate in this modification [25, 27, 30]. The additional epimerization can create different distributions of M and G monomers in the *A. vinelandii* alginates generating stretches or blocks of consecutive mannuronic acid residues, or consecutive guluronic acids residues, or alternating mannuronic-guluronic residues [27, 31].

Most of the genes coding for the alginate biosynthetic enzymes are found clustered. The cluster *algD-8-44-K-J-X-L-I-V-F-A* contains all but one (*algC*) of the genes needed for the synthesis of the activated monomer GDP-mannuronic acid (*algD* and *algA*), those for the polymerization of polymannuronate (*alg8* and *alg44*), the genes for the formation of a periplasmic scaffold (*alg44, algG, algX,* and *algK*), those coding for the enzymes that acetylate (*algI, algF, algV,* and *algX*), epimerize alginate (*algG*), the alginate lyase (*algL*), and the gene that codes for the porin (*algJ*), which is located in the outer membrane [24-26]. In *A. vinelandii*, three promoters upstream *algD* have been identified and partially characterized, but additional promoters upstream *algA, algG,* and *alg8* have also been mapped [25, 32, 33]. The other genes needed for the synthesis and additional modification of alginate are found in different locations in the genome, like *algC* [34], the mannuronan C-5 epimerases genes *algE1-6*, the genes for alginate lyases *alyA1-3* and *alyB*, and *algE7*, the gene for the bifunctional epimerase-lyase [25, 27, 30]. The expression of these genes is regulated by several mechanisms. The use of alternative sigma factors is one of them. The three promoters of *algD* are important targets because the corresponding enzyme catalyzes an irreversible step in alginate synthesis [33]. One of the promoters is recognized by RpoS, the sigma factor of the stationary phase. A second promoter is recognized by the extracytoplasmic function sigma factor AlgU, also named σE, whereas the third promoter has not been fully characterized [32, 33, 35, 36]. The promoter of *algC* is also dependent on AlgU and in this case, is essential for its expression [34]. Thus, AlgU is central in regulating alginate synthesis.

The activity of AlgU is negatively controlled by interaction with the membrane-bound anti-sigma factor MucA [35, 37, 38], which sequesters AlgU, preventing its interaction with RNA polymerase and therefore its target promoters [39]. MucA also interacts with the periplasmic protein MucB, providing an additional mechanism of environmental sensing. AlgU is activated through proteolysis of MucA, releasing AlgU and inducing alginate synthesis [40], so mutations in *mucA* increase the alginate production [35, 37]. Bacterial envelope stress and misfold outer member proteins [24], and cell wall recycling impairment [39] activate the proteases.

Besides the participation of RpoS in the transcription of one of the promoters of *algD*, it is also needed for a full expression of the epimerases genes algE1-7 and of its secretory system eexDEF, thus regulating proportions of G residues in alginate [33]. Besides sigma factors, the expression of the alginate genes is also controlled by other regulators. AmrZ (alginate and motility regulator) is needed for full alginate synthesis, and although the molecular mechanism has not been studied, it is probably similar to *Pseudomonas aeruginosa*, which is required for normal transcription algD [40].

Post-transcriptional control is also present in the regulation of alginate synthesis. The two-component GacS/GacA system regulates *algD* expression [36, 41], and this control works through the Rsm system, which is formed by the RsmA protein, a translational repressor of *algD* mRNA and is counteracted by the small RNAs RsmZ1–7 and RsmY, which titrate it. GacA activates transcription of the genes of these small RNAs [42, 43]. The two-component system CbrA/CbrB negatively regulates *algD* translation, and this is probably related to the need for this system for optimal rsmA expression. CbrA/CbrB was also required for normal accumulation of RpoS [44].

The second messenger cyclic dimeric GMP (c-di-GMP) post-translationally controls alginate synthesis, because binding of this molecule to the PilZ domain of Alg44 activates the polymerase activity of the complex Alg8-Alg44. The diguanylate cyclase protein AvGReg provides the c-di-GMP for Alg44, and the inner membrane protein MucG is probably the phosphodiesterase degrading c-di-GMP to inhibit alginate synthesis [28, 45]. In addition, the c-di-GMP has a role in the regulation of the content of guluronic residues of alginate during cyst formation. Under this physiological condition, the content of c-di-GMP is increased by the alternative diguanylate cyclase MucR, which in *P. aeruginosa* provides the c-di-GMP for Alg44. This increase is needed for the expression of the epimerases genes *algE1* to *algE6*, which modify the alginate to form the different layers of the cyst capsule. The expression of *mucR* is in turn activated by the response regulator AlgR [46].

18.4 Production of Bacterial Alginate on a Bioreactor Scale

18.4.1 Cultivation Modality for Alginate Production

In a particular bioprocess, a proper cultivation strategy will provide controlled kinetics parameters and productivities for optimal growth and bioproduct formation. In this section, we summarize the effects of cultivation modalities on alginate production in *A. vinelandii* cultures, which has been widely studied in shake flasks to small fermenters at the laboratory level. Recompilation of different studies (n = 29) about alginate production under different cultivation modalities is presented in Figure 18.2.

In shake flasks many experiments at a small scale can be conducted simultaneously [3, 47], usually used for optimizing culture conditions (e.g. medium composition), screening microorganisms, and establishing basic process conditions; however, using shake flasks important process variables cannot be controlled (e.g. pH or dissolved oxygen) [14, 48, 49]. Goméz-Pazarín et al. [49] demonstrated that by changing the filling volumes, the alginate production by *A. vinelandii* ATCC 9046 varied between 1.6 and 5.8 g/L. Later, García *et al.* [14] reported the alginate production using shake flask cultures and different strains of *A. vinelandii*: ATCC 9046, AT9 (ATCC 9046 derivative carrying a *mucG*::Tn5 mutation), CG9 (AEIV derivative carrying a *mucG*::Tn5 mutation), and OPAlgU+ (OP derivative carrying an *algU*::Km^r), reaching an alginate concentration between 2.5 and 3.6 g/L. One of the factors that importantly influence the alginate concentration in shake flasks is the viscosity, which is directly associated with cell growth in *A. vinelandii* [3, 47]. Peña *et al.* [3] demonstrated the relationship between alginate production and viscosity in shake flasks. These authors using alginate 3 g/L reported that the viscosity of cultures conducted at 100 rpm was twofold more with respect to those obtained at 200 rpm (~17 MPa s).

Oxygen has an important role in alginate production in *A. vinelandii* [1, 15, 50]. The cultivation operational conditions such as agitation rate and aeration or process variables such as dissolved oxygen tension (DOT), oxygen transfer rate (OTR), and pH can be controlled

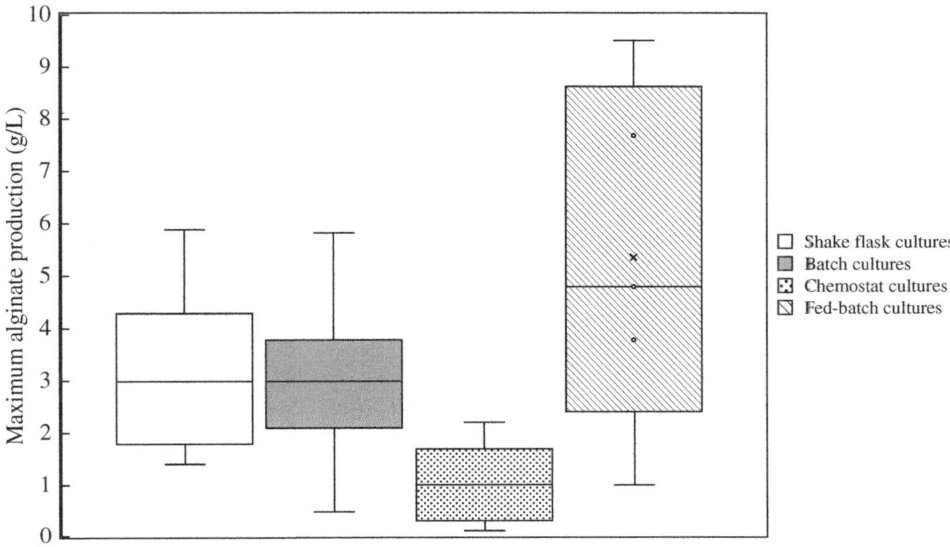

Figure 18.2 Production of bacterial alginate by *A. vinelandii* sp. obtained under different cultivation modalities.

or variated in bioreactor systems [1, 15, 51]. In cultures developed in a bioreactor (stirred tank), it has been possible to produce alginates with a specific composition by manipulating the operational conditions during the cultivation. Usually, these processes are simple and result in high substrate to product yields [52–55].

In batch cultures the alginate production has been widely studied using *A. vinelandii* ATCC 9046, where the alginate concentration can vary between 0.5 and 4.8 g/L (Figure 18.2), depending on cultivation conditions. Flores et al. [50] obtained alginates between 1.5 and 1.8 g/L from batch cultures of *A. vinelandii* ATCC 9046 grown under DOT control (1% and 5%) using gas blending and non-diazotrophic conditions. Díaz-Barrera *et al.* [54] demonstrated that increasing the agitation rate (from 300 to 700 rpm) in cultures without DOT control increased the alginate concentration from 2.1 to 3.3 g/L. In order to enhance alginate production, a two-stage or exponential fermentation strategy has been evaluated. Mejía *et al.* [56] using two-stage in fed-batch culture of *A. vinelandii* AT6 (ATCC 9046 derivative carrying a *phbB*::Tn5-lacZ mutation) reached a maximum concentration of alginate of up to 9.0 g/L (Figure 18.3). More recently, under this modality (batch mode), the studies have focused on optimizing alginate concentration without adding more nutrients to the medium culture (as in fed-batch mode), just controlled by process variables (such as pH or DOT). Our group [55] has studied the controlled OTR during the cell growth in *A. vinelandii* cultures as an appropriate strategy to enhance alginate production (up to 5.5 g/L).

On the other hand, chemostat cultures provide an ideal system for characterizing and studying biological systems, i.e. provide a steady state for different metabolic studies. The advantage of using this type of cultivation strategy is to allow (at steady state operated at the fixed dilution rate) a cellular growth and alginate concentration controlled and defined that

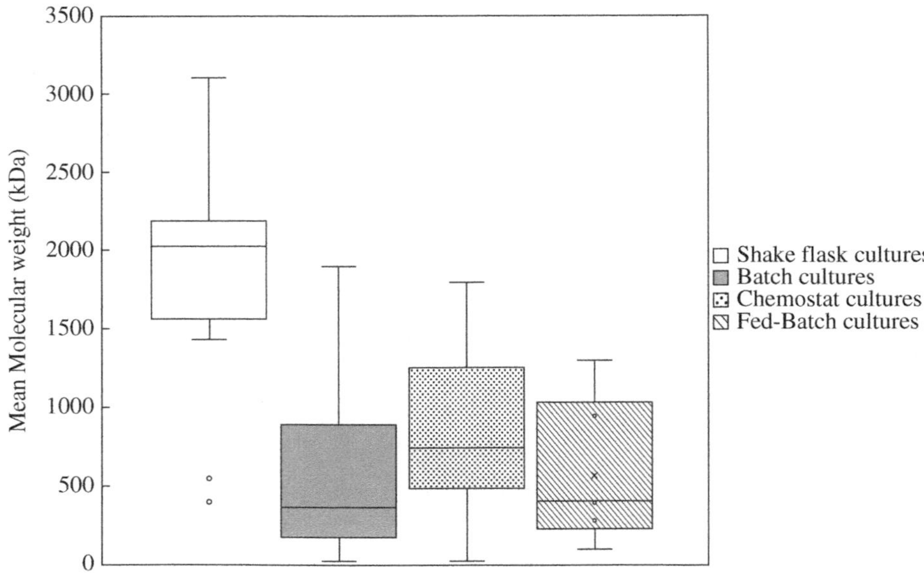

Figure 18.3 Mean molecular weight of alginate obtained under different cultivation modalities.

does not change over time [57–62]. In steady state, the dilution rate and OTR can be used to vary the alginate composition [61, 63]. The data collected from the chemostat cultures (Figure 18.2) shows that it is possible to reach an alginate concentration up to 2.2 g/L. Díaz-Barrera *et al.* [61] reported that in steady state ($D = 0.06$ 1/h), the alginate concentration (1.1 g/L) was not affected by the change in DOT control (1% and 8%) in the cultures.

18.4.2 Influence of Oxygen on Alginate Production

As aforementioned, DOT and oxygen availability are two key parameters in the alginate synthesis by *A. vinelandii*. Several studies in the literature on alginate production have focused on optimizing polymer yield and process productivity at high OTR conditions [15, 64, 65]. In a fermentation process, oxygen is continuously supplied by a gas phase during the cultivation operation, and thus the OTR can be used for bioreactor design at the laboratory and pilot plant level [50, 66, 67].

In the cultivation process of *A. vinelandii*, in the cell growth, the DOT is nearly zero, and the OTR can be used for indicating that the cultures are limited by oxygen [3, 64, 65, 68]. During the period of maximum and constant OTR in the culture, the dissolved oxygen does not accumulate ($dO_2/dt \approx 0$), and it is possible to consider the OTR equal to the appropriate oxygen demand (OUR) for the microorganism [55, 64, 67]. Therefore, OTR values depend on the operation conditions in the fermenter, such as the DOT, agitation, or aeration rate [51, 59, 61, 62].

To compare the alginate concentration and maximum specific alginate production rate (q_p) in *A. vinelandii*, we present a summary of shake flasks and batch cultures performed at different OTRs. The collected data (Table 18.2) highlights that in shaken flasks, changes in

Table 18.2 Alginate production under different cultivation strategies in *A. vinelandii* sp.

Mode of cultivation	Strain	Oxygen transfer strategy	Agitation rate (rpm)	OTR_{max} (mmol/L/h)	Alginate (g/L)	q_p (g/g/h)	Reference
Shake flasks	ATCC 9046	100 mL of filling vol. /500 mL vol. total	200	3.1	1.3	0.03	[69]
	AT6	100 mL of filling vol. /500 mL vol. total	200	3.4	1.7	0.09	[69]
	GG9	50 mL of filling vol. /250 mL vol. total	200	4.8	2.5	0.037	[14]
	ATCC 9046	50 mL of filling vol. /250 mL vol. total	200	4.8	3.6	0.05	[14]
	OPAlgU+	50 mL of filling vol. /250 mL vol. total	200	5.2	2.5	0.046	[14]
	AT9	50 mL of filling vol. /250 mL vol. total	200	5.2	3.8	0.052	[14]
	OPAlgU+	10 mL of filling vol. /250 mL vol. total	200	24.0	4.2	0.6	[70]
	ATCC 9046	10 mL of filling vol. /250 mL vol. total	200	29.4	4.1	0.5	[70]
	ATCC 9046	100 mL of filling vol. /500 mL vol. total	100	2.6	3.1	nr.	[3]
	ATCC 9046	200 mL of filling vol. /500 mL vol. total	200	6.0	4.3	nr.	[3]
Batch culture (bioreactor)	ATCC 9046	Oxygen at 21% in inlet gas (air)	300	5.0	2.1	0.014	[54]
	ATCC 9046	Oxygen at 21% in inlet gas (air)	500	10.4	2.9	0.017	[54]
	ATCC 9046	Oxygen at 21% in inlet gas (air)	700	19.2	3.3	0.014	[54]
	ATCC 9046	Non-controlled OTR	500	16.8	2.9	0.011	[55]
	ATCC 9046	Controlled OTR	500	20.3	5.5	0.042	[55]
	ATCC 9046	DOT = 0.5%	700	70	2.1	0.025	[51]
	ATCC 9046	DOT = 5%	700	100	3.1	0.031	[51]
	ATCC 9046	Oxygen at 21% in inlet gas (air)	200	18	3.8	0.021	[60]
	ATCC 9046	Oxygen at 21% in inlet gas (air)	600	19	3.8	0.093	[60]

nr., not reported; vol., volume.

the OTR (for example, manipulating the filling volume) and depending on the microorganism employed it is possible to obtain alginates with a high concentration (between 1.3 and 4.3 g/L) and higher specific alginate production rate (from 0.03 to 0.6 g/g/h). Peña *et al.* [3] in shaken flasks of *A. vinelandii* ATCC 9046 reported that by modifying the shaking frequency from 100 to 200 rpm, the OTR increased from 2.6 to 6.0 mmol/L/h and alginate production from 3.1 to 4.3 g/L. Jiménez *et al.* [69] demonstrated that the q_P was high (0.09 g/g/h) for the *A. vinelandii* AT6, compared to ATCC 9046 strain (0.03 g/g/h), both in shaken flasks cultures with OTR similar (~3.0 mmol/L/h). On the other hand, some studies [51, 54, 71] in batch culture have demonstrated the importance of the DOT and agitation rate on alginate production and productivity. Díaz-Barrera *et al.* [54] in batch culture of *A. vinelandii* ATCC 9046 demonstrated that by modifying the agitation rate from 300 to 500 rpm, the OTR increases from 5.0 to 10.4 mmol/L/h, increased the alginate productivity (0.014 to 0.017 g/g/h). Under 0.5% and 5% DOT, Lozano *et al.* [51] in batch cultures at 700 rpm of *A. vinelandii* ATCC 9046 demonstrated that DOT control caused an increase in the OTR from 70 to 100 mmol/L/h, enhancing alginate production, as well as, the q_P.

18.4.3 Influence of Cultivation Modality on the Molecular Weight of Alginate

Figure 18.3 shows a recompilation of the mean molecular weight of alginate obtained under different cultivation modalities. In shake flasks [14, 48, 49, 72] it has been reported that alginates with a wide molecular weight range from 400 to 3112 kDa – an advantage in terms of their viscosifying ability required for different industrial applications [73]. For example, García *et al.* [14] in shake flasks reported that is possible to obtain alginates with a higher molecular weight using different *A. vinelandii* strain, between 2100 and 3112 kDa. In batch culture [50, 54, 74, 75] and fed-batch culture [56, 73], it is possible to obtain alginates having molecular weights between 20 and 1900 kDa, varying the operational conditions in the bioreactor. For example, Trujillo-Roldan *et al.* [75] observed in batch culture (at 700 rpm) of *A. vinelandii* SML2 (ATCC 9046 derivative carrying a nonpolar mutation within *algL*) that the molecular weight increased from 250 to 985 kDa under an increase of DOT control from 1 % to 3 %. In chemostat culture, the dilution rate, DOT, or OTR (by adjusting the agitation rate) set at steady state influences the molecular weight of alginate. The data collected (Figure 18.3) shows a recompilation of molecular weight from 480 to 1250 kDa [57, 60, 61, 63]. The advantage of using this type of cultivation strategy is that it permits maintaining the molecular weight during the steady state of culture [1, 61].

18.5 Chemical Characterization of Alginate Quality

The bacterial alginate has a wide range of biotechnological applications as a biomaterial (e.g. hydrogels, encapsulation of drugs or foods). It is well known that the alginate composition depends on the guluronic/mannuronic (G/M) ratio, acetylation level, and molecular weight that varies according to the microorganism of study and cultivation strategy, which determines their rheological properties [1, 13, 26, 64, 76, 77]. Unlike brown algae (which serve as an alginate source in industrial applications), the bacterial alginate is often

O-acetylated (at C_2 and/or C_3 positions of mannuronic acid residues), and acetylation level affects water holding and viscosity of alginate [15, 64, 78, 79], expanding the application commercial of this polysaccharide. To compare the alginate quality that determines their rheological properties in *A. vinelandii*, we present a summary of its chemical composition in different modes of cultivation and operational conditions at the laboratory level (Table 18.3); however, the evidence indicates that the number of studies of G/M ratio and acetylation degree is scarce.

The collected data (Table 18.3) highlights that in shaken flasks and batch cultures it is possible to obtain alginates with the widest range of molecular weight (from 193 to 4000 kDa) and G/M ratio (between 0.37 and 4.50) [53, 54, 80]. So, the G/M ratio and molecular weight are affected by the nutritional conditions and changing the cultivation parameters. Gaytán *et al.* [48] reported that the alginate quality (molecular weight, G/M ratio, and acetylation degree) produced by *A. vinelandii* ATCN4 (ATCC 9046 derivative carrying a *nqrE*::Tn5) in shake flasks could be altered under non- or diazotrophic conditions. More recently, Díaz-Barrera *et al.* [54] in batch cultures of *A. vinelandii* ATCC 9046 demonstrated that the molecular weight of alginate decreased from 520 to 370 kDa when the OTR increased from 10.4 to 19.2 mmol/L/h, but no differences were observed in the G/M ratio (0.83).

On the other hand, the acetylation degree can be controlled by the cultivation conditions in the bioreactor, such as DOT control, and microorganism of study [8, 81]. Peña *et al.* [81] reported in batch culture of *A. vinelandii* ATCC 9046, keeping DOT control (3%) and 300 rpm, an acetylation degree of 3.6%, whereas under similar conditions using *A. vinelandii* AT268 (ATCC 9046 derivative carrying a *phbR*::mini-Tn5 of mutant strain DS268) only reached 3.3%. Subsequent studies reported that small changes in the degree of acetylation positively affect alginate viscosity and are independent of the molecular weight obtained [3]. Even if the acetylation degree has not been studied extensively, only a study has demonstrated that the addition of sodium salt of MOPS (as a buffer for maintaining the pH in the medium culture) has a strong influence on the acetylation degree [13]. Peña *et al.* [13] reported that shaken flasks of *A. vinelandii* ATCC 9046 supplemented with MOPS (13.6 mmol/L) increased twofold the acetylation degree, in comparison with cultures without MOPS.

18.5.1 Scale-up of Alginate Production

In the cultures where high viscosities are achieved, such as in polysaccharide production, the scaling faces several problems [82]. Such is the case of alginate production, where the viscosities can increase at the end of the culture to values close to 600 MPa s [15, 52, 72]. However, the scale-up of alginate production has not been studied extensively. Maybe because currently there is no industrial process reported on the bacterial production of this polysaccharide.

To our knowledge, the commercial alginate that is currently used is of algal origin and not bacterial. Most of the information in the literature is focused on the scale-up from shake flasks to small bioreactors and bioreactors at the pilot plant level. The first approach to this subject is the study by Trujillo-Roldán *et al.* [83], who carried out an analysis of scaling-down of the alginate production. In that study, the authors simulated conditions

Table 18.3 Chemical composition of alginates in *A. vinelandii* sp. under different conditions of cultivation.

Mode of cultivation	Strain	Carbon source (g/L)	Nitrogen source (g/L)	Operational conditions: DOT (% air saturation)	Operational conditions: Agitation rate (rpm)	MMW (kDa)	G/M ratio	Acetylation (%)	Reference
Shake flasks	ATCN4	Suc 20	Diazotrophy	Non-controlled	200	2200	2.0	4.2	[42]
	ATCN4	Suc 20	Yeast extract[a]	Non-controlled	200	1500	4.5	0.0	[48]
	ATCC 9046	Glu 20	Diazotrophy	Non-controlled	200	nr.	nr.	4.7	[69]
	AT6	Glu 20	Diazotrophy	Non-controlled	200	nr.	nr.	4.4	[69]
	ATCC 9046	Suc 20	Yeast extract[a]	Non-controlled	200	1430	0.78	0.7	[13]
	ATCC 9046	Suc 20	Yeast extract[a]	Non-controlled	200	550		1.0	[72]
Batch culture (bioreactor)	DSM 576	Glu 20	$(NH_4)_2SO_4$ 0.6	nr.	400	193	0.27	29.2	[80]
	ATCC 9046	Suc 20	Yeast extract[a]	5%	700	160	3.63	nr.	[53]
	ATCC 9046	Suc 20	Diazotrophy	Non-controlled	300	406	0.73	nr.	[54]
	ATCC 9046	Suc 20	Diazotrophy	Non-controlled	500	520	0.83	nr.	[54]
	ATCC 9046	Suc 20	Diazotrophy	Non-controlled	700	370	0.83	nr.	[54]
Batch culture (bioreactor)	ATCC 9046	Suc 20	Yeast extract[a]	3%	300	800	nr.	3.6	[81]
	AT268	Suc 20	Yeast extract[a]	3%	300	850	nr.	3.3	[81]
	DM	Suc 20	Yeast extract[a]	3%	300	4000	nr.	2.6	[81]
	AT6	Glu 6	Diazotrophy	1%	300	nr.	nr.	1.5	[69]

Glu, glucose; MMW, mean molecular weight; nr, not reported; Suc, sucrose.
[a] Yeast extract concentration = 3 g/L.

occurring in large-scale bioreactors in 1.0 L laboratory bioreactors. When *A. vinelandii* was grown under oscillating DOT (similar to larger-scale fermenters), the molecular weight of the alginate was negatively affected and therefore the quality of the product [83]. The results highlight the importance of maintaining strict control of DOT in order to produce alginates with a constant molecular weight during cultivation, particularly in large fermenters, where very extreme oxygen concentration gradients occur, caused by the high viscosity of the broth and insufficient mixing. Later, Reyes *et al.* [52], in studies of scaling-up from shake flasks to bioreactors, found that initial power input, as a scale-up criterion, did not reproduce culture behavior, in terms of broth viscosity, alginate concentration, and molecular weight. These authors proposed that the evolution of power input during the cultures developed in shake flasks and bioreactors was very different. They found that using a low initial P/V in the bioreactor allowed to obtain alginates with a molecular weight very similar to that obtained in shake flasks.

The design of new equipment for measurement of the power input and the OTR in shaken flasks [84, 85] has allowed characterizing (online) the changes in both parameters, in cultures where broth viscosities are very high. Taking into account the above has characterized the evolution of P/V and OTR [3, 86], in cultures of *A. vinelandii* producers of alginate under different conditions [47]. Due to an increase in viscosity in the broth culture, an exponential increase in the power consumption was observed during the course of the fermentation (up to 1.4 kW/m^3). A slight decrease in power consumption was observed at the end of the cultivation when the viscosity and alginate concentration reached a maximum [3, 47]. Taking into account that information, our group reported [86] that by simulating the evolution of the power input (determined in shake flasks), in 14 L bioreactors, it was possible to reproduce both the concentration and molecular weight of the alginates synthesized by *A. vinelandii*. Using this strategy, a maximal production of 3.6 ± 0.5 g/L and a molecular weight of 1700 ± 200 kDa were obtained, with very similar values to those achieved in shaken flasks [86].

Another culture parameter widely used as a scale-up criterion is the OTR [71]. Díaz-Barrera et al. [71] reported that by maintaining the OTR_{max} (19 mmol/L/h.) in bioreactors of 3 and 14 L, the same concentration of alginate (3.8 g/L) was reached, during the cell growth on both scales. In contrast, the maximal molecular weight of alginate was 1250 kDa in a 3 L bioreactor and 590 kDa in a 14 L bioreactor. The authors explain these discrepancies based on differences in the specific oxygen consumption rate (qO2) between the two fermenters, reaching 8.3 mmol/g/h in a 3 L bioreactor and 10.6 mmol/g/h in a 14 L bioreactor. Based on this information, the authors propose the use of qO2 instead of OTR as a scale-up criterion to produce polymers with similar molecular weights during cultivation [71].

Subsequently, Gómez-Pazarín et al. [49] found in cultures of *A. vinelandii* in shaken flasks that the use of very low power input (0.02–0.68 kW/m^3) and an OTR_{max} of 0.85–2.8 mmol/L/h negatively affected both the growth of the bacteria and the production of alginate. In contrast, the viscosifying capacity and the molecular weight of the alginate increased significantly, with respect to that obtained when using higher powers (1.6 kW/m^3) [49]. In that same study, the evolution of input power, measured in shake flasks, was simulated in a 3 L bioreactor. Using this strategy, it was possible to reproduce the same trends in both alginate production and polymer viscosifying capacity. This study offers useful information to implement new scale-up strategies for the production of alginates, exhibiting chemical characteristics and viscosifying properties superior to commercial alginates.

In order to reach a viable cell concentration of *Azotobacter chroococcum*, a scaling-up strategy for a liquid fermentation process was validated and adjusted on a laboratory and pilot scale, using two strains of *A. chroococcum*. For this purpose, a batch fermentation process was developed under the previously defined conditions in a 3.5 L bioreactor. In order to optimize the process, the influence of decreasing the agitation rate was analyzed in combination with a fed-batch type culture modality. In order to reproduce the parameters obtained on a smaller scale, the geometric and fluid dynamic behavior was kept constant. By using this strategy, it was possible to simulate cell concentration in pilot and laboratory cultures [87] and in both scales, the kinetic parameters were not affected.

More recently, Díaz-Barrera et al. [84] studied the effect of scale-up on alginate production, especially evaluating the molecular weight and G/M ratio in *A. vinelandii* cultures of 3–30 L scale in different OTRs under diazotrophic conditions. These authors found that the OTR as a scaling criterion for the production of alginate from 3 to 30 L bioreactors allowed reproducing the concentration and productivity of alginate; however, the quality (molecular weight and G/M ratio) was not reproduced. It was found that the molecular weight, in both scales, could be reproduced when the same specific oxygen consumption rates were applied in the bioreactors, which indicates that this parameter may be more appropriate to scale the molecular weight of alginate. However, by using both scale-up criteria (OTR and specific oxygen consumption rate), it was not possible to reproduce the G/M ratio of the alginates. The authors propose that the molecular weight and G/M ratio of alginate are affected by cellular respiration and that the design of a process for the production of alginates on a larger scale and with a defined composition could be based on the oxygen consumption of the cells, more than in the OTR. In general, the earlier data reveal that depending on the objective function (concentration, molecular weight, or G/M ratio) is the type of scaling criteria to be used in the development of the bioprocess. The compilation of different studies on the scale-up of alginate production is presented in Table 18.4.

18.6 Prospects and Conclusions

Alginates are polymers that have ideal chemical and geological characteristics for their application as viscosifiers, thickeners, and gelling agents. In the past decades, these materials have been attracting attention because of their potential use in the medical field as tissue support and scaffold, drug delivery systems, and soft robots among others. Because the performance of alginates for particular applications is determined by their chemical and physical properties, and these are affected by the molecular weight and mannuronate to guluronate composition, the use of *A. vinelandii* is interesting due to the diversity of alginate modifying enzymes present in this bacterium: six epimerases, four alginate lyases, and a bifunctional epimerase-lyase, in addition to the enzymes usually present in other bacteria [27]. The different epimerases produce different distributions of M and G [88]; thus, the selective genetic manipulation of the expression of these enzymes could enable the production of new alginates with controlled properties. A few examples of this kind of

Table 18.4 Scale-up of bacterial alginate production using *A. vinelandii* sp.

Scale	Scale-up criterion	Main results	Reference
Scale-down in 2 L bioreactor	Oscillating DOT	Cultures of cells under oscillating DOT promoting a decrease in the quality of the alginate, in terms of its molecular weight, were observed.	[83]
Scale-up from shake flasks to 3 L bioreactor	Initial power input	Initial P/V did not reproduce culture behavior, in terms of broth viscosity, alginate concentration, and molecular weight.	[52]
Scale-up from shake flasks to 14 L bioreactor	Power input evolution	Simulating the evolution of P/V during the cultivation it was possible to reproduce both the production and the molecular weight of the alginates.	[86]
Scale-up from 3 to 14 L bioreactor	Oxygen transfer rate	Keeping constant the OTR_{max} the same alginate concentration was obtained in both scales. However, the maximum molecular weight of the alginate was very different.	[71]
Scale-up from shake flasks to 3 L bioreactor	Power input evolution	Simulating the profiles of P/V was possible to reproduce the same trends in both the alginate production and the polymer viscosifying power.	[49]
Scale-up from 4 to 70 L bioreactor	Manipulation of power input	Keeping the same geometric configuration and fluid dynamic behavior was possible to reach the same cellular concentration in both scales.	[87]
Scale-up from 3 to 30 L bioreactor	Oxygen transfer rate/oxygen consumption rate	A similar alginate molecular weight was reached when the cultures were conducted at similar specific oxygen uptake rate.	[54]

genetic manipulation have been reported [28, 30, 89], so this is a field with opportunities for investigation.

The chemical composition of alginates can also be tailor-made by the growth conditions. It is known that the culture modality will affect both the production and the molecular weight of the alginate. For example, in batch cultures, both in shake flasks and bioreactors using *A. vinelandii* ATCC 9046, the alginate concentration can vary between 0.5 and 5.8 g/L. On the other hand, in order to improve alginate production, a two-stage or exponential cultivation strategy has been evaluated, where a maximal concentration of alginate of up to 9.0 g/L was reached. Finally, the chemostat cultures provide an ideal system for characterizing and studying biological systems, i.e. provide a steady state for different metabolic studies. The advantage of using this type of cultivation strategy is to allow (at steady state operated at the fixed dilution rate) a cellular growth and alginate concentration controlled and defined that does not change over time.

There are a large number of reports in literature related to the effect of culture conditions on the synthesis and chemical characteristics of alginates. Among these parameters, oxygen plays a relevant role. It is known that increasing the DOT or OTR promotes the synthesis of alginate and the acetylation degree of the polymer. However, the molecular weight of the alginate is affected at high levels of oxygenation in the cultures. It is for that reason that a compromise must be established in the culture conditions to generate alginates with a high yield and desirable chemical characteristics. In this context, multistage cultures, such as fed-batch, turn out to be viable strategies.

Several strategies for scaling up the alginate production process, most of them based on the use of power consumption and OTR as scaling-up criteria, have been reported. Most of the results have shown that it is possible to replicate both the alginate yields and the chemical characteristics, such as the molecular weight and G/M ratio of the polymer at different scales, from shake flask level to pilot scale bioreactors. However, the commercial production of bacterial alginate has not yet materialized, mainly due to the low yields obtained so far. In the case of fed-batch cultures, maximum concentrations of alginate close to 10 g/L have been reached; however, the cost of production limits the commercial production of the bacterial polymer. Taking into account the above, it would be interesting to implement new production strategies, such as multistage fermentation, which promote better growth and alginate production, in order to take advantage of the higher specific alginate production capacities of strains overproducing of alginate. On the other hand, the use of low-cost raw materials, such as agricultural waste or industrial byproducts, such as glycerol, whey, or even molasses, will make it possible to propose more attractive processes from the economic point of view.

Acknowledgment

VU is grateful to FONDECYT postdoctoral project 3180406, ANID-Chile.

References

1 Urtuvia, V., Maturana, N., Acevedo, F. et al. (2017). Bacterial alginate production: an overview of its biosynthesis and potential industrial production. *World J. Microbiol. Biotechnol.* 33: 198.

2 Franklin, M.J., Nivens, D.E., and Weadge, J.T. (2011). Biosynthesis of the *Pseudomonas aeruginosa* extracellular polysaccharides, alginate, Pel and Psl. *Front. Microbiol.* 167: 1–15.

3 Peña, C., Galindo, E., and Büchs, J. (2011). The viscosifying power, degree of acetylation and molecular mass of the alginate produced by *Azotobacter vinelandii* in shake flasks are determined by the oxygen transfer rate. *Process Biochem.* 46: 290–297.

4 Brownlee, I.A., Allen, A., Pearson, J.P. et al. (2005). Alginate as a dietary fiber. *Crit. Rev. Food Sci. Nutr.* 45: 497–510.

5 Singh, S.P. and Upadhyay, S.K. (eds.) (2021). *Bioprospecting of Microorganism-Based Industrial Molecules*. John Wiley & Sons Ltd. doi:10.1002/9781119717317.

6 Upadhyay, S.K. and Singh, S.P. (eds.) (2021). *Bioprospecting of Plant Biodiversity for Industrial Molecules*. John Wiley & Sons Ltd. doi:10.1002/9781119718017.

7 Sun, J. and Tan, H. (2013). Alginate-based biomaterials for regenerative medicine applications. *Materials* 6: 1285–1309.

8 Castillo, T., Galindo, E., and Peña, C. (2013). The acetylation degree of alginates in *Azotobacter vinelandii* ATCC 9046 is determined by dissolved oxygen and specific growth rate: studies in glucose-limited chemostat cultivations. *J. Ind. Microbiol. Biotechnol.* 40: 715–723.

9 Skjåk-Bræk, G., Zanetti, F., and Paoletti, S. (1989). Effect of acetylation on some solution and gelling properties of alginates. *Carbohydr. Res.* 185: 131–138.

10 Liparoti, S., Speranza, V., and Marra, F. (2021). Alginate hydrogel: the influence of the hardening on the rheological behaviour. *J. Mech. Behav. Biomed. Mater.* 116: 104341.

11 Soleimanpour, M., Mirhaji, S.S., Jafari, S. et al. (2022). Designing a new alginate-fibrinogen biomaterial composite hydrogel for wound healing. *Sci. Rep.* 12 (1): 7213.

12 Liu, W., Madry, H., and Cucchiarini, M. (2022). Application of alginate hydrogels for next-generation articular cartilage regeneration. *Int. J. Mol. Sci.* 23 (3): 1147.

13 Peña, C., Hernández, L., and Galindo, E. (2006). Manipulation of the acetylation degree of *Azotobacter vinelandii* alginate by supplementing the culture medium with 3-(N-morpholino)-propane-sulfonic acid. *Lett. Appl. Microbiol.* 43: 200–204.

14 García, A., Castillo, T., Ramos, D. et al. (2020). Molecular weight and viscosifying power of alginates produced by mutant strains of *Azotobacter vinelandii* under microaerophilic conditions. *Biotech. Rep.* 26: e00436.

15 Galindo, E., Peña, C., Núñez, C. et al. (2007). Molecular and bioengineering strategies to improve alginate and polyhydroxyalkanaote production by *Azotobacter vinelandii*. *Microb. Cell Factories* 6: 7.

16 Chen, X., Yang, Y., Guan, Y. et al. (2022). A solar-heated antibacterial sodium alginate aerogel for highly efficient cleanup of viscous oil spills. *J. Colloid Interface Sci.* 621: 241–253.

17 Zhang, S., He, F., Fang, X. et al. (2022). Enhancing soil aggregation and acetamiprid adsorption by ecofriendly polysaccharides hydrogel based on Ca2+- amphiphilic sodium alginate. *J. Environ. Sci.* 113: 55–63.

18 Yang, P., Yu, F., Yang, Z. et al. (2022). Graphene oxide modified κ-carrageenan/sodium alginate double-network hydrogel for effective adsorption of antibiotics in a batch and fixed-bed column system. *Sci. Total Environ.* 837: 155662.

19 Thivya, P., Bhosale, Y.K., Anandakumar, S. et al. (2022). Study on the characteristics of gluten/alginate-cellulose/onion waste extracts composite film and its food packaging application. *Food Chem.* 390: 133221.

20 Rizzotto, F., Vasiljevic, Z.Z., Stanojevic, G. et al. (2022). Antioxidant and cell-friendly Fe_2TiO_5 nanoparticles for food packaging application. *Food Chem.* 90: 133198.

21 Li, Y., Lu, J., Tian Xiaolu, X. et al. (2021). Alginate with citrus pectin and pterostilbene as healthy food packaging with antioxidant property. *Int. J. Biol. Macromol.* 193, Part B: 2093–2102.

22 Jeon, O., Bin, L.Y., Lee, S.J. et al. (2022). Stem cell-laden hydrogel bioink for generation of high resolution and fidelity engineered tissues with complex geometries. *Bioact. Mater.* 15: 185–193.

23 Rezaei, A. and Ehtesabi, H. (2022). Fabrication of alginate/chitosan nanocomposite sponges using green synthesized carbon dots as potential wound dressing. *Mater. Today Chem.* 24: 100910.

24 Hay, I.D., Wang, Y., Moradali, M.F. et al. (2014). Genetics and regulation of bacterial alginate production. *Environ. Microbiol.* 16: 2997–3011.

25 Núñez, C., López-Pliego, L., Ahumada-Manuel, C.L., and Castañeda, M. (2022). Genetic regulation of alginate production in *Azotobacter vinelandii* a bacterium of biotechnological interest: a mini-review. *Front. Microbiol.* 13: 845473.

26 Hay, I., Rehman, Z., Moradali, F. et al. (2013). Microbial alginate production, modification and its applications. *Microb. Biotechnol.* 6: 637–650.

27 Ertesvåg, H. (2015). Alginate-modifying enzymes: biological roles and biotechnological uses. *Front. Microbiol.* 6: 523.

28 Ahumada-Manuel, C.L., Martínez-Ortíz, I.C., Hsueh, B.Y. et al. (2020). Increased c-di-GMP levels lead to the production of alginates of high molecular mass in *Azotobacter vinelandii*. *J. Bacteriol.* 202: e00134–e00120.

29 Rehm, B.H. (2009). Alginate production: precursor biosynthesis, polymerization and secretion. In: *Alginates: Biology and Applications*, Microbiol. Monographs, vol. 13 (ed. B. Rehm). Berlin, Heidelberg: Springer.

30 Gimmestad, M., Ertesvåg, H., Heggeset, T.M.B. et al. (2009). Characterization of three new *Azotobacter vinelandii* alginate lyases, one of which is involved in cyst germination. *J. Bacteriol.* 191: 4845–4853.

31 Donati, I. and Paoletti, S. (2009). Material properties of alginates. In: *Alginates: Biology and Applications* (ed. B.H.A. Rehm), 1–53. Berlin, Heidelberg, Genmany: Springer Berlin Heidelberg.

32 Campos, M., Martínez-Salazar, J.M., Lloret, L. et al. (1996). Characterization of the gene coding for GDP-mannose dehydrogenase (algD) from *Azotobacter vinelandii*. *J. Bacteriol.* 178: 1793–1799.

33 Moreno, S., Nájera, R., Guzmán, J. et al. (1998). Role of alternative sigma factor AlgU in encystment of *Azotobacter vinelandii*. *J. Bacteriol.* 180: 2766–2769.

34 Gaona, G., Núñez, C., Goldberg, J.B. et al. (2004). Characterization of the *Azotobacter vinelandii algC* gene involved in alginate and lipopolysaccharide production. *FEMS Microbiol. Lett.* 238: 199–206.

35 Núñez, C., León, R., Guzmán, J. et al. (2000). Role of *Azotobacter vinelandii mucA* and *mucC* gene products in alginate production. *J. Bacteriol.* 182: 6550–6556.

36 Castañeda, M., Sánchez, J., Moreno, S. et al. (2001). The global regulators GacA and σS form part of a Cascade that controls alginate production in *Azotobacter vinelandii*. *J. Bacteriol.* 183: 6787–6793.

37 Martínez-Salazar, J.M., Moreno, S., Nájera, R. et al. (1996). Characterization of the genes coding for the putative sigma factor AlgU and its regulators MucA, MucB, MucC, and MucD in *Azotobacter vinelandii* and evaluation of their roles in alginate biosynthesis. *J. Bacteriol.* 178: 1800–1808.

38 Wang, H., Yang, Z., Swingle, B., and Kvitko, B.H. (2021). AlgU, a conserved sigma factor regulating abiotic stress tolerance and promoting virulence in *Pseudomonas syringae*. *Mol. Plant-Microbe Interact.* 34 (4): 326–336.

39 Núñez, C., Moreno, S., Cárdenas, L. et al. (2000). Inactivation of the *ampDE* operon increases transcription of *algD* and affects morphology and encystment of *Azotobacter vinelandii*. *J. Bacteriol.* 182: 4829–4835.

40 Mærk, M., Jakobsen, Ø.M., Sletta, H. et al. (2020). Identification of regulatory genes and metabolic processes important for alginate biosynthesis in *Azotobacter vinelandii* by screening of a transposon insertion mutant library. *Front. Bioeng. Biotechnol.* 7: 475.

41 Castañeda, M., Guzmán, J., Moreno, S., and Espín, G. (2000). The GacS sensor kinase regulates alginate and poly-β-hydroxybutyrate production in *Azotobacter vinelandii*. *J. Bacteriol.* 182: 2624–2628.

42 Manzo, J., Cocotl-Yañez, M., Tzontecomani, T. et al. (2011). Post-transcriptional regulation of the alginate biosynthetic gene *algD* by the Gac/Rsm system in *Azotobacter vinelandii*. *Microb. Physiol.* 21: 147–159.

43 Hernández-Eligio, A., Moreno, S., Castellanos, M. et al. (2012). RsmA post-transcriptionally controls PhbR expression and polyhydroxybutyrate biosynthesis in *Azotobacter vinelandii*. *Microbiology* 158: 1953–1963.

44 Quiroz-Rocha, E., Bonilla-Badía, F., García-Aguilar, V. et al. (2017). Two-component system CbrA/CbrB controls alginate production in *Azotobacter vinelandii*. *Microbiology* 163: 1105–1115.

45 Ahumada-Manuel, C.L., Guzmán, J., Peña, C. et al. (2017). The signaling protein MucG negatively affects the production and the molecular mass of alginate in *Azotobacter vinelandii*. *Appl. Microbiol. Biotechnol.* 101: 1521–1534.

46 Martínez-Ortiz, I.C., Ahumada-Manuel, C.L., Hsueh, B.Y. et al. (2020). Cyclic-di-GMP-mediated regulation of extracellular mannuronan C-5 epimerases is essential for cyst formation in *Azotobacter vinelandii*. *J. Bacteriol.* 202: e00135-20.

47 Peña, C., Peter, C., Büchs, J., and Galindo, E. (2007). Evolution of the specific power consumption and oxygen transfer rate in alginate-producing cultures of *Azotobacter vinelandii* conducted in shake flasks. *Biochem. Eng. J.* 36: 73–80.

48 Gaytán, I., Peña, C., Núñez, C. et al. (2012). *Azotobacter vinelandii* lacking the Na^+-NQR activity: a potential source for producing alginates with improved properties and at high yield. *World J. Microbiol. Biotechnol.* 28: 22731–22740.

49 Gómez-Pazarín, K., Flores, C., Castillo, T. et al. (2016). Molecular weight and viscosifying power of alginates produced in *Azotobacter vinelandii* cultures in shake flasks under low power input. *J. Chem. Technol. Biotechnol.* 91: 1485–1492.

50 Flores, C., Moreno, S., Espín, G. et al. (2013). Expression of alginases and alginate polymerase genes in response to oxygen, and their relationship with the alginate molecular weight in *Azotobacter vinelandii*. *Enzym. Microb. Technol.* 53: 85–91.

51 Lozano, E., Galindo, E., and Peña, C. (2011). Oxygen transfer rate during the production of alginate by *Azotobacter vinelandii* under oxygen limited and non oxygen-limited conditions. *Microb. Cell Factories* 10: 13.

52 Reyes, C., Peña, C., and Galindo, E. (2003). Reproducing shake flasks performance in stirred fermentors: production of alginates by *Azotobacter vinelandii*. *J. Biotechnol.* 105: 189–198.

53 Moral, C.K. and Sanin, F.D. (2012). An investigation of agitation speed as a factor affecting the quantity and monomer distribution of alginate from *Azotobacter vinelandii* ATCC 9046. *J. Ind. Microbiol. Biotechnol.* 39: 513–519.

54 Díaz-Barrera, A., Sanchez-Rosales, F., Padilla-Córdova, C. et al. (2021). Molecular weight and guluronic/mannuronic ratio of alginate produced by *Azotobacter vinelandii* at two bioreactor scales under diazotrophic conditions. *Bioprocess Biosyst. Eng.* 44: 1275–1287.

55 Ponce, B., Urtuvia, V., Maturana, N. et al. (2021). Increases in alginate production and transcription levels of alginate lyase (*alyA1*) by control of the oxygen transfer rate in *Azotobacter vinelandii* cultures under diazotrophic conditions. *Electron. J. Biotechnol.* 52: 35–44.

56 Mejía, M.A., Segura, D., Espín, G. et al. (2010). Two-stage fermentation process for alginate production by *Azotobacter vinelandii* mutant altered in poly-b-hydroxybutyrate (PHB) syntesis. *J. Appl. Microbiol.* 108: 55–61.

57 Sabra, W., Zeng, A.P., Lünsdorf, H., and Deckwer, W.D. (2000). Effect of oxygen on formation and structure of *Azotobacter vinelandii* alginate and its role in protecting nitrogenase. *Appl. Environ. Microbiol.* 66: 4037–4044.

58 Hoskisson, P. and Hobbs, G. (2005). Continuous culture – making a comeback? *Microbiology* 151: 3153–3159.

59 Díaz-Barrera, A., Silva, P., Avalos, R., and Acevedo, F. (2009). Alginate molecular mass produced by *Azotobacter vinelandii* in response to changes of the O_2 transfer rate in chemostat cultures. *Biotechnol. Lett.* 31: 825–829.

60 Díaz-Barrera, A., Martínez, F., Guevara Pezoa, F., and Acevedo, F. (2014). Evaluation of gene expression and alginate production in response to oxygen transfer in continuous culture of *Azotobacter vinelandii*. *PLoS One* 9 (8): e105993.

61 Díaz-Barrera, A., Maturana, N., Pacheco-Leyva, I. et al. (2017). Different responses in the expression of alginases, alginate polymerase and acetylation genes during alginate production by *Azotobacter vinelandii* under oxygen-controlled conditions. *J. Ind. Microbiol. Biotechnol.* 44: 1041–1051.

62 García, A., Ferrer, P., Albiol, J. et al. (2018). Metabolic flux analysis and the NAD(P)H/NAD(P) ratios in chemostat cultures of *Azotobacter vinelandii*. *Microb. Cell Factories* 17: 10.

63 Díaz-Barrera, A., Silva, P., Berrios, J., and Acevedo, F. (2010). Manipulating the molecular weight of alginate produced by *Azotobacter vinelandii* in continuous cultures. *Bioresour. Technol.* 101: 9405–9408.

64 Flores, C., Díaz-Barrera, A., Martínez, F. et al. (2015). Role of oxygen in the polymerization and de-polymerization of alginate produced by *Azotobacter vinelandii*. *J. Chem. Technol. Biotechnol.* 90: 356–365.

65 Castillo, T., García, A., Padilla-Cordova, C. et al. (2020). Respiration in *Azotobacter vinelandii* and its relationship with the synthesis of biopolymers. *Electron. J. Biotechnol.* 48: 36–45.

66 Garcia-Ochoa, F., Gomez, E., Santos, V.E., and Merchuk, J.C. (2010). Oxygen uptake rate in microbial processes: an overview. *Biochem. Eng. J.* 49: 289–307.

67 Garcia-Ochoa, F., Gomez, E., and Santos, V.E. (2020). Fluid dynamic conditions and oxygen availability effects on microbial cultures in STBR: an overview. *Biochem. Eng. J.* 164: 107803.

68 Maier, U. and Büchs, J. (2001). Characterization of the gas–liquid mass transfer in shaking bioreactors. *Biochem. Eng. J.* 7: 99–106.

69 Jiménez, L., Castillo, T., Flores, C. et al. (2016). Analysis of respiratory activity and carbon usage of a mutant of *Azotobacter vinelandii* impaired in poly-beta-hydroxybutyrate synthesis. *J. Ind. Microbiol. Biotechnol.* 43: 1167–1174.

70 Castillo, T., López, I., Flores, C. et al. (2018). Oxygen uptake rate in alginate producer (algU+) and no producer (algU-) strains of *Azotobacter vinelandii* under nitrogen-fixation conditions. *J. Appl. Microbiol.* 125: 181–189.

71 Díaz-Barrera, A., Gutierrez, J., Martínez, F., and Altamirano, C. (2014). Production of alginate by *Azotobacter vinelandii* grown at two bioreactor scales under oxygen-limited conditions. *Bioprocess Biosyst. Eng.* 37: 1133–1140.

72 Peña, C., Campos, N., and Galindo, E. (1997). Changes in alginate molecular mass distribution, broth viscosity and morphology of *Azotobacter vinelandii* cultured in shake flasks. *Appl. Microbiol. Biotechnol.* 48: 510–515.

73 Priego-Jimenéz, R., Peña, C., Ramírez, O.T., and Galindo, E. (2005). Specific growth rate determines the molecular mass of the alginate produced by *Azotobacter vinelandii. Biochem. Eng. J.* 25: 187–193.

74 Peña, C., Trujillo-Roldán, M.A., and Galindo, E. (2000). Influence of dissolved oxygen tension and agitation speed on alginate production and its molecular weight in cultures of *Azotobacter vinelandii. Enzym. Microb. Technol.* 27: 390–398.

75 Trujillo-Roldán, M., Moreno, S., Espín, G., and Galindo, E. (2004). The roles of oxygen and alginate-lyase in determining the molecular weight of alginate produced by *Azotobacter vinelandii. Appl. Microbiol. Biotechnol.* 63: 742–747.

76 Rehm, B.H. (2010). Bacterial polymers: biosynthesis, modifications and applications. *Nat. Rev. Microbiol.* 8: 578–559.

77 Ramos, P.E., Silva, P., Alario, M.M. et al. (2018). Effect of alginate molecular weight and M/G ratio in beads properties foreseeing the protection of probiotics. *Food Hydrocoll.* 77: 8–16.

78 Gacesa, P. (1998). Bacterial alginate biosynthesis – recent progress and future prospects. *Microbiology* 114: 1133–1143.

79 Noar, J.D. and Bruno-Bárcena, J.M. (2018). *Azotobacter vinelandii*: the source of 100 years of discoveries and many more to come. *Microbiology* 164: 421–436.

80 Clementi, F., Crudele, M.A., Parente, E. et al. (1999). Production and characterization of alginate from *Azotobacter vinelandii. J. Sci. Food Agric.* 79: 602–610.

81 Peña, C., Miranda, L., Segura, D. et al. (2002). Alginate production by *Azotobacter vinelandii* mutant altered in poly-beta-hydroxybutyrate and alginate biosynthesis. *J. Ind. Microbiol. Biotechnol.* 29 (5): 209–213.

82 Hempel, D. and Dziallas, H. (1999). Scale-up, stirred-tank reactors. In: *Encyclopedia of Bioprocess Technology. Fermentation, Biocatalysis and Bioseparation, Biotechnology Encyclopedias*, vol. 4 (ed. M. Flickinger and S. Drew), 2314–2332. Wiley.

83 Trujillo-Roldán, M., Peña, C., Ramírez, O.T., and Galindo, E. (2001). The effect of oscillating dissolved tension upon the kinetics of growth, alginate production and molecular weight in cultures of *Azotobacter vinelandii. Biotechnol. Prog.* 17: 1042–1048.

84 Büchs, J., Maier, U., Milbradt, C., and Zoels, B. (2000). Power consumption in shaking flasks on rotary machines: I. Power consumption measurements in unbaffled flasks at low viscosity. *Biotechnol. Bioeng.* 68: 589–593.

85 Büchs, J. (2001). Introduction to advantages and problems of shaken cultures. *Biochem. Eng. J.* 7 (91): 98.

86 Peña, C., Millán, M., and Galindo, E. (2008). Production of alginate by *Azotobacter vinelandii* in a stirred fermentor simulating the evolution of power input observed in shake flasks. *Process Biochem.* 43: 775–778.

87 Quiroga-Cubides, G., Díaz, A., and Gómez, M. (2017). Adjustment and scale-up strategy of pilot liquid fermentation process of *Azotobacter sp. Int. J. Sci. Res. Innov.* 11 (4): 322–330.

88 Høidal, H.K., Svanem, B.I.G., Gimmestad, M., and Valla, S. (2000). Mannuronan C-5 epimerases and cellular differentiation of *Azotobacter vinelandii*. *Environ. Microbiol.* 2: 27–38.

89 Trujillo-Roldán, M., Moreno, S., Segura, D. et al. (2003). Alginate production by an *Azotobacter vinelandii* mutant unable to produce alginate lyase. *Appl. Microbiol. Biotechnol.* 60: 733–737.

19

Environment-Friendly Microbial Bioremediation

Areej Shahbaz[1], Nazim Hussain[1], Tehreem Mahmood[2], Mubeen Ashraf[3], and Nida Khaliq[3]

[1] *Center for Applied Molecular Biology (CAMB), University of the Punjab, Lahore, Pakistan*
[2] *Department of Biotechnology, Quaid-i-Azam University, Islamabad, Pakistan*
[3] *Department of Microbiology, University of Central Punjab, Lahore, Pakistan*

19.1 Introduction

There are a wide variety of microorganisms that are spread over the whole biosphere. This distribution is basically based on the fact that their metabolic capacities are different and unique and allows them to show growth in a greater range of ecological conditions. For the process of biodegradation of various forms of pollutants, the versatility of their nutritional requirements can also undergo exploitation. The process is sustained though on the basis of the capacity of the microorganism for converting, modifying, and utilizing the deadly chemicals or pollutants for the purpose of gaining energy and the production of biomass [1]. As compared to the simple process of pollutants accumulation and storage, bioremediation is a microbiology-based and highly systematized process that is used for breaking down and transforming toxic compounds into least toxic elements or compounds with zero toxicity. Those agents that are required for this process of bioremediation for the purpose of cleaning up polluted areas are called bioremediators. Primary type of bioreactors includes archaea, fungi, and bacteria [2]. As a claim to be considered as a process of biotechnological nature, bioremediation is a method of using various microbes that are able to solve and remove the dangers caused by different toxic agents from the surroundings over and done with the process of biodegradation. The words "biodegradation" and "bioremediation" can be used interchangeably. Microorganisms have an advantage over other kinds of protocol that are used for remediation process and thus are considered a potential tool for removing the lethal pollutants from water, sediments, and soil. Microorganisms are gaining attention because they are helpful not only for the restoration of the original and usual environmental conditions but also for the protection of surroundings from any additional pollutant accumulation [3, 4, 5]. The target of the review is to describe the ongoing

Microbial Bioreactors for Industrial Molecules, First Edition.
Edited by Sudhir P. Singh and Santosh Kumar Upadhyay.

trends in bioremediation protocols carried out by the exploitation of microorganisms and to include the appropriate context that acknowledged the gaps in this particular area. Currently, it is a hot topic in the research field because microbes are environmentally friendly and constitute a favorable and treasured genome for solving ecological fears.

Resources that were once accessible have now been overutilized because of industrialization and overpopulation. Such advancements have also led to enhanced air, water, and land pollution. Currently, because of the great economic importance of heavy metals in industrial area, it has posed a number of ecological difficulties over the globe [6, 7]. Within an ecosystem, the pollution of the environment caused by heavy metals has now become the main cause of threatening conditions for living beings [8, 9]. Referring to the definition posed by Environmental Protection Agency in 2010 [10], the process of parameterization is a process that is carried out spontaneously in which the microbiological protocol is required for breaking down and transforming the precarious contaminating agents into forms that are either least toxic or having no toxicity. As a result, the contaminants are removed, eliminated, or remedied from the environment. Chemical contaminants are exploited as a major source of energy by microbes. The microbes utilize their metabolic abilities to carry out the whole microbiological action. Nevertheless, the bioremediation mediating activity of the microbes can be inhibited by the presence in the soil of large quantities of inorganic nutrients [11]. Microorganisms, notably, have the capacity for detoxifying, degrading, and even accumulating lethal compounds that can be both organic or inorganic. Several different sources contribute to the accumulation of toxic heavy metals in the surroundings, which include natural, agricultural, atmospheric, inland effluent, and sources of industrial solid wastes. A diverse range of areas throughout the world has been polluted because of different activities, which include the usage of pesticides and fertilizers, mining processes such as metallurgical smelting activities, and electroporation [12].

The production of crops and the quality of food crops have been reduced to a great extent because of the increasing pollution posed by heavy metals. The reason underlies the use of different agricultural inputs, such as the application of pesticides or fertilizers, as a consequence of which soils are heavily contaminated by heavy metals [13]. Organic compounds are most commonly applied as pesticides. However, some inorganic compounds, as well as pure minerals, can also be used as pesticides. Moreover, a few heavy metals such as Zn, Hg, As, and Cu can also be employed as heavy metals [14]. As compared with the contaminants of organic nature, metals are unable to degrade and can be retained in the environment for a longer time period. Because of their accessibility for a long time and because of being available in large quantities, heavy metals can have destructive effects on the metabolism of the plant [15]. There is a dire need to develop inventive management strategies to remove heavy metal ions from the water bodies or the soil. A diverse range of microbes have been considered as efficient and cost-effective substitutes for eliminating heavy metals that are present in soil or water bodies [11] (Figure 19.1).

The microbes are first discovered on a biochemical basis, and then their capacity to resist the presence of heavy metals is described. A number of advancements to carry out the bioremediation process have been made in the past two decades, the main purpose of which is the restoration of natural environments from polluted ecosystems at an

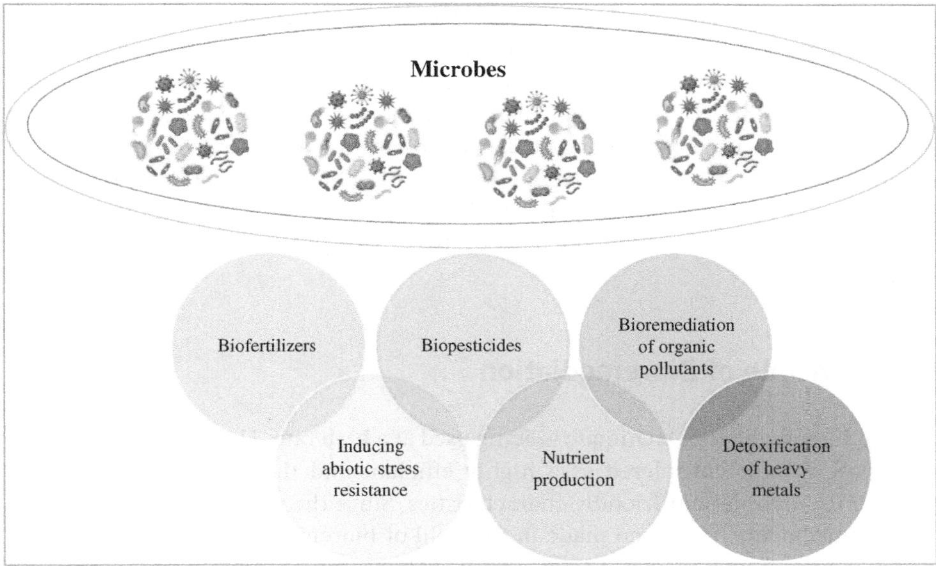

Figure 19.1 Uses of soil microbes in bioremediation.

effective cost. A number of different bioremediation techniques have been modeled and established. However, on the basis of the type and nature of the pollutant that needs to be removed, there is not even a single strategy that we can say acts as a "silver bullet" for the restoration of the environment that has been polluted. Key problems that are encountered during the biodegradation or bioremediation processes are most likely to be solved by the presence of autochthonous (indigenous) microbes in the contaminated surroundings [16] provided that the circumstances of the environment are appropriate for the metabolism and growth of these microorganisms. Unlike the physical and chemical methods, the most important benefit provided by the bioremediation process is the cost-effectiveness and the ecofriendly nature of the process. In some cases, however, bioremediation and biodegradation are interchangeable terms, the latter term referring to the process under the former. In this review, bioremediation is described as a method that depends on biological means in order to achieve the aim of degrading, mineralizing, detoxifying, or transforming hazardous amounts of pollutants into a form that is nonpoisonous.

Depending on the type of the pollutant, whether it is a chlorinated compound, an agrochemical, any greenhouse gas, heavy metals, plastics, or any sewage material, pollutant removal takes place. There are two categories of the bioremediation process on the basis of the site of the protocol. The process can be carried out in both *ex situ* or *in situ* environment. In order to choose a better and appropriate strategy for bioremediation, some factors must be taken into account, such as the extent of pollution, types of contaminant, the environment that is polluted, the site of action, and the expense of the whole process [17]. In addition to this criterion on the basis of which bioremediation technique is selected, other criteria for the performance must also be given some important considerations before carrying out the whole project of bioremediation. This performance criteria is the one that regulates the extent to which the bioremediation technique would be successful [18]. Advancements in the field of bioremediation technology and the attention of current research are now focused on microbe-mediated bioremediation, which is considered as a novel and effective strategy for solving the problem of polluted environments that retain high amounts of accumulated heavy metals. The following review addresses the types of microorganisms that have been reported to be a potential root for carrying out the process of bioremediation.

19.2 Principle of Bioremediation

Currently, bioremediation technique is being used at a high rate. The use of microbes for this purpose is now considered as a highly efficient and the most reliable method because of its environment-friendly characteristics. Since the past two decades, a number of advancements have been made in the field of bioremediation, and the ultimate goal of these developments was to successfully achieve the restoration of the contaminated surroundings keeping the expenses low and utilizing the approach that should be environment friendly. Some of the researchers succeeded in developing different strategies that carried out bioremediation in a suitable manner. Both indigenous and non-indigenous types of microbes can be exploited for this purpose. The microbes holding the key to removing the problems that are related to bioremediation process are mostly of indigenous type [19]. The most important benefits that are provided by microbe-mediated bioremediation involve the cost-effectiveness and the ecofriendly features it offers.

The principle of bioremediation involves the reduction, detoxification, degradation, mineralization, or transformation of the toxic agents into a form that shows the least toxicity (Figure 19.2). The whole pathway of removing the pollutant relies on the type of pollutant. The pollutant can be of any nature, such as pesticides, chlorinated

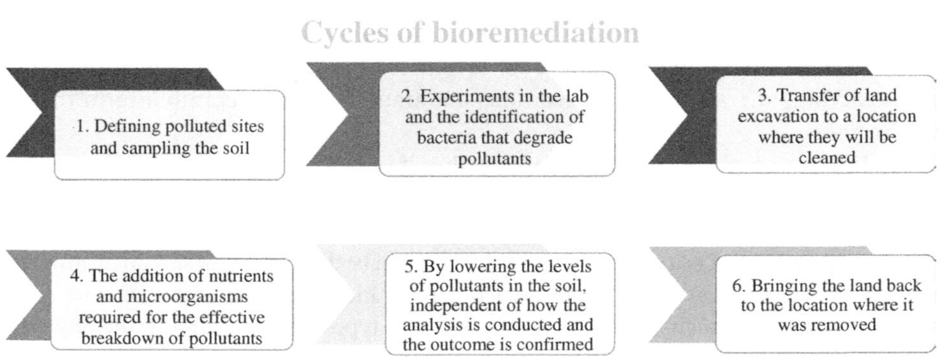

Figure 19.2 Cycles of bioremediation.

components, heavy metals, compounds having xenobiotic nature, agricultural compounds, hydrocarbons, waste from nuclear industries, plastics, and untreated sludge. Cleaning approaches are employed for removing the polluting wastes from the contaminated surroundings. Bioremediation greatly focuses on the possibility of degrading, immobilizing, detoxifying, and completely eradicating the wide range of toxic chemical and physical agents from the environment by means of all-encompassing activity of microbes [20].

The process of removing, detoxifying, or reducing the amounts of a harmful xenobiotic polluting agent from the surroundings by means of biological living beings is described as bioremediation. The living beings employed can be microbes or their enzymes. Plants can also be used, in case of which the process is termed "phytoremediation." The process of bioremediation proposes the probability of using microbial [21, 22], fungal [23], or plant-based actions in order to remove, degrade, alter, immobilize, or detoxify the toxic and pollution causing chemicals that exist in the biosphere. The action of microbes in bioremediation ends up in the cleanup of harmful polluting agents from the surroundings by detoxifying, breaking, mineralizing, removing, or sequestering them.

The process by which structurally large and chemically intricate groups of polluting agents are crumbled or decomposed into a minute and simple compound that are nontoxic is termed as "degradation." Some byproducts are also produced during the process, which may include water, carbon dioxide, or any other nontoxic component. The process of sequestration refers to the one by which the toxic agents present in the surroundings are either confined or manipulated in such a way that they become harmless or inaccessible to living beings. In the method of removing pollutants, though the harmful polluting agent is not essentially degraded, toxic agents are physically removed from the surroundings so that they can be carefully extracted and disposed of. Normally, when toxic compounds get retained in the surroundings for a very long time, bioremediation technique takes place in a series of steps by means of the microbial enzymes or by various microbial members existing in the polluted location. The process is typically applied at different sites such as wastewater, agricultural soil, groundwater or surface water, water bodies, air, or sediments, which can get polluted by the release of harmful polluting agents or chemicals [24, 25].

One of the most useful and environment-friendly strategy involved in removing persistent toxic agents from the surroundings is the degradation of the xenobiotic by means of microorganisms. The capability of microbes to degrade, metabolizing and transforming the xenobiotic components has been renowned as a proficient method of eliminating toxic and unfavorable trashes [26, 27]. Microbes are preferably appropriate for the job of pollutant annihilation and elimination because they have an enzyme system, permitting them to utilize ecologically poisonous contaminants as food and energy. The maximum of the developments in bioremediation discipline has been accredited to the discrete and interdisciplinary input provided by technical fields of molecular biology, biochemistry, environmental engineering, analytical chemistry, and microbiology [28]. The bioremediation method encompasses mineralization and detoxification, where the waste is transformed into inorganic complexes, for instance, methane, carbon dioxide, and water [29].

19.3 Types of Bioremediations

There are diverse forms of management approaches or methods under bioremediation process and the types (Figure 19.3).

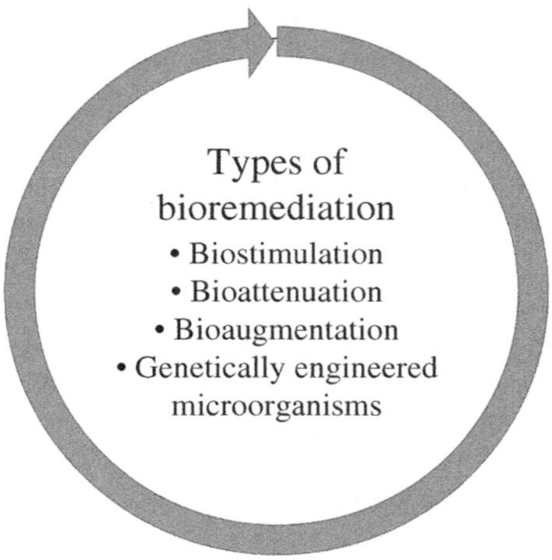

Figure 19.3 Types of bioremediations.

19.3.1 Biostimulation

The strategy of biostimulation involves the inoculation of particular nutrients at the location of groundwater or soil. This injection is done for the purpose of stimulating the action of indigenous microbes. Biostimulation is focused on prompting the activity of indigenous microbes as well as the bacterial and fungal community that exists naturally in the soil. This can be done first by adding trace elements and growth supplements or by employing fertilizers. Further, the metabolic activities and pathways of the microbes can be enhanced by fulfilling their ecological necessities, such as oxygen, temperature, or pH [30, 31]. The operons coding for the enzymes involved in bioremediation can be turned on by the availability of minute quantities of a pollutant that has the ability to act as a stimulant. Most frequently, this kind of pathway is followed by adding nutrients and oxygen to aid local microorganisms. Microorganisms are capable of producing basic needs by taking help from the nutrients, which are considered as the elementary building blocks of life. Carbon, nitrogen, and phosphorous are required by most microbes. These basic necessities include cell biomass, energy, or the enzymes required for the degradation of polluting agents [32].

19.3.2 Bioattenuation

The process of eradicating the pollutants from the environment is termed as "bioattenuation" or "natural attenuation." It is done by means of a biological method, including uptake by plants or animals and biodegradation carried out in either an aerobic or an anaerobic manner. Volatilization, diffusion, advection, dilution, dispersion, and desorption/sorption comprise the physical methods, while chemical ones constitute the reactions involving abiotic transformation, complexation, and the process of ion exchange. Biotransformation and intrinsic remediation are the terms that are encompassed in the more common definition of natural attenuation [33].

There are four methods that nature can adopt in order to remove the toxic and pollution-causing agents from the surroundings [33]: (i) some of the chemicals are utilized as food by a number of microorganisms inhabiting the soil. Chemicals, after being fully digested, are transformed into H_2O and gases that are not harmful at all. (ii) Soil makes the chemicals stick or sorb over the soil by holding them in the abode. This method does not help in removing the chemicals, but it surely helps in protecting the groundwater from getting polluted. (iii) As the pollution flows over the soil and groundwater and can get mixed with pure water, this method can help in either diluting or reducing pollution. (iv) The fourth way in which the environment can be cleaned is by the evaporation of chemicals. For instance, in the soil, the conversion of solvents or oils into gases, which might get escaped in the air and get extinguished by the sunlight. If the process of natural process solely is not done rapidly, the bioremediation can be further augmented by the methods of bioaugmentation as well as biostimulation.

19.3.3 Bioaugmentation

It is one of the strategies adopted for carrying out the process of degradation. Bioaugmentation refers to the process by which the ability of the indigenous microbial community to degrade the polluting agents is augmented or enhanced in the polluted environment. With the aim of achieving the degradation, the microbial population increased that relies on feeding over the polluting agents in the polluted space. Microbes are isolated from the bioremediation site, followed by their individual culturing and genetic modifications. After that, the microbes are placed back at the site. It has been recognized that almost every necessary microbe exists in the soil or groundwater, which have become polluted because of the presence of chlorinated ethenes, for instance, trichloroethylene or tetrachloroethylene. Thus, it has been made sure that the contaminants like these can be efficiently eliminated or modified by *in situ* microbes into forms like ethylene or chloride, which have no toxicity [34]. During the process of bioaugmentation, genetically modified microbes are counted into the system, which then shows the actions of bioremediators for the purpose of eliminating the polluting agents in a rapid manner. Relying on the fact that microbial communities comprise varied metabolism capacity for producing the least harmful end compounds [35]. Recombinant DNA technology is exploited in order to manipulate the microorganisms for effective degradation of the polluting agents because the naturally existing microbial species do not perform this action in a quick manner. Moreover, genetically modified microorganisms greatly compete with naturally existing microbes and predators for abiotic factors. A number of different sites such as contaminated soil, activated sludge, and water bodies have been shown to effectively undergo the process of bioremediation by genetically modified microbes because these microbial communities offer better degradation abilities leading to the removal of a wide range of physical and chemical polluting agents [36].

19.3.4 Genetically Engineered Microorganisms (GEMs)

A microorganism having its genetic material altered by means of different genetic engineering approaches is referred to as genetically engineered microorganism (GEM). This type of creative work and technical approach falls under the field of recombinant DNA technology. A number of harmful and toxic pollutant agents have been completely utilized or removed by means of developing genetically engineered microbes attributing to the science of genetic engineering [37]. Genetically modified microbes can be obtained either by exploiting the approaches of recombinant DNA technology or by the exchange of genetic components between the microorganisms that exist naturally. Currently, it is done by the insertion of a suitable gene that codes for a specific enzyme involved in the degradation of a diverse range of polluting substances [38].

Currently, a variety of opportunities advance toward the enhancement of the degradation capability of microorganisms by using genetic engineering techniques. For instance, rate-limiting steps in well-known metabolic processes can be genetically altered to produce faster rates of degradation, or brand-new metabolic pathways might be added to bacterial strains to speed up the degradation of chemicals that have previously shown resistance. Four kinds of actions are done while developing genetically modified organisms. These strategies are: (i) altering the affinity and particularity of the targeted enzyme; (ii) building and regulating the mechanism pathway; (iii) establishing, monitoring, and regulating the bioprocess; (iv) developing the bioreporter sensor employments for sensing chemicals, eliminating toxins, and making end point inquiries. The necessary bacterial genes are usually present on an individual chromosome, but those genes that code for the enzymes involved in catabolizing the rare pollutant substrates may be present on the extrachromosomal DNA that is plasmid. The catabolism has been linked to plasmids. As a result, GEMs can be effectively exploited for biodegradation purposes, which represents or indicates a research frontier with significant future consequences [39].

19.4 Factors Affecting Microbial Bioremediation

The lack of contact between bacteria and contaminants is the main factor affecting the rate of deterioration. Additionally, the distribution of bacteria and contaminants in the environment is not constant. Due to a variety of circumstances, regulation and optimization of bioremediation processes are complex systems. The presence of a microbial population able to degrade pollutants, the accessibility of contaminants to the microbial population, and environmental conditions are all considered here (pH, temperature, type of soil, nutrients, presence of oxygen, and other electron acceptors) (Figure 19.4).

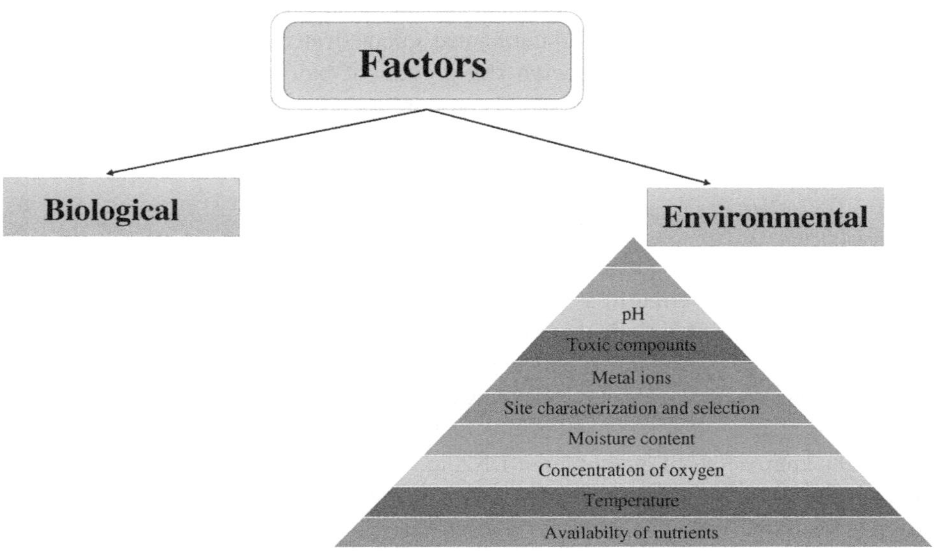

Figure 19.4 Factors affecting microbial bioremediation.

19.4.1 Biological Factors

Microbes involved in bioremediation compete with each other for inadequate carbon supplies. Moreover, on the basis of the antagonistic communication between microbes and predators, biotic factors can influence the rate at which the organic substrates are degraded. This rate at which the contaminants are degraded is usually reliant on the amounts of existing contaminant and catalysts. Here, the catalyst concentration refers to the quantity of the microbial organisms that have the ability to metabolize the polluting agents as well as the sum of enzymes made by every cell. The rate of pollutant degradation also depends on the upregulation or downregulation of the gene expression that codes for such degrading enzymes. In addition, the access of the contaminant to enzymes, the affinity of the enzymes, and the availability of a particular enzyme also influence the degree to which the contaminants are metabolized. The size of the population; gene transfer by horizontal manner; action of the enzymes involved; interactions such as predation, succession, or competition; critical biomass obtained; and the presence of mutagenesis are some of the significant biological aspects that influence the degree of bioremediation [32].

19.4.2 Environmental Factors

How the microbes will possibly communicate during the process is determined by the metabolizing features of the microbes as well as the physiochemical characteristics of the pollutants that need to be removed. However, an actual effective collaboration is usually reliant on the circumstances of the polluted site. The pH, temperature, humidity, soil composition, water solubility, nutrients, site features, redox potential, oxygen levels, shortage of competent human resources in this field, and physiochemical accessibility of contaminants are all factors that affect microorganism growth and activity. The kinetics of biodegradation also relies on the aforementioned aspects [30, 32]. A diverse range of pH can be employed during the biodegradation process, but an optimal biodegradation process occurring in water bodies or land ecosystems is usually carried out at pH of 6.5–8.5. Since it impacts the type and supply of soluble materials, and also the turgor stress and pH of land and freshwater systems, moisture influences the speed of contaminant metabolism [40]. Following is a list of a majority of environmental factors.

19.4.2.1 Availability of Nutrients

The growing and reproductive capacity of the microbes require a vital balance of nutrient supply. This is the reason why nutrients are supplied for adjusting this balance. Moreover, the nutrient supply also influences the frequency and efficacy of the biodegradation process. The effectiveness of the process can be enhanced by the optimization of C: N: P ratio provided by an enriched supply of nutrients, particularly nitrogen and phosphorus. Moreover, a diverse range of nutrients, especially phosphorus, carbon, and nitrogen, are required by microbes in order to survive and maintain their microbial actions. The degradation of the hydrocarbons is also reduced if the nutrient supply is provided in small amounts. The metabolizing action of microbes and consequently the degree of bioremediation process in cold surroundings is enhanced by providing an adequate amount of nutrients [41, 42]. The process of biodegradation is reduced by the accessibility of nutrients in water bodies [43]. Apart from fulfilling the nutritional requirement of other organisms, optimal growth and development of the oil-eating microorganisms also need nutrient supply. In the natural surroundings, these nutrients exist but in small amounts [44].

19.4.2.2 Temperature and pH

Several physical aspects influence the ability of microorganisms to survive, among which temperature is the most vital one. The configuration of hydrocarbons is also reliant on temperature [45]. The degradation of oil by means of a natural approach proceeds at a low rate in cold surroundings, such as those of the Arctic region. It puts the microbial community under a higher burden of cleaning up the dropped petroleum. Inactivation of the metabolizing activity of the microbial cells results from the below zero temperature of water in such areas, which inhibit the transport pathways in the microbial cells or may even cause freezing of the whole cytoplasm. The reason that lies under this phenomenon is that all biologically coded enzymes work in order to degrade only at an optimal temperature, and once the temperature rises or falls below the optimum range, the degradation and metabolic capacity of the enzyme will not remain the same. In addition, a particular temperature is required in order to degrade a particular substrate. The physiochemical features of the microbial community are greatly impacted by the temperature leading to the upregulation or downregulation of the bioremediation protocol. Temperature affects the rate of microbial activity, which rises with temperature and peaks at the ideal temperature. As the temperature continued to rise or fall, it abruptly started to plummet and eventually stopped when it reached a certain value [46].

The metabolic potential of microorganisms is likely to be affected by the nature of the pollutant, whether it is a base, acid, or alkali. Moreover, the pH of a substrate also has the potential to enhance or limit the removal process. Measuring the pH of the soil can help us to identify the possibility of the growth of microbial communities [47]. However, pH values that are really high or really low may result in low-grade consequences as the metabolic actions are greatly vulnerable to even minute alterations in pH [48].

19.4.2.3 Concentration of Oxygen and Moisture Content

The requirement of oxygen varies between different microbes, and the basis of this oxygen requirement assists the biodegradation extent in an improved manner. As oxygen acts as a gaseous need for a huge number of living beings, its presence in a number of cases can improve the degradation of hydrocarbons [44]. The growth of microbes is achieved in the presence of a sufficient quantity of water. The amount of water present in the soil also adversely impacts the agents required for bioremediation process.

19.4.2.4 Site Characterization and Selection

Before proposing an appropriate bioremediation approach, a satisfactory remedial exploration effort must be made in order to sufficiently describe the scale and degree of pollution. This task should be done at a minimal level and include the following elements: fully defining the lateral and vertical magnitude of pollutants, reviewing the variables and sites to be sampled and the explanation for their selection, and defining the steps to be taken for sample collection and analysis.

19.4.2.5 Metal Ions and Toxic Compounds

When present in trace quantities, metal ions are vital for the fungal and bacterial community, but their excess amounts may block the metabolizing action of the cells. The extent of the degradation process is directly or indirectly influenced by metal ions. When the polluting agents having contaminating features are present in an excess amount, they can lead to the development of toxicity leading to the downregulation of bioremediation process. The amounts and presence of the particular pollutants and accessibility of the microbes determine the extent and pathway of toxicity development. Living beings may be affected by the toxicity of some organic or inorganic substrates [32].

19.5 Bioremediation Techniques

Both *ex situ* and *in situ* systems can be employed for carrying out the process of bioremediation. When choosing an appropriate bioremediation system, several aspects must be considered, such as the type of contaminant, concentration of the polluting agent, the type of contaminated surroundings, expenses, and the ecological policies. The rate at which the process is successfully done is reliant on the performance standards such as pH, oxygen and nutrient supply, temperature, and several added abiotic elements [17, 49].

The process of bioremediation in which the polluting agents are first dug up from the contaminated environment followed by their successful transport toward the other management sites is referred to as *ex situ* bioremediation techniques. These kinds of bioremediation approaches are usually considered on the basis of the extent of pollution, the kind of polluting agent, the expense of the process, and the place at which the contaminated site is located. The selection of an appropriate *ex situ* bioremediation strategy is also regulated by the performance standards [20]

In an *ex situ* microbe-mediated bioremediation method, the polluted soil needs to be evacuated, and the contaminated groundwater needs to be pumped for the facilitation of the degrading process. However, this method has more drawbacks than benefits. The expenses for this kind of bioremediation technique are usually higher due to the fact that the polluted samples are needed to be evacuated. Moreover, in contrast to the *in situ* bioremediation approaches, the frequency and constancy of the *ex situ* bioremediation method results can be changed. The solid phase and the slurry phase methods are the two classes of *ex situ* bioremediation on the basis of the state of the polluted agent that needs to be evacuated. A number of different solids, for instance, municipal wastes, wastes from the chemical and agricultural industries, sludge, organic wastes, and waste from water bodies are all encompassed in the solid phase treatment. The methods of soil biopile, composting, and land farming are all involved in the solid phase system. In land farming, a complete breakdown or conversion of the pollutants is done by stimulating the naturally existing microbes having the ability to degrade. This is a simple approach for remediation in which the contaminated soil is dug and spread on a created bed and occasionally turned over to encourage the microbes. However, when only the superior 0.5 m of the soil is contaminated, the process can only be beneficial. Since it can be considered a good option for clean disposal, land farming has gained huge attention [50]. One other form of surface management that has been effectively utilized for the degradation, removal, or transformation of

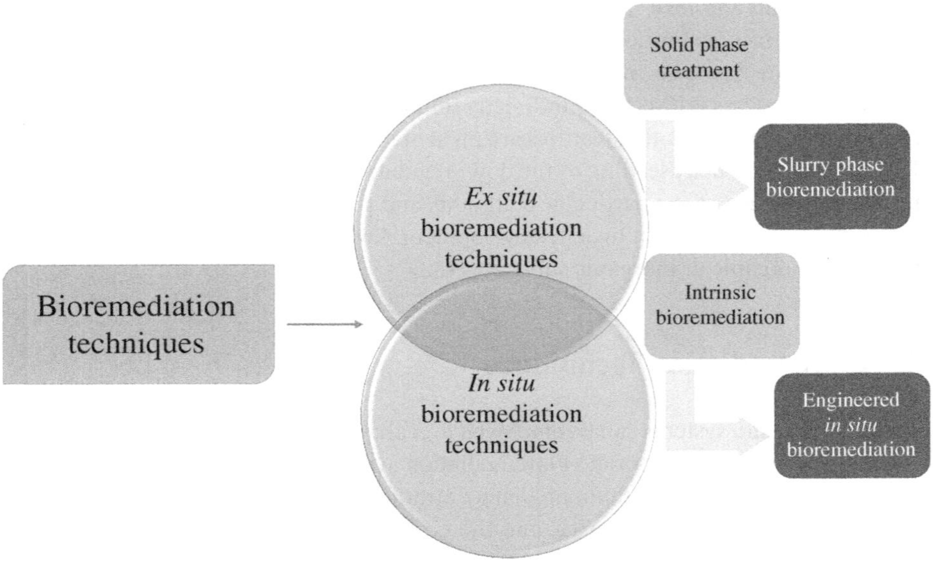

Figure 19.5 Techniques of bioremediation.

toxic agents is composting. In this method, the process is the speed by improving the establishment of a rich microbial community. This developmental process is usually enhanced by mixing the polluted soil with nontoxic organic wastes [51].

When the techniques of land farming and composting get combined, they lead to a new bioremediation approach that is called biopiles. The main purpose of this technique is to limit the physical loss of the pollutants that occur as a result of leaching and volatilization. It is a revised form of land farming technique, and the sites that have become polluted with petroleum hydrocarbons are usually treated with biopiles. Suitable conditions for both aerobic and anaerobic microflora are offered by this process of biopiles [48]. In contrast to other treatment techniques, the slurry phase system offers a faster approach. It involves mixing the polluted soil with water and naturally existing microbes in a bioreactor. Moreover, some of the appropriate oxygen and nutrient supply is provided to the bioreactor to regulate the optimal conditions required to carry out the bioremediation process. Once the process has been done, the liquid part of the soil is extracted and carefully disposed [51]. The frequency and degree of the bioremediation process are higher when performed in a bioreactor or fermenter as compared to the one carried out in *in situ* or solid phase systems because a bioreactor offers a more predictive and controllable process (Figure 19.5).

19.6 Methods for *Ex Situ* Bioremediation

19.6.1 Solid Phase Treatment

Solid phase treatment falls under the category of an *ex situ* technique and involves the evacuation and piling up of the soil, which is polluted by a number of pollutants. A number of different organic wastes are also encompassed in this technology, such as the leaves.

Moreover, wastes from industries, agricultural productions, domestic places, and animals are also included. A number of piles have diverged from the pipes through which the bacterial community is moved. In order to make the respiration of the microorganisms and for proper ventilation, it is vital to aerate the pipes. Unlike the slurry phase method, this is an extensive process that needs a large space. Methods for solid phase management encompass composting, land farming, windrows, biopiles, and more [52].

19.6.1.1 Slurry Phase Bioremediation

In contrast to other approaches employed for treatment, the method of slurry phase bioremediation occurs more quickly. Inside a bioreactor, soil that has been polluted with a variety of contaminants is mixed with oxygen, nutrients, and water for the purpose of developing an optimized condition that will further facilitate the degradation of the pollutants by means of microbes. This method also helps in separating the grits and other debris from the polluted soil. The rate at which the biodegradation takes place and the amounts of pollutants along with the physiochemical features of the soil determine the concentration of water that needs to be added. Once the process has been completely done, the soil is removed, followed up by drying by the use of centrifugation, vacuum filtration, or pressure filtration. Subsequently, the resulting fluids are treated in an advanced manner, and the soil is disposed.

19.6.1.2 *In Situ* Bioremediation

In contrast to other methods, the *in situ* bioremediation strategy does not need the extraction and elimination of polluted soil or water with the aim of achieving bioremediation. Generally, this process involves the flow of aqueous suspensions through the contaminated soil for the excitation of the originally existing microbes, thus enhancing the degradation of the toxic polluting agents. This circulation is done for the purpose of providing the source of electron acceptors, oxygen, and some other nutrients. *In situ* bioremediation involves exploitation of microorganism-based inoculum and the enzymes that are at liberty from the cell. This usually treats saturated soils and water at the surface of ground being contaminated by employing this strategy [53]. This approach of cleaning up the surroundings is considered to be better than other strategies because it is cost-effective and exploits the naturally existing and not-so-toxic microbes in order to achieve the degradation of the contaminants. Moreover, this process is a safe method and can help in managing the huge amounts of polluted soil and water bodies with the discharge of the polluting agents in minute amounts. Two groups that fall under the *in situ* bioremediation category are engineered bioremediation and intrinsic bioremediation.

19.6.2 Engineered Bioremediation

The provision of particular engineered microbes to the polluted site, thus enhancing the microbial growth by improving their physiochemical changes, is termed as "engineered bioremediation." This method results in the acceleration of the bioremediation process.

19.6.3 Intrinsic Bioremediation

On the other hand, the intrinsic bioremediation strategy involves the stimulation of the naturally existing microbes and their metabolizing action by nourishing them with stimulants, which are usually oxygen and some other nutrients. Natural reduction, another name for intrinsic bioremediation, is an *in situ* bioremediation process that involves passive remediation of polluted places without the need for any outside force (human intervention). Electron acceptors, oxygen, phosphorus, and nitrogen are usually employed for enhancing the growth of microbes as well as for speeding up the bioremediation process [53]. Chlorinated hydrocarbons, nitriles, nitrobenzenes, and plasticizers are generally degraded by this strategy of *in situ* bioremediation [54].

19.7 Bioremediation Using Microbial Enzymes

Bioremediation is primarily dependent on degrading enzymes to implement a system. It is an emerging approach in the modern technological age use of enzymes as an ecofriendly, sustainable approach for soil and water cleanup. Enzyme-based bioremediation is a fusion of biological and chemical remediation technology. Because of current innovations in biology, it is now possible to isolate and purify these enzymes. These enzymes are then extracted and infused into contaminated water or soil. Enzyme remediation has been successfully tested in other countries for the cleanup of sites contaminated with a variety of organic contaminants, such as petroleum, PCBs, and nitrophenols. Since no waste is produced or disposed of during this enzymatic bioremediation, there is no long-term environmental consequence. It has been revealed that a vast number of enzymes secreted by bacteria and fungi are crucial in the biodegradation of hazardous organic contaminants. Bioremediation, which is powered by microbial enzymes, is an innovative bioprocess technology. It is economical and environmentally friendly. Different classes of enzymes used in bioremediation have been identified (Figure 19.6).

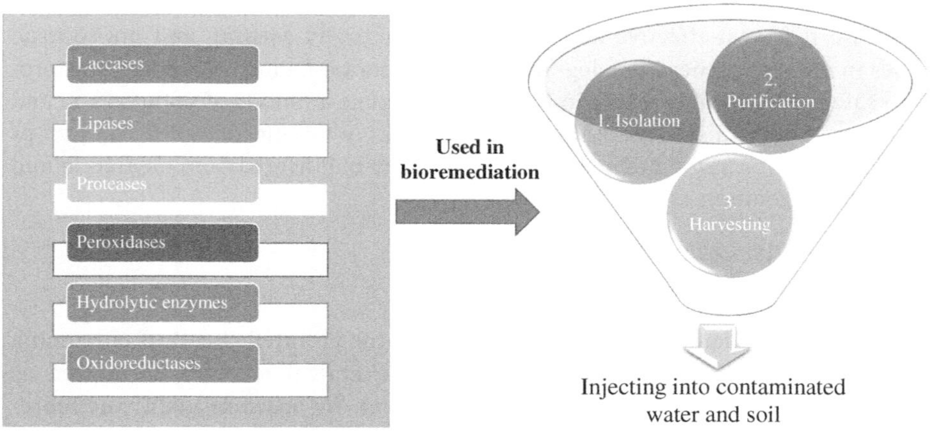

Figure 19.6 Enzymes used in bioremediation.

19.7.1 Laccases

Laccases are a group of multicopper oxidases that are made by bacteria, fungi, and insects. These laccases stimulate the oxidation of a wide spectrum of reduced phenolic and aromatic compounds while concurrently reducing molecular oxygen to water [55]. According to Rodriguez Couto and Toca Herrera [55], laccases are produced both inside and outside of cells and have the ability to catalyze the oxidation of ortho- and para-diphenyls, aminophenols, polyphenols, polyamines, lignins, and aryl diamines as well as some inorganic ions. They are also responsible for the depolymerization of lignin, making them appear to be a universal set of enzymes with a lot of potential for biotechnological and bioremediation applications [56].

19.7.2 Lipases

The biodegradation of soil-based organic contaminants is closely linked to lipases. They operate on a variety of lipids produced by microbes, animals, and plants and breakdown them. Lipases, which may be extracted from bacteria, actinomycetes, and mammalian cells, have been shown to significantly lower the overall hydrocarbon content of polluted soil [57]. Because of their significant contribution to the remediation of oil spills, industrial wastes, triglycerides [58], and hydrocarbon pollutants, microbial lipases are more adaptable [58]. They have been discovered to be the most helpful by catalyzing a variety of processes, including hydrolysis, interesterification, esterification, alcoholysis, and amylolysis [59]. A total of 157 factors for measuring hydrocarbon breakdown in the soil are included in the paper "Microbe-Mediated Bioremediation: An Eco-friendly Sustainable Approach." Although they are used for bioremediation diagnostic purposes, the cost of manufacture has limited their industrial utilization [58].

19.7.3 Proteases

A lot of proteinaceous material enters the environment as a byproduct of some industries, such as poultry, fishing, and textiles, as well as from the withering and molting of animal appendages. The proteases, which are classified into endopeptidase and exopeptidase groups based on where they are active on the substrate, hydrolyze these proteinaceous molecules. With their numerous uses in food, medicinal, textile, and chemical industries, proteases play both a direct and an indirect role in bioremediation [60, 61].

19.7.4 Peroxidases

Another group of universal enzymes generated by fungi and prokaryotes is the peroxidase family. Chlorinated phenolic are eliminated by peroxidase from polluted surroundings [62]. At the expense of hydrogen peroxide, peroxidases also stimulate the oxidation of lignin and other phenolic substances (H_2O_2). They are essential for auxin metabolism, lignin and suberin synthesis, cross-linking of cell wall constituents, protection toward pathogens, and cell elongation in plants. They can be heme or non-heme peptides [63]. Peroxidases are classified as lignin peroxidase (LiP), manganese-dependent peroxidase (MnP), and versatile peroxidase based on their origin and action (VP). Due to their strong capacity to break down harmful chemicals in nature, all of these have been extensively examined.

19.7.5 Hydrolytic Enzymes

Hydrolytic enzymes are a significant threat as harmful pollutants in many aquatic and terrestrial environments because of the widespread use of industrial chemicals and petroleum hydrocarbons. An efficient route for the microbial degradation of oil spills and organophosphate and carbamate pesticides is provided by hydrolytic enzymes, which break important chemical interactions in hazardous chemicals. They are crucial in the degradation process of organochlorine pesticides like DDT and heptachlor that are durable in soil with good air circulation but quickly disintegrate in anaerobic conditions [64]. Due to their use in the decomposition of biomass, hemicellulose, cellulose, and glycosidase, all have significant potential applications [65].

19.7.6 Oxidoreductases

According to [62], oxidoreductases are a group of microbial enzymes that perform oxidation-reduction processes and also help in the detoxification of a variety of hazardous chemical molecules. These oxidative coupling processes change the contaminants into nontoxic molecules. Numerous microbial oxidoreductases have been used to detoxify various harmful xenobiotics, including phenolic or anilinic chemicals, by polymerization, copolymerization with some other targets, or coupling to compounds as well as degrading azo dyes [66]. Oxidoreductases are crucial for the biodegradation of a number of phenolic compounds that are formed during the decomposition of lignin in a terrestrial ecosystem.

19.8 Bioremediation Prospects

Bioremediation techniques are diverse and ought to be established effectively in reinstating contaminated places. Microorganisms show a central part in bioremediation; thus, their variety, profusion, and communal assembly in contaminated surroundings propose a vision into the chance of some bioremediation method as long as additional ecological influences can inhibit the action of microbes.

"Omics" encompasses the fields of transcriptomic, metabolomics, proteomics, and genomics and is considered a promising advancement in the science of molecular biology. It offers a significant contribution to the process of identifying microbes and their functions. Moreover, their metabolic and catabolic potential is also determined by using this science. Inconsistency in the nutrient supply and no presence of required amounts of microbes having the potential to degrade may lead to delayed accomplishment of the bioremediation process. As the process of bioremediation is microbe mediated, the strategies to carry out bioremediation and bioaugmentation help in speeding up the metabolizing actions of the microbial community in the contaminated surroundings. Biostimulation offers the provision of nutrient stimulants in order to enhance the degradation capacities of the microbes at the polluted site. An excess amount of microbes usually inhabits a diverse range of environments; thus, it is perceptible that microorganisms having the ability to degrade the polluting agents already exist in the contaminated location. The type of pollutant that needs to be removed and its concentration may have a significant impact on the growth and metabolizing action of the microbes. The nutrient supply may provide the polluted sites by means of the wastes obtained from agricultural industries. As compared to pure isolates, an association of microbes has been described to achieve biodegradation in a more effective manner [67].

When such isolates are combined, this activity may result in complete and quick degradation of pollutants due to the metabolic diversity of individual isolates, which potency is created by their isolation source, adaption process, pollutant composition, and synergistic effects [68]. Additionally, compared to a non-amended setup (control), both bioaugmentation and biostimulation have been reported to be successful in eliminating pollutants such polyaromatic hydrocarbons (PAHs) from a substantially polluted sample [69].

Although the process of biodegradation is considered to be an efficient strategy, thereby enhancing degradation of many compounds, if adequate microbes do not exist in the polluted surroundings or if the pollutants reduce the amounts of the microbial community, particular microbes can be provided to enhance the existing microbial community and the likelihood that the injected microbes may not endure the new surroundings; this may lead to the explanation of this process as an uncertain approach. Bioaugmentation process refers to a bioremediation approach that involves the use of microbes having particular metabolic features and either existing naturally or engineered genetically. There are certain hurdles that are related to the process of bioaugmentation, which involves the utilization of carrier substances such as gelatin, agar, alginate, polyurethane, and gellan gum [70].

Biosurfactants are considered to be equivalent to the chemical compounds exhibiting a number of environment-friendly and recyclable features. Nevertheless, high expenses associated with the construction and applicability of the biosurfactants to the contaminated surroundings at a low scale make this process economically unsuitable. During fermentation, combining the wastes from the agricultural industries acts as a nutrient supply for the establishment of biosurfactant manufacturers [71]. The employment of the genetically modified microbes for the purpose of improving the bioremediation potential is considered a promising strategy.

Nevertheless, gene transferring in a parallel manner and the reproduction of the genetically engineered microbes in a surrounding application are also considered as a promising strategy. A process that involves the containment of bacterial community encompasses the escaping of any genetically modified microorganism into the surroundings for the purpose of rebuilding the environment that has become contaminated. Furthermore, the efficacy of the bioremediation approach can be enhanced by employing the derivative mechanisms of GEM with a directed polluting agent. The time to complete the bioremediation process and the associated expenses can be reduced by the use of nanomaterials. This is due to the fact that nanoparticles have the high surface area to volume ratio, have low energy of activation, and can decrease the toxicity that the polluting agents cause to microbes [72]. The merits, demerits, and limitations of bioremediation are mentioned in Table 19.1.

Table 19.1 Merits, demerits, and possible limitations of bioremediation.

Merits	Demerits	Limitations
1) Offers the chance for organic contaminants to completely break down, degrade, or mineralize into other harmless substances in a natural ecosystem.	The formation of harmful intermediate molecules with greater mobility than the original pollutants due to incomplete or partial breakdown of organic contaminants.	The biological processes are very particular. The availability of metabolically competent microbial populations, optimal environmental growth conditions, and appropriate quantities of nutrients and pollutants are crucial site elements that are essential for success.
2) Depending on the circumstances, it can be used as both an *in situ* and an *ex situ* approach.	Tracking the development of the organic pollutants' biodegradation requires outdoor monitoring.	Scaling up the bioremediation process from batch and pilot scale investigations to large-scale field operations is challenging.
3) When compared to other remediation techniques, the cost of treatment per unit volume of soil or groundwater is quite low.	Process takes longer than other remediation technologies and typically needs longer treatment times.	For sites with composite combinations of toxins that are not evenly distributed in the environment, modern engineered bioremediation systems must still be developed. It might exist in the form of solids, liquids, or gases.
4) Low environmental effect and minimal site disruption, making it simple for the public to view positively.	In contaminated areas, residual levels of dangerous intermediates can occasionally become too high (exceeding regulatory requirements), persistent, and toxic.	In comparison to alternative treatment options like excavation and soil removal from contaminated sites, bioremediation requires more time.
5) Need low-technology equipment or equipment that is easily accessible.	Performance evaluations are challenging because there is no set standard for a "clean" site because there are no regulations governing performance criteria.	There is no acceptable goal for bioremediation therapies since it is difficult to evaluate their effectiveness.

Source: Adapted from Sharma [20]; Sangwan and Dukare [73].

19.9 Future Prospective

When it comes to remediating, cleaning, maintaining, and recovering methods for resolving a polluted environment via microbial activity, biodegradation is a very profitable and appealing alternative. The competition between biological agents like fungi, bacteria, and algae as well as unfavorable external abiotic factors (aeration, moisture, pH, and temperature) and limited bioavailability dictate how quickly undesired waste materials degrade. The effectiveness of bioremediation depends on a number of parameters, including but not limited to cost, site features, and the kind and quantity of pollutants. The site description is the first step in successful bioremediation since it aids in the creation of the most effective bioremediation technique (*ex situ* or *in situ*). Due to excavation and transportation from the archaeological site, bioremediation processes are typically more expensive. They can, however, be utilized to remediate a variety of contaminants. Contrarily, *in situ* techniques do not incur additional costs for excavation, yet some inefficient *in situ* bioremediation approaches can be reduced by the onsite installation equipment costs, connected efficiently, and controlling the bottom of a contaminated environment. When choosing the most effective bioremediation method to successfully treat polluted sites, geological features of the contaminated environment, particularly soil; pollutant kind and depth; human settlements on site; and performance of each bioremediation approach should be considered [11].

19.10 Conclusion

Environmental pollution is the major problem of the twenty-first century, and research communities are paying close attention to it. Because microbes exhibit the capability of adapting to new and potentially toxic environments, bioremediation using microbes is a useful method for reducing pollution by boosting innate biodegradation systems. Acknowledging the microbial populations as well as how they react to the natural surroundings and in the existence of contaminants is essential for developing environmentally sustainable, new, and potentially helpful bioremediation strategies.

References

1 Tang, C.Y., Fu, Q.S., Criddle, C.S., and Leckie, J.O. (2007). Effect of flux (transmembrane pressure) and membrane properties on fouling and rejection of reverse osmosis and nanofiltration membranes treating perfluorooctane sulfonate containing wastewater. *Environmental Science & Technology* 41 (6): 2008–2014.

2 Strong, P.J. and Burgess, J.E. (2008). Treatment methods for wine-related and distillery wastewaters: a review. *Bioremediation Journal* 12 (2): 70–87.

3 Demnerová, K., Mackova, M., Spevákova, V. et al. (2005). Two approaches to biological decontamination of groundwater and soil polluted by aromatics – characterization of microbial populations. *International Microbiology* 8 (3): 205–211.

4 Singh, S.P. and Upadhyay, S.K. (eds.) (2021). *Bioprospecting of Microorganism-Based Industrial Molecules*. John Wiley & Sons Ltd. doi:10.1002/9781119717317.

5 Shintani, T., Upadhyay, S.K., Singh, S.P. (2021). An introduction to microbial biodiversity and bioprospection. In: *Bioprospecting of Microorganism-Based Industrial Molecules* (ed. S.P. Singh and S.K. Upadhyay). John Wiley & Sons Ltd. https://doi.org/10.1002/9781119717317.ch1.

6 Igiri, B.E., Okoduwa, S.I.R., Idoko, G.O. et al. (2018). Toxicity and bioremediation of heavy metals contaminated ecosystem from tannery wastewater: a review. *Journal of Toxicology* 2018: 2568038.

7 Siddiquee, S., Rovina, K., Al Azad, S. et al. (2015). Heavy metal contaminants removal from wastewater using the potential filamentous fungi biomass: a review. *Journal of Microbial and Biochemical Technology* 7 (6): 384–395.

8 Okolo, V.N., Olowolafe, E.A., Akawu, I., and Okoduwa, S.I.R. (2016). Effects of industrial effluents on soil resources in Challawa industrial area, Kano, Nigeria. *Journal of Global Ecology and Environment* 5 (1): 1–10.

9 Deepa, C. and Suresha, S. (2014). Biosorption of lead (II) from aqueous solution and industrial effluent by using leaves of *Araucaria cookii*: application of response surface methodology. *IOSR Journal of Environmental Science, Toxicology and Food Technology* 8 (7): 67–79.

10 Fox, R. and Tuchman, M. (1996). The assessment and remediation of contaminated sediments (ARCS) program. *Journal of Great Lakes Research* 3 (22): 493–494.

11 Ahirwar, N.K., Gupta, G., Singh, R., and Singh, V. (2016). Isolation, identification and characterization of heavy metal resistant bacteria from industrial affected soil in central India. *International Journal of Pure & Applied Bioscience* 4 (6): 88–93.

12 Zhang, W., Jiang, F., and Ou, J. (2011). Global pesticide consumption and pollution: with China as a focus. *Proceedings of the International Academy of Ecology and Environmental Sciences* 1 (2): 125.

13 Su, C. (2014). A review on heavy metal contamination in the soil worldwide: situation, impact and remediation techniques. *Environmental Skeptics and Critics* 3 (2): 24.

14 Arao, T., Ishikawa, S., Murakami, M. et al. (2010). Heavy metal contamination of agricultural soil and countermeasures in Japan. *Paddy and Water Environment* 8 (3): 247–257.

15 Ferraz, P., Fidalgo, F., Almeida, A., and Teixeira, J. (2012). Phytostabilization of nickel by the zinc and cadmium hyperaccumulator *Solanum nigrum* L. are metallothioneins involved? *Plant Physiology and Biochemistry* 57: 254–260.

16 Azubuike, C., Chikere, C., and Okpokwasili, G. (2016). Bioremediation techniques–classification based on site of application: principles, advantages, limitations and prospects. *World Journal of Microbiology and Biotechnology* 32 (11): 180.

17 Smith, E., Palanisami, T., Ramadass, K. et al. (2015). Remediation trials for hydrocarbon-contaminated soils in arid environments: evaluation of bioslurry and biopiling techniques. *International Biodeterioration & Biodegradation* 101: 56–65.

18 Medfu Tarekegn, M., Salilih, F.Z., and Ishetu, A.I. (2020). Microbes used as a tool for bioremediation of heavy metal from the environment. *Cogent Food & Agriculture* 6 (1): 1783174.

19 Khan, M.Y., Swapna, T., Hameeda, B., and Reddy, G. (2015). Bioremediation of heavy metals using biosurfactants. In: *Advances in Biodegradation and Bioremediation of Industrial Waste*, 20. CRC Press.

20 Sharma, I. (2020). Bioremediation techniques for polluted environment: concept, advantages, limitations, and prospects. In: *Trace Metals in the Environment-New Approaches and Recent Advances* (ed. M. Alfonso Murillo-Tovar, H. Saldarriaga-Noreña, and A. Saeid). IntechOpen.

21 Chowdhury, A., Pradhan, S., Saha, M., and Sanyal, N. (2008). Impact of pesticides on soil microbiological parameters and possible bioremediation strategies. *Indian Journal of Microbiology* 48 (1): 114–127.

22 Jha, A.K., Singh, K., Sharma, C. et al. (2011). Assessment of methane and nitrous oxide emissions from livestock in India. *Journal of Earth Science and Climatic Change* 2 (1): 107.

23 Kishore Kumar, K., M. Krishna Prasad, G. V. S. Sarma, Ch. V. R. Murthy, Removal of Cd (II) from aqueous solution using immobilized Rhizomucor tauricus. *Journal of Microbial and Biochemical Technology*, 2009. 1(1): p. 15–21.

24 Elredaisy, S.M.A. (2010). Ecological benefits of bioremediation of oil contaminated water in rich Savannah of Palogue, Upper Nile area-southern Sudan. *Journal of Bioremediation & Biodegradation* 1 (1): 103.

25 Aghamiri, S., Kabiri, K., and Emtiazi, G. (2011). A novel approach for optimization of crude oil bioremediation in soil by the Taguchi method. *Journal of Petroleum & Environmental Biotechnology* 2: https://doi.org/10.4172/2157-7463.1000110.

26 Sridevi, V., Lakshmi, M.V.V.C., Swamy, A.V.N., and Rao, M.N. (2011). Implementation of response surface methodology for phenol degradation using Pseudomonas putida (NCIM 2102). *Journal of Bioremediation & Biodegradation* 2 (2).

27 Agarry, S. and Solomon, B. (2008). Kinetics of batch microbial degradation of phenols by indigenous *Pseudomonas fluorescence. International Journal of Environmental Science and Technology* 5 (2): 223–232.

28 Sheehan, D. (1997). *Bioremediation Protocols*, Methods in Biotechnology. Totowa, NJ: Humana Press.

29 Reshma, S., Spandana, S., and Sowmya, M. (2011). *Bioremediation Technologies*. India: World Congress of Biotechnology.

30 Adams, G.O., Tawari-Fufeyin, P., Okoro, S., and Ehinomen, I. (2015). Bioremediation, biostimulation and bioaugmention: a review. *International Journal of Environmental Bioremediation & Biodegradation* 3 (1): 28–39.

31 Kumar, A., Bisht, B.S., Joshi, V.D., and Dhewa, T. (2011). Review on bioremediation of polluted environment: a management tool. *International Journal of Environmental Sciences* 1 (6): 1079.

32 Naik, M. and Duraphe, M. (2012). Review paper on-parameters affecting bioremediation. *Advance Research in Pharmaceuticals and Biologicals* 2 (3).

33 Mulligan, C. and Yong, R. (2004). Natural attenuation of contaminated soils. *Environment International* 30: 587–601.

34 Niu, G.-L., Zhang, J.-J., Zhao, S. et al. (2009). Bioaugmentation of a 4-chloronitrobenzene contaminated soil with Pseudomonas putida ZWL73. *Environmental Pollution* 157 (3): 763–771.

35 Gomez, F. and Sartaj, M. (2014). Optimization of field scale biopiles for bioremediation of petroleum hydrocarbon contaminated soil at low temperature conditions by response surface methodology (RSM). *International Biodeterioration & Biodegradation* 89: 103–109.

36 Thapa, B., Kc, A.K., and Ghimire, A. (2012). A review on bioremediation of petroleum hydrocarbon contaminants in soil. *Kathmandu University Journal of Science, Engineering and Technology* 8 (1): 164–170.

37 Jain, P., Gupta, V.K., Bajpal, V. et al. (2011). GMO's: perspective of bioremediation. In: *Recent Advances in Environmental Biotechnology*, 6–23. Germany: LAP Lambert Academic Publishing AG and Co. KG.

38 Jain, P., Gupta, V.K., Gaur, R.K. et al. (2010). Fungal enzymes: potential tools of environmental processes. In: *Fungal Biochemistry and Biotechnology*, 44–56. Germany: LAP Lambert Academic Publishing AG and Co. KG.

39 Kulshreshtha, S. (2013). Genetically engineered microorganisms: a problem solving approach for bioremediation. *Journal of Bioremediation & Biodegradation* 4 (4): 1–2.

40 Cases, I. and Lorenzo, V.d. (2005). Genetically modified organisms for the environment: stories of success and failure and what we have learned from them. *International Microbiology* 8: 213–222.

41 Couto, M. N., Fritt-Rasmussen, Janne; Jensen, Pernille; Højrup, Mads; Rodrigo, Ana; Ribeiro, Alexandra, Suitability of oil bioremediation in an Artic soil using surplus heating from an incineration facility. *Environmental Science and Pollution Research*, 2014. 21 (9): p. 6221–6227.

42 Phulia, V. et al. (2013). Technologies in aquatic bioremediation. In: *Freshwater Ecosystem and Xenobiotics*, 65–91. New Delhi, India: Discovery Publishing House PVT. Ltd.

43 Thavasi, R., Jayalakshmi, S., and Banat, I.M. (2011). Application of biosurfactant produced from peanut oil cake by Lactobacillus delbrueckii in biodegradation of crude oil. *Bioresource Technology* 102 (3): 3366–3372.

44 Macaulay, B. (2015). Understanding the behaviour of oil-degrading micro-organisms to enhance the microbial remediation of spilled petroleum. *Applied Ecology and Environmental Research* 13 (1): 247–262.

45 Das, N. and Chandran, P. (2011). Microbial degradation of petroleum hydrocarbon contaminants: an overview. *Biotechnology Research International* 2011.

46 Si-Zhong, Y., Jin, H., Wei, Z. et al. (2009). Bioremediation of oil spills in cold environments: a review. *Pedosphere* 19 (3): 371–381.

47 Asira, E.E. (2013). Factors that determine bioremediation of organic compounds in the soil. *Academic Journal of Interdisciplinary Studies* 2 (13): 125–125.

48 Wang, Q., Zhang, S., and Klassen, W. (2011). Potential approaches to improving biodegradation of hydrocarbons for bioremediation of crude oil pollution. *Journal of Environmental Protection* 2 (01): 47.

49 Frutos, F.G., Pérez, R., Escolano, O. et al. (2012). Remediation trials for hydrocarbon-contaminated sludge from a soil washing process: evaluation of bioremediation technologies. *Journal of Hazardous Materials* 199: 262–271.

50 Shaw, D. (2003). EPA's Report on the Environment (2003 Draft).

51 Cunningham, C. and Philip, J. (2000). *Comparison of Bioaugmentation and Biostimulation in Treatment of Diesel Contaminated Soil, Land Contamination and Reclamation*. Edinburgh: University of Edinburgh.

52 Kulshreshtha, A., Agrawal, R., Barar, M., and Saxena, S. (2014). A review on bioremediation of heavy metals in contaminated water. *IOSR Journal of Environmental Science, Toxicology and Food Technology* 8 (7): 44–50.

53 Evans, G.M. and Furlong, J.C. *Theory and Application*. Wiley.

54 Wang, L., Barrington, S., and Kim, J.-W. (2007). Biodegradation of pentyl amine and aniline from petrochemical wastewater. *Journal of Environmental Management* 83 (2): 191–197.

55 Mai, C., Schormann, W., Milstein, O., and Hüttermann, A. (2000). Enhanced stability of laccase in the presence of phenolic compounds. *Applied Microbiology and Biotechnology* 54 (4): 510–514.

56 Narnoliya, L.K., Agarwal, N., Patel, S.N. et al. (2019). Kinetic characterization of laccasefrom *Bacillus atrophaeus*, and its potential in juice clarification in free and immobilized forms. *Journal of Microbiology* 57: 900–909.

57 Phukon, L.C., Chourasia, R., Padhi, S., et al. (2022). Cold-adaptive traits identified by comparative genomic analysis of a lipase-producing *Pseudomonas* sp. HS6 isolated from snow-covered soil of Sikkim Himalaya and molecular simulation of lipase for wide substrate specificity. *Current Genetics* 68 (3–4): 375–391.

58 Sharma, D., Sharma, B., and Shukla, A. (2011). Biotechnological approach of microbial lipase: a review. *Biotechnology* 10 (1): 23–40.

59 Prasad, M. and Manjunath, K. (2011). Comparative study on biodegradation of lipid-rich wastewater using lipase producing bacterial species. *Indian Journal of Biotechnology* 10: 121–124.

60 Beena, A. and Geevarghese, P. (2010). A solvent tolerant thermostable protease from a psychrotrophic isolate obtained from pasteurized milk. *Developmental Microbiology and Molecular Biology* 1: 113–119.

61 Singh, A.K., Kumari, M., Sharma, N. et al. (2022). Metagenomic views on taxonomic and functional profiles of the Himalayan Tsomgo cold lake and unveiling its deterzome potential. *Current Genetics* 68 (5–6): 565–579.

62 Rubilar, O., Diez, M.C., and Gianfreda, L. (2008). Transformation of chlorinated phenolic compounds by white rot fungi. *Critical Reviews in Environmental Science and Technology* 38 (4): 227–268.

63 Koua, D., Cerutti, L., Falquet, L. et al. (2009). PeroxiBase: a database with new tools for peroxidase family classification. *Nucleic Acids Research* 37 (suppl_1): D261–D266.

64 Vasileva-Tonkova, E. and Galabova, D. (2003). Hydrolytic enzymes and surfactants of bacterial isolates from lubricant-contaminated wastewater. *Zeitschrift für Naturforschung. Section C* 58 (1–2): 87–92.

65 Schmidt, O. (2006). *Wood and Tree Fungi*. Springer.

66 Husain, Q. (2006). Potential applications of the oxidoreductive enzymes in the decolorization and detoxification of textile and other synthetic dyes from polluted water: a review. *Critical Reviews in Biotechnology* 26 (4): 201–221.

67 Silva-Castro, G., Uad, I., Gonzalez-Lopez, J. et al. (2012). Application of selected microbial consortia combined with inorganic and oleophilic fertilizers to recuperate oil-polluted soil using land farming technology. *Clean Technologies and Environmental Policy* 14 (4): 719–726.

68 Bhattacharya, M., Guchhait, S., Biswas, D., and Datta, S. (2015). Waste lubricating oil removal in a batch reactor by mixed bacterial consortium: a kinetic study. *Bioprocess and Biosystems Engineering* 38 (11): 2095–2106.

69 Sun, G.-D., Xu, Y., Jin, J.-H. et al. (2012). Pilot scale ex-situ bioremediation of heavily PAHs-contaminated soil by indigenous microorganisms and bioaugmentation by a PAHs-degrading and bioemulsifier-producing strain. *Journal of Hazardous Materials* 233: 72–78.

70 Tyagi, M., da Fonseca, M.M.R., and de Carvalho, C.C. (2011). Bioaugmentation and biostimulation strategies to improve the effectiveness of bioremediation processes. *Biodegradation* 22 (2): 231–241.

71 Cassidy, D.P., Srivastava, V.J., Dombrowski, F.J., and Lingle, J.W. (2015). Combining in situ chemical oxidation, stabilization, and anaerobic bioremediation in a single application to reduce contaminant mass and leachability in soil. *Journal of Hazardous Materials* 297: 347–355.

72 Singh, M. and Morve, R. (2014). Ecofriendly application of nanomaterials: nanobioremediation. *Journal of Nanoparticles* 2014: 431787.

73 Sangwan, S. and Dukare, A. (2018). Microbe-mediated bioremediation: an eco-friendly sustainable approach for environmental clean-up. In: *Advances in Soil Microbiology: Recent Trends and Future Prospects*, 145–163. Springer.

20

Microbial Bioresource for Plastic-Degrading Enzymes

Ayodeji Amobonye[1], *Christiana Eleojo Aruwa*[1,2], *and Santhosh Pillai*[1]

[1] *Department of Biotechnology and Food Science, Faculty of Applied Sciences, Durban University of Technology, Durban, South Africa*
[2] *Department of Microbiology, School of Sciences, Federal University of Technology, Akure, Nigeria*

20.1 Introduction

The invention of plastics and their unprecedented global acceptance have encouraged the fabrication of different conventional plastic polymers, which include polyethylene (PE), polypropylene (PP), polyvinyl chloride (PVC), polyethylene terephthalate (PET), polyurethane, and polystyrene (PS), from nonrenewable sources like petroleum, coal, and natural gas. Although conventional plastics have provided humans with enormous potential and an array of diverse functions, they are practically nondegradable and persist in the environment for centuries [1]. Subsequently, the bioplastic industry has emerged as a promising solution to alleviate some of the challenges posed by fossil fuel-based polymers. Interestingly, the building blocks of these bioplastics are sourced from plant biomass making them more renewable and relatively more degradable. Common examples of bioplastics include polybutylene succinate (PBS), polyhydroxyalkanoates (PHAs), polylactic acid (PLA), and thermoplastic starch (TPS). Others include bio-polyethylene (bioPE) and bio-polyethylene terephthalate (bioPET) that have similar properties to their petroleum-derived counterpart polymers [2, 3]. However, despite the progress gained due to the emergence of bioplastics, there are still many gaps that need to be filled to ensure the resounding effects of these advancements. For instance, the bioplastic proportion of the total plastic market is still less than one percent [4], which makes their supposed advantages less impactful. Furthermore, although the degradation of bioplastics is faster than their conventional counterparts, both types of plastics still accumulate in the environment, especially in aquatic environments. For instance, Lebreton et al. [5] observed that approximately 80 metric tons of plastic were floating in the Great Pacific Garbage Patch, an approximately 1.6 million km^2 area, formed in subtropical waters between California and Hawaii. Numerous marine species have been found to be entangled by or ingested with plastic litter, which is likely to have

Microbial Bioreactors for Industrial Molecules, First Edition.
Edited by Sudhir P. Singh and Santosh Kumar Upadhyay.

unpredictable consequences for the ecosystem [6]. Furthermore, the miniaturized forms of plastics, viz., microplastics and nanoplastics, have been revealed to be more likely to be taken up by marine organisms such as bivalves, copepods, echinoderms, and polychaetes at different stages of their life cycle [7].

Considering the persistent increase in plastic production and plastic waste generation, the economic cost associated with plastic litter in the marine environment was estimated at $21.3 billion in 2020 and projected to rise to $229 billion and $731 billion by 2030 and 2050, respectively [8]. However, current approaches aimed at mitigating this deleterious environmental threat, viz., incineration, recycling, and landfilling, all come at a massive cost, are unsustainable, and are more likely to increase the burden on the different ecosystems [9]. Incineration of various plastic polymers, for example, has been described to end in the generation of more poisonous wastes such as dioxins, furans, heavy metals, and sulfides [10]. Plastic recycling has also been found economically nonviable due to the aggregated costs of the technology and labor involved, especially the cost of segregation, processing, and byproduct disposal, which cumulatively render recycled plastics costlier than virgin plastics [11]. Hence, in recent decades, much emphasis has been placed on the application of biological systems as ecofriendly alternatives to degrade defiant plastic polymers. It is hoped that these microbes, in their natural states, in consortiums or as microbial bioreactors will be more effective and efficient in fighting the plastic waste menace on a long-term basis.

Generally, a bioreactor is any system that involves organisms or biochemically active substances sourced from organisms, and they have been useful in the manufacturing of biopharmaceuticals, chemicals, food, and food additives. In addition, significant progress has been made in the application of microbial bioreactors in the bioremediation of various environmental pollutants such as dyes [12, 13], heavy metals [14], organic contaminants [15], petroleum hydrocarbons, and, to a lesser degree, microplastics [16]. Thus, it is believed that designing and developing microbial bioreactors to facilitate the degradation of plastic, both in bulk and miniaturized form, are essential steps for the large-scale applications of plastic bioremediation.

Studies have shown that organisms, including higher and lower ones, are capable of metabolizing and transforming plastic polymers into less complex compounds such as CO_2 and H_2O. For instance, different insect larvae including waxworm, mealworm, and superworm have been investigated for their ability to ingest, breakdown, and mineralize a wide variety of plastic materials, with the aid of their gut microbiome [17, 18]. In addition to the symbiotic microorganisms in the larvae's gut, remarkable microbes isolated from various ecosystems have also been shown with notable biodegradative potential [6, 19, 20]. In this regard, actinomycetes [21], algae [22], bacteria [23], and fungi [24], functioning through their enzymatic systems, have been highlighted to possess remarkable potential in plastic degradation and can serve as a repository for plastic-degrading enzymes. Although the plastic biodegradative system is relatively slow, various efforts are being made, through protein engineering and recombinant technology, to improve the efficiency of degradation. This chapter thus sheds more light on the key areas regarding the production, improvement, and potential of microbial enzymes for the biodegradation of plastic polymers in the increasing efforts against plastic bioaccumulation.

20.2 Classification of Plastics: Biobased, Biodegradable, and Fossil-Based Plastics

Approximately 99% of all plastic materials are petroleum based, while the remaining one percent is biobased and derived from natural raw materials such as corn, cane sugar, potatoes, wheat, and vegetable oil [4]. Despite being produced from natural or renewable resources, bioplastics are not the sole solution to plastics bioaccumulation, since they are not readily biodegradable [25]. Furthermore, it is critical to distinguish between biodegradable plastics, bioplastics, and biobased plastics, despite the terms being commonly used, interchangeably.

20.2.1 Fossil-Based Plastics

The majority of plastic products are petroleum-based or fossil fuel-based plastic polymers, derived from nonrenewable fossil fuels as their feedstock. These widely exploited fossil-based polymers include PE, PET, PP, PVC, and PS [26]. PET, one of the most popular among plastics, is synthesized using ethylene glycol and terephthalic acid, both derived from the cracking process of fossil fuels [27]. Similarly, PVC is produced from the polymerization of its monomer, vinyl chloride, which is synthesized from ethylene sourced from natural gas [28]. The global plastic production of fossil-based plastics has tripled in the past 25 years. Thus, recent estimates revealed that 8.3 billion tons of virgin plastic have been produced so far, with more than two-thirds of them ending up being discarded indiscriminately in different habitats [29]. This has prompted the search for environmentally sustainable alternatives to fossil-based plastics.

20.2.2 Biobased Plastics

Biobased polymers, more popularly referred to as bioplastics, utilize renewable biological products as the major feedstock. According to the United States Department of Agriculture (USDA), under its Biopreferred Program, the ASTM's (American Society for Testing and Materials) D6866 standard is the yardstick for measuring and certifying biobased products such as bioplastics [30]. This standard evaluates the percentage of a material's biobased content by measuring the presence and the proportion of carbon-14 in the test material, hence, differentiating between carbon-sourced renewable (bioplastics) and carbon sourced from nonrenewable petroleum (fossil-based plastic). There are many bioplastics currently on the market such as PLA, Bio-PET polycaprolactone (PCL), PBS, and poly(butylene adipate-co-terephthalate) (PBAT) [25]. By design, some bioplastics are both functionally and chemically similar to some fossil-based polymers and are commonly referred to as "drop-ins," whereas others are entirely different polymers [31]. Examples include PLA, which may serve as an alternative to PS; biobased polybutylene succinate (Bio-PBS), which exhibits similar properties to PP; and biobased polyethylene terephthalate (Bio-PET), a potential substitute for fossil-based PET. Unlike its conventional counterpart, Bio-PET is manufactured from ethylene glycol and terephthalic acid that are sourced from the fermentation of sugarcane and the oxidization of *p*-xylene from lignocellulosic materials, respectively [32].

20.2.3 Biodegradable Plastics

Confusion exists between the two terms "biodegradable plastics" and "biobased plastics," with many assuming they are interchangeable. In the long run, all organic matter-based materials undergo biodegradation including the most recalcitrant conventional petroleum-based plastics. However, it should be noted that the biodegradation rate of different plastic polymers can vary exponentially with the degradation environment, acting as a significant factor. Hence, this warrants the need to specify the timeframe and the environment, while defining the biodegradability of a plastic polymer. Typically, plastics referred to as "biodegradable" are created to degrade under set environments and/or conditions, which include the soil and aquatic environments, under sunlight, industrial, home composting facilities, etc. [33]. Thus, plastics labeled as biodegradable or compostable are designed to degrade in an industrial composting environment at a comparable rate to other well-known compostable substances with no visual or toxic residue. However, certain countries have specific requirements for plastic to be considered biodegradable. For example, in the United States, biodegradable plastic must pass the ASTM's D6400 test protocols [34]. In this test, the polymer, which is in a controlled environment, must have disintegrated to less or approximately equal to 10% of its original dry weight after 84 days and 90% of its organic carbon must have been converted to CO_2 within 180 days, among many other criteria [35]. In addition, the rate of biodegradation is also dependent on the thickness of the product as biobased plastics, like their conventional alternatives such as fossil-based plastics, are available in many grades with a wide variety of properties [36].

20.3 General Mechanism of Plastic Biodegradation

During plastic degradation, microplastics (less than 5 mm) are produced, which are toxic to the ecosystem [7]. In addition, the degradation of bulk plastic materials also leads to the production of inhalable fibrous microplastics and additives such as plasticisers and dyes, which are allergenic, mutagenic, and carcinogenic [37]. Hence, elaborating the underlying mechanisms involved in the microbial-plastic association could contribute immensely to initiating, maintaining, and enhancing the complete environmental degradation of plastics [38].

Generally, four mechanisms are associated with plastic breakdown: hydrolytic degradation, photo (light)-degradation, thermo-oxidative degradation, and biodegradation [39]. Moreover, the plastic polymer degradation rate can be achieved and/or enhanced by moisture as well as by photodegrading and thermo-oxidant agents. However, microorganisms are perceived to be more effective and more environmentally sustainable with regards to plastic polymers breakdown as they have been reported to achieve further degradation of plastic beyond carbonyl compounds' generation, to partial or complete plastic monomers that could be bioassimilated and could also lead to end products that can be recycled into various geochemical cycles [38]. The microbial biodegradative pathway is majorly dependent on the enzymatic machinery, especially extracellular enzymes that catalyze the breakdown of the bulk plastic materials to lower molecular weight and reduced chain-length compounds. In summary, microbial biodegradation occurs through different stages of biodeterioration, biofragmentation, bioassimilation, and mineralization [40], as depicted in Figure 20.1.

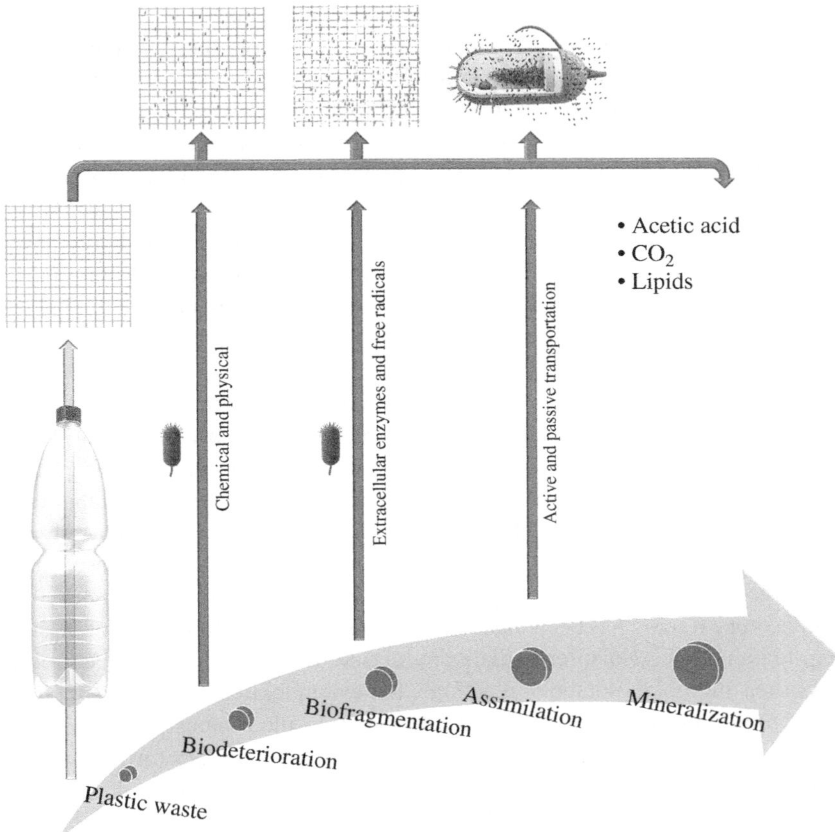

Figure 20.1 Steps involved in plastic biodegradation. *Source:* Amobonye et al. [6] / with permission of Elsevier.

At the initial stage of biodegradation, biodeterioration occurs and it is typically a superficial activity that translates into significant mechanochemical changes in the structure of the polymer. During biodeterioration, microorganisms alongside other biotic factors or agents deteriorate plastic polymers via physicochemical reactions resulting in peripheral changes in polymer properties [6]. Microbes attach to and colonize plastic surfaces, thus reducing the plastic's resistance and durability. These structural changes are also enhanced by the exposure of the materials to abiotic factors, such as chemicals in the environment, light, and temperature [41].

Biofragmentation is the second stage and it involves a loss of stability in the long carbon chains that were initially impacted by microbial enzymes but may also be enhanced by abiotic factors. This is the major depolymerization step that is based on the enzymatic depolymerization of biodeteriorated polymers into lower molecular weight units, and the process is enhanced by microbe-derived reactive species [42]. Biofragmentation targets polymer weight reduction and oxidation of derived less weighted compounds, both of which facilitated by enzymes cleavage [43]. The degrading enzymes carry out hydrolytic and nucleophilic attacks on the carbonyl carbon bonds and groups, that is, the peptide, ester, and glycosidic bonds in the polymer structure. While the exo-hydrolytic attacks result in monomeric or oligomeric compounds like terephthalic acid that are assimilated by microbes, endo-attacks result in products that still need to undergo further degradation before microbial assimilation [44].

Bioassimilation involves the intake of shorter chained carbon molecules into the cell across the cell membrane as starter substrates in catabolism for the generation of biomass and energy [45]. Plastics bioassimilation is posited to involve both active and passive transportation across the membrane of microbial cells [6]. Furthermore, different microbes have been demonstrated to assimilate via facilitated passive or active transport systems [46]. In this regard, the genes for different transporters have been observed to be upregulated in microbes while assimilating plastic biodegradation intermediates, especially those of the ATP binding cassette protein family [47].

The final stage, mineralization, ensures that end products of the plastic degradation process are inorganic molecules, and this occurs after the successful transportation of the polymer derivatives into the cells [38]. The assimilated compounds further undergo a stepwise reaction involving enzymes resulting in the total breakdown to oxidized metabolites, including water, CH_4, N_2, and CO_2 [48]. Mineralization takes place both anaerobically or aerobically and it requires synergy among participating enzymes such as peroxidase, cutinases, lipases, esterases, and laccases [49]. The utilization of Strum's method of quantifying released CO_2 and isotope tracing has been useful in demonstrating this final stage of plastic polymer biodegradation [50]. Also, intermediate products at the final stage could be useful as substrates in other microbial metabolic reactions. For example, in *Ideonella sakaiensis*, the internalized terephthalic acid is transformed into the molecule protocatechuic acid by the action of terephthalate 1,2-dioxygenase. Protocatechuic acid undergoes further catalysis into compounds (oxaloacetate and pyruvate) that feed the citric Kreb's cycle and are eventually mineralized to water and carbon (IV) oxide [51].

20.4 Microbial Sources of Plastic-Degrading Enzymes

Despite the various forms and chemical properties of plastic materials, microbial plastic degraders seem to continually evolve to tackle plastic wastes and other emerging environmental contaminants, albeit slowly and over prolonged exposure [38]. In this regard, various microbial plastic degraders have been isolated from diverse environments, primarily bacteria, algae, actinomycetes, and fungi [52, 53]. Some of the microbial sources of plastic-degrading enzymes are discussed in detail in this section.

20.4.1 Actinomycetes

Though actinomycetes are mainly credited with the secretion of secondary metabolites with antibiotic properties, some of their bioproducts have also been linked to plastic depolymerization. The polymer-degrading *Thermoactinomyces* (a thermophile), *Actinomadura* spp., *Rhodococcus ruber*, and *Streptomyces* spp. are some of the most studied members of the actinomycete group [54]. Earlier studies by Adhi et al. [55] showed that extracellular enzymes that degraded plastics were secreted during the submerged cultivation of

Streptomyces setoni and *Streptomyces badicus*. Furthermore, actinomycetes have been noted to degrade different conventional plastics, for example, PP film treated with UV and nitric acid was degraded by *Actinomycetes* sp. [56], while polyesters were degraded by *Micromonospora* and *Streptomyces* species [57, 58]. In addition, a highly stable 25-kDa suberinase, sourced from *Streptomyces scabies,* was demonstrated to possess a significant ability to degrade PET to terephthalic acid (TCA) [59].

This group of microbes has also been shown to break down biobased plastics; for example, polyhydroxybutyrate-co-3-hydroxyvalerate (PHBV) was degraded by *Streptoverticillium kashmirense* AF1 with the aid of its extracellular PHBV depolymerase [60]. The filamentous *Streptomyces lydicus* MM10 secreted PHB depolymerase during its breakdown of PHB [61]. *Moritella* sp., *Shewanella* sp., and *Psychrobacter* sp. have also been demonstrated to breakdown PCL [62, 63], while the significant degradation of PCL, PHB, and polyethylene succinate (PES) by the heat-loving *Streptomyces*, *Microbispora*, *Thermoactinomyces*, and *Actinomadura* species have also been reported [64].

20.4.2 Algae

It has been established that algae are able to utilize plastic polymers as carbon sources, which can be measured in terms of enhanced growth and cellular metabolism [65]. Plastic biodegradation by algal organisms has been recorded to be mediated by their ligninolytic enzymes together with exopolysaccharide [22]. Algae break down plastic polymer through fouling, corrosion, hydrolysis, penetration, leached components breakdown, and pigment diffusion into the polymers [22]. *Oscillatoria subbrevis* and *Phormidium lucidum*, both fast growing, freshwater algae, have been shown to efficiently attack and break down PE/LDPE (low-density polyethylene) [22, 65]. More recently, *Uronema africanum,* isolated from waste plastic bags, was also shown to degrade LDPE film within a month of incubation [66].

Although not much has been achieved in the isolation and characterization of plastic-degrading enzymes from algae, significant research has been made using this class of organisms as hosts for plastic-degrading enzymes [67, 68]. This progress has been remarkable as algae do not require organic carbon sources under photoautotrophic conditions, are easy to cultivate, are capable of fast growth, and typically do not produce endotoxins [69]. For example, *Phaeodactylum tricornutum*, a diatom was demonstrated as an efficient host of an engineered version of PETase [68]. The *Chlamydomonas reinhardtii* green alga was also demonstrated to express the PETase enzyme [67].

20.4.3 Bacteria

The bacteria group make up a significant part of the world's nutrient cycles, ensuring the liberation of simple molecules from complex ones and their natural recycling. Furthermore, bacteria possess immense potential as plastic degraders, utilizing a cohort of enzymes and forming bioactive biofilms [41]. Studies have shown that bacterial plastic degraders can be found in various ecosystems and niches like insect gut [70], cold marine environments [23], landfill, and dumpsites [71]. In addition, thermophilic bacteria including *Hydrogenobacter*, *Aquifex*, *Rhodothermus*, *Thermus*, *Thermotoga*, *Anoxybacillus*, *Geobacillus*, and *Brevibacillus* have been noted with plastic-degrading potentials [19]. Similarly, psychrophiles have also been reported in plastic degradation, including *Arthrobacter*, *Psychrobacter*, *Pseudomonas*, *Colwellia*, *Pseudoalteromonas*, among others [19].

Bacteria's ability to break down long-chained fatty acids (LCFA) is closely associated with their capacity to cleave plastic polymers. The *Pseudomonas* genus is mostly investigated in relation plastics and LCFA biodegradation [72]. For example, *Pseudomonas* sp. E4 expresses an alkane hydroxylase which had significant activity in PE breakdown; furthermore, the gene of the enzyme has also been successfully cloned into *Escherichia coli* and its activity confirmed [73]. Furthermore, *Pseudomonas aeruginosa* also showed the expression of alkane hydroxylase, rubredoxin, and rubredoxin reductase [74]. In addition, polymerases from *Pseudomonas* as well as *Bacillus* species are linked to PS breakdown [75]. However, various bacterial genera have also demonstrated enzyme machineries that are able to degrade an array of plastic polymers. These include *Micrococcus*, *Streptococcus*, *Pseudomonas*, and *Staphylococcus*. [76]. In addition, other plastic-degrading bacteria have also been shown to include *Ideonella*, *Comamonas*, *Escherichia*, *Flavobacterium*, *Nocardia*, *Azotobacter*, *Rhodococcus*, *Streptomyces*, *Mycobacterium*, *Thermoactinomycetes*, *Micromonospora*, and *Alcaligenes* species [77, 78]. It has been reported that some of these plastic-degrading bacteria sequester up to 90% of plastic polymer within their cells [78].

The enzyme, PETase, is a well investigated polyesterase expressed in *I. sakaiensis* that can take up PET and transform it to mono(2-hydroxyethyl) terephthalate (MHET). MHET is further broken down to ethylene glycol (EG) and TPA using the MHETase accessory enzyme [79]. In another study, the Gram-negative *Klebsiella pneumoniae* secreted lipase, tyrosinase, peroxidase, and laccase, which were all associated with PE degradation [38].

20.4.4 Fungi

Fungi, a microbial class made up of yeasts and moulds, are also revered for their key roles in nature's nutrient cycles and are also capable of breaking down plastic polymers. Hence, different fungal species have been identified to be able to cleave various polymers and use them and/or their products as carbon sources. It is noteworthy, that these fungi come from a wide variety of habitats and belong to different classes and forms. For example, popular industrial fungi such as *Penicillium*, *Aspergillus*, and *Fusarium* isolated from agricultural soils have been shown to utilize LDPE as a nutrient source [80]. Similarly, *Aspergillus niger* and *Aspergillus japonicus* were demonstrated to break down LDPE and PE following incubation for a month [81]. In another study, *Aspergillus nomius* RH06 and *Trichoderma viride* RH03 degraded LDPE films up to 6% after 45 days of incubation, with a decreased film tensile strength of 40% and 58%, respectively [82]. *Fusarium* sp. was also shown to depolymerize PE following pretreatment with UV and/or heat [83]. The high-density polyethylene (HDPE) and LDPE plastic films were also cleaved by *Penicillium* species, viz., *P. oxalicum* and *P. chrysogenum* to about 56% and 34%, respectively, following a three-month incubation [84]. Similarly, *Mucor circinilloides* and *Aspergillus flavus* isolated from landfills demonstrated potential for LDPE degradation [85].

Interestingly, fungal depolymerases in particular are among the most studied fungal enzymes because they have a broad spectrum of activity that includes polymers breakdown [86]. For example, *Thermobifida fusca* hydrolases were described to be responsible for the enzymatic hydrolysis of PET polymers [87]. In another study, laccases were also expressed in *Pleurotus ostreatus* mushroom [88] and were implicated in the significant degradation of PE. Earlier, lignin-degrading manganese peroxidase expressed in *Trametes versicolor* and *Phanerochaete chrysosporium* was associated with PE cleavage [89]. Similarly, the *Lentinus tigrinus* mushroom esterase was demonstrated to break down PS [90]. Like bacteria, certain fungi also produce polyhydroxybutyrate (PHB) depolymerase [91] and PHB-decarboxylase [92]. The former breaks PHB into mono(3-hydroxybutyrate) and short-chained oligomers that are further cleaved and bioassimilated to H_2O and CO_2 [91].

20.5 Biotechnological Strategies for Identifying/Improving Microbial Enzymes and Their Sources for Plastic Biodegradation

In the quest for a greener planet, the past five decades have witnessed increased application of biocatalysts in various industrial processes, replacing a wide range of chemically based processes. Despite the challenges faced in identifying specific enzymes for the metabolism of various anthropogenic materials, significant progress has been made in this regard. The major tools continually being explored in this field are the conventional culturing approach, the application of metagenomic techniques, recombinant technology, and protein engineering. Scientists worldwide are fully relying on the full complements of these tools to identify novel microbial strains as well as their enzymes and other bioproducts, with the sole aim of improving plastic biodegradation.

20.5.1 Conventional Culturing Approach

The conventional method of obtaining novel industrially important microbes, enzymes, and metabolites by culturing and screening of pure microbial strains is probably the oldest and still the most widely used procedure. However, microorganisms that can be traditionally cultured are only an insignificant proportion of the total genomic diversity in any specific environment, and thus, focusing solely on these would restrict the potential of discovering a variety of new bioproducts necessary for different industrial bioprocesses [93]. Specifically, the culture-dependent approaches are limited in the scope of finding novel enzymes since only 1% of the total microorganisms on the Earth can be cultured [94].

Despite this being a very efficient and less expensive method of sourcing plastic-degrading enzymes, this approach is very laborious when there is the need to screen a large number of microorganisms, thus failing to leverage microbial diversity and versatility. However, numerous microbes including bacteria, fungi, actinomycetes, and algae with varying degrees of plastic-degrading abilities have been identified in various environments based on the conventional approach [95]. Many of these organisms have been isolated from both extreme and moderate environments and in a consortium of different microbes [6]. The strategy to maximize the chances of isolating microbes with adaptation to metabolize plastic polymers is to collect samples from different environmental habitats rich in plastic waste. Worthy of note is *I. sakaiensis*, which was isolated following an extensive screening by microbiologists at the Keio University and Kyoto Institute of Technology. The PET-degrading bacteria were identified from a consortium of microorganisms derived from environmental samples (activated sludge, wastewater, soil, and sediments) from a PET recycling site located in Sakai, Japan [51]. The consortia were recorded to achieve depolymerization of approximately 75% and the eventual mineralization of the PET degradation products into water and carbon (IV) oxide [51]. The use of enriched media that consists of basal salt media and the specific plastic polymer, as the sole or main carbon source, has also been identified as another effective approach. In a recent study by Dey et al. [96], both LDPE beads and films were utilized as the carbon sources for the isolation of PE-degrading *Stenotrophomonas* sp. and *Achromobacter* sp. Similarly, *Bacillus* sp., *Brevibacterium* sp., while *Pseudomonas* sp. with PS-degrading ability were isolated using Styrofoam-enriched liquid carbon-free basal medium [97].

20.5.2 Metagenomics

Metagenomics has appeared as a feasible alternative technique to the traditional microbial screening approach for enzyme discovery. This innovative approach has accelerated the exploration of unique and important microbial genes by directly cloning the metagenome (environmental DNA) from diverse environmental sources in surrogate hosts [94]. This has resulted into the emergence of various new enzymes with unique activities and/or sequences including many genes encoding proteins with the capability of depolymerizing various plastic polymers [98]. Furthermore, this approach is independent of the culturability of the source organisms as metagenome screening is mainly carried out based on the sequence or the activity (function) [99].

Sequence-based screening leverages the comparison of similar sequences and annotation of the functional gene through the relevant database searches. This approach has found a lot of applications in the discovery of various plastic metabolizing biocatalysts as it is relatively less time consuming and more cost-effective [98]. For example, using the well-researched PETase enzyme from *I. sakaiensis* as the template, 27 putative PETases from 24 environmental metagenomics samples were identified in the JGI Integrated Microbial Genomes and Microbiomes database [100]. Similarly, many genes responsible for established enzymes, putative enzymes, and metabolic pathways involved in the biodegradation of polyurethane and other xenobiotic additives were identified in a landfill microbial community via the proximity ligation-based metagenomic approach [101]. However, it is important to note that the sequence-based approach is limited by the content and gene annotation quality of currently available databases of plastic-degrading enzymes. Furthermore, the approach does not guarantee plastic-degrading activity, and hence, subsequent biochemical characterization and validation of enzyme functionality are required [102].

Function-based screening or functional metagenomics on the other hand makes use of activity screening to narrow down to the appropriate phenotypes from metagenomic libraries. Furthermore, it has the added advantage of efficiently mining unique enzymes with sequences that are more divergent from existing homologous ones [103]. Thus, it can also serve as an efficient complement to sequence-based metagenomics. It is usually achieved via traditional plate-based assays and/or the construction of a comprehensive gene expression library using a heterologous expression system. These expression systems could include *E. coli*, which is most widely used, as well as *Pichia pastoris*, *Saccharomyces cerevisiae*, cosmids, and phages [104]. In this regard, novel polyesterases with activity on PLA,

PCL, and PBS were identified using function-based plate assays from environmental metagenomes [105]. It is believed that the comparative metagenomic evaluation of different "plastisphere" of diverse ecology will go a long way in identifying the "core" microbial population that is predominant in "plastispheres" across various geographical regions.

20.5.3 Recombinant Technology

Typically, recombinant DNA technology involves manipulating and isolating DNA segments of interest to obtain improved and desired characteristics in the organism or their bioproducts. It is based on the insertion of DNA fragments from various sources, having a desirable gene sequence via the appropriate vector. Thus, several genes encoding various biocatalysts have been successfully cloned and expressed in host systems to attain levels that are multiple times higher than the wild organism. For example, more than half of the enzymes used in the food, detergent, and starch industry are heterologous proteins [106]. In addition, the production of heterologous enzymes cuts costs in the long run as it removes the need for expensive protein purification and environmental conditioning of pure enzymes [107]. Furthermore, genetic engineering tools have been very useful in manipulating microbial genetics to enhance their ability toward the biodegradation of plastic polymers [108, 109].

The well-studied MHETase gene from *I. sakaensis* was recently constructed in pUCIDT plasmids with its original signal peptide and a constitutive promoter and subsequently expressed in *E. coli* BL21-DE3 [108]. Similarly, *Pseudozyma antarctica* esterase, with the ability to degrade different bioplastics including PBS, was cloned using pUC19 plasmid for the construction of a gene expression cassette in another *P. antarctica* strain, and the production level of the recombinant strain was approximately 10 times that of the native strain [110]. More recently, cutinase from a phytopathogen, *Moniliophthora roreri*, was overexpressed in *E. coli* with significantly improved specific activity; in addition, the heterologous enzyme was also discovered to function without induction, unlike the native enzyme [106]. Furthermore, the cutinase was demonstrated to efficiently degrade plastic films made from aliphatic polyesters such as PES and polycaprolactone as well as aromatic polyesters such as PET [109].

Significant progress has also been made using higher organisms such as algae as hosts for plastic-degrading enzymes [67, 68]. In this regard, the diatom *P. tricornutum* was found efficient as a host of an engineered version of PETase [68]. The same enzyme was also demonstrated to be expressed in a green alga, *C. reinhardtii* [67].

20.5.4 Protein Engineering

Despite the significant progress achieved in enzyme catalysis with the emergence of recombinant DNA technology, the issues of producing enzymes with the desired biocatalytic properties such as thermostability, extreme pH stability, robustness, and stability in organic solvents remain a huge challenge [111]. In more recent times, protein engineering has offered many researchers the option of establishing the catalytic pathways for the synthesis of compounds that are highly competitive with organic synthesis. Protein engineering is basically the design of novel enzymes or proteins with desirable characteristics by modifying the natural protein sequence through deletion, substitution, or insertion of the nucleotides in the encoding gene. This approach to enzyme modification is achieved via two strategies: directed evolution and rational design [112].

While rational design involves the use of structural and systematized data, as well as molecular modeling for the prediction of modifications in the structure of proteins with the aim of altering or inducing the desired characteristics, directed evolution involves the creation of mutant libraries via random changes, the screening of the library for the desired property [113]. However, the dearth of efficient high-throughput screening techniques has been identified as a major limitation against the application of directed evolution to engineer plastic-degrading enzymes [114]. For example, the major attempt at applying direct evolution in the engineering of *Ralstonia pickettii* PHB depolymerase did not achieve the desired result as the mutants produced had no improved activity [115]. As a result, most attempts to engineer plastic-degrading enzymes have been conducted using the rational design approach.

Thermostability is believed to be one of the most desirable biochemical characteristics of industrial enzymes, particularly as plastic polymers possess high glass transition temperature. In this regard, mutations at two different positions of the calcium binding site of an esterase from *T. fusca* introduced a new disulfide bond into the protein, which remarkably increased its thermostability and plastic-degrading ability [116]. Furthermore, the substitution of threonine for proline at the 235 amino acid residue of a cutinase from *Thermobifida alba* was also observed to increase the melting temperature of the enzyme significantly as well as its plastic hydrolytic activity [117]. Similar results were also shown by Son et al. [118].

Significant results were also recorded by tailoring the surface properties of a plastic-degrading enzyme, in one instance, rendering the surface of *Thermobifida cellulosilytica* cutinase neutral [119]. This consequently reduced the existing electrostatic repulsion of the enzyme to its corresponding plastic substrate, thus improving the binding and efficiency of PET degradation [119]. Furthermore, efforts at increasing the hydrolytic capabilities of various plastic-degrading enzymes have also been attempted by reducing product inhibition effects, creating multifunctional biocatalysts, and/or enabling the enzyme catalytic promiscuity [94, 120]. Recently, a mutation in the substrate-binding site of *T. fusca* esterase caused a ~500% decrease in the binding constant for mono(2-hydroxyethyl) terephthalate (MHET), its inhibitory hydrolysis product, consequently enhancing its PET-degrading activity [94]. Furthermore, in the study by Biundo et al. [120], plastic-degrading polyesterases were modified to develop enhanced amidase activity toward depolymerizing artificial polyamides.

20.6 Conclusion and Future Perspectives

This chapter has highlighted that microbes, especially bacteria and fungi, have great potential for ameliorating the damaging impacts of plastic pollution due to their robust enzymatic system that possesses the capacity for pollutant detoxification, their diverse substrate specificity, and their ability to colonize solid substrates. Although some progress has been made in the scientific investigation of microbes as bioresources for plastic-degrading enzymes, the same cannot be said about their real-life technological applications and environmental effects. However, the future looks promising based on the vast amount of untapped resources at our disposal. It has been acknowledged by many scientists that the diversity of known microbes and enzymes involved in plastic degradation is still limited. It is believed that the full deployment of metagenomics and the other -omic tools – genomics, metabolomics, proteomics, and transcriptomics – will greatly enhance our knowledge of the complex biological interactions happening between genes, metabolites, proteins, transcripts, as well as external environmental conditions during the initial colonization and the subsequent biodegradation of plastic materials. This in turn will facilitate the application and usefulness of the different microorganisms, with the complement of their enzymes, in the biodegradation of plastic materials. For example, the development of more advanced algorithms for mining metagenome data sets will be a promising direction to follow. Furthermore, the application of various microorganisms in consortiums, thus exploring their synergistic effect, is expected to result in heightened efficiency in plastic biodegradation as it is the true picture in nature. Likewise, the discovery of efficient function-based assays for the identification of plastic-active biocatalysts is as important. It was observed that some of the current microbial screening and enzyme assays are designed in such a way that they lead to false positives. As a result, the different methodologies applied in the analysis of microbial plastic biodegradation should be further optimized and standardized to high-throughput screening methods.

Genome editing has been demonstrated to be a very effective technology in a wide diversity of applications, and hence, the utilization of its most promising approach, viz., CRISPR/Cas9 technology, should be investigated for the optimized engineering of plastic polymer biodegradation in microbes. Furthermore, a deeper knowledge of the biochemical and structural characteristics of plastic-degrading enzymes is key to further unravelling the mechanism of recalcitrant plastic biodegradation. If attained, all these will ultimately lead to the scalable production and more efficient deployment of plastic-degrading biocatalysts in real-time industrial plastic processing.

In the short term, other areas that could be explored include the application of synthetic biology in discovering microbes with the ability to produce high-value compounds from various plastic wastes, thus, contributing to the efficient circular use of plastic polymers. In this regard, the monomeric and oligomeric derivatives produced after the biodegradation process would be used to synthesize value-added products for various industries as well as biodegradable polymers. More regulatory efforts must also be placed on replacing the more popular inert plastic polymers with the relatively more readily biodegradable plastics specifically tailored for improved biodegradability. Furthermore, a drastic reduction in the indiscriminate use, disposal, and contamination of the planet with conventional and biodegradable plastic polymers will remarkably lead to reduced plastic loads and the resultant environmental burden. This can be achieved by the effective implementation of regulatory measures that could include statutory bans and monetary incentives, as well as increased awareness and educational campaigns on the ecological impacts of increasing plastic waste in the environment.

References

1 Chamas, A., Moon, H., Zheng, J. et al. (2020). Degradation rates of plastics in the environment. *ACS Sustainable Chemistry & Engineering* 8: 3494–3511.

2 Fredi, G. and Dorigato, A. (2021). Recycling of bioplastic waste: a review. *Advanced Industrial and Engineering Polymer Research* 4: 159–177.

3 Singh, S.P. and Upadhyay, S.K. (eds.) (2021). *Bioprospecting of Microorganism-Based Industrial Molecules*. John Wiley & Sons Ltd. doi:10.1002/9781119717317.

4 Leal Filho, W., Salvia, A.L., Bonoli, A. et al. (2021). An assessment of attitudes towards plastics and bioplastics in Europe. *Science of the Total Environment* 755: 142732.

5 Lebreton, L., Slat, B., Ferrari, F. et al. (2018). Evidence that the Great Pacific Garbage Patch is rapidly accumulating plastic. *Scientific Reports* 8: 4666.

6 Amobonye, A., Bhagwat, P., Singh, S., and Pillai, S. (2021). Plastic biodegradation: frontline microbes and their enzymes. *Science of the Total Environment* 759: 143536.

7 Amobonye, A., Bhagwat, P., Raveendran, S. et al. (2021). Environmental impacts of microplastics and nanoplastics: a current overview. *Frontiers in Microbiology* 12: 768297.

8 McIlgorm, A., Raubenheimer, K., McIlgorm, D.E., and Nichols, R. (2022). The cost of marine litter damage to the global marine economy: insights from the Asia-Pacific into prevention and the cost of inaction. *Marine Pollution Bulletin* 174: 113167.

9 Evode, N., Qamar, S.A., Bilal, M. et al. (2021). Plastic waste and its management strategies for environmental sustainability. *Case Studies in Chemical and Environmental Engineering* 4: 100142.

10 Chen, Y. and Selvinsimpson, S. (2022). Current trends, challenges, and opportunities for plastic recycling. In: *Plastic and Microplastic in the Environment: Management and Health Risks* (ed. A. Ahamad, P. Singh, and D. Tiwary), 205–221. Wiley.

11 Manickavelan, K., Ahmed, S., Mithun, K. et al. (2022). A review on transforming plastic wastes into fuel. *Journal of the Nigerian Society of Physical Sciences* 64–74.

12 Talha, M.A., Goswami, M., Giri, B. et al. (2018). Bioremediation of Congo red dye in immobilized batch and continuous packed bed bioreactor by *Brevibacillus parabrevis* using coconut shell bio-char. *Bioresource Technology* 252: 37–43.

13 Darwesh, O.M., Matter, I.A., and Eida, M.F. (2019). Development of peroxidase enzyme immobilized magnetic nanoparticles for bioremediation of textile wastewater dye. *Journal of Environmental Chemical Engineering* 7: 102805.

14 Gola, D., Chawla, P., Malik, A., and Ahammad, S.Z. (2020). Development and performance evaluation of native microbial consortium for multi metal removal in lab scale aerobic and anaerobic bioreactor. *Environmental Technology and Innovation* 18: 100714.

15 Srivastava, A.K., Singh, R.K., and Singh, D. (2021). Microbe-based bioreactor system for bioremediation of organic contaminants: present and future perspective. In: *Microbe Mediated Remediation of Environmental Contaminants* (ed. A. Kumar, V.K. Singh, P. Singh, and V.K. Mishra), 241–253. Woodhead Publishing.

16 Liu, S.Y., Leung, M.M.-L., Fang, J.K.-H., and Chua, S.L. (2021). Engineering a microbial "trap and release" mechanism for microplastics removal. *Chemical Engineering Journal* 404: 127079.

17 Wu, W.-M. and Criddle, C.S. (2021). Characterization of biodegradation of plastics in insect larvae. In: *Methods in Enzymology*, vol. 648 (ed. G. Weber, U.T. Bornscheuer, and R. Wei), 95–120. Academic Press.

18 Brandon, A.M., Garcia, A.M., Khlystov, N.A. et al. (2021). Enhanced bioavailability and microbial biodegradation of polystyrene in an enrichment derived from the gut microbiome of *Tenebrio molitor* (mealworm larvae). *Environmental Science & Technology* 55: 2027–2036.

19 Atanasova, N., Stoitsova, S., Paunova-Krasteva, T., and Kambourova, M. (2021). Plastic degradation by extremophilic bacteria. *International Journal of Molecular Sciences* 22: 5610.

20 Shintani, T., Upadhyay, S.K., Singh, S.P. (2021). An introduction to microbial biodiversity and bioprospection. In: *Bioprospecting of Microorganism-Based Industrial Molecules* (ed. S.P. Singh and S.K. Upadhyay). John Wiley & Sons Ltd. https://doi.org/10.1002/9781119717317.ch1.

21 Hari, S. (2019). Screening of enzymes from actinomycetes and fungi isolated from plastic dumped soil. *Research Journal of Pharmacy and Technology* 12: 2261–2266.

22 Chia, W.Y., Tang, D.Y.Y., Khoo, K.S. et al. (2020). Nature's fight against plastic pollution: algae for plastic biodegradation and bioplastics production. *Environmental Science and Ecotechnology* 4: 100065.

23 Urbanek, A.K., Rymowicz, W., and Mirończuk, A.M. (2018). Degradation of plastics and plastic-degrading bacteria in cold marine habitats. *Applied Microbiology and Biotechnology* 102: 7669–7678.

24 Sánchez, C. (2020). Fungal potential for the degradation of petroleum-based polymers: an overview of macro-and microplastics biodegradation. *Biotechnology Advances* 40: 107501.

25 Nandakumar, A., Chuah, J.-A., and Sudesh, K. (2021). Bioplastics: a boon or bane? *Renewable and Sustainable Energy Reviews* 147: 111237.

26 Asgher, M., Qamar, S.A., Bilal, M., and Iqbal, H.M. (2020). Bio-based active food packaging materials: sustainable alternative to conventional petrochemical-based packaging materials. *Food Research International* 137: 109625.

27 Chen, L., Pelton, R.E., and Smith, T.M. (2016). Comparative life cycle assessment of fossil and bio-based polyethylene terephthalate (PET) bottles. *Journal of Cleaner Production* 137: 667–676.

28 Zheng, R., Liu, Z., and Xie, Z. (2022). Reaction: process changes of PVC manufacture driven by ethane chlorination. *Chem* 8: 888–889.

29 Nielsen, T.D., Hasselbalch, J., Holmberg, K., and Stripple, J. (2020). Politics and the plastic crisis: a review throughout the plastic life cycle. *Wiley Interdisciplinary Reviews: Energy and Environment* 9: e360.

30 Kunioka, M., Ninomiya, F., and Funabashi, M. (2007). Biobased contents of organic fillers and polycaprolactone composites with cellulose fillers measured by accelerator mass spectrometry based on ASTM D6866. *Journal of Polymers and the Environment* 15: 281–287.

31 De Almeida Oroski, F., Chaves Alves, F., and Bomtempo, J.V. (2014). Bioplastics tipping point: drop-in or non-drop-in? *Journal of Business Chemistry* 11: 43–50.

32 Xiao, B., Zheng, M., Pang, J. et al. (2015). Synthesis and characterization of poly (ethylene terephthalate) from biomass-based ethylene glycol: effects of miscellaneous diols. *Industrial & Engineering Chemistry Research* 54: 5862–5869.

33 Haarstrick, A., Hempel, D., Ostermann, L. et al. (2001). Modelling of the biodegradation of organic matter in municipal landfills. *Waste Management & Research* 19: 320–331.

34 Kijchavengkul, T., Auras, R., Rubino, M. et al. (2006). Development of an automatic laboratory-scale respirometric system to measure polymer biodegradability. *Polymer Testing* 25: 1006–1016.

35 Bhagwat, G., Gray, K., Wilson, S.P. et al. (2020). Benchmarking bioplastics: a natural step towards a sustainable future. *Journal of Polymers and the Environment* 28: 3055–3075.

36 Van den Oever, M., Molenveld, K., van der Zee, M., and Bos, H. (2017). *Bio-based and Biodegradable Plastics: Facts and Figures: Focus on Food Packaging in the Netherlands.* Wageningen Food & Biobased Research.

37 Wong, J.K.H., Lee, K.K., Tang, K.H.D., and Yap, P.-S. (2020). Microplastics in the freshwater and terrestrial environments: prevalence, fates, impacts and sustainable solutions. *Science of the Total Environment* 719: 137512.

38 Elahi, A., Bukhari, D.A., Shamim, S., and Rehman, A. (2021). Plastics degradation by microbes: a sustainable approach. *Journal of King Saud University-Science* 33: 101538.

39 Webb, H., Arnott, J., Crawford, R., and Ivanova, E. (2013). Plastic degradation and its environmental implications with special reference to poly(ethylene terephthalate). *Polymers* 5: 1–18.

40 Lucas, N., Bienaime, C., Belloy, C. et al. (2008). Polymer biodegradation: mechanisms and estimation techniques. *Chemosphere* 73: 429–442.

41 Devi, R.S., Kannan, V.R., Natarajan, K. et al. (2016). The role of microbes in plastic degradation. In: *Environmental Waste Manage* (ed. R. Chandra), 341. CRC Press.

42 Jenkins, S., Quer, A.M., Fonseca, C., and Varrone, C. (2019). Microbial degradation of plastics: new plastic degraders, mixed cultures and engineering strategies. In: *Soil Microenvironment for Bioremediation and Polymer Production* (ed. N. Jamil, P. Kumar, and R. Batool), 213–238. Scrivener Publishing LLC.

43 Restrepo-Flórez, J.-M., Bassi, A., and Thompson, M.R. (2014). Microbial degradation and deterioration of polyethylene – a review. *International Biodeterioration & Biodegradation* 88: 83–90.

42 Pathak, V.M. (2017). Review on the current status of polymer degradation: a microbial approach. *Bioresources and Bioprocessing* 4: 1–31.

45 Barone, G.D., Ferizović, D., Biundo, A., and Lindblad, P. (2020). Hints at the applicability of microalgae and Cyanobacteria for the biodegradation of plastics. *Sustainability* 12: 10449.

46 Hua, F., Wang, H.Q., Li, Y., and Zhao, Y.C. (2013). Trans-membrane transport of n-octadecane by *Pseudomonas* sp. DG17. *Journal of Microbiology* 51: 791–799.

47 Gravouil, K., Ferru-Clément, R., Colas, S. et al. (2017). Transcriptomics and lipidomics of the environmental strain *Rhodococcus ruber* point out consumption pathways and potential metabolic bottlenecks for polyethylene degradation. *Environmental Science and Technology* 51: 5172–5181.

48 Ho, B.T., Roberts, T.K., and Lucas, S. (2018). An overview on biodegradation of polystyrene and modified polystyrene: the microbial approach. *Critical Reviews in Biotechnology* 38: 308–320.

49 Alshehrei, F. (2017). Biodegradation of synthetic and natural plastic by microorganisms. *Journal of Applied & Environmental Microbiology* 5: 8–19.

50 Yang, Y., Wang, J., and Xia, M. (2020). Biodegradation and mineralization of polystyrene by plastic-eating super worms *Zophobas atratus*. *Science of the Total Environment* 708: 135233.

51 Yoshida, S., Hiraga, K., Takehana, T. et al. (2016). A bacterium that degrades and assimilates poly (ethylene terephthalate). *Science* 351: 1196–1199.

52 Glaser, J.A. (2019). *Biological Degradation of Polymers in the Environment*, vol. 1, 13. London, UK: IntechOpen.

53 Jadaun, J.S., Bansal, S., Sonthalia, A. (2022). Biodegradation of plastics for sustainable environment. *Bioresource Technology* 347: 126697.

54 Sriyapai, P., Chansiri, K., and Sriyapai, T. (2018). Isolation and characterization of polyester-based plastics-degrading bacteria from compost soils. *Microbiology* 87: 290–300.

55 Adhi, T.P., Korus, R.A., and Crawford, D.L. (1989). Production of major extracellular enzymes during lignocellulose degradation by two *Streptomyces* in agitated submerged culture. *Applied and Environmental Microbiology* 55: 1165–1168.

56 Sepperumal, U. and Markandan, M. (2014). Growth of Actinomycetes and *Pseudomonas* sp., biofilms on abiotically pre-treated polypropylene surface. *European Journal of Zoological Research* 3: 6–17.

57 Hoang, K.C., Lee, C., Chung, Y. et al. (2007). Polyester-degrading actinomycetes isolated from the Touchien river of Taiwan. *World Journal of Microbiology and Biotechnology* 23: 201–205.

58 Devanshi, S., Shah, K.R., Arora, S., and Saxena, S. (2021). Actinomycetes as an environmental scrubber. In: *Crude Oil-New Technologies and Recent Approaches* (ed. M.E. Abdel-Raouf and M.H. El-Keshawy). IntechOpen.

59 Jabloune, R., Khalil, M., Moussa, I.E.B. et al. (2020). Enzymatic degradation of p-nitrophenyl esters, polyethylene terephthalate, cutin, and suberin by sub1, a suberinase encoded by the plant pathogen *Streptomyces scabies*. *Microbes and Environments* 35: ME19086.

60 Shah, A.A., Hasan, F., Hameed, A., and Ahmed, S. (2007). Isolation and characterisation of poly (3-hydroxybutyrate-co-3-hydroxyvalerate) degrading actinomycetes and purification of PHBV depolymerase from newly isolated *Streptoverticillium kashmirense* AF1. *Annals of Microbiology* 57: 583–588.

61 Aly, M.M., Tork, S., Qari, H.A., and Al-Seenim, M.N. (2015). Poly-β-hydroxy butyrate Depolymerase from *Streptomyces lydicus* MM10, Isolated from Wastewater Sample. *International Journal of Agriculture and Biology* 17: 891–900.

62 Sekiguchi, T., Sato, T., Enoki, M. et al. (2011). Isolation and characterization of biodegradable plastic degrading bacteria from deep-sea environments. *JAMSTEC Report of Research and Development* 11: 33–41.

63 Asiandu, A.P., Wahyudi, A., and Sari, S.W. (2021). A review: plastics waste biodegradation using plastics-degrading bacteria. *Journal of Environmental Treatment Techniques* 9: 148–157.

64 Jarerat, A. and Tokiwa, Y. (2001). Degradation of poly (tetramethylene succinate) by thermophilic actinomycetes. *Biotechnology Letters* 23: 647–651.

65 Sarmah, P. and Rout, J. (2018). Efficient biodegradation of low-density polyethylene by cyanobacteria isolated from submerged polyethylene surface in domestic sewage water. *Environmental Science and Pollution Research* 25: 33508–33520.

66 Sanniyasi, E., Gopal, R.K., Gunasekar, D.K., and Raj, P.P. (2021). Biodegradation of low-density polyethylene (LDPE) sheet by microalga, *Uronema africanum* Borge. *Scientific Reports* 11: 1–33.

67 Kim, J.W., Park, S.-B., Tran, Q.-G. et al. (2020). Functional expression of polyethylene terephthalate-degrading enzyme (PETase) in green microalgae. *Microbial Cell Factories* 19: 1–9.

68 Moog, D., Schmitt, J., Senger, J. et al. (2019). Using a marine microalga as a chassis for polyethylene terephthalate (PET) degradation. *Microbial Cell Factories* 18: 1–15.

69 Yan, N., Fan, C., Chen, Y., and Hu, Z. (2016). The potential for microalgae as bioreactors to produce pharmaceuticals. *International Journal of Molecular Sciences* 17: 962.

70 Ren, L., Men, L., Zhang, Z. et al. (2019). Biodegradation of polyethylene by *Enterobacter* sp. D1 from the guts of wax moth *Galleria mellonella*. *International Journal of Environmental Research and Public Health* 16: 1941.

71 Muhonja, C.N., Makonde, H., Magoma, G., and Imbuga, M. (2018). Biodegradability of polyethylene by bacteria and fungi from Dandora dumpsite Nairobi-Kenya. *PLoS One* 13: e0198446.

72 Wilkes, R.A. and Aristilde, L. (2017). Degradation and metabolism of synthetic plastics and associated products by *Pseudomonas* sp.: capabilities and challenges. *Journal of Applied Microbiology* 123: 582–593.

73 Yoon, M.G., Jeon, H.J., and Kim, M.N. (2012). Biodegradation of polyethylene by a soil bacterium and AlkB cloned recombinant cell. *Journal of Bioremediation & Biodegradation* 3: 1–8.

74 Jeon, H.J. and Kim, M.N. (2015). Functional analysis of alkane hydroxylase system derived from *Pseudomonas aeruginosa* E7 for low molecular weight polyethylene biodegradation. *International Biodeterioration & Biodegradation* 103: 141–146.

75 Mohanan, N., Montazer, Z., Sharma, P.K., and Levin, D.B. (2020). Microbial and enzymatic degradation of synthetic plastics. *Frontiers in Microbiology* 11: 2837.

76 Das, M.P. and Kumar, S. (2015). An approach to low-density polyethylene biodegradation by *Bacillus amyloliquefaciens*. *Biotechnology* 5: 81–86.

77 Leja, K. and Lewandowicz, G. (2010). Polymer biodegradation and biodegradable polymers – a review. *Polish Journal of Environmental Studies* 19: 255–266.

78 Joo, S., Cho, I.J., Seo, H. et al. (2018). Structural insight into molecular mechanism of poly (ethylene terephthalate) degradation. *Nature Communications* 9: 382.

79 Austin, H.P., Allen, M.D., Donohoe, B.S. et al. (2018). Characterization and engineering of a plastic-degrading aromatic polyesterase. *Proceedings of the National Academy of Sciences* 115: E4350–E4357.

80 Shah, A.A., Hasan, F., Hameed, A., and Ahmed, S. (2008). Biological degradation of plastics: a comprehensive review. *Biotechnology Advances* 26: 246–265.

81 Raaman, N., Rajitha, N., Jayshree, A., and Jegadeesh, R. (2012). Biodegradation of plastic by *Aspergillus* spp. isolated from polythene polluted sites around chennai. *Journal of Academia and Industrial Research* 1: 313–316.

82 Munir, E., Harefa, R.S.M., Priyani, N., and Suryanto, D. (2018). Plastic degrading fungi *Trichoderma viride* and *Aspergillus nomius* isolated from local landfill soil in Medan. *IOP Conference Series: Earth and Environmental Science* 126: 012145.

83 Ammala, A., Bateman, S., Deana, K. et al. (2011). An overview of degradable and biodegradable polyolefins. *Progress in Polymer Science* 36: 1015–1049.

84 Gilan, I., Hadar, Y., and Sivan, A. (2004). Colonization, biofilm formation and biodegradation of polyethylene by a strain of *Rhodococcus ruber*. *Applied Microbiology and Biotechnology* 65: 97–104.

85 Pramila, R. and Ramesh, K.V. (2011). Biodegradation of low density polyethylene (LDPE) by fungi isolated from marine water a SEM analysis. *African Journal of Microbiology Research* 5: 5013–5018.

86 da Luz, J.M.R., da Silva, M.d.C.S., dos Santos, L.F., and Kasuya, M.C.M. (2019). Plastics polymers degradation by fungi. In: *Microorganisms* (ed. M. Blumenberg, M. Shaaban, and A. Elgaml). IntechOpen.

87 Müller, R.J., Schrader, H., Profe, J. et al. (2005). Enzymatic degradation of poly (ethylene terephthalate): rapid hydrolyse using a hydrolase from *T. fusca*. *Macromolecular Rapid Communications* 26: 1400–1405.

88 Gomez-Mendez, L.D., Moreno-Bayona, D.A., Poutou-Pinales, R.A. et al. (2018). Biodeterioration of plasma pretreated LDPE sheets by *Pleurotus ostreatus*. *PLoS One* 13: e0203786.

89 Iiyoshi, Y., Tsutsumi, Y., and Nishida, T. (1998). Polyethylene degradation by lignin-degrading fungi and manganese peroxidase. *Journal of Wood Science* 44: 222–229.

90 Tahir, L., Ali, M.I., Zia, M. et al. (2013). Production and characterization of esterase in *Lantinus tigrinus* for degradation of polystyrene. *Polish Journal of Microbiology* 62: 101–108.

91 Ojha, N., Pradhan, N., Singh, S. et al. (2017). Evaluation of HDPE and LDPE degradation by fungus, implemented by statistical optimization. *Scientific Reports* 7: 39515.

92 Panagiotidou, E., Konidaris, C., Baklavaridis, A. et al. (2014). A simple route for purifying extracellular poly (3-hydroxybutyrate)-depolymerase from *Penicillium pinophilum*. *Enzyme Research* 2014: 159809.

93 Xing, M.-N., Zhang, X.-Z., and Huang, H. (2012). Application of metagenomic techniques in mining enzymes from microbial communities for biofuel synthesis. *Biotechnology Advances* 30: 920–929.

94 Zhu, B., Wang, D., and Wei, N. (2022). Enzyme discovery and engineering for sustainable plastic recycling. *Trends in Biotechnology* 40: 22–37.

95 Shahnawaz, M., Sangale, M.K., and Ade, A.B. (ed.) (2019). Case studies and recent update of plastic waste degradation. In: *Bioremediation Technology for Plastic Waste*, 31–43. Singapore: Springer.

96 Dey, A.S., Bose, H., Mohapatra, B., and Sar, P. (2020). Biodegradation of unpretreated low-density polyethylene (LDPE) by *Stenotrophomonas* sp. and *Achromobacter* sp., isolated from waste dumpsite and drilling fluid. *Frontiers in Microbiology* 11: 3095.

97 Arunrattiyakorn, P., Ponprateep, S., Kaennonsang, N. et al. (2022). Biodegradation of polystyrene by three bacterial strains isolated from the gut of super worms (*Zophobas atratus* larvae). *Journal of Applied Microbiology* 132: 2823–2831.

98 Purohit, J., Chattopadhyay, A., and Teli, B. (2020). Metagenomic exploration of plastic degrading microbes for biotechnological application. *Current Genomics* 21: 253–270.

99 Madhavan, A., Sindhu, R., Parameswaran, B. et al. (2017). Metagenome analysis: a powerful tool for enzyme bioprospecting. *Applied Biochemistry and Biotechnology* 183: 636–651.

100 Karunatillaka, I., Jaroszewski, L., and Godzik, A. (2022). Novel putative polyethylene terephthalate (PET) plastic degrading enzymes from the environmental metagenome. *Proteins: Structure, Function, and Bioinformatics* 90: 504–511.

101 Gaytán, I., Sánchez-Reyes, A., Burelo, M. et al. (2020). Degradation of recalcitrant polyurethane and xenobiotic additives by a selected landfill microbial community and its biodegradative potential revealed by proximity ligation-based metagenomic analysis. *Frontiers in Microbiology* 10: 2986.

102 Müller, C.A., Perz, V., Provasnek, C. et al. (2017). Discovery of polyesterases from moss-associated microorganisms. *Applied and Environmental Microbiology* 83: e02641–e02616.

103 Lam, K.N., Cheng, J., Engel, K. et al. (2015). Current and future resources for functional metagenomics. *Frontiers in Microbiology* 6: 1196.

104 Contesini, F.J., Davanço, M.G., Borin, G.P. et al. (2020). Advances in recombinant lipases: production, engineering, immobilization and application in the pharmaceutical industry. *Catalysts* 10: 1032.

105 Hajighasemi, M., Tchigvintsev, A., Nocek, B. et al. (2018). Screening and characterization of novel polyesterases from environmental metagenomes with high hydrolytic activity against synthetic polyesters. *Environmental Science & Technology* 52: 12388–12401.

106 Adrio, J.-L. and Demain, A.L. (2010). Recombinant organisms for production of industrial products. *Bioengineered Bugs* 1: 116–131.

107 Wood, D.W. (2014). New trends and affinity tag designs for recombinant protein purification. *Current Opinion in Structural Biology* 26: 54–61.

108 Janatunaim, R.Z. and Fibriani, A. (2020). Construction and cloning of plastic-degrading recombinant enzymes (MHETase). *Recent Patents on Biotechnology* 14: 229–234.

109 Vázquez-Alcántara, L., Oliart-Ros, R.M., García-Bórquez, A., and Peña-Montes, C. (2021). Expression of a cutinase of *Moniliophthora roreri* with Polyester and PET-plastic residues degradation activity. *Microbiology Spectrum* 9: e00976–e00921.

110 Sameshima-Yamashita, Y., Watanabe, T., Tanaka, T. et al. (2019). Construction of a *Pseudozyma antarctica* strain without foreign DNA sequences (self-cloning strain) for high yield production of a biodegradable plastic-degrading enzyme. *Bioscience, Biotechnology, and Biochemistry* 83: 1547–1556.

111 Lutz, S. (2010). Beyond directed evolution – semi-rational protein engineering and design. *Current Opinion in Biotechnology* 21: 734–743.

112 Arslan, M., Karadağ, D., and Kalyoncu, S. (2019). Protein engineering approaches for antibody fragments: directed evolution and rational design approaches. *Turkish Journal of Biology* 43: 1–12.

113 Bornscheuer, U.T. and Pohl, M. (2001). Improved biocatalysts by directed evolution and rational protein design. *Current Opinion in Chemical Biology* 5: 137–143.

114 Tan, L.-T., Hiraishi, T., Sudesh, K., and Maeda, M. (2013). Directed evolution of poly [(R)-3-hydroxybutyrate] depolymerase using cell surface display system: functional importance of asparagine at position 285. *Applied Microbiology and Biotechnology* 97: 4859–4871.

115 Hiraishi, T., Hirahara, Y., Doi, Y. et al. (2006). Effects of mutations in the substrate-binding domain of poly [(R)-3-hydroxybutyrate] (PHB) depolymerase from *Ralstonia pickettii* T1 on PHB degradation. *Applied and Environmental Microbiology* 72: 7331–7338.

116 Then, J., Wei, R., Oeser, T. et al. (2016). A disulfide bridge in the calcium binding site of a polyester hydrolase increases its thermal stability and activity against polyethylene terephthalate. *FEBS Open Bio* 6: 425–432.

117 Thumarat, U., Kawabata, T., Nakajima, M. et al. (2015). Comparison of genetic structures and biochemical properties of tandem cutinase-type polyesterases from *Thermobifida alba* AHK119. *Journal of Bioscience and Bioengineering* 120: 491–497.

118 Son, H.F., Joo, S., Seo, H. et al. (2020). Structural bioinformatics-based protein engineering of thermo-stable PETase from *Ideonella sakaiensis*. *Enzyme and Microbial Technology* 141: 109656.

119 Herrero Acero, E., Ribitsch, D., Dellacher, A. et al. (2013). Surface engineering of a cutinase from *Thermobifida cellulosilytica* for improved polyester hydrolysis. *Biotechnology and Bioengineering* 110: 2581–2590.

120 Biundo, A., Subagia, R., Maurer, M. et al. (2019). Switched reaction specificity in polyesterases towards amide bond hydrolysis by enzyme engineering. *RSC Advances* 9: 36217–36226.

21

Strategies, Trends, and Technological Advancements in Microbial Bioreactor System for Probiotic Products

Soubhagya Tripathy[1], Ami R. Patel[2], Deepak Kumar Verma[1], Smita Singh[3], Gemilang Lara Utama[4,5], Mamta Thakur[6], Alaa Kareem Niamah[7], Nihir Shah[2], Shayma Thyab Gddoa Al-Sahlany[7], Prem Prakash Srivastav[1], Mónica L. Chávez-González[8], and Cristobal Noe Aguilar[8]

[1] Agricultural and Food Engineering Department, Indian Institute of Technology Kharagpur, Kharagpur, West Bengal, India
[2] Division of Dairy Microbiology, Mansinhbhai Institute of Dairy and Food Technology-MIDFT, Dudhsagar Dairy Campus, Mehsana, Gujarat, India
[3] Department of Allied Health Sciences, Chitkara School of Health Sciences, Chitkara University, Rajpura, Punjab, India
[4] Faculty of Agro-Industrial Technology, Universitas Padjadjaran, Sumedang, Indonesia
[5] Center for Environment and Sustainability Science, Universitas Padjadjaran, Bandung, Indonesia
[6] Department of Food Technology, School of Sciences, ITM University, Gwalior, Madhya Pradesh, India
[7] Department of Food Science, College of Agriculture, University of Basrah, Basra City, Iraq
[8] Bioprocesses and Bioproducts Research Group, Food Research Department, School of Chemistry, Autonomous University of Coahuila, Saltillo, Coahuila, Mexico

21.1 Introduction

Recently, there has been a lot of buzz about adding probiotic microbes to all sorts of foods. As a result, our understandings of the gut microbiota and the mechanisms that modulate it have improved. By definition, a probiotic is a "living microbe" or a "product containing viable microbial cells in sufficient quantities to affect the microflora," both of which, when provided at adequate levels, impart a health benefit to the host [1, 2]. Large numbers of people take probiotics from the bacterial families *Lactobacillus*, *Bifidobacterium*, and *Bacillus*, as well as a small number of yeasts. Considered to be advantageous to human health, lactic acid bacteria (LAB) may be found in both naturally fermented foods and dietary supplements [2, 3, 4, 5]. However, a major breakthrough has been the introduction of fortified foods containing a combination of probiotics and prebiotics that, when ingested as part of a regular diet, increase beneficial bacteria in the gut [2, 6–9].

Humans have a diverse array of microorganisms in their gastrointestinal (GI) tracts, with estimates placing the number of microbial cells there at roughly 1013 [10]. Beneficial microbes improve the host's immune system and metabolic capacity, alter gene expression, and increase the body's use of prebiotics (non-digestible oligosaccharides) [9, 11–16]. Dysbiosis, a microbial imbalance between the host and the population of gut microflora, likely results from any

Microbial Bioreactors for Industrial Molecules, First Edition.
Edited by Sudhir P. Singh and Santosh Kumar Upadhyay.

change in the abundance or composition of the gut microbes' community due to external influences [10]. In the treatment of several GI illnesses, including antibiotic-associated diarrhea, *Helicobacter pylori* infection, and viral infections, probiotics have been recommended in the literature as a potential alternative to antibiotics [2, 17]. The effectiveness of probiotics in combating multiple sclerosis and in enhancing a wide range of mental health indicators has also been suggested by a recent study by Rahimlou and colleagues [18]. Despite these findings, it is yet unknown whether commercially available probiotics are useful in treating severe GI problems [19]. It is directly tied to the total number of living cells (colony forming unit per mL) present in the labeled product, which in turn is related to claims made about the efficacy of probiotic microbial strains. Further reduction in viable cells may occur during transit through the GI tract as a result of the complicated GI environment (such as acidic conditions) probiotics must endure before they reach the colon [20].

Because of the prevalence of the various issues, the vast majority of companies that produce probiotics have switched their attention away from products that include a single probiotic strain and instead are concentrating on developing multi-strain solutions [2, 7, 9]. Products or preparations that contain multiple strains of probiotics are thought to be superior to single-culture probiotics in terms of their ability to alleviate the symptoms of gastric disorders. Additionally, multi-strain probiotics are thought to be more effective at colonizing the intricate ecosystem that exists within the human gut [21, 22]. Encapsulating probiotics into a specific carrier material is one of the potential ways that may be used to boost the probiotics' survival [2, 23].

This chapter covers the most recent advancements that have been made in bioreactor systems for the production of probiotics, as well as prospective novel methods for boosting the performance of probiotics while they are being fermented and processed further down the production chain. In this regard, methods that incorporate applications of sublethal stress during the synthesis of probiotics and innovative bioreactor technologies such as immobilized cell fermentations are techniques that show promise as potential solutions.

21.2 Bioreactors and Production of Probiotics

The process of fermentation is an essential part of bioengineering and has enormous ramifications for human civilization [1, 7, 13, 14, 24, 25–27]. The bioreactor or fermenter is the production unit that is utilized most frequently for this method of operation [24, 27–29]. The simplicity of its construction is the primary benefit it offers. It is beneficial for bioreactors to be outfitted with a control system since proper management of the fermentation process may considerably raise both the product's quality and the overall efficiency of the production process [24, 27, 28, 30–32].

One way to think of the human digestive tract is as a fermenter, as it is responsible for producing and maintaining complex ecosystems of both aerobic and anaerobic bacteria [33]. Fermentation is the most typical approach used in the manufacturing of commercial probiotics. Fermented dairy products are where we will find the most frequent kinds of probiotics in their natural state [2, 9, 24]. In addition, fermented foods such as kimchi, kombucha, sourdough, pickles, and sauerkraut have been utilized by humans for a significant amount of time as a source of probiotics [2, 7, 13, 14, 34, 35–39]. Living bacteria are known as probiotics, and it is thought that they have a health-promoting action in addition to having biofunctional impacts on humans. The types of bacteria belonging to the genera *Bifidobacterium* and *Lactobacillus* are the ones that have received the most attention and research out of all the other probiotic bacteria [9, 40, 41, 42]. They are used rather regularly to increase the quantity of desirable and helpful bacteria in the colon as well as to rebuild the intestinal flora, for example following treatments with antibiotics. Because of its therapeutic significance in the treatment of a variety of GI illnesses, the value of the probiotics market, as well as the market for probiotic nutritional supplements, has significantly expanded. Because of their potential use in health care, *Lactobacillus* species have been the subject of a significant amount of research as starter cultures for the fermentation of dairy products or probiotics [9, 24, 42].

Numerous LABs from various genera are found in commercially marketed probiotics. These include *Limosilactobacillus reuteri, Lacticaseibacillus rhamnosus, Lacticaseibacillus casei, Lactobacillus acidophilus* group, bifidobacteria, and *Bacillus coagulans*. However, commercially manufactured probiotics include LAB from the genera *Propionibacterium, Pediococcus, Lactococcus, Streptococcus, Leuconostoc, Enterococcus* (especially specific enterococci like *Enterococcus faecium* SF68), and the yeast *Saccharomyces boulardii* [3, 40, 43]. Table 21.1 summarizes the strains of commercially available probiotics that have been the subject of at least one thorough randomized control experiment to assess their safety and efficacy in GI circumstances. However, a new generation of probiotic bacteria is now being produced with the use of -omics technology. Recent advances in -omics

Table 21.1 Commercialized probiotic strains effective against specific gastrointestinal disorders with their relevant dosage.

Probiotic strain	Use against gastrointestinal disorder	Dosage value (in CFU)	Reference
Akkermansia muciniphila ATCC BAA-835	Inflammatory bowel disease	2×10^8	[44]
Bacillus subtilis PXN 21	Immunity	2×10^9	[45]
Saccharomyces boulardii CNCM I-745	Diarrhea	5×10^9	[46]
Bifidobacterium animalis	Eradication of *Helicobacter pylori*	1×10^9	[47]
Streptococcus thermophilus KCTC 11870BP	Constipation	2.5×10^8	[48]
Pediococcus acidilactici CECT 7483	Irritable bowel syndrome	$3–6 \times 10^9$	[49]
Bifidobacterium bifidum MIMBb75	Irritable bowel syndrome	1×10^9	[50]
Lacticaseibacillus paracasei B21060 and *L. rhamnosus* GG	Acute diarrhea	1×10^{10}	[51]
Escherichia coli Nissle 1917	Inflammatory bowel disease, ulcerative colitis	5×10^{10}	[52]
Lactobacillus delbrueckii subsp. *bulgaricus* LBY-27	Irritable bowel syndrome	4×10^9	[52]

technologies have not allowed for a full characterization of the microbial population in the gut. In spite of this, computer studies have been conducted to try to figure out which microbial communities are actually present in the gut and what function each strain plays [53]. However, it is vital to recognize that our complete understanding of the gut microflora is far from complete and that new bacteria are continuously being discovered and defined.

Recent years have seen a rise in interest in the use of probiotic blends containing both aerobic and anaerobic strains for nutritional supplementation and medical purposes. The number of live cells present in the GI system is one factor that determines how effective probiotic combinations are. Probiotics are traditionally manufactured in separate bioreactors and then enhanced with various prebiotics and carrier agents. This procedure is done in order to maximize its effectiveness [33]. An illustration of some important bioreactors used for probiotic production has been indicated in Figure 21.1. For the purpose of the efficient and cost-effective industrial production of probiotic biomass, it is important to develop a novel medium that has a low operating cost. The agricultural and food sectors generate a lot of waste, which might be turned into useful components for low-cost medium development. Furthermore, industrial wastes such as cheese or paneer whey and corn steep liquor are exceptional examples of dependable nitrogen (N_2) sources for the development of probiotic biomass [54].

Fermentation, media composition, cell harvesting process, cell drying methods (spray-drying or freeze-drying, depicted in Figure 21.2), and conditions during storage like humidity, temperature, and pH are some of a wider array of manufacturing parameters that can affect the growth, viability, and ultimately survival of probiotic strains. The quality of the

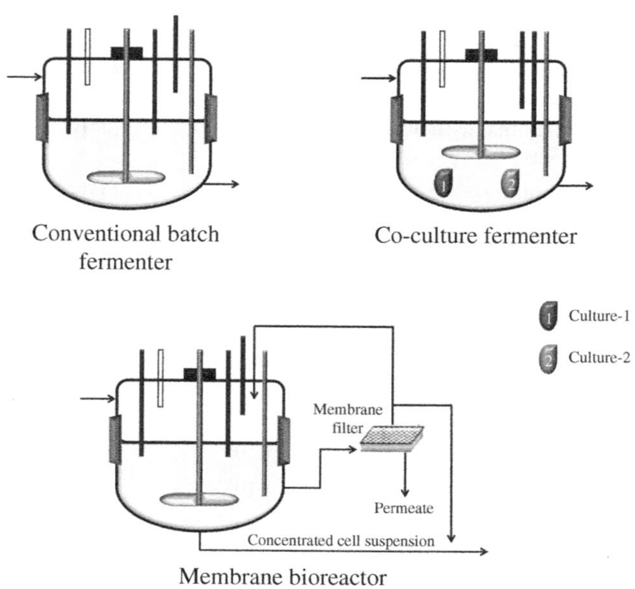

Figure 21.1 Diagrammatic presentation of the numerous bioreactors that are utilized in the production of probiotics. *Source:* Adapted from Hathi et al. [33].

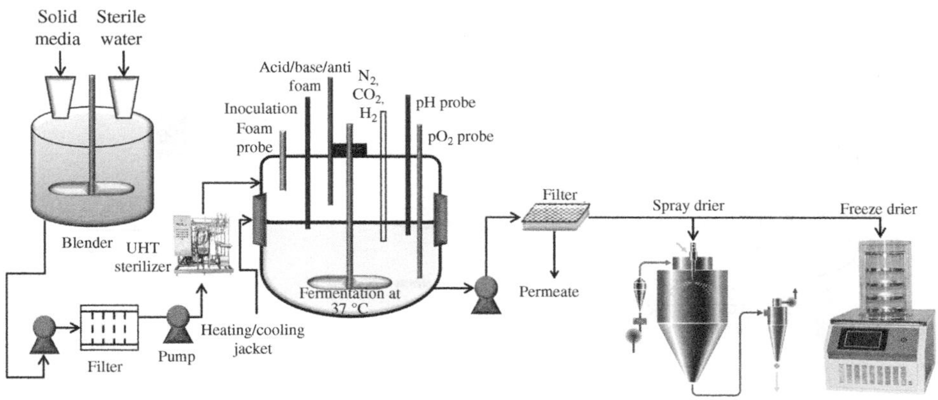

Figure 21.2 A schematic depiction of the production process for probiotics, employing either spray-drying or freeze-drying. *Source:* Hathi et al. [33] / with permission from Elsevier.

final probiotic preparation is primarily dependent on the procedures that are followed during the manufacturing process [55].

21.2.1 Conventional Batch Bioreactor System

Traditional fermentation methods for single-strain probiotics are shown in Figure 21.1. To prevent the bacteria from sinking to the bottom of the bioreactor, continual stirring is used to grow single-strain probiotics in a particular liquid broth within a huge pH-controlled fermenter tank. It is important to note that throughout the process of producing anaerobic probiotics, dissolved gases such as N_2, hydrogen (H_2), and carbon dioxide (CO_2) are frequently subjected to stringent regulation and circulation. Figure 21.3 is a representation of the aseptic process, which consists of combining a solid growth medium with sterile water in a mixing tank, filtering the mixture to remove any leftover particles, and then pumping the mixture into a fermenter vessel that has been sanitized.

When it comes to inoculating with single-strain probiotics, there are two typical ways that are utilized [56]. The preparation of cultures on a smaller scale that may later be increased in size is made possible through the use of freeze-dried stocks. An alternate technique in which a significant number of frozen, concentrated cells are delivered to the fermenter is called direct vat inoculation/direct vat set (DIV/DVS). When employing the first method, there is a possibility of genetic drift occurring due to the multiple generations of bacteria that are produced while raising the culture volume; however, using DVI/DVS removes the possibility of this happening [56].

In the fed-batch fermentation approach, which is one of the older methods, the utilization of concentrated whey and waste from the processing of mussels was emphasized as a means of achieving high-volume production of the probiotic strain *Lactococcus lactis* CECT 539. This was done as a means to achieve high-quality production of the strain. The following are the circumstances that were present during the batch fermentation: 2% inoculum (v/v), 30 °C temperature, and agitation at 200 revolutions per minute (rpm). According to the findings, the amount of biomass was raised to 5.49 g/L, and a greater

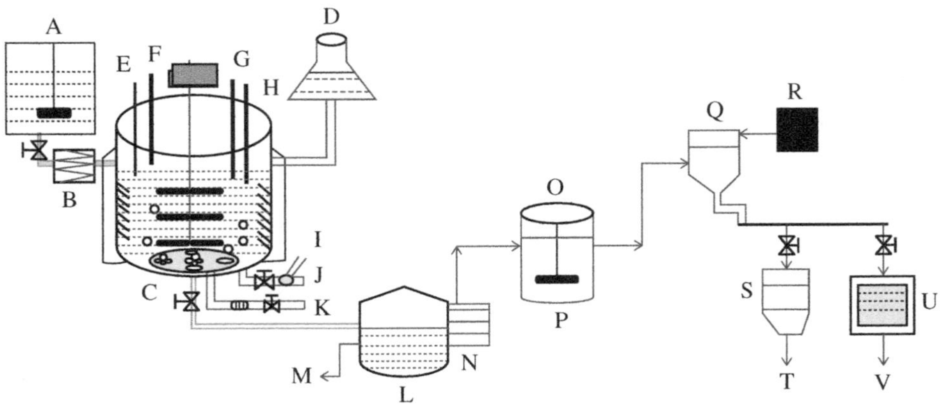

Figure 21.3 Schematic representation of various components of bioreactor systems for the production of probiotics. (A) Water and solid media mix; (B) sterilizer; (C) fermenter; (D) nutrient media containing vessel; (E) inoculation port; (F) gas exhaust port; (G) oxygen probe; (H) pH probe; (I) steam valve; (J) sampling port; (K) air inlet; (L) centrifuge/filtration unit; (M) supernatant; (N) filtrate; (O) cryo/lyo-protectant vessel; (P) cryo/lyo-protectant and cell concentrate mix tank; (Q) pelletizer; (R) liquid N_2 bath; (S) spray drier; (T) spray-dried powder; (U) freeze drier; (V) freeze-dried powder.

number of viable cells (2.33 × 1010 CFU/mL) were achieved [57]. The study effectively proposes a technique to utilize the culture medium made by effluents of the food industry using a fed-batch method for better production of probiotic bacteria, which is one of the most important contributions that can be taken away from the study [57]. In a study that was quite similar to this one, researchers evaluated many different types of dietary strategies that may modulate the gut microbiota in the circumstance of dysbiosis. Researchers in this work employed batch fermentation to learn how different *Bifidobacterium* strains symbiotically exploited prebiotics (inulin and short-chain fructo-oligosaccharides) as a carbon source. The researchers were interested in how inulin and short-chain fructo-oligosaccharides were utilized by the *Bifidobacterium* strains. In particular, the researchers were interested in the manner in which the *Bifidobacterium* strains metabolized inulin and short-chain fructo-oligosaccharides. The use of short-chain fructo-oligosaccharides by strains of *Bifidobacterium longum* was shown to be much more successful than the utilization of inulin, as evidenced by the findings of Malvido and coworkers [57]. In addition, the use of short-chain fructo-oligosaccharides as a source of carbon was found to modify the synthesis of metabolites [58]. Both the studies that were described earlier in this section came to the conclusion that batch fermentation techniques may be employed effectively for increased probiotic production when the appropriate carbon source is utilized.

In a recent investigation on the gut probiotic *Prevotella copri* DSM18205T, done by Huang and coworkers [59], the researchers reported effective batch cultivations using stirred tank bioreactors. The primary objective of the research was to conduct a comparative growth analysis between two distinct kinds of peptone yeast medium, each of which could either employ glucose or xylose as its source of carbon. In addition, this paper included a more in-depth investigation into the technique by which succinic acid is produced throughout the fermentation process. Fermentation took place in a 1.4 L bioreactor with a pH of 7.2 maintained by adding 1 M NaOH to the medium. The reactor was heated to 37 °C and agitated at a rate of 80 rpm. According to the findings, the organism exhibited a rather consistent cell dry weight of more than 2 g/L when the pH was measured at 7.2 [59]. Nevertheless, the organism that produces succinic acid preferred conditions containing xylose since it resulted in 0.8 g of succinic acid being formed for every gram of xylose that was used. This was a significantly greater amount in comparison to the 0.6 g of succinic acid that Huang and coworkers found to be created per gram of glucose [59]. It is possible to monitor the growth of probiotics in the fermenter through the use of cell density measurements, and both optical and fluorescence microscopes are utilized in order to examine the surrounding environment for any indications of contamination. The temperature that is typically maintained during fermentation is 37 °C; however, this temperature may change depending on the probiotic strain being used.

Following the completion of the batch fermentation process, a filtered and concentrated suspension of cells is either spray-dried or freeze-dried to generate a dry powder probiotic. This powder may then be combined with a variety of food products or marketed in capsule form. For the fermentation process to be profitable on a commercial scale, it must have a high cell density in addition to its viability. In order to avoid the loss of bacterial viability, cryoprotectants and lyoprotectants are often introduced just before to the freeze-drying and spray-drying processes, respectively. It has been found that freeze-dried probiotic bacteria have a greater survival rate than spray-dried bacteria because of the lower temperatures involved in the freeze-drying method [60]. Freeze-drying in the presence of lactulose as a cryoprotectant was shown to be an effective method for the preservation of the probiotic strain *Enterococcus durans* in a study that was carried out by Estilarte and colleagues [61]. The cell pellets were centrifuged and then resuspended in 0.15 mL of a 0.85% (w/v) NaCl solution, followed by 0.15 mL of a solution containing either 200 mg/L of sucrose, lactulose, or maltodextrin. After being frozen (at −80 °C for 24 hours), the cell suspension was freeze-dried for 48 hours at −50 °C and 0.006 mbar correspondingly. However, a number of studies that focused on the administration of probiotics show that the spray drying technique is just as successful as other ways because it provides a barrier of protection for cells against the harsh conditions of the intestinal environment and storage methods [62, 63]. On the basis of the molecular weight of the constituent components, cryoprotectants may be divided into two distinct categories [64], and they are as follows: (i) low-molecular weight (LMW) cryoprotectants and (ii) high-molecular weight (HMW) cryoprotectants. Sugars such as trehalose, lactose, glucose, lactose, sorbitol, and mannose are examples of LMW cryoprotectants that are utilized often. Sugars are assumed to prevent cellular damage by reacting with phospholipids that are already present in the cell wall during LMW cryopreservation, despite the fact that this process is not completely understood. Cryoprotectants with an HMW, such as polysaccharides and polypeptides (such as whey proteins, casein, alginate, gelatin, maltodextrin, etc.), coat the surface of the microorganisms with a thick, viscous film that is impermeable to ice crystals [64]. Moreover, it has been found that the probiotic cells are able to maintain higher viability rates (greater than 4 log CFU/mL) when the encapsulation conditions are adequate and sufficient lyoprotectants are used [65].

21.2.2 Membrane Bioreactor System

The use of cross-flow membrane ultra/microfiltration into the batch fermentation vessel is one method that may be utilized to boost the output rate (Figures 21.1 and 21.3). When the required cell density has been attained, the broth is passed through the membrane unit of the filtration system to get rid of any potentially dangerous metabolites and excess acids, and then the retentate is transferred back into the fermenter vessel. After then, the new medium is continually fed into the fermenter, and the flow rates are adjusted as necessary in order to keep the overall volume from fluctuating. After that, the concentrated cell suspension can be removed either in batches or in a continuous fashion. When grown in membrane bioreactors, probiotics have been shown in studies to provide a higher cell output and overall productivity than when produced in conventional bioreactors [1, 26, 56]. When compared to traditional batch fermentation, the research conducted by Anandharaj and coworkers [26] found that the amount of biomass produced by continuous fermentation was 7–10 times higher than that of batch fermentation. High cell yield and a 15-fold increase in volumetric productivity were reported by Corre and coworkers [66] for *Bifidobacterium bifidum* in liquid culture. This was in comparison to previous findings that there was a high cell yield. On the other hand, it has been noted that membrane fouling and a greater cost input are associated with membrane-based bioreactors in comparison to conventional batch fermentation (Table 21.2). In spite of this, Fan and colleagues who concentrated their research on the problem of membrane fouling found that employing a combination of fed-batch fermentation technology and *B. coagulans* boosts the final concentration as well as the purity of the target product [71]. In this particular instance, fructo-oligosaccharides are the product of interest. These fructo-oligosaccharides are produced by enzymes known as fructosyltransferases or levansucrases [72]. In the beginning, a semi-continuous manufacturing method using a membrane bioreactor was established. This method was then followed by fermentation using *B. coagulans*. Studies involving fermentation were conducted in a 3 L bioreactor at 40 °C and pH 6.8 with 1 vvm aeration. After the fermentation had progressed to the point where it had reached the stationary phase, the mode of operation was changed to fed-batch, and the fructo-oligosaccharide solution was constantly diluted so as to avoid substrate inhibition. According to the findings of Fan and colleagues [71], the most significant contributor to membrane fouling is the aggregation of proteins, which might be responsible for more than 50% of the resistance encountered during the filtering process. During the fermentation process, the fructo-oligosaccharide solution was appropriately diluted with water so as to avoid substrate inhibition. This was done in order to prevent substrate inhibition. After the elimination of all inhibitory metabolites, the final product was the fructo-oligosaccharide, which had a very high purity of 96.6% [71].

Immobilized cell fermentation is an additional method that may be utilized to increase the production rates of probiotic bacteria (Figure 21.3). The development of novel fermentation technologies has resulted in a rise in the level of attention that is being paid to the use of novel nanoparticles, polymers, and compounds that are derived from plants in current research. For immobilized cell fermentation, the biopolymers that can most effectively operate as an immobilization matrix are chosen. This helps to enhance the overall growth rate, maintains cell viability, and makes it possible to more easily collect the value-added product. Because of this, there is a possibility that the probiotic strains' activity can be enhanced to a greater degree [73]. In immobilized cell culture, the cells are initially

Table 21.2 Different fermentation systems employed for producing probiotic cultures with specific pros and cons.

Type of fermentation	Batch fermentation	Fed-batch fermentation	Continuous system	Membrane bioreactor system	Immobilized culture system
Probiotic strain(s) studied	*Lactococcus lactis*, *Bifidobacterium longum*, *Bifidobacterium bifidum*	*Lactococcus lactis* CECT 539	*Bifidobacterium longum*	*Bifidobacterium longum* *Bacillus coagulans* PS5	*Bifidobacterium longum*, *Lactococcus lactis*, *Lactobacillus delbrueckii* ssp. *bulgaricus*
Advantages	• Simple control and easy to sterilize • Low chance of contamination • Low operating cost	• High viability of probiotic strains • Higher yield	• Under controlled conditions, a higher growth rate is noticed • Usually used in adaptation and evolution studies	• Improves productivity • Diminishes effects of inhibitory metabolites on probiotic strains produced during the growth	• Cost-effective approach • Greater viability of the probiotic culture • Continuous and controlled delivery of probiotics to the gut
Drawbacks	• New seed culture is necessitated for every batch • Fresh carbon source is required for each batch • Hard to culture anaerobic strains	• A build-up of inhibitory metabolites including toxins may occur	• A constant supply of nutrients is an obligation • Aggregation of microbial cells can limit optimal growth	• High-cost approach • Membrane fouling is the biggest issue	• Biocompatibility of some immobilizing agents restricts their use • Production processes are very complex
Reference(s)	[58, 67]	[57]	[68, 69]	[26, 27, 70]	[30, 68]

immobilized in polysaccharide hydrogels. This allows the cells to be cultured more effectively. The immobilized bacterium gels are then placed inside of a normal fermenter vessel, where they are subjected to routine medium additions and periodic cell harvesting in order to guarantee that the required concentration of bacteria is always present.

Maximum productivity and long-term survival may be achieved by careful regulation of metabolic activity, gene expression, and growth rate during immobilized cell fermentation. There has not been much research done on *B. longum*, but the ones that have been done have found that strains created through immobilized cell fermentation had higher productivity, improved transcriptional rates, and superior long-term stability [74, 75]. The phase of the production process in which bacterial cells are collected is primarily responsible for the observed increase in metabolic activity and stability of the probiotic strains that are produced. When bacterial cells are collected, it is either during the exponential phase of growth or the early stationary phase. This helps to ensure that the cells will maintain a higher level of viability throughout storage and subsequent operations. Making probiotic bacterial strains that can be used in commercial food products costs a lot of money, takes a long time, and presents a big problem when it comes to making sure they can be used for a long time. As a result, it is challenging to obtain the bare minimum required number of viable cells after the fermentation process [26].

Immobilized cell fermentation had been associated with conventional batch fermentation in the method that Desmond and colleagues developed [76]. The goal of this method was to increase survival rates and tolerance toward a variety of stress-producing factors that were the result of nutritional, environmental, and physicochemical limitations [76]. However, further research is required to determine whether the implementation of these methodologies could have an effect on the capability of downstream processing to include carrier agents and compounds. It is possible that in the decades to come, producers of probiotics may utilize a combination of these tactics in order to establish a system that is both easy and economical for the manufacturing of probiotics.

21.2.3 Co-culture Fermentation

Combinations of several probiotic strains can be found in a number of commercially marketed probiotic supplements. At the moment, practically all commercial multi-strain probiotic solutions are made by fermenting and drying the bacteria in separate batches, and then combining the bacteria in different proportions or mixes in order to achieve the desired result [28]. In the past few years, there has been a significant increase in interest in the application of multi-strain fermentation techniques for the production of probiotic bacteria on a large scale. A successful co-culture strategy (Figure 21.1) may produce viable microorganisms in appropriate quantities, and it can also act as a cost-effective alternative to conventional methods. Probiotics are expensive to produce, but a number of studies have suggested that growing several types of bacteria in the same bioreactor might bring down manufacturing costs and make treatment of severe stomach problems accessible to people in developing nations [77]. Nevertheless, in order to establish such an approach for co-culture fermentation, one must first have an in-depth knowledge of the dynamics, synergism, and growth rates of the probiotic strains during the initial exponential phase. This is a precondition for constructing such a strategy. Some examples of the types of interactions that might take

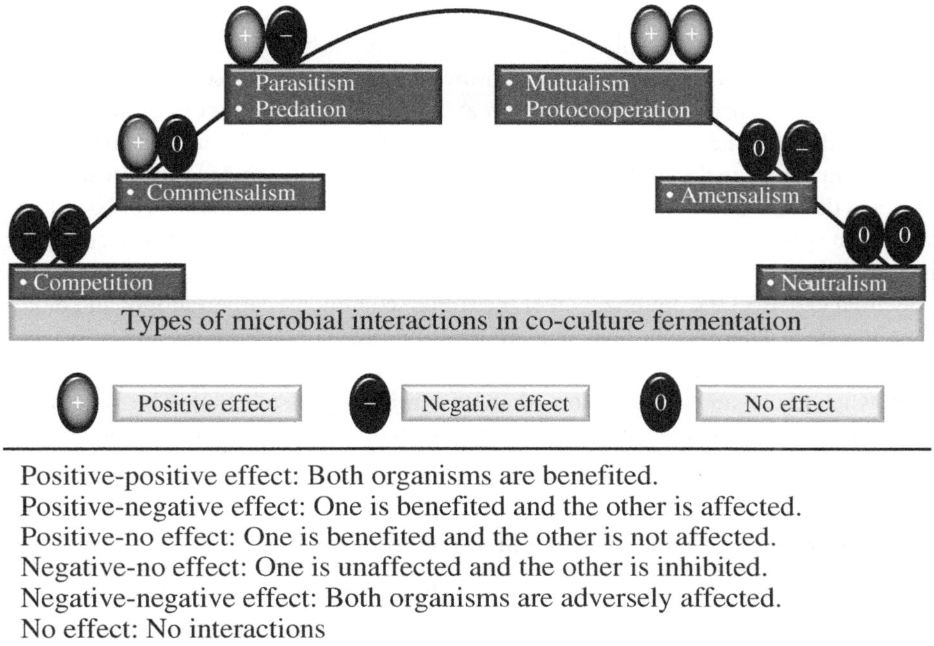

Figure 21.4 A graphical presentation of the various kinds of interactions that are related to co-culture fermentation. *Source:* Hathi et al. [33]; Sarsan et al. [78].

place are indicated in Figure 21.4. Therefore, having a grasp of the interactions that might occur between two or more bacterial strains is a vital component in the process of establishing protocols for co-culture fermentation. These complexities have led to a reliance on traditional or batch procedures employing pure cultures produced from freeze-dried powder or previously held stocks rather than co-culture fermentation, which has its benefits. However, co-culture fermentation does have its advantages. After the cultures have been prepared individually, they are dried and then combined in a variety of different quantities. Nevertheless, there are a few rare instances in which two or more than two cultures can be produced in a bioreactor at the same time inside an environment that is under regulated conditions [25]. Batch fermentation has a number of benefits, but typical ways of producing probiotics have a poor productivity rate. This is caused by a number of factors, including competition for nutrients and biological space, inhibitory metabolic end products, and a lack of synergism [79]. Cell immobilization technologies, on the other hand, have the potential to provide a viable solution to the difficulties associated with the production of multi-strain cultures. In point of fact, new research on cell immobilization techniques has shown that it is possible to employ them in conjunction with co-culture fermentation in order to enhance the number of probiotic cultures produced [25, 42]. In addition, as opposed to the batch fermentation method, the conjunction of co-culture fermentation and immobilization approaches has resulted in enhanced cell density, productivity, and greater resistance to contamination [77, 80]. This is all because of the approaches' synergistic effect. According to the findings of these investigations, it may be feasible to maintain stable circumstances for an extended length of time while cultivating a diverse microbiome. The key challenges faced by industries seeking to commercially produce probiotics are the high pace of growth and acidification experienced by traditional starter cultures. In addition, it is essential to choose the most appropriate choice of growing medium and fermentation method in order to maximize cell production and maintain its integrity during storage. Dinkçi et al. [81] and Fenster et al. [56] contributed to the studies in which these findings were included. Notwithstanding the benefits that these approaches offer, one of the most significant limitations of co-culture fermentation systems is that the microorganisms that are co-cultured continue to engage in competitive interactions with one another. In order to have a successful co-culture process, it is essential to perfect the optimization of the culture conditions, carbon supply, and bioreactor architecture. There can be several microorganisms that compete for resources and generate secondary metabolites that inhibit the growth of other microorganisms. This can happen when there are two or more than two microorganisms [82]. The process of co-culturing necessitates substantial prior screening in order to identify which microorganism species may be co-cultured in a way that is beneficial.

21.2.4 Recent Methods for Producing Multiple Probiotic Strains

In majority instances, the synthesis of probiotic cells is performed by using the strain's own growing medium, as this is the most common method. The most common medium for the culture of LAB or bifidobacteria is called MRS broth, which stands for de Man Rogosa Sharpe broth. In addition, redox-reducing chemicals such as cysteine were added to the bifidobacteria growth medium in order to stimulate their proliferation [41]. Researchers also employed a medium composed of whey in a few of the cases in order to boost cell counts at the end of the fermentation process [83]. For instance, a method that was developed not too long ago demonstrated that starchy acid hydrolysates, particularly those that were obtained from byproducts of agro-industrial processes, are effective substrates for the production of yeast biomass (*Saccharomyces cerevisiae*) and other derivative products at a low cost [83]. In addition to this, it may be able to contribute to the environmentally responsible development of biorefinery technologies.

The human digestive tract is a highly effective bioreactor that receives food as its input, digests that food, absorbs nutrients from that food, and then expels surplus biomass. The intestinal tract is capable of maintaining pH and oxygen gradients in the stomach and the rectum, which provides habitats that are conducive to the growth of both aerobic and anaerobic microorganisms in appropriate sites [84]. It is feasible to utilize bioreactors to cultivate several aerobic and anaerobic gut microorganism strains at the same time by simulating the human digestive tract by varying the pH and oxygen levels in the bioreactors in the same way that the human gut does. For instance, one approach of immobilization based on hydrogel may be utilized to bring about the desired spatial gradients. There is evidence that cellulose-based hydrogels are appropriate alternatives for the immobilization of probiotics. These hydrogels are both biocompatible and biodegradable, and they have already been utilized in a variety of bioengineering procedures [33, 41, 77]. In addition, it has been demonstrated that cellulose hydrogels exhibit a considerable degree of bio-elasticity. This property has the potential to be studied as a technique to produce pH and oxygen gradients and to replicate the ecology of the intestinal tract. However, the development of such a bioreactor is difficult from a technological standpoint [77]. Despite recent advancements in hydrogel-based nano- and micro-structured materials, comprehensive multidisciplinary research is necessary to confirm the effectiveness of bioreactors that mimic the human digestive tract. Producing probiotic biotherapeutics and nutraceuticals might be accomplished with the use of a bioreactor that mimics the digestive tract and allows for the simultaneous production of a variety of aerobic and anaerobic bacteria [77]. Furthermore, in order to bring these innovative materials and engineering to market, it will first be necessary to conduct exhaustive clinical studies and comprehensive lifecycle assessment evaluations of the process.

21.3 Strategies Employed for Harvesting and Drying Probiotic Cells

The subsequent critical stage in the synthesis of probiotics is cell harvesting, and this step is essential regardless of the bioreactors technology that was employed. Centrifugation and filtering are two typical procedures that are used for harvesting produced cells from the liquid medium. These methods are comparable to cross-flow filtration or membrane technology. The effectiveness of a multi-strain probiotic product that is commercially useful is greatly reliant on the kind of strains that are utilized in combination and the amount of viable bacterial cells that are produced once the product is made. As a result, it is of utmost importance to determine whether the strains being utilized have a high level of production, a high concentration of spores, and are simple to recover during the harvesting procedure [85]. In addition, the processes that are utilized in order to harvest the spores for the multi-strain probiotics need to be productive enough to ensure that greater than 80% of the viable spores are obtained [85]. A spore count of more than 1×10^{10} spores per dose is considered optimal for all. This is essential in order to make up for the reduction in the number of viable cells that occurs during the subsequent processes and product development [86]. At this time, the most common method of harvesting is centrifugation performed at a cool temperature (about 4 °C). It is generally agreed upon that this approach yields the best results when it comes to concentrating probiotic cells. An investigation was carried out with the purpose of evaluating the effect that different centrifugation conditions, such as speed, pH, temperature, and time, had on starting cultures [87]. In this study, the impacts on the acidification activity of probiotic cells were analyzed. According to the findings, the particular acidification activity that took place during the phase of cell concentration was not significantly affected by the conditions in which the cells were harvested. This was demonstrated by the absence of a substantial impact. However, the cells' capacity to tolerate cold storage was altered both by the length of time that was spent centrifuging the cells and by the speed at which the process was carried out. In addition, when centrifugation was performed at temperatures ranging from 4 to 15 °C, there was not a discernible difference in the results. When the proper centrifugation circumstances were paired with the acidification of the cells in the fermented broth, a relative increase in the cells' ability to withstand freezing temperatures was seen [87]. According to the findings of another study, which focused on *Propionibacterium freudenreichii* and *L. casei*, microbial cells in high solid (20–30%) containing medium have a better growth rate, greater viability, and greater ability to retain nutrients [88]. The increased tolerance that is produced as a result of the high osmolality that is present in a medium that contains more than 20% of total solid content is responsible for enhanced cell survival [88]. In addition, the cells should be harvested during the stationary phase, and the pH should be neutralized since these are the ideal techniques for increasing the vitality of the probiotic cells [88, 89].

Physically separating bacterial cells from the growing medium using membrane filtration increases expenses and resource usage [56, 90]. The retention and processing of the growing media for spray drying of cells is recommended for most large-scale manufacturing plants. When cells are harvested immediately and then spray-dried, valuable metabolites including short-chain fatty acids, vitamins, and bacteriocin are preserved. The recovery process has been improved by the use of state-of-the-art techniques such as vacuum filtration, microfiltration, and ultracentrifugation to isolate cells from fermented liquor [91]. After the cells have been harvested in the best way possible, they need to be mixed with protective agents like cryo (which prevents ice crystal formation) and lyo (which stabilizes the lipid bilayer of the cell membrane) to prevent too much damage from occurring during subsequent steps in the process, such as drying (Figure 21.2). This cryogenically preserved concentrate is freeze-dried utilizing a number of different methods. The process that may be summarized as "pouring of cryoprotected concentrate into cans, closing it, then immersing it in liquid N_2 bath, and finally storing it" is the easiest method. A second method involves the dripping of cryoprotected concentrate through calibrated pores in a bath of liquid N_2 in order to make pellets. The pellets that are retrieved from the bottom of the bath are either treated to freeze-drying or packaged and kept at temperatures between −45 and −55 °C. However, the most popular method involves filling trays with cryoprotected concentrate and placing them on top of temperature-controlled shelves. The shelves are then chilled to freezing under air pressure and heat in the presence of a vacuum. The vacuum and temperature ranges are 100–1000 mTorr and −40 to +40 °C, respectively [56]. Despite these developments in harvesting methods, centrifugation looks to be a suitable alternative, and with further refinement in technology, this might serve as a more trustworthy alternative when compared to the other ways for the downstream processing of probiotics. To be both ecologically and economically sustainable, as well as straightforward and doable, a harvesting method must first and foremost aim to minimize the number of steps required to collect the desired output. This crucial component is at the very core of manufacturing facilities aiming to formulate multi-strain probiotic products.

In addition to the many systems that have been addressed, it demonstrates that exposing probiotic cells to proper environmental factors such as temperature, pH, osmotic pressure, and so on has the potential to lead to an increase in the viability of probiotics. In addition, probiotic bacterial strains respond to environmental conditions by producing certain chemicals such as exopolysaccharides and proteins. These molecules assist the probiotic strains in maintaining their viability during the subsequent processing steps as well.

21.4 Final Remarks and Possible Directions for the Future

Over the course of the past several years, there has been a significant rise in the use of probiotic foods that have functional purposes. However, due to the poor survival rate of many probiotic strains, in vitro production of these microorganisms continues to be a difficult task. Nevertheless, novel fermentation methodologies, such as immobilized cell technologies, batch methods, membrane bioreactors, and continuous fermentation, are extremely suited and encourage the growth of probiotic strains due to the simple management of pH gradients and dispersed O_2 levels. Compared to batch procedures, fed-batch and continuous fermentation systems might be more cost-effective due to their ability to maximize lactic acid productivity. Furthermore, the utilization of non-hazardous nano-cellulose hydrogels that are biodegradable and recyclable in membrane fermenters has the potential to make probiotic research substantially more successful.

Complex degrees of planning and engineering is required in order to design a method that may successfully grow probiotics in an environment that is a close analog to their natural habitat. The utilization of agricultural and/or food waste resources for the purpose of the growth of probiotics appears to be a viable strategy with the potential to minimize the total cost of the production process by utilizing any form of the bioreactor system. Even while large-scale culture using continuous production methodologies has been described as effective, continuous fermentation always presents a danger of contamination and, more significantly, modifying the features of probiotics in the process owing to genetic drift. The process has the potential to change the characteristics of the probiotics themselves as a result of genetic drift. This is because continuous fermentation involves the use of the same culture for an extended period of time. The growing popularity of functional foods has redirected the attention of researchers in this field toward determining the probiotic strains that are most effective in certain applications. The development of multi-strain probiotic products will make it possible to develop a strategy for treating GI problems and for alleviating the symptoms of malnutrition, at least to some extent. Unfortunately, traditional techniques are still relied on for the generation of probiotic cultures in a significant majority. It should come as no surprise that many of these approaches require a significant investment of both time and money. Cellulose hydrogels appear to be an appropriate choice for the production of multi-strain probiotics and hold the promise of being a cost-effective alternative that requires very minimal processing in the downstream stages. These characteristics are essential for establishing a probiotic product in the marketplace that is viable from a marketing standpoint moving forward.

Abbreviations

DVI	direct vat inoculation
GI	gastrointestinal
IBD	inflammatory bowel disease
IBS	irritable bowel syndrome
LAB	lactic acid bacteria

References

1 Lacroix, C. and Yildirim, S. (2007). Fermentation technologies for the production of probiotics with high viability and functionality. *Current Opinion in Biotechnology* 18 (2): 176–183.

2 Niamah, A.K., Al-Sahlany, S.T.G., Ibrahim, S.A. et al. (2021). Electro-hydrodynamic processing for encapsulation of probiotics: a review on recent trends, technological development, challenges and future prospect. *Food Bioscience* 44: 101458. https://doi.org/10.1016/j.fbio.2021.101458.

3 Verma, D.K., Thakur, M., Singh, S. et al. (2022). Bacteriocins as antimicrobial and preservative agents in food: biosynthesis, separation and application. *Food Bioscience* 48: 101594. https://doi.org/10.1016/j.fbio.2022.101594.

4 Chourasia, R., Kumari, R., Singh, S.P. et al. (2022). Characterization of native lactic acid bacteria from traditionally fermented chhurpi of Sikkim Himalayan region for the production of chhurpi cheese with enhanced antioxidant effect. *LWT – Food Science and Technology* 154: 112801.

5 Kumar, J., Sharma, N., Kaushal, G. et al. (2019). Metagenomic insights into the taxonomic and functional features of Kinema, a traditional fermented soybean product of Sikkim Himalaya. *Frontiers in Microbiology* 10: 1744.

6 Hasler, C.M. (2002). Functional foods: benefits, concerns and challenges – a position paper from the American Council on Science and Health. *The Journal of Nutrition* 132 (12): 3772–3781.

7 Verma, D.K., Patel, A.R., and Srivastav, P.P. (2018). *Bioprocess Technology in Food and Health: Potential Applications and Emerging Scope*. As part of book series on *Innovation in Agricultural Microbiology*. Palm Bay, FL: Apple Academic Press https://doi.org/10.1201/9781351167888.

8 Verma, D.K., Niamah, K.A., Patel, A.R. et al. (2020). Chemistry and microbial sources of curdlan with potential application and safety regulations as prebiotic in food and health. *Food Research International* https://doi.org/10.1016/j.foodres.2020.109136.

9 Verma, D.K., Patel, A., Prajapati, J.B. et al. (2020). Starter culture and probiotic bacteria in dairy food products. In: *Microbiology for Food and Health Technological Developments and Advances* (ed. D.K. Verma, A.R. Patel, P.P. Srivastav, et al.), 3–49. Palm Bay, FL: Apple Academic Press.

10 Kho, Z.Y. and Lal, S.K. (2018). The human gut microbiome–a potential controller of wellness and disease. *Frontiers in Microbiology* 9: 1835.

11 Sharma, M., Sangwan, R.S., Khatkar, B.S., and Singh, S.P. (2019). Alginate–pectin co-encapsulation of dextransucrase and dextranase for oligosaccharide production from sucrose feedstocks. *Bioprocess and Biosystems Engineering* 42 (10): 1681–1693.

12 Zmora, N., Suez, J., and Elinav, E. (2019). You are what you eat: diet, health and the gut microbiota. *Nature Reviews Gastroenterology & Hepatology* 16 (1): 35–56.

13 Verma, D.K., Patel, A.R., Srivastav, P.P. et al. (2020). *Microbiology for Food and Health Technological Developments and Advances*. Palm Bay, FL: Apple Academic Press 10.1201/9780429276170.

14 Verma, D.K., Patel, A.R., Billoria, S. et al. (2021). *Microbial Biotechnology in Food Processing and Health Advances, Challenges, and Potential*. Palm Bay, FL: Apple Academic Press.

15 Sharma, M., Patel, S.N., Lata, K. et al. (2016). A novel approach of integrated bioprocessing of cane molasses for production of prebiotic and functional bioproducts. *Bioresource Technology* 219: 311–318. https://doi.org/10.1016/j.biortech.2016.07.131.

16 Jadaun, J.S., Narnoliya, L.K., Agarwal, N. et al. (2019). Catalytic biosynthesis of levan and short-chain fructooligosaccharides from sucrose-containing feedstocks by employing the levansucrase from *Leuconostoc mesenteroides* MTCC10508. *International Journal of Biological Macromolecules* 127: 486–495.

17 Maladkar, M., Yadav, A., and Save, N. (2020). LGGTM-A promising therapy in gastro-intestinal infections. *Indian Journal of Microbiology Research (IJMR)* 7: 302–312.

18 Rahimlou, M., Hosseini, S.A., Majdinasab, N. et al. (2022). Effects of long-term administration of multi-strain probiotic on circulating levels of BDNF, NGF, IL-6 and mental health in patients with multiple sclerosis: a randomized, double-blind, placebo-controlled trial. *Nutritional Neuroscience* 25 (2): 411–422.

19 Zawada, A., Rychter, A.M., Ratajczak, A.E. et al. (2020). Does gut-microbiome interaction protect against obesity and obesity-associated metabolic disorders? *Microorganisms* 9 (1): 18.

20 Aldawsari, F.S., Bin Helel, B.S., Alharbi, Y.T., and Abudahash, M.A. (2020). Probiotics and their quality-related concerns: highlights from the Saudi Arabian market. *Therapeutic Innovation & Regulatory Science* 54 (2): 365–369.

21 Ishaque, S.M., Khosruzzaman, S.M., Ahmed, D.S., and Sah, M.P. (2018). A randomized placebo-controlled clinical trial of a multi-strain probiotic formulation (Bio-Kult®) in the management of diarrhea-predominant irritable bowel syndrome. *BMC Gastroenterology* 18 (1): 1–12.

22 Davoodvandi, A., Marzban, H., Goleij, P. et al. (2021). Effects of therapeutic probiotics on modulation of microRNAs. *Cell Communication and Signaling* 19 (1): 1–22.

23 Sudheendra, C.V. and Patel, A. (2022). Carrier materials for encapsulation. In: *Encapsulation in Food Processing and Fermentation* (ed. S. Lević, V. Nedović, and B. Bugarski), 21. CRC Press.

24 Verma, D.K., Mahato, D.K., Billoria, S. et al. (2017). Microbial approaches in fermentations for production and preservation of different food. In: *Microorganisms in Sustainable Agriculture, Food and the Environment* (ed. D.K. Verma and P.P. Srivastav). Volume-1, as part of book series on *Innovations in Agricultural Microbiology*, 105–142. Palm Bay, FL: Apple Academic Press.

25 Smid, E.J. and Lacroix, C. (2013). Microbe–microbe interactions in mixed culture food fermentations. *Current Opinion in Biotechnology* 24 (2): 148–154.

26 Anandharaj, M., Rani, R.P., Swain, M.R. et al. (2017). Production of high-quality probiotics by fermentation. In: *Microbial Functional Foods and Nutraceuticals* (ed. V.K. Gupta, H. Treichel, V. (Olga) Shapaval, et al.), 235–266. Wiley.

27 Fan, R., Ebrahimi, M., and Czermak, P. (2017). Anaerobic membrane bioreactor for continuous lactic acid fermentation. *Membranes* 7 (2): 26.

28 Jangra, M., Belur, P.D., Oriabinska, L.B., and Dugan, O.M. (2016). Multistrain probiotic production by co-culture fermentation in a lab-scale bioreactor. *Engineering in Life Sciences* 16 (3): 247–253.

29 Verma, D.K., Thakur, M., Tripathy, S. et al. (2022). Emerging biosensor technology and its potential application in food. In: *Innovations in Fermentation and Phytopharmaceutical Technologies* (ed. H. Thatoi, S. Mohapatra, and S.K. Das), 127–164. Elsevier Inc.

30 Faschian, R., De, S., and Pörtner, R. (2016). Multi-fixed-bed bioreactor system applied for bioprocess development of immobilized lactic acid bacteria. *The Open Biotechnology Journal* 10 (1): https://doi.org/10.2174/1874070701610010001.

31 Ritonja, J. (2019). Design of identification based adaptive control for fermentation process in bioreactor. *International Journal of Electrical and Computer Engineering* 13 (2): 65–68.

32 Ritonja, J., Goršek, A., and Pečar, D. (2021). Use of a heating system to control the probiotic beverage production in batch bioreactor. *Applied Sciences* 11 (1): 84.

33 Hathi, Z., Mettu, S., Priya, A. et al. (2021). Methodological advances and challenges in probiotic bacteria production: ongoing strategies and future perspectives. *Biochemical Engineering Journal* 176: 108199. https://doi.org/10.1016/j.bej.2021.108199.

34 Thakkar, P.N., Patel, A., Modi, H.A., and Prajapati, J.B. (2020). Hypocholesterolemic effect of potential probiotic *Lactobacillus fermentum* strains isolated from traditional fermented foods in Wistar rats. *Probiotics and Antimicrobial Proteins* 12 (3): 1002–1011.

35 Upadhyay, S.K. and Singh, S.P. (eds.) (2021). *Bioprospecting of Plant Biodiversity for Industrial Molecules*. John Wiley & Sons Ltd. doi:10.1002/9781119718017.

36 Dixit, S., Shukla, A., Singh, V., Upadhyay, S.K. (2021). Bioprospecting of natural compounds for industrial and medical applications; current scenario and bottleneck. In: *Bioprospecting of Plant Biodiversity for Industrial Molecules* (ed. S.K. Upadhyay and S.P. Singh), 53–57. John Wiley & Sons Ltd. https://doi.org/10.1002/9781119718017.ch3.

37 Upadhyay, S.K. and Singh, S.P. (eds.) (2023). *Plants as Bioreactors for Industrial Molecules*. John Wiley & Sons Ltd. doi:10.1002/9781119875116.

38 Singh, S.P. and Upadhyay, S.K. (eds.) (2021). *Bioprospecting of Microorganism-Based Industrial Molecules*. John Wiley & Sons Ltd. doi:10.1002/9781119717317.

39 Sharma, S., Shukla, M.K., Sharma, K.C. et al. (2023). Revisiting the therapeutic potential of gingerols against different pharmacological activities. *Naunyn-Schmiedeberg's Arch Pharmacol* 396: 633–647. https://doi.org/10.1007/s00210-022-02372-7.

40 Patel, A., Shah, N., and Verma, D.K. (2017). Lactic acid bacteria (LAB) bacteriocins: an ecological and sustainable biopreservative approach to improve the safety and shelf-life of foods. In: *Microorganisms in Sustainable Agriculture, Food and the Environment* (ed. D.K. Verma and P.P. Srivastav). Volume-1, as part of book series on *Innovations in Agricultural Microbiology*, 197–258. Palm Bay, FL: Apple Academic Press.

41 Doleyres, Y. and Lacroix, C.J.I.D.J. (2005). Technologies with free and immobilised cells for probiotic bifidobacteria production and protection. *International Dairy Journal* 15 (10): 973–988.

42 Frakolaki, G., Giannou, V., Kekos, D., and Tzia, C. (2021). A review of the microencapsulation techniques for the incorporation of probiotic bacteria in functional foods. *Critical Reviews in Food Science and Nutrition* 61 (9): 1515–1536.

43 Mishra, S.S., Behera, P.K., Kar, B., and Ray, R.C. (2018). Advances in probiotics, prebiotics and nutraceuticals. In: *Innovations in Technologies for Fermented Food and Beverage Industries* (ed. S.K. Panda and P.H. Shetty), 121–141. Cham: Springer.

44 Zhai, R., Xue, X., Zhang, L. et al. (2019). Strain-specific anti-inflammatory properties of two *Akkermansia muciniphila* strains on chronic colitis in mice. *Frontiers in Cellular and Infection Microbiology* 9: 239.

45 Goldenberg, J.Z., Yap, C., Lytvyn, L. et al. (2017). Probiotics for the prevention of Clostridium difficile-associated diarrhea in adults and children. *Cochrane Database of Systematic Reviews* 12: CD006095.

46 Elshaghabee, F.M., Rokana, N., Gulhane, R.D. et al. (2017). *Bacillus* as potential probiotics: status, concerns, and future perspectives. *Frontiers in Microbiology* 8: 1490.

47 Hauser, G., Salkic, N., Vukelic, K. et al. (2015). Probiotics for standard triple *Helicobacter pylori* eradication: a randomized, double-blind, placebo-controlled trial. *Medicine* 94 (17): e685.

48 Yeun, Y. and Lee, J. (2015). Effect of a double-coated probiotic formulation on functional constipation in the elderly: a randomized, double blind, controlled study. *Archives of Pharmacal Research* 38 (7): 1345–1350.

49 Lorenzo-Zúñiga, V., Llop, E., Suárez, C. et al. (2014). I. 31, a new combination of probiotics, improves irritable bowel syndrome-related quality of life. *World Journal of Gastroenterology: WJG* 20 (26): 8709.

50 Guglielmetti, S., Mora, D., Gschwender, M., and Popp, K. (2011). Randomised clinical trial: *Bifidobacterium bifidum* MIMBb75 significantly alleviates irritable bowel syndrome and improves quality of life – a double-blind, placebo-controlled study. *Alimentary Pharmacology & Therapeutics* 33 (10): 1123–1132.

51 Grossi, E., Buresta, R., Abbiati, R. et al. (2010). Clinical trial on the efficacy of a new symbiotic formulation, Flortec, in patients with acute diarrhea: a multicenter, randomized study in primary care. *Journal of Clinical Gastroenterology* 44: S35–S41.

52 Kruis, W., Frič, P., Pokrotnieks, J. et al. (2004). Maintaining remission of ulcerative colitis with the probiotic *Escherichia coli* Nissle 1917 is as effective as with standard mesalazine. *Gut* 53 (11): 1617–1623.

53 Almeida, A., Mitchell, A.L., Boland, M. et al. (2019). A new genomic blueprint of the human gut microbiota. *Nature* 568 (7753): 499–504.

54 Kumar, V., Naik, B., Kumar, A. et al. (2022). Probiotics media: significance, challenges, and future perspective – a mini review. *Food Production, Processing and Nutrition* 4 (1): 1–13. https://doi.org/10.1186/s43014-022-00098-w.

55 Alvarez-Calatayud, G. and Margolles, A. (2016). Dual-coated lactic acid bacteria: an emerging innovative technology in the field of probiotics. *Future Microbiology* 11 (3): 467–475. https://doi.org/10.2217/fmb.15.150.

56 Fenster, K., Freeburg, B., Hollard, C. et al. (2019). The production and delivery of probiotics: a review of a practical approach. *Microorganisms* 7 (3): 83.

57 Malvido, M.C., González, E.A., Bazán Tantaleán, D.L. et al. (2019). Batch and fed-batch production of probiotic biomass and nisin in nutrient-supplemented whey media. *Brazilian Journal of Microbiology* 50 (4): 915–925.

58 Valdés-Varela, L., Ruas-Madiedo, P., and Gueimonde, M. (2017). in vitro fermentation of different fructo-oligosaccharides by *Bifidobacterium* strains for the selection of synbiotic combinations. *International Journal of Food Microbiology* 242: 19–23.

59 Huang, F., Sardari, R.R., Jasilionis, A. et al. (2021). Cultivation of the gut bacterium *Prevotella copri* DSM 18205T using glucose and xylose as carbon sources. *MicrobiologyOpen* 10 (3): e1213.

60 Rashidinejad, A., Bahrami, A., Rehman, A. et al. (2022). Co-encapsulation of probiotics with prebiotics and their application in functional/synbiotic dairy products. *Critical Reviews in Food Science and Nutrition* 62 (9): 2470–2494.

61 Estilarte, M.L., Tymczyszyn, E.E., de los Ángeles Serradell, M., and Carasi, P. (2021). Freeze-drying of *Enterococcus durans*: effect on their probiotics and biopreservative properties. *LWT-Food Science and Technology* 137: 110496.

62 Dimitrellou, D., Kandylis, P., Petrović, T. et al. (2016). Survival of spray dried microencapsulated *Lactobacillus casei* ATCC 393 in simulated gastrointestinal conditions and fermented milk. *LWT-Food Science and Technology* 71: 169–174.

63 Jokicevic, K., Kiekens, S., Byl, E. et al. (2021). Probiotic nasal spray development by spray drying. *European Journal of Pharmaceutics and Biopharmaceutics* 159: 211–220.

64 Liu, H., Cui, S.W., Chen, M. et al. (2019). Protective approaches and mechanisms of microencapsulation to the survival of probiotic bacteria during processing, storage and gastrointestinal digestion: a review. *Critical Reviews in Food Science and Nutrition* 59 (17): 2863–2878.

65 Kanmani, P., Kumar, R.S., Yuvaraj, N. et al. (2011). Effect of cryopreservation and microencapsulation of lactic acid bacterium *Enterococcus faecium* MC13 for long-term storage. *Biochemical Engineering Journal* 58: 140–147.

66 Corre, C., Madec, M.N., and Boyaval, P. (1992). Production of concentrated *Bifidobacterium bifidum*. *Journal of Chemical Technology & Biotechnology* 53 (2): 189–194.

67 Malvido, M.C., Alonso González, E., Outeiriño, D., and Pérez Guerra, N. (2018). Production of a highly concentrated probiotic culture of *Lactococcus lactis* CECT 539 containing high amounts of nisin. *3 Biotech* 8 (7): 1–15.

68 Mozzetti, V., Grattepanche, F., Moine, D. et al. (2010). New method for selection of hydrogen peroxide adapted bifidobacteria cells using continuous culture and immobilized cell technology. *Microbial Cell Factories* 9 (1): 1–9.

69 Cha, K.H., Lee, E.H., Yoon, H.S. et al. (2018). Effects of fermented milk treatment on microbial population and metabolomic outcomes in a three-stage semi-continuous culture system. *Food Chemistry* 263: 216–224.

70 Jung, I.S., Oh, M.K., Cho, Y.C., and Kong, I.S. (2011). The viability to a wall shear stress and propagation of *Bifidobacterium longum* in the intensive membrane bioreactor. *Applied Microbiology and Biotechnology* 92 (5): 939–949.

71 Fan, R., Burghardt, J.P., Prell, F. et al. (2020). Production and purification of fructo-oligosaccharides using an enzyme membrane bioreactor and subsequent fermentation with probiotic *Bacillus coagulans*. *Separation and Purification Technology* 251: 117291.

72 Jadaun, J.S., Narnoliya, L.K., Agarwal, N., and Singh, S.P. (2019). Catalytic biosynthesis of levan and short-chain fructooligosaccharides from sucrose-containing feedstocks by employing the levansucrase from *Leuconostoc mesenteroides* MTCC10508. *International Journal of Biological Macromolecules* 127: 486–495.

73 Schoina, V., Terpou, A., Papadaki, A. et al. (2019). Enhanced aromatic profile and functionality of cheese whey beverages by incorporation of probiotic cells immobilized on *Pistacia terebinthus* resin. *Food* 9 (1): 13.

74 Indrasti, D. and Hidaya, H.C. (2018). Improvement of viability of *Lactobacillus casei* and *Bifidobacterium longum* with several encapsulating materials using extrusion method. *Jurnal Ilmu Ternak dan Veteriner* 23 (4): 189–201.

75 Ji, R., Wu, J., Zhang, J. et al. (2019). Extending viability of *Bifidobacterium longum* in chitosan-coated alginate microcapsules using emulsification and internal gelation encapsulation technology. *Frontiers in Microbiology* 10: 1389.

76 Desmond, C., Corcoran, B.M., Coakley, M., and Fitzgerald, G.F. (2005). Development of dairy-based functional foods containing probiotics and prebiotics. *Australian Journal of Dairy Technology* 60 (2): 121.

77 Mettu, S., Hathi, Z., Athukoralalage, S. et al. (2021). Perspective on constructing cellulose-hydrogel-based gut-like bioreactors for growth and delivery of multiple-strain probiotic bacteria. *Journal of Agricultural and Food Chemistry* 69 (17): 4946–4959.

78 Sarsan, S., Pandiyan, A., Rodhe, A.V., and Jagavati, S. (2021). Synergistic interactions among microbial communities. In: *Microbes in Microbial Communities* (ed. R.P. Singh, G. Manchanda, K. Bhattacharjee, and H. Panosyan), 1–37. Singapore: Springer.

79 Othman, M., Ariff, A.B., Wasoh, H. et al. (2017). Strategies for improving production performance of probiotic *Pediococcus acidilactici* viable cell by overcoming lactic acid inhibition. *AMB Express* 7 (1): 1–14.

80 Fehlbaum, S., Chassard, C., Haug, M.C. et al. (2015). Design and investigation of PolyFermS in vitro continuous fermentation models inoculated with immobilized fecal microbiota mimicking the elderly colon. *PLoS One* 10 (11): e0142793.

81 Dinkçi, N., Akdeniz, V., and Akalin, A.S. (2019). Survival of probiotics in functional foods during shelf life. In: *Food Quality and Shelf Life* (ed. C. Galanakis), 201–233. Academic Press.

82 do Amaral Santos, C.C.A., da Silva Libeck, B., and Schwan, R.F. (2014). Co-culture fermentation of peanut-soy milk for the development of a novel functional beverage. *International Journal of Food Microbiology* 186: 32–41.

83 Martiniano, S.E., Fernandes, L.A., Alba, E.M. et al. (2020). A new approach for the production of selenium-enriched and probiotic yeast biomass from agro-industrial by-products in a stirred-tank bioreactor. *Metabolites* 10 (12): 508.

84 Kalantar-Zadeh, K., Berean, K.J., Burgell, R.E. et al. (2019). Intestinal gases: influence on gut disorders and the role of dietary manipulations. *Nature Reviews Gastroenterology & Hepatology* 16 (12): 733–747.

85 Ramlucken, U., Ramchuran, S.O., Moonsamy, G. et al. (2021). Production and stability of a multi-strain *Bacillus* based probiotic product for commercial use in poultry. *Biotechnology Reports* 29: e00575.

86 Ramlucken, U., Lalloo, R., Roets, Y. et al. (2020). Advantages of *Bacillus*-based probiotics in poultry production. *Livestock Science* 241: 104215.

87 Streit, F., Corrieu, G., and Béal, C. (2010). Effect of centrifugation conditions on the cryotolerance of *Lactobacillus bulgaricus* CFL1. *Food and Bioprocess Technology* 3 (1): 36–42.

88 Huang, S., Cauty, C., Dolivet, A. et al. (2016). Double use of highly concentrated sweet whey to improve the biomass production and viability of spray-dried probiotic bacteria. *Journal of Functional Foods* 23: 453–463.

89 Terpou, A., Papadaki, A., Lappa, I.K. et al. (2019). Probiotics in food systems: significance and emerging strategies towards improved viability and delivery of enhanced beneficial value. *Nutrients* 11 (7): 1591.

90 Huang, S., Vignolles, M.L., Chen, X.D. et al. (2017). Spray drying of probiotics and other food-grade bacteria: a review. *Trends in Food Science & Technology* 63: 1–17.

91 Dushkova, M., Kodinova, S., Denkova, Z. et al. (2021). Physicochemical, microbiological, and sensory characteristics of probiotic Bulgarian yoghurts obtained by ultrafiltration of goat's milk. *Zeitschrift für Naturforschung. Section C* 76 (11–12): 481–489.

22

Microbial Bioproduction of Antiaging Molecules

Ankita Dua[1], Aeshna Nigam[1], Anjali Saxena[2], Gauri Garg Dhingra[3], and Roshan Kumar[4]

[1] *Department of Zoology, Shivaji College, University of Delhi, Raja Garden, New Delhi, India*
[2] *Department of Zoology, Bhaskaracharya College of Applied Sciences, University of Delhi, Dwarka, New Delhi, India*
[3] *Department of Zoology, Kirori Mal College, University of Delhi, New Delhi, India*
[4] *Post-Graduate Department of Zoology, Magadh University, Bodh Gaya, Bihar, India*

22.1 Introduction

Microbial community is known to produce specialized metabolites, secondary products, pigments, toxins, etc., which are not involved in primary metabolism but are beneficial to the organism. Some of these products are potential medical agents with therapeutic applications [1, 2]. The ever-increasing demand for pharmaceuticals from natural resources has been largely met by microbial bioproduction and has proven to be a boon to the growing population. In 1978, about 23,500 pigs and cattle were sacrificed to obtain one pound of insulin (https://www.gene.com/stories/cloning-insulin). Today, microbial bioproduction of synthetic "human" insulin has been achieved in *Escherichia coli* and *Saccharomyces cerevisiae* on a large scale [3] and approved by FDA for therapeutical applications. Many other natural compounds and metabolites produced by microbes and plants could be potential drug candidates; however, they are produced in very low quantities. This problem necessitates the development of microbial bioproduction of these compounds for consumption in drug industry. For example, ergothioneine, commonly known as longevity vitamin, is a vitamin-like antioxidant synthesized by mushrooms and a few microbes but at very low concentrations [4]. The same compound has been reported to be produced in large scale in *E. coli* with a high-cysteine production system [5].

Aging is a multifactorial process that includes both extrinsic and intrinsic factors. Three popular research aspects on aging have been outlined as: *compressed morbidity*, *decelerated aging*, and *arrested aging*. Compressed morbidity aims to prevent ailments of old age by intervening at the molecular level to increase average human lifespan. Decelerated aging seeks to stall the processes of aging, and arrested aging has a long-term goal of restoring vitality and body functions [6]. Arrested aging negates the damaging effects of senescence by controlling the aging process.

Microbial Bioreactors for Industrial Molecules, First Edition.
Edited by Sudhir P. Singh and Santosh Kumar Upadhyay.

In recent years, economic growth, lifestyles, and health awareness has led to increased usage of antiaging products. Calorie/dietary restriction, genetic manipulation, and antiaging molecule treatments are the most effective proven methods to reverse aging and increase the life expectancy [7, 8]. Diet is the principal source of calorie intake and contributes to many lifestyles associated diseases and aging process. Calorie Restriction Mimetics (CRM) are the molecules that can increase longevity by altering aging-related pathways [9], and consumption of these mimetics can even help in overcoming the challenges posed by a calorie restriction diet. Some of these CRM include hydroxy-citric acid, inhibitors of glycolysis (e.g. D-allulose and D-glucosamine), and polyamines (e.g. curcumin, resveratrol, quercetin, di-methoxychalcones, salicylic acid, trientine, and rapamycin) that are freely available in nature [10]. These can facilitate metabolism by activating autophagy and recycling of cellular components. CRM are known to act on four biological pathways namely insulin signaling pathway, mammalian target of rapamycin (mTOR) pathway, sirtuin-1 (SIRT1) pathway, and adenosine monophosphate-activated kinase (AMPK) pathway. Understanding of these pathways is required for extending human health span [11]. There is no alternative to calorie restriction to extend life expectancy in higher organisms except the discovery of antiaging molecules and their mass production. A possibility to achieve this would be to express the genes in microbial hosts such as bacteria or fungi through biotechnological means.

Since most of these compounds are chemical based and are bound to have unknown side effects, demand for ecofriendly, naturally safe, and harmless molecules in the market continues to grow [12]. New products being used are cosmeceuticals and nutricosmetics: former being used to define safe and active products that benefit skin maintenance and treat dermatological conditions produced by the cosmetic industry, and the latter are products that function as nutritional supplements for skin, hair, and nails [13].

Furthermore, using the latest advances in gut-microbial research, antiaging interventions are being developed by changing dietary habits, exercising, and taking certain specific drugs [14]. The gut microbiome tends to change with age, diet, physical activity, lifestyle, etc. and restoring dysbiosis in the gut entails harnessing those beneficial bacteria that have been depleted to achieve minimal damage to internal milieu [15]. Keeping in view the current research trends in microbes, this chapter focuses on the source of antiaging compounds from microbial world and their bioproduction and association with human health.

22.2 The Aging Process: An Overview

Aging can be classified into two categories: intrinsic aging (due to senescence in cells caused by oxidative stress and reactive oxygen species [ROS]) and extrinsic aging (due to environmental factors such as UV light, particulate matter, and pollution [16]). Lifestyle with diet, stress, consumption of alcohol/drugs, disorders, and medications exert a great influence on physical aging [17]. Byproducts of normal metabolism as well as environmental stress and UV radiation lead to the buildup of ROS and reactive nitrogen species (RNS). The buildup of these radicals in oxidative stress leads to pathological conditions and untimely cell death. The phenomenon of aging is initiated at the time of conception and goes on in the entire lifespan of an individual. Various molecular and cellular events add on to the damage in any cell and result in aging, also causing risk of onset of diseases and ultimately death (Figure 22.1). Multiple factors affect aging processes in the human body; for example, ROS that includes superoxide anions (O_2^-), hydroxyl radicals (OH), hydrogen peroxide (H_2O_2), and hypocholorous acid (HOCl) have been implicated with inflammation, aging-related disorders, cardiovascular disorders, cancer, and atherosclerosis [18]. Particulate matter, soot and nitrogen dioxide, that are resultants of traffic-related air pollution have been also shown to be associated with premature aging of skin [19]. The cardiovascular system, nervous system, and skin (the human body's biggest defense system) are affected the most by aging-related processes. The most prevalent factor affecting the skin is exposure to radiations, especially ultraviolet causing damage to the dermal and epidermal layers and release of ROS [20]. Thus, there is an increasing demand in the usage of natural products that can help in extended vitality and rejuvenation of skin and general health. Multiple such products have been explored from bacterial and fungal reservoirs for preparation of commercial antiaging products for human usage [21]. Nitric oxide ($NO^•$) is a molecule produced in the body by nitric oxide synthases (NOSs) (three types: neuronal NOS, endothelial NOS, and inducible NOS) that acts as a second messenger and functions in the smooth muscle relaxation in blood vessels [22]. NO is toxic to bacteria and viruses that may invade the body but is also detrimental to host cells and may inactivate vital enzymes resulting in cellular death. Repeated infections in the body produce changes that lead to aging via release of high quantities of $NO^•$. It has been reported that melatonin, vitamin C, and vitamin E are capable of combatting the ill effects of damage caused by NO [23].

All autotrophic marine organisms such as bacteria, archaea, fungi, and algae contain carotenoids, xanthophylls, which play roles of antioxidants that shield the skin from ROS that are dispersed in the cell after natural oxidation induced by aging-related activities. Examples of carotenes are lycopene and α/β carotene made of carbon and hydrogen; xanthophylls on the other hand contain carbon, hydrogen, and oxygen (examples such as astaxanthin, flucoxanthin, and lutein) [20]. Carotenoids play a major role in defense against these radiations, including lycopene, lutein, and α/γ/β carotenes. Studies have shown effect

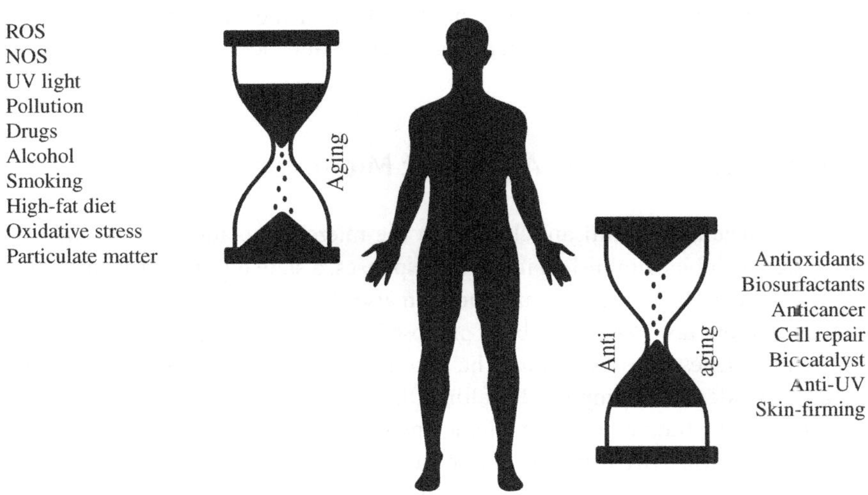

Figure 22.1 Aging and antiaging molecules.

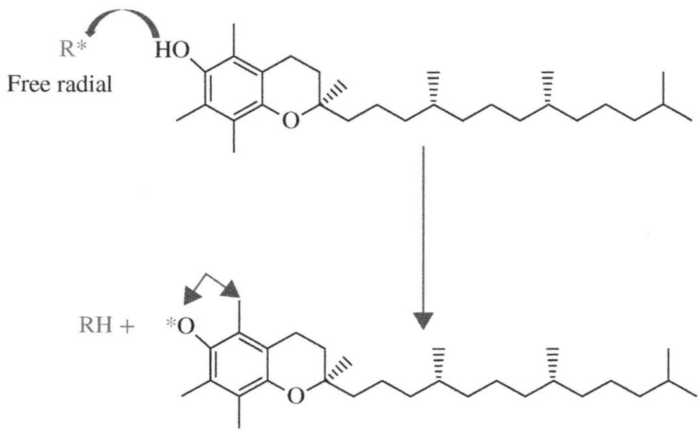

Figure 22.2 Vitamin E quenches free radicals by donating its H-atom from the phenolic group and thus preventing oxidative damage.

of β-carotene coupled with vitamins E and C can protect cells from toxic damage by peroxynitrite anion and the nitrogen dioxide radical (NO˙) [24]. Carotenoids are a group of more than 700 pigments produced by plants, algae, and photosynthetic bacteria and are known to counter harmful effects of ROS. Pathogenesis of atherosclerosis, diabetes mellitus, inflammations, and many types of cancers are linked to the presence of increasing levels of ROS. It has become imperative to look for natural products that can reduce disease burden and aging due to the presence of oxidative stress. Detailed work has been carried out on the green microalga *Tetraselmis suecica* (class Chlorophyceae) that is abundant in vitamin E, carotenoids, chlorophyll, and tocopherols. Pigment extract of this algal species has been implicated in strong antioxidant effects and cell repairing capabilities [25]. Imbalance between antioxidant defense systems and free radicals circulating in our bodies results in oxidative stress. Polyphenols obtained from plant extracts have a strong free radical-scavenging activity and possess antiaging potential [26]. Their mechanism of action is of scavenging free radicals by H-atom transfer (Figure 22.2) and hence reduction of toxic effects of oxidative stress [27].

22.3 Human Health and the Aging Gut Microbiome

Physiological aging results in significant changes in the microbiome and immune system. The immune system exhibits diminished immune responses, a state referred to as immuno-senescence [28], while bacterial diversity and function also shifts due to the reduction in the heterogeneity of different taxa [29]. It has been proposed that by reducing low-grade inflammation and immuno-senescence, which are the two main causes of age-related dysbiosis, human lifespan may extend with improved health [30]. A variety of factors affect the composition and function of the human gut microbiome, including host genetics, lifestyle, dietary factors, extent of physical activity, drug use, and social environment [31]. Among the leading causes of mortality worldwide are age-related conditions that often accompany gut dysbiosis. The gut microbiome is dynamic to keep up with a constantly changing environment, continuously changing its species composition, abundance, and functional potential [32].

According to studies, the human gut microbiome develops rapidly between the ages of 0 and 3 years, followed by a relatively stable microbiome for a longer period if not subjected to dysbiosis or medication, and then slowly changes as we approach the old age [31].

Biagi et al. investigated the enrichment of age-related gut microbial taxa using four age categories, i.e. semi-supercentenarians (105–109 years old), centenarians (99–104 years old), elderly (65–75 years old), and adults (22–48 years old), and reported the shrinkage of core microbial taxa in gut microbiome [33]. In each age group, three families namely *Bacteroidaceae*, *Lachnospiraceae*, and *Ruminococcaceae* dominated, but showed decline in their relative abundance with increasing age [33]. Moreover, the genera *Coprococcus*, *Roseburia*, and *Faecalibacterium*, members of the family *Lachnospiraceae* and *Ruminococcaceae*, showed a negative correlation, while *Oscillospira* showed positive correlation with increasing age [33]. Few genera and families, such as *Akkermansia*, *Eggerthella*, *Anaerotruncus*, *Bilophila*, *Synergistaceae*, and *Christensenellaceae* showed an increasing trajectory, which became even more prominent in semi-supercentenarians, suggesting their role in longevity [33]. Similarly, Xu et al. reported 35 bacterial genera associated with age progression in human subjects [34]. The genera that showed increasing pattern include *Odoribacter*, *Butyricimonas*, *Desulfovibrio*, *Bilophila*, *Corynebacterium*, *Parvimonas*, etc. Few of the genera that showed random pattern include *Lactobacillus*, *Oscillospira*, *Oxalobacter*, *Butyrivibrio*, etc. With age, these beneficial species increased in number, but sharply decreased in elderly subjects [34]. A shift in the composition of the human microbiome characterized by enrichment of *Proteobacteria* and the loss of *Clostridiales* and *Bifidobacterium* along with an increase in pathobionts, i.e. potentially pathogenic microbes such as *Enterobacteriaceae*, is indicative of microbial aging [34]. The healthier gastrointestinal microbiome has been associated with lower levels of low-density lipoprotein (LDL) cholesterol and increased levels of vitamin D and many other useful metabolites circulating in the blood, which are known to reduce inflammation and thereby suggestive of extended lifespan. These studies indicate that the presence of beneficial microbial species contributes largely toward disease-free health and good metabolism.

22.4 The Antiaging Bioproducts from Microbes

Numerous microbial products have been widely used in cosmetic industries for their antiaging properties, biocompatibility, simplified extraction, and usage (Figures 22.3 and 22.4). In subsequent paragraphs, each section covers a specific microbe and its ability to produce antiaging molecules.

22.4.1 Bacteria

Dextran, an exopolysaccharide obtained from *Leuconostoc mesenteriodes* and *Streptococcus mutans*, has found extensive use as a skin care agent. It also shows anti-inflammatory properties and helps in brightening and smoothening of the skin [21]. **Microbial biosurfactants**, surface-active molecules produced by bacteria and yeast, have the advantages of being ecofriendly and biocompatible and are generously used for cosmetic applications [35]. **Glycolipid** microbial biosurfactants are extensively used in the cosmetic industry, such as mannosylerythritol lipids produced by yeast. Biosurfactants have a

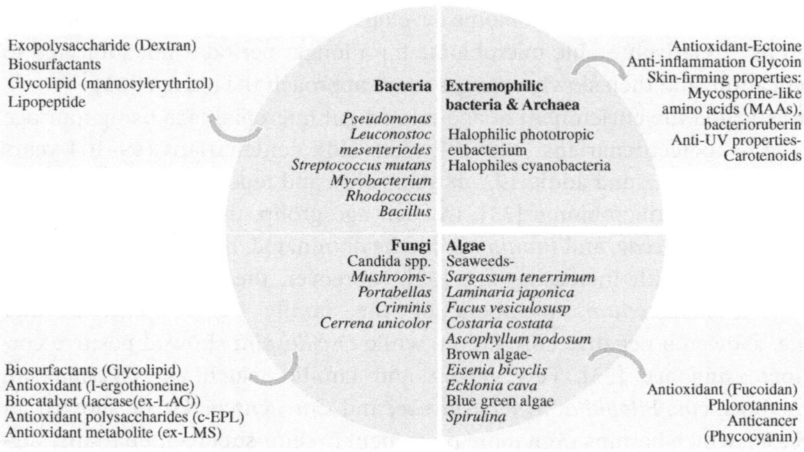

Figure 22.3 Microbes and their antiaging compounds.

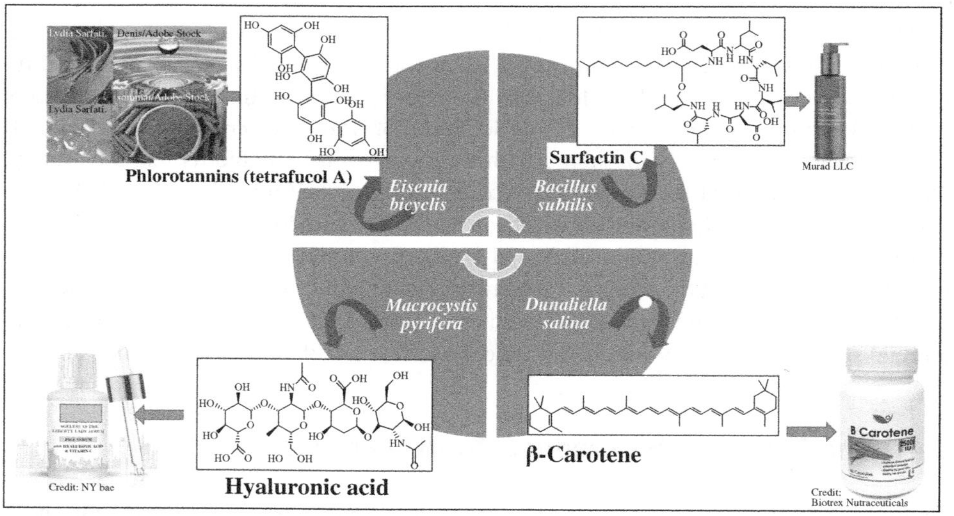

Figure 22.4 Examples of microbial products used in cosmetic industry.

potential in being active ingredients of cosmetic formulations as they exhibit a prebiotic character and are biodegradable in nature [36, 37]. Examples of glycolipids from bacteria include rhamnolipids (mono- and di-rhamnolipids) of *Pseudomonas* spp., sophorolipids (lactonic and acidic) and mannosylerythritol lipids (MELs) (MELs A-C) of *Candida* spp., and trehalose lipids of *Mycobacterium* and *Rhodococcus* spp.

Many studies have delineated bacterial species that constitute major population of microbes on the skin. Using the 16S rRNA sequencing technique, many workers determined that the maximum population belongs to the phyla: *Actinobacteria*, *Firmicutes*, *Proteobacteria*, and *Bacteroides* [38]. *Staphylococci* spp. and *Corynebacteria* spp. were predominantly found in sebaceous and moist skin environments. Several of these biosurfactants have also been implied in protecting the skin microbiome against harmful bacteria as well as *Pseudomonas aeruginosa*, *Staphylococcus aureus*, *Streptococcus pyogenes*, and

Cutibacterium acnes [39]. Biosurfactants have been implied in moisturization of dry skin, hair care as well as repair, and antioxidant functions. *Bacillus* spp. are major producers of lipopeptides, and many of these biosurfactants have anti-wrinkle, moisturizing, and cleansing properties. Their low toxicity and improved efficacy make them efficient candidates to be used instead of chemical surfactants in current cosmetic and personal skincare pharmaceutical formulations. Rincón-Fontán et al. formulated a sunscreen solution based on mica powder and a lipopeptide microbial extract from corn steep liquor, and it was concluded that they provided protection against the harmful effects of sun [40].

22.4.2 Fungi

One of the best studied model organisms for the process of aging is yeast. Yeast cells are easily manipulatable due to their short lifespan, their clearly defined cell structure, and the vast amount of information on their evolutionary conserved genes and signaling pathways. Lifespan of a yeast cell is calculated as chronological lifespan (CLS; duration for which the nondividing cells are viable) and replicative lifespan (RLS; counting the number of divisions a mother cell can undergo to produce new daughter cells) [41]. Budding yeast, *S. cerevisiae*, has been a favorite to quantify longevity and identification of various factors associated with it. Several compounds have been investigated in detail that have been implicated in increasing longevity such as plant-derived chemicals, phenols, and flavonoids. Their major function is to act as an antioxidant that prevents buildup of ROS and loss of mitochondrial activity.

As a source of antiaging molecules, fungi have proven to be a valuable resource. As for example, the antioxidant L-ergothioneine has been isolated from mushrooms namely *Portabellas* and *Criminis*. It is implicated in protecting the skin from oxidative and DNA damage. It is frequently used in antiaging creams for the same benefit [42]. Similarly, white rot fungus *Cerrena unicolor* (Polyporaceae family) has been implicated in production of three bioactive molecules: extracellular laccase (ex-LAC), intracellular nonpurified polysaccharides (c-EPL), and a low molecular weight subfraction of extracellular metabolites (ex-LMS) [43]. *Cerrena unicolor* has already been known to produce the enzyme laccase that functions as an efficient biocatalyst. The molecules mentioned above have been suggested to be a source of antioxidative molecules and hence can be of potential use in antiaging therapies.

22.4.3 Algae

Marine algae have also been proposed as abundant sources of products beneficial for human health. Seaweeds have been postulated to be the richest sources of biologically active ingredients and harbor a large number of proteins, carbohydrates, fatty acids, and amino acids and hence find varied applications in pharmaceutical and cosmeceutical industries as the demand for natural products is increasing by the day versus the synthetic alternatives in the market [44]. These species offer skin beneficial activities that delay aging and display anti-inflammatory, antioxidant, antimicrobial, and anticancer properties.

Fucoidan, a sulfated polysaccharide of long branched chain of sugars with high content of fucose, has been shown to have high antioxidant capacity. It has been isolated from various species of algae such as brown seaweed *Sargassum tenerrimum* [45], *Laminaria japonica* [46], and *Fucus vesiculosusp* [47] and shown positive results of superoxide radical scavenging and antioxidant assays. *Costaria costata* produce fucoidan that showed antiaging activity in human foreskin fibroblast HS68 cells and from *Ascophyllum nodosum* demonstrated antiaging potential in human skin by preventing the degradation of elastic fiber and leukocyte elastase activity [48]. Species of brown algae *Eisenia bicyclis* and *Ecklonia cava* are known to produce **phlorotannins** that play a role in reduction of elastase activity. Elastin fibers along with collagen help in maintaining skin texture intact and slowing down aging [49]. Brown seaweed extracts from *A. nodosum* are rich in phlorotannins and are being explored for their inhibitory activity against oxidative stress, inflammation, and advanced aging effects [50].

Blue green algae, *Spirulina*, has also many antioxidant properties, and **phycocyanin** isolated from it has been shown as an immunity enhancer and shows cancer resistance; its superoxide dismutase inactivates free radicals that slow down aging of cells as well [51].

22.5 The Impact of Microbial Bioproducts on Gut Diversity

Streptomyces hygroscopicus produces **rapamycin** and is well known for its antiaging potential [52]. It not only suppresses senescence and delays the onset of aging-related disorders but also impacts the gut microbial diversity. In case of *Drosophila*, it has been reported to reduce the CFU count significantly for the members of family *Enterobacteriaceae*, *Lactobacillae*, and *Acetobacteriaceae* [53], suggesting that it can delay the microbial expansion in the aging guts. On the contrary, in case of mice, rapamycin regimen has induced significant changes in the gut microbial diversity and to be precise, it has shown to increase the prevalence of *Candidatus arthromitus* sp., a segmented filamentous bacterium, which is generally absent in aging mice [54]. This bacterium is known to induce the postnatal maturation of homeostatic innate and adaptive immune responses [55]. **Acarbose** is another microbial product produced by members of the *Actinoplanes* genus, a family of *Micromonosporaceae* [56]. This molecule has proven antiaging effects in geriatric subjects (https://clinicaltrials.gov/ct2/show/NCT02865499). Following treatment, members of the order *Clostridiale* significantly increased while the *Ruminococcaceae* and *Lachnospiraceae* decreased. Acarbose is known to modulate the gut microbiota to enhance the short chain fatty acid (SCFA) production and fermentation [57]. Similarly, **spermidine** is known to increase the *Lachnospiraceae*, a well-known family of SCFA-producing bacteria [58]. These compounds have considerable gut modulating potential, and as a result in the upcoming years, their use will be explored in even more depth.

22.6 Microbial Bioproduction of Extremolytes

Extremolytes are small organic molecules with low molecular weight present in extremophilic bacteria and archaea. Extremophilic organisms are present in virtually inhabitable extreme environments such as hot springs, polar ice, and volcanic areas. Studies have shown the presence of accumulated extremolytes (either synthetized de novo or taken up from the environment) inside extremophilic bacteria and archaea that play crucial role in their survival and protect cells from harsh conditions [59]. They mostly consist of sugars, amino acids, polyols, heterosides, and their derivatives [60, 61]. These molecules stabilize membranes, proteins, or nucleic acids and protect the extremophiles against multiple stress as they can act as multifunctional agents, a notable property [59]. **Ectoine, hydroxyectoine, proline, mannitol**, and **glycine-betaine**, all of which are often present in halophiles, are notable examples. Glucosyl-glycerol (GG, glycoin), glucosyl-glycerate (GGA), mannosyl-glycerate

(MG, firoin), and mannosyl-glyceramide (MG) are all heterosides found in hyperthermophilic microorganisms [59, 61]. Furthermore, complex molecular UV light-scavenging substances such as scytonemin, carotenoids, mycosporine-like amino acids (MAAs), bacterioruberin, and melanin have been identified from radiation-resistant bacteria [59–61]. Many extremolytes build and stabilize protective water layers surrounding extremophile macromolecules and cell structures to shield them from their harsh environments.

There are some instances of use of extremolytes as active antiaging ingredients in the cosmetic industry. Ectoine, produced by halophilic phototropic eubacterium, is one of the most important molecules being used. Its cosmotropic properties make it a desired ingredient in skin and hair care products [59]. Owing to its immense water binding capacity, application of ectoine prevents skin dehydration and skin aging [62]. Studies have shown that ectoine also protects skin from UV-induced damage by suppressing the expression of UV-induced secondary messenger release, AP2 transcription factor, and ICAM1 expression and by preventing mutation in mitochondria [63].

The 2-*O*-alpha-D-glucopyranosylglycerin (**glycoin**) found in halophilic cyanobacteria and in desert resurrection plant *Myrothamnus flabellifolia* is a potent antiaging agent. Studies have shown that glycoin protects cell membrane from stress and increases cell vitality and renewal by stimulating ATP production in aged cells. Further, it induces antioxidant properties in aged skin cells by increasing the production of superoxide dismutase 1 (SOD1). Additionally, glycoin is known for increasing elasticity and recovery of skin by inducing the expression of transforming growth factor beta 1 (TGF β) and fibroblast growth factor 7 (FGF7) [64]. Certain microbes have started producing extremolytes (natural sunscreen) due to exposure to extensive radiations (chemical scavengers). **Mycosporine-like amino acids (MAA)** are prominent examples with antiaging effects comprising antioxidant, anti-inflammation, and skin-firming properties [65]. They are also known to be more efficient than currently existing sunscreens with inorganic (zinc/titanium oxide) and organic filters (aminobenzoic acid, avobenzone, etc.) [66]. Studies have shown that MAAs are extremely stable at high temperature and pH. They are produced in response to UV radiation that causes sunburn, premature aging, and skin cancer. MAAs provide proper protection against UVA radiation [67] while providing comparatively less protection against UVB radiation [66], and hence primarily is used as a natural bioactive ingredient in antiaging cosmetic products [67]. **Carotenoids** too have anti-UV properties and are natural antioxidants that help in reducing the skin's photodamage by reducing the production of free radicals in cells [68]. The health industry has hailed extremolytes as "untapped gold mines" with enormous potential in the beauty, medicinal, and food industries. Thus, protection of these delicate biomolecules and biostructures is critical to the success of these industries.

22.7 The Role of Antiaging and Antioxidant Molecules

The microbial diversity on the Earth is a reservoir of biomolecules that have antiaging potential through their antioxidant properties. Natural antioxidants have been largely reported from plants and our kitchen spices [69], but microbial products remain largely unexplored [70]. Some biomolecules like hyaluronic acid [71], phlorotannins [21], porphyra-334 [72], and surfactin C [73] are being used in the cosmetic industry. Flavonoids like kaempferol [74] and quercetin [26] reduce oxidative damage. Though these flavonoids are produced by plants, their upscale production is carried out in genetically engineered microbes [75]. Table 22.1 and Figures 22.3 and 22.4 list a few of these microbial biomolecules with antioxidant potential.

Table 22.1 Microbial compounds with antioxidant potential.

S. no.	Antiaging compound	Microbial source	Function studied/mechanism of action	Structure
1	Astaxanthin	*Agrobacterium aurantiacum, Paracoccus carotinifaciens*	Acts as an antioxidant because of its enhanced capacity to capture free radicals and ROS released as a byproduct of metabolism in biological systems [76]	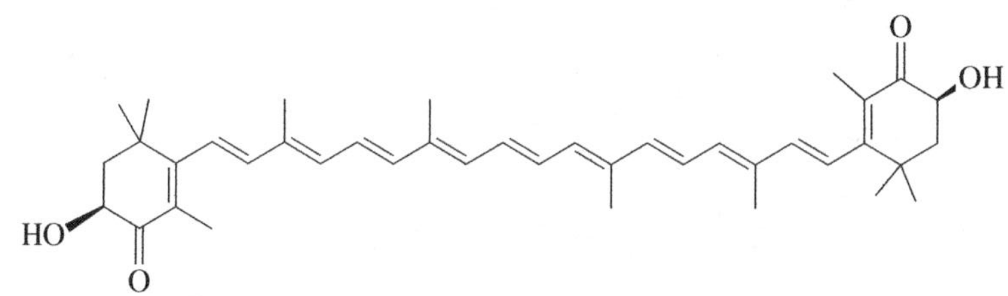
2	β-Carotene	*Mucor circinelloides*	Antioxidant [77]	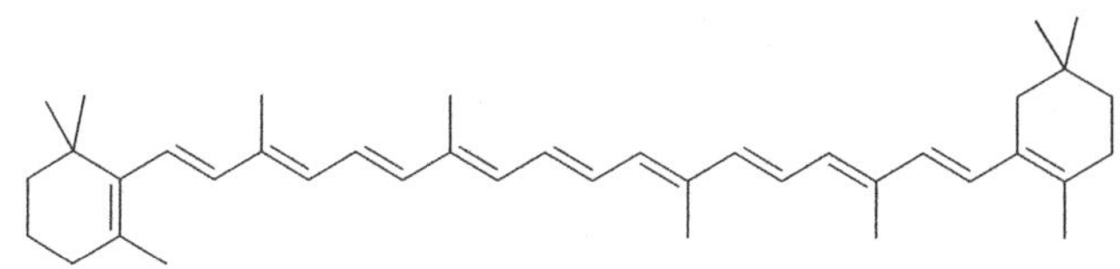
3	Canthaxanthin	*Bradyrhizobium* sp., *Lactobacillus pluvalis*	Alters the pro-oxidation and antioxidation balance and suppresses oxidative stress induced by cholesterol via modulation of antioxidant system and cholesterol metabolism in rats [78]	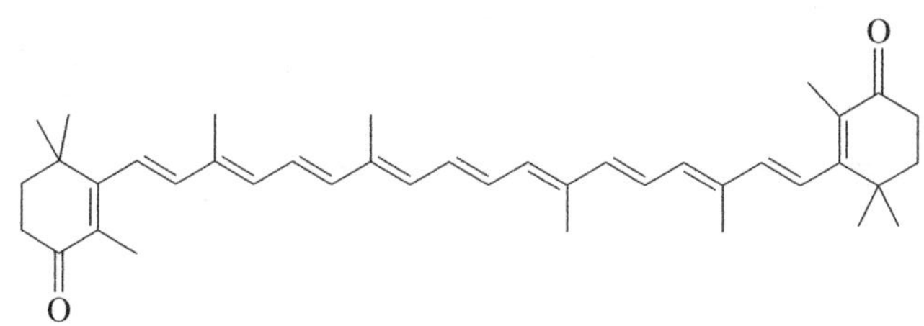
4	Chaetopyranin	*Chaetomium globosum*	Acts as an antioxidant by scavenging 1, 1-diphenyl-2-picrylhydrazyl radical [79]	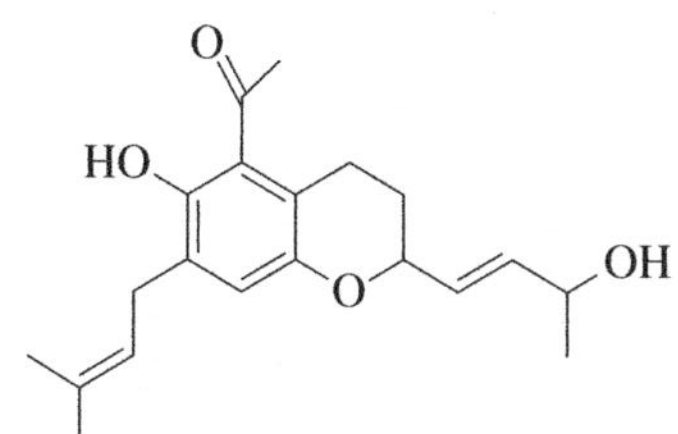

5	Citric acid	*Aspergillus niger*	Prevents the oxidation of biomolecules by scavenging ROS that are produced as a result of metabolism [80]	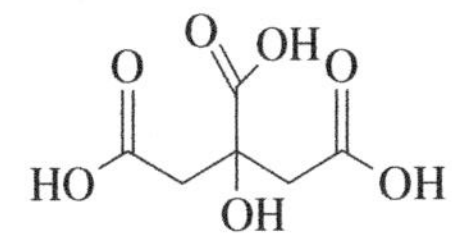
6	Curcumin	*Curcuma longa* L. The introduction of curcumin synthase genes in engineered strain of *E. coli* resulted in upscale production of curcuminoids [81]	It increases the lifespan of *Drosophila melanogaster* by 25.8% by upregulation of Mn-SOD and CuZn-SOD genes and the downregulation of dInR, ATTD, Def, CecB, and DptB genes [82]	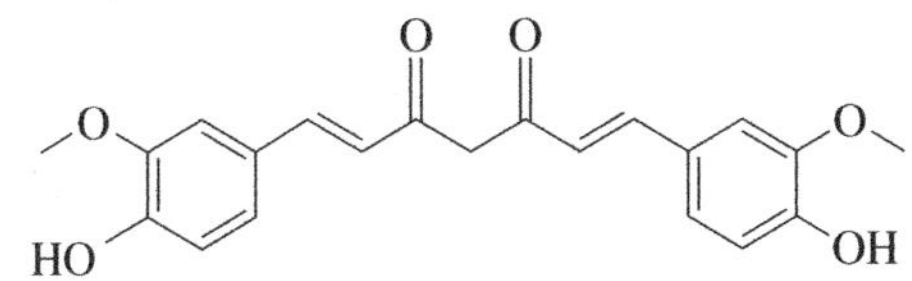
7	Dimethyl sulfide (DMS)	Main metabolite of marine algae	Promotes methionine sulfoxide reductase A activity and was found to increase the longevity in *Caenorhabditis elegans* and *Drosophila* [83]	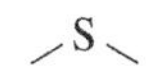
8	Flavipin	*Aspergillus flavipes* *Aspergillus terrus*	Can scavenge five free radicals at a time via three phenolic hydroxyl and two aldehyde groups [84]	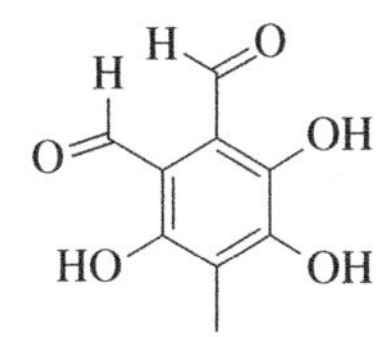
9	Hyaluronic acid	*Streptococcus* spp. *Bacillus* spp.	It binds and retains water molecules in skin and protects it from aging [71]	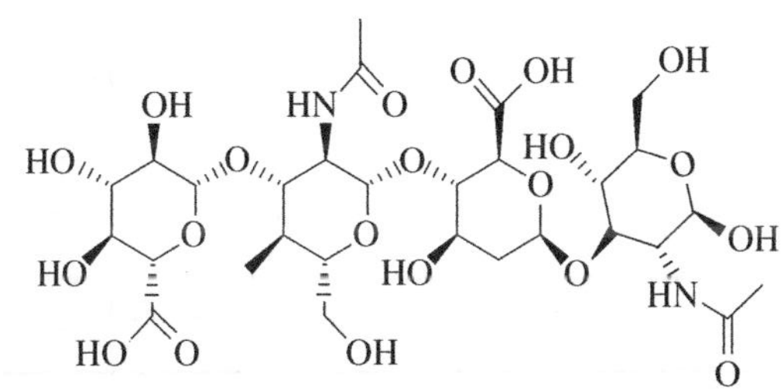

(*Continued*)

Table 22.1 (Continued)

S. no.	Antiaging compound	Microbial source	Function studied/mechanism of action	Structure
10	Isopestacin	*Pestalotiopsis microspora*	Antioxidant activity [85]	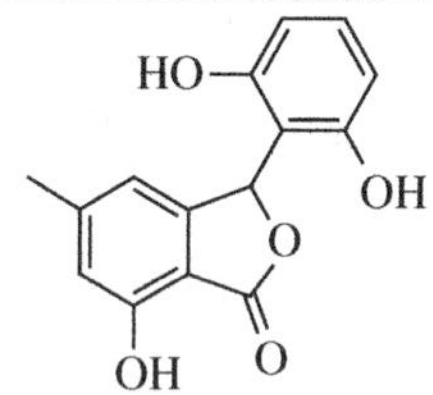
11	Kaempferol	Broccoli and tea Microbial synthesis of kaempferol by introducing F3H and FLS into a naringenin-producing *S. cerevisiae* strain Y-22, which rendered the recombinant strain the ability to generate kaempferol at titer level 86 mg/L [86]	Reduces oxidative cell damage induced by glucose and dysfunction in HIT-T15 pancreatic β cells [74]	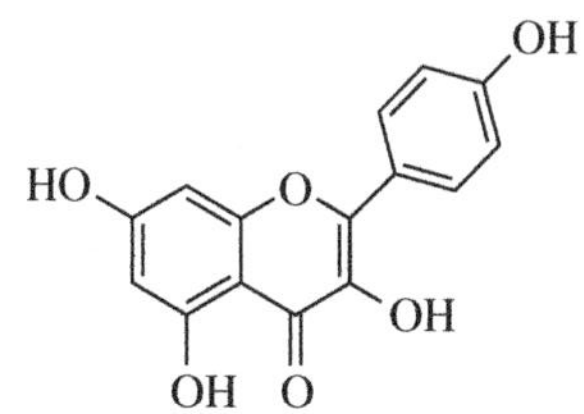
12	Lutein	*Chlorella sorokiniana*	Protects skin from UV-induced damage [87]	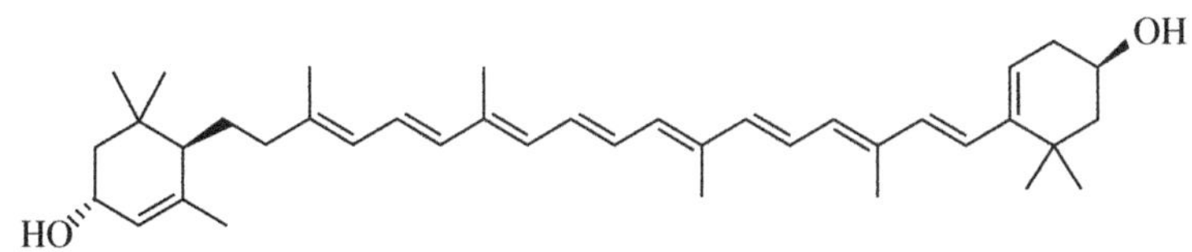
13	Lycopene	*Fusarium sporotrichioides*	Potent antioxidant and a O_2 quencher [88]	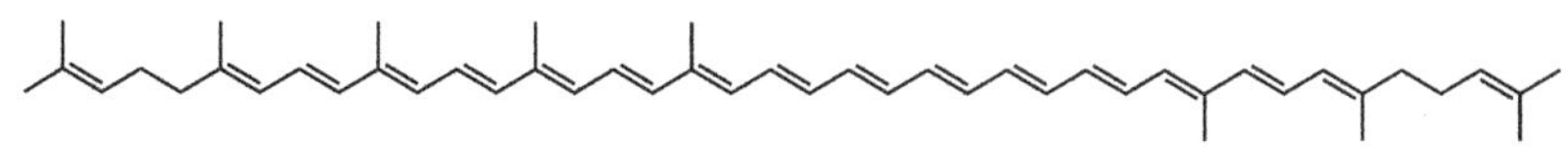
14	Melanin	*Saccharomyces, Streptomyces* spp.	The unpaired electrons in melanin can scavenge free radicals making it a potent antioxidant [89]	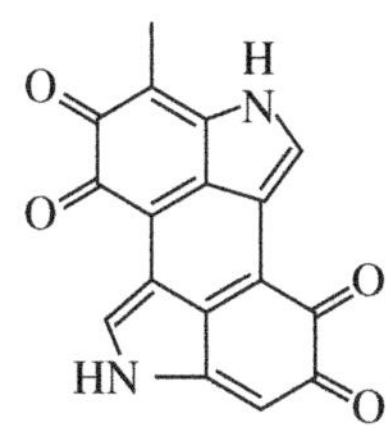

15	Phlorotannins	*Eisenia bicyclis* *Ecklonia cava*	Reduces elastase activity of the skin [21]	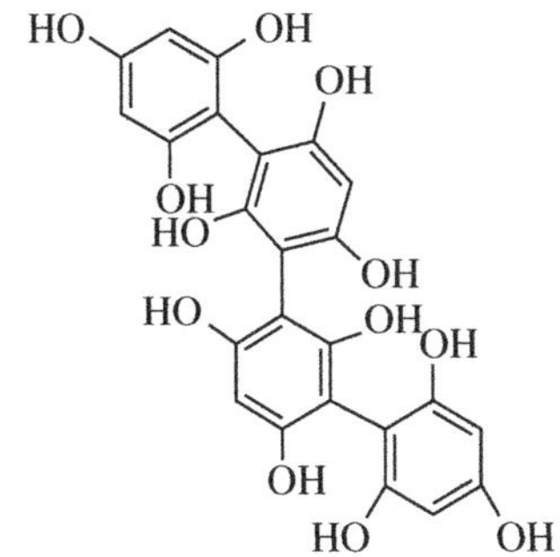
16	Porphyra-334 (P-334)	*Porphyra rosengurttii*	The topical application of P-334 protected against UV-induced skin damage by preventing sunburn cell formation in mice and helped in maintaining the antioxidant defense system of the skin as well as expression of Hsp70 protein [72]	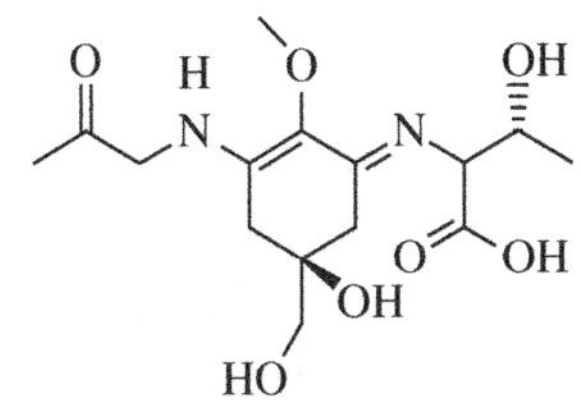
17	Quercetin	Flavonoid found in plant extracts. An artificial pathway in actinomycete containing the TAL from *Rhodobacter capsulatus*, 4CL from *Streptomyces coelicolor*, CHS and CHI from *Glycine max*, naringenin 3-dioxygenase from *Pichia crispum*, and FLS and F3′H from *Arabidopsis thaliana* was constructed and introduced into *Streptomyces albus* and *S. coelicolor*. The recombinant *S. albus* produced the higher quercetin titer of 0.1 mg/L [75]	Stress resistance, reduction in the levels of reactive oxygen species (ROS) in *Saccharomyces cerevisiae* [26]	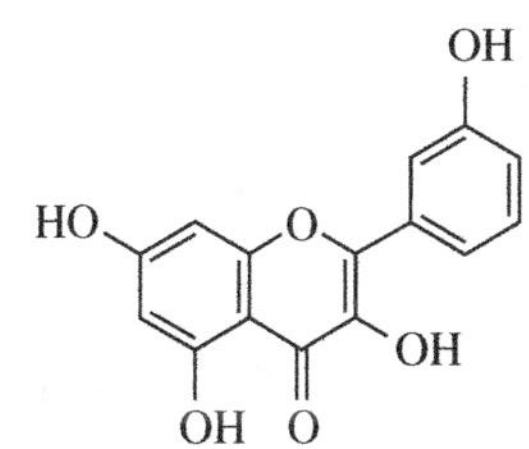

(*Continued*)

Table 22.1 (Continued)

S. no.	Antiaging compound	Microbial source	Function studied/mechanism of action	Structure
18	Resveratrol	Non-flavonoid natural phenol found in small quantities in grape skin, berries, peanuts, and red wine. A study showed that DJ526 rice seeds when treated with yeast extract greatly enhanced the production and antioxidant property of resveratrol [90]	Shown to counteract aging factors in fish, nematodes, flies, mice, and human cells; increases RLS in yeast [91]	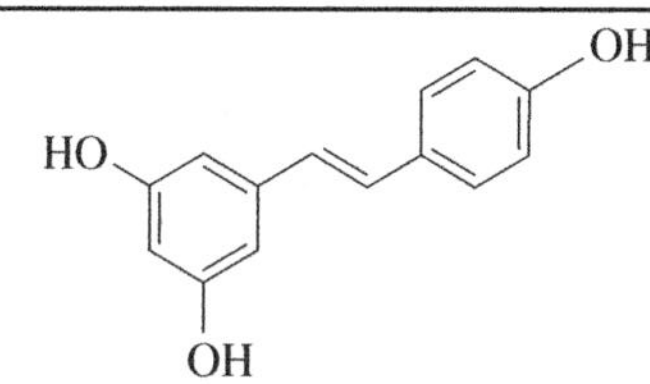
19	Riboflavin	*Ashbya gossypi*	Increases the lifespan and strengthened reproduction in *Drosophila* by enhancing the activity of antioxidant enzymes [86, 92]	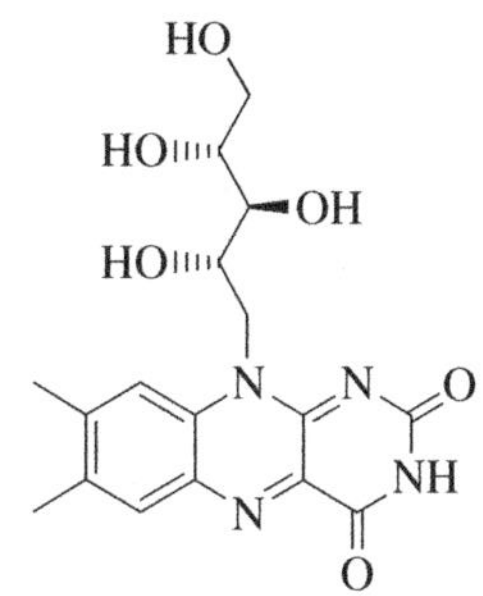
20	Surfactin C	*Bacillus subtilis*	Antiaging function for skin (https://patents.google.com/patent/US9364413B2/en)	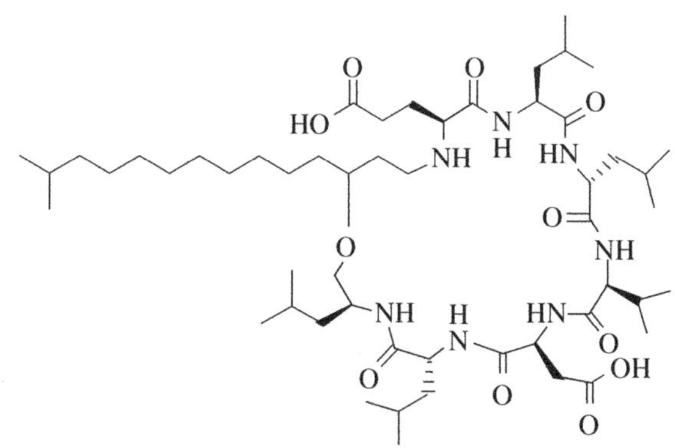

21	Scytonemin	Cyanobacteria	Has the ability to absorb UV-radians (UV-C, UV-B, and UV-A) and prevent skin aging [93]	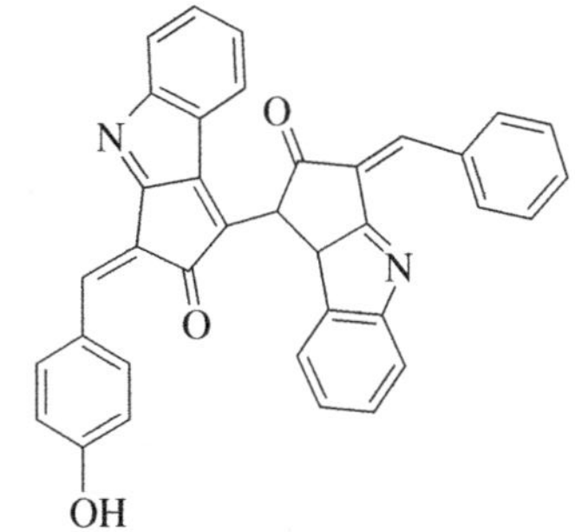
22	Tyrosol	*Diaporthe helianthin, Glomerella cingulata*	Prevents lipid peroxidation [94]	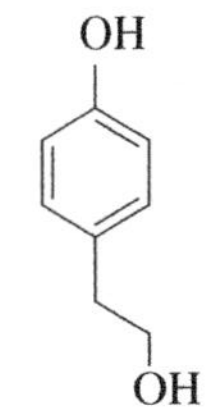
23	Valine	*Bacillus subtilis, Bacillus licheniformis, Corynebacterium glutamicum* [95]	Increases the lifespan of *C. elegans* [96]	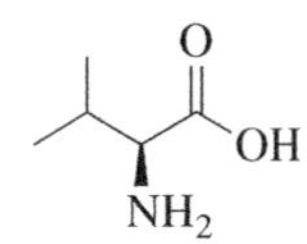
24	Vitamin E	In nature synthesized exclusively by photosynthetic organisms. An attempt has been made for large-scale production in *S. cerevisiae* via fermentation by introducing artificially created pathway in it and subjecting it to cold-shock-triggered temperature control to produce a titer level of 4.1 mg/L [97]	Acts as an antioxidant (Figure 22.2)	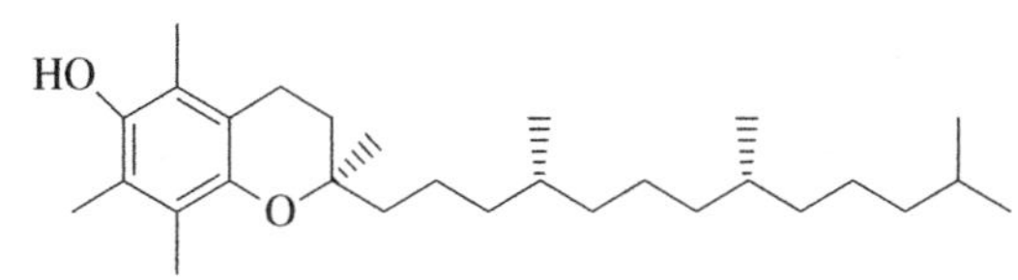
25	Zeaxanthin	*Flavobacterium* sp.	Inhibits lipid peroxidation [97]	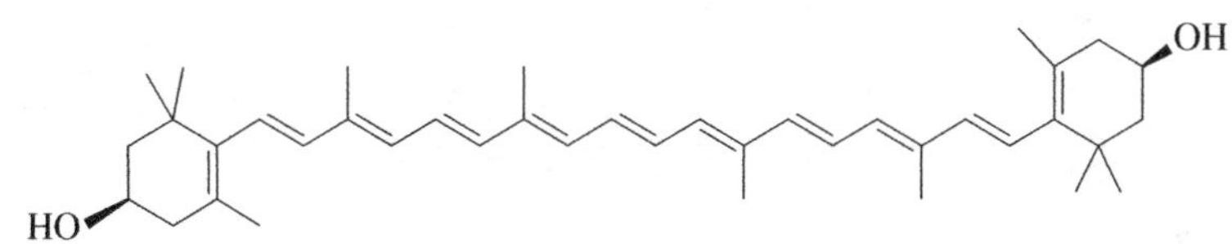

22.8 Conclusions

Research in the field of "biogerontology," i.e. study of the biological basis of aging and age-related diseases, is promising. It offers opportunities for development of technologies, formulations, and preparations for maintaining and improving metabolism and health and also ways to retard degenerative processes in living organisms. Environment friendly natural products are in great demand today, and microbes are the source of untapped reservoirs that can be used in abundance for production of antiaging compounds on an industrial scale. Moreover, extremophiles subjected to extreme environmental stress produce extremolytes with exceptional and unexplored properties that make them good candidates for health promoting and therapeutic properties. The opportunities thus presented by natural products owing to their novelty, diversity of structures, and being amenable to genetic modifications to create useful hybrids holds great promise in the future of medicine and global personal care markets.

References

1 Singh, S.P. and Upadhyay, S.K. (eds.) (2021). *Bioprospecting of Microorganism-Based Industrial Molecules*. John Wiley & Sons Ltd. doi:10.1002/9781119717317.

2 Shintani, T., Upadhyay, S.K., Singh, S.P. (2021). An introduction to microbial biodiversity and bioprospection. In: *Bioprospecting of Microorganism-Based Industrial Molecules* (ed. S.P. Singh and S.K. Upadhyay). John Wiley & Sons Ltd. https://doi.org/10.1002/9781119717317.ch1.

3 Baeshen, N.A., Baeshen, M.N., Sheikh, A. et al. (2014). Cell factories for insulin production. *Microb. Cell Factories* 13: 141.

4 Beelman, R.B., Kalaras, M.D., Phillips, A.T., and Richie, J.P. Jr. (2020). Is ergothioneine a "longevity vitamin" limited in the American diet? *J. Nutr. Sci.* 9: e52.

5 Tanaka, N., Kawano, Y., Satoh, Y. et al. (2019). Gram-scale fermentative production of ergothioneine driven by overproduction of cysteine in *Escherichia coli*. *Sci. Rep.* 9 (1): 1895.

6 Stefánsson, H. (2005). The science of ageing and anti-ageing. *EMBO Rep.* 6 (Suppl 1): S1–S3.

7 Kapahi, P., Kaeberlein, M., and Hansen, M. (2017). Dietary restriction and lifespan: lessons from invertebrate models. *Ageing Res. Rev.* 39: 3–14.

8 Tosato, M., Zamboni, V., Ferrini, A., and Cesari, M. (2007). The aging process and potential interventions to extend life expectancy. *Clin. Interv. Aging* 2 (3): 401–412.

9 Martel, J., Chang, S.H., Wu, C.Y. et al. (2021). Recent advances in the field of caloric restriction mimetics and anti-aging molecules. *Ageing Res. Rev.* 66: 101240.

10 Hofer, S.J., Davinelli, S., Bergmann, M. et al. (2021). Caloric restriction mimetics in nutrition and clinical trials. *Front. Nutr.* 8: 717343.

11 Yu, M., Zhang, H., Wang, B. et al. (2021). Key signaling pathways in aging and potential interventions for healthy aging. *Cell* 10 (3): 660.

12 Lopez-Hortas, L., Flórez-Fernández, N., Torres, M.D. et al. (2021). Applying seaweed compounds in cosmetics, cosmeceuticals and nutricosmetics. *Mar. Drugs* 19 (10): 552.

13 Dini, I. and Laneri, S. (2019). Nutricosmetics: a brief overview. *Phytother. Res.* 33 (12): 3054–3063.

14 de Cabo, R., Carmona-Gutierrez, D., Bernier, M. et al. (2014). The search for antiaging interventions: from elixirs to fasting regimens. *Cell* 157 (7): 1515–1526.

15 Kumar, R., Sood, U., Gupta, V. et al. (2020). Recent advancements in the development of modern probiotics for restoring human gut microbiome dysbiosis. *Indian J. Microbiol.* 60 (1): 12–25.

16 Hernandez, D.F., Cervantes, E.L., Luna-Vital, D.A., and Mojica, L. (2021). Food-derived bioactive compounds with anti-aging potential for nutricosmetic and cosmeceutical products. *Crit. Rev. Food Sci. Nutr.* 61 (22): 3740–3755.

17 Antell, D.E. and Taczanowski, E.M. (1999). How environment and lifestyle choices influence the aging process. *Ann. Plast. Surg.* 43 (6): 585–588.

18 Ames, B.N., Shigenaga, M.K., and Hagen, T.M. (1993). Oxidants, antioxidants, and the degenerative diseases of aging. *Proc. Natl. Acad. Sci. U. S. A.* 90 (17): 7915–7922.

19 Schikowski, T. and Huls, A. (2020). Air pollution and skin aging. *Curr. Environ. Health Rep.* 7 (1): 58–64.

20 Galasso, C., Corinaldesi, C., and Sansone, C. (2017). Carotenoids from marine organisms: biological functions and industrial applications. *Antioxidants (Basel)* 6 (4): 96.

21 Gupta, P.L., Rajput, M., Oza, T. et al. (2019). Eminence of microbial products in cosmetic industry. *Nat. Prod. Bioprospecting* 9 (4): 267–278.

22 Phaniendra, A., Jestadi, D.B., and Periyasamy, L. (2015). Free radicals: properties, sources, targets, and their implication in various diseases. *Indian J. Clin. Biochem.* 30 (1): 11–26.

23 McCann, S.M., Licinio, J., Wong, M.L. et al. (1998). The nitric oxide hypothesis of aging. *Exp. Gerontol.* 33 (7–8): 813–826.

24 Poljsak, B. and Fink, R. (2014). The protective role of antioxidants in the defence against ROS/RNS-mediated environmental pollution. *Oxidative Med. Cell. Longev.* 2014: 671539.

25 Sansone, C., Galasso, C., Orefice, I. et al. (2017). The green microalga *Tetraselmis suecica* reduces oxidative stress and induces repairing mechanisms in human cells. *Sci. Rep.* 7 (1): 41215.

26 Alugoju, P., Periyasamy, L., and Dyavaiah, M. (2020). Protective effect of quercetin in combination with caloric restriction against oxidative stress-induced cell death of *Saccharomyces cerevisiae* cells. *Lett. Appl. Microbiol.* 71 (3): 272–279.

27 Di Meo, F., Lemaur, V., Cornil, J. et al. (2013). Free radical scavenging by natural polyphenols: atom versus electron transfer. *J. Phys. Chem. A* 117 (10): 2082–2092.

28 Crooke, S.N., Ovsyannikova, I.G., Poland, G.A., and Kennedy, R.B. (2019). Immunosenescence and human vaccine immune responses. *Immun. Ageing* 16: 25.

29 Badal, V.D., Vaccariello, E.D., Murray, E.R. et al. (2020). The gut microbiome, aging, and longevity: a systematic review. *Nutrients* 12 (12): 3759.

30 Thevaranjan, N., Puchta, A., Schulz, C. et al. (2018). Age-associated microbial dysbiosis promotes intestinal permeability, systemic inflammation, and macrophage dysfunction. *Cell Host Microbe* 23 (4): 570.

31 Yatsunenko, T., Rey, F., Manary, M. et al. (2012). Human gut microbiome viewed across age and geography. *Nature* 486 (7402): 222–227.

32 Bosco, N. and Noti, M. (2021). The aging gut microbiome and its impact on host immunity. *Genes Immun.* 22 (5–6): 289–303.

33 Biagi, E., Franceschi, C., Rampelli, S. et al. (2016). Gut microbiota and extreme longevity. *Curr. Biol.* 26 (11): 1480–1485.

34 Xu, C., Zhu, H., and Qiu, P. (2019). Aging progression of human gut microbiota. *BMC Microbiol.* 19 (1): 236.

35 Moldes, A.B., Rodríguez-López, L., Rincón-Fontán, M. et al. (2021). Synthetic and bio-derived surfactants versus microbial biosurfactants in the cosmetic industry: an overview. *Int. J. Mol. Sci.* 22 (5): 2371.

36 Vecino, X., Cruz, J.M., Moldes, A.B., and Rodrigues, L.R. (2017). Biosurfactants in cosmetic formulations: trends and challenges. *Crit. Rev. Biotechnol.* 37 (7): 911–923.

37 Bezerra, K.G.O., Rufino, R.D., Luna, J.M., and Sarubbo, L.A. (2018). Saponins and microbial biosurfactants: potential raw materials for the formulation of cosmetics. *Biotechnol. Prog.* 34 (6): 1482–1493.

38 Grice, E.A., Kong, H.H., Conlan, S. et al. (2009). Topographical and temporal diversity of the human skin microbiome. *Science* 324 (5931): 1190–1192.

39 Findley, K. and Grice, E.A. (2014). The skin microbiome: a focus on pathogens and their association with skin disease. *PLoS Pathog.* 10 (10): e1004436.

40 Rincon-Fontan, M., Rodríguez-López, L., Vecino, X. et al. (2020). Novel multifunctional biosurfactant obtained from corn as a stabilizing agent for antidandruff formulations based on Zn pyrithione powder. *ACS Omega* 5 (11): 5704–5712.

41 Postnikoff, S.D. and Harkness, T.A. (2014). Replicative and chronological life-span assays. *Methods Mol. Biol.* 1163: 223–227.

42 Dubost, N., Beelman, R., and Royse, D. (2007). Influence of selected cultural factors and postharvest storage on ergothioneine content of common button mushroom agaricus bisporus (J. Lge) Imbach (Agaricomycetideae). *Int. J. Med. Mushrooms* 9: 163–176.

43 Jaszek, M., Osińska-Jaroszuk, M., Janusz, G. et al. (2013). New bioactive fungal molecules with high antioxidant and antimicrobial capacity isolated from *Cerrena unicolor* idiophasic cultures. *Biomed. Res. Int.* 2013: 497492.

44 Kalasariya, H.S., Yadav, V.K., Yadav, K.K. et al. (2021). Seaweed-based molecules and their potential biological activities: an eco-sustainable cosmetics. *Molecules* 26 (17): 5313.

45 Marudhupandi, T., Kumar, T.T., Senthil, S.L., and Devi, K.N. (2014). in vitro antioxidant properties of fucoidan fractions from *Sargassum tenerrimum*. *Pak. J. Biol. Sci.* 17 (3): 402–407.

46 Wang, J., Zhang, Q., Zhang, Z., and Li, Z. (2008). Antioxidant activity of sulfated polysaccharide fractions extracted from *Laminaria japonica*. *Int. J. Biol. Macromol.* 42 (2): 127–132.

47 Ruperez, P., Ahrazem, O., and Leal, J.A. (2002). Potential antioxidant capacity of sulfated polysaccharides from the edible marine brown seaweed *Fucus vesiculosus*. *J. Agric. Food Chem.* 50 (4): 840–845.

48 Kim, J.H., Lee, J.E., Kim, K.H., and Kang, N.J. (2018). Beneficial effects of marine algae-derived carbohydrates for skin health. *Mar. Drugs* 16 (11): 459.

49 Wijesekara, I., Yoon, N.Y., and Kim, S.K. (2010). Phlorotannins from *Ecklonia cava* (Phaeophyceae): biological activities and potential health benefits. *Biofactors* 36 (6): 408–414.

50 Dutot, M., Fagon, R., Hemon, M., and Rat, P. (2012). Antioxidant, anti-inflammatory, and anti-senescence activities of a phlorotannin-rich natural extract from brown seaweed *Ascophyllum nodosum*. *Appl. Biochem. Biotechnol.* 167 (8): 2234–2240.

51 Babich, O., Sukhikh, S., Larina, V. et al. (2022). Algae: study of edible and biologically active fractions, their properties and applications. *Plants (Basel)* 11 (6): 780.

52 Blagosklonny, M.V. (2019). Rapamycin for longevity: opinion article. *Aging (Albany NY)* 11 (19): 8048–8067.

53 Fan, X., Liang, Q., Lian, T. et al. (2015). Rapamycin preserves gut homeostasis during *Drosophila* aging. *Oncotarget* 6 (34): 35274–35283.

54 Bitto, A., Ito, T.K., Pineda, V.V. et al. (2016). Transient rapamycin treatment can increase lifespan and healthspan in middle-aged mice. *elife* 5: e16351.

55 Bolotin, A., de Wouters, T., Schnupf, P. et al. (2014). Genome sequence of "*Candidatus arthromitus*" sp. strain SFB-Mouse-NL, a commensal bacterium with a key role in postnatal maturation of gut immune functions. *Genome Announc.* 2 (4): e00705–e00714.

56 Harrison, D.E., Strong, R., Alavez, S. et al. (2019). Acarbose improves health and lifespan in aging HET3 mice. *Aging Cell* 18 (2): e12898.

57 Zhang, X., Fang, Z., Zhang, C. et al. (2017). Effects of acarbose on the gut microbiota of prediabetic patients: a randomized, double-blind, controlled crossover trial. *Diabetes Ther.* 8 (2): 293–307.

58 Ma, L., Ni, Y., Wang, Z. et al. (2020). Spermidine improves gut barrier integrity and gut microbiota function in diet-induced obese mice. *Gut Microbes* 12 (1): 1–19.

59 Becker, J. and Wittmann, C. (2020). Microbial production of extremolytes – high-value active ingredients for nutrition, health care, and well-being. *Curr. Opin. Biotechnol.* 65: 118–128.

60 McParland, E.L., Alexander, H., and Johnson, W. (2021). The osmolyte ties that bind: genomic insights into synthesis and breakdown of organic osmolytes in marine microbes. *Front. Mar. Sci.* 8: https://doi.org/10.3389/fmars.2021.689306.

61 Rastogi, R.P. and Incharoensakdi, A. (2014). Characterization of UV-screening compounds, mycosporine-like amino acids, and scytonemin in the cyanobacterium Lyngbya sp. CU2555. *FEMS Microbiol. Ecol.* 87 (1): 244–256.

62 Graf, R., Anzali, S., Buenger, J. et al. (2008). The multifunctional role of ectoine as a natural cell protectant. *Clin. Dermatol.* 26 (4): 326–333.

63 Buenger, J. and Driller, H. (2004). Ectoin: an effective natural substance to prevent UVA-induced premature photoaging. *Skin Pharmacol. Physiol.* 17 (5): 23sss2–23sss7.

64 Schagen, S., Overhagen, S., and Bilstein, A. (2017). New data confirm skin revitalizing and stress protection by glycoin® natural. *Euro Cosmet.* 1: 24–27.

65 Rosic, N.N. (2019). Mycosporine-like amino acids: making the foundation for organic personalised sunscreens. *Mar. Drugs* 17 (11): 638.

66 Geraldes, V. and Pinto, E. (2021). Mycosporine-like amino acids (MAAs): biology, chemistry and identification features. *Pharmaceuticals (Basel)* 14 (1): 63.

67 Corinaldesi, C., Barone, G., Marcellini, F. et al. (2017). Marine microbial-derived molecules and their potential use in cosmeceutical and cosmetic products. *Mar. drugs* 15: E118.

68 Mendes-Silva, T., da Silva Andrade, R.F., Ootani, M.A. et al. (2020). Biotechnological potential of carotenoids produced by extremophilic microorganisms and application prospects for the cosmetics industry. *Adv. microbiol* 10: 397–410.

69 Yashin, A., Yashin, Y., Xia, X., and Nemzer, B. (2017). Antioxidant activity of spices and their impact on human health: a review. *Antioxidants (Basel)* 6 (3): 70.

70 Anand, S., Hallsworth, J.E., Timmis, J., Verstraete, W., et al. (2023). Weaponising microbes for peace. *Microbial Biotechnology.* https://doi.org/10.1111/1751-7915.14224.

71 Papakonstantinou, E., Roth, M., and Karakiulakis, G. (2012). Hyaluronic acid: a key molecule in skin aging. *Dermato-Endocrinology* 4 (3): 253–258.

72 de la Coba, F., Aguilera, J., de Gálvez, M.V. et al. (2009). Prevention of the ultraviolet effects on clinical and histopathological changes, as well as the heat shock protein-70 expression in mouse skin by topical application of algal UV-absorbing compounds. *J. Dermatol. Sci.* 55 (3): 161–169.

73 Lewinska, A., Domzal- Kedzia, M., Jaromin, A., and Lukaszewics, M. (2020). Nanoemulsion stabilized by safe surfactin from *Bacillus Subtilis* as a multifunctional, custom-designed smart delivery system. *Pharmaceutics* 12 (10): 953.

74 Lee, Y.J., Suh, K.S., Choi, M.C. et al. (2010). Kaempferol protects HIT-T15 pancreatic beta cells from 2-deoxy-D-ribose-induced oxidative damage. *Phytother. Res.* 24 (3): 419–423.

75 Marin, L., Gutiérrez-Del-Río, I., Entrialgo-Cadierno, R. et al. (2018). De novo biosynthesis of myricetin, kaempferol and quercetin in *Streptomyces albus* and *Streptomyces coelicolor*. *PLoS One* 13 (11): e0207278.

76 Pertiwi, H., Mahendra, M.Y.N., and Kamaludeen, J. (2022). Astaxanthin as a potential antioxidant to improve health and production performance of broiler chicken. *Vet. Med. Int.* 2022: 4919442.

77 Hameed, A., Hussain, S.A., Yang, J. et al. (2017). Antioxidants potential of the filamentous fungi (Mucor circinelloides). *Nutrients* 9 (10): 1101.

78 Shih, C.K., Chang, J.H., Yang, S.H. et al. (2008). Beta-carotene and canthaxanthin alter the pro-oxidation and antioxidation balance in rats fed a high-cholesterol and high-fat diet. *Br. J. Nutr.* 99 (1): 59–66.

79 Vitale, G.A., Coppola, D., Palma Esposito, F. et al. (2020). Antioxidant molecules from marine fungi: methodologies and perspectives. *Antioxidants (Basel)* 9 (12): 1183.

80 Ryan, E.M., Duryee, M.J., Hollins, A. et al. (2019). Antioxidant properties of citric acid interfere with the uricase-based measurement of circulating uric acid. *J. Pharm. Biomed. Anal.* 164: 460–466.

81 Kim, E.J., Cha, M.N., Kim, B.G., and Ahn, J.H. (2017). Production of curcuminoids in engineered *Escherichia coli*. *J. Microbiol. Biotechnol.* 27 (5): 975–982.

82 Shen, L.R., Xiao, F., Yuan, P. et al. (2013). Curcumin-supplemented diets increase superoxide dismutase activity and mean lifespan in *Drosophila*. *Age (Dordr.)* 35 (4): 1133–1142.

83 Guan, X.L., Wu, P.F., Wang, S. et al. (2017). Dimethyl sulfide protects against oxidative stress and extends lifespan via a methionine sulfoxide reductase A-dependent catalytic mechanism. *Aging Cell* 16 (2): 226–236.

84 Hubka, V., Nováková, A., Kolařík, M. et al. (2015). Revision of *Aspergillus* section *Flavipedes*: seven new species and proposal of section Jani sect. nov. *Mycologia* 107 (1): 169–208.

85 Strobel, G., Ford, E., Worapong, J. et al. (2002). Isopestacin, an isobenzofuranone from *Pestalotiopsis microspora*, possessing antifungal and antioxidant activities. *Phytochemistry* 60 (2): 179–183.

86 Ledesma-Amaro, R., Serrano-Amatriain, C., Jiménez, A. et al. (2015). Metabolic engineering of riboflavin production in *Ashbya gossypii* through pathway optimization. *Microb. Cell Factories* 14 (1): 163.

87 Choi, M.J., Kim, B.K., Park, K.Y. et al. (2010). Anti-aging effects of cyanidin under a stress-induced premature senescence cellular system. *Biol. Pharm. Bull.* 33 (3): 421–426.

88 Imran, M., Ghorat, F., Ul-Haq, I. et al. (2020). Lycopene as a natural antioxidant used to prevent human health disorders. *Antioxidants (Basel)* 9 (8): 706.

89 El-Naggar, N.E. and El-Ewasy, S.M. (2017). Bioproduction, characterization, anticancer and antioxidant activities of extracellular melanin pigment produced by newly isolated microbial cell factories *Streptomyces glaucescens* NEAE-H. *Sci. Rep.* 7: 42129.

90 Kantayos, V., Kim, J.-S., and Baek, S.-H. (2022). Enhanced anti-skin-aging activity of yeast extract-treated resveratrol rice DJ526. *Molecules* 27 (6): 1951.

91 Kaeberlein, M. (2010). Lessons on longevity from budding yeast. *Nature* 464 (7288): 513–519.

92 Suwannasom, N., Kao, I., Pruß, A. et al. (2020). Riboflavin: the health benefits of a forgotten natural vitamin. *Int. J. Mol. Sci.* 21 (3): 950.

93 Rastogi, R.P., Sinha, R.P., Moh, S.H. et al. (2014). Ultraviolet radiation and cyanobacteria. *J Photochem Photobiol B* 141: 154–169.

94 Covas, M.I., Miró-Casas, E., Fitó, M. et al. (2003). Bioavailability of tyrosol, an antioxidant phenolic compound present in wine and olive oil, in humans. *Drugs Exp. Clin. Res.* 29 (5–6): 203–206.

95 Gao, H., Tuyishime, P., Zhang, X. et al. (2021). Engineering of microbial cells for L-valine production: challenges and opportunities. *Microb. Cell Factories* 20 (1): 172.

96 Edwards, C., Canfield, J., Copes, N. et al. (2015). Mechanisms of amino acid-mediated lifespan extension in *Caenorhabditis elegans*. *BMC Genet.* 16: 8.

97 Prabhu, S., Rekha, P.D., Young, C.C. et al. (2013). Zeaxanthin production by novel marine isolates from coastal sand of India and its antioxidant properties. *Appl. Biochem. Biotechnol.* 171 (4): 817–831.

Index

Microbial Bioreactors for Industrial Molecules, First Edition.
Edited by Sudhir P. Singh and Santosh Kumar Upadhyay.